LAPLACE TRANSFORM
OPERATIONAL PROPERTIES

$f(t)$	$\hat{f}(s)$
1. $C_1 f_1(t) + C_2 f_2(t)$	$C_1 \hat{f}_1(s) + C_2 \hat{f}_2(s)$
2. $f(at)$	$(1/a)\hat{f}(s/a)$
3. $f'(t)$	$s\hat{f}(s) - f(0)$
$f''(t)$	$s^2\hat{f}(s) - sf(0) - f'(0)$
4. $tf(t)$	$-d/ds\,\hat{f}(s)$
$t^2 f(t)$	$d^2/ds^2\,\hat{f}(s)$
5. $e^{at}f(t)$	$\hat{f}(s-a)$
6. $H(t-a)f(t-a)$	$e^{-as}\hat{f}(s)\,a > 0$
7. $\int_0^t f(\tau)\,d\tau$	$(1/s)\hat{f}(s)$
8. $(1/t)f(t)$	$\int_s^\infty \hat{f}(z)\,dz$
9. $f*g(t)$	$\hat{f}(s)\hat{g}(s)$

Here C_1, C_2, and a denote nonzero constants.

LAPLACE TRANSFORM PAIRS

$f(t)$ $(t>0)$	$\hat{f}(s)$
1. $H(t-0) = 1$	$1/s$
2. t^n	$n!/s^{n+1}$
3. $\exp[kt]$	$1/(s-k)$
4. $\sin at$	$a/(s^2+a^2)$
5. $\cos at$	$s/(s^2+a^2)$
6. $(\pi t)^{-1/2}\exp[-k^2/4t]$	$s^{-1/2}\exp[-ks^{1/2}],\ k \geq 0$
7. $k(4\pi t^3)^{-1/2}\exp[-k^2/4t]$	$\exp[-ks^{1/2}],\ k > 0$
8. $\text{erfc}[k(4t)^{-1/2}]$	$(1/s)\exp[-ks^{-1/2}],\ k > 0$

where

$$\text{erfc}(z) = 2/\pi^{1/2}\int_z^\infty \exp[-s^2]\,ds \text{ and } \text{erf}(z) = 2/\pi^{1/2}\int_0^z e^{-s^2}ds$$
$$\text{erfc}(z) = 1 - \text{erf}(z)$$

Here k and a denote constants.

APPLIED PARTIAL DIFFERENTIAL EQUATIONS

Paul DuChateau & David Zachmann
Colorado State University

DOVER PUBLICATIONS, INC.
Mineola, New York

Bibliographical Note

This Dover edition, first published in 2002, is a corrected republication of the work originally published by Harper & Row, Publishers, Inc., New York, in 1989.

Library of Congress Cataloging-in-Publication Data

DuChateau, Paul.
    Applied partial differential equations / Paul DuChateau, David Zachmann.—Dover ed.
        p. cm.
    Originally published: New York : Harper & Row, 1989.
    Includes index.
    ISBN 0-486-41976-2 (pbk.)
        1. Differential equations, Partial. I. Zachmann, David W. II. Title.

QA374 .D8 2002
515'.353—dc21

2001047180

Manufactured in the United States of America
Dover Publications, Inc., 31 East 2nd Street, Mineola, N.Y. 11501

Contents

Preface

The importance of partial differential equations as a component of applied mathematics has long been recognized, and the increasing complexity of today's technology requires engineers and applied scientists to understand this subject on a level previously attained only by specialists. In addition the accelerating development of computers, both super-computers and powerful personal computers, has had a significant impact on the way in which we approach problems in partial differential equations. In spite of all this, the teaching of the elementary courses in this subject has changed very little in three or four decades.

In writing this book we have tried to reflect some of the modern attitudes toward the subject of partial differential equations without losing sight of the fact that this is intended to be an introductory text. The only prerequisites we assume are a good foundation in calculus and an introductory course in ordinary differential equations. Some background in linear algebra would be helpful, but we recognize that it is likely that many of the students in a course on partial differential equations have little knowledge of linear algebra. For that reason we have included a linear algebra appendix and integrated additional linear algebra facts into the text when they are needed.

EMPHASIS

Throughout we have tried to emphasize the development of only a few mathematical tools, but we concentrate on learning to use them very well. In this way, with only a modest mathematical background we are able to consider a number of interesting physical applications including flow in a porous medium, dispersive and nondispersive wave propagation, advection with a random coefficient, traffic flow, and waves in a ripple tank. We know from experience that not only can students understand these examples, they appreciate them more than the traditional but bland examples like the vibrating string.

This book differs from other elementary books on partial differential equations in a number of ways. Most obvious is the fact that approximately one half of the book is devoted to numerical methods for solving partial differential equations. In part this is in recognition of the need to train applied mathematicians in computational mathematics [see A National Computing Initiative, the agenda for leadership, SIAM workshop, Feb. 1987]. In addition it is a reflection of the belief that, at an elementary level, it no longer makes sense to study the classical aspects of partial differential equations disjoint from the numerical side of the subject. We prefer to think of the two faces of a problem in partial differential equations as the continuous model and discrete model, respectively, for some underlying physical system. It is one of the principal themes of this book that intelligent use of numerical solution to the discrete version of a partial differential equation is not possible without a good understanding of the behavior of the solution of the associated continuous problem. Conversely it is often the case that insight into the workings of the continuous problem can be obtained by studying the associated discrete model. This is particularly evident in Chapter 1, in which we formulate the discrete model for some example physical systems. In the discrete setting it is quite clear what sort of auxiliary data in the form of initial and boundary conditions must be imposed in addition to the basic equation in order to obtain a well-posed problem. This motivates the selection of auxiliary conditions in formulating the associated continuous problem.

CONTENT

We begin in Chapter 1 and continue throughout the book to promote the point of view that in the modeling of physical systems there are three entities to be considered and that these should be accorded equivalent status. The three entities are the physical system, the continuous model, and the discrete model. In order to illustrate the relations that exist between the continuous and discrete model for a system, we very often will solve a continuous problem in the first half of the book by means of an eigenfunction expansion or integral transform and then solve the associated discrete problem in the second part of the book by an entirely parallel method. Each model provides information of some sort about the physical system and each model requires a specific set of tools to uncover that information. The tools of linear algebra are used to study the discrete model while the continuous model is analyzed by methods that have their roots in calculus. We hope to create in the students the impression that linear algebra and analysis (calculus) each has its place in the toolbox of the applied mathematician.

Other differences between this text and most other elementary books on partial differential equations are more subtle. For example, we do not think that the student should leave this course with the feeling that once he or she has constructed a solution to a continuous problem in the form of an eigenfunction expansion or an integral representation the job is finished; in truth it is only the routine part of the problem that is finished. At that point begins the task of extracting information about the behavior of the system being modeled. Examples of the types of information available and how that information may be obtained include the following:

a boundary-value problem for the Laplace equation that shows how the shape of the region can affect the smoothness of the solution

initial-boundary-value problems for the heat equation that illustrate the influence of the lower-order terms in the equation (i.e., conduction with convection and conduction with dissipation)

initial-boundary-value problems for the wave equation that illustrate the influence of lower-order terms (i.e., dispersive and dissipative wave propagation)

initial-boundary-value problems for the heat equation that illustrate both time-dependent and time-independent steady-state solutions

initial-boundary-value problems for the heat equation and wave equation that illustrate the difference between diffusionlike and wavelike evolution

an advection equation with a random coefficient that induces diffusive behavior in the solution

a traffic-flow model to illustrate the solution of first-order conservation laws

a model of waves in a ripple tank to illustrate systems of first-order hyperbolic equations

The treatment of first-order partial differential equations is found in Chapter 7, following the material on second-order problems. We have found that this material can be more easily grasped after an introduction to second-order partial differential equations. Chapter 7 is more than just a cursory introduction to first-order equations and includes discussions of shocks, fans, and generalized solutions. In addition to a thorough treatment of first-order equations and conservation laws, this chapter contains many more examples of the applications of first-order equations than is usual in a text at this level.

ARRANGEMENT OF MATERIAL

There are other ways in which the arrangement of material in this book differs from most other texts. In Chapter 1 on modeling of physical systems, the discrete model is derived before the associated continuous model. The reasons for this have already been mentioned. Chapter 2 develops the classical theory of Fourier series followed immediately by a discussion of generalized Fourier series and Sturm–Liouville problems. Here we generalize the usual pointwise notion of a function slightly and introduce the class of square integrable functions. The chapter discusses in detail the convergence of Fourier series but does not include a proof of the Fourier convergence theorem. We feel that this proof belongs more appropriately in a course on real analysis. We also omit any mention of special functions from this text. Although special functions have a place in solving partial differential equations, to include anything more than a superficial treatment would force us to exclude more relevant material.

We begin solving partial differential equations in Chapter 3. This chapter focuses on problems on bounded regions so that the apropriate solution method is then eigenfunction expansion. We begin with static problems described by Laplace's equation and proceed to time-dependent problems governed by the heat and wave equations. Chapter 4 presents the essentials of the Fourier and Laplace transforms, and in Chapter 5 these are applied to the solving of partial differential equations on unbounded sets. The problems in Chapter 5 are covered in the same order as in Chapter 3, and whenever possible we draw attention to similarities and differences between problems on bounded regions and their analogues on regions that are not bounded.

Chapter 6 is devoted to some of the qualitative aspects of the problems we have studied in Chapters 1 through 5. In particular energy integral methods and max–min principles are introduced and used to expose various facts about the solutions to some of the examples considered in previous chapters. Chapter 7 presents a thorough treatment of first-order equations including a discussion of weak solutions, fans, and shocks.

Chapters 8, 9, and 10 cover the solution of discrete models of parabolic, hyperbolic, and elliptic types, respectively. Parabolic problems are the focus of Chapter 8, in which a number of the standard topics associated with finite difference methods are introduced. Some of the less usual topics covered in this chapter include discrete Fourier methods and discrete conservation laws. Discrete material balance laws are the basis for developing conservative difference methods, a notion of considerable importance in many applications.

Hyperbolic equations are treated in Chapter 9, which contains much material that is not to be found in any other text at this level. Examples include conservation law equations, numerical dispersion and dissipation, and the numerical method of characteristics.

Chapter 10 concludes the book with a presentation of numerical methods for elliptic problems. As in the previous two chapters, the methods are illustrated with physical applications and an effort is made to emphasize the parallels between the discrete problem being discussed and the analogous continuous problem. Algorithms in the form of pseudocode are included in each of the last three chapters, and we strongly urge users of this book to engage in extensive computer experimentation. We have done our best to provide exercises to leave the student with the impression that the subject of numerical methods for solving partial differential equations is something far more subtle than just "crunching numbers."

There is more material in this book than can be covered in a one-semester or even full-year course in partial differential equations. However, by selecting certain portions of the book there are various courses and sequences of courses that can be offered using this book as a text. Some of these are indicated in flowcharts following the preface. Note that there is sufficient material in each chapter that, by selecting or omitting examples, an instructor can adapt the level of the course to suit the class.

ACKNOWLEDGMENTS

We should like to express our appreciation to Professor C. W. Groetsch of the University of Cincinnati, Professor Stephen Krantz of Washington University, Professor John Palmer of the University of Arizona, Professor Ed Landesman of the University of California, Santa Cruz, Professor Gilbert N. Lewis of Michigan Technological University, Professor Gary Walls of the University of Southern Mississippi, and Professor Bernard Marshall of McGill University. Each of them read various versions of the manuscript and made suggestions that led to improvements. We should also like to thank Professor John Hunter and Professor Jim Thomas of Colorado State University for many helpful conversations and Professor Ralph Niemann also of Colorado State for testing a preliminary version of this text in his junior/senior-level partial differential equations class. A number of rough spots were made smooth as a result of his input. Of course we are also grateful to the many students who diligently found and reported errors typographical and otherwise. Finally, we should like to express our appreciation to Peter Coveney and the staff at Harper & Row for their cooperation and support during this project.

Paul DuChateau
David Zachmann

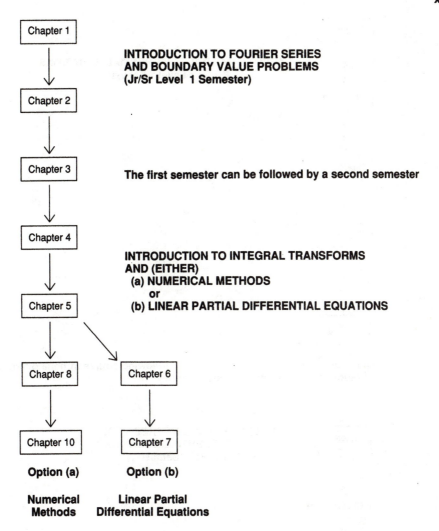

Chapter 1

**INTRODUCTION TO FOURIER SERIES
AND BOUNDARY VALUE PROBLEMS
(Jr/Sr Level 1 Semester)**

Chapter 2

Chapter 3

The first semester can be followed by a second semester

Chapter 4

**INTRODUCTION TO INTEGRAL TRANSFORMS
AND (EITHER)
 (a) NUMERICAL METHODS
  or
 (b) LINEAR PARTIAL DIFFERENTIAL EQUATIONS**

Chapter 5

Chapter 8   Chapter 6

Chapter 10   Chapter 7

Option (a)   **Option (b)**

**Numerical
Methods**   **Linear Partial
Differential Equations**

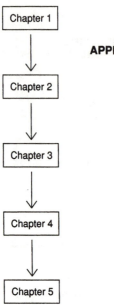

APPLIED PARTIAL DIFFERENTIAL EQUATIONS
(Sr/Gr 1 Level 1 Semester)

The above applied PDE course and the following course in finite differences can be taught as a sequence or independently.

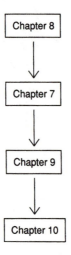

FINITE DIFFERENCE METHODS FOR
PARTIAL DIFFERENTIAL EQUATIONS
(Sr/Gr 1 Level 1 Semester)

PART ONE

EXACT METHODS

Chapter 1

Mathematical Modeling and Partial Differential Equations

This is a book about solving partial differential equations. More specifically, it is a book about solving physically motivated problems in partial differential equations. Before we can proceed to solve such problems, we should first know something about how the problems arise, how they are properly formulated, and the notation and terminology used to discuss them. This is the subject matter of Chapter 1.

Mathematical modeling at an elementary level is illustrated for three different physical systems. Examples of properly posed problems for the discrete model are given for each system, and these problems are used to motivate the formulation of a corresponding well-posed problem for the continuous model. In Chapter 1 we are only trying to demonstrate by example the meaning of well-posedness for problems in partial differential equations, and none of the examples in this chapter are *proved* to be well posed; this is done in Chapter 6 for some of the examples.

The chapter concludes with a brief discussion of the notions of classification of equations.

1.1 MATHEMATICAL MODELING OF PHYSICAL SYSTEMS

The term *mathematical model* refers to a mathematical problem whose solution allows us to describe or predict the behavior of an associated physical system as it responds to a given set of inputs. The physical system is governed by a well-defined set of physical principles that are then translated into corresponding mathematical statements. These statements often take the form of equations in which the "state" of the physical system plays the role of the unknown in the problem. The mathematical model is considered to be "well formulated" if the output or response of the physical system is uniquely determined by the input for the problem.

We intend to demonstrate the process of developing a mathematical model for a physical system by means of some examples. In each example we describe the physical system and the "state variables" that characterize its behavior. In addition, we explain how the governing physical principles may be expressed as mathematical equations. Finally, we try to illustrate in these examples the role that simplifying assumptions play in the development of a mathematical model.

Continuous and Discrete Models of Physical System

In each of the examples we present, we develop two distinct versions of a mathematical model for the physical system. One of the versions, the continuous model, treats the physical system as a continuous medium. A consequence of this point of view is that the governing physical principles translate into differential equations in the mathematical setting; if the number of independent variables is more than one, then these differential equations are *partial* differential equations.

In addition to the continuous model, we discuss a discrete version of the mathematical model for each of the physical systems we consider. This model arises from viewing the physical system as composed of discrete entities each of finite rather than infinitesimal size. In this case, the mathematical statement of the governing physical principles takes the form of a system of algebraic equations.

Relationship of Models to Each Other

The continuous and discrete models for a given physical system are related to each other in the following way. The continuous model may be obtained from the discrete model by allowing the size of the discrete entities comprising the system to shrink to zero. Properly applied limiting procedures then cause the algebraic expressions to become differential expressions. Conversely, the discrete model may be obtained from the continuous model by approximating the derivatives in the continuous model by suitable difference quotients. If this is done correctly, the differential equations in which the unknowns are functions are replaced by algebraic equations in which the unknowns are the values of these same functions evaluated at discrete points in the domain of interest.

The continuous and the discrete models each contribute to the understanding of a physical system and each can contribute to the understanding of the other. Currently it is more usual to rely on the continuous model for *qualitative* information about the physical system and to resort to the discrete model for *quantitative* (numerical) results. The motivation for this point of view is practical. The large-scale algebraic problems arising in connection with the discrete model are well suited to treatment by computer where numerical results are readily generated. On the other hand, even when no solution can be explicitly constructed, the continuous-model problem will often yield information of a qualitative nature using methods that have their roots in the calculus.

Practical though it might be, this point of view propagates the following *false* impressions about the discrete model:

(a) The discrete model is just an approximation of the continuous model (this reduces the discrete model to the status of an approximation to an approximation of the physical system).

(b) The discrete model is obtained from the continuous model by simply replacing derivatives in the continuous-model equations by finite-difference expressions based on Taylor series expansions of the unknown functions.

The impression created is that the continuous model stands *between* the physical system and the discrete model. We aim to illustrate in the pages to follow that the discrete and continuous models are equally valid *alternative* descriptions for a physical system and that neither should be viewed as any more "approximate" than the other. Moreover, while the continuous model can always be obtained from the discrete model by a passing to the limit, it often happens that there is no approximation of derivatives by Taylor series expansions that will lead from the continuous model to the discrete. More precisely, we should say that replacing derivatives in the continuous-model equation by finite-difference expressions based on Taylor series expansions does not always lead to *correct* discrete models. In order to ensure that the discrete model is a correct one, the derivation should be based on discrete versions of the physical principles used to develop the continuous model. A detailed discussion of the process of obtaining discrete models from continuous ones is found in Section 8.1.

Qualitative versus Quantitative Information

Another of the goals of the presentation in this text will be to alter the perception that the discrete model must serve a purely quantitative role in the study of a physical system while the continuous model is to be used only for qualitative purposes. In the chapters to come we provide examples of how the continuous-model solution can be used for quantitative purposes. These examples show that there are certain limitations in the responses (outputs) that can be modeled discretely. For example, in any discrete model, all responses of sufficiently high frequency are modeled in an ambiguous way. This is the phenomenon of "aliasing" mentioned in Chapter 2 in connection with discrete Fourier series. Deficiencies of this sort, resulting from the very discreteness of the discrete model are, of course, not present in the continuous model. This is one example then of a situation where the continuous model is to be preferred as a source of quantitative information about the physical system.

On the other hand, particularly in this first chapter, we use the discrete model as a source of qualitative information about the continuous model. Here, and again in the later chapters where the methods for constructing the solutions to the discrete model problem are developed, we make extensive use of a few basic principles from linear algebra. For those whose background in this area is lacking, we have collected most of the results we need in an appendix at the end of this text. Throughout the development we strive to emphasize the parallel between the treatment of the continuous problem by methods having their roots in the calculus (analysis) and the treatment of the discrete problem by the techniques of linear algebra. As a by-product of this approach, we hope to create the impression that for an applied mathematician, strong foundations in both analysis and linear algebra are essential.

Summarizing what we have said so far, we have described three distinct entities: the physical system, the discrete model, and the continuous model (see Figure 1.1.1). Each of these entities provides information about the other two. In this text we are primarily concerned with extracting information about the physical system from the two

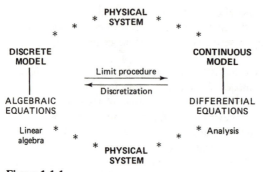

Figure 1.1.1

mathematical models. Chapters 2–7 are devoted to applying the tools of analysis to solving the partial differential equations of the continuous model and to using those solutions to provide information about the physical system. Chapters 8–10 describe how the techniques of linear algebra can draw out from the discrete model information about the behavior of a physical system. A secondary theme throughout this text is the exchange of information between the continuous and discrete models.

The Role of Simplifying Assumptions

We must also recognize that whether we are pursuing the continuous or the discrete approach, the final form of the mathematical model we develop depends on the exact nature of the simplifying assumptions we make. For example, although the state of a physical system under consideration may depend on a very large number of factors, we generally decide to take into account only those factors we consider to be "of primary importance." Other simplifying assumptions may take the form of omitting certain terms from the equations that represent the mathematical expression of the system's governing principles. These terms can be omitted for reasons of expediency (this is often done when the equation is more easily solvable when they are not present) or some terms in the equation may actually be "negligible" with respect to the other terms in the equation.

In any case, since there may be several ways in which a given model may be simplified, it can happen that a single physical system is described by more than a single mathematical model. This does not necessarily represent an ambiguous situation. It may be that different models can be associated with different levels of refinement, and the solutions that result from solving these models can be thought of as analogous to viewing the physical system under differing degrees of magnification. On the other hand, a single physical system may be accurately represented by one model under one set of conditions while another set of conditions requires that a different model be used.

We now illustrate the meaning of these remarks with some examples. Each of the examples leads to a partial differential equation that is of second order. Since so many physical problems do lead to equations of second order, and since the treatment of second-order problems is somewhat more systematic than it is for first-order equations, we discuss the solution of equations of order two before discussing equations of order one. Chapters 3 and 5 are primarily devoted to second-order problems while the first-order problems are considered in Chapter 7.

1.2 EQUATION OF HEAT CONDUCTION

We begin by considering a three-dimensional region Ω that we imagine to be filled with a heat-conducting material. Let $u = u(x, y, z, t)$ denote the temperature at the position (x, y, z) in Ω at time t, and our aim is to be able to compute the unknown function $u(x, y, z, t)$. The conduction of heat is governed by certain physical principles that we translate into mathematical equations. Since these equations are statements relating to the rate of heat flow, it will be convenient to define a quantity Φ that we call the heat flux. The heat flux $\Phi(x, y, z, t)$ is a vector quantity whose magnitude equals the rate of heat flow at the point (x, y, z) at the time t and whose direction indicates the direction of heat flow. We have then

$$u = u(x, y, z, t) = \text{temperature at } (x, y, z, t)$$
$$\Phi = \Phi_1(x, y, z, t)\mathbf{i} + \Phi_2(x, y, z, t)\mathbf{j} + \Phi_3(x, y, z, t)\mathbf{k}$$
$$\Phi = \text{heat flux (vector) at } (x, y, z, t)$$

where $\mathbf{i}$, $\mathbf{j}$, $\mathbf{k}$ denote unit vectors in the x, y, z coordinate directions. Imagine now a small cubical cell in Ω situated so that each of its faces is perpendicular to one of the coordinate axes and its center is at the generic point (x, y, z). Denote the side length of each side of the cell by ϵ and number the faces 1–6 such that if $\mathbf{N}_i$ denotes the unit outward normal to face number i, then

$$\mathbf{N}_1 = \mathbf{i} \qquad \mathbf{N}_3 = \mathbf{j} \qquad \mathbf{N}_5 = \mathbf{k}$$
$$\mathbf{N}_2 = -\mathbf{i} \qquad \mathbf{N}_4 = -\mathbf{j} \qquad \mathbf{N}_6 = -\mathbf{k}$$

Figure 1.2.1 shows Ω viewed along the z axis. Now if we consider a generic interval of time (t_0, t_1), the amount of heat that flows out of face number i during this period of time can be expressed as

$$\Delta Q_i = \Phi \cdot \mathbf{N}_i \, \Delta A \, \Delta t = \Phi \cdot \mathbf{N}_i \, \epsilon^2(t_1 - t_0), \qquad i = 1, \ldots, 6$$

Here ΔQ is expressed in units of heat since Φ is expressed in units of heat per unit time

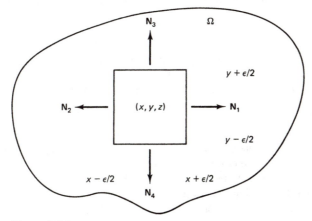

Figure 1.2.1

per unit area. The total amount of heat flowing out of the cubical cell during this time interval is then the sum of these six terms:

$$\Delta Q = \sum_{i=1}^{6} \Delta Q_i = \epsilon^2 \, \Delta t \, [\Phi_1(x + \tfrac{1}{2}\epsilon, y, z, t) - \Phi_1(x - \tfrac{1}{2}\epsilon, y, z, t)$$
$$+ \, \Phi_2(x, y + \tfrac{1}{2}\epsilon, z, t) - \Phi_2(x, y - \tfrac{1}{2}\epsilon, z, t)$$
$$+ \, \Phi_3(x, y, z + \tfrac{1}{2}\epsilon, t) - \Phi_3(x, y, z - \tfrac{1}{2}\epsilon, t)] \qquad (1.2.1)$$

Here we are making use of the fact that

$$\mathbf{\Phi} \cdot \mathbf{N}_1 = \Phi_1(x + \tfrac{1}{2}\epsilon, y, z, t), \qquad \mathbf{\Phi} \cdot \mathbf{N}_2 = -\Phi_1(x - \tfrac{1}{2}\epsilon, y, z, t), \qquad \text{etc.}$$

Now we bring to bear the first of the principles governing the physical system. This principle states that in the absence of internal heat sources or heat sinks, when an amount of heat is added to or taken from a heat-conducting body, a temperature change occurs. More precisely, the relationship between the quantity of heat, ΔQ, and the corresponding temperature change, Δu, is given by

$$\Delta Q = C \, \Delta m \, \Delta u = C\sigma\epsilon^3[u(x, y, z, t_0) - u(x, y, z, t_1)] \qquad (1.2.2)$$

That is, ΔQ is proportional to Δu with the proportionality factor equal to the product $(C \, \Delta m)$. Here C denotes a material-dependent parameter called the *heat capacity* and Δm denotes the mass of the conductor. Then Δm equals the density of the material, σ, times the volume (ϵ^3 for the cubical cell). We make the simplifying assumptions here that both C and σ are constants.

We can express (1.2.1) and (1.2.2) more compactly if we introduce the notation

$$\delta_p f(p) = [f(p + \tfrac{1}{2}\epsilon) - f(p - \tfrac{1}{2}\epsilon)] \qquad (1.2.3)$$

for the *differences* in the flux components appearing in (1.2.1). Then equating the two expressions for ΔQ and using the notation (1.2.3), we obtain the equation

$$C\sigma\epsilon^3[u(x, y, z, t_0) - u(x, y, z, t_1)]$$
$$= \epsilon^2 \, \Delta t \, [\delta_x\Phi_1(x, y, z, t) + \delta_y\Phi_2(x, y, z, t) + \delta_z\Phi_3(x, y, z, t)]$$

If we cancel the factor ϵ^2 from the equation, this reduces to the *discrete heat balance equation*

$$C\sigma\epsilon[u(x, y, z, t_0) - u(x, y, z, t_1)] = \Delta t \, [\delta_x\Phi_1 + \delta_y\Phi_2 + \delta_z\Phi_3](x, y, z, t) \qquad (1.2.4)$$

The value of the time argument t appearing on the right side of (1.2.4) is indeterminate. It follows from the mean-value theorem that equality in (1.2.4) occurs for *some* t value between t_0 and t_1. For simplicity, we assume this value to be t_0.

Discrete Heat Equation

Equation (1.2.4) is a statement (in a discrete setting) of the principle that the physical process of heat conduction proceeds so as to conserve thermal energy. This principle has been expressed in terms of the temperature and the heat flux (the state variables). This is one equation for the four unknown functions u, Φ_1, Φ_2, Φ_3. What we are seeking, then, is an equation involving only the temperature function, and in order to achieve this, we need an additional equation relating heat flux to temperature. This equation is based on an empirical physical principle known as *Fourier's law of heat conduction.* Fourier's law is named for Joseph Fourier, a French mathematician of the Napoleanic era who carried out some of the early work on the mathematical modeling of heat conduction. The law states that heat flows from hot regions in a conductor to cool ones and that the rate of heat flow is proportional to the temperature difference between the hot and cold regions. Expressed in mathematical terms, this becomes

$$\Phi(x, y, z, t) = -(K/\epsilon)[\delta_x u \, \mathbf{i} + \delta_y u \, \mathbf{j} + \delta_z u \, \mathbf{k}](x, y, z, t) \qquad (1.2.5)$$

where K denotes another material-dependent parameter called *thermal conductivity.* We assume that, as for density and heat capacity, K is a constant. Using (1.2.5) in (1.2.4) leads to

$$C\sigma\epsilon^2[u(x, y, z, t_1) - u(x, y, z, t_0)]$$
$$= \Delta t \, [\delta_x(K \, \delta_x u) + \delta_y(K \, \delta_y u) + \delta_z(K \, \delta_z u)](x, y, z, t_0) \qquad (1.2.6)$$

We refer to (1.2.6) as the discrete version of the heat conduction equation. The temperature function $u(x, y, z, t)$ must satisfy this equation at each point (x, y, z) in the conductor and for all times t_0 and t_1 during the process of heat conduction.

Equation (1.2.6) is not, by itself, enough to allow us to compute the time evolution of the temperature in the conductor. Additional information is required, and the discrete version of the heat conduction equation is particularly well suited to illustrating by example what this information should be and how it is used.

A Special Case: Discrete Model in One Dimension For ease of illustration we suppose that we are dealing with a one-dimensional conductor, a thin rod whose lateral surface has been insulated against the flow of heat (Figure 1.2.2). Insulating the lateral surface of the rod ensures that the heat flows axially along the rod, and if the rod is sufficiently thin, it is an acceptable approximation of reality to suppose that the temperature does not vary across the thickness of the rod.

If the rod lies with its center line along the x axis, the temperature in the rod depends only on x and t and (1.2.6) reduces to

$$C\sigma\epsilon^2[u(x, t_1) - u(x, t_0)] = K \, \Delta t \, [\delta_x \, \delta_x u(x, t_0)] \qquad (1.2.7)$$

This can be rewritten in the more explicit form

$$u(x, t_1) = u(x, t_0) + \beta[u(x + \epsilon, t_0) - 2u(x, t_0) + u(x - \epsilon, t_0)] \qquad (1.2.8)$$

where $\beta = K \, \Delta t/(C\sigma\epsilon^2)$.

Now in order to be able to relate Equation (1.2.8) to the thin rod, we imagine that

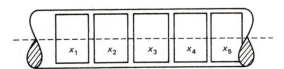

Figure 1.2.2

the rod is composed of an array of one-dimensional "cells" each of length ϵ. If we introduce the notation

$$x_n = n\epsilon, \qquad n = \text{integer}$$
$$t^j = t_0 + j(t_1 - t_0), \qquad j = 0, 1, \ldots$$
$$u_n^j = u(x_n, t^j)$$

then each of the cells can be thought of as being centered at one of the points x_n and (1.2.8) can be written as

$$u_n^{j+1} = \beta u_{n+1}^j + (1 - 2\beta)u_n^j + \beta u_{n-1}^j \qquad (1.2.8^*)$$

This equation is satisfied in each cell (i.e., at each of the discrete points x_n) in the one-dimensional conductor for every one of the discrete times t^j. Equation (1.2.8*) is an example of the type of equation treated in more detail in Chapter 8. We now consider two example problems for the rod in order to see what sort of additional information will be required in order to be able to compute the temperatures in the rod from Equation (1.2.8*). In the first example we consider a rod of infinite length. Of course, practically speaking, there is no such thing, but if the rod is long, there will be a portion of the rod near the middle where for a finite amount of time the influence of the ends of the rod may be neglected. Then for this finite amount of time, the middle portion of the rod may be thought of as a rod of infinite length. This artifice provides a way of treating the problem of finding the temperature in the rod when we do not wish to deal with the effects of the ends of the rod.

EXAMPLE 1.2.1 _____

An Infinitely Long Rod

Suppose the temperature in the rod is everywhere known at $t^0 = 0$; that is, u_n^0 is known for every n. Then we can compute u_n^1 for every n from (1.2.8*). Once u_n^1 is known, we can proceed to compute u_n^2 in the same way, and the procedure may (theoretically) be continued for as long as we like (see Problem 1).

Plotting u_n^j versus n for successive values of k then provides a "time lapse picture" of the spatial distribution of temperature in the rod. The graph of u_n^j versus n is often referred to as a "*temperature profile*" at time t^j. A detailed description of how this procedure may be carried out is the subject of Sections 8.2 and 8.6.   ■ ■

EXAMPLE 1.2.2 _____

A Rod of Finite Length

Suppose the rod occupies the interval $(0, L)$ on the x axis and let $x_0 = 0$ and $x_N = N\epsilon = L$. Suppose that the initial distribution of temperature u_n^0 is known for $n = 0, 1, \ldots, N$ and that in addition we know the end temperatures u_0^j and u_N^j, for $j = 0, 1, \ldots$. Then it is easy to verify that u_n^1 may be calculated for $n = 1, 2, \ldots, N$ from (1.2.8*). Of course, once u_n^1 is known, u_n^2 may be computed in the same way (for $n = 1, \ldots, N$), and the process may (again theoretically) be continued for as long as we like in order to predict the evolution of the temperature in the rod at the discrete points x_n. Just as in the case of the infinite rod, we can plot u_n^j versus n for successive values of j in order to obtain a time lapse picture of the evolution of the temperature in the rod with time.

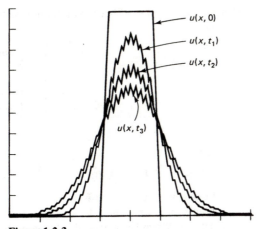

Figure 1.2.3

In particular, for a rod of finite length $L = 10\epsilon$, suppose that $u_0^j = 0$ and $u_{10}^j = 0$ for all j. Suppose also that $u_n^0 = 1$ for $n = 4, 5, 6$. Then we can compute u_n^j from (1.2.8*) and plot it against n for several values of j in order to obtain Figure 1.2.3.

Specifying the temperature everywhere in the rod at $t = 0$ is referred to as imposing an *initial condition* on the temperature. Evidently, Equation (1.2.8*) together with an initial condition are sufficient to determine subsequent states of the temperature in an infinitely long rod. In a rod of finite length we must impose *boundary conditions* as well as an initial condition in order to be able to use (1.2.8*) to calculate subsequent temperature states. Conditions on $u(x, t)$ imposed at the ends (i.e., the boundaries) of the rod are referred to as boundary conditions. ∎ ∎

Semidiscrete Heat Equation

The mean-value theorem for derivatives can be applied to Equation (1.2.7) in order to rewrite it in the form

$$\frac{\partial}{\partial t} u(x, \tau) = \mu[u(x + \epsilon, t_0) - 2u(x, t_0) + u(x - \epsilon, t_0)]$$

for $\mu = K/(\sigma C \epsilon^2)$ and some τ, $t_0 < \tau < t_1$.

Then letting t_1 approach t_0, we have the "semidiscrete" version of (1.2.8),

$$u_n'(t) = \mu[u_{n-1}(t) - 2u_n(t) + u_{n+1}(t)] \tag{1.2.8'}$$

where $u_n(t) = u(x_n, t)$ for $x_n = n\epsilon$ (n an integer) and $t > 0$.

We may apply (1.2.8') in the case of a rod of finite length L with $N\epsilon = L$. Suppose

$$u_0(t) = p(t), \qquad u_N(t) = q(t), \quad \text{for } t > 0$$

for known functions $p(t)$, $q(t)$ and suppose

$$u_n(0) = C_n \quad \text{for } n = 0, 1, \ldots, N$$

for specified constants C_n. Then we have a system of linear ordinary differential equations for the $N - 1$ unknown functions $u_1(t), \ldots, u_{N-1}(t)$,

$$\frac{d}{dt}\begin{bmatrix} u_1 \\ u_2 \\ \vdots \\ u_{N-1} \end{bmatrix}(t) = \mu \begin{bmatrix} -2 & 1 & 0 & \cdots \\ 1 & -2 & 1 & \cdots \\ & & \cdots & \\ & & 1 & -2 \end{bmatrix}\begin{bmatrix} u_1 \\ u_2 \\ \vdots \\ u_{N-1} \end{bmatrix} + \begin{bmatrix} \mu p(t) \\ 0 \\ \vdots \\ \mu q(t) \end{bmatrix}$$

$$\mathbf{U}(0) = [C_1, C_2, \ldots, C_{N-1}]^T$$

In matrix notation this is written as

$$\mathbf{U}'(t) = [A]\mathbf{U}(t) + \mathbf{F}(t), \quad \mathbf{U}(0) = \mathbf{C} \tag{1.2.9}$$

where $\mathbf{U}(t)$ denotes the vector of unknown functions and $\mathbf{F}, \mathbf{C}$ denote vectors containing the data of the problem. In the simpler case where $\mathbf{F} = \mathbf{0}$, we know from the theory of ordinary differential equations that this system has a general solution of the form

$$\mathbf{U}(t) = \Gamma_1 \exp[\alpha_1 t]\mathbf{W}_1 + \cdots + \Gamma_{N-1}\exp[\alpha_{N-1}t]\mathbf{W}_{N-1} \tag{1.2.10}$$

where $\alpha_1, \ldots, \alpha_{N-1}$ denote the eigenvalues of the matrix $[A]$ and $\mathbf{W}_1, \ldots, \mathbf{W}_{N-1}$ denote the corresponding eigenvectors. The Γ's denote arbitrary constants. See Theorem 8.4.1 for expressions for the eigenvalues α_n and eigenvectors $\mathbf{W}_n$.

Our purpose in discussing the semidiscrete equation (1.2.9) here is to show that information about the qualitative character of the solution to the continuous problem can be deduced from analysis of the discrete or semidiscrete problems. Before we pursue this point further with this example, we introduce the continuous version of this problem.

Partial Differential Equation of Heat Conduction

The continuous model may be obtained from the discrete model, in general, by applying an appropriate limit procedure to the discrete equation as the size of the discrete elements in the model is allowed to decrease to zero. We have already allowed the time interval (t_0, t_1) to decrease to zero in going from (1.2.8) to (1.2.8′). Passing to the limit as ϵ decreases to zero in (1.2.8′) leads to the partial differential equation

$$u_t(x, t) = k u_{xx}(x, t), \quad k = K/\sigma C \tag{1.2.11}$$

which is often referred to as the one-dimensional *heat equation*. Here we use the notation

$$u_t = \frac{\partial u}{\partial t} \quad \text{and} \quad u_{xx} = \frac{\partial^2 u}{\partial x^2}$$

to indicate the partial derivatives in this equation.

Equation (1.2.11) is the continuous version of the heat conduction equation in just one dimension. Applying similar limit procedures to (1.2.6) leads to

$$C\sigma u_t(x, y, z, t) = [(Ku_x)_x + (Ku_y)_y + (Ku_z)_z] \tag{1.2.12}$$

This is the full three-dimensional continuous version of the heat conduction equation.

Auxiliary Conditions for Continuous Problem   In Example 1.2.1 we saw that in the case of a discrete model of heat conduction in an infinitely long rod it was necessary to know the initial state of the temperature in the rod in order to compute all subsequent temperature states. In Example 1.2.2 we saw that when the rod is of finite length, in addition to the initial temperature, temperatures at the ends of the rod must be known in order to uniquely determine all subsequent temperature states in the rod. Based on this experience with the discrete problems, we guess that similar specifications of auxiliary information will be required in the continuous problems.

> **An Infinite Rod.** If the initial temperatures $u(x, 0)$ are known for all x, $-\infty < x < \infty$, then $u(x, t)$ can be computed for all $t > 0$ from (1.2.11).
>
> **A Rod of Finite Length.** Suppose the rod occupies the interval $(0, L)$ of the x axis. If the end temperatures $u(0, t)$ and $u(L, t)$ are known for all $t > 0$, in addition to knowing $u(x, 0)$ for $0 < x < L$, then $u(x, t)$ can be computed for all $t > 0$ from (1.2.11).

The first problem here is referred to as a *pure initial-value problem* while the second is called a *mixed initial-boundary-value problem*. Later in this chapter we state these problems with more precision, and in Chapters 3 and 5 we solve several examples of such problems. In Chapter 6 we show how to prove there can be no more than one solution to a problem for which the proper auxiliary conditions have been imposed.

Qualitative Properties of Solutions   We shall see that the solutions to initial-value problems for the heat equation exhibit certain characteristic qualitative behavior, and it is interesting to note that we can anticipate some of this behavior by applying results from linear algebra to the simpler semidiscrete problem. These linear algebra results are collected in the appendix. Their application to this problem is described in detail in Section 8.4. Here we simply summarize the conclusions.

The matrix $[A]$ appearing in (1.2.9) is *symmetric*. This implies that the eigenvalues α_n are all real and that the corresponding eigenvectors are an orthogonal basis for R_{N-1}. In addition, the symmetric matrix $[A]$ can be shown to be *negative definite*, a property that implies that the eigenvalues α_n are all negative. Then it follows from (1.2.10) that the solution to the semidiscrete heat equation decreases exponentially to zero with time (in the case that $\mathbf{F} = \mathbf{0}$). We shall see later that the solution to the analogous problem for the continuous heat equation exhibits this same asymptotic behavior. In addition, results apply that are analogous to the negativity of eigenvalues and the existence of an orthogonal basis of eigenvectors.

In this example we have chosen to derive the discrete model for the conduction process first and to then pass to the semidiscrete and finally the continuous model by applying a limiting procedure as the size of the elements in the discrete model decreases to zero. We could have as well derived the continuous model first, using methods similar to those leading to the discrete model. This is in fact what is usually done in a text of this sort. The discrete model is then obtained by replacing the derivatives in the continuous model by finite-difference expressions resulting from a Taylor series development for the unknown function. As we have remarked previously, deriving the discrete model from the continuous model in this way can lead to discrete models that are inaccurate or even incorrect unless considerable care is taken. It is safer to derive the discrete model directly from discrete versions of the applicable conservation principles.

Exercises: Heat Conduction Equation

1. Consider an infinitely long rod for which the parameters K, ϵ, σ, C are such that $\beta = 0.1$. Then Equation (1.2.8*) becomes

$$u_n^{j+1} = 0.1u_{n+1}^j + 0.8u_n^j + 0.1u_{n-1}^j$$

Suppose

$$u_n^0 = \begin{cases} 1 & \text{for } n = 4, 5, 6 \\ 0 & \text{for all other } n \end{cases}$$

Then use (1.2.8*) and this initial condition to compute u_n^j for $n = -5, \ldots, 15$ for $j = 1, \ldots, 5$. For each value of j, for how many n is u_n^j different from zero?

2. Repeat Exercise 1 for the situation in which the rod is of finite length L with $10\epsilon = L$. Suppose

$$u_0^j = 1 \quad \text{and} \quad u_{10}^j = -1 \quad \text{for all } j > 0$$

and

$$u_n^0 = 0 \quad \text{for all } n$$

Then use (1.2.8*) to compute u_n^j for $n = 1, \ldots, 9$ and $j = 1, \ldots, 5$.

3. Repeat Exercise 2 in the case that

$$u_0^j = u_{10}^j = 0 \quad \text{for all } j > 0$$
$$u_9^0 = u_1^0 = 1, \qquad u_8^0 = u_2^0 = 2, \qquad u_7^0 = u_3^0 = 3$$
$$u_6^0 = u_4^0 = 4, \qquad u_5^0 = 5$$

4. Let the rod be as in Exercise 2 but suppose that at the ends of the rod we have the conditions

$$u_1^j = u_0^j \quad \text{and} \quad u_{10}^j = u_9^j \quad \text{for all } j > 0$$

Show that these conditions are consistent with the condition of no heat flow near the ends of the rod; the heat flux is zero at $x = \epsilon$ and $x = L - \epsilon$. Solve for u_n^j for $n = 0, 1, \ldots, 10$ for $j = 1, 2, \ldots, 5$ if

$$u_n^0 = 1 - n\epsilon/L, \qquad n = 1, \ldots, 9$$

5. Repeat Exercise 4 for a rod of length $L = \pi$. Use the values $u(x, 0) = \sin x$ evaluated at the nodes x_n as the initial values,

$$u_n^0 = \sin(x_n) = \sin(n\pi/10), \qquad n = 0, 1, \ldots, 10$$

Then use (1.2.8*) with $\beta = 0.1$ to compute u_n^1 for $n = 1, \ldots, 9$.

6. Let k denote a positive integer and let

$$u_k(x, t) = \exp[-k^2t]\sin(kx)$$

Then verify that $u_k(x, t)$ satisfies

$$u_t(x, t) = u_{xx}(x, t), \qquad 0 < x < \pi, \qquad t > 0$$
$$u(x, 0) = \sin(kx), \qquad 0 < x < \pi$$
$$u(0, t) = u(\pi, t) = 0, \qquad t > 0$$

7. Let $u_k(x, t)$ be as in Exercise 6. Then what equation, what initial condition, and what boundary conditions are satisfied by the function $v(x, t) = u_1(x, t) + 2u_2(x, t)$?

8. Use the values of the function $u(x, 0) = \sin(10x)$ evaluated at the nodes x_n as the initial values for the discrete model for the problem in Exercise 5. Let $10\epsilon = \pi$; that is, $u_n^0 = \sin(10x_n) = \sin(n\pi)$, $n = 0, 1, \ldots, 10$. Then use (1.2.8*) with $\beta = 0.1$ to compute u_n^1 for $n = 1, \ldots, 9$. Compare these results with the exact solution

$$u(x, t) = e^{-100t}\sin(10x)$$

9. For all positive integers n, show that the functions

$$u_1(x, t) = \exp[-(n\pi)^2 t]\sin n\pi x$$
$$u_2(x, t) = \exp[-(n\pi)^2 t]\cos n\pi x$$
$$u_3(x, t) = \exp[-(n + \tfrac{1}{2})^2\pi^2 t]\cos(n + \tfrac{1}{2})\pi x$$
$$u_4(x, t) = \exp[-(n + \tfrac{1}{2})^2\pi^2 t]\sin(n + \tfrac{1}{2})\pi x$$

all satisfy

$$u_t(x, t) = u_{xx}(x, t)$$

In addition, each satisfies at least one of the following boundary conditions:
(a) $u(0, t) = 0$ (b) $u(1, t) = 0$
(c) $u_x(0, t) = 0$ (d) $u_x(1, t) = 0$
Match each function with all of the conditions it satisfies.

10. Consider the semidiscrete heat equation (1.2.9) in the case of a rod of finite length. Let the parameters K, C, σ, ϵ be such that $\mu = 1$. Suppose $N = 2$ and that

$$u_0(t) = u_2(t) = 0 \quad \text{for } t > 0$$
$$u_1(0) = 1$$

Then write and solve the initial-value problem (1.2.9) for $u_1(t)$.

1.3 STEADY-STATE CONDUCTION OF HEAT

Equations (1.2.6) and (1.2.11) are, respectively, the discrete and continuous models for an *evolution process* and as such are called *evolution equations*. This name derives from the fact that these equations describe a physical process that is evolving with time. Physical processes that are stationary with respect to time lead to somewhat different problems in partial differential equations. For example, suppose the temperature $u(x, y, z, t)$ in (1.2.6) is independent of time. This would be the case if the temperature were observed after all transient behavior had died out and the system had reached a steady state, that is, a state of thermal equilibrium. Then (1.2.6) reduces to

$$\delta_x(\delta_x u) + \delta_y(\delta_y u) + \delta_z(\delta_z u) = 0 \tag{1.3.1}$$

where we continue to assume that K is a constant. Equation (1.3.1) is known as the *discrete Laplace equation* for the unknown function $u = u(x, y, z)$. Problems of this type are taken up again in Chapter 10 where they are considered in more detail.

EXAMPLE 1.3.1 ──

Two-dimensional Conduction Problem

Here we only want to illustrate the type of auxiliary data needed in connection with Equation (1.3.1), and for that purpose we consider, for simplicity, an example in which the temperature does not depend on z. Then in determining $u = u(x, y)$, it will be sufficient to consider not all of the region Ω but rather just a planar cross section taken such that the plane is normal to the z axis. Let the plane cross section be denoted by D. Since u is a function only of x and y, Equation (1.3.1) reduces to

$$\delta_x\delta_x u(x, y) + \delta_y\delta_y u(x, y) = 0 \quad \text{in } D \tag{1.3.2}$$

where

$$\delta_x\delta_x u(x, y) = \delta_x[u(x + \tfrac{1}{2}\epsilon, y) - u(x - \tfrac{1}{2}\epsilon, y)]$$
$$= [u(x + \epsilon, y) - 2u(x, y) + u(x - \epsilon, y)]$$

and $\delta_y\delta_y u(x, y)$ is similarly defined. Then Equation (1.3.2) may be rearranged to read

$$4u(x, y) = u(x + \epsilon, y) + u(x - \epsilon, y) + u(x, y + \epsilon) + u(x, y - \epsilon) \tag{1.3.2*}$$

For the purpose of relating Equation (1.3.2*) to the region D, we must imagine that D is covered by an array of square cells, each of side length ϵ, and that the cells are oriented so that their sides lie parallel to the coordinate axes. Each cell has four neighboring cells with which it shares a side, and Equation (1.3.2) is just the statement that the net flow of heat between any cell and its four neighbors is equal to zero. The array of cells is called a *grid* and the center points of the cells in the grid are called *nodes*. Suppose the nodes in the array covering D are arranged and numbered as in Figure 1.3.1.

Let us assume that the function $u = u(x, y)$ is known at all points on the boundary of D, which means then that u is known at the nodes that lie on (or approximately on) the boundary (i.e., nodes 1–10). Then we can seek to determine u at each of the nodes interior to D (i.e., nodes 11–16). It is evident from the figure that placement of the grid on D is somewhat arbitrary and that the nodes considered to be boundary nodes may not actually lie precisely on the boundary. A node is considered to be a boundary node if it

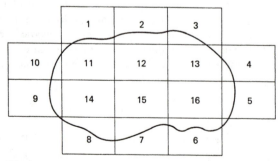

Figure 1.3.1

lies outside of D but is the center of a cell that shares a side with a cell whose center is an interior node of D.

We apply Equation (1.3.2*) with (x, y) located at each of the nodes 11–16; that is, (1.3.2*) must be satisfied for each of the six cells centered at an interior node of D. For example, at node 11 the equation states

$$4u_{11} = u_{12} + u_{14} + u_1 + u_{10}$$

Similarly,

$$4u_{12} = u_{13} + u_{15} + u_{11} + u_2$$

$$\vdots \qquad\qquad \vdots$$

$$4u_{16} = u_5 + u_6 + u_{15} + u_{13}$$

We can rearrange these equations so that all the unknown quantities (u values corresponding to interior nodes) are on the left side of the equation and all known quantities (u values corresponding to boundary nodes) are on the right. In matrix notation this becomes

$$\begin{bmatrix} 4 & -1 & 0 & -1 & 0 & 0 \\ -1 & 4 & -1 & 0 & -1 & 0 \\ 0 & -1 & 4 & 0 & 0 & -1 \\ -1 & 0 & 0 & 4 & -1 & 0 \\ 0 & -1 & 0 & -1 & 4 & -1 \\ 0 & 0 & -1 & 0 & -1 & 4 \end{bmatrix} \begin{bmatrix} u_{11} \\ u_{12} \\ u_{13} \\ u_{14} \\ u_{15} \\ u_{16} \end{bmatrix} = \begin{bmatrix} u_1 + u_{10} \\ u_2 \\ u_4 + u_3 \\ u_8 + u_9 \\ u_7 \\ u_5 + u_6 \end{bmatrix}$$

This is a system of linear algebraic equations of the form

$$[L_N]\mathbf{u}_N = \mathbf{f}_N \qquad\qquad (1.3.2')$$

where N denotes the number of interior nodes (in this example $N = 6$), $\mathbf{u}_N$ denotes the vector of unknowns (containing the unknown values of u at the interior nodes), and $\mathbf{f}_N$ denotes the data vector (containing the known values of u at the boundary nodes). Finally, $[L_N]$ denotes the $N \times N$ coefficient matrix. This matrix can be seen to have some special structure that is the result of the form of the equation (1.3.2). For example, the matrix entries are symmetric about the diagonal and the largest entry in each row is the number 4 that appears on the diagonal. In Chapter 10 we show the properties of $[L_N]$ are sufficient to imply that the system (1.3.2') has a unique solution $\mathbf{u}_N$ for each data vector $\mathbf{f}_N$. We conclude that if u is known at each of the *boundary* nodes of the grid covering D, then the equation (1.3.2) uniquely determines u at the *interior* nodes of the grid. This is a discrete version of a *well-posed boundary value problem*. Chapter 10 contains a more complete discussion of discrete boundary value problems. In that chapter we not only learn how to solve these problems, we will begin to see how the physical properties of the modeled system are reflected in the algebraic properties of the matrix $[L_N]$.

We conclude that if u is known at each of the boundary nodes of the grid covering D, Equation (1.3.2) uniquely determines u at the interior nodes of the grid. Then this is a discrete version of a *well-posed boundary value problem*. Section 10.3 contains additional information on discrete boundary-value problems.　　　　　　　　■ ■

Qualitative Properties of Solution

An interesting property of the solution to the discrete Laplace equation becomes evident when we write (1.3.2*) in the form

$$u(x, y) = \tfrac{1}{4}[u(x + \epsilon, y) + u(x - \epsilon, y) + u(x, y + \epsilon) + u(x, y - \epsilon)]$$

This says that at each node in D, $u(x, y)$ equals the arithmetic average of the values of u at the four nodes adjacent to the node (x, y). Then u at (x, y) cannot exceed the largest of these four values nor can it be smaller than the smallest of the four values. Since this equation applies at each of the nodes in the grid that covers D, it follows that at each interior node the value of u is not greater than the maximum u value attained at a boundary node nor is it smaller than the minimum u value attained at a boundary node. We state this result formally as a property that holds for any solution of the discrete Laplace equation.

Discrete Max–Min Principle.   Let $u(x, y)$ satisfy the discrete Laplace equation through-out the region D and let M and m denote, respectively, the maximum and minimum values attained by u at a boundary node of the grid covering D. Then

$$m \leq u(x, y) \leq M$$

at each interior node (x, y) in D.

One particularly significant consequence of the discrete max–min principle is that if $u(x, y)$ is zero at all the boundary nodes of a grid, then $u(x, y)$ must be zero at all interior nodes as well. It follows almost immediately then that if the discrete boundary-value problem we described has a solution, that solution is unique. (See Exercise 4 at the end of this section.) In Section 6.3 we consider analogues of the discrete max–min principle for solutions of the continuous Laplace equation.

Continuous Laplace Equation

The continuous version of (1.3.2) can be obtained by taking the limit in (1.3.2) as ϵ decreases to zero or by taking $u(x, y, z, t)$ in (1.2.11) to be independent of z and t. In any case, we have the continuous Laplace equation (in two variables)

$$u_{xx}(x, y) + u_{yy}(x, y) = 0 \tag{1.3.3}$$

Here we continue to use the following notation for partial derivatives:

$$u_{xx} = \frac{\partial^2 u}{\partial x^2}, \qquad u_{xy} = \frac{\partial^2 u}{\partial x\, \partial y}, \qquad u_{yy} = \frac{\partial^2 u}{\partial y^2}, \qquad \text{etc.}$$

We also use the notation

$$\nabla^2 u = u_{xx} + u_{yy}$$

for the so-called *Laplacian* of $u = u(x, y)$. Then the Laplace equation (1.3.3) can be written

$$\nabla^2 u(x, y) = 0 \quad \text{in } D$$

In the case of the discrete model we found that if the unknown function $u(x, y)$ is known on the boundary of the grid, $u(x, y)$ is uniquely determined by the discrete Laplace equation on the interior of the grid. Reasoning by analogy, we conjecture that if $u(x, y)$ satisfies the continuous version of Laplace's equation throughout the region D and if $u(x, y)$ is known over the entire boundary S of D, $u(x, y)$ is uniquely determined inside D. This is in fact correct as we demonstrate in subsequent chapters. Any problem in which an unknown function is to be determined throughout some region D from a partial differential equation used in conjuction with certain boundary information is known as a *boundary-value problem*. In the present example, a complete statement of the continuous version of the boundary-value problem reads as follows: Given $f(x, y)$ defined on S, the boundary of D, find $u = u(x, y)$ defined in D such that

$$\nabla^2 u(x, y) = 0 \quad \text{in } D$$
$$u = f \quad \text{on } S$$

Applying the methods of linear algebra to the discrete problem (1.3.2') leads to the following results:

Discrete Model

$$[L_N]\mathbf{u}_N = \mathbf{f}_N$$

Results
1. $\mathbf{f}_N = 0$ implies $\mathbf{u}_N = 0$.
2. For each $\mathbf{f}_N$ there exists a unique $\mathbf{u}_N$.
3. $[L_N]$ has real eigenvalues, all of one sign.
4. $[L_N]$ has a complete set of orthogonal eigenvectors.
5. $\mathbf{u}_N$ satisfies the discrete max–min principle.

Results 1–4 follow from the fact that $[L_N]$ is symmetric and positive definite.

Later in this book, using techniques of analysis, we uncover the following parallel properties for the continuous problem:

Continuous Model

$$\nabla^2 u(x, y) = 0 \quad \text{in } D$$
$$u = f \quad \text{on } S$$

Results
1. $f(x, y) = 0$ on S implies $u(x, y) = 0$ in D.
2. For each f there exists a unique $u = u(x, y)$.
3. Associated with the continuous problem is a set of eigenvalues, all real and of one sign.
4. Associated with the continuous problem is a set of eigenfunctions. These functions are orthogonal and form a complete set in a sense to be defined later.
5. $u(x, y)$ satisfies a continuous version of the discrete max–min principle.

Exercises: Steady-State Conduction of Heat

1. Suppose the region D is covered by the following rectangular grid:

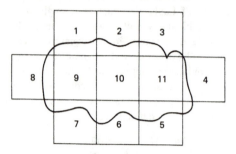

Suppose $u(x, y)$ satisfies the discrete Laplace equation on D and that u is given at the boundary nodes as

$$u_1 = u_2 = u_3 = 1, \qquad u_8 = u_4 = 2, \qquad u_5 = u_7 = 4, \qquad u_6 = 5$$

Then find $u(x, y)$ at the interior nodes 11, 10, and 9 and verify that $u(x, y)$ satisfies the discrete max–min principle.

2. If the value u_2 in Exercise 1 is changed from 1 to 3, use the discrete Laplace equation to compute the new values of u at the interior nodes. Do all the interior u values change, or just u_{10}, the one closest to u_2?

3. In Exercise 1, suppose that u is not specified over the entire boundary of D. In particular, suppose u_1 is not specified. Then at which interior nodes can u be determined using the discrete Laplace equation?

4. Suppose that two solutions are obtained for Exercise 1, that is, each solution is based on the boundary values $u_1, \ldots, u_8$ but two solutions are obtained for the values at the interior nodes. Denote these solutions by u_{11}, u_{10}, u_9 and v_{11}, v_{10}, v_9. Let $w_k = u_k - v_k$, $k = 1$, $\ldots, 11$, and note that $w_k = 0$ for $k = 1, \ldots, 8$ by definition. Use the discrete max–min principle to conclude that $w_k = 0$ for $k = 11, 10, 9$. Conclude from this that $u_k = v_k$ for $k = 11, 10, 9$, which implies that if Exercise 1 has a solution, this solution is the only solution.

5. Suppose the rectangular grid of Exercise 1 is replaced by the following grid:

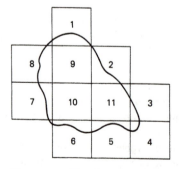

Then answer Exercises 1–4 as they pertain to this grid.

6. Reconsider Exercise 1 with the same conditions on boundary nodes 1–7 but where we impose the following condition at the boundary node 8: $u_8 = u_9$. This condition has the effect of making the x derivative of the temperature u equal to zero at the face shared by the cells centered at nodes 8 and 9. Since the heat flux through a cell face is proportional to the derivative of temperature in the direction normal to the face, this boundary condition is equivalent to supposing that there is no heat flow through the cell face between nodes 8 and 9; that is, the face is insulated. Solve for the temperatures at the three interior nodes under these conditions. How is the matrix $[L_N]$ affected by this new boundary condition?

7. Suppose D is the unit square $(0, 1) \times (0, 1)$. Then verify that the function $u(x, y) = xy$ satisfies the following boundary-value problem:

$$u_{xx}(x, y) + u_{yy}(x, y) = 0 \quad \text{in } D$$
$$u(x, 0) = 0, \quad u(x, 1) = x, \quad 0 < x < 1$$
$$u(0, y) = 0, \quad u(1, y) = y, \quad 0 < y < 1$$

At what point in D do the max and min values of $u(x, y)$ occur and what are they?

8. Verify that the functions

$$u_1(x, y) = \sin(\pi x)\sinh(\pi y), \quad u_2(x, y) = \sin(2\pi x)\sinh(2\pi y)$$

each satisfy all of the conditions

$$\nabla^2 u(x, y) = 0 \quad \text{in } D = (0, 1) \times (0, 1)$$
$$u(0, y) = u(1, y) = 0, \quad 0 < y < 1$$
$$u(x, 0) = 0, 0 < x < 1$$

Is this a well-posed problem for $u(x, y)$?

9. Verify that for that for all positive integers m, the functions

$$U_1(x, y) = \cos m\pi x \sinh m\pi y$$
$$U_2(x, y) = \sin m\pi x \cosh m\pi y$$
$$U_3(x, y) = \cos m\pi x \sinh m\pi(1 - y)$$
$$U_4(x, y) = \sin(m + \tfrac{1}{2})\pi x \sinh(m + \tfrac{1}{2})\pi y$$

each satisfy the equation $\nabla^2 U(x, y) = 0$ in R^2.

10. Each of the functions in the previous exercise satisfies at least one of the following boundary conditions:

(a) $U(0, y) = 0$ (b) $U(1, y) = 0$
(c) $U_x(0, y) = 0$ (d) $U_x(1, y) = 0$
(e) $U(x, 0) = 0$ (f) $U(x, 1) = 0$
(g) $U_y(x, 0) = 0$ (h) $U_y(x, 1) = 0$

Match each function to all of the boundary conditions it satisfies.

1.4 TRANSMISSION LINE EQUATIONS

We have seen, in Sections 1.2 and 1.3, two examples of partial differential equations and the auxiliary conditions required for each in order that there exists one and only one function that satisfies both the equation and the auxiliary conditions. We see in this section one more such example, intended to illustrate the role of the auxiliary conditions in the problem.

Consider a one-dimensional electrical conductor lying along the x axis:

$$\overline{\qquad\quad\underset{x_1}{\cdot}\underline{\qquad\quad}\underset{x_2}{\cdot}\underline{\qquad\quad}}$$

We assume that at each point x in the line and at each time t we have

$$i(x,\,t) = \text{current at } (x,\,t)$$
$$v(x,\,t) = \text{voltage at } (x,\,t)$$

It is our goal to compute for all times the current and the voltage at every point in the line.

The relevent physical properties of the transmission line are denoted,

$$R = \text{resistance} \qquad L = \text{inductance}$$
$$C = \text{capacitance} \qquad G = \text{leakage}$$

These properties are treated as constant distributed properties, which means that they are given in units per unit length.

If we consider an arbitrary segment $(x_1,\,x_2)$ of the line, we have four empirical principles that govern the interaction of current and voltage in the line:

(a) In measuring the voltage at two points in the line, we observe that there is a voltage loss that is proportional to the current level in the line. The proportionality factor is called the "resistance." There is an additional loss of voltage that is proportional to the time rate of change of the current in the line and here the proportionality factor is referred to as the "inductance." We express these relationships as follows:

$$R \int_{x_1}^{x_2} i(x,\,t)\,dx = \text{voltage drop from } x_1 \text{ to } x_2 \text{ due to resistance in the line}$$

$$L \int_{x_1}^{x_2} i_t(x,\,t)\,dx = \text{voltage drop from } x_1 \text{ to } x_2 \text{ due to inductance in the line}$$

(b) If we measure the current flow at two distinct points in the line, we find a current loss in an amount that is proportional to the voltage level in that part of the line. The proportionality factor is called the "leakage." There is an additional observable current loss that is proportional to the time rate of change of the voltage in this portion of the line, and we refer to the factor of proportionality in this equation as the "capacitance." These two relationships are expressed as follows:

$$C \int_{x_1}^{x_2} v_t(x,\,t)\,dx = \text{current needed to charge section of line from } x_1 \text{ to } x_2$$

$$G \int_{x_1}^{x_2} v(x,\,t)\,dx = \text{current lost due to leakage from section of line from } x_1 \text{ to } x_2$$

Conservation of charge requires that

$$i(x_1,\,t) - i(x_2,\,t) = \int_{x_1}^{x_2} Cv_t(x,\,t)\,dx + \int_{x_1}^{x_2} Gv(x,\,t)\,dx$$

That is, any decrease in the current flow between x_1, and x_2 must be due to the charging of or leakage from the section of line from x_1 to x_2.

We write

$$i(x_1, t) - i(x_2, t) = -\int_{x_1}^{x_2} i_x(x, t) \, dx$$

and hence the conservation-of-charge statement reduces to

$$\int_{x_1}^{x_2} [i_x + Cv_t + Gv] \, dx = 0 \tag{1.4.1a}$$

In addition to the principle of conservation of charge, we have an empirical principle known as Ohm's law, which states that any voltage loss in the line must be the result of the sum of resistance loss and inductance loss; that is,

$$v(x_1, t) - v(x_2, t) = \int_{x_1}^{x_2} Ri(x, t) \, dx + \int_{x_1}^{x_2} Li_t(x, t) \, dx$$

Then since

$$v(x_1, t) - v(x_2, t) = -\int_{x_1}^{x_2} v_x(x, t) \, dx$$

Ohm's law reduces to

$$\int_{x_1}^{x_2} [v_x + Li_t + Ri] \, dx = 0 \tag{1.4.1b}$$

Since (x_1, x_2) denotes a completely arbitrary segment of the transmission line and the quantities in brackets in (1.4.1) are assumed to be continuous functions of x and t, we conclude that

$$i_x(x, t) + Cv_t(x, t) + Gv(x, t) = 0 \tag{1.4.2a}$$
$$v_x(x, t) + Li_t(x, t) + Ri(x, t) = 0 \tag{1.4.2b}$$

These equations are satisfied at each x and all t and govern the behavior of the function $i(x, t)$, $v(x, t)$ describing the current and voltage in the line. The system of equations (1.4.2) is known as the transmission line equations. Further discussion of these equations can be found in Section 7.6.

Telegraph and Wave Equations

Note that if we form the combination

$$L \frac{\partial}{\partial t} [\text{Eq. (1.4.2a)}] - \frac{\partial}{\partial x} [\text{Eq. (1.4.2b)}]$$

equations (1.4.2a) and (1.4.2b) reduce to the single equation

$$LCv_{tt}(x, t) + [LG + RC]v_t(x, t) + RGv(x, t) - v_{xx}(x, t) = 0$$

Similarly, the combination

$$\frac{\partial}{\partial x} \text{[Eq. (1.4.2a)]} - C \frac{\partial}{\partial t} \text{[Eq. (1.4.2b)]}$$

leads to

$$LCi_{tt}(x, t) + [LG + RC]i_t(x, t) + RGi(x, t) - i_{xx}(x, t) = 0$$

Evidently, both $i(x, t)$ and $v(x, t)$ satisfy the so-called *telegraph equation*

$$LCu_{tt}(x, t) + [LG + RC]u_t(x, t) + RGu(x, t) - u_{xx}(x, t) = 0 \qquad (1.4.3)$$

The system of two first-order equations (1.4.2) and the single second-order equation (1.4.3) are equivalent continuous models for the transmission line. Equations (1.4.2) are called first-order equations because the highest order derivative that appears is of order 1. Equation (1.4.3) contains derivatives of order 2 (as well as of order 1) and so is called a second-order equation.

We consider the model in the form (1.4.3) and, for simplicity, we consider the special case $R = G = 0$. If $a^2 = 1/LC$, (1.4.3) reduces to a special case of the telegraph equation called the *wave equation*,

$$u_{tt}(x, t) - a^2 u_{xx}(x, t) = 0 \qquad (1.4.4)$$

This equation applies to both the current and the voltage in a line for which there is no leakage and no resistance. In order to discover what auxiliary conditions must accompany the wave equation, we now develop the discrete version of the wave equation. It will be easy to see what extra information is needed in order to determine the solution to the discrete problem, and this will provide some guidance in formulating the continuous problem.

Discrete Wave Equation

In the previous examples we derived the discrete model first and then passed to the continuous model by means of a limit process. In this example we start with the continuous model and make the transition to the discrete model by approximating the derivatives in Equation (1.4.4) by difference quotients. We hope in this way to emphasize the point that neither model should take precedence over the other. We can use conservation statements to derive either of the two models and then obtain the other by either discretizing or passing to the limit.

We use the following discrete approximations for the derivatives in (1.4.4):

$$u_{xx}(x, t) \approx [u(x + h, t) - 2u(x, t) + u(x - h, t)]/h^2$$
$$u_{tt}(x, t) \approx [u(x, t + \tau) - 2u(x, t) + u(x, t - \tau)]/\tau^2 \qquad (1.4.5)$$

Then using the notation,

$$u_n^k = u(nh, k\tau) \quad \text{for } n, k \text{ integers}$$

and replacing the derivatives in (1.4.4) by their discrete approximants from (1.4.5), we obtain the discrete wave equation

$$u_n^{k+1} = 2u_n^k - u_n^{k-1} + \alpha^2 [u_{n+1} - 2u_n + u_{n-1}]^k \qquad (1.4.6)$$

where

$$\alpha^2 = a^2\tau^2/h^2$$

We now show by example how this equation, together with certain additional information, can be used to determine the function $u(x, t)$ at the discrete points $(nh, k\tau)$.

EXAMPLE 1.4.1 ───

An Infinitely Long Line (Discrete Model)
   Suppose u_n^0 and u_n^1 are each known for all values of n. Then u_n^2 can be found for every n by applying (1.4.6) in the case $k = 1$. We can continue, finding u_n^k for $k = 3, 4, \ldots$, for all n (see Exercise 1). ■■

EXAMPLE 1.4.2 ───

A Line of Finite Length (Discrete Model)
   Suppose the line is of finite length L and that $Nh = L$. Suppose that u_n^0 and u_n^1 are each known for all values of n and in addition that

$$u_0^k = f(k\tau) = f^k \quad \text{and} \quad u_N^k = g(k\tau) = g^k$$

for $k = 0, 1, \ldots$, where f and g denote given functions. Then u can be computed for $n = 1, 2, \ldots, N - 1$ from (1.4.6) (see Exercise 4). ■■

   With respect to Example 1.4.1, note that both u_n^0 and u_n^1 are needed in order to compute u_n^2 from (1.4.6) since this equation expresses the values of u_n at level $k + 1$ in terms of the two previous t levels, namely, levels k and $k - 1$. Example 1.4.2 shows that when the line is of finite length, additional information is needed. In particular, knowing the values of u_n^k at the ends of the line for all values of k will allow the values of u inside the line to be computed for all $k > 1$ from u_n^0 and u_n^1 and (1.4.6).
   Specifying u_n^0 and u_n^1 for all values of n in Example 1.4.1 amounts to specifying two *initial conditions* on the solution. Two conditions are needed for the wave equation as opposed to only one initial condition required in connection with the conduction equation of the previous example. This is a consequence of the fact that the heat equation involves only a first-order derivative with respect to the time variable while the wave equation is second order in time. Roughly speaking, for each differentiation that appears in an equation, one auxiliary condition is required in order for the solution to the problem to be uniquely determined.

Auxiliary Conditions for Continuous Problem

Based on these examples for the discrete model, we make the following guesses about the auxiliary conditions required in the continuous model:

   (i) **An Infinitely Long Line** (Continuous Model). Suppose that $u(x, t)$ satisfies the wave equation (1.4.4) and, in addition, $u(x, 0)$ and $u_t(x, 0)$ are known for all values of x. Then $u(x, t)$ can be found for all values of x and all $t > 0$.
   (ii) **A Line of Finite Length** (Continuous Model). In a line of finite length L,

suppose that $u(x, t)$ satisfies (1.4.4) and that $u(x, 0)$ and $u_t(x, 0)$ are both known for $0 < x < L$. Suppose, in addition, that $u(0, t)$ and $u(L, t)$ are both specified for $t > 0$. Then $u(x, t)$ can be found for all $t > 0$ and all x, $0 < x < L$.

Problem (i) is referred to as a *pure initial-value problem* while (ii) is called a *mixed initial-boundary-value problem*. In Chapters 3 and 5 we construct solutions for these problems, and in Chapter 6 we show that the solutions we construct are the *only* solutions. This will then prove that the problems are well formulated.

Semidiscrete Wave Equation

If we discretize only the spatial derivatives in (1.4.4), we obtain the semidiscrete version of the wave equation,

$$u_n''(t) = \beta^2[u_{n+1}(t) - 2u_n(t) + u_{n-1}(t)] \tag{1.4.7}$$

where

$$\beta^2 = a^2/h^2$$

In matrix notation this becomes

$$\frac{d^2}{dt^2}\begin{bmatrix} u_1 \\ u_2 \\ \vdots \\ u_{N-1} \end{bmatrix} = \beta^2 \begin{bmatrix} -2 & 1 & 0 & \cdots \\ 1 & -2 & 1 & \cdots \\ & & \ddots & \cdots \\ & & 1 & -2 \end{bmatrix} \begin{bmatrix} u_1 \\ u_2 \\ \vdots \\ u_{N-1} \end{bmatrix} + \begin{bmatrix} F(t) \\ 0 \\ \vdots \\ G(t) \end{bmatrix}$$

Here we are assuming that the line is of finite length $L = Nh$, and

$$u_0(t) = F(t), \qquad u_N(t) = G(t)$$

are given. In the special case that $F(t) = G(t) = 0$, this equation is of the form

$$U''(t) = \beta^2[A]U(t) \tag{1.4.8}$$

where the matrix $[A]$ is the same one that appears in the semidiscrete model for the heat-conducting rod and $U(t)$ denotes the vector whose $N - 1$ entries are the functions $u_1(t), \ldots, u_{N-1}(t)$. The general solution for this second-order system of differential equations is

$$U(t) = W_1[C_1\exp[i\theta_1\beta t] + D_1\exp[-i\theta_1\beta t]] + \cdots$$
$$+ W_{N-1}[C_{N-1}\exp[i\theta_{N-1}\beta t] + D_{N-1}\exp[-i\theta_{N-1}\beta t]] \tag{1.4.9}$$

Here, $\alpha_1, \ldots, \alpha_{N-1}$ denote the $N - 1$ *negative* eigenvalues of the matrix $[A]$ and $-(\theta_j^2) = \alpha_j$ for $j = 1, \ldots, N - 1$. Also W_j for $j = 1, \ldots, N - 1$ denotes the $N - 1$ eigenvectors of $[A]$.

Since $\exp[i\theta t] = \cos(\theta t) + i\sin(\theta t)$, it appears that the solution to the semidiscrete wave equation is a periodic, (oscillating) function of time. We shall see later that the solution to the continuous wave equation exhibits this same evolutionary behavior, which is quite different from the evolutionary behavior of the solution to the conduction equation. This is one further illustration of the way in which the essential behavior of the physical system is contained in and can be discovered from each of the models: discrete, semidiscrete, and continuous.

Exercises: Transmission Line Model

Consider an infinitely long transmission line for which $R = G = 0$ and $LC = 1$. Use the discrete wave equation (1.4.6) in the case that $\tau = h$ to answer the following questions.

1. Suppose

$$u_n^0 = \begin{cases} 1 & \text{for } n = 10, \ldots, 15 \\ 0 & \text{for all other } n \end{cases}$$
$$u_n^1 = u_n^0, \quad \text{all } n$$

Then compute u_n^2 for $n = 0, \ldots, 25$.

2. What is the first value of k such that u_0^k is different from zero? Is u_0^k then different from zero for all larger values of k or is there some value of k where u_0^k again equals zero? If so, does u_0^k then remain zero for all larger values of k or does u_0^k become different from zero for some still larger value of k?

3. Continuing with Exercise 1, is there any positive integer k such that u_n^k is zero when $n = 10$ and when $n = 15$? Is there more than one such value for k? Is there any integer value of n such that u_n^k is different from zero for all positive integers k? Is there any integer value of n such that u_n^k is equal to zero for all positive integers k?

4. Consider Exercise 1 in the case of a transmission line of finite length $L = 20h = L$. Suppose

$$u_0^k = u_{20}^k = 0 \quad \text{for all } k$$

What is the smallest positive integer k such that u_5^k is different from zero? Is u_5^k different from zero for all larger integers k?

5. Suppose in Exercise 4 that only the initial conditions of Exercise 1 are imposed and that u_0^k and u_{20}^k are not specified for $k > 1$. Compute all the values of u_n^k that are computable from (1.4.6) and this data.

6. If the initial conditions in Exercise 1 are changed to

$$u_n^1 = 0 \quad \text{for all } n$$

how does this affect the answers to Exercises 1–5?

7. For all positive integers m, show that each of the functions

$$u_1(x, t) = \sin m\pi x \sin m\pi a t$$
$$u_2(x, t) = \sin m\pi x \cos m\pi a t$$
$$u_3(x, t) = \cos m\pi x \sin m\pi a t$$
$$u_4(x, t) = \cos m\pi x \cos m\pi a t$$
$$u_5(x, t) = \sin(m + \tfrac{1}{2})\pi x \cos(m + \tfrac{1}{2})\pi a t$$
$$u_6(x, t) = \cos(m + \tfrac{1}{2})\pi x \sin(m + \tfrac{1}{2})\pi a t$$

satisfies

$$u_n(x, t) = a^2 u_{xx}(x, t) \quad \text{for all } x, t$$

8. Each of the preceding functions satisfies at least one of the following boundary and initial conditions. Match *each* function with *all* of the conditions it satisfies.

(a) $u(0, t) = 0$ (b) $u(1, t) = 0$
(c) $u_x(0, t) = 0$ (d) $u_x(1, t) = 0$
(e) $u(x, 0) = 0$ (f) $u_t(x, 0) = 0$

1.5 WELL-POSED PROBLEMS

In the previous sections, we gave three examples of mathematical models for physical systems:

(i) time-dependent conduction of heat (heat equation),
(ii) steady-state conduction of heat (Laplace equation), and
(iii) transmission line model (wave equation).

In each case, the discrete model for the system was used to illustrate that in addition to a governing equation, it is necessary to impose certain auxiliary conditions on the unknown function in order for the overall problem to have a unique solution. This fact is best demonstrated in the setting of the discrete model where the governing equations are a system of algebraic equations and a lack of sufficient information shows up as a problem in which the number of unknowns exceeds the number of equations. Similarly, imposing too many auxiliary conditions will lead to a problem in which the number of equations exceeds the number of unknowns. In such a case, the problem will in general have no solution at all.

A similar situation exists with regard to the continuous model, although it is usually more difficult to demonstrate the interplay between the auxiliary conditions and the existence of a unique solution to the overall problem. The partial differential equation that governs the model generally will have infinitely many solutions, and in order to select from this infinite family of functions the one that describes a particular physical system, it is necessary to impose additional conditions on the solution. These additional conditions, called auxiliary conditions, serve to further characterize the system being modeled. Auxiliary conditions of two types will be considered: boundary and initial conditions.

Boundary Conditions

Boundary conditions are conditions that are satisfied by the solution of the partial differential equation at points on the boundary S of the region D where the equation is satisfied. The three common forms of boundary conditions are:

1. **Dirichlet Conditions.** The solution values are specified on S; for example, $u = g$ on S.
2. **Neumann Conditions.** The values of the directional derivative of the solution in the direction normal to S are specified on S; for example, $\partial_N u = g$ on S. Here $\partial_N u = \mathbf{grad}\ u \cdot \mathbf{N} = $ normal derivative of u where $\mathbf{N}$ denotes the unit outward normal to S.

3. Robin Conditions. A combination of the solution values and the values of the normal derivative of the solution are specified on S; for example, $a \, \partial_N u + bu = g$ on S.

Initial Conditions

Initial conditions are usually prescribed only in the case of evolution equations. These conditions are satisfied by the solution of the partial differential equation throughout the region D at the instant when consideration of the physical system commences. Typical initial conditions prescribe the value of the solution together with some of its time derivatives at the initial instant (which is generally labeled $t = 0$).

A problem in which an unknown function is required to satisfy a partial differential equation throughout some region D and to satisfy some boundary condition over the boundary S of D is called a *boundary-value problem*. The example in Section 1.3 involving steady-state conduction of heat led to a boundary-value problem.

A problem in which an unknown function is required to satisfy an evolution equation in an unbounded region and to satisfy some initial conditions throughout the region at $t = 0$ is called a *pure initial-value problem*. Modeling time-dependent heat conduction in an infinite rod or modeling an infinitely long transmission line leads to a pure initial-value problem. If the evolution equation is to be satisfied throughout a bounded region and if in addition to initial conditions there are conditions that the solution must satisfy on the boundary of the region, the problem is called a *mixed initial-boundary-value problem*. Problems of this type arose in the examples where we modeled heat conduction in a rod of finite length or transmission in a line of finite length.

In any problem, the prescribed functions appearing in the auxiliary conditions together with any coefficients and inhomogeneous terms in the partial differential equation are said to comprise the *data* in the problem. The solution of the problem is said to *depend continuously on the data* if small changes in the data produce correspondingly small changes in the solution. Then we say that a problem is *well posed* if

1. a solution exists,
2. the solution is unique, and
3. the solution depends continuously on the data.

If any of these conditions is not satisfied, the problem is said to be *ill-posed*.

The question of whether a given problem is well posed is generally difficult to answer. Roughly speaking, the following statements apply:

(a) The auxiliary conditions imposed must not be *too many* or the solution will not exist.

(b) The auxiliary conditions imposed must not be *too few* or the solution will not be unique.

(c) The auxiliary conditions must be of the "correct type" to go with the partial differential equation or the solution will not depend continuously on the data.

We shall not attempt to give a precise characterization of what constitutes a well-posed problem for a partial differential equation. However, in Chapter 6 we provide a list of several examples of well-posed problems of various kinds. Looking at that list now may help to give some idea of what is meant by the "correct" number and type of auxiliary conditions needed to form a well-posed problem. For example, it is clear from this list that for a problem on an unbounded domain the fact that the boundary is infinitely distant does not necessarily remove the need for some sort of condition on that boundary. For some problems on unbounded domains it is necessary to impose some sort of "behavior-at-infinity" condition in order for the problem to be well posed. In Chapters 3 and 5 we construct solutions to many problems in partial differential equations. All of the problems we solve in those chapters will be well-posed problems, and in Chapter 6 we prove the well-posedness for some of the examples.

1.6 CLASSIFICATION OF EQUATIONS

In the preceding sections we discussed various aspects of the following three partial differential equations:

(i) $u_{xx}(x, y) + u_{yy}(x, y) = 0$ (Laplace equation)
(ii) $u_{xx}(x, y) - u_y(x, y) = 0$ (heat equation) (1.6.1)
(iii) $u_{xx}(x, y) - u_{yy}(x, y) = 0$ (wave equation)

Each of these equations contains partial derivatives of order 2 and no derivatives of order higher than 2. For this reason we refer to these as *second-order partial differential equations*. In general, an equation containing derivatives of order N and no derivatives of order higher than N is referred to as a differential equation of order N or as an an Nth-order differential equation.

Each of the equations in (1.6.1) is an example of a *linear* differential equation. A partial differential equation is said to be *linear* if the equation is algebraically linear in the unknown function u and its derivatives. This means that the equation contains no terms involving products of u and/or its derivatives and the equation contains no terms where u or its derivatives appear as the argument of a function whose graph is not a straight line.

Each of the equations in (1.6.1) is a special case of the equation

$$a(x, y)u_{xx} + 2b(x, y)u_{xy} + c(x, y)u_{yy} + d(x, y)u_x + e(x, y)u_y + f(x, y)u = 0 \quad (1.6.2)$$

This is the most general linear partial differential equation of order 2 in two independent variables. It is interesting that at least in the special case of constant coefficients, the character of the solution of Equation (1.6.2) is, to a large extent, determined by the leading terms in the equation, that is, by the terms containing the derivatives of highest order.

This part of the equation (we refer to it as the *principal part* of the equation) can be written as follows in matrix notation:

$$a \, \partial_{xx}u + 2b \, \partial_{xy}u + c \, \partial_{yy}u = (\partial_x, \partial_y) \begin{bmatrix} a & b \\ b & c \end{bmatrix} \begin{bmatrix} \partial_x \\ \partial_y \end{bmatrix} [u]$$

$$= \partial^T[A] \, \partial[u]; \qquad \partial^T = (\partial/\partial x, \partial/\partial y) \quad (1.6.3)$$

Equation (1.6.2) can be characterized in terms of the algebraic properties of the matrix [A] in the representation (1.6.3) for the principal part of Equation (1.6.2). In particular, we classify equations of the form (1.6.2) in terms of the eigenvalues of the matrix [A]. This is not the only way in which the classification may be carried out, but it is the one that generalizes most naturally to the case of equations involving more than two independent variables. In Chapter 7 we discuss the classification of systems of first-order partial differential equations.

Since the matrix [A] is symmetric, its two eigenvalues are both real. In fact, they are the two roots of the equation

$$\det[A - \mu I] = \mu^2 - (a + c)\mu - (b^2 - ac) = 0 \qquad (1.6.4)$$

Then it is not difficult to show (see Exercise 2) that at each point the character of the eigenvalues is controlled by the sign of the *discriminant* $d(x, y) = (b^2 - ac)(x, y)$. That is, at a given (fixed) point (x, y), the following holds:

1. If $b^2 - ac < 0$, the two roots of (1.6.4) are of the same sign. In this case we say Equation (1.6.2) is of *elliptic type* at (x, y).
2. If $b^2 - ac = 0$, one of the two roots of (1.6.4) is zero and (1.6.2) is said to be of *parabolic type* at (x, y).
3. If $b^2 - ac > 0$, one of the roots of (1.6.4) is negative and the other is positive. We say that (1.6.2) is of *hyperbolic type* at (x, y).

Since exactly one of these three alternatives must apply, it follows that every partial differential equation of the form (1.6.2) can be classified at each point (x, y) as one of the three types: elliptic, parabolic, or hyperbolic. In particular, in (1.6.1) (i), (ii), and (iii), $b^2 - ac$ is -1, 0, and $+1$, respectively, so that Laplace's equation, the heat equation, and the wave equation are, respectively, of elliptic, parabolic, and hyperbolic types. Since these equations have constant coefficients, these classifications do not vary from point to point. On the other hand, the equation

$$yu_{xx}(x, y) + u_{yy}(x, y) = 0$$

is of elliptic, parabolic, or hyperbolic type at (x, y) according to whether y is positive, zero, or negative.

In the next several chapters we develop methods for constructing the solution to various well-posed problems for partial differential equations. You will notice that most problems involve one or another of the equations in (1.6.1), and it may seem therefore that we are taking an unnecessarily narrow view in our selection of examples. In fact, there is no loss in generality in considering only the equations in (1.6.1) instead of considering other cases of Equation (1.6.2). We can show that any equation of the form (1.6.2) can be transformed by a change of independent variable into an equation whose principal part is one of the three "canonical forms" appearing in (1.6.1).

This rather striking result suggests that the canonical forms in (1.6.1) may be viewed as the "prototypes" of the second-order equations of elliptic, parabolic, and hyperbolic type, respectively. We mean by this that what is found to be true about solutions to Laplace's equation is liable to hold true for solutions to elliptic partial differential equations in general. Similar statements can be made about the heat and wave equations with respect to parabolic and hyperbolic equations. Thus, by studying the three equations in (1.6.1), we discover the properties of solutions to all equations of the form (1.6.2).

Exercises: Classification

1. Determine the set of points (x, y) where each of the following equations is elliptic, parabolic, and hyperbolic:
   (a) $x^2 u_{xx}(x, y) - y^2 u_{yy}(x, y) = 0$
   (b) $xy u_{xy}(x, y) - u_y(x, y) + u = 0$
   (c) $2u_{xx}(x, y) - 4u_{xy}(x, y) + u_x(x, y) = 0$
   (d) $2y u_{xx}(x, y) + (x + y)u_{xy}(x, y) + 2xu_{yy}(x, y) = 0$
   (e) $\sin(xy)u_{xy}(x, y) = 0$

2. Consider the equation

$$F(\mu) = \mu^2 - (a + c)\mu - (b^2 - ac) = 0$$

Show that the following hold:
   (a) The graph of $F(\mu) = 0$ is a parabola opening upward.
   (b) If $b^2 - ac < 0$, then $F(\mu) = 0$ has two real roots of the same sign. In this case, if $(a + c) > 0$, then the roots are both negative, and if $(a + c) < 0$, they are both positive.
   (c) If $b^2 - ac = 0$, then $\mu = 0$ is a root of $F(\mu) = 0$. In this case the other root is positive or negative according to whether $a + c$ is negative or positive.
   (d) If $b^2 - ac > 0$, then $F(\mu) = 0$ has distinct roots of opposite sign.

3. Which of the following differential operators is linear?
   (a) $L[u] = u_{xx}(x, y) + x^2 u_x(x, y) - u_{yy}(x, y)$
   (b) $L[u] = u_{xx}(x, y) - u(x, y)u_x(x, y) + u_{yy}(x, y)$
   (c) $L[u] = u_y(x, y)u_x(x, y) - u_{yy}(x, y)$
   (d) $L[u] = \sin(xy)u_{xx}(x, y) - \cos(xy)u_{xy}(x, y)$

Chapter 2

Fourier Series and Eigenfunction Expansions

INTRODUCTION

In this chapter we introduce the first of the two principal tools we shall be using to solve problems in partial differential equations: Fourier series, or more generally, eigenfunction expansions.

It is easy to check that for $n = 1, 2, \ldots, N$ and $\mu_n = n\pi/L$, the function

$$u_n(x, t) = \exp[-\mu_n^2 t]\sin(\mu_n x)$$

satisfies

$$u_t(x, t) = u_{xx}(x, t), \qquad 0 < x < L, t > 0$$
$$u(0, t) = u(L, t) = 0$$

It is easily checked that for arbitrary constants $C_1, \ldots, C_N$ the function $U(x, t)$ given by

$$U(x, t) = \sum_{n=1}^{N} C_n u_n(x, t)$$

satisfies the partial differential equation and boundary conditions above. Clearly, at $t = 0$ $U(x, t)$ reduces to

$$U(x, 0) = \sum_{n=1}^{N} C_n \sin(\mu_n x)$$

and hence in order to satisfy an initial condition of the form

$$U(x, 0) = F(x), \qquad 0 < x < L$$

it is necessary to be able to find values for the constants C_n such that

$$\sum_{n=1}^{N} C_n \sin(\mu_n x) = F(x), \qquad 0 < x < L$$

If such constants can be found, then $U(x, t)$ is seen to be the solution to the initial-boundary-value problem consisting of the partial differential equation together with the inhomogeneous initial condition and the two homogeneous boundary conditions. We show in Chapter 6 that if this problem has *any* solution, then that is the *only* solution. Evidently, we will have constructed the unique solution to this initial-boundary-value problem once the appropriate constants C_n are found.

For an arbitrary function $F(x)$ it is not, in general, possible for any *finite* value of N to find constants C_n such that $U(x, 0) = F(x)$. However, when N is allowed to become infinite, then for a large class of functions $F(x)$, the constants C_n can be chosen so that the (now infinite) series not only converges but converges to $F(x)$.

Although other mathematicians, including Euler, were working in the area as early as 1750, this discovery is generally attributed to the previously mentioned Joseph Fourier, and series of the sort to be discussed in this chapter are called Fourier series. In a classic work published about 1820, *Theorie Analytique de la Chaleur*, Fourier presented the mathematical techniques on which this chapter and the next are based. Although his ideas were essentially correct, they were not rigorously set forth, and Fourier had considerable difficulty in convincing other mathematicians of his time that it is possible to represent an arbitrary function as a series of periodic functions. Part of the confusion was no doubt a result of confusion about the precise meaning of the term *function*. In Sections 2.1 and 2.2 we introduce the notion of a *square integrable function*, and we shall see that this interpretation of *function* is particularly suitable for discussing Fourier series. A more substantial discussion of the space of square integrable functions is the subject of Section 2.5.

In section 2.2 we shall see that a Fourier series in terms of the sine and cosine functions is only a special case of a more general series expansion. Section 2.3 explains how to find families of functions that are suitable for use in these *generalized Fourier expansions*. Section 2.4 is a digression into the subject of *discrete Fourier series*, which is of interest in connection with the second part of the book, where discrete solution methods are discussed. Finally, in Section 2.6 we briefly describe the use of Fourier expansions for functions of several variables.

2.1 FOURIER SERIES

Let $f(x)$ denote an arbitrary function of x defined on the interval $(-L, L)$ and write

$$f(x) \sim \tfrac{1}{2}a_0 + \sum_{n=1}^{\infty} a_n \cos \frac{n\pi x}{L} + b_n \sin \frac{n\pi x}{L} \tag{2.1.1}$$

A series of this form, where the constants a_n and b_n are yet to be determined, is called a trigonometric infinite series. The series may or may not converge, and for those values of x where it does converge, it may or may not converge to the value $f(x)$. Of course, for the series to converge to $f(x)$, it is necessary that the constants a_n and b_n depend in some way on the function $f(x)$. It will be our first task to show that (2.1.5) below expresses the necessary dependence of the constants a_n and b_n on $f(x)$.

Elementary integration formulas show that for arbitrary integers m, n

$$\int_{-L}^{L} \cos \frac{m\pi x}{L} \cos \frac{n\pi x}{L} \, dx = \begin{cases} 0, & m \neq n \\ L, & m = n > 0 \\ 2L, & m = n = 0 \end{cases}$$

$$\int_{-L}^{L} \sin \frac{m\pi x}{L} \sin \frac{n\pi x}{L} \, dx = \begin{cases} 0, & m \neq n \\ L, & m = n > 0 \end{cases}$$

$$\int_{-L}^{L} \sin \frac{n\pi x}{L} \cos \frac{m\pi x}{L} \, dx = 0, \quad \text{all integers } m, n \tag{2.1.2}$$

The set of relations (2.1.2) are referred to as *orthogonality relations*, and the family of functions

$$\{1, \cos \pi x/L, \cos 2\pi x/L, \ldots; \sin \pi x/L, \sin 2\pi x/L, \ldots\}$$

is called an *orthogonal family* on the interval $(-L, L)$.

Proceeding formally, if we multiply both sides of (2.1.1) by $\cos M\pi x/L$ for a fixed integer M and then integrate from $-L$ to L, (2.1.1) becomes

$$\left(f, \cos \frac{M\pi x}{L} \right) \sim \tfrac{1}{2} a_0 \left(1, \cos \frac{M\pi x}{L} \right) + \sum_{n=1}^{\infty} a_n \left(\cos \frac{n\pi x}{L}, \cos \frac{M\pi x}{L} \right)$$

$$+ \sum_{n=1}^{\infty} b_n \left(\sin \frac{n\pi x}{L}, \cos \frac{M\pi x}{L} \right) \tag{2.1.3}$$

Here we are using the notation

$$(f, g) = \int_{-L}^{L} f(x)g(x) \, dx \tag{2.1.4}$$

for arbitrary functions $f(x)$ and $g(x)$ defined on $(-L, L)$. It follows from the orthogonality relations (2.1.2) that for every $n \neq M$,

$$(\sin n\pi x/L, \cos M\pi x/L) = 0$$
$$(\cos n\pi x/L, \cos M\pi x/L) = 0$$

Note that for $n = 0$, $\cos n\pi x/L = 1$.

Then if (2.1.1) is to be an equality, it must be the case that for each integer $M = 0, 1, \ldots$,

$$(f, \cos M\pi x/L) = a_M L$$

That is,

$$a_M = 1/L(f, \cos M\pi x/L) \quad \text{for } M = 0, 1, \ldots$$

Similarly, multiplying on both sides of (2.1.1) by $\sin M\pi x/L$ and integrating from $-L$ to L leads to

$$\left(f, \sin \frac{M\pi x}{L}\right) \sim \tfrac{1}{2}a_0 \left(1, \sin \frac{M\pi x}{L}\right) + \sum_{n=1}^{\infty} a_n \left(\cos \frac{n\pi x}{L}, \sin \frac{M\pi x}{L}\right)$$
$$+ \sum_{n=1}^{\infty} b_n \left(\sin \frac{n\pi x}{L}, \sin \frac{M\pi x}{L}\right)$$

Then the orthogonality relations (2.1.2) imply that in order for (2.1.1) to be an equation, we must have

$$(f, \sin M\pi x/L) = b_M L \quad \text{for each } M > 0$$

Evidently, a necessary condition for (2.1.1) to be an equality is that the coefficients a_n and b_n in the series are related to $f(x)$ as follows:

$$a_n = \frac{1}{L} \int_{-L}^{L} f(x)\cos \frac{n\pi x}{L}\, dx, \qquad n = 0, 1, \ldots$$

$$b_n = \frac{1}{L} \int_{-L}^{L} f(x)\sin \frac{n\pi x}{L}\, dx, \qquad n = 1, 2, \ldots \qquad (2.1.5)$$

When the coefficients a_n and b_n in (2.1.1) are given by (2.1.5), they are called the *Fourier coefficients* for the function $f(x)$ and the series (2.1.1) is called the *Fourier series* for the function $f(x)$.

Up to this point we have shown only that *if* the series (2.1.1) is convergent to $f(x)$, then the coefficients a_n and b_n in the series *must* be given by (2.1.5). Of course, it was to be expected that the coefficients would depend on $f(x)$; (2.1.5) shows explicitly *how* a_n and b_n are determined from $f(x)$. Our next task will be to find conditions sufficient to ensure convergence of the series (2.1.1) with coefficients a_n and b_n given by (2.1.5). This development may be more meaningful if we first compute the coefficients a_n and b_n in a few simple examples and examine the resulting Fourier series. For convenience in these examples, we take $L = \pi$.

EXAMPLE 2.1.1 ──

Consider

$$f(x) = \begin{cases} 0 & \text{if } -\pi < x < 0 \\ 1 & \text{if } 0 < x < \pi \end{cases} \qquad (2.1.6)$$

Note that *no value* has been specified for $f(x)$ at the three points $x = -\pi, 0, \pi$. Since the Fourier coefficients are calculated by integrating $f(x)$, this will have no effect on the Fourier coefficients (changing the value of the integrand at finitely many points does not affect the value of an integral), but it suggests an interesting point: Functions that differ at finitely many points have the same Fourier coefficients. We return to this point later.
For $n = 0, 1, \ldots$

$$a_n = \frac{1}{\pi} \int_{-\pi}^{\pi} f(x)\cos(nx)\, dx = \frac{1}{\pi} \int_{0}^{\pi} \cos(nx)\, dx$$

After integrating, this reduces to

$$a_0 = 1$$
$$a_n = 1/(n\pi)\sin(nx)\big|_0^\pi = 0 \quad \text{for } n = 1, 2, \ldots$$

Note that the integration formula used to compute a_n in the case $n = 1, 2, \ldots$ was not the same one used to compute a_0.

Similarly,

$$b_n = \frac{1}{\pi} \int_0^\pi \sin(nx)\, dx \quad \text{for } n = 1, 2, \ldots$$

$$= \left. \frac{-1}{n\pi} \cos(nx) \right|_0^\pi = \frac{1 - \cos(n\pi)}{n\pi}$$

Note that $\cos(n\pi) = (-1)^n$ for $n = $ integer. Then

$$b_n = \begin{cases} 0 & \text{if } n = \text{even integer} \\ 2/n\pi & \text{if } n = \text{odd integer} \end{cases}$$

This produces the following Fourier series for $f(x)$:

$$f(x) \sim \frac{1}{2} + \frac{2}{\pi} \sin x + \frac{2}{3\pi} \sin 3x + \frac{2}{5\pi} \sin 5x + \cdots$$

$$\sim \frac{1}{2} + \frac{2}{\pi} \sum_{m=1}^\infty \frac{\sin[(2m-1)x]}{2m-1} \tag{2.1.7}$$

If we define the partial sums of this series by

$$S_0(x) = \tfrac{1}{2}$$
$$S_1(x) = \tfrac{1}{2} + (2/\pi)\sin x$$
$$S_2(x) = \tfrac{1}{2} + (2/\pi)\sin x + (2/3\pi)\sin 3x$$

$$\vdots$$

then plotting $S_0, S_1, S_2. \ldots$ versus x shows that with each additional term, the graphs of the partial sums of this Fourier series draw closer to the graph of $f(x)$ (Figure 2.1.1). Note in particular that each of these three partial sums assumes the value $\tfrac{1}{2}$ at each of the points $x = -\pi, 0, \pi$ where $f(x)$ was left undefined. ∎ ∎

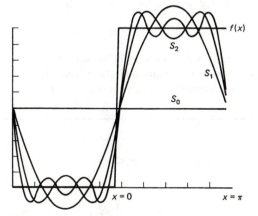

Figure 2.1.1

EXAMPLE 2.1.2 _____

Consider the function

$$f(x) = x, \qquad -\pi < x < \pi$$

Using (2.1.2), we compute, for $n = 1, 2, \ldots$,

$$a_n = \frac{1}{\pi} \int_{-\pi}^{\pi} x \cos(nx) \, dx = \frac{1}{\pi} \left[\frac{x \sin nx}{n} - \frac{(-\cos nx)}{n^2} \right] \Big|_{-\pi}^{\pi}$$
$$= 0$$

and for $n = 0$

$$a_0 = \int_{-\pi}^{\pi} x \, dx = 0$$

Similarly, for $n = 1, 2, \ldots$

$$b_n = \frac{1}{\pi} \int_{-\pi}^{\pi} x \sin(nx) \, dx = \frac{1}{\pi} \left[\frac{-x \cos nx}{n} + \frac{\sin nx}{n^2} \right] \Big|_{-\pi}^{\pi}$$
$$= \frac{-2 \cos(n\pi)}{n} = \frac{2(-1)^{n+1}}{n}$$

Then the Fourier series for $f(x)$ in this example is

$$f(x) \sim 2 \sin x - \sin 2x + \frac{2}{3} \sin 3x - \frac{2}{4} \sin 4x \ldots$$
$$\sim 2 \sum_{n=1}^{\infty} \frac{(-1)^{n+1}}{n} \sin nx \tag{2.1.8}$$

If we define

$$S_1(x) = 2 \sin x$$
$$S_2(x) = 2 \sin x - \sin 2x$$
$$S_3(x) = 2 \sin x - \sin 2x + \tfrac{2}{3} \sin 3x$$
$$\vdots$$

then plotting $S_1, S_2, S_3 \ldots$ versus x shows that including additional terms of the Fourier series improves the approximation to the function $f(x)$ (Figure 2.1.2). Note also that each of the partial sums $S_k(x)$ assumes the value zero at $x = -\pi, \pi$. ■ ■

EXAMPLE 2.1.3 _____

Consider the function

$$f(x) = |x|, \qquad -\pi < x < \pi$$

Then

$$a_0 = \frac{1}{\pi} \int_{-\pi}^{0} -x \, dx + \frac{1}{\pi} \int_{0}^{\pi} x \, dx = \pi$$

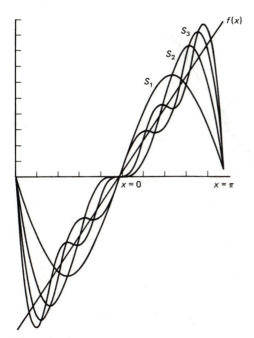

Figure 2.1.2

and for $n = 1, 2, \ldots$

$$a_n = \frac{1}{\pi} \int_{-\pi}^{0} - x \cos nx \, dx + \frac{1}{\pi} \int_{0}^{\pi} x \cos nx \, dx$$

$$= -\frac{1}{\pi} \left[\frac{x \sin nx}{n} + \frac{\cos nx}{n^2} \right] \Bigg|_{-\pi}^{0}$$

$$+ \frac{1}{\pi} \left[\frac{x \sin nx}{n} + \frac{\cos nx}{n^2} \right] \Bigg|_{0}^{\pi}$$

$$= \frac{2}{\pi} \frac{(\cos n\pi - 1)}{n^2}$$

We also have, for $n = 1, 2, \ldots$,

$$b_n = \frac{1}{\pi} \int_{-\pi}^{0} - x \sin nx \, dx + \frac{1}{\pi} \int_{0}^{\pi} x \sin nx \, dx$$

$$= \frac{1}{\pi} \left[\frac{x \cos nx}{n} - \frac{\sin nx}{n^2} \right] \Bigg|_{-\pi}^{0}$$

$$+ \frac{1}{\pi} \left[\frac{-x \cos nx}{n} + \frac{\sin nx}{n^2} \right] \Bigg|_{0}^{\pi} = 0$$

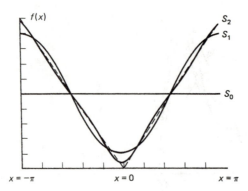

Figure 2.1.3

Then the Fourier series for $f(x)$ is

$$f(x) = \frac{\pi}{2} - \frac{4}{\pi} \cos x - \frac{4}{9\pi} \cos 3x - \cdots$$

$$= \frac{\pi}{2} - \frac{4}{\pi} \sum_{m=1}^{\infty} [2m - 1]^{-2} \cos(2m - 1)x \qquad (2.1.9)$$

In this case, the partial sums of the Fourier series,

$$S_0(x) = \pi/2$$
$$S_1(x) = \pi/2 - (4/\pi)\cos x$$
$$S_2(x) = \pi/2 - (4/\pi)\cos x - (4/9\pi)\cos 3x$$

approach $f(x)$ "uniformly." That is, they have the property that for $k = 1, 2, \ldots$ the maximum difference $|S_k(x) - f(x)|$ for x between $-\pi$ and π is a decreasing function of k (see Figure 2.1.3) ■ ■

We now begin to develop the notions that will allow us to precisely describe and understand some of the things we have noticed about these examples.

Periodic Functions and Periodic Extensions

A function $f(x)$ defined for all values of x is said to be periodic with period P if

$$f(x) = f(x + P) \quad \text{for all } x \qquad (2.1.10)$$

For example:

$\sin x$ and $\cos x$ are periodic with period 2π.

$\sin nx$ and $\cos nx$ are periodic with period 2π for all integers n.

$\sin(n\pi x/L)$ and $\cos(n\pi x/L)$ are periodic with period $2L$ for all integers n.

Any constant function is periodic with period P for every value P.

Note that if $f(x)$ is periodic with period P, then

$$f(x) = f(x + P) = f(x + 2P) = \cdots \quad \text{for all } x$$

Evidently, if $f(x)$ is periodic with period P, then $f(x)$ is also periodic with period Q for Q equal to any integer multiple of P. If it is necessary to speak of "the period" of $f(x)$ unambiguously, then the period of $f(x)$ is defined to be the *smallest* positive number P such that (2.1.10) holds.

For example, we have said that for each integer n the function $\sin nx$ is periodic with period 2π. This is so, but the smallest number P such that (2.1.10) holds for $f(x) = \sin nx$ is $P = 2\pi/n$. Thus,

$\sin x$ is periodic with period 2π.

$\sin 2x$ is periodic with period π.

$\sin 3x$ is a periodic with period $2\pi/3$.

And so on.

However, it is still correct to say that each of these functions is periodic with period 2π since (2.1.10) holds for each of these functions with $P = 2\pi$.

If $f_1(x), f_2(x), \ldots, f_N(x)$ are all periodic functions having a common period P, then for any integer N, the function

$$F(x) = \sum_{n=1}^{N} f_n(x)$$

is also periodic with period P. In particular, each of the partial sums $S_N(x)$ associated with the Fourier series (2.1.7), (2.1.8), and (2.1.9) is periodic with period 2π. Clearly, this period is determined by the first nonconstant term in the Fourier series.

A function $f(x)$ defined only on $(-L, L)$ can be extended to $(-\infty, \infty)$ as a periodic function of period $2L$ in the following way. For each number z, $-\infty < z < \infty$, there is a unique integer M such that $-L \le z - 2ML \le L$. Then we define $\tilde{f}(x)$ by

$$\tilde{f}(z) = f(z - 2ML) \quad \text{for } -\infty < z < \infty \tag{2.1.11}$$

The function $\tilde{f}(x)$ is everywhere defined and is periodic with period $2L$, and $\tilde{f}(x)$ agrees with $f(x)$ on $(-L, L)$. The function $\tilde{f}(x)$ is called the *2L-periodic extension of* $f(x)$.

The functions of Examples 2.1.1, 2.1.2, and 2.1.3 are each defined on $(-\pi, \pi)$. The following sketches show the 2π-periodic extensions of each of these functions on the interval $(-2\pi, 2\pi)$:

Example 2.1.1:

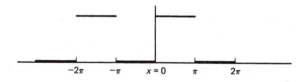

Example 2.1.2:

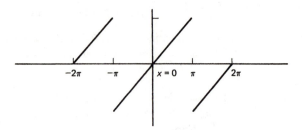

Example 2.1.3:

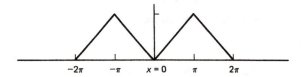

Note the following facts about these extensions:

Example 2.1.1: the function $f(x)$ is not continuous on $(-\pi, \pi)$ and hence $\tilde{f}(x)$ is not continuous on $(-\infty, \infty)$.

Example 2.1.2: $f(x)$ is continuous on $(-\pi, \pi)$ but $\tilde{f}(x)$ is not continuous on $(-\infty, \infty)$.

Example 2.1.3: $f(x)$ is continuous and its extension is continuous. From these examples it is clear that the conditions under which the periodic extension is continuous are as follows:

Lemma 2.1.1. $\tilde{f}(x)$ is continuous on $(-\infty, \infty)$ if and only if

(a) $f(x)$ is continuous on $[-L, L]$ and
(b) $f(-L) = f(L)$.

There is something else we should notice about the examples. The Fourier series (2.1.8) contains only sine terms while the series (2.1.9) contains only cosine terms. To see why this is the case, note that for each integer n,

$\cos n\pi x/L$ satisfies

$$f(x) = f(-x) \quad \text{for all } x \tag{2.1.12}$$

$\sin n\pi x/L$ satisfies

$$f(x) = -f(-x) \quad \text{for all } x \tag{2.1.13}$$

Any function having property (2.1.12) is said to be an *even function of x* while any function having property (2.1.13) is said to be an *odd function of x*. Most functions are neither even nor odd, but *any function* can be written as the sum of an even and an odd function as follows:

$$f(x) = f_E(x) + f_0(x) \tag{2.1.14}$$

where

$$f_E(x) = \tfrac{1}{2}[f(x) + f(-x)] = f_E(-x)$$

denotes the "even part of $f(x)$" and

$$f_0(x) = \tfrac{1}{2}[f(x) - f(-x)] = -f_0(-x)$$

denotes the "odd part of $f(x)$."

If

$$f(x) = \tfrac{1}{2}a_0 + \sum_{n=1}^{\infty} a_n\cos\frac{n\pi x}{L} + b_n\sin\frac{n\pi x}{L}$$

then since $\cos n\pi x/L$ is *even* for $n = 0, 1, \ldots$ and $\sin n\pi x/L$ is *odd* for $n = 1, 2, \ldots$, it follows from (2.1.14) that

$$f_E(x) = \tfrac{1}{2}a_0 + \sum_{n=1}^{\infty} a_n\cos\frac{n\pi x}{L}$$

$$f_0(x) = \sum_{n=1}^{\infty} b_n\sin\frac{n\pi x}{L}$$

If $f(x)$ is known to be an even function on $(-L, L)$, then $f_0(x)$, the odd part of $f(x)$, is identically zero and hence $b_n = 0$ for every n. This is the case in Example 2.1.3. On the other hand, if $f(x)$ is known to be an odd function on $(-L, L)$, as in Example 2.1.2, then the even part of $f(x)$, $f_E(x)$, is identically zero and a_n vanishes for $n = 0, 1, \ldots$. When $f(x)$ is neither even nor odd, as in Example 2.1.1, the Fourier series for $f(x)$ will have both sine and cosine terms in it.

For every *even* function, not only can we anticipate that b_n is zero for $n = 1, 2, \ldots$ (without having to calculate the b_n's) but in addition it is the case that

$$a_n = \frac{2}{L}\int_0^L f(x)\cos\frac{n\pi x}{L}\, dx, \qquad n = 0, 1, \ldots \tag{2.1.15}$$

Likewise, for every *odd* function, in addition to anticipating that $a_n = 0$ for $n = 0, 1, \ldots$, we have

$$b_n = \frac{2}{L}\int_0^L f(x)\sin\frac{n\pi x}{L}\, dx, \qquad n = 1, 2, \ldots \tag{2.1.16}$$

Moreover, formulas (2.1.15) and (2.1.16) suggest the following additional consequence of these observations. Suppose we are given a function $f(x)$ defined only on the "half-interval" $(0, L)$. We have then two alternatives:

(a) We can *define* $b_n = 0$ for $n = 1, 2, \ldots$ and compute the Fourier coefficients a_n for $f(x)$ from (2.1.15). Then the Fourier series for $f(x)$ will have only cosine terms in it and must therefore represent an even function. We refer to this as the *half-range Fourier cosine series for f(x)*.

(b) We can *define $a_n = 0$* for $n = 0, 1, 2, \ldots$ and compute the Fourier coefficients b_n for $f(x)$ from (2.1.16). In this case the Fourier series for $f(x)$ will have only sine terms in it and must then represent an odd function. We refer to this as the *half-range Fourier sine series for f(x)*.

Alternative (a) is equivalent to extending the function $f(x)$ to the interval $(-L, 0)$ as an *even function* and then computing the Fourier coefficients in the usual way from (2.1.6). Alternative (b) is equivalent to extending $f(x)$ to the interval $(-L, 0)$ as an *odd function*. We refer to these extensions as the even and odd extensions of $f(x)$. If these extensions are themselves extended to the whole real line as $2L$-periodic functions, we obtain what we call the even $2L$-periodic and odd $2L$-periodic extensions for $f(x)$. In the next section we establish the fact that the half-range Fourier cosine–sine series for a function converge respectively to the even–odd $2L$-periodic extension of that function. In the next chapter we shall see some applications of half-range series.

We have observed that if $f(x)$ defined on $(-L, L)$ is symmetric about the y axis, then the Fourier series for $f(x)$ contains no sine terms, and if $f(x)$ is symmetric about the origin, then the Fourier series contains no cosine terms. Other types of symmetry cause other terms to be absent from the Fourier series for $f(x)$. We shall not pursue this here, but some of the problems at the end of this section explore the connection between symmetries and the Fourier coefficients.

One point we do wish to make note of, however, is that the initial term in any Fourier series is the term

$$\tfrac{1}{2}a_0 = \frac{1}{2L} \int_{-L}^{L} f(x) \, dx$$

Clearly this is just the average value of the function $f(x)$ on the interval $(-L, L)$. Then we can think of the Fourier series for $f(x)$ as composed of the (constant) average-value component plus all the periodic components that oscillate about this average value with amplitudes equal to the Fourier coefficients.

Convergence of Fourier Series

We begin our discussion of convergence for Fourier series by recalling a few facts about convergence of infinite series of functions. Consider then an infinite family of functions $u_1(x), u_2(x), \ldots$, all defined on a common interval I. Associated with this family of functions consider the infinite series

$$\sum_{k=1}^{\infty} u_k(x) \tag{2.1.17}$$

If each of the functions $u_k(x)$ is of the form $u_k(x) = a_k(x - x_0)^k$, then the infinite series is called a *power series*. If the terms $u_k(x)$ are all trigonometric functions, then the series is called a *trigonometric series*. Fourier series are a special type of trigonometric series.

The series (2.1.17) defines a function $S(x)$ that we refer to as the "sum" of the series. The domain of $S(x)$ is the set of points x in I where the series converges and the value of $S(x)$ at a point of convergence is the value to which the series converges at the point. Of course, this set of points may be empty, but as we shall see, under certain conditions on the functions $u_k(x)$ in (2.1.17), the domain of $S(x)$ is not empty.

For $N = 1, 2, \ldots$ let $S_N(x)$ be defined by

$$S_N(x) = \sum_{k=1}^{N} u_k(x) \quad \text{for } x \text{ in } I$$

Then the functions $S_1(x)$, $S_2(x)$, . . . form what is called the *sequence of partial sums* for the infinite series (2.1.17). The convergence of the infinite series can now be defined in terms of the convergence of the sequence of partial sums. In fact, there is more than one "mode" of convergence we need to define.

Definition.   The infinite series (2.1.17) is said to converge to the sum $S(x)$ on the interval I,

(a) in the *mean-square sense* if

$$\int_I |S_N(x) - S(x)|^2 \, dx \to 0 \quad \text{as } N \to \infty$$

(b) in the *pointwise sense* if

For each x in I, $|S_N(x) - S(x)| \to 0$ as $N \to \infty$

(c) *uniformly on I* if

$$\max_I |S_N(x) - S(x)| \to 0 \quad \text{as } N \to \infty$$

Each of these modes of convergence provides a way of describing how the graph of $S_N(x)$ draws close to that of $S(x)$ as N tends to infinity. For example, when the series (2.1.17) converges *uniformly* on I to $S(x)$, then for each $\epsilon > 0$ there is an integer $N_\epsilon > 0$ and an "ϵ-strip,"

$$E_\epsilon = \{(x, y) : S(x) - \epsilon < y < S(x) + \epsilon; x \text{ in } I\}$$

such that for every $N > N_\epsilon$ the graph of $S_N(x)$ is entirely contained in E_ϵ. This is the case, for example, in Figure 2.1.3.

If the series (2.1.17) converges pointwise to $S(x)$, this means only that for each x in I, $S_N(x)$ converges to $S(x)$ but the *rate* of convergence may vary with x. In Figure 2.1.2 we see an example where $S_N(x)$ appears to be converging to $S(x)$ fairly rapidly for x between -0.9π and 0.9π, but the convergence for $x = 0.99\pi$ is evidently slower. This is an example of pointwise convergence that is not uniform.

Finally, mean-square convergence of $S_N(x)$ to $S(x)$ ensures only that the area contained between the graphs of $S_N(x)$ and $S(x)$ decreases to zero as N tends to infinity. This does not then necessarily mean that $S_N(x)$ tends to the value $S(x)$ at every x in I. In many practical situations where our knowledge of the functions involved is not sufficiently precise to support pointwise evaluations, it is natural to think of the mean-square mode of convergence.

Every series that is uniformly convergent is also convergent in the mean-square and pointwise senses. However, there are mean-square convergent series that converge neither pointwise nor uniformly, and there are pointwise convergent series that do not converge in the mean-square sense and do not converge uniformly. A "litmus test" for distinguishing uniform from nonuniform convergence is contained in the following.

Proposition 2.1.1. Suppose the series (2.1.17) converges uniformly on I to the sum $S(x)$. If each of the functions $u_n(x)$ is continuous on I, then $S(x)$ is necessarily continuous on I.

It follows from this proposition that the series in Examples 2.1.1 and 2.1.2 cannot converge uniformly on any interval I of length longer than 2π since any such interval will contain a discontinuity of the limit function $S(x)$.

A useful test for determining if a given series does converge uniformly is the so-called Weierstrass M-test. A more complete discussion of this and the previous result can be found in most advanced calculus texts. A very good nontechnical discussion can be found in *Foundations of Applied Mathematics* by Michael D. Greenberg (Prentice-Hall, 1978).

Proposition 2.1.2 (Weierstrass M-test). A sufficient condition for the uniform convergence on I of the series (2.1.17) is the existence of a convergent series of positive constants ΣM_n such that for all x in I, $|u_n(x)| \le M_n$ for $n = 1, 2, \ldots$.

Note that according to this proposition, the series in (2.1.9) is uniformly convergent on any interval I of finite length.

Consider now the following special case of the series (2.1.17),

$$\frac{1}{2} a_0 + \sum_{n=1}^{\infty} a_n \cos \frac{n\pi x}{L} + b_n \sin \frac{n\pi x}{L} \qquad (2.1.18)$$

Whether or not this is a Fourier series for some function $f(x)$, we can use tests such as the Weierstrass M-test to determine whether the series converges in one mode or another. Conditions sufficient to ensure convergence of the series must then bear on the coefficients a_n and b_n and can imply nothing about the sum $S(x)$ to which the series converges. When (2.1.18) *is* a Fourier series for a given $f(x)$, then it is our aim to determine conditions which are sufficient to ensure that the Fourier series *converges to* $f(x)$. It is clear that such conditions must then bear on the function $f(x)$.

In order to state these conditions on $f(x)$, it will be convenient to define the notion of a sectionally continuous function. First we introduce the notation $f(x^+)$ and $f(x^-)$ to denote the limit of $f(x)$ as we approach x from the right and left, respectively. Note that any point x where $f(x^+) = f(x^-)$ is a point of continuity for $f(x)$. Then:

Definition. $f(x)$ is said to be *sectionally continuous* if

  (a) at each point x the limits $f(x^+)$ and $f(x^-)$ both exist and are finite and
  (b) in any interval of finite length there are at most finitely many x such that $f(x^+)$ is not equal to $f(x^-)$.

Obviously every continuous function is sectionally continuous. The functions in Examples 2.1.1 and 2.1.2 are examples of functions that are sectionally continuous but not continuous. The function $f(x) = x^{-1}$ is an example of a function that is not sectionally continuous on any interval that contains $x = 0$ since neither $f(0^+)$ nor $f(0^-)$ are finite in this case.

The class of sectionally continuous functions is a class of functions that is strictly larger than the class of continuous functions. A still larger class of functions is the class of square-integrable functions. The function $f(x)$ is said to be *square integrable* on $(-L, L)$ if

$$\int_{-L}^{L} f(x)^2 \, dx < \infty$$

Every sectionally continuous function is square integrable on $(-L, L)$, but the converse is false. For example, $f(x) = x^{-1/4}$ is square integrable but is not sectionally continuous on $(-L, L)$. We shall have more to say about square-integrable functions later in the chapter.

Now we can state some conditions under which the Fourier series for $f(x)$ converges to $f(x)$.

Theorem 2.1.1. Suppose that $f(x)$ is defined on the interval $(-L, L)$ and let $\tilde{f}(x)$ denote the $2L$-periodic extension of $f(x)$.

(a) If $f(x)$ is square integrable on $(-L, L)$, then the Fourier series for f converges to $f(x)$ in the mean-square sense on $(-L, L)$.

(b) If $\tilde{f}(x)$ and its derivative $d/dx[\tilde{f}(x)]$ are both sectionally continuous, then at each x the Fourier series for $f(x)$ converges pointwise to the value

$$\tfrac{1}{2}[\tilde{f}(x^+) + \tilde{f}(x^-)]$$

(c) If $\tilde{f}(x)$ is continuous and $[\tilde{f}(x)]'$ is sectionally continuous, then the Fourier series for $f(x)$ converges uniformly to $\tilde{f}(x)$.

The theorem has the following corollary describing the conditions under which we can differentiate the Fourier series for $f(x)$ and expect the differentiated series to converge to the derivative of $f(x)$.

Corollary. With the notation of the theorem we have:

(d) If $\tilde{f}(x)$ is continuous and if $[\tilde{f}(x)]'$ and $[\tilde{f}(x)]''$ are both sectionally continuous, then the Fourier series for $f(x)$ converges uniformly to $\tilde{f}(x)$. In addition, the Fourier series for $f(x)$ may be differentiated, term by term, and at each x the differentiated series converges pointwise to the value

$$\tfrac{1}{2}[\tilde{f}(x^+)]' + \tfrac{1}{2}[\tilde{f}(x^-)]'$$

(e) If $\tilde{f}(x)$ and $[\tilde{f}(x)]'$ are both continuous and if the second derivative, $[\tilde{f}(x)]''$, is sectionally continuous, then the Fourier series for $f(x)$ converges uniformly to $\tilde{f}(x)$. In addition, the Fourier series for $f(x)$ may be differentiated, term by term, and this differentiated series converges uniformly to $[\tilde{f}(x)]'$.

Note that with the exception of part (a) of the theorem, the convergence properties of the Fourier series for $f(x)$ are determined by the smoothness properties of the periodic extension $\tilde{f}(x)$; the smoother the function $\tilde{f}(x)$, the stronger is the sense in which the Fourier series for $f(x)$ converges. Observe also that we do not mention differentiating the Fourier series for $f(x)$ unless $\tilde{f}(x)$ is at least continuous with a sectionally continuous derivative. Without at least this much smoothness the derivative of $\tilde{f}(x)$ may fail to exist at some points and the differentiated Fourier series could not converge.

Proving the Fourier convergence theorem is a difficult exercise in real analyis. Such an effort would be out of place in this book, whose focus is intended to be the solution of partial differential equations. Instead we illustrate the meaning and use of the theorem with some examples.

EXAMPLE 2.1.4 ───

Consider the function

$$g(x) = \begin{cases} -1 & \text{for } -\pi < x < 0 \\ 1 & \text{for } 0 < x < \pi \end{cases}$$

Then $g(x)$ is related to the function $f(x)$ of Example 2.1.1 by the equation

$$g(x) = 2[f(x) - \tfrac{1}{2}]$$

and the Fourier coefficients for $g(x)$ are related to the Fourier coefficients for $f(x)$ in the same way. That is,

$$g(x) = \frac{4}{\pi} \sum_{m=1}^{\infty} (2m - 1)^{-1}\sin(2m - 1)x$$

The 2π-periodic extension of $g(x)$ is the function $\tilde{g}(x)$ whose graph on the interval $[-3\pi, 3\pi]$ is shown in Figure 2.1.4. It is evident from the graph that $\tilde{g}(x)$ is sectionally continuous but not continuous. In particular,

$$\tilde{g}(0^+) = +1 \quad \text{does not equal} \quad -1 = \tilde{g}(0^-)$$
$$\tilde{g}(\pi^+) = -1 \quad \text{does not equal} \quad +1 = \tilde{g}(\pi^-), \quad \text{etc.}$$

Thus $\tilde{g}(x)$ has a "jump discontinuity" at every integer multiple of π, but on any interval of finite length, there are only finitely many of these jumps. The derivative $[\tilde{g}(x)]'$ is also sectionally continuous; in fact, the derivative of $\tilde{g}(x)$ is zero at each point where it is defined. At the integer multiples of π, the derivative is not defined, but the left and right limiting values for $[\tilde{g}(x)]'$ at these points do exist.

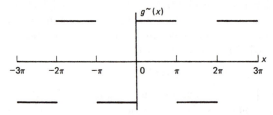

Figure 2.1.4

It follows then from part (b) of the Fourier convergence theorem that the Fourier series for $g(x)$ converges pointwise to the value

$$\tfrac{1}{2}[\tilde{g}(x^+) + \tilde{g}(x^-)] = \begin{cases} \tilde{g}(x) & \text{at each } x \text{ where } \tilde{g} \text{ is continuous} \\ 0 & \text{at integer multiples of } \pi \end{cases}$$

Note that since $\tilde{g}(x)$ is not continuous, the convergence of this Fourier series could not be uniform as this would violate Proposition 2.1.1   ■ ■

Note also that the differentiated Fourier series does not converge since the nth term of the differentiated series fails to go to zero. This is consistent with the fact that $\tilde{g}(x)$ is not continuous and hence lacks the minimal amount of smoothness required for differentiating its Fourier series.

EXAMPLE 2.1.5 _____

Consider the function

$$G(x) = |x|, \qquad -\pi < x < \pi$$

This is just the function of Example 2.1.3, where we found the Fourier series to be

$$\frac{\pi}{2} - \frac{4}{\pi} \sum_{m=1}^{\infty} (2m - 1)^{-2}\cos(2m - 1)x$$

The graph of Example 2.1.3 of the periodic extension $\tilde{G}(x)$ shows that $\tilde{G}(x)$ is continuous for all values of x. The derivative $[\tilde{G}(x)]'$ is just the function $\tilde{g}(x)$ from Example 2.1.4 and hence $[\tilde{G}(x)]'$ is sectionally continuous. Then part (c) of the Fourier convergence theorem implies that the Fourier series for $\tilde{G}(x)$ converges uniformly to $\tilde{G}(x)$ on $(-\infty, \infty)$.

Compare this with the function

$$F(x) = x, \qquad -\pi < x < \pi$$

whose Fourier series was found in Example 2.1.2 to be given by (2.1.8). The graph of the periodic extension $\tilde{F}(x)$ was also shown previously, and it is clear that $\tilde{F}(x)$ is sectionally continuous but not continuous. The derivative $[\tilde{F}(x)]'$ is also sectionally continuous; $[\tilde{F}(x)]' = 1$ at each point where it is defined. At odd-integer multiples of π, $\tilde{F}(x)$ has a finite jump discontinuity, and so the derivative $[\tilde{F}(x)]'$ is not defined, although its left and right limits at these points do exist. Then part (b) of the Fourier convergence theorem implies that the Fourier series for $F(x)$ converges pointwise to the value

$$\tfrac{1}{2}[\tilde{F}(x^+) + \tilde{F}(x^-)] = \begin{cases} \tilde{F}(x) & \text{at points of continuity} \\ 0 & \text{at odd-integer multiples of } \pi \end{cases}$$

It is striking that although both $F(x)$ and $G(x)$ are continuous on the interval $(-\pi, \pi)$ and, in fact, $F(x)$ is the smoother of the two functions there, the Fourier series for $G(x)$ converges more strongly than the Fourier series for $F(x)$. Evidently the convergence of the Fourier series for a function is controlled by the smoothness of the periodic extension of the function. As Lemma 2.1.1 implies, smoothness of the function does not necessarily imply smoothness of its periodic extension.

Note that the function $\tilde{G}(x)$ is continuous and $\tilde{G}(x)$ has as its derivative the sectionally continuous function $\tilde{g}(x)$, which also has a sectionally continuous derivative [which is then the second derivative of $\tilde{G}(x)$]. Then part (d) of the corollary to Theorem 2.1.1 implies that the Fourier series for $G(x)$ may be differentiated, term by term, and the resulting series must then converge pointwise to the value

$$\tfrac{1}{2}[\tilde{G}(x^+)]' + \tfrac{1}{2}[\tilde{G}(x^-)]' = \tfrac{1}{2}[\tilde{g}(x^+) + \tilde{g}(x^-)]$$

This conclusion can be verified by differentiating the Fourier series for $G(x)$, term by term, and observing that one does in fact get the Fourier series for $g(x)$. On the other hand, $\bar{F}(x)$ does not satisfy the conditions of the corollary to Theorem 2.1.1, and one sees that differentiating the Fourier series (2.1.8) for $F(x)$ leads to the series

$$2 \sum_{n=1}^{\infty} (-1)^{n+1}\cos nx$$

Not only does this series not converge to the derivative of $F(x)$, but the series is obviously divergent since the nth term does not go to zero. As with the function $\tilde{g}(x)$ in the previous example, $\bar{F}(x)$ lacks the minimal smoothness required for differentiating the Fourier series.   ■ ■

Exponential Form of Fourier Series

We have seen that under certain conditions a function $f(x)$ defined on the interval $(-L, L)$ can be represented as an infinite series of the form (2.1.1). In view of the Euler identity, $\exp[i\theta] = \cos \theta + i \sin \theta$ (i denotes $\sqrt{-1}$), the series has the alternative form

$$f(x) \sim \sum_{n=-\infty}^{\infty} c_n\exp \frac{in\pi x}{L} \tag{2.1.19}$$

where the coefficients c_n can be expressed in terms of the a_n and b_n by substituting the Euler identity into (2.1.19), collecting terms, and comparing with (2.1.1).

Instead we will find the coefficients c_n directly in terms of $f(x)$ as follows. For n equal to any integer, let

$$E_n(x) = \exp[in\pi x/L] \tag{2.1.20}$$

and introduce the notation

$$«F, G» = \int_{-L}^{L} F(x)\overline{G}(x)\, dx \tag{2.1.21}$$

Here $\overline{G}(x)$ in the integrand denotes the complex conjugate of $G(x)$. In particular,

$$\overline{E}_n(x) = \exp[-in\pi x/L]$$

Then it is easily shown that

$$«E_m, E_n» = \begin{cases} 0, & m \neq n \\ 2L, & m = n \end{cases} \tag{2.1.22}$$

These are the orthogonality relations for the family of functions $\{E_n(x) : n = \text{integer}\}$ on the interval $(-L, L)$. We can use the relations (2.1.22) to find the coefficients c_n in the same way we used the orthogonality relations (2.1.2) to find the a_n and b_n. That is, if (2.1.19) is an equation, then (formally at least) we have, for fixed integer M,

$$\langle\!\langle f, E_M \rangle\!\rangle = \sum_{-\infty}^{\infty} c_n \langle\!\langle E_n, E_M \rangle\!\rangle = c_M 2L$$

This leads to

$$c_n = \frac{1}{2L} \int_{-L}^{L} f(x)\overline{E}_n(x)\, dx \qquad (2.1.23)$$

The series (2.1.19) with the coefficients c_n given by (2.1.23) is called the *exponential form of the Fourier series*. The exponential form of the Fourier series is completely equivalent to the form (2.1.1), and in fact, substituting the Euler identity into (2.1.23) shows immediately that

$$c_n = \begin{cases} \frac{1}{2}[a_n - ib_n] & \text{for } n = 0, 1, \ldots \\ \frac{1}{2}[a_n + ib_n] & \text{for } n = -1, -2, \ldots \end{cases} \qquad (2.1.24)$$

In view of this equivalence, the Fourier convergence theorem applies without change to the exponential form of the series.

EXAMPLE 2.1.6

Consider the function defined on the interval $(-\pi, \pi)$ by

$$f(x) = \begin{cases} 1 & \text{if } |x| < a \\ 0 & \text{if } -\pi < x < -a \ \text{ or } \ a < x < \pi \end{cases}$$

Here a denotes a positive constant less than π.
Then

$$\begin{aligned} c_n &= \frac{1}{2\pi} \int_{-a}^{a} 1 \exp[-inx]\, dx \\ &= \frac{1}{-2n\pi i} \exp[-inx] \Big|_{-a}^{a} \\ &= \frac{1}{n\pi} \sin na, \qquad n = -1, 1, -2, 2, \ldots \end{aligned}$$

and $c_0 = a/\pi$.
Then

$$f(x) = \frac{a}{\pi} + \sum_{n<0} \frac{1}{n\pi} (\sin na)e^{inx} + \sum_{n>0} \frac{1}{n\pi} (\sin na)e^{inx}$$

This is the exponential form of the Fourier series for this function $f(x)$. Note that

$$\sin(na)/n = \sin(-na)/(-n)$$

and hence the sum over positive n and the sum over negative n can be combined to give

$$f(x) = \frac{a}{\pi} + \sum_{n>0} \frac{1}{n\pi} \sin na[e^{inx} + e^{-inx}]$$

$$= \frac{a}{\pi} + \sum_{n>0} \frac{2}{n\pi} \sin na \cos nx$$

That is, this is the trigonometric form of the Fourier series for this $f(x)$. This example clearly illustrates the equivalence of the exponential and trigonometric forms of the Fourier series. ■ ■

Our discussion of Fourier series up to this point has dealt with the so-called classical aspects of the subject. We have seen that when the coefficients a_n and b_n are related to the function $f(x)$ as described in (2.1.5), then the series (2.1.1) converges to the function $f(x)$. The mode of convergence is dependent on certain properties of $f(x)$ explained in Theorem 2.1.1. The conditions sufficient to ensure convergence for the Fourier series are mild compared to what is required to ensure convergence of other forms of infinite series. For example, power series converge to a function in a neighborhood of a point only if that function is analytic at the point.

We have seen also that the Fourier series may be expressed in the alternative form (2.1.19). Of course, this form of the series is completely equivalent to the form (2.1.1); in fact, the coefficients c_n and the coefficients a_n and b_n are related by (2.1.24). In addition, for functions $f(x)$ defined only on the "half-interval" $(0, L)$, we have the so-called half-range sine or cosine series, which converge to $f(x)$ on $(0, L)$ and converge, respectively, to the odd or even extension of $f(x)$ on the interval $(-L, 0)$.

All of this is part of a more general theory we develop in the next section.

Exercises: Fourier Series

The following functions are defined on $(-L, L)$.

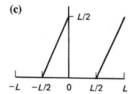

(a)

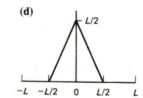

(b)

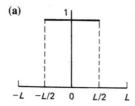

(c)

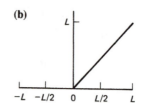

(d)

1. Compute the Fourier coefficients a_n and b_n for each of these functions and sketch the graph on $(-2L, 2L)$ of the function to which the Fourier series converges.

2. Compute the coefficients c_n for the exponential form of the Fourier series for each of the functions (a)–(d).

3. Tell whether the Fourier series for the functions $f(x)$ in (a)–(d) converge pointwise or uniformly. Tell whether the differentiated series converges to the derivative of $f(x)$.

4. Compute the Fourier coefficients a_n and b_n for each of the following functions:

   (a) $f(x) = \sin x,$ $-\pi < x < \pi$

   (b) $f(x) = \sin \tfrac{1}{2}x,$ $-\pi < x < \pi$

   (c) $f(x) = |\sin x|,$ $-\pi < x < \pi$

   (d) $f(x) = \begin{cases} \cos x, & -\pi/2 < x < \pi/2 \\ 0, & -\pi < x < -\pi/2, \ \pi/2 < x < \pi \end{cases}$

   (e) $f(x) = \begin{cases} \sin x, & -\pi/2 < x < \pi/2 \\ 0, & -\pi < x < -\pi/2, \ \ \pi/2 < x < \pi \end{cases}$

5. In parts (a), (b), and (c) of Exercise 4, each of these Fourier series is missing some terms; that is, some of the coefficients a_n and b_n are zero. In each case, explain the reason for the missing terms. For example, the $f(x)$ in (c) is symmetric about the lines $x = 0$ and $x = \pi/2$. Such symmetry will cause some of the Fourier coefficients to vanish. In each of the five cases, calculate the coefficients, see which, if any, vanish, and examine the graph of $f(x)$ for symmetries. Then try to connect the missing terms with the symmetries.

6. Compute the Fourier coefficients a_n and b_n for the following functions:

   (a) $f(x) = x^2,$ $-1 < x < 1$

   (b) $f(x) = |x|,$ $-1 < x < 1$

   (c) $f(x) = \begin{cases} x + 1 & \text{if } -1 < x < 0 \\ 1 - x & \text{if } \ \ 0 < x < 1 \end{cases}$

   (d) $f(x) = 1 - x^2,$ $-1 < x < 1$

7. For each of the functions $f(x)$ in Exercise 6, sketch the graph on the interval $(-2, 2)$ of the periodic extension $\bar{f}(x)$. Also sketch the graph of the derivative of the periodic extension. Does the differentiated Fourier series converge to this derivative in any of the four cases?

8. Write the Fourier series for the functions:

   (a) $f(x) = \begin{cases} x^2 & \text{if } -\pi/2 < x < \pi/2 \\ 0 & \text{if } -\pi < x < -\pi/2, \ \pi/2 < x < \pi \end{cases}$

   (b) $g(x) = \begin{cases} x^2 & \text{if } -\pi/2 < x < \pi/2 \\ \pi^2/4 & \text{if } -\pi < x < -\pi/2, \ \pi/2 < x < \pi \end{cases}$

   (c) $h(x) = \begin{cases} 2x & \text{if } -\pi/2 < x < \pi/2 \\ 0 & \text{if } -\pi < x < -\pi/2, \ \pi/2 < x < \pi \end{cases}$

9. Use the Fourier series of Exercise 8 to discuss the following: Which of the functions $f(x)$ or $g(x)$ has as its derivative the function $h(x)$? Explain what is wrong with the following statement: "Since $f(x)$ and $g(x)$ differ by a constant, they have the same derivative." Each of the functions f and g is an antiderivative of h. Which of the two has as its Fourier series an antiderivative of the Fourier series for h?

10. Each of the following functions is defined on the interval $(0, 1)$.

    (a) $f(x) = \begin{cases} 0 & \text{if } a < x < 1, \ a = \text{positive const} \\ 1 & \text{if } 0 < x < a \end{cases}$

(b) $f(x) = \begin{cases} 1 & \text{if } \frac{1}{2} - s < x < \frac{1}{2} + s, \ s = \text{const}, \ 0 < s < \frac{1}{2} \\ 0 & \text{if } 0 < x < \frac{1}{2} - s, \ s + \frac{1}{2} < x < 1 \end{cases}$

For each of the functions (a) and (b) write a Fourier cosine series. Sketch the graph on the interval $(-2, 2)$ to which the series converges. Does the cosine series converge pointwise or uniformly? Discuss whether the differentiated Fourier series converges to the derivative of $f(x)$.

11. For each of the functions (a) and (b) in Exercise 10 write a Fourier sine series. Sketch the graph on the interval $(-2, 2)$ to which the series converges. Does the sine series converge pointwise or uniformly? Discuss whether the differentiated Fourier series converges to the derivative of $f(x)$.

12. Use the Fourier half-range series to discuss the following statements:

(a) The derivative of an even/odd function is odd/even.
(b) If $f(x)$ is defined on $[-L, L]$ and is odd, then $f(x)$ vanishes at $x = 0, L, -L$.
(c) If $f(x)$ is defined on $[-L, L]$ and is even, then the derivative $f'(x)$ vanishes at the points $x = 0, L, -L$.

Tell whether each statement is true or false. If the statement is true only under special conditions on $f(x)$, describe those conditions.

13. The following functions are defined on the interval $(0, \pi)$:

$$f_0(x) = x, \qquad 0 < x < \pi$$

$$f_1(x) = \begin{cases} x & \text{if } 0 < x < \pi/2 \\ \pi - x & \text{if } \pi/2 < x < \pi \end{cases}$$

$$f_2(x) = \begin{cases} x & \text{if } 0 < x < \pi/2 \\ x - \pi & \text{if } \pi/2 < x < \pi \end{cases}$$

Write a Fourier sine series and Fourier cosine series for each of the three functions. Then answer the following questions:

(a) For each of the six series, sketch the graph on $(-\pi, \pi)$ of the function to which the series converges.
(b) For which function does the sine series contain no terms of the form (i) $\sin 2nx$ or (ii) $\sin(2n - 1)x$.
(c) For which function does the cosine series contain no terms of the form (i) $\cos 2nx$ or (ii) $\cos(2n - 1)x$.

14. In order that $f(x)$ defined on $[-\pi, \pi]$ has a continuous, 2π-periodic extension, it is necessary and sufficient that (i) $f(x)$ is continuous on $[-\pi, \pi]$ and (ii) $f(\pi) = f(-\pi)$. Under what conditions is the 2π-periodic extension continuously differentiable (i.e., $\tilde{f}$ is continuous, together with its first derivative)? Carry this further and find conditions on f sufficient to ensure that $\tilde{f}$ is continuous, together with each of its derivatives up to order M.

15. Suppose that $f(x)$ defined on $[-\pi, \pi]$ has a continuously differentiable 2π-periodic extension. Then show that the Fourier coefficients for $f(x)$ must satisfy

$$|a_n| \le C/n \quad \text{and} \quad |b_n| \le C/n, \qquad n = 1, 2, \ldots$$

where $C > 0$ denotes a constant that is independent of n. Show that if $\tilde{f}$ is continuous, together with its derivatives up to order M, then there exists a constant $C > 0$ such that

$$n^M |a_n| \le C \quad \text{and} \quad n^M |b_n| \le C, \qquad n = 1, 2, \ldots$$

[*Hint:* Integrate by parts.]

16. The function $f(x) = -x$ is defined for $0 < x < 1$. Show that the Fourier sine series for $f(x)$ is

$$f(x) = 2 \sum_{n=1}^{\infty} (n\pi)^{-1} \sin n\pi x$$

Integrate this series term by term to obtain the series

$$g(x) = C - 2 \sum_{n=1}^{\infty} (n\pi)^{-2} \cos n\pi x$$

where C denotes a constant of integration. Find C such that $g(x) = x - x^2/2$. *Hint:*

$$\sum_{n=1}^{\infty} n^{-2} = \tfrac{1}{6}\pi^2$$

17. Compute the Fourier cosine series coefficients for the function $g(x) = x - x^2/2$ defined on $0 < x < 1$ directly and compare with the results of Exercise 16.

18. Compute the Fourier cosine series coefficients for the function

$$h(x) = 2/\pi \sin \pi x/2, \qquad 0 < x < 1$$

If this series is differentiated term by term, does the resulting series converge to $h'(x) = \cos \pi x/2$?

2.2 GENERALIZED FOURIER SERIES

We show in this section that the Fourier series expansion of a square-integrable function in terms of sines and cosines is really an abstract version of an operation that is already familiar to many of the readers of this book. We are thinking of the operation of writing an arbitrary vector in R^2 or R^3 in terms of a basis. Engineering texts frequently refer to the basis vectors in R^3 as $\mathbf{i}$, $\mathbf{j}$, and $\mathbf{k}$. Then the vector

$$\mathbf{v} = a\mathbf{i} + b\mathbf{j} + c\mathbf{k}$$

denotes the directed line segment from $(0, 0, 0)$ to (a, b, c). We begin by recalling a few salient facts about R^N.

Vector Space R^N

Very early in the study of the applications of mathematics, one encounters the techniques of vector algebra. Often vectors are first presented as directed line segments, and the operations of forming linear combinations of vectors are carried out geometrically by means of the so-called parallelogram law. Such methods are inconvenient for computational purposes and do not readily generalize. This motivates the introduction of an orthonormal basis and the description of vectors in terms of components. In this setting, vectors are generally written as N-tuples of numbers $\mathbf{x} = (x_1, \ldots, x_N)$. We may interpret this to mean the following:

$$\mathbf{x} = x_1\mathbf{e}_1 + x_2\mathbf{e}_2 + \cdots + x_N\mathbf{e}_N$$

where $\mathbf{e}_i$ denotes the vector or N-tuple with a 1 in the ith entry and zeros in every other entry (Figure 2.2.1). Then the operations of vector algebra, namely, vector addition and scalar multiplication, can be carried out componentwise as follows:

$$\alpha\mathbf{x} + \beta\mathbf{y} = (\alpha x_1 + \beta y_1)\mathbf{e}_1 + \cdots + (\alpha x_N + \beta y_N)\mathbf{e}_N$$

Here α, β denote scalars (numbers) and $\mathbf{x}$, $\mathbf{y}$ denote vectors.

Mathematicians refer to a system consisting of two types of "objects," vectors and scalars, together with two "operations," vector addition and scalar multiplication, as a *vector space*. In order to describe a vector space one has only to describe the vectors and the scalars and to make clear how the operations of vector addition and scalar multiplication are to be carried out. The system is then a vector space if the following conditions are satisfied:

$$\mathbf{u} + \mathbf{v} = \mathbf{v} + \mathbf{u}$$
$$(\mathbf{u} + \mathbf{v}) + \mathbf{w} = \mathbf{u} + (\mathbf{v} + \mathbf{w})$$
$$\mathbf{u} + \mathbf{0} = \mathbf{0} + \mathbf{u}$$
$$\mathbf{u} + (-\mathbf{u}) = \mathbf{0}$$
$$c(d\mathbf{u}) = (cd)\mathbf{u}$$
$$(c + d)\mathbf{u} = c\mathbf{u} + d\mathbf{u}$$
$$c(\mathbf{u} + \mathbf{v}) = c\mathbf{u} + c\mathbf{v}$$
$$1(\mathbf{u}) = \mathbf{u}, \qquad 0(\mathbf{u}) = \mathbf{0}$$

The system that consists of N-tuples of numbers as vectors with (real) numbers as scalars and the operations of vector addition and scalar multiplication defined componentwise as in the preceding satisfies all of these conditions and is referred to as the vector space R^N.

An additional operation, called the inner product (or "dot" product), is often defined on a vector space. For $\mathbf{x}$, $\mathbf{y}$ arbitrary vectors in R^N, the inner product is denoted $(\mathbf{x}, \mathbf{y})$ and is defined as follows:

$$(\mathbf{x}, \mathbf{y}) = x_1 y_1 + \cdots + x_N y_N \tag{2.2.1}$$

The product defined in (2.2.1) is easily seen to satisfy the conditions

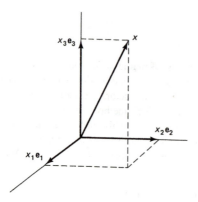

Figure 2.2.1

$$(\mathbf{x}, \mathbf{y}) = (\mathbf{y}, \mathbf{x})$$
$$(\alpha\mathbf{x} + \beta\mathbf{y}, \mathbf{z}) = \alpha(\mathbf{x}, \mathbf{z}) + \beta(\mathbf{y}, \mathbf{z})$$
$$(\mathbf{x}, \mathbf{x}) \geq 0 \quad \text{and} \quad (\mathbf{x}, \mathbf{x}) = 0 \quad \text{if and only if } \mathbf{x} = \mathbf{0} \qquad (2.2.2)$$

Having defined the inner product, we can say that two vectors in R^N are *orthogonal* if their inner product is zero, and a family of vectors in R^N, $\mathbf{x}^1, \mathbf{x}^2, \ldots, \mathbf{x}^M$, is said to form an *orthogonal family* if

$$(\mathbf{x}^i, \mathbf{x}^j) = 0 \quad \text{for } i \text{ different from } j \quad \text{for } 1 \leq i, j \leq M \qquad (2.2.3)$$

When $N = 2, 3$, vectors in R^N that are orthogonal are geometrically perpendicular. If $\mathbf{x}^1, \ldots, \mathbf{x}^M$ is an orthogonal family in R^N and, in addition,

$$(\mathbf{x}^i, \mathbf{x}^i) = 1 \quad \text{for } i = 1, \ldots, M \qquad (2.2.4)$$

then the family is said to be an *orthonormal family* in R^N. The vectors $\mathbf{e}_i$, $i = 1, \ldots, N$, mentioned previously are an example of an orthonormal family. Any orthogonal family can, of course, be made into an orthonormal family by dividing each vector in the orthogonal family by its *norm*; that is, by the scalar

$$\|\mathbf{x}^i\| = [(\mathbf{x}^i, \mathbf{x}^i)]^{1/2}$$

When $N = 2, 3$, the norm of a vector can be interpreted as the length of the vector. When $N > 3$, neither the vector nor its norm need have any geometric significance, although we often continue to think of the norm as being some kind of generalized length. In particular, the only vector whose norm/length is zero is the zero vector.

It is a fundamental result in linear algebra that if $\mathbf{x}^1, \ldots, \mathbf{x}^M$ is an orthogonal family of nonzero vectors in R^N, then necessarily $M \leq N$. If $\mathbf{x}^1, \ldots, \mathbf{x}^M$ is an orthogonal family in R^N and, in addition, $M = N$, then the family is said to be *complete*. We can prove the following facts about orthogonal families in R^N:

Lemma 2.2.1. If $\mathbf{x}^1, \ldots, \mathbf{x}^N$ is a complete orthogonal family in R^N and $\mathbf{v}$ is a vector in R^N such that

$$(\mathbf{v}, \mathbf{x}^i) = 0 \quad \text{for } i = 1, \ldots, N$$

then it follows that $\mathbf{v} = \mathbf{0}$. Conversely, if there is a $\mathbf{v}$ in R^N that is not zero and yet

$$(\mathbf{v}, \mathbf{x}^i) = 0 \quad \text{for } i = 1, \ldots, M$$

then it follows that $\mathbf{x}^1, \ldots, \mathbf{x}^M$ is *not* a complete family; that is, $M < N$

A complete orthonormal family in R^N is also called an *orthonormal basis*. The importance of orthonormal bases lies in their convenience for computational purposes. If $\mathbf{x}^1, \ldots, \mathbf{x}^N$ is an orthonormal basis in R^N and $\mathbf{v}$ denotes an arbitrary vector in R^N, then there exist unique constants v_n, $n = 1, \ldots, N$ such that

$$\mathbf{v} = v_1\mathbf{x}^1 + \cdots + v_N\mathbf{x}^N = \sum_{n=1}^{N} v_n\mathbf{x}^n$$

In general (for a family that is complete but not orthogonal), to find the constants v_n

requires solution of a system of linear algebraic equations. For an orthonormal basis, however, it follows from (2.2.3) and (2.2.4) that

$$(\mathbf{v}, \mathbf{x}^k) = \sum_{n=1}^{N} v_n(\mathbf{x}^n, \mathbf{x}^k) = v_k \quad \text{for } k = 1, \ldots, N$$

That is,

$$v_k = (\mathbf{v}, \mathbf{x}^k) \quad \text{for } k = 1, \ldots, N$$

Then

$$\mathbf{v} = \sum_{n=1}^{N} (\mathbf{v}, \mathbf{x}^n)\mathbf{x}^n \tag{2.2.5}$$

We emphasize that (2.2.5) is an equality for all vectors $\mathbf{v}$ in R^N only if the orthonormal family is *complete*. For if it is not complete, then there are nonzero vectors $\mathbf{v}$ in R^N for which $(\mathbf{v}, \mathbf{x}^n) = 0$ for all n so the right side of (2.2.5) vanishes, while the $\mathbf{v}$ on the left side is not zero.

Now we extend these notions to a vector space where functions of one variable play the role of the vectors. In this vector space there can be an unlimited number of linearly independent vectors, and so we shall be facing a situation that is more complicated than the one in R^N. However, in superficial ways at least, these two vector spaces will appear to be very similar.

$L^2(a, b)$ A Vector Space of Functions

Let $f(x)$ denote a function defined on the interval (a, b) and that satisfies

$$\int_a^b f(x)^2 \, dx < \infty \tag{2.2.6}$$

The collection of all such functions will be denoted by $L^2(a, b)$, or if there is no danger of confusion, simply by L^2. We examine in more detail later exactly what sort of functions are in this collection, but note that L^2 is a vector space in which the functions $f(x)$ play the role of the vectors and real numbers are the scalars. The operations of vector addition and multiplication by a scalar are defined as follows:

$$[\alpha f + \beta g](x) = \alpha f(x) + \beta g(x) \quad \text{for } x \text{ in } (a, b) \tag{2.2.7}$$

That is, for f and g in L^2 and arbitrary scalars α, β, the function $\alpha f + \beta g$ in L^2 is that function whose value at x in (a, b) is given by (2.2.7).

We can also define an inner product in $L^2(a, b)$ by

$$(f, g) = \int_a^b f(x)g(x) \, dx \quad \text{for } f, g \text{ in } L^2 \tag{2.2.8}$$

It is easy to check that similar to the product (2.2.1), the product defined by (2.2.8) satisfies all of the conditions of (2.2.2).

In the event that we wish to take the complex numbers for the scalars [in which case the functions $f(x)$ in L^2 are allowed to assume complex values], we modify the definition of the inner product as follows:

$$\langle\!\langle f, g \rangle\!\rangle = \int_a^b f(x)\overline{g}(x)\ dx \tag{2.2.8'}$$

Here the $\overline{g}(x)$ in the integrand denotes the complex conjugate of $g(x)$, and we use the notation $\langle\!\langle\ ,\ \rangle\!\rangle$ in place of $(\ ,\ )$ to emphasize when we are using the *complex* inner product.

We define the *norm* of a function $f(x)$ in $L^2(a, b)$ to be the number

$$\|f\| = (f, f)^{1/2} \quad (\text{or } \langle\!\langle f, f \rangle\!\rangle^{1/2})$$

Note that

$$\|f\| \geq 0 \quad \text{for all } f \text{ in } L^2(a, b)$$

and any function f in $L^2(a, b)$ for which $\|f\| = 0$ is said to be zero in the sense of L^2. However, this is not the same as saying $f(x) = 0$ for every x in (a, b). The function $f(x)$ may differ from zero on a discrete set of points (i.e., any set of points containing no interval) and still have $\|f\| = 0$. This affects the way in which we may regard the functions in L^2. Two functions f and g in $L^2(a, b)$ are equal in $L^2(a, b)$ if $\|f - g\| = 0$. As we have just seen, this does not mean that $f(x) = g(x)$ at each point x in (a, b) but only that

$$\int_a^b [f(x) - g(x)]^2\ dx = 0 \tag{2.2.9}$$

Then $L^2(a, b)$ is composed of "equivalence classes" of pointwise functions; pointwise functions f and g belong to the same equivalence class in $L^2(a, b)$ if $\|f - g\| = 0$. In computing with functions in $L^2(a, b)$, we still treat the functions as pointwise functions, but we are to understand that altering the values of the function or leaving values unspecified on any discrete subset of (a, b) does not in any way affect the function as an element of $L^2(a, b)$. We shall see later that this has the important consequence of causing the correspondence between a "function" f in L^2 and the Fourier coefficients for f to be a one-to-one correspondence. We have already observed that the Fourier coefficients for a pointwise function $F(x)$ are unaffected by altering the values of $F(x)$ on a discrete set of points. It follows that the correspondence between *pointwise* functions and the associated Fourier coefficients is then many-to-one.

Functions $f(x)$ and $g(x)$ in $L^2(a, b)$ are said to be orthogonal on (a, b) if $(f, g) = 0$. While orthogonality in R^2 or R^3 had geometric significance (orthogonality meant perpendicularity), in $L^2(a, b)$ there is no such interpretation. In particular, orthogonality of the functions f and g does not imply any relationship between the graphs of the functions.

A family of functions $\{f_1, f_2, \ldots\}$ in $L^2(a, b)$ is said to be *orthogonal* if

$$(f_i, f_j) = 0 \quad \text{if } i \neq j$$

The family is said to be an *orthonormal* family if

$$(f_i, f_j) = \begin{cases} 0 & \text{if } i \neq j \\ 1 & \text{if } i = j \end{cases}$$

Clearly, these definitions carry over directly from R^N. However, in contrast to the situation in R^N, an orthogonal family in $L^2(a, b)$ may have infinitely many members.

EXAMPLE 2.2.1

1. For $k = 1, 2, \ldots$, let

$$u_k(x) = \sin kx, \qquad 0 < x < \pi$$

Then,

$$(u_k, u_j) = \begin{cases} 0 & \text{if } k \neq j \\ \pi/2 & \text{if } k = j \end{cases}$$

So this family is orthogonal but not orthonormal in $L^2(0, \pi)$. Of course, the family

$$U_k(x) = (2/\pi)^{1/2} \sin kx, \qquad k = 1, 2, \ldots$$

is an orthonormal family in $L^2(0, \pi)$ obtained by dividing each of the functions $u_k(x)$ by its norm.

2. For $n = 0, 1, 2, \ldots$, let

$$v_n(x) = \cos nx, \qquad 0 < x < \pi$$

Then,

$$(v_n, v_m) = \begin{cases} 0 & \text{if } m \neq n \\ \pi & \text{if } m = n = 0 \\ \pi/2 & \text{if } m = n > 0 \end{cases}$$

and this family is orthogonal but not orthonormal in the vector space $L^2(0, \pi)$. Note that $v_0(x) = 1$. The family of "normalized" functions

$$V_0(x) = (1/\pi)^{1/2}, \qquad V_n(x) = (2/\pi)^{1/2} \cos nx, \qquad n = 1, 2, \ldots$$

is then orthonormal in $L^2(0, \pi)$. ∎∎

Let $u_1(x), u_2(x), \ldots$ denote an orthonormal family in $L^2(a, b)$ and let $g(x)$ denote an arbitrary function in $L^2(a, b)$. By analogy with (2.2.5), we can form the series

$$\sum_{n=1}^{\infty} (g, u_n)u_n(x)$$

and ask under what conditions and in what sense the series can be said to represent $g(x)$.
We define convergence in $L^2(a, b)$ as follows. Associated with the infinite series

$$\sum_{n=1}^{\infty} u_n(x) \qquad\qquad (2.2.10)$$

is the sequence of partial sums

$$S_N(x) = \sum_{n=1}^{N} u_n(x)$$

Then the infinite series (2.2.10) is said to converge to the sum $S(x)$ in $L^2(a, b)$ if

$$\int_a^b |S(x) - S_N(x)|^2 \, dx \to 0 \quad \text{as } N \to \infty$$

Evidently, convergence in $L^2(a, b)$ is identical to the previously defined mean-square convergence on (a, b). In discussing $L^2(a, b)$ convergence, it will be convenient to have the following result.

Proposition 2.2.1. Let $\{u_k : k = 1, 2, \ldots\}$ denote an orthonormal family in $L^2(a, b)$, and for arbitrary constants $g_1, g_2, \ldots$, let $\hat{g}(x)$ denote the formal sum

$$\hat{g}(x) = \sum_{k=1}^{\infty} g_k u_k(x)$$

Then the expression

$$J[\hat{g}] = \int_a^b |g(x) - \hat{g}(x)|^2 \, dx$$

is minimized by choosing the constants g_k in the sum such that $g_k = (g, u_k)$. Moreover, for arbitrary g in $L^2(a, b)$,

$$\sum_{k=1}^{\infty} |(g, u_k)|^2 \le \|g\|^2 = \int_a^b |g(x)|^2 \, dx \tag{2.2.11}$$

Proof. For positive integer N, let

$$\hat{g}_N(x) = \sum_{k=1}^{N} g_k u_k(x)$$

and define

$$J_N(g_1, \ldots, g_N) = \int_a^b |g(x) - \hat{g}_N(x)|^2 \, dx$$

Then for each N, J_N is a quadratic function of the N real variables $g_1, \ldots, g_N$. Since J_N is nonnegative for all choices of $g_1, \ldots, g_N$, it follows that J_N has a unique minimum at the point g^* in R_N where

$$\frac{\partial J_N}{\partial g_k} = 0 \quad \text{for } k = 1, \ldots, N \tag{2.2.12}$$

Now the chain rule implies

$$\frac{\partial J_N}{\partial g_k} = 2 \int_a^b [g(x) - \hat{g}_N(x)] \frac{\partial \hat{g}_N}{\partial g_k} \, dx$$

and

$$\frac{\partial \hat{g}_N}{\partial g_k} = u_k \quad \text{for } k = 1, \ldots, N$$

Then, using these in connection with the definition of the inner product in $L^2(a, b)$, we get

$$\frac{\partial J_N}{\partial g_k} = 2[(g, u_k) - (\hat{g}_N, u_k)] = 2[(g, u_k) - g_k] \tag{2.2.13}$$

From (2.2.12) and (2.2.13) together, it follows that for each $N = 1, 2, \ldots, J_n(\hat{g}_n)$ is minimized when

$$\hat{g}_N(x) = \sum_{k=1}^{N} (g, u_k)u_k(x) \tag{2.2.14}$$

Note that, with $\hat{g}_N$ given by (2.2.14), we have

$$J_N(\hat{g}_N) = \int_a^b \left| g(x) - \sum_{k=1}^{N} (g, u_k)u_k(x) \right|^2 dx$$

$$= (g, g) - 2 \sum_{k=1}^{N} |(g, u_k)|^2 + \sum_{k=1}^{N} |(g, u_k)|^2$$

$$= (g, g) - \sum_{k=1}^{N} |(g, u_k)|^2 \geq 0$$

and it follows that for each N,

$$\|g\|^2 \geq \sum_{k=1}^{N} |(g, u_k)|^2 \geq 0$$

Since the left side of this inequality does not depend on N, we are entitled to let N go to infinity on the right side, proving (2.2.11). In fact, this proves both parts of the proposition together since it implies that N can be allowed to go to infinity in (2.2.14).

The inequality (2.2.11) is called *Bessel's inequality*, and it ensures that the infinite series

$$\sum_{k=1}^{\infty} (g, u_k)u_k(x) \tag{2.2.15}$$

converges in $L^2(a, b)$. However, the Bessel inequality does not ensure that the series converges to $g(x)$. In order that the series converge to g for arbitrary g in $L^2(a, b)$, it is necessary that the orthonormal family have a property known as "completeness."

Definition. The orthonormal family $\{u_k : k = 1, 2, \ldots\}$ in $L^2(a, b)$ is said to be *complete* if, for arbitrary g in $L^2(a, b)$, the series (2.2.15) converges to g.

In R_N we saw that an orthonormal set was said to be complete if it contained the maximal number of orthogonal vectors (namely N). In L^2 completeness cannot be characterized in terms of the *number* of members in the orthonormal set. In L^2 every complete orthonormal family has infinitely many members but not every infinite orthonormal family is complete. We can prove the following facts about completeness of orthonormal families in L^2.

Lemma 2.2.2. If $\{u_k : k = 1, 2, \ldots\}$ is a complete orthonormal family in L^2 and g in L^2 satisfies

$$(g, u_k) = 0 \quad \text{for } k = 1, 2, \ldots$$

then it follows that $g = 0$. Alternatively, if $\{u_k : k = 1, 2, \ldots\}$ is an orthonormal family in L^2 and if there exists a g in L^2 that is not zero and yet

$$(g, u_k) = 0 \quad \text{for } k = 1, 2, \ldots$$

then it follows that $\{u_k : k = 1, 2, \ldots\}$ is not complete.

EXAMPLE 2.2.2

1. We have seen in Example 2.2.1 that the family

$$U_k(x) = (2/\pi)^{1/2} \sin kx, \qquad k = 1, 2, \ldots$$

is an orthonormal family in $L^2(0, \pi)$. Then the subfamily consisting of the functions

$$U_{2k}(x) = (2/\pi)^{1/2} \sin 2kx, \qquad k = 1, 2, \ldots$$

is also orthonormal but cannot be complete since the nonzero function $\sin x$ is orthogonal to each of the functions $U_{2k}(x)$, $k = 1, 2, \ldots$. We shall see later that the full family of functions $\{U_k : k = 1, 2, \ldots\}$ *is* complete in $L^2(0, \pi)$.

2. The family $\{V_k : k = 1, 2, \ldots\}$

$$V_k(x) = (2/\pi)^{1/2} \cos kx, \qquad k = 1, 2, \ldots$$

is also an orthonormal family in $L^2(0, \pi)$. The constant function, $v_0(x) = 1$, is orthogonal to $V_k(x)$ for $k = 1, 2, \ldots$, and it follows that the family $\{V_k : k = 1, 2, \ldots\}$ is *not* complete. We shall see later that the family $\{V_k : k = 1, 2, \ldots\}$ together with the function $V_0(x) = (1/\pi)^{1/2}$ is a complete orthonormal family in $L^2(0, \pi)$. ∎ ∎

Exercises: Generalized Fourier Series

1. For $j = 0, 1, \ldots, 10$, let $x_j = j/10$ and let

$$\Phi_j(x) = \begin{cases} 1 & \text{if } x_{j-1} < x < x_j, \\ 0 & \text{otherwise,} \end{cases} \qquad j = 1, \ldots, 10$$

Show that the family $\{\Phi_1, \ldots, \Phi_{10}\}$ is an orthogonal family in $L^2(0, 1)$. Is this family complete? Compute $\|\Phi_j\|$ for each j, and then normalize the family. For $f(x) = x^2$, compute

$$(f, \phi_j) = \int_0^1 f(x)\phi_j(x)\, dx, \qquad j = 1, \ldots, 10$$

where $\phi_j(x)$ denotes the normalized function $\Phi_j/\|\Phi_j\|$. Graph the function

$$f^{\#}(x) = \sum_{j=1}^{10} (f, \phi_j)\phi_j(x)$$

and compare the graph with that of $f(x)$ on $(0, 1)$.

2. For $j = 0, 1, \ldots, 10$, let $x_j = j/10$, and let $\theta_j(x)$ denote the continuous and piecewise-linear function on $(0, 1)$ that is equal to 1 at x_j and 0 at x_k for k not equal to j. The family $\theta_0(x)$, $\ldots, \theta_{10}(x)$ are so-called "hat" functions. Sketch the graph of $\theta_0(x)$, $\theta_5(x)$, and $\theta_{10}(x)$. Is this family orthogonal? Is it complete? For $f(x) = x^2$, compute (f, θ) for $j = 0, \ldots, 10$ and graph the function

$$f^{\#}(x) = \sum_{j=1}^{10} (f, \theta_j)\theta_j(x)$$

3. For $j = 0, 1, \ldots, 10$, let $\sigma_j(x) = 2^{-1/2}\cos j\pi x$ for $0 < x < 1$. Is this an orthonormal family in $L^2(0, 1)$? Is the family complete?

4. Give an example of a set of vectors in R^4 that is
   (a) orthonormal but not complete,
   (b) orthonormal and complete,
   (c) complete but not orthogonal,
   (d) orthogonal but not orthonormal.

5. Using the vectors (denote them by $\mathbf{u}_j$) from Exercise 4(a), give an example of a vector $\mathbf{V}$ in R^4 such that

$$\mathbf{V} = \sum_j (\mathbf{V}, \mathbf{u}_j)\mathbf{u}_j$$

Give an example of a vector $\mathbf{W}$ that cannot be expressed as a linear combination of the $\mathbf{u}_j$'s.

6. Let $\sigma_j(x)$ denote the family of functions from Exercise 3. For $g(x) = x$, compute $g_j = (g, \sigma_j)$, $j = 0, 1, \ldots, 10$. Compute

$$\|g\|^2 = \int_0^1 x^2 \, dx$$

and compare this with

$$\sum_{j=0}^{10} g_j^2$$

2.3 STURM–LIOUVILLE PROBLEMS

An infinite series of the form (2.2.15) is called a *generalized Fourier series* for the function $g(x)$ in terms of the orthonormal family $\{u_k : k = 1, 2, \ldots\}$. According to the definition of completeness, the generalized Fourier series converges to $g(x)$ in $L^2(a, b)$ if the orthonormal family is complete. Unfortunately, the definition does not suggest any way of *finding* complete orthonormal families in $L^2(a, b)$. As we shall see, such families arise naturally as the solutions to certain kinds of boundary-value problems for ordinary differential equations.

Consider the following boundary-value problem for the unknown function $u(x)$ on the interval (a, b):

$$\begin{aligned}
-[p(x)u'(x)]' + q(x)u(x) &= \mu r(x)u(x), \quad a < x < b \\
A_1 u(a) + A_2 u'(a) &= 0 \\
B_1 u(b) + B_2 u'(b) &= 0
\end{aligned} \qquad (2.3.1)$$

Here μ denotes a parameter. We assume that

> $p(x)$, $p'(x)$, $q(x)$, $r(x)$ are continuous on (a, b)
> $p(x)$ and $r(x)$ are strictly positive on $[a, b]$
> A_1, A_2 do not both vanish nor do B_1, B_2 (2.3.2)

When all of the conditions (2.3.2) are satisfied, problem (2.3.1) is called a Sturm–Liouville problem. More precisely, it is called a Sturm–Liouville problem with "separated boundary conditions." This refers to the fact that each of the boundary conditions in (2.3.1) imposes a condition on the value of the solution $u(x)$ and/or its derivative $u'(x)$ at just one of the endpoints of the interval where the differential equation is to be satisfied. Another type of boundary condition we shall want to consider is the so-called periodic boundary condition, which requires the solution to satisfy

$$u(a) = u(b) \quad \text{and} \quad u'(a) = u'(b) \tag{2.3.3}$$

Note that the first condition in (2.3.3) involves the value of $u(x)$ at both endpoints of the interval, and the second condition in (2.3.3) involves the values of $u'(x)$ at the two ends of the interval.

It is easy to see that for any choice of the parameter μ, one solution of (2.3.1) is the "trivial solution," $u(x)$ identically zero on (a, b). However, for certain special values of μ, there exist nontrivial solutions for (2.3.1). Each value of μ for which (2.3.1) has a nontrivial solution is called an *eigenvalue* of the problem. The corresponding nontrivial solution is then called an *eigenfunction*. We state the following theorem without proof.

Theorem 2.3.1.   Under the assumptions (2.3.2), the eigenvalues and eigenfunctions for (2.3.1) have the following properties:

1. All eigenvalues μ of (2.3.1) are real numbers. The eigenvalues form a (countably) infinite collection μ_1, μ_2, . . . satisfying

$$\mu_1 < \mu_2 < \cdots < \mu_n \to \infty$$

2. To each eigenvalue μ_n there corresponds a single independent eigenfunction $u_n(x)$.
3. The "weighted eigenfunctions"

$$w_n(x) \equiv \frac{\sqrt{r(x)}\, u_n(x)}{\|\sqrt{r}\, u_n\|}, \qquad n = 1, 2, \ldots$$

are a *complete* orthonormal family in $L^2(a, b)$.

Note that property 1 implies that there can be at most a finite number of negative eigenvalues for a Sturm–Liouville problem. Note also that if $u(x)$ is an eigenfunction for (2.3.1), then for any constant C, $v(x) = Cu(x)$ is also an eigenfunction for (2.3.1). However, $u(x)$ and $v(x)$ are not considered to be "independent" eigenfunctions; property 2 implies that for each eigenvalue there is just one *independent* eigenfunction. Finally, property 3 implies that for any g in $L^2(a, b)$ the generalized Fourier series for g in terms of the orthonormal family of weighted eigenfunctions will converge in at least the L^2 sense to g. For stronger convergence statements we have the following theorem.

Theorem 2.3.2. For $f(x)$ defined on (a, b), consider the generalized Fourier series for f in terms of the family of weighted eigenfunctions for (2.3.1).

1. If $f(x)$ belongs to $L^2(a, b)$, then the series converges in the L^2 sense.
2. If f and f' are both sectionally continuous on (a, b), then the series converges pointwise to

$$\tfrac{1}{2}[f(x^+) + f(x^-)] \quad \text{at each } x \text{ in } (a, b)$$

3. If $f(x)$ is continuous and $f'(x)$ is sectionally continuous on (a, b) and if $f(x)$ satisfies

$$A_1 f(a) + A_2 f'(a) = 0 \quad \text{and} \quad B_1 f(b) + B_2 f'(b) = 0$$

then the series converges uniformly to $f(x)$ on (a, b).

We now consider some examples of Sturm–Liouville problems and the complete families of eigenfunctions they generate.

EXAMPLE 2.3.1 _____

Consider the problem

$$-u''(x) = \mu u(x), \quad 0 < x < L$$
$$u(0) = u(L) = 0 \tag{2.3.4}$$

This is a problem of the form (2.3.1) in the special case that

$$p(x) = 1, \quad r(x) = 1, \quad \text{and } q(x) = 0 \quad \text{for all } x \text{ in } (0, L)$$
$$a = 0, \quad b = L > 0 \quad \text{and} \quad A_1 = B_1 = 1, \quad A_2 = B_2 = 0$$

Then the assumptions (2.3.2) are all fulfilled, and the conclusions of Theorem 2.3.1 must apply. In particular, all the eigenvalues are real, which means we need only consider the following three cases: $\mu < 0$, $\mu = 0$, and $\mu > 0$.

(a) $\mu = 0$ (Zero Eigenvalue)
In this case the differential equation (2.3.4) reduces to

$$u''(x) = 0$$

The general solution to this equation is $u(x) = Ax + B$. Then the boundary conditions in (2.3.4) imply that

$$u(0) = B = 0 \quad \text{and} \quad u(L) = AL + B = 0$$

that is, $A = B = 0$. Evidently the only solution to the problem when $\mu = 0$ is the trivial solution, and it follows that $\mu = 0$ is not an eigenvalue for this Sturm–Liouville problem.

(b) $\mu < 0$ (Negative Eigenvalues)
We write $\mu = -\beta^2$ for β real and not zero. Then $u(x)$ must satisfy

$$-\mathbf{u}''(x) = -\beta^2 u(x), \quad 0 < x < L$$

The general solution to this equation can be written as a linear combination of any *two* of the following six linearly independent functions:

$$e^{\beta x}, \; e^{-\beta x}, \; \sinh \beta x, \; \cosh \beta x, \; \sinh \beta(L - x), \; \cosh \beta(L - x)$$

If we choose to write

$$u(x) = A \sinh \beta x + B \sinh \beta(L - x)$$

then the boundary conditions imply

$$u(0) = B \sinh \beta L = 0$$
$$u(L) = A \sinh \beta L = 0$$

Since $\sinh \beta L$ is *not* zero (recall $L > 0$), it follows that the trivial solution ($A = B = 0$) is the *only* solution to the problem in the case μ is negative; that is, there are no negative eigenvalues.

Note that forming the general solution from the independent functions $\sinh \beta x$ and $\sinh \beta(L - x)$ made the determination of A and B from the boundary conditions particularly straightforward. For other sorts of boundary conditions, other choices for the functions comprising the general solution will be more convenient. We illustrate this point further in the examples to come.

(c) $\mu > 0$ (Positive Eigenvalues)

We write $\mu = \beta^2$ for β real and not zero. Then the differential equation reduces to

$$-u''(x) = \beta^2 u(x)$$

The general solution to this equation is most conveniently expressed in the form

$$u(x) = A \cos \beta x + B \sin \beta x \tag{2.3.5}$$

Then the boundary conditions reduce to

$$u(0) = A = 0$$
$$u(L) = B \sin \beta L = 0$$

The first condition clearly requires that $A = 0$. However, the second condition is satisfied for $B \neq 0$ provided that β satisfies

$$\sin \beta L = 0 \tag{2.3.6}$$

The zeros of the sine function are well known (they are the integer multiples of π), and it follows that the solutions of (2.3.6) are the values

$$\beta_n = n\pi/L, \quad n = \text{nonzero integer}$$

There are infinitely many roots of Equation (2.3.6), and they occur in positive–negative pairs; for each positive root β_n, there is a corresponding negative root $-\beta_n$. Since we began by writing $\mu = \beta^2$, the eigenvalues of (2.3.4) are the *squares* of the roots β_n; that is,

$$\mu_n = (n\pi/L)^2, \quad n = 1, 2, \ldots$$

Then each *pair* of roots β_n, $-\beta_n$ leads to just a *single* eigenvalue $\mu_n = \beta_n^2$. We refer to Equation (2.3.6) as the "eigenvalue equation" (even though the roots of the equation are not themselves the eigenvalues). The eigenfunction corresponding to the eigenvalue μ_n is then

$$u_n(x) = \sin \beta_n x, \qquad n = 1, 2, \ldots$$

Recall that the boundary condition at $x = 0$ required that the coefficient A in (2.3.5) must vanish while the boundary condition at $x = L$ was satisfied as a result of (2.3.6) (without B having to be zero).

Note that the eigenvalues are all real and positive as predicted by Theorem 2.3.1. The theorem also ensures that the eigenfunctions $u_n(x)$ are a complete orthogonal family in $L^2(0, L)$. Note that since $r(x) = 1$ in (2.3.4), the weighted eigenfunctions $w_n(x)$ of the theorem are identical to the normalized eigenfunctions

$$U_n(x) = (2/L)^{1/2} u_n(x)$$

Then for arbitrary $f(x)$ in $L^2(0, L)$, the generalized Fourier series

$$\sum_{n=1}^{\infty} (f, U_n) U_n(x)$$

converges in accordance with the conclusions of Theorem 2.3.2. Note that this generalized Fourier series is just what we have previously called the half-range Fourier sine series. Evidently, then, the series must converge to the odd $2L$-periodic extension of the function $f(x)$. ∎ ∎

EXAMPLE 2.3.2 ───

Consider the problem

$$-u''(x) = \mu u(x), \qquad 0 < x < L$$
$$u'(0) = 0, \qquad u'(L) = 0$$

This boundary-value problem differs from the previous example only in the boundary conditions. It is readily seen to be a problem of the Sturm–Liouville type so that all the conclusions of Theorems 2.3.1 and 2.3.2 will apply. In particular, the eigenvalues μ must all be real so that μ must be negative, zero, or positive. Each of these three possibilities for μ leads to a different form of the general solution to the differential equation, and we must then check to see which solution(s) can satisfy the homogeneous boundary conditions without reducing to the trivial solution.

(a) $\mu < 0$ (Negative Eigenvalues)

Write $\mu = -\beta^2$ for β real and not zero. Then the general solution to the differential equation can be written as

$$u(x) = A \cosh \beta x + B \cosh \beta(L - x)$$

Note that we have chosen a different pair of independent solutions from the list of six solutions for the equation listed previously. The advantage of this choice becomes evident when we write the boundary conditions:

$$u'(0) = -\beta B \sinh \beta L = 0$$
$$u'(L) = \beta A \sinh \beta L = 0$$

Since $\sinh \beta L \neq 0$, it follows from the boundary conditions that both A and B must be zero. Then there are no negative eigenvalues.

(b) $\mu = 0$ (Zero Eigenvalue)

In this case the general solution to the differential equation is just the linear function

$$u(x) = Ax + B$$

The boundary conditions

$$u'(0) = A = 0 \quad \text{and} \quad u'(L) = A = 0$$

are satisfied by $u(x) = B$ for $B \neq 0$. Then $\mu_0 = 0$ is an eigenvalue with the corresponding eigenfunction $u_0(x) = B$ ($B = $ const).

(c) $\mu > 0$ (Positive Eigenvalues)

We write $\mu = \beta^2$ in this case for β real and not zero. Then the general solution to the differential equation may be expressed in the form

$$u(x) = A \cos \beta x + B \sin \beta x$$

The boundary conditions lead to the following pair of homogeneous equations for the unknowns A and B:

$$u'(0) = \beta B = 0$$
$$u'(L) = \beta A \sin \beta L = 0$$

The first condition obviously requires that $B = 0$, but the second condition can be satisfied without A having to vanish if β satisfies

$$\sin \beta L = 0$$

This equation appeared in the previous example and leads to the eigenvalues

$$\mu_n = (n\pi/L)^2, \qquad n = 1, 2, \ldots$$

Since it is B that vanishes here, the corresponding eigenfunctions are

$$u_n(x) = \cos \beta_n x, \qquad n = 1, 2, \ldots$$

The full set of eigenvalues and normalized eigenfunctions for this problem is then

$$\mu_0 = 0, \quad \mu_1 = (\pi/L)^2, \quad \mu_2 = (2\pi/L)^2, \ldots$$
$$U_0(x) = (1/L)^{1/2}$$
$$U_1(x) = (2/L)^{1/2}\cos(\pi x/L)$$
$$U_2(x) = (2/L)^{1/2}\cos(2\pi x/L)$$
$$\vdots$$

Theorem 2.3.1 ensures that these eigenfunctions form a complete orthonormal set in $L^2(0, L)$, and hence, for any $f(x)$ in $L^2(0, L)$ we can write

$$\sum_{n=0}^{\infty} (f, U_n)U_n(x)$$

and expect this generalized Fourier series to converge in accordance with the statements in Theorem 2.3.2. Notice that this series is what we called (in Section 2.1) the half-range Fourier cosine series. Then this generalized Fourier series converges to the even $2L$-periodic extension of $f(x)$. ■ ■

EXAMPLE 2.3.3 _____

In the previous examples the eigenvalues were obtained by solving the equation

$$\sin \beta L = 0$$

for the unknown β. This equation is referred to as the *eigenvalue equation*, and its roots, when squared, give the eigenvalues. The real zeros of the sine function are well known, and so this equation can be solved by inspection. We now consider a problem in which the eigenvalue equation is not so easily solved.

Consider the problem

$$-u''(x) = \mu u(x), \qquad 0 < x < L$$
$$u(0) + au'(0) = 0$$
$$u(L) + bu'(L) = 0$$

where a and b denote constants such that $a > b > 0$. This is once again a problem of the Sturm–Liouville type with separated boundary conditions so that all of the conditions of Theorems 2.3.1 and 2.3.2 apply here. In particular, all of the eigenvalues of this problem are real numbers.

(a) $\mu = 0$

The general solution of the differential equation in this case is the function

$$u(x) = Ax + B$$

and the boundary conditions imply

$$B + aA = 0 \quad \text{and} \quad B + (L + b)A = 0$$

This is a set of two homogeneous linear algebraic equations in two unknowns (A and B), and we know that in order for this system to have a nontrivial solution, it is necessary and sufficient for the determinant of this system to be zero. The determinant of this system is

$$\det = L + b - a$$

and it follows that $\mu = 0$ is not an eigenvalue of the Sturm–Liouville problem unless the parameters L, a, and b in the problem are such that $L = a - b$. If this condition is satisfied, then $\mu = 0$ is an eigenvalue with the corresponding eigenfunction

$$u_0(x) = A(x - a)$$

Let us suppose in this example that the $L \neq a - b$ and that $\mu = 0$ is therefore not an eigenvalue.

(b) $\mu < 0$

Write $\mu = -\beta^2$ for β real and not zero. In this case, we choose to write the general solution to the differential equation in the form

$$u(x) = Ae^{\beta x} + Be^{-\beta x}$$

Then the boundary conditions take the form

$$A(1 + a\beta) + B(1 - a\beta) = 0$$
$$Ae^{L\beta}(1 + b\beta) + Be^{-L\beta}(1 - b\beta) = 0$$

Once again we make use of the fact that this system of two equations in the two unknowns A and B has only the trivial solution $A = B = 0$ unless the determinant of the system vanishes. We can easily show that the determinant vanishes if and only if β satisfies

$$e^{2\beta L} = \frac{(1 + a\beta)(1 - b\beta)}{(1 - a\beta)(1 + b\beta)}$$

This is an example of a transcendental equation for which the roots cannot be found by inspection. However, we can gain some insight into the existence and location of roots for this equation by graphing on the same set of axes the two functions

$$F(\beta) = e^{2\beta L} \quad \text{and} \quad G(\beta) = \frac{(1 + a\beta)(1 - b\beta)}{(1 - a\beta)(1 + b\beta)}$$

From Figures 2.3.1 and 2.3.2 we see that the two graphs intersect at $\beta = 0$. Whether there are any *other* intersections depends on the slopes of the tangent lines to the graphs of F and G at the point $\beta = 0$. We compute the derivatives,

$$F'(0) = 2L \quad \text{and} \quad G'(0) = 2(a - b)$$

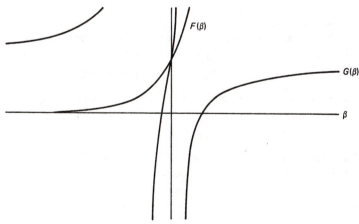

$F(\beta)$

$G(\beta)$

β

Figure 2.3.1

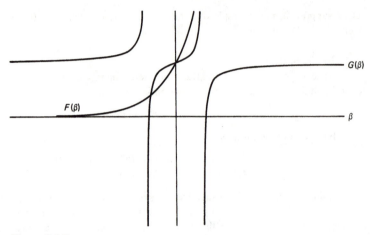

Figure 2.3.2

Evidently $F'(0) < G'(0)$ if $L < a - b$. In this case there is only one intersection at $\beta = 0$. This is the case sketched in Figure 2.3.1. If $L > a - b$, then $F'(0) > G'(0)$ and there are *three* intersections. This is the situation sketched in Figure 2.3.2. The two nonzero roots in this case can be shown to have the same square (see Exercise 9), and hence they produce the same (negative) eigenvalue $\mu_{-1} = -\beta^2$. The eigenfunction corresponding to this negative eigenvalue may be written in the form

$$u_{-1}(x) = A[(a\beta - 1)e^{\beta x} - (a\beta + 1)e^{-\beta x}]$$

where β here denotes the positive root of the eigenvalue equation $F(\beta) = G(\beta)$. We mention in passing that solutions of this equation must be computed by some numerical algorithm.

(c) $\mu > 0$

In this case, we write $\mu = \beta^2$ for β real and not zero. The general solution to the differential equation in this case is of the form

$$u(x) = A \cos \beta x + B \sin \beta x$$

and then the boundary conditions lead to the following pair of equations for A and B:

$$A + a\beta B = 0$$
$$[\cos \beta L - b\beta \sin \beta L]A + [\sin \beta L + b\beta \cos \beta L]B = 0$$

Once again, this system has a nontrivial solution only if the determinant of the system vanishes. Setting the determinant equal to zero leads then to the following equation for β:

$$\tan \beta L = \frac{(a - b)\beta}{1 + ab\beta^2}$$

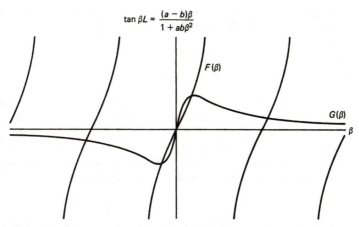

$$\tan \beta L = \frac{(a - b)\beta}{1 + ab\beta^2}$$

Figure 2.3.3

This is again a transcendental equation whose roots can only be found by means of some numerical algorithm. However, by sketching (qualitatively) graphs of

$$F(\beta) = \tan \beta L \quad \text{and} \quad G(\beta) = (a - b)\beta/[1 + ab\beta^2]$$

on one set of axes, we can see that the eigenvalue equation has an infinite set of symmetrically located roots. That is, for each positive root of $F(\beta) = G(\beta)$ there is a negative root of equal magnitude but opposite sign. This, of course, follows from the fact that both F and G are odd functions of β. A qualitative sketch of F and G versus β is shown in Figure 2.3.3.

If we denote the roots of $F(\beta) = G(\beta)$ by β_n, then the corresponding family of eigenfunctions may be written in the form

$$u_n(x) = \sin \beta_n x - a\beta_n \cos \beta_n x, \quad n = 1, 2, \ldots$$

The family of eigenfunctions

$$u_{-1}(x), u_1(x), u_2(x), \ldots$$

form a complete orthogonal family in L^2 on the interval $(0, L)$. This is guaranteed by Theorem 2.3.1. Dividing each of the functions $u_n(x)$ by the constants

$$C_n = \|u_n\| = (u_n, u_n)^{1/2}, \quad n = 1, 1, 2, \ldots$$

makes this an orthonormal family of eigenfunctions. Note, however, that C_n in this case depends on n (see Exercise 10). ∎ ∎

EXAMPLE 2.3.4 _____

Each of the previous examples involved a Sturm–Liouville problem with *separated* boundary conditions; that is, each boundary condition involved the value of the unknown function and/or its derivative at just *one* of the endpoints of the interval on which the problem is posed. In this example we consider a problem with boundary conditions that are not separated, the so-called *periodic* boundary conditions:

$$-u''(x) = \mu u(x), \qquad -L < x < L$$
$$u(-L) = u(L)$$
$$u'(-L) = u'(L)$$

Even though the boundary conditions in this Sturm–Liouville problem are not separated, all the conclusions of Theorems 2.3.1 and 2.3.2 continue to hold with just one exception. Conclusion 2 of Theorem 2.3.1 does not apply to this problem. As we shall see, for each nonzero eigenvalue there are two independent eigenfunctions. All of the other conclusions of Theorem 2.3.1 continue to hold; in particular, the eigenvalues are all real so that we need only consider the usual three possibilities for μ.

(a) $\mu < 0$

We write $\mu = -\beta^2$ for β real and not zero. Then the general solution to the differential equation may be written in the form

$$u(x) = A \cosh \beta x + B \sinh \beta x$$

and the boundary conditions then imply

$$A \cosh \beta L - B \sinh \beta L = A \cosh \beta L + B \sinh \beta L$$
$$\beta[-A \sinh \beta L + B \cosh \beta L] = \beta[A \sinh \beta L + B \cosh \beta L]$$

From the first boundary condition we get $B = 0$, and from the second we see that $A = 0$. Then there can be no negative eigenvalues.

(b) $\mu = 0$

Then the general solution to the differential equation is the linear function

$$u(x) = Ax + B$$

and

$$u(-L) - u(L) = -AL + B - AL - B = -2AL = 0$$
$$u'(-L) - u'(L) = A - A = 0$$

Evidently the boundary conditions are satisfied by the nontrivial function $u(x) = B$. Then $\mu = 0$ is an eigenvalue with the corresponding eigenfunction $u_0(x) = 1$. (The choice $B = 1$ is entirely arbitrary.)

(c) $\mu > 0$

We write $\mu = \beta^2$ for β real and not zero. Then the general solution for the differential equation can be written as

$$u(x) = A \cos \beta x + B \sin \beta x$$

and the boundary conditions lead to

$$2A \sin \beta L = 0$$
$$2B \sin \beta L = 0$$

Evidently, if β satisfies $\sin \beta L = 0$, then the boundary conditions are both satisfied without *either A or B* having to vanish. Then the positive eigenvalues are the numbers

$$\mu_n = (n\pi/L)^2, \qquad n = 1, 2, \ldots$$

and to each eigenvalue corresponds the *two* independent eigenfunctions

$$u_n(x) = \cos \mu_n x \quad \text{and} \quad v_n(x) = \sin \mu_n x$$

In each of the previous examples, the boundary conditions implied that one or the other of the constants A or B must be zero or else some relationship existed between A and B. This removed one degree of freedom from the general solution for the differential equation, leaving one independent eigenfunction for each eigenvalue. In this example, neither A nor B vanishes, and there need be no relation between A and B in order that the boundary conditions be satisfied. The full set of eigenvalues and eigenfunctions in this example is then

$$\mu_0 = 0, \qquad u_0(x) = 1$$
$$\mu_1 = (\pi/L)^2, \qquad u_1(x) = \cos(\pi x/L), \qquad v_1(x) = \sin(\pi x/L)$$
$$\mu_2 = (2\pi/L)^2, \qquad u_2(x) = \cos(2\pi x/L) \qquad v_2(x) = \sin(2\pi x/L)$$
$$\vdots \qquad\qquad \vdots \qquad\qquad\qquad \vdots$$

Note that these functions are just the functions used in the classical form of the Fourier series as it was developed in Section 2.1. Thus the classical form of the Fourier series is a "generalized" Fourier series associated with this Sturm–Liouville problem with periodic boundary conditions. Note also that we *could* have written the general solution to the differential equation here in part (c) of this example as

$$u(x) = Ae^{i\beta x} + Be^{-i\beta x}$$

This leads to the same set of positive eigenvalues $\mu_n = (n\pi/L)^2$ to each of which then corresponds the pair of independent eigenfunctions

$$u_n(x) = \exp[i\beta_n x] \quad \text{and} \quad v_n(x) = \exp[-i\beta_n x]$$

These are just the eigenfunctions used to form the exponential form of the Fourier series. ■ ■

Exercises: Sturm–Liouville Problems

1. Compute all the eigenvalues and corresponding eigenfunctions for the following Sturm–Liouville problems:
   (a) $-u''(x) = \mu u(x)$, $u(0) = u'(1) = 0$
   (b) $-v''(x) = \mu v(x)$, $v(0) = v(2) = 0$
   (c) $-w''(x) = \mu w(x)$, $w'(0) = w(1) = 0$
   (d) $-z''(x) = \mu z(x)$, $z'(0) = z'(2) = 0$

2. Let $f(x) = 1$ for $0 < x < 1$. Write a generalized Fourier series for $f(x)$ in terms of the eigenfunctions of Exercise 1(a). Discuss the convergence of this series.

3. Repeat Exercise 2 for the function $f(x) = x - \frac{1}{2}x^2$, $0 < x < 1$.

4. Repeat Exercise 2, this time using the eigenfunctions from Exercise 1(c).

5. For $g(x) = x^2$, $0 < x < 1$, write a generalized Fourier series in terms of the eigenfunctions of Exercise 1(c). Discuss the convergence of this series.

6. Under what conditions on the parameter α does the following Sturm–Liouville problem have a negative eigenvalue?

$$-U''(x) = \mu U(x), \qquad U'(0) + \alpha U(0) = 0, \qquad U'(1) = 0$$

7. Find all the eigenvalues and corresponding eigenfunctions for Exercise 6 in the following cases:
   (i) $\alpha < 0$ (ii) $\alpha = 0$ (iii) $\alpha > 0$

8. Under what conditions on the parameter L does the following Sturm–Liouville problem have a negative eigenvalue?

$$-u''(x) = \mu u(x), \qquad 0 < x < L$$
$$u(0) - 1u'(0) = 0$$
$$u(L) - 3u'(L) = 0$$

9. Show that the determinant of the system of two equations in the two unknowns A and B,

$$A(1 + a\beta) + B(1 - a\beta) = 0$$
$$Ae^{L\beta}(1 + b\beta) + Be^{-L\beta}(1 - b\beta) = 0$$

vanishes if and only if β satisfies

$$e^{2\beta L} = \frac{(1 + a\beta)(1 - b\beta)}{(1 - a\beta)(1 + b\beta)}$$

Show that the solutions of this equation are symmetrically located; that is, if there is a positive solution β, then $-\beta$ is also a root.

10. Consider the family of eigenfunctions generated by the problem in Example 2.2.3, namely

$$u_{-1}(x) = A[(a\beta - 1)e^{\beta x} - (a\beta + 1)e^{-\beta x}]$$

where β satisfies

$$e^{2\beta L} = \frac{(1 + a\beta)(1 - b\beta)}{(1 - a\beta)(1 + b\beta)}$$

together with the functions

$$u_n(x) = \sin \beta_n x - a\beta_n \cos \beta_n x, \qquad n = 1, 2, \ldots$$

where β satisfies

$$\tan \beta L = \frac{(a - b)\beta}{1 + ab\beta^2}$$

Verify by integrating that this family is indeed orthogonal on the interval $(0, L)$. Compute the norms of the eigenfunctions, $\|u_n\|$, for $n = -1, 1, 2, \ldots$.

11. Suppose $f(x)$ denotes a function defined on the interval $(-L, L)$. Using the notation (2.1.4) for the inner product, we make various assumptions about the Fourier coefficients for $f(x)$. In each of the following three cases, state all conclusions you can draw about the function $f(x)$:
   (a) $(f, \cos n\pi x/L) = 0$ for $n = 0, 1, \ldots$
   (b) $(f, \sin n\pi x/L) = 0$ for $n = 1, 2, \ldots$
   (c) $(f \cos n\pi x/L) = (f, \sin n\pi x/L) = 0$ for $n = 1, 2, \ldots$

12. Let $g(x)$ denote a function defined on the interval $(0, L)$. What conclusions can you draw about $g(x)$ in the following situations?

(a) $\displaystyle\int_0^L g(x) \sin \frac{n\pi x}{L} \, dx = 0$ for $n = 1, 2, \ldots$

(b) $\displaystyle\int_0^L g(x) \cos \frac{n\pi x}{L} \, dx = 0$ for $n = 0, 1, \ldots$

2.4 DISCRETE FOURIER SERIES

The topics discussed in this section are a digression from the main theme of this chapter. The material does not appear again until Section 8.4 but is presented here to introduce the discrete aspects of Fourier series.

In representing a given function $f(x)$ by its Fourier series, one generally does not sum the series. Instead, the series is truncated after some *finite* number of terms, and the function is then approximated by a trigonometric series of finite length,

$$f(x) \approx f_N(x) = \tfrac{1}{2}a_0 + \sum_{n=1}^{M} a_n \cos nx + b_n \sin nx \qquad (2.4.1)$$

For convenience here we are taking $L = \pi$. In addition, we find it more convenient to work with a truncated series of the form

$$f_N(x) = \sum_{n=0}^{N-1} c_n \exp[inx] \qquad (2.4.2)$$

with N equal to a power of 2. Using Euler's formula, we can show that (2.4.1) and (2.4.2) are equivalent with $M = N/2$ (see Exercise 2).

Instead of computing the coefficients c_n in (2.4.2) from the formula (2.1.23) for the exponential form of the Fourier coefficients, we require instead that the c_n satisfy the condition

$$f_N(x_k) = \sum_{n=0}^{N-1} c_n \exp[inx_k] = f(x_k) \quad \text{for } k = 0, 1, \ldots, N - 1 \qquad (2.4.3)$$

where $x_k = (2\pi/N)k$, $k = 0, 1, \ldots, N - 1$. In other words, we are requiring that $f_N(x)$ interpolate the function $f(x)$ at the N uniformly spaced points x_k. The system of equations (2.4.3) is a set of N equations in the N unknowns $c_0, c_1, \ldots, c_{N-1}$. We write this in matrix notation as

$$[A_N]\mathbf{C} = \mathbf{F} \qquad (2.4.4)$$

EXAMPLE 2.4.1 _____

For $N = 4$ we have $\mathbf{X} = (0, \pi/2, \pi, 3\pi/2)$, and if we write out Equations (2.4.4), we find the matrix $[A_4]$ to be

$$[A_4] = \begin{bmatrix} 1 & 1 & 1 & 1 \\ 1 & i & i^2 & i^3 \\ 1 & i^2 & i^4 & i^6 \\ 1 & i^3 & i^6 & i^9 \end{bmatrix}, \qquad i = (-1)^{1/2}$$

Reducing this to lowest terms, we get

$$[A_4] = \begin{bmatrix} 1 & 1 & 1 & 1 \\ 1 & i & -1 & -i \\ 1 & -1 & 1 & -1 \\ 1 & -i & -1 & i \end{bmatrix}$$

This matrix $[A_4]$ is symmetric but not Hermitian. In fact, the structure of this matrix is really very special. If we denote by $[A_4^*]$ the conjugate matrix obtained by replacing every element of $[A_4]$ by its complex conjugate, then it is easy to verify that

$$[A_4][A_4^*] = [A_4^*][A_4] = 4I_4$$

That is,

$$[A_4]^{-1} = \tfrac{1}{4}[A_4^*]$$

and it follows then that

$$\mathbf{C} = \tfrac{1}{4}[A_4^*]\mathbf{F}$$

In the general case, N is equal to a power of 2, if we let

$$w = \exp[i2\pi/N], \qquad w^k = \exp[i(2\pi/N)k] = \exp[ix_k] \qquad (2.4.5)$$

for $k = 0, 1, \ldots, N - 1$, then

$$[A_N] = \begin{bmatrix} 1 & 1 & 1 & \cdots & 1 \\ 1 & w & w^2 & \cdots & w^{N-1} \\ 1 & w^2 & w^4 & \cdots & w^{2(N-1)} \\ \vdots & \vdots & \vdots & & \vdots \\ 1 & w^{N-1} & w^{2(n-1)} & \cdots & w^{(N^2-1)} \end{bmatrix} \qquad (2.4.6)$$

and

$$[A_N]^{-1} = (1/N)[A_N^*] \qquad (2.4.7)$$

If we adopt the notation

$$[A_N]^{(k)} = k\text{th column of } [A_N]$$

then (2.4.7) is equivalent to the statement

$$\langle\!\langle [A_N]^{(k)}, [A_N]^{(j)} \rangle\!\rangle = \begin{cases} 0 & \text{for } k \neq j \\ N & \text{for } k = j \end{cases} \qquad (2.4.8)$$

Note that we are using the complex inner product

$$\langle\!\langle \mathbf{U}, \mathbf{V} \rangle\!\rangle = U_1\overline{V}_1 + \cdots + U_N\overline{V}_N$$

Then the solution of (2.4.4) is given by

$$\mathbf{C} = [A_N]^{-1}\mathbf{F}$$

That is,

$$C_n = \frac{1}{N} \sum_{k=0}^{N-1} (\overline{w})^{nk} F_k$$

Since $(\overline{w})^{nk} = w^{-nk} = \exp[-inx_k]$, we can write this as

$$C_n = \frac{1}{2\pi} \sum_{k=0}^{N-1} \exp[-inx_k] f(x_k) \frac{2\pi}{N} \tag{2.4.9}$$

Evidently, the sum on the right side of (2.4.9) is just the approximation of the following integral by a trapezoidal integration rule:

$$\frac{1}{2\pi} \int_0^{2\pi} e^{-inx} f(x)\, dx$$

But this is just the expression given by (2.1.23) for c_n, the nth Fourier coefficient of $f(x)$ in the exponential form of the Fourier series. We conclude that interpolating $f(x)$ on the interval $(0, 2\pi)$ with a trigonometric polynomial over N uniformly spaced points is equivalent to approximating the integral in (2.1.23) by a trapezoidal sum on N equal intervals.

Going further, we note that (2.4.4) is equivalent to

$$\mathbf{F}_N = C_0 \mathbf{A}_N^{(1)} + \cdots + C_{N-1} \mathbf{A}_N^{(N)} \tag{2.4.10}$$

where the columns $\mathbf{A}_N^{(k)}$ of $[A]$ are just the vectors

$$A_N^{(k)} = \begin{bmatrix} \exp[ikx_0] \\ \exp[ikx_1] \\ \vdots \\ \exp[ikx_{N-1}] \end{bmatrix} = \begin{bmatrix} 1 \\ \exp[ik(2\pi/N)] \\ \vdots \\ \exp[ik(N-1)2\pi/N] \end{bmatrix} \tag{2.4.11}$$

We have already observed that these vectors are (complex) orthogonal, and since there are N of them, they form a basis. Since the family of functions $\{\exp[ikx] : k = 0, 1, \ldots\}$ form a (complex) orthogonal basis in the vector space $L^2(0, 2\pi)$, we are tempted to look for a correspondence

$$e^{ikx} \leftrightarrow [e^{ikx_0}, e^{ikx_1}, \ldots, e^{ikx_{N-1}}]^T = A_N^{(k)} \tag{2.4.12}$$

between the basis vectors in L^2 and the apparently related basis vectors. However, since there are infinitely many of the function vectors e^{ikx} and only N of the column vectors $A_N^{(k)}$, the correspondence cannot be one-to-one. To see what happens, consider again the example $N = 4$ and consider the function e^{i5x}. Now,

$$e^{i5x_k} = e^{ix_k} e^{i4x_k}, \qquad k = 0, 1, 2, 3$$

But

$$e^{i4x_k} = 1 \quad \text{for } k = 0, 1, 2, 3$$

That is,

$$\exp[i4x_0] = e^0 = 1$$
$$\exp[i4x_1] = \exp[i4(\pi/2)] = e^{i2\pi} = 1$$
$$\vdots$$

Evidently, then, on the interpolation points x_k,

$$\exp[imx_k] = \exp[inx_k] \quad \text{for } k = 0, 1, \ldots, N - 1 \quad \text{if } m = n(\text{mod } N) \quad (2.4.13)$$

Here, $m = n(\text{mod } N)$ if $m - n$ is an integer multiple of N. We conclude that the correspondence in (2.4.12) is many-to-one with each column vector $A_N^{(k)}$ corresponding to every one of the function vectors $\exp[imx]$ for which $m = k(\text{mod } N)$. Numerical analysts refer to this phenomenon as "aliasing"; that is, all the functions $\exp[imx]$ such that $m = k(\text{mod } N)$ for a given k "share the same alias," namely, $A_N^{(k)}$.

For any smooth function $f(x)$, the Fourier coefficients c_n decrease at least as fast as $1/n$ as n increases; in fact, the smoother the function $f(x)$, the faster the coefficients c_n decrease with n. (see Exercise 15 at the end of Section 2.1.) It follows that approximating $f(x)$ by a Fourier expansion that uses only the lowest frequency terms is liable to be a good approximation if $f(x)$ is smooth. It only remains then to execute the procedure for solving (2.4.4). In view of the considerable amount of structure and symmetry in the matrix $[A_N]$, it should come as no surprise that there exist algorithms that solve (2.4.4) very much faster and more efficiently than algorithms based on Gaussian elimination or multiplying by $[A_N]^{-1}$. These algorithms are referred to as *fast Fourier transforms*. The best of these algorithms involve sophisticated programming techniques that we are not in a position to discuss, but further information about such algorithms is available from several sources including J. Stoer and R. Bulirsch, *Introduction to Numerical Analysis*, Springer-Verlag, Berlin, 1980, and W. Kaplan, *Advanced Engineering Mathematics*, Addison-Wesley, Reading, Mass., 1981.   ■ ■

Exercises: Discrete Fourier Series

1. For $x_k = (2\pi/N)k$, $k = 0, 1, \ldots, N - 1$, show that

$$\exp[-inx_k] = \exp[(N - n)ix_k], \qquad n = 0, 1, \ldots, N$$

and hence that

$$\cos nx_k = [\exp\{nix_k\} + \exp\{(N - n)ix_k\}]/2$$
$$\sin nx_k = [\exp\{nix_k\} - \exp\{(N - n)ix_k\}]/2i$$

for $n = 0, 1, \ldots, N$.

2. Use the results of Exercise 1 to show that (2.4.2) is equivalent to a trigonometric polynomial of the form

$$\tfrac{1}{2}a_0 + \sum_{n=1}^{M-1} [a_n\cos nx + b_n\sin nx] + \tfrac{1}{2}a_M\cos Mx$$

with $M = N/2$, $c_0 = \tfrac{1}{2}a_0$, $c_M = \tfrac{1}{2}a_M$, and $c_j = \tfrac{1}{2}(a_j - ib_j)$, $c_{N-j} = \tfrac{1}{2}(a_j + ib_j)$ for $j = 1, 2, \ldots, M - 1$.

3. Compute the entries in the matrix $[A_N]$ for $N = 8$. Verify that the columns of this matrix satisfy (2.4.8).

4. Solve (2.4.4) in the case $N = 4$ if F is generated by the function
   (a) $F(x) = \cos 5x$
   (b) $F(x) = \cos 3x$
   (c) $F(x) = \cos 4x$
   (d) $F(x) = \cos 5x + 100 \sin 4x$

5. Solve (2.4.4) with $N = 6$ if F is generated by the function
   (a) $F(x) = (x \sin x)/\pi$, $\quad 0 < x < 2\pi$
   (b) $F(x) = x(2\pi - x)/\pi^2$, $\quad 0 < x < 2\pi$

6. Given a positive integer N, can a corresponding integer M be found such that

$$\sum_{n=-N}^{N} c_n e^{inx} = \sum_{n=0}^{M} c_n e^{inx} \quad \text{for all } x$$

7. Let $x_k = 2\pi k/(2N + 1)$ for $k = 0, 1, \ldots, 2N$. Then for positive integers m and n less than or equal to N, show

$$\sum_{k=0}^{2N} \cos mx_k \cos nx_k = \begin{cases} 0 & \text{for } m \text{ not equal } n \\ (2N + 1)/2 & \text{for } m = n \end{cases}$$

$$\sum_{k=0}^{2N} \sin mx_k \sin nx_k = \begin{cases} 0 & \text{for } m \text{ not equal } n \\ (2N + 1)/2 & \text{for } m = n \end{cases}$$

$$\sum_{k=0}^{2N} \sin mx_k \cos nx_k = 0 \quad \text{for all } m, n$$

2.5 FUNCTION SPACE L^2

In Section 2.2 we introduced the notion of $L^2(a, b)$, the vector space in which functions play the role of vectors. In particular, $L^2(a, b)$ is composed of functions defined on (a, b) that satisfy the condition

$$\int_a^b |f(x)|^2 \, dx \quad \text{is finite} \tag{2.5.1}$$

Stated in terms of the inner product or the norm on $L^2(a, b)$, (2.5.1) is just the condition

$$\|f\|^2 = (f, f) \quad \text{is finite}$$

We have already alluded to the fact in Section 2.2 that the functions in L^2 are not functions in the pointwise sense but are instead equivalence classes of pointwise functions. This fact has some important consequences with regard to solutions to partial differential equations. There are other properties of L^2 with significant implications, and in this section we examine a few of these, paying particular attention to the relationship between an L^2 function f and its Fourier series.

Throughout this chapter, when we speak of L^2, we always mean $L^2(a, b)$ for (a, b) a *bounded* interval. In Chapter 4 we consider L^2 on an *unbounded* interval. We begin our examination of $L^2(a, b)$ by listing a number of examples of functions defined on (a, b), some of which belong to L^2 and some of which do not.

EXAMPLE 2.5.1 _____

For convenience, we take $(a, b) = (0, 1)$.

1. Consider the function $f(x) = x^p$ for p real. Then

$$\int_0^1 f(x)^2\, dx = \int_0^1 x^{2p}\, dx = \begin{cases} \log x & \text{for } 2p = -1 \\ x^{2p+1}/(2p + 1) & \text{for } 2p + 1 \neq 0 \end{cases}$$

It follows that

$$\|f\|^2 = \begin{cases} (2p + 1)^{-1} & \text{if } 2p + 1 > 0 \\ \infty & \text{if } 2p + 1 \leq 0 \end{cases}$$

Then the function $f(x) = x^{-1/3}$ belongs to $L^2(0, 1)$, but the function $g(x) = x^{-1/2}$ does not. For both of these functions, the limit as x approaches zero fails to exist (the functions both grow without bound as x tends to zero). Evidently it is the "rate" at which the function grows that determines whether it belongs to L^2. Note that *neither* $f(x)$ nor $g(x)$ is sectionally continuous on $(0, 1)$.

2. If $f(x)$ is defined and *continuous* on the closed interval $[0, 1]$, then f must belong to $L^2(0, 1)$. This follows immediately from the fact that if $f(x)$ is continuous on the closed interval $[0, 1]$, then $f(x)$ is bounded on $[0, 1]$; that is, for some finite constant M we have

$$\max_{0 \leq x \leq 1} |f(x)| \leq M$$

Then

$$\|f\|^2 \leq \int_0^1 M^2\, dx = M^2 < \infty$$

It is not difficult to extend this reasoning to conclude that if $f(x)$ is sectionally continuous on (a, b), then f must belong to $L^2(a, b)$. However, the previous example shows that not every function that is in $L^2(a, b)$ need be sectionally continuous on (a, b).   ■ ■

As we have said several times now, the set of all functions that satisfy (2.5.1) is more than just a collection of functions. The function space $L^2(a, b)$ is a *vector space* where the functions play the role of the vectors and the vector space operations are as defined in (2.2.7) and (2.2.8). The norm of a function,

$$\|f\| = (f, f)^{1/2}, \quad f \text{ in } L^2(a, b)$$

is the generalization of the notion of length for (geometric) vectors. We have the following important inequalities relating the norm and inner product.

Theorem 2.5.1 For f and g in $L^2(a, b)$,

$$|(f, g)| \le \|f\| \, \|g\| \tag{2.5.2}$$
$$\|f + g\| \le \|f\| + \|g\| \tag{2.5.3}$$

These inequalities are known, respectively, as the *Cauchy–Schwarz* and *triangle* inequalities.

Proof. We begin by observing that since

$$[f(x) - g(x)]^2 \ge 0$$

then it follows that

$$\tfrac{1}{2}[f(x)^2 + g(x)^2] \ge f(x)g(x)$$

Therefore, for f and g in $L^2(a, b)$,

$$\int_a^b f(x)g(x) \, dx \le \tfrac{1}{2}[\|f\|^2 + \|g\|^2] < \infty$$

Next observe that for all values of the real parameter α,

$$P(\alpha) = \int_a^b [f(x) + \alpha g(x)]^2 \, dx \ge 0$$

$$= \int_a^b f(x)^2 \, dx + 2\alpha \int_a^b f(x)g(x) \, dx + \alpha^2 \int_a^b g(x)^2 \, dx$$

That is,

$$P(\alpha) = A + 2B\alpha + C\alpha^2 \ge 0$$

where

$$A = \int_a^b f(x)^2 \, dx, \qquad B = \int_a^b f(x)g(x) \, dx, \qquad C = \int_a^b g(x)^2 \, dx$$

Since $P(\alpha)$ is nonnegative, it follows that $P(\alpha)$ has at most one real zero, and it must be the case that $B^2 - AC \le 0$; that is,

$$\left[\int_a^b f(x)g(x) \, dx\right]^2 \le \int_a^b f(x)^2 \, dx \int_a^b g(x)^2 \, dx$$

Then, taking the square root of both sides of this inequality leads to (2.5.2). To prove (2.5.3), note that (2.5.2) implies

$$\int_a^b [f(x) + g(x)]^2 \, dx = \int_a^b [f(x)^2 + 2f(x)g(x) + g(x)^2] \, dx$$

$$\le \int_a^b f^2(x) \, dx + \left|2 \int_a^b f(x)g(x) \, dx\right| + \int_a^b g^2(x) \, dx$$

$$\le \|f\|^2 + 2\|f\| \, \|g\| + \|g\|^2$$

$$\int_a^b [f(x) + g(x)]^2 \, dx \le [\|f\| + \|g\|]^2$$

Taking the square root on each side of this inequality gives (2.5.3).

Along with the notion of an inner product come the notions of orthogonality, orthogonal families, and complete orthogonal families. We have the following equivalent ways of characterizing completeness for an orthogonal family.

Theorem 2.5.2. Let $\{u_1, u_2, \ldots\}$ denote an orthogonal family in $L^2(a, b)$. Then the following are equivalent statements about the family. If any of them apply to the family $\{u_1, u_2, \ldots\}$, they all apply, and we say then that the family is *complete*.

1. For each g in $L^2(a, b)$, the generalized Fourier series for g in terms of the family $\{u_1, u_2, \ldots\}$ converges to g in $L^2(a, b)$; that is,

$$g(x) = \sum_{n=1}^{\infty} \frac{(g, u_n)}{(u_n, u_n)} u_n(x)$$

2. For g in $L^2(a, b)$

$$(g, u_n) = 0 \quad \text{for every } n \quad \text{if and only if} \quad \|g\| = 0$$

3. For g in $L^2(a, b)$

$$\|g\|^2 = \sum_{n=1}^{\infty} \frac{(g, u_n)^2}{(u_n, u_n)}$$

The equation in part 3 of this theorem is often referred to as the *Parseval relation*. It is just the L^2 version of the Pythagorean theorem.

We can show now that the correspondence between functions in $L^2(a, b)$ and a Fourier series representation for the function is one-to-one.

Suppose that $f(x)$ and $g(x)$ are two pointwise functions corresponding to the *same* element in $L^2(a, b)$. That is, $\|f - g\| = 0$. If $\{u_1, u_2, \ldots\}$ is a complete orthogonal family in $L^2(a, b)$, then generalized Fourier coefficients for f and g relative to the family $\{u_1, u_2, \ldots\}$ are, respectively, (f, u_n) and (g, u_n) for $n = 1, 2, \ldots$ Then, for $n = 1, 2, \ldots$

$$|(f, u_n) - (g, u_n)| \le |(f - g, u_n)| \le \|f - g\| \|u_n\| = 0$$

That is, *each* function in $L^2(a, b)$ has just *one* set of Fourier coefficients. Conversely, if two functions f and g in $L^2(a, b)$ have the *same* Fourier coefficients c_n relative to the basis $\{u_1, u_2, \ldots\}$, then

$$0 = c_n - c_n = (f, u_n) - (g, u_n) = (f - g, u_n) \quad \text{for } n = 1, 2, \ldots$$

Then it follows from Theorem 2.5.2 that $\|f - g\| = 0$; that is, $f = g$. Note that this one-to-one correspondence *does not* exist between sectionally continuous functions and their Fourier series. This result is important enough to be stated as a theorem.

Theorem 2.5.3. Let $\{u_1, u_2, \ldots\}$ denote a complete orthogonal family in $L^2(a, b)$. Then there is a one-to-one correspondence between functions f in $L^2(a, b)$ and the generalized Fourier series

$$\sum_{n=1}^{\infty} (f, u_n) u_n(x)$$

A complete orthogonal family in $L^2(a, b)$ is a basis; that is, the functions in the family play the same role in $L^2(a, b)$ that the vectors **i**, **j**, **k** play in R^3. To illustrate how the use of an orthogonal basis of functions provides the same sort of computational convenience in L^2 that the use of the vectors **i**, **j**, **k** provides in R^3, consider the following example.

EXAMPLE 2.5.2 ——

We want to consider a physical system that is subject to an input $F(t)$ that is a periodic function of time. We expect that after all transient effects have died out, the response consists of an output $u(t)$ that is also a periodic function of time:

Periodic input $F(t) \rightarrow$ physical system $\rightarrow u(t)$ periodic output

We suppose that the physical system is modeled by a linear second-order ordinary differential equation with constant coefficients,

$$Au''(t) + Bu'(t) + Cu(t) = F(t) \qquad (2.5.4)$$

Physical systems modeled by an equation of the form (2.5.4) include

(a) vibrating spring-mass-dashpot system driven by a periodic external force and
(b) an electrical circuit containing resistance, inductance, and capacitance driven by a periodic electromotive force.

Equation (2.5.4) has a solution consisting of two parts,

$$u(t) = u_H(t) + u_P(t)$$

where $u_H(t)$ satisfies $Au'' + Bu' + Cu = 0$ and initial condition $u_P(t)$ satisfies $Au'' + Bu' + Cu = F$. The component $u_H(t)$ is called the *homogeneous solution*, while $u_P(t)$ is referred to as the *particular solution*. The homogeneous solution is of the form

$$u_H(t) = C_1\exp[\alpha_1 t] + C_2\exp[\alpha_2 t]$$

where the α's are the roots of the algebraic equation,

$$P(\alpha) = A\alpha^2 + B\alpha + C = 0$$

In general, the roots of $P(\alpha) = 0$ are complex numbers. If Re $\alpha < 0$ for $\alpha = \alpha_1, \alpha_2$, then $u_H(t)$ tends to zero as t tends to infinity; that is, the transient part of the response dies out exponentially with increasing time. We say then that the system modeled by (2.5.4) is *stable*. For a stable system

$$u(t) \rightarrow u_P(t), \quad t \rightarrow \infty$$

Then $u_P(t)$ represents the so-called *steady-state* or *equilibrium solution*. It is this part of the solution that we expect to be periodic in time if the input $F(t)$ is periodic in time.

We suppose that $F(t)$ is periodic with period T,

$$F(t + T) = F(t) \quad \text{for all } t$$

In addition, we assume that $F(t)$ belongs to $L^2(0, T)$. Now we choose an orthogonal basis for $L^2(0, T)$ and use it in order to represent both the input $F(t)$ and the steady-state output $u_P(t)$.

We use the following orthogonal basis for $L^2(0, T)$:

$$\{1, \cos \Omega t, \cos 2\Omega t, \ldots ; \sin \Omega t, \ldots\}, \qquad \Omega = 2\pi/T$$

Then if we write

$$F(t) = \tfrac{1}{2}F_0 + \sum_{n=1}^{\infty} F_n\cos n\Omega t + G_n\sin n\Omega t$$

it is evident that the coefficients F_n, G_n are the usual Fourier coefficients for $F(t)$; that is,

$$F_n = \frac{1}{2T} \int_0^T F(t)\cos n\Omega t \, dt, \qquad n = 0, 1, \ldots$$

$$G_n = \frac{1}{2T} \int_0^T F(t)\sin n\Omega t \, dt, \qquad n = 1, 2, \ldots$$

Next we write the steady-state output $u_P(t)$ in terms of the same basis,

$$u_P(t) = \tfrac{1}{2}u_0 + \sum_{n=1}^{\infty} u_n\cos n\Omega t + v_n\sin n\Omega t$$

and seek to determine the unknown coefficients u_n and v_n.

Formally, we have

$$u_P'(t) = \sum_{n=1}^{\infty} (n\Omega)[-u_n\sin n\Omega t + v_n\cos n\Omega t]$$

$$u_P''(t) = -\sum_{n=1}^{\infty} (n\Omega)^2[u_n\cos n\Omega t + v_n\sin n\Omega t]$$

If we substitute these series into (2.5.4) and then equate coefficients of $\cos n\Omega t$ and $\sin n\Omega t$ on the two sides of the equation, we obtain, for $n = 1, 2, \ldots,$

$$[C - (n\Omega)^2A]u_n + (n\Omega)Bv_n = F_n$$
$$-(n\Omega)Bu_n + [C - (n\Omega)^2A]v_n = G_n$$

That is,

$$\begin{bmatrix} C - (n\Omega)^2A & n\Omega B \\ -n\Omega B & C - (n\Omega)^2A \end{bmatrix}\begin{bmatrix} u_n \\ v_n \end{bmatrix} = \begin{bmatrix} F_n \\ G_n \end{bmatrix}$$

Computing the inverse of this 2×2 matrix leads to

$$\begin{bmatrix} u_n \\ v_n \end{bmatrix} = \frac{1}{D_n}\begin{bmatrix} C - (n\Omega)^2A & -n\Omega B \\ n\Omega B & C - (n\Omega)^2A \end{bmatrix}\begin{bmatrix} F_n \\ G_n \end{bmatrix} \qquad (2.5.5)$$

where $D_n = [C - (n\Omega)^2A]^2 + (n\Omega)^2B^2$.

Equation (2.5.5) gives the Fourier coefficients of the response in terms of the Fourier coefficients of the input; that is, (2.5.5) defines the output $u_P(t)$ in terms of the input $F(t)$. But Equation (2.5.5) in fact tells us much *more* than that. For $n = 1, 2, \ldots,$ consider the two 2-vectors

$$\Phi_n[u_P](t) = (u_n, v_n) = u_n\cos n\Omega t + v_n\sin n\Omega t$$
$$\Phi_n[F](t) = (F_n, G_n) = F_n\cos n\Omega t + G_n\sin n\Omega t$$

Evidently, $\Phi_n[u_P]$ and $\Phi_n[F]$ represent the nth harmonic component, respectively, of the output and input. Then Equation (2.5.5) can be written in terms of $\Phi_n[u_P]$ and $\Phi_n[F]$ as follows:

$$\Phi_n[u_P] = [Y_n]\Phi_n[F] \tag{2.5.6}$$

where $[Y_n]$ denotes the following 2×2 matrix:

$$\frac{1}{D_n}\begin{bmatrix} C - (n\Omega)^2 A & -n\Omega B \\ n\Omega B & C - (n\Omega)^2 A \end{bmatrix}$$

The matrix $[Y_n]$ represents the effect of the physical system on the nth harmonic component of the input. From (2.5.6) we can compute the magnitude of the output in terms of the magnitude of the input,

$$\|\Phi_n[u_P]\|^2 = \Phi_n[F]^T[Y_n]^T[Y_n]\Phi_n[F] \tag{2.5.7}$$
$$[u_n^2 + v_n^2] = (1/D_n)[F_n^2 + G_n^2]$$

Here we have made use of the fact that $[Y_n]^T[Y_n] = 1/D_n[I]$, where $[I]$ denotes the 2×2 identity matrix.

It is clear from (2.5.7) that $1/D_n$ represents the magnification factor that the system applies to the nth harmonic component of the input to produce the nth harmonic component of the system output. Note that for a fixed frequency Ω, D_n^{-1} is a function of n, and it is then a simple matter to determine for which integer n this factor is largest. Thus not only do we know how the system responds to a given periodic input, but also we know how it responds to *every component* of the input, and we can find which component of the input produces the most dramatic response in the system output. All of this information is brought to light by the use of the orthogonal basis to represent the input and the output in this problem.

Exercises: Function Space L^2

1. Which of the following functions belong to $L^2(0, 1)$?
   (a) $f(x) = [2x - 1]^{2/3}$
   (b) $g(x) = \log(4x - 3)$
   (c) $h(x) = \sum_{n=1}^{\infty} u_n(x)$, where $u_n(x) = \begin{cases} n^{-1/2}, & 1/n < x < 1/(n-1) \\ 0, & \text{otherwise} \end{cases}$
   (d) $p(x) = [\sin x]/x$

2. The spring-mass-dashpot system pictured here

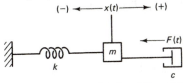

   is modeled by the following differential equation:
   $$mx''(t) = -kx(t) - cx'(t) + F(t)$$

   If $F(t) = A\cos\Omega t$ for A and Ω fixed constants, find the steady response $x(t)$.

3. Suppose the constants in Exercise 2 are as follows:

$$m = 1, \qquad k = 360, \qquad c = 0.1 \qquad \Omega = 6$$

Is this system stable? For which n is D_n^{-1} largest?

4. If k, c, and Ω are as in Exercise 3, for what values of n is it the case that
   (a) $D_n^{-1/2} \le 1$
   (b) $D_n^{-1/2} \ge 1$

2.6 MULTIPLE FOURIER SERIES

We have seen that an arbitrary function in $L^2(a, b)$ can be represented by a generalized Fourier series and that this representation amounts to writing the function in terms of an orthogonal basis for the vector space $L^2(a, b)$.

Consider now a function of *two* variables $F(x, y)$ defined for $a < x < b$, $c < y < d$, and satisfying

$$\int_a^b \int_c^d F(x, y)^2 \, dx \, dy < \infty \tag{2.6.1}$$

Then we say that F belongs to $L^2[(a, b) \times (c, d)]$.

Suppose now that

$\{u_1(x), u_2(x), \ldots\}$ denotes an orthogonal basis for $L^2(a, b)$
$\{v_1(y), v_2(y), \ldots\}$ denotes an orthogonal basis for $L^2(c, d)$

Then it can be shown that

$$\phi_{mn}(x, y) = u_m(x)v_n(y), \qquad m, n = 1, 2, \ldots$$

is an orthogonal basis for $L^2[(a, b) \times (c, d)]$. This means that we can write

$$F(x, y) = \sum_{m=1}^{\infty} \sum_{n=1}^{\infty} F_{mn} u_m(x) v_n(y) \tag{2.6.2}$$

where

$$F_{mn} = \frac{1}{U_m V_n} \int_c^d \int_a^b F(x, y) u_m(x) v_n(y) \, dx \, dy \tag{2.6.3}$$

and

$$U_m = (u_m, u_m), \qquad V_n = (v_n, v_n)$$

The series (2.6.2) is called a *multiple* Fourier series. This notion may be extended to functions of more than two variables.

EXAMPLE 2.6.1 ———

1. Consider the special case that $F(x, y) = f(x)g(y)$ for f in $L^2(a, b)$ and g in $L^2(c, d)$. Then if $\{u_1(x), u_2(x), \ldots\}$ and $\{v_1(y), v_2(y), \ldots\}$ denote orthogonal bases for $L^2(a, b)$ and $L^2(c, d)$, respectively,

$$F_{mn} = \frac{1}{U_m V_n} \int_c^d \int_a^b f(x)g(y)u_m(x)v_n(y) \, dx \, dy$$

$$F_{mn} = \frac{1}{U_m} \int_a^b f(x)u_m(x) \, dx \, \frac{1}{V_n} \int_c^d g(y)v_n(y) \, dx$$

$$= \frac{(f, u_m)}{(u_m, u_m)} \frac{(g, v_n)}{(v_n, v_n)}$$

Then

$$F(x, y) = \sum_{m=1}^{\infty} \frac{(f, u_m)}{(u_m, u_m)} u_m(x) \sum_{n=1}^{\infty} \frac{(g, v_n)}{(v_n, v_n)} v_n(y)$$

2. Consider the function

$$F(x, y) = 6 - 2x - 3y, \qquad 0 < x < 3, \qquad 0 < y < 2$$

This function $F(x, y)$ is continuous on the closed set $[0, 3] \times [0, 2]$ and so F is clearly in $L^2[(0, 3) \times (0, 2)]$. However, F is not of the form $f(x)g(y)$ as in the previous example. We know that

$\{\sin(m\pi x/3) : m = 1, 2, \ldots)$ is an orthogonal basis in $L^2(0, 3)$

$(\sin(n\pi y/2) : n = 1, 2, \ldots\}$ is an orthogonal basis in $L^2(0, 2)$

and hence we can compute F_{mn} such that

$$F(x, y) = \sum_{m=1}^{\infty} \sum_{n=1}^{\infty} F_{mn} \sin \alpha_m x \sin \beta_n y$$

where $\alpha_m = (m\pi/3)$, $\beta_n = (n\pi/2)$. According to (2.6.3),

$$F_{mn} = \frac{1}{U_m V_n} \int_0^2 \int_0^3 (6 - 2x - 3y) \sin \alpha_m x \sin \beta_n y \, dx \, dy$$

$$= \frac{1}{U_n V_m} \int_0^2 \int_0^3 [(6 - 3y) - 2x] \sin \alpha_m x \, dx \sin \beta_n y \, dy$$

$$= \frac{1}{U_m V_n} \int_0^2 \frac{3}{m\pi} [6 - (1 - (-1)^m)3y] \sin \beta_n y \, dy$$

$$= \frac{1}{U_m V_n} \frac{(36/\pi^2)[1 - (-1)^{m+n}]}{mn}$$

But $U_m = (\sin \alpha_m x, \sin \alpha_m x) = \frac{3}{2}$, $V_n = (\sin \beta_n y, \sin \beta_n y) = 1$, and therefore,

$$F_{mn} = (24/\pi^2)[1 - (-1)^{m+n}]/mn, \qquad m, n = 1, 2, \ldots \qquad \blacksquare \blacksquare$$

Exercises: Multiple Fourier Series

1. Write a double Fourier series for the following functions in terms of the orthogonal basis:

$$\{\sin kx : k = 1, 2, \ldots\} \times \{\cos ny : n = 0, 1, \ldots\}$$

for $L^2[(0, \pi) \times (0, \pi)]$.

(a) $f(x, y) = \begin{cases} 1 & \text{for } 0 < x < \pi, 0 < y < \pi/2 \\ 0 & \text{for } 0 < x < \pi, \pi/2 < y < \pi \end{cases}$

(b) $g(x, y) = xy$ for $0 < x, y < \pi$

(c) $h(x, y) = \begin{cases} 1 & \text{if } x < y < \pi - x, 0 < x < \pi/2 \\ -1 & \text{if } \pi - x < y < x, \pi/2 < x < \pi \\ 0 & \text{otherwise} \end{cases}$

(d) $p(x, y) = \begin{cases} x - xy & \text{if } 0 < x < \pi/2, \pi/2 < y < \pi \\ 0 & \text{if } 0 < x < \pi/2, 0 < y < \pi/2, \\ & \text{or if } \pi/2 < x < \pi, \pi/2 < y < \pi \\ y - xy & \text{if } \pi/2 < x < \pi, 0 < y < \pi/2 \end{cases}$

2.7 SUMMARY

In this chapter we promoted the notion of viewing the space L^2 of square-integrable functions as a vector space in which orthogonal families of eigenfunctions play the role played by orthogonal bases in R^N. In order to emphasize the parallels between the vector space L^2 and the more familiar vector space R^N, we list here for both of these spaces, the definitions of the vectors, the linear combinations, the inner products, the norms, orthonormal sets, and completeness.

R^N	$L^2(a, b)$
1. The vectors: $\mathbf{X} = (x_1, \ldots, x_N)$	$f(x)$, defined on (a, b) satisfying $\int_a^b \|f(x)\|^2 \, dx < \infty$
2. Linear combinations: $\alpha \mathbf{X} + \beta \mathbf{Y} = (\alpha x_1 + \beta y_1, \ldots, \alpha x_N + \beta y_N)$	$(\alpha f + \beta g)(x) = \alpha f(x) + \beta g(x)$
3. Inner (dot) product: $(\mathbf{X}, \mathbf{Y}) = \mathbf{X} \cdot \mathbf{Y} = x_1 y_1 + \cdots x_N y_N$	$(f, g) = \int_a^b f(x)g(x) \, dx$
4. Norm: $\|\mathbf{X}\|^2 = (\mathbf{X}, \mathbf{X})$	$\|f\|^2 = (f, f)$
5. Orthonormal family: $(\mathbf{X}_j, \mathbf{X}_k) = \delta_{kj}, j, k = 1, \ldots, N$	$(f_k, f_j) = \delta_{kj}, j, k = 1, 2, \ldots, \infty$
6. Completeness: $$\mathbf{V} = \sum_{k=1}^M (\mathbf{V}, \mathbf{X}_k)\mathbf{X}_k$$ Equality holds for all $\mathbf{V}$ if and only if $M = N$	$$G(x) = \sum_{k=1}^\infty (G, f_k)f_k(x)$$ Equality holds for all G if and only if $\{f_k\}$ is complete
$$\|\mathbf{V}\|^2 \geq \sum_{k=1}^M (\mathbf{V}, \mathbf{X}_k)^2$$	$$\|G\|^2 \geq \sum_{k=1}^\infty (G, f_k)^2$$

Boundary-Value Problems and Initial-Boundary-Value Problems on Spatially Bounded Domains

In this chapter we begin solving problems in partial differential equations. Each of the problems of this chapter will have in common the feature that the spatial domain of the problem is a bounded set. We find that on bounded domains, the eigenfunction expansion techniques of the previous chapter can be brought to bear in order to find the solution of the problem. When the spatial domain of the problem is not bounded, the eigenfunction expansion methods do not generally apply. In the next chapters we introduce the techniques that must be used when the spatial domain of the problem is not a bounded set.

In order to apply the method of eigenfunction expansion it will be necessary to make extensive use of a principle known as the "principle of superposition."

The Principle of Superposition   Note that for $u = u(x, y)$ and $v = v(x, y)$ smooth functions of x and y, and for arbitrary constants α, β,

$$\frac{\partial}{\partial x} [\alpha u + \beta v] = \alpha u_x + \beta v_x$$

Identical results hold when $\partial/\partial x$ is replaced by any of the derivatives $\partial/\partial y$, $\partial^2/\partial x^2$, $\partial^2/\partial y^2$, and so on. In fact, for smooth function $u(x, y)$, let $L[u]$ denote a combination of derivatives of the form

$$L[u] = a(x, y)u_{xx} + 2b(x, y)u_{xy} + c(x, y)u_{yy}$$
$$+ d(x, y)u_x + e(x, y)u_y + f(x, y)u$$

It is not hard to see that for a pair of smooth functions $u(x, y)$, $v(x, y)$ and arbitrary constants α and β,

$$L[\alpha u + \beta v] = \alpha L[u] + \beta L[v]$$

We say in this case that $L[\ ]$ is a *linear differential operator*. If $L[\ ]$ is a linear differential operator and if $u_1(x, y)$ and $u_2(x, y)$ each satisfy the linear homogeneous partial differential equation $L[u] = 0$, it is easy to see that $\alpha u_1 + \beta u_2$ must also then satisfy the same homogeneous equation, whatever the constants α and β. This is the simplest statement of the principle of superposition. More generally, we have

Principle. For linear differential operator $L[\ ]$ and smooth functions $u_1, \ldots, u_N$ satisfying $L[u_k] = 0$ for $k = 1, \ldots, N$,

$$L\left[\sum_{k=1}^{N} C_k u_k \right] = 0 \quad \text{for arbitrary constants } C_1, \ldots, C_N$$

When N, the number of solutions to the homogeneous equation, is allowed to become infinite, the principle remains valid so long as the series

$$\sum_{k=1}^{\infty} C_k u_k \quad \text{and} \quad L\left[\sum_{k=1}^{\infty} C_k u_k \right]$$

both converge.

The principle of superposition applies to linear boundary conditions and initial conditions as well as to linear partial differential equations. That is, if each one of a set of functions satisfies a given homogeneous auxiliary condition, then arbitrary linear combinations of these functions will also satisfy the same homogeneous condition.

3.1 BOUNDARY-VALUE PROBLEMS FOR LAPLACE AND POISSON EQUATIONS

In this section we consider several examples of boundary-value problems on bounded domains. In all of them the governing equation will be either the homogeneous Laplace equation or its inhomogeneous version, the Poisson equation.

EXAMPLE 3.1.1 _____

Dirichlet Problem for Laplace Equation
    Let Ω denote the unit square $(0, 1) \times (0, 1)$ and suppose that $u = u(x, y)$ satisfies

$$\nabla^2 u(x, y) = 0 \quad \text{in } \Omega$$
$$u(0, y) = 0, \qquad u(1, y) = f(y), \qquad 0 < y < 1$$
$$u(x, 0) = 0, \qquad u(x, 1) = 0, \qquad 0 < x < 1$$

where $f(y)$ denotes a given function of y belonging, say, to $L^2(0, 1)$. We refer to this as a ''Dirichlet problem'' because Dirichlet-type boundary conditions are imposed around the entire boundary of Ω; that is, the value of the unknown function is specified over the boundary of Ω. A problem of this sort arises in connection with determining the steady-state distribution of temperature in a square plate, three of whose edges are maintained at temperature zero while the temperature of the fourth edge is given by the prescribed function $f(y)$.

Separation of Variables   To find the unknown function $u(x, y)$, we assume that

$$u(x, y) = X(x)Y(y) \tag{3.1.1}$$

Admittedly, there is nothing about the problem to suggest ahead of time that the dependence of u on x and y should be of this special form. However, the assumption is made in the hopes that it will simplify the problem in some way that will allow the solution to be found. As we shall see, that is what happens.

For $u(x, y)$ as in (3.1.1), we have

$$u_{xx}(x, y) = X''(x)Y(y), \qquad u_{yy}(x, y) = X(x)Y''(y)$$

where the primes denote differentiation of a function of one variable with respect to its argument. Substituting these expressions for the derivatives into the Laplace equation leads to

$$X''(x)Y(y) + X(x)Y''(y) = 0$$

that is,

$$X''(x)/X(x) = -Y''(y)/Y(y)$$

The left side of this equation depends only on the variable x, and the right side depends only on the variable y. The only way the two sides of the equation can remain equal to each other as x and y range over the values from 0 to 1 is for each side to equal the same constant, say, μ. Then the assumption (3.1.1) reduces the Laplace equation in the two independent variables x and y to the two ordinary differential equations

$$X''(x) = \mu X(x) \tag{3.1.2a}$$
$$-Y''(y) = \mu Y(y) \tag{3.1.2b}$$

Substituting (3.1.1) into the boundary conditions of the problem leads to

$$X(x)Y(0) = 0, \qquad X(x)Y(1) = 0, \qquad 0 < x < 1 \tag{3.1.3a}$$
$$X(0)Y(y) = 0, \qquad X(1)Y(y) = f(y), \qquad 0 < y < 1 \tag{3.1.3b}$$

One way for (3.1.3a) to be satisfied is for $X(x)$ to vanish for all x, but this is not consistent with the second part of (3.1.3b). A second way in which (3.1.3a) can be satisfied is for $Y(0)$ and $Y(1)$ to vanish, and since this does not lead to any evident inconsistencies, we accept this possibility. Then we have

$$-Y''(y) = \mu Y(y), \qquad Y(0) = Y(1) = 0 \tag{3.1.4}$$

We observe that (3.1.4) is a Sturm–Liouville problem; in fact, it is exactly the Sturm–Liouville problem of Example 2.2.1 (with $L = 1$). We found, in Example 2.2.1 that the eigenvalues for problem (3.1.4) are the following:

$$\mu_n = (n\pi)^2, \qquad n = 1, 2, \ldots \tag{3.1.5}$$

The corresponding eigenfunctions were found to be

$$Y_n(y) = \sin n\pi y, \qquad n = 1, 2, \ldots \tag{3.1.6}$$

Reasoning as we did for $Y(y)$, we conclude that $X(0)$ must be zero. For the time being, we cannot make use of the inhomogeneous condition in (3.1.3b) [we may be tempted to believe that $Y(y)$ is equal to $f(y)/X(1)$, but this will, in general, conflict with (3.1.2b)]. Then we have, for $X(x)$,

$$X''(x) = \mu X(x), \qquad X(0) = 0 \tag{3.1.7}$$

Equations (3.1.4) and (3.1.7) are two separate ordinary differential equations for the unknown functions $X(x)$ and $Y(y)$, but they are linked together by the common separation constant μ. For μ equal to one of the values μ_n from (3.1.5), a general solution of (3.1.7) can be written in the form

$$X_n(x) = A_n \sinh n\pi x + B_n \sinh n\pi(1 - x)$$

The homogeneous boundary condition

$$X_n(0) = B_n \sinh n\pi = 0$$

implies that $B_n = 0$. Then the solution of (3.1.7) corresponding to the value $\mu = \mu_n = (n\pi)^2$ for the separation constant is

$$X_n(x) = A_n \sinh n\pi x, \qquad n = 1, 2, \ldots \tag{3.1.8}$$

Here A_n denotes an arbitrary constant of integration whose value may depend on the index n.

Inhomogeneous Boundary Condition   At this point, our assumption (3.1.1) has led us to a family of functions

$$u_n(x, y) = A_n \sinh n\pi x \sin n\pi y, \qquad n = 1, 2, \ldots \tag{3.1.9}$$

with the property that for each n, $u_n(x, y)$ satisfies the Laplace equation and all of the homogeneous boundary conditions of the problem we are trying to solve. Then by the principle of superposition, the function

$$u(x, y) = \sum_{n=1}^{\infty} A_n \sinh n\pi x \sin n\pi y \tag{3.1.10}$$

has all of these same properties whatever choice we make for the constants A_n. Of course, this statement is only formal in the sense that we are temporarily ignoring any questions relating to the convergence of the infinite series involved. If we subject the function (3.1.10) to the inhomogeneous boundary condition, we obtain

$$u(1, y) = \sum_{n=1}^{\infty} A_n \sinh n\pi \sin n\pi y = f(y) \tag{3.1.11}$$

If values can be found for the constants A_n such that (3.1.11) holds, then with this choice for the A_n, (3.1.10) will satisfy all of the conditions of the Dirichlet problem, and the problem will be solved. Finding values for A_n such that (3.1.11) holds is made possible by the fact that the family of functions

$$\{\sin n\pi y : n = 1, 2, \ldots\}$$

are just the eigenfunctions listed in (3.1.6), and therefore they form an orthogonal basis in $L^2(0, 1)$. If we take the inner product with $Y_k(y) = \sin k\pi y$ on both sides of (3.1.11), then the orthogonality of the eigenfunctions implies

$$\sum_{n=1}^{\infty} A_n(\sinh n\pi)(Y_n, Y_k) = (f, \sin k\pi y)$$

$$\tfrac{1}{2}A_k\sinh k\pi = (f, \sin k\pi y)$$

since

$$(Y_n, Y_k) = \begin{cases} \tfrac{1}{2} & \text{if } n = k \\ 0 & \text{if } n \neq k \end{cases}$$

Evidently, if we choose, for $n = 1, 2, \ldots$,

$$A_n = 2(f, \sin n\pi y)/\sinh n\pi$$

$$= \frac{2}{\sinh n\pi} \int_0^1 f(y)\sin n\pi y \, dy \tag{3.1.12}$$

then the completeness of the family of eigenfunctions implies that (3.1.11) holds, and it follows that (3.1.10) [with A_n given by (3.1.12)] is the solution to the Dirichlet problem. That is, if

$$u(x, y) = \sum_{n=1}^{\infty} f^n \frac{\sinh n\pi x}{\sinh n\pi} \sin n\pi y \tag{3.1.13}$$

where

$$f^n = 2 \int_0^1 f(y)\sin n\pi y \, dy, \qquad n = 1, 2, \ldots \tag{3.1.14}$$

then $u(x, y)$ satisfies all of the conditions of our Dirichlet problem. This solution is just formal since we have not verified that the series in (3.1.13) converges. Although we do not prove this now, it can be shown that the infinite series in (3.1.13) converges absolutely and uniformly on closed subsets of Ω. Moreover, this series can be differentiated any number of times with respect to either x or y, and it can be shown that this differentiated series will also converge absolutely and uniformly to the appropriate derivative of $u(x, y)$. These convergence proofs are the basis of a rigorous proof of the fact that $u(x, y)$ given by (3.1.13) satisfies Laplace's equation in Ω. We do not pursue at this time a rigorous proof of the fact that $u(x, y)$ is a solution of the partial differential equation.

Discussion of Solution   The homogeneous boundary conditions in the problem are easily seen to be satisfied by the function $u(x, y)$ of (3.1.13) (without engaging in any discussion of convergence) since $\sin n\pi y$ vanishes, for every n, at $y = 0$ and $y = 1$, and $\sinh n\pi x$ vanishes at $x = 0$ for every n.

Note that constants f^n are just the generalized Fourier coefficients for the function $f(y)$ relative to the family of eigenfunctions given by (3.1.6). Then Theorem 2.2.3 implies that the inhomogeneous boundary condition at $x = 1$ is satisfied in at least the mean-square sense. Note also that the quotient

$$\sinh n\pi x/\sinh n\pi \tag{3.1.15}$$

acts as a "modifier" on the Fourier coefficients f^n. That is, at $x = 1$

$$f^n \sinh n\pi x/\sinh n\pi = f^n$$

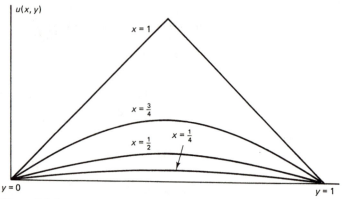

Figure 3.1.1

and it follows from (3.1.13) that $u(1, y) = f(y)$; at $x = 0$

$$f^n \text{sinh } n\pi x / \text{sinh } n\pi = 0$$

and then (3.1.13) implies that $u(0, y) = 0$. For x between 0 and 1 the quotient in (3.1.15) causes the coefficients of $\sin n\pi y$ in (3.1.13) to vary between these two extremes in such a way as to ensure that $u(x, y)$ satisfies the Laplace equation for $0 < y < 1$ for each x, $0 < x < 1$; that is,

$$0 \leq f^n \text{sinh } n\pi x / \text{sinh } n\pi \leq f^n$$

Consider, for example,

$$f(y) = \begin{cases} 2y & \text{for } 0 < y < \tfrac{1}{2} \\ 2(1 - y) & \text{for } \tfrac{1}{2} < y < 1 \end{cases}$$

Then,

$$f^n = 4(-1)^{n+1}[\pi(2n - 1)]^{-2}, \qquad n = 1, 2, \ldots$$

and if we plot $u(x, y)$ versus y for discrete values of x, we obtain Figure 3.1.1.

At $x = 1$, $u(x, y)$ equals the piecewise linear function $f(y)$. Note that $f(y)$ has a discontinuity in its derivative at $y = \tfrac{1}{2}$. At $x = \tfrac{3}{4}, \tfrac{1}{2}, \tfrac{1}{4}$ the profiles of u versus y are smooth, even at $y = \tfrac{1}{2}$, and we can see that in addition to this smoothing effect, the modifier (3.1.15) causes the profiles (plots of u vs. y) to decrease uniformly with increasing x. We can give a physical interpretation for Figure 3.1.1 also. Suppose a thin, square heat-conducting plate has the temperature distribution described by the piecewise linear function $f(y)$ imposed along the right edge of the plate. Then the heat diffuses through the plate, and the temperature decreases with increasing distance from the edge at $x = 1$, as is pictured in Figure 3.1.1.

We can also view the solution as it is expressed in (3.1.13) in the following way. If we write

$$u^n(x) = f^n \text{sinh } n\pi x / \text{sinh } n\pi, \qquad n = 1, 2, \ldots \tag{3.1.16}$$

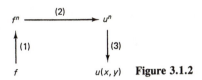

Figure 3.1.2

then the $u^n(x)$ represent the x-dependent generalized Fourier coefficients for the solution u relative to the eigenfunctions of this problem. The solution procedure can be thought of as consisting of the following steps:

1. The *data function* $f(y)$ is converted to a generalized Fourier series in terms of the eigenfunctions of this problem via (3.1.14).
2. The generalized Fourier coefficients of the solution u relative to the appropriate eigenfunctions are found by applying the separation-of-variables procedure. These coefficients are the ones given in (3.1.16).
3. The solution $u(x, y)$ can be evaluated for (x, y) in Ω by summing the Fourier series (3.1.13). More precisely, we can approximate the value of $u(x, y)$ by summing finitely many terms of the series (3.1.13).

The composition of these three steps forms a mapping that carries the *data coefficients* f^n into the solution coefficients $u^n(x)$. This is one way of representing the *solution operator* for this problem. Note that step 2 of the procedure is the actual solution step while steps 1 and 3 are conversion to and from the Fourier series representation. If step 1 in Figure 3.1.2 is replaced by a discrete Fourier series similar to what is described in Section 2.3 and then a correspondingly truncated series is used in step 3, the result is an approximation to the exact solution. An approximation scheme of this sort is referred to as a *spectral method*. ■ ■

EXAMPLE 3.1.2 _____

Neumann Problem for Laplace Equation
We continue to let Ω denote the unit square and consider the problem

$$\nabla^2 u(x, y) = 0 \quad \text{in } \Omega$$
$$u_y(x, 0) = f(x), \qquad u_y(x, 1) = 0, \qquad 0 < x < 1$$
$$u_x(0, y) = 0, \qquad u_x(1, y) = 0, \qquad 0 < y < 1$$

Since the unknown function $u(x, y)$ in this problem is subjected to Neumann-type boundary conditions (the normal derivative of u is specified at each point on the boundary), we refer to this as a Neumann problem. In the context of steady-state heat conduction, the boundary conditions here can be given the following interpretation. The heat flux vector field satisfies the continuous Fourier law of heat conduction, namely,

$$\Phi(x, y) = -K \operatorname{grad} \mathbf{u}(x, y)$$

On the vertical sides of the square, the unit outward normal vector is $\mathbf{i}$ or $-\mathbf{i}$ and hence the heat flux $\Phi \cdot \mathbf{N}$ through those edges of the square plate is proportional to u_x. Similarly, the normal to the horizontal edges of the plate is $\mathbf{j}$ or $-\mathbf{j}$, and the heat flux through the top and bottom edges of the square plate is proportional to u_y. Thus the boundary conditions in this Neumann problem can be interpreted as requiring that three edges of the plate are insulated against the flow of heat while the flow of heat through the fourth edge is equal to a prescribed function.

Separation of Variables   Using the separation-of-variables assumption (3.1.1) in the equation and boundary conditions of this problem leads to the following separated problems:

$$-X''(x) = \mu X(x), \qquad X'(0) = X'(1) = 0 \qquad\qquad (3.1.17a)$$
$$Y''(y) = \mu Y(y), \qquad Y'(1) = 0 \qquad\qquad\qquad (3.1.17b)$$

In this problem it is the X part of the solution, $X(x)$, which satisfies the Sturm–Liouville problem, namely, (3.1.17a). Note that we have chosen the sign of the separation constant μ so that the X problem (3.1.17a) conforms to the notation we introduced in Chapter 2 in connection with Sturm–Liouville problems. In particular, (3.1.17a) is exactly the Sturm–Liouville problem of Example 2.2.2 (with $L = 1$).

It should be evident from this example and the previous one that at least one of the two separated ordinary differential equations that results from using assumption (3.1.1) in the boundary-value problem will be a Sturm–Liouville problem. Moreover, the Sturm–Liouville problem can be identified as the problem containing two homogeneous boundary conditions. In this boundary-value problem, the homogeneous boundary conditions occur at $x = 0$ and $x = 1$, and the Sturm–Liouville problem is then the X problem. In the previous example, the homogeneous conditions occurred at $y = 0$ and $y = 1$, and the Y problem was the Sturm–Liouville problem. We show in the next example how to deal with a boundary-value problem in which there are no homogeneous boundary conditions.

Continuing with this example, we recall from Example 2.2.2 that the eigenvalues and eigenfunctions associated with problem (3.1.17a) are

$$\mu_n = (n\pi)^2 \qquad\qquad\qquad\qquad (3.1.18)$$
$$X_n(x) = \cos n\pi x, \qquad n = 0, 1, 2, \ldots \qquad\qquad (3.1.19)$$

Using the eigenvalues (3.1.18) in the Y equation (3.1.17b), we can write the general solution of that equation in the form

$$Y_n(y) = A_n\cosh n\pi y + B_n\cosh n\pi(1 - y) \qquad\qquad (3.1.20)$$

Here we have chosen to use this pair of independent solutions for Equation (3.1.17b) because of the derivative boundary conditions in this boundary-value problem. In future problems it may be convenient to recall that any two of the following six functions are a pair of independent solutions for (3.1.17b):

(i) $\exp[\beta y]$, $\exp[-\beta y]$
(ii) $\sinh \beta y$, $\sinh \beta(L - y)$
(iii) $\cosh \beta y$, $\cosh \beta(L - y)$, $\beta^2 = \mu$

Note that the pair of functions in (i) vanish as y tends to minus and plus infinity, respectively. The functions in pair (ii) vanish at $y = 0$ and L, respectively, while the first derivative of the solutions in pair (iii) vanish at $y = 0$ and L, respectively.

From (3.1.20) and (3.1.17b), we have

$$Y_n'(1) = n\pi[A_n\sinh n\pi - 0] = 0$$

from which it follows immediately that $A_n = 0$ for $n = 0, 1, \ldots$. Then for each n,

$$u_n(x, y) = Y_n(y)X_n(x) = B_n\cosh n\pi(1 - y)\cos n\pi x$$

satisfies the Laplace equation and all three of the homogeneous boundary conditions in this boundary-value problem for any choice of the (at this point arbitrary) constants B_n. Then by the principle of superposition, the same can be said (at least formally) for the series

$$u(x, y) = \sum_{n=1}^{\infty} B_n\cosh n\pi(1 - y)\cos n\pi x + B_0 \qquad (3.1.21)$$

Inhomogeneous Boundary Condition   The final step in the process of solving this boundary-value problem is to try to find values for the unspecified constants B_n such that the inhomogeneous boundary condition in the problem can be satisfied. We begin by writing

$$u_y(x, 0) = -\sum_{n=1}^{\infty} n\pi B_n\sinh n\pi \cos n\pi x = f(x) \qquad (3.1.22)$$

Using the orthogonality of the eigenfunctions, $\cos n\pi x$, we have, for $n = 1, 2, \ldots$,

$$-\tfrac{1}{2}n\pi B_n\sinh n\pi = \int_0^1 f(x)\cos n\pi x \, dx$$

Then,

$$B_n = -f^n/\{n\pi \sinh n\pi\}, \qquad n = 1, 2, \ldots \qquad (3.1.23)$$

where

$$f^n = 2\int_0^1 f(x)\cos n\pi x \, dx, \qquad n = 1, 2, \ldots$$

The numbers f^n are just the generalized Fourier coefficients in the eigenfunction expansion of $f(x)$.

Discussion of Solution   For $n = 0$, it follows from (3.1.22) that

$$0 = \int_0^1 f(x) \, dx \qquad (3.1.24)$$

There are two pieces of information implied by (3.1.22) for $n = 0$. First, the constant B_0 is indeterminate; that is, the solution to this boundary-value problem is not unique since the series (3.1.21) satisfies all the conditions of the boundary-value problem whatever value we give for B_0. Second, the boundary-value problem has no solution at all unless the condition (3.1.24) is satisfied. Here (3.1.24) is a condition of compatibility on the data function $f(x)$ originating in the following observation:

$$0 = \int_0^1 \int_0^1 [u_{xx}(x, y) + u_{yy}(x, y)]\, dx\, dy$$

$$0 = \int_0^1 [u_x(1, y) - u_x(0, y)]\, dy + \int_0^1 [u_y(x, 1) - u_y(x, 0)]\, dx$$

$$0 = 0 - \int_0^1 f(x)\, dx$$

Here we began with the condition that $u(x, y)$ satisfies Laplace's equation in Ω, integrated by parts, and used the boundary conditions to arrive at the condition (3.1.24). This is a special case of the following application of the divergence theorem (see Chapter 6):

$$\int_\Omega \text{div}[\text{grad } u]\, dx\, dy = \int_S \text{grad } u \cdot \mathbf{N}\, d\sigma \tag{3.1.25}$$

where S denotes the smooth boundary of the closed, bounded region Ω and N denotes the unit outward normal vector to S. It follows from (3.1.25) that a necessary condition for $u = u(x, y)$ to satisfy

$$\text{div}[\text{grad } u] = \nabla^2 u(x, y) = F(x, y) \quad \text{in } \Omega$$
$$\partial_N u(x, y) = f \quad \text{on } S$$

is then

$$\int_\Omega F\, dx\, dy = \int_S f\, d\sigma$$

The Dirichlet-type boundary-value problem has no such compatibility condition associated with it.

If $f(x)$ satisfies the compatibility condition (3.1.24), then a nonunique solution of the Neumann problem is given by

$$u(x, y) = B_0 + \sum_{n=1}^\infty f^n \frac{\cosh n\pi(1 - y)}{n\pi \sinh n\pi} \cos n\pi x \tag{3.1.26}$$

for arbitrary constant B_0.

The solution (3.1.26) is formal in the sense that convergence of the series has not been established. However, as in the previous example, this convergence can be established, and in fact, it can be shown that on closed subsets of Ω the series can be differentiated arbitrarily often with respect to either x or y, and the differentiated series will converge uniformly to the corresponding derivative of $u(x,y)$. This is true for $f(x)$ an arbitrary element of $L^2(0, 1)$.

The solution procedure for this boundary-value problem is represented schematically in Figure 3.1.2. In this problem, however, the generalized Fourier coefficients of the solution are given by

$$u^n(y) = f^n \frac{\cosh n\pi(1 - y)}{n\pi \sinh n\pi}$$

The scheme of Figure 3.1.2 also suggests a method for approximating the solution to this Neumann problem by a spectral method. That is, step 1 of the scheme could be carried out approximately by means of a discrete Fourier series similar to what was described in Section 2.3. Then step 3 of the scheme could be replaced by a suitably truncated series. ■ ■

In each of the two examples considered so far, there has been just one inhomogeneous ingredient in the problem, that is, one inhomogeneous boundary condition. These examples were selected for the purpose of illustrating how the separation assumption (3.1.1) causes the Laplace equation to separate into two ordinary differential equations coupled by the separation constant μ. In each of the examples, one of the ordinary differential equations carries with it the pair of homogeneous, two-point boundary conditions needed to form a Sturm–Liouville problem. This Sturm–Liouville problem is then the source of the eigenfunctions that ultimately are used in the eigenfunction expansion of the solution. It is the completeness of this family of functions that ensures that the inhomogeneous boundary condition can be satisfied by appropriately choosing the free constants in the eigenfunction expansion of the solution. We now show how a boundary-value problem with no homogeneous boundary conditions can be reduced to a series of simpler subproblems of the type we have already considered.

EXAMPLE 3.1.3

Mixed Boundary-Value Problem for Poisson Equation
We continue to let Ω denote the unit square and consider the problem of finding an unknown function $u = u(x, y)$ satisfying

$$\nabla^2 u(x, y) = F(x, y) \quad \text{in } \Omega$$
$$u(0, y) = f(y), \quad u(1, y) = g(y), \quad 0 < y < 1$$
$$u_y(x, 0) = p(x), \quad u_y(x, 1) = q(x), \quad 0 < x < 1$$

Here, F, f, g, p, q all denote given data functions that are at least square-integrable functions of their respective arguments. We refer to this as a mixed boundary-value problem because the boundary conditions are a mixture of Dirichlet and Neumann boundary conditions. In the context of steady-state heat conduction, the boundary conditions can be interpreted to mean that on two edges of the square plate, the temperature is specified, while on the remaining two edges of the plate, the flux of heat is specified. The forcing function $F(x, y)$ can be interpreted as the result of a time-independent source or sink of heat inside Ω. This could be due to some sort of chemical or fission reaction, for example.

Reducing Problem to Subproblems We shall split this problem into five subproblems, one for each of the five inhomogeneous terms in the main problem. Then it will be seen that the sum of the five subproblem solutions is a solution of the main problem. We proceed now to state and solve each of the five subproblems in succession.

Subproblem A

$$\nabla^2 u^A(x, y) = 0 \quad \text{in } \Omega$$
$$u^A(0, y) = f(y)$$
$$u^A(1, y) = 0, \quad 0 < y < 1$$
$$u_y^A(x, 0) = u_y^A(x, 1) = 0, \quad 0 < x < 1$$

$u_y^A = 0$

$u^A = f \quad \boxed{\nabla^2 u^A = 0} \quad u^A = 0$

$u_y^A = 0$

Separation of Variables (Subproblem A)   The separation assumption (3.1.1) will cause the Laplace equation to separate, as usual, into two ordinary differential equations. Moreover the homogeneous boundary conditions will lead to the following homogeneous conditions for the separated solutions $X(x)$ and $Y(y)$:

$$Y'(0) = Y'(1) = 0 \quad \text{and} \quad X(1) = 0$$

Evidently it is the $Y(y)$ problem here that is the Sturm–Liouville problem. This is foreseeable from the fact that two homogeneous boundary conditions involving the same independant variable are, in this subproblem, the homogeneous conditions on $u_y(x, y)$ at $y = 0$ and $y = 1$. The inhomogeneous condition in this subproblem is the boundary condition at $x = 0$, and it follows that the $X(x)$ separated problem is not the Sturm–Liouville problem.

The Sturm–Liouville problem for $Y(y)$ is the following:

$$-Y''(y) = \mu Y(y), \qquad Y'(0) = Y'(1) = 0$$

As we have seen in the previous example, the eigenvalues and eigenfunctions for this problem are

$$\mu_n = (n\pi)^2, \qquad Y_n(y) = \cos n\pi y, \qquad n = 0, 1, \dots$$

The coupled $X(x)$ problem for the case $n > 0$ reads

$$X_n''(x) = (n\pi)^2 X_n(x), \qquad X_n(1) = 0$$

The solution to this problem is expressed in the form

$$X_n(x) = A_n \sinh n\pi x + B_n \sinh n\pi(1 - x)$$

and then

$$X_n(1) = A_n \sinh n\pi + 0 = 0$$

It follows that A_n vanishes, and for $n = 1, 2, \dots$ we have

$$X_n(x) = B_n \sinh n\pi(1 - x)$$

where B_n is, at this point, still arbitrary.

For $n = 0$ the $X(x)$ problem reduces to

$$X_0''(x) = 0, \qquad X_0(1) = 0$$

which has the solution $X_0(x) = B_0(1 - x)$ for B_0 arbitrary.

Then combining the X and Y solutions, we have

$$u^A(x, y) = B_0(1 - x) + \sum_{n=1}^{\infty} B_n \sinh n\pi(1 - x)\cos n\pi y$$

and this function satisfies the Laplace equation and the three homogeneous boundary conditions. The inhomogeneous boundary condition is satisfied as well if the constants B_n satisfy

$$f(y) = B_0 + \sum_{n=1}^{\infty} B_n \sinh n\pi \cos n\pi y, \qquad 0 < y < 1$$

The orthogonality of the eigenfunctions implies that for fixed integer m, if we multiply both sides of this expression by $\cos(m\pi x)$ and integrate from 0 to 1, then only one term will remain on the right side of the equation. Then we easily find

$$B_0 = \int_0^1 f(y)\, dy, \qquad B_m = \frac{2}{\sinh(m\pi)} \int_0^1 f(y) \cos m\pi y\, dy$$

That is,

$$B_0 = f^0, \qquad B_m = f^m / \sinh m\pi$$

Finally, the completeness of the eigenfunctions implies that with this choice for the constants B_n, the inhomogeneous boundary condition is satisfied, at least in the mean-square sense. Then the solution to subproblem A is given by

$$u^A(x, y) = f^0(1 - x) + \sum_{n=1}^{\infty} f^n \frac{\sinh n\pi(1 - x)}{\sinh n\pi} \cos n\pi y \qquad (3.1.27)$$

It can be shown that this formal solution is in fact the unique solution to subproblem A.

Subproblem B

$$\nabla^2 u^B(x, y) = 0 \quad \text{in } \Omega$$
$$u^B(0, y) = 0$$
$$u^B(1, y) = g(y), \qquad 0 < y < 1,$$
$$u_y^B(x, 0) = u_y^B(x, 1) = 0, \qquad 0 < x < 1$$

$$u_y^B = 0$$
$$u^B = 0 \quad \boxed{\nabla^2 u^B = 0} \quad u^B = g$$
$$u_y^B = 0$$

If we let $z = 1 - x$, then $u_x^B = -u_z^B$ and $u_{xx}^B = u_{zz}^B$. It follows that $u^B = u^B(z, y)$ satisfies

$$u_{zz}^B + u_{yy}^B = 0 \quad \text{in } \Omega$$
$$u^B(1, y) = 0, \qquad u^B(0, y) = g(y), \qquad 0 < Y < 1$$
$$u_y^B(z, 0) = u_y^B(z, 1) = 0, \qquad 0 < x < 1$$

Having just solved subproblem A, we recognize that $u(z, y)$ is given by

$$u^B(z, y) = g^0(1 - z) + \sum_{n=1}^{\infty} g^n \frac{\sinh n\pi(1 - z)}{\sinh n\pi} \cos n\pi y$$

Then, since $z = 1 - x$, this reduces to

$$u^B(x, y) = g^0 x + \sum_{n=1}^{\infty} g^n \frac{\sinh n\pi x}{\sinh n\pi} \cos n\pi y \qquad (3.1.28)$$

This formal solution can be shown to be the unique solution of subproblem B.

Subproblem C

$$\nabla^2 u^C(x, y) = 0 \quad \text{in } \Omega$$
$$u_y^C(x, 0) = p(x)$$
$$u_y^C(x, 1) = 0, \qquad 0 < x < 1$$
$$u^C(0, y) = u^C(1, y) = 0, \qquad 0 < y < 1$$

$$u_y^C = 0$$
$$u^C = 0 \quad \boxed{\nabla^2 u^C = 0} \quad u^C = 0$$
$$u_y^C = p$$

The separation assumption (3.1.1) applied to this subproblem leads to two coupled ordinary differential equations. The homogeneous boundary conditions in this subproblem induce the following homogeneous boundary conditions for solutions of the separated ordinary differential equations:

$$X(0) = X(1) = 0 \quad \text{and} \quad Y'(1) = 0$$

Evidently, the $X(x)$ problem is the Sturm–Liouville problem

$$-X''(x) = \mu X(x), \qquad X(0) = X(1) = 0$$

with eigenvalues

$$\mu_n = (n\pi)^2, \qquad n = 1, 2, \ldots$$

and eigenfunctions

$$X_n(x) = \sin n\pi x, \qquad n = 1, 2, \ldots$$

We write the general solution to the $Y(y)$ problem in the form

$$Y_n(y) = A_n \cosh n\pi y + B_n \cosh n\pi(1 - y)$$

and then conclude from the boundary condition $Y'(1) = 0$ that $A_n = 0$. Then,

$$Y_n(y) = B_n \cosh n\pi(1 - y)$$

and

$$u^C(x, y) = \sum_{n=1}^{\infty} B_n \cosh n\pi(1 - y) \sin n\pi x$$

Finally, the inhomogeneous boundary condition is satisfied if the free constants B_n can be chosen so that

$$u_y^C(x, 0) = \sum_{n=1}^{\infty} -n\pi B_n \sinh n\pi \sin n\pi x = p(x)$$

Using the orthogonality of the eigenfunctions $\sin n\pi x$ in the usual way, we find that in order for this boundary condition to be satisfied, it is necessary that

$$B_n = -p^n/(n\pi \sinh n\pi), \qquad n = 1, 2, \ldots$$

where

$$p^n = 2 \int_0^1 p(x) \sin n\pi x \, dx, \qquad n = 1, 2, \ldots$$

The completeness of the eigenfunctions implies that this condition is also sufficient. Then the solution to subproblem C is given by

$$u^C(x, y) = \sum_{n=1}^{\infty} p^n \frac{\cosh n\pi(1 - y)}{-n\pi \sinh n\pi} \sin n\pi x \qquad (3.1.29)$$

Subproblem D

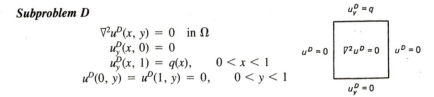

$$\nabla^2 u^D(x, y) = 0 \quad \text{in } \Omega$$
$$u^D_y(x, 0) = 0$$
$$u^D_y(x, 1) = q(x), \qquad 0 < x < 1$$
$$u^D(0, y) = u^D(1, y) = 0, \qquad 0 < y < 1$$

We introduce the following change of independent variable: $z = 1 - y$, so that $u_z = -u_y$ and $u_{zz} = u_{yy}$. This transforms subproblem D to

$$u_{xx}(x, z) + u_{zz}(x, z) = 0 \quad \text{in } \Omega$$
$$u_z(x, 1) = 0, \qquad u_z(x, 0) = -q(x)$$
$$u(0, z) = 0, \qquad u(1, z) = 0$$

Using the solution for subproblem C, we can write the solution for this problem as

$$u(x, z) = \sum_{n=1}^{\infty} q^n \frac{\cosh n\pi(1 - z)}{n\pi \sinh n\pi} \sin n\pi x$$

and then, since $z = 1 - y$, we have

$$u^D(x, y) = \sum_{n=1}^{\infty} q^n \frac{\cosh n\pi y}{n\pi \sinh n\pi} \sin n\pi x \qquad (3.1.30)$$

This formal solution can be shown to be the unique solution of subproblem D.

We have at this point solved the four subproblems for which the partial differential equation is homogeneous. In each case, the separation assumption (3.1.1) reduced the partial differential equation in two independent variables to two ordinary differential equations coupled by the separation constant. In each of the four subproblems, one of the two separated ordinary differential equations carries with it the homogeneous boundary conditions needed to form a Sturm–Liouville problem. The Sturm–Liouville problem then produces the complete family of orthogonal eigenfunctions that make it possible to satisfy the inhomogeneous boundary condition.

Now we proceed to solve the last subproblem, the problem involving the inhomogeneous equation.

Subproblem E

$$\nabla^2 u^E(x, y) = F(x, y) \quad \text{in } \Omega$$
$$u^E_y(x, 0) = 0$$
$$u^E_y(x, 1) = 0, \qquad 0 < x < 1$$
$$u^E(0, y) = u^E(1, y) = 0, \qquad 0 < y < 1$$

As in the previous subproblems, the key to the solution of this subproblem is the eigenfunctions. We can represent the solution in terms of the x eigenfunctions,

$$u^E(x, y) = \sum_{n=1}^{\infty} V_n(y) \sin n\pi x$$

where the unknown functions $V_n(y)$ must be found from the equation. We can also write the solution in terms of the y eigenfunctions,

$$u^E(x, y) = \sum_{m=0}^{\infty} U_m(x)\cos m\pi y$$

where, in this case, it is the unknown functions $U_m(x)$ that are to be found from the partial differential equation. Finally, we can choose to write the solution using both families of eigenfunctions,

$$u^E(x, y) = \sum_{n=1}^{\infty} \sum_{m=0}^{\infty} u_{mn}\sin n\pi x \cos m\pi y \qquad (3.1.31)$$

and in this instance the constants u_{mn} are the unknowns that are to be found from the equation. In the first approach we write $u(x, y)$ as a function of y with values in $L^2(0, 1)$, and we use the basis $\{\sin n\pi x : n = 1, 2, \ldots\}$ to represent elements of $L^2(0, 1)$. In the second approach, we treat $u(x, y)$ as a function of x, which takes its values in $L^2(0, 1)$, and here we use the eigenfunctions $\{\cos m\pi y : m = 0, 1, 2, \ldots)$ as a basis in $L^2(0, 1)$. In the third approach $u(x, y)$ is viewed as a function in $L^2(\Omega)$, and we use the basis $\{\sin n\pi x \cos m\pi y : n = 1, 2, \ldots, m = 0, 1, \ldots\}$ for $L^2(\Omega)$. Although these three ways of viewing $u(x, y)$ are equivalent and any one of the three approaches will lead eventually to the solution of the subproblem, we choose to use the third method.

If we suppose that the forcing function $F(x, y)$ is an element of $L^2(\Omega)$, then it can be represented in terms of the basis $\{\sin n\pi x \cos m\pi y : n = 1, 2, \ldots, m = 0, 1, \ldots\}$ as follows:

$$F(x, y) = \sum_{n=1}^{\infty} \sum_{m=0}^{\infty} F_{mn}\sin n\pi x \cos m\pi y$$

where, for $m = 0, 1, \ldots, n = 1, 2, \ldots,$

$$F_{mn} = 4 \int_0^1 \int_0^1 F(x, y) \sin n\pi x \cos m\pi y \, dx \, dy$$

Here we are using Equations (2.6.2) and (2.6.3).

From (3.1.31) we compute

$$u_{xx}^E(x, y) = \sum_m \sum_n -(n\pi)^2 u_{mn}\sin n\pi x \cos m\pi y$$

$$u_{yy}^E(x, y) = \sum_m \sum_n -(m\pi)^2 u_{mn}\sin n\pi x \cos m\pi y$$

Then substituting the expansions for $u^E(x, y)$ and $F(x, y)$ into the Poisson equation of subproblem E leads to

$$\sum_m \sum_n [F_{mn} + (m^2 + n^2)\pi^2 u_{mn}]\sin n\pi x \cos m\pi y = 0$$

Since the eigenfunctions form a basis for $L^2(\Omega)$, it follows that, for all m and n,

$$F_{mn} + (m^2 + n^2)\pi^2 u_{mn} = 0$$

that is,

$$u_{mn} = -F_{mn}/[(m^2 + n^2)\pi^2]$$

Using this last result in (3.1.31), we have the so-called full eigenfunction expansion of the solution to subproblem E,

$$u^E(x, y) = \sum_{n=1}^{\infty} \sum_{m=0}^{\infty} \frac{-F_{mn}}{(m^2 + n^2)\pi^2} \sin n\pi x \cos m\pi y \qquad (3.1.32)$$

The sum of the solutions to the five subproblems is the solution to the original mixed boundary-value problem. To see this, observe that this sum satisfies each of the following conditions:

The equation:

$$\begin{aligned}
\nabla^2 u(x, y) &= \nabla^2[u^A + u^B + u^C + u^D + u^E] \\
&= \nabla^2 u^A + \nabla^2 u^B + \nabla^2 u^C + \nabla^2 u^D + \nabla^2 u^E \\
&= 0 + 0 + 0 + 0 + F(x, y)
\end{aligned}$$

The boundary conditions:

At $x = 0$: $\begin{aligned}[t] u(0, y) &= u^A(0, y) + u^B(0, y) + u^C(0, y) + u^D(0, y) + u^E(0, y) \\ &= f(y) + 0 + 0 + 0 + 0 \end{aligned}$

At $x = 1$: $\begin{aligned}[t] u(1, y) &= u^A(1, y) + u^B(1, y) + u^C(1, y) + u^D(1, y) + u^E(1, y) \\ &= 0 + g(y) + 0 + 0 + 0 \end{aligned}$

At $y = 0$: $\begin{aligned}[t] u_y(x, 0) &= u_y^A(x, 0) + u_y^B(x, 0) + u_y^C(x, 0) + u_y^D(x, 0) + u_y^E(x, 0) \\ &= 0 + 0 + p(x) + 0 + 0 \end{aligned}$

At $y = 1$: $\begin{aligned}[t] u_y(x, 1) &= u_y^A(x, 1) + u_y^B(x, 1) + u_y^C(x, 1) + u_y^D(x, 1) + u_y^E(x, 1) \\ &= 0 + 0 + 0 + q(x) + 0 \end{aligned}$

This technique of dividing a complex problem into several simpler subproblems is carried to an extreme in this example but is frequently useful. ■ ■

EXAMPLE 3.1.4 ──

An Eigenvalue Problem

Let Ω be as in the previous example and consider the eigenvalue problem

$$\nabla^2 u(x, y) = \mu u(x, y) \quad \text{in } \Omega$$
$$u(x, 0) = u(x, 1) = 0, \qquad 0 < x < 1$$
$$u(0, y) = u(1, y) = 0, \qquad 0 < y < 1$$

Here we are looking for values of the parameter μ such that this Dirichlet boundary-value problem has nontrivial solutions. These values are then the eigenvalues of this problem, and the corresponding nontrivial solutions are the eigenfunctions.

A problem of this sort could arise in a number of situations, among them the investigation of the transverse vibrations of a thin, square membrane that is clamped at the edges. If the membrane in its equilibrium position lies in a plane, then the out-of-plane deflections of the membrane satisfy the two-dimensional wave equation

$$U_{tt}(x, y, t) = a^2[U_{xx}(x, y, t) + U_{yy}(x, y, t)]$$

Suppose the out-of-plane deflections $U(x, y, t)$ separate as follows:

$$U(x, y, t) = u(x, y)T(t)$$

Then the spatial part of the solution, $u(x, y)$, will satisfy the preceding eigenvalue problem. The homogeneous boundary conditions there are a reflection of the fact that the edges of the membrane are constrained to remain in the equilibrium plane; that is, they are clamped. The eigenvalues of the problem in this case have the interpretation of being the fundamental frequencies of the membrane and the corresponding eigenfunctions then represent the associated mode shapes assumed by the membrane when vibrating at a fundamental frequency.

Separation of Variables We now solve the eigenvalue problem by a further separation of variables. When the assumption that $u(x, y) = X(x)Y(y)$ is substituted into the partial differential equation, we obtain

$$X''(x)/X(x) = \mu - Y''(y)/Y(y)$$

Since the left side of the equation depends only on the variable x and the right side is a function of y alone, it follows that each side must be equal to a constant τ. That is,

$$
\begin{aligned}
X''(x) &= \tau X(x), & X(0) &= X(1) = 0 & \text{(3.1.33a)} \\
-Y''(y) &= (\tau - \mu)Y(y), & Y(0) &= Y(1) = 0 & \text{(3.1.33b)}
\end{aligned}
$$

Each of these is a Sturm–Liouville problem of the form of Example 2.3.1. The eigenvalues and eigenfunctions of problem (3.1.33a) are then

$$\tau_n = -(n\pi)^2, \quad X_n(x) = \sin n\pi x, \quad n = 1, 2, \ldots$$

and for problem (3.1.33b) we have

$$(\tau_n - \mu_{mn}) = (m\pi)^2, \quad Y_m(y) = \sin m\pi y, \quad m = 1, 2, \ldots$$

These are the eigenvalues and eigenfunctions for the separated problems. For the *original* problem we have

$$
\begin{aligned}
\mu_{mn} &= \tau_n - (m\pi)^2 = -\pi^2(m^2 + n^2) \\
G_{mn}(x, y) &= \sin(n\pi x)\sin(m\pi y), \quad m, n = 1, 2, \ldots & \text{(3.1.34)}
\end{aligned}
$$

Evidently, the eigenvalues associated with the Dirichlet problem for Laplace's equation are all real and negative. Moreover, since the eigenfunctions generated by the two separated equations each form a complete orthogonal family in $L^2(0, 1)$, it follows that the eigenfunctions $G_{mn}(x, y)$ are a complete orthogonal family in $L^2(\Omega)$.

Application: Inhomogeneous Problem Consider now the following inhomogeneous Dirichlet problem:

$$
\begin{aligned}
\nabla^2\phi(x, y) &= 0 \quad \text{in } \Omega \\
\phi(0, y) &= f(y) \\
\phi(1, y) &= g(y), \quad 0 < y < 1 \\
\phi(x, 0) &= p(x) \\
\phi(x, 1) &= q(x), \quad 0 < x < 1
\end{aligned}
\qquad \text{(3.1.35)}
$$

If we let

$$h(x, y) = (1 - x)f(y) + xg(y) + (1 - y)p(x) + yq(x)$$

where, for convenience, we suppose that

$$f(0) = g(0) = p(0) = q(0) = 0$$
$$f(1) = g(1) = p(1) = q(1) = 0 \qquad (3.1.36)$$

then $\Phi(x, y) = \phi(x, y) - h(x, y)$ satisfies

$$-\nabla^2\Phi(x, y) = F(x, y) \quad \text{in } \Omega$$
$$\Phi(x, 0) = \Phi(x, 1) = 0, \quad 0 < x < 1$$
$$\Phi(0, y) = \Phi(1, y) = 0, \quad 0 < y < 1 \qquad (3.1.37)$$

where

$$F(x, y) = \nabla^2 h(x, y) = f''(y)(1 - x) + g''(y)x + p''(x)(1 - y) + q''(x)y$$

Now we *could* solve this inhomogeneous boundary-value problem (3.1.37) in much the same way we solved subproblem E of Example 3.1.3, by expanding $F(x, y)$ in terms of the eigenfunctions G_{mn}. Then, adding $h(x, y)$ to $\Phi(x, y)$ yields $\phi(x, y)$, the solution to the original Dirichlet problem (3.1 35). However, our purpose here is not just to *solve* the Dirichlet problem (we could have done that directly as we did in Example 3.1.1). We have transformed problem (3.1.35) to the problem (3.1.37) in order to move the inhomogeneous terms in the problem from the boundary to the equation. We can then illustrate more clearly the analogy between the continuous problem considered here and the discrete Dirichlet problem (1.3.2'). We list now some of the similarities between these two problems.

Comparison of Continuous and Discrete Dirichlet Problem   In Equation (1.3.2') the Dirichlet boundary data is contained in the vector $\mathbf{f}_N$, while in (3.1.37) the function $F(x, y)$ contains the boundary information. In both cases, then, the system response is related to the boundary input by means of an "operator"; in (1.3.2') it is the matrix $[L_N]$ and in (3.1.37) it is the Laplacian ∇^2.

The matrix $[L_N]$ of (1.3.2') has N real, distinct, and negative eigenvalues, and from (3.1.34) we see that the continuous Dirichlet problem has a countably infinite number of distinct eigenvalues, all real and negative. For $[L_N]$ the fact that the eigenvalues are all negative implies that the matrix is "negative definite," which in turn implies that $[L_N]\mathbf{z} = \mathbf{0}$ if and only if $\mathbf{z} = \mathbf{0}$; that is, the null space of the matrix $[L_N]$ consists of just the zero vector. The Laplacian operator ∇^2 with the Dirichlet boundary conditions of problem (3.1.37) is similarly negative definite with a trivial null space (see Exercises 9 and 11).

The N distinct eigenvalues of the matrix $[L_N]$ generate N mutually orthogonal eigenvectors that then form a basis for R_N. Similarly, the eigenfunctions $G_{mn}(x, y)$ of the continuous problem are a complete orthogonal family in $L^2(\Omega)$.

In addition, $\mathbf{u}_N$, the solution to (1.3.2'), satisfies the discrete max–min principle while $\phi(x, y)$, the solution to (3.1.35), satisfies a continuous version of this principle (see Chapter 6).

Discussion of Solution: Satisfying Boundary Conditions We point out that the condition (3.1.36) is not necessary in order to be able to carry out the transformation of (3.1.35) into (3.1.37). If we suppose that

$$f(1) = q(0) = A, \qquad q(1) = g(0) = B$$
$$f(0) = p(0) = D, \qquad p(1) = g(0) = C \qquad (3.1.38)$$

for A, B, C, D all constants, then the function

$$H(x, y) = h(x, y) - Ay(1 - x) - Bxy - Cx(1 - y) - D(1 - x)(1 - y)$$

is such that $\phi(x, y) - H(x, y)$ is zero on the boundary of Ω. Then $\Phi(x, y) = \phi(x, y) - H(x, y)$ satisfies the inhomogeneous Poisson equation together with homogeneous Dirichlet conditions, and $\Phi(x, y)$ can be found by the methods of subproblem E of Example 3.1.3.

The condition (3.1.38) is just the condition that the boundary data is compatible with a continuous solution to the Dirichlet problem. If (3.1.38) is *not* satisfied, it implies that the solution $\phi(x, y)$ is being forced to assume two different values at a single point on the boundary, that is, at any of the corners of Ω where (3.1.38) is violated. In such a case the solution of the problem will be smooth on the interior of the region Ω but is discontinuous on the closed region consisting of the interior together with the boundary of Ω.

On the other hand, the fact that (3.1.38) is not satisfied does not prevent us from formally carrying out the solution of the problem (3.1.35) by the method of eigenfunction expansion. In such a case, however, we cannot expect that the solution to the problem will satisfy the boundary conditions in a pointwise sense (particularly not at a corner where the solution is discontinuous!). Instead, it turns out that the boundary conditions are satisfied in a mean-square sense, and we refer to this solution as a "generalized solution" to the Dirichlet problem. In this problem, the lack of smoothness in the solution is induced by the lack of smoothness in the boundary data. We will see an example soon where the irregular shape of the region Ω introduces some irregularity into the solution to the Dirichlet problem. We lead up to this example by considering first a Dirichlet problem on a region that is extremely smooth. There we see another of the behaviors typical of solutions to Laplace's equation, the so-called organic behavior. ■ ■

Exercises: Boundary-Value Problems

In Exercises 1 through 4, solve for $u(x, y)$.

1. $u_{xx}(x, y) + u_{yy}(x, y) = 0$ for $0 < x, y < 1$
   $u_x(0, y) = 0, \qquad u_x(1, y) = 0, \qquad 0 < y < 1$
   $u_y(x, 0) = 1, \qquad u_y(x, 1) = 1, \qquad 0 < x < 1$

2. $u_{xx}(x, y) + u_{yy}(x, y) = 0$ for $0 < x, y < 1$
   $u_x(0, y) = \frac{1}{2} - y, \qquad u_x(1, y) = 0, \qquad 0 < y < 1$
   $u_y(x, 0) = 0, \qquad u_y(x, 1) = 0, \qquad 0 < x < 1$

3. $u_{xx}(x, y) + u_{yy}(x, y) = 0$ for $0 < x, y < 1$
   $u(0, y) = \sin \pi y, \qquad u(1, y) = 0, \qquad 0 < y < 1$
   $u(x, 0) = 0, \qquad u(x, 1) = 0, \qquad 0 < x < 1$

4. $u_{xx}(x, y) + u_{yy}(x, y) = 0$ for $0 < x, y < 1$
$u_x(0, y) = 0,$ $u(1, y) = 0,$ $0 < y < 1$
$u(x, 0) = 1,$ $u_y(x, 1) = 0,$ $0 < x < 1$

5. Let Ω denote the unit square $(0, 1) \times (0, 1)$. Then find $u(x, y)$ satisfying

$$\nabla^2 u(x, y) = 0 \quad \text{in } \Omega$$
$$u(0, y) = \sin \pi(\tfrac{1}{2} - y), \quad u(1, y) = 0, \quad 0 < y < 1$$
$$u_y(x, 0) = x(1 - x), \quad u_y(x, 1) = 0, \quad 0 < x < 1$$

6. Let Ω be as in Exercise 5. Then find the eigenvalues and eigenfunctions for the problem

$$\nabla^2 u(x, y) = \mu u(x, y) \quad \text{in } \Omega$$
$$u(0, y) = u(1, y) = 0, \quad 0 < y < 1$$
$$u_y(x, 0) = u_y(x, 1) = 0, \quad 0 < x < 1$$

7. Let Ω be as in the previous two exercises and use the eigenfunctions from Exercise 6 to solve the inhomogeneous problem

$$\nabla^2 U(x, y) = F(x, y) \quad \text{in } \Omega$$
$$U(0, y) = U(1, y) = 0, \quad 0 < y < 1$$
$$U_y(x, 0) = U_y(x, 1) = 0, \quad 0 < x < 1$$

In particular, what is $U(x, y)$ when $F(x, y)$ is the eigenfunction associated with the smallest eigenvalue?

8. Let Ω be as in Exercise 5. Then find the eigenvalues and eigenfunctions for the problem

$$\nabla^2 u(x, y) = \mu u(x, y) \quad \text{in } \Omega$$
$$u(0, y) = u_x(1, y) = 0, \quad 0 < y < 1$$
$$u_y(x, 0) = u(x, 1) = 0, \quad 0 < x < 1$$

Expand the function $F(x, y) = x^2 + y^2$ in a double Fourier series in terms of these eigenfunctions.

9. Let Ω denote the unit square and suppose that $u(x, y)$ is two times continuously differentiable with respect to both x and y on Ω. In addition, suppose that $u(x, y)$ is continuous on the closed square and that $u(x, y)$ is zero on the boundary of Ω. Then show that

$$\int_0^1 \int_0^1 u u_{xx} \, dx \, dy = -\int_0^1 \int_0^1 u_x^2 \, dx \, dy$$

$$\int_0^1 \int_0^1 u u_{yy} \, dx \, dy = -\int_0^1 \int_0^1 u_y^2 \, dx \, dy$$

Use these results to show that the only smooth function that satisfies

$$\nabla^2 u(x, y) = 0 \quad \text{in } \Omega$$
$$u(x, y) = 0 \quad \text{on } \Gamma, \text{ the boundary of } \Omega$$

is the zero function $u(x, y) = 0$.

10. Let Ω be as in the previous exercise and suppose that $u(x, y), v(x, y)$ are two functions that are two times continuously differentiable in Ω and are continuous on the closed square. Suppose they both satisfy

$$\nabla^2 u(x, y) = F(x, y) \quad \text{in } \Omega$$
$$u(x, y) = f(x, y) \quad \text{on } \Gamma, \text{ the boundary of } \Omega$$

where F and f denote given functions defined on Ω and Γ, respectively. Then use Exercise 9 to show that $u(x, y) = v(x, y)$ at each point of the closed square.

11. Let Ω denote the unit square and suppose that $u(x, y)$ is two times continuously differentiable with respect to both x and y on Ω. In addition, suppose that $u(x, y)$ is continuous on the closed square and that $u(x, y)$ satisfies

$$\nabla^2 u(x, y) = \mu u(x, y) \quad \text{in } \Omega$$
$$u(x, y) = 0 \qquad\qquad \text{on } \Gamma, \text{ the boundary of } \Omega$$

for some constant μ. Show that μ satisfies

$$\mu = -\frac{\displaystyle\int_\Omega \nabla u(x, y)^2 \, dx \, dy}{\displaystyle\int_\Omega u(x, y)^2 \, dx \, dy}$$

(i.e., this shows that the eigenvalues for the Dirichlet problem for Laplace's equation are necessarily real and negative).

EXAMPLE 3.1.5 _____

A Dirichlet Problem on a Disk

Let Ω denote the disk $\{(x, y) : x^2 + y^2 < R^2\}$ and let Γ denote the circular boundary of the disk. Then consider the problem of finding the unknown function $u = u(r, \theta)$ that satisfies

$$\nabla^2 u(r, \theta) = 0 \quad \text{in } \Omega$$
$$u(R, \theta) = f(\theta) \quad \text{on } \Gamma$$

$$(3.1.39)$$

Here $f(\theta)$ denotes a given function of θ.

This problem could arise in connection with determining the steady-state temperature distribution in a heat-conducting disk whose circumference is maintained at a prescribed temperature described by $f(\theta)$.

Separation of Variables in Polar Coordinates   Problem (3.1.39) is most naturally expressed in polar coordinates where it assumes the form

$$r^{-1}[ru_r(r, \theta)]_r + r^{-2}u_{\theta\theta}(r, \theta) = 0, \qquad 0 < r < R, \qquad -\pi \le \theta < \pi$$
$$u(R, \theta) = f(\theta), \qquad\qquad -\pi \le \theta < \pi \qquad (3.1.40)$$

In addition to the Dirichlet boundary conditions in this problem, we have the *implied* conditions

$$u(r, -\pi) = u(r, \pi), \qquad u_\theta(r, -\pi) = u_\theta(r, \pi), \qquad 0 < r < R \quad (3.1.41)$$

These boundary conditions are made necessary by the fact that the coordinates (r, θ) and $(r, \theta + 2n\pi)$ for all integers n describe the same point in the plane. Then the conditions (3.1.41) cause $u(r, \theta)$ to be 2π periodic in θ so that this multiple addressing of points does not cause the function $u(r, \theta)$ to be multiple valued. The conditions (3.1.41) are called "periodic" boundary conditions.

If we suppose that $u(r, \theta) = R(r)T(\theta)$, then the partial differential equation reduces to

$$r[rR']'/R(r) = -T''(\theta)/T(\theta)$$

and by the usual arguments, we are led to the two separated problems

$$-T''(\theta) = \mu T(\theta), \qquad T(-\pi) = T(\pi), \qquad T'(-\pi) = T'(\pi) \qquad (3.1.42a)$$
$$r[rR'(r)]' = \mu R(r) \qquad\qquad\qquad\qquad (3.1.42b)$$

Problem (3.1.42) is a Sturm–Liouville problem. In fact, it is the problem of Example 2.3.4 (with $L = \pi$). There we found that periodic boundary conditions led to *two* independent eigenfunctions for *each* eigenvalue, and we computed the following eigenvalues and eigenfunctions:

Eigenvalues	*Eigenfunctions*
$\mu_n = n^2$	$e^{-in\theta}, e^{in\theta}, \quad n = 1, 2, \ldots$
$\mu_0 = 0$	1

Setting the separation constant μ equal to an eigenvalue in the $R(r)$ equation leads to

$$r^2 R''(r) + r R'(r) - n^2 R(r) = 0$$

This equation has as its general solution the function

$$R(r) = Ar^n + Br^{-n}$$

and if we wish to select the component of the solution that remains bounded on the disk Ω, we set $B = 0$. Then the function

$$u(r, \theta) = \sum_{n=-\infty}^{\infty} A_n r^{|n|} e^{in\theta}$$

satisfies (formally) the equation and the periodic boundary conditions for all choices of the constants A_n.

The inhomogeneous boundary condition will be satisfied if the constants A_n can be chosen such that

$$u(R, \theta) = \sum_{n=-\infty}^{\infty} A_n R^{|n|} e^{in\theta} = f(\theta) \qquad (3.1.43)$$

Multiplying both sides of this expression by $e^{-ik\theta}$ and integrating from $-\pi$ to π leads to the following necessary condition on the A_n if the boundary condition is to be satisfied:

$$A_k R^{|k|} 2\pi = \int_{-\pi}^{\pi} f(\theta) e^{-ik\theta} \, d\theta$$

The completeness of the eigenfunctions ensures that this condition is also sufficient for (3.1.43) to hold. Then

$$u(r, \theta) = \sum_{n=-\infty}^{\infty} f^n \left(\frac{r}{R}\right)^{|n|} e^{in\theta} \qquad (3.1.44)$$

where

$$f^n = \frac{1}{2\pi} \int_{-\pi}^{\pi} f(\theta) e^{-in\theta} \, d\theta \tag{3.1.45}$$

must be the unique bounded solution of the Dirichlet problem on the disk, Ω.

Poisson Integral Formula   An interesting *alternative* to the representation (3.1.44) for the solution to this problem is obtained if we substitute (3.1.45) into (3.1.44) to get

$$u(r, \theta) = \frac{1}{2\pi} \int_{-\pi}^{\pi} \sum_{n=-\infty}^{\infty} \left(\frac{r}{R}\right)^{|n|} e^{in(\theta - \phi)} f(\phi) \, d\phi \tag{3.1.46}$$

Now the formula for the sum of a geometric series can be applied to show that for $r < R$,

$$\sum_{n=-\infty}^{\infty} \left(\frac{r}{R}\right)^{|n|} e^{in(\theta - \phi)} = \sum_{n=1}^{\infty} \left(\frac{r}{R}\right)^{n} e^{-in(\theta - \phi)} + \sum_{n=0}^{\infty} \left(\frac{r}{R}\right)^{n} e^{in(\theta - \phi)}$$

$$= \frac{(r/R)e^{-i(\theta - \phi)}}{1 - (r/R)e^{-i(\theta - \phi)}} + \frac{1}{1 - (r/R)e^{i(\theta - \phi)}} \quad \text{for } r < R$$

"Rationalizing the denominators" on the right side of this expression leads to

$$\frac{[1 - (r/R)e^{i(\theta - \phi)}](r/R)e^{-i(\theta - \phi)} + [1 - (r/R)e^{-i(\theta - \phi)}]}{1 - (r/R)[e^{-i(\theta - \phi)} + e^{i(\theta - \phi)}] + (r/R)^2}$$

$$= \frac{1 - (r/R)^2}{1 - 2(r/R)\cos(\theta - \phi) + (r/R)^2}$$

$$= \sum_{n=-\infty}^{\infty} \left(\frac{r}{R}\right)^{|n|} e^{in(\theta - \phi)}$$

Then using this replacement for the series in (3.1.46) produces the result

$$u(r, \theta) = \int_{-\pi}^{\pi} P(r, \theta - \phi) f(\phi) \, d\phi \tag{3.1.47}$$

where

$$P(r, \theta) = \frac{1}{2\pi} \frac{R^2 - r^2}{R^2 - 2Rr \cos \theta + r^2}$$

The function $P(r, \theta)$ is called the "*Poisson kernel*," and (3.1.47) is called the *Poisson integral* representation for the solution to the Dirichlet problem for the disk Ω.

Discussion of Solution: Organic Behavior   We are particularly interested in the representation (3.1.47) because it clearly shows that at each point (r, θ) inside Ω, the value $u(r, \theta)$ of the solution depends on the values of the data function f at *all* the points on the boundary Γ of Ω. For example, if we were to increase the values of $f(\theta)$ on a small segment σ of Γ, this would produce a corresponding increase in the value of $u(r, \theta)$ at every point (r, θ) in Ω (not just at those points that are near the segment σ). It is as if any stimulus applied (even very locally) to the "skin" Γ of the region Ω is felt throughout the "body" of Ω. We refer to this phenomena as the "organic behavior" of solutions to Laplace's equation. This behavior is related to the fact that Laplace's equation is descriptive of physical systems that are in a state of equilibrium. An increase on the boundary of such a system (even a very localized increase) requires the interior of the region to go through a complete rearrangement in order to bring the system into a new condition of equilibrium. In particular, in the setting of steady-state heat conduction, a small positive disturbance in the boundary temperature will result in some change in the temperature at *every* point in the interior of the region Ω.

It is not hard to show, using the Weierstrass M-test for the uniform convergence of an infinite series of functions, that the series representation (3.1.44) for the solution converges uniformly on every closed disk inside Ω. Furthermore, the series (3.1.44) can be differentiated any number of times with respect to either r or θ, and the differentiated series will also converge uniformly on every disk inside Ω (see Exercise 6). These facts imply that the solution of this Dirichlet problem is a very smooth function of the variables r and θ. Such smoothness is typical of solutions to Laplace's equation and elliptic partial differential equations in general. However, there are circumstances that can interfere with the smoothness of the solution to an elliptic boundary-value problem. We shall see now an example of one such circumstance.   ■ ■

EXAMPLE 3.1.6 ───

Irregularity Induced by Boundary

Let Ω denote the disk of radius 1, centered at the origin, from which a wedge-shaped segment has been removed; that is,

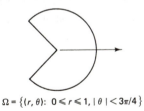

$$\Omega = \{(r, \theta): \ 0 \leqslant r \leqslant 1, |\theta| < 3\pi/4\}$$

Then consider the following boundary-value problem in Ω;

$$\nabla^2 u(r, \theta) = 0 \quad \text{in } \Omega$$
$$u(1, \theta) = f(\theta) \quad \text{for } |\theta| < 3\pi/4$$
$$u(r, -3\pi/4) = u(r, 3\pi/4) = 0, \qquad 0 < r < 1$$

We can think of this as a problem in which we seek to determine the steady-state distribution of temperature in Ω resulting from a temperature $f(\theta)$ on the curved part of the boundary and a zero temperature along the edges of the missing wedge. We find that this solution exhibits a lack of smoothness near the point of the wedge.

The usual separation assumption $u(r, \theta) = R(r)T(\theta)$ leads to the following Sturm–Liouville problem:

$$-T''(\theta) = \mu T(\theta), \qquad T(-3\pi/4) = T(3\pi/4) = 0$$

Proceeding as in previous examples, we find that zero is not an eigenvalue nor are there any negative eigenvalues. To find the positive eigenvalues, we write $\mu = \alpha^2$, and then

$$T(\theta) = A \sin \alpha\theta + B \cos \alpha\theta$$

is the general solution to the differential equation. The boundary conditions then require that A and B satisfy

$$T(-3\pi/4) = -A \sin 3\alpha\pi/4 + B \cos 3\alpha\pi/4 = 0$$
$$T(3\pi/4) = A \sin 3\alpha\pi/4 + B \cos 3\alpha\pi/4 = 0$$

A nontrivial solution for A and B exists if and only if the determinant of this system vanishes; that is, if

$$\det = -2 \sin 3\alpha\pi/4 \cos 3\alpha\pi/4 = -\sin 3\alpha\pi/2 = 0$$

Evidently, the determinant vanishes for α equal to

$$\alpha_n = 2n/3, \qquad n = 1, 2, \ldots$$

Then the eigenvalues of the problem are the numbers

$$\mu_n = (2n/3)^2, \qquad n = 1, 2, \ldots$$

and the corresponding eigenfunctions are

$$T_n(\theta) = A_n \sin 2n\theta/3 + B_n \cos 2n\theta/3$$

The boundary conditions imply that A_n and B_n are related by

$$A_n \sin n\pi/2 = \pm B_n \cos n\pi/2, \qquad n = 1, 2, \ldots$$

and hence

$$T_n(\theta) = \sin 2n\theta/3 \cos n\pi/2 \pm \cos 2n\theta/3 \sin n\pi/2$$
$$= \sin[n(2\theta/3 + \pi/2)]$$

Setting the separation constant equal to an eigenvalue in the differential equation for $R(r)$ leads to

$$r^2 R''(r) + rR'(r) - (2n/3)^2 R(r) = 0$$

The general solution to this equation is found to be

$$R_n(r) = C_n r^{2n/3} + D_n r^{-2n/3}$$

and if we are interested in the solution that remains bounded on Ω, then D_n must vanish. This leaves us with the following formal solution to the boundary-value problem:

$$u(r, \theta) = \sum_{n=1}^{\infty} C_n r^{2n/3} \sin\left[n\left(\frac{2\theta}{3} + \frac{\pi}{2}\right)\right]$$

This function satisfies (formally) the partial differential equation and the homogeneous boundary conditions on the two straight edges of the boundary of Ω. In order to satisfy the inhomogeneous boundary condition along the curved edge of the boundary, we make use of the fact that the eigenfunctions $T_n(\theta)$ are a complete orthogonal family in $L^2(-3\pi/4, 3\pi/4)$. The inhomogeneous boundary condition is satisfied if the constants C_n satisfy

$$\sum_{n=1}^{\infty} C_n \sin\left[n\left(\frac{2\theta}{3} + \frac{\pi}{2}\right)\right] = f(\theta)$$

That is, if

$$C_n = \frac{4}{3\pi} \int_{-3\pi/4}^{3\pi/4} \sin\left[n\left(\frac{2\theta}{3} + \frac{\pi}{2}\right)\right] f(\theta)\, d\theta$$
$$= f^n$$

Here we are using the notation f^n to indicate the generalized Fourier coefficients for the function $f(\theta)$ with respect to the eigenfunctions $T_n(\theta)$. Then the solution to the boundary-value problem is given by

$$u(r, \theta) = \sum_{n=1}^{\infty} f^n r^{2n/3} \sin\left[n\left(\frac{2\theta}{3} + \frac{\pi}{2}\right)\right] \tag{3.1.48}$$

Discussion of Solution   This series can be shown to converge absolutely and uniformly on closed subsets of Ω (see Exercise 7). Since the terms of the infinite series are themselves continuous functions, it follows from Proposition 2.1.1 that the function $u(r, \theta)$ must be continuous on the interior of Ω.

Note, however, that differentiating this series with respect to r leads to the series

$$u_r(r, \theta) = \sum_{n=1}^{\infty} f^n \frac{2n}{3} r^{2n/3 - 1} T_n(\theta)$$
$$= \tfrac{2}{3} f^1 r^{-1/3} T_1(\theta) + \tfrac{4}{3} f^2 r^{1/3} T_2(\theta) + \cdots$$

The first term of this series is *not bounded* on Ω; that is, as r approaches zero, the first term in the series goes to infinity. Thus, unless f^1 should happen to vanish, $u_r(r, \theta)$ is not continuous on Ω. Evidently the source of this irregular behavior is the angle formed by removing the wedge from the disk Ω. That this is so can be seen by solving the following boundary-value problem for the unknown function $u(r, \theta)$ on the region:

$$\Omega = \{(r, \theta) : 0 < r < 1, |\theta| < \Phi\}$$
$$\nabla^2 u(r, \theta) = 0 \qquad \text{in } \Omega$$
$$u(1, \theta) = f(\theta) \qquad \text{for } |\theta| < \Phi$$
$$u(r, -\Phi) = u(r, \Phi) = 0 \quad \text{for } 0 < r < 1$$

Here Φ denotes a fixed (given) angle between 0 and π. (The case $\Phi = 3\pi/4$ is the boundary-value problem we have just solved.) Then we can show that

   (i) $u(r, \theta)$ is continuous on Ω,
   (ii) $u_r(r, \theta)$ is continuous on Ω if $\Phi \leq \pi/2$, and
   (iii) $u_r(r, \theta)$ goes to infinity as r goes to zero if $\Phi > \pi/2$.

Apparently, when Ω is a half-disk (i.e., $\Phi = \pi/2$) the solution $u(r, \theta)$ of the boundary-value problem is continuous together with its derivative $u_r(r, \theta)$, but as soon as Φ exceeds $\pi/2$, the derivative develops an infinite discontinuity at $r = 0$. Since all of the other ingredients in the problem are unchanged, it must be that it is the *region Ω itself* that induces this irregularity in the solution.   ■ ■

Exercises: Problems in Polar Coordinates

1. Let Ω denote the semiannular region, $1 < r < 2, 0 < \theta < \pi$. Then consider the problem

$$\nabla^2 u(r, \theta) = 0 \quad \text{in } \Omega$$
$$u_r(1, \theta) = 0, \qquad u(2, \theta) = f(\theta), \qquad 0 < \theta < \pi$$
$$u(r, 0) = u(r, \pi) = 0, \qquad 1 < r < 2$$

Find $u(r, \theta)$ if $f(\theta) = \sin \theta + \frac{1}{16} \sin 5\theta$. Find $u(r, \theta)$ if

$$f(\theta) = \begin{cases} 1 & \text{if } \pi/4 < \theta < 3\pi/4 \\ 0 & \text{if } \theta < \pi/4 \text{ or } 3\pi/4 < \theta \end{cases}$$

2. Find $u(r, \theta)$ satisfying

$$\nabla^2 u(r, \theta) = 0 \quad \text{in } 0 < r < 2, 0 < \theta < \pi/2$$
$$u(2, \theta) = f(\theta), \qquad 0 < \theta < \pi/2$$
$$u(r, 0) = u(r, \pi/2) = 0, \qquad 0 < r < 2$$

3. How is the solution to Exercise 2 related to the solution of the following problem?

$$\nabla^2 u(r, \theta) = 0 \quad \text{in } 0 < r < 2, |\theta| < \pi/4$$
$$u(2, \theta) = f(\theta), \qquad |\theta| < \pi/4$$
$$u(r, -\pi/4) = u(r, \pi/4) = 0, \quad 0 < r < 2$$

4. Solve the problem

$$\nabla^2 u(r, \theta) = 0 \quad \text{in } 0 < r < 2, |\theta| < \pi/4$$
$$u_r(2, \theta) = f(\theta), \qquad |\theta| < \pi/4$$
$$u(r, -\pi/4) = u(r, \pi/4) = 0, \quad 0 < r < 2$$

5. Solve the problem

$$\nabla^2 u(r, \theta) = 0 \quad \text{in } 0 < r < 2, |\theta| < \pi/4$$
$$u(2, \theta) = f(\theta), \qquad |\theta| < \pi/4$$
$$u_\theta(r, -\pi/4) = u_\theta(r, \pi/4) = 0, \qquad 0 < r < 2$$

6. Use the Weierstrass M-test to show that the series (3.1.44) converges uniformly on each closed disk of radius $r < R$. Conclude that the series converges to a continuous function $u(r, \theta)$ on each closed disk in Ω. Apply the same argument to the series obtained by differentiating the series (3.1.44) any number of times with respect to either r or θ.

7. Apply the reasoning of the previous exercise to the series (3.1.48). How does the argument fail when applied to the differentiated series?

Remarks on Method of Eigenfunction Expansion

In this section we solved several examples using the technique of eigenfunction expansion. This is a very powerful method for solving problems in partial differential equations, but we do not wish to create the false impression that *every* boundary-value problem can be solved in this way. Each of the examples of this section has the following two properties: The partial differential equation is separable and the domain where the equation is to be satisfied is a bounded set that is a coordinate cell for some coordinate system.

In each of the examples of this section, the Laplace equation in two variables separates into a pair of coupled ordinary differential equations under the assumption that the solution is a product of two functions of one variable. However, *not all* partial differential equations can be separated in this way. For example, the assumption $u(x, y) = X(x)Y(y)$ reduces the equation

$$u_{xx}(x, y) + u_{xy}(x, y) + u_{yy}(x, y) = 0$$

to

$$X''(x)Y(y) + X'(x)Y'(y) + X(x)Y''(y) = 0$$

However, there is no way to rearrange this equation so that one side of the equation depends only on x and the other side depends only on y, and hence this equation could not be solved by separation of variables.

A bounded set is a *coordinate cell* in the $\alpha\beta$ coordinate system if it is of the form $\{(\alpha, \beta) : a < \alpha < b, c < \beta < d\}$, where a, b, c, and d denote *finite* constants. The following are examples of coordinate cells in Cartesian and polar coordinates,

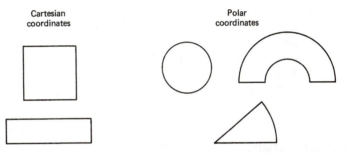

| Cartesian coordinates | Polar coordinates |

When the domain of the problem is not a coordinate cell, the boundary conditions associated with the solution to the partial differential equation do not lead to Sturm–Liouville boundary conditions for any of the separated problems. In such a case there is then no Sturm–Liouville problem to provide the eigenfunctions needed for building the solution. The following are examples of domains that are not coordinate cells in either Cartesian or polar coordinates.

Exercises: Method of Eigenfunction Expansion

In each of the following problems decide whether the method of eigenfunction expansion can be successfully carried out. For each problem where it cannot, isolate and explain the reason for the failure of the method.

1. $u_{xyy}(x, y) + u_{xxy}(x, y) = 0,$ $0 < x, y < 1$
$u(x, 0) = u(x, 1) = 0,$ $0 < x < 1$
$u(0, y) = f(y),$ $u(1, y) = 0,$ $0 < y < 1$

2. $\nabla^2 u(r, \theta) = 0,$ $0 < r < 1,$ $-\pi/3 < \theta < \pi/3$
$u(r, \pi/3) = u(r, -\pi/3) = 0$
$u(1, \theta) = f(\theta),$ $|\theta| < \pi/3$

3. $u_{xx}(x, y) + u_{yy}(x, y) = 0,$ $0 < x^2 + y^2 < 1,$ $x, y > 0$
$u(x, 0) = u(0, y) = 0$
$u(x, y) = f(x, y)$ on $x^2 + y^2 = 1$

4. $u_{xx}(x, y) + u_{yy}(x, y) = 0,$ $x, y > 0$
$u(x, 0) = f(x),$ $u(0, y) = 0$
$u(x, y)$ bounded for all $x, y > 0$

5. $u_{xx}(x, y) + u_{yy}(x, y) = 0,$ $0 < x, y < 1,$ $y < x$
$u(x, 0) = 0,$ $u(0, y) = 0$
$u(x, x) = f(x),$ $0 < x < 1$

6. $x^2 u_{xx}(x, y) + (1 + y^2)u_{yy}(x, y) = 0,$ $0 < x, y < 1$
$u(x, 0) = u(x, 1) = 0,$ $0 < x < 1$
$u(0, y) = f(y),$ $u(1, y) = 0,$ $0 < y < 1$

7. $\nabla^2 u(x, y) = 0,$ $0 < x, y < 1$
$(1 + x^2)u(x, 0) = x,$ $u(x, 1) = 0,$ $0 < x < 1$
$u_x(0, y) = 0,$ $u_x(1, y) = 0,$ $0 < y < 1$

Problems on Partially Bounded Sets

In general, a boundary-value problem on a region that is unbounded will not be solvable by the method of eigenfunction expansion. Even if the partial differential equation separates, the resulting ordinary differential equations problems will be posed on unbounded intervals and hence cannot be Sturm–Liouville problems. However, if the unbounded region is bounded in at least one direction, it may be that an eigenfunction expansion method will apply. For example, consider the following boundary-value problem on a semi-infinite strip.

EXAMPLE 3.1.7 _____

Problem on Semi-infinite Strip

$$\nabla^2 u(x, y) = 0 \quad \text{for } 0 < y < 1, \quad x > 0$$
$$u(x, 0) = u(x, 1) = 0, \quad 0 < y < 1$$
$$u(0, y) = f(y), \quad 0 < x \tag{3.1.49}$$
$$u(x, y) \text{ bounded for } 0 < y < 1 \text{ and } x > 0$$

Here $u(x, y)$ could represent the equilibrium temperature distribution in a heat-conducting strip that is very long in comparison with its width. In practical terms it is as if the remote end is treated as having negligible effect on the temperatures near the end $x = 0$, where a specified temperature distribution is maintained. The sides of the strip are maintained at the constant temperature zero.

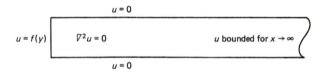

It will be shown in Chapter 6 that we must specify we are looking for the bounded solution in order that this boundary-value problem be well posed. In solving the problem, we shall see that the condition of boundedness plays the role of a boundary condition "at infinity."

Separation of Variables The assumption that $u(x, y) = X(x)Y(y)$ leads to the following ordinary differential equations for $X(x)$ and $Y(y)$:

$$\text{(i)} \quad -Y''(y) = \mu Y(y), \quad Y(0) = Y(1) = 0, \qquad\qquad (3.1.50)$$
$$\text{(ii)} \quad X''(x) = \mu X(x), \quad X(x) \text{ bounded for } x \to \infty$$

Clearly (3.1.50) (i) is a Sturm–Liouville problem for which the eigenvalues and eigenfunctions are by now familiar. They are

$$\mu_n = (n\pi)^2 \quad \text{and} \quad Y_n(y) = \sin n\pi y, \qquad n = 1, 2, \ldots$$

Since the variable x ranges over the unbounded interval $(0, \infty)$, it will be convenient to express the general solution of the problem (3.1.50) (ii) (where μ has been replaced by the eigenvalue μ_n) as follows:

$$X_n(x) = A_n e^{n\pi x} + B_n e^{-n\pi x}, \qquad x > 0$$

The general solution is thus expressed in terms of two linearly independent solutions of Equation (3.1.50) (ii) where it is clear that one component remains bounded for all $x > 0$, whereas the other solution grows without bound for $x \to \infty$. In order for the boundedness condition to be satisfied then, we choose $A_n = 0$. Note that the boundedness condition acts similar to a "boundary condition at infinity" in the sense that it selects that component of the general solution that vanishes at infinity while rejecting the component that grows without bound at infinity. This leads to

$$u(x, y) = \sum_{n=1}^{\infty} B_n e^{-n\pi x} \sin n\pi y$$

which (formally) satisfies all of the conditions of the problem except the inhomogeneous boundary condition at $x = 0$. This condition will be satisfied as well if the arbitrary constants B_n are chosen so as to satisfy

$$u(0, y) = \sum_{n=1}^{\infty} B_n \sin n\pi y = f(y), \qquad 0 < y < 1$$

As usual, this is accomplished by the choice

$$B_n = f^n = 2 \int_0^1 f(y) \sin n\pi y \, dy, \qquad n = 1, 2, \ldots$$

Then the (formal) solution of the boundary-value problem (3.1.49) is

$$u(x, y) = \sum_{n=1}^{\infty} f^n e^{-n\pi x} \sin n\pi y \qquad (3.1.51)$$

Evidently the method of eigenfunction expansion requires only one of the independent variables to be confined to a bounded interval so long as the separated equation in that variable is a Sturm–Liouville equation and the corresponding boundary conditions are homogeneous conditions of the Sturm–Liouville type. Other examples of boundary-value problems on unbounded domains where the method of eigenfunction expansion can be successfully carried out are found in the exercises.

Discussion of Solution   Note that just as for each of the previous examples of boundary-value problems involving Laplace's equation, this problem has been solved in three steps:

1. The data $f(y)$ is converted to a sequence f^n of generalized Fourier coefficients.
2. The corresponding sequence u^n of Fourier coefficients for the solution are found. In this example,

$$u^n = f^n e^{-n\pi x}, \qquad n = 1, 2, \ldots$$

3. Evaluation of the solution $u(x, y)$ is carried out by summing the generalized Fourier series whose coefficients are u^n.

If we denote the operation of converting the data function $f(y)$ into a sequence of Fourier coefficients in terms of the eigenfunctions of this problem by $\mathcal{F}_n[f]$, then

$$\mathcal{F}_n[f] = f^n,$$

and $\mathcal{F}_n^{-1}$ denotes the inverse operation of summing the series to convert the sequence of coefficients back to a function. In particular,

$$\mathcal{F}_n^{-1}\{u^n\} = u(x, y)$$

Finally, let σ denote the operation that transforms the Fourier coefficients of the data $f(y)$ into the Fourier coefficients of the solution. In this example,

$$\sigma\{f^n\} = u^n = e^{-n\pi x} f^n$$

In Example 3.1.1 we had

$$\sigma\{f^n\} = \frac{\sinh n\pi x}{\sinh n\pi} f^n$$

and in Example 3.1.5, it was

$$\sigma\{f^n\} = (r/R)^{|n|} f^n$$

In any case, the solution can be expressed as the composition of these three operations as follows:

$$u(x, y) = \mathcal{F}_n^{-1}[\sigma\{\mathcal{F}_n[f]\}]$$

Exercises: Boundary-Value Problems on Partially Bounded Sets

1. Solve the following boundary-value problem on the unbounded domain $\Omega = \{(x, y) : 0 < x < 1, y > 0\}$:

$$\nabla^2 u(x, y) = 0 \quad \text{on } \Omega$$
$$u(x, 0) = x - x^2/2, \quad 0 < x < 1$$
$$u(0, y) = u_x(1, y) = 0, \quad y > 0$$
$$u(x, y) \text{ bounded for all } (x, y) \text{ in } \Omega$$

2. Let Ω denote the exterior of the unit disk; that is,

$$\Omega = \{(r, \theta) : r > 1, |\theta| < \pi\}.$$

Show that for all real values of the constant C,

$$u(r, \theta) = 1 + C \log r$$

satisfies

$$\nabla^2 u(r, \theta) = 0 \quad \text{in } \Omega$$
$$u(1, \theta) = 1$$

Find the unique value for C when we add the condition that $u(r, \theta)$ must remain bounded on Ω.

3. Let Ω be as in the previous problem. Find the unique solution to the boundary-value problem

$$\nabla^2 u(r, \theta) = 0 \quad \text{in } \Omega$$
$$u(1, \theta) = f(\theta) \quad \text{and} \quad u(r, \theta) \text{ bounded on } \Omega$$

Assume here

$$f(\theta) = \begin{cases} 1 & \text{if } 0 < \theta < \pi \\ 0 & \text{if } -\pi < \theta < 0 \end{cases}$$

4. Let Ω denote the region $\{(r, \theta) : r > 1, 0 < \theta < \pi\}$. Find the bounded solution to the problem

$$\nabla^2 u(r, \theta) = 0 \quad \text{in } \Omega$$
$$u(r, \pi) = u(r, 0) = 0, \quad r > 1$$
$$u(1, \theta) = 1 \quad \text{for } 0 < \theta < \pi$$

5. Let Ω denote the region $\{(r, \theta) : r > R, 0 < \theta < \pi\}$. Find the bounded solution to the problem

$$\nabla^2 u(r, \theta) = 0 \quad \text{in } \Omega$$
$$u_\theta(r, \pi) = 0, \quad r > R$$
$$u(r, 0) = 0, \quad r > R$$
$$u(R, \theta) = f(\theta) \quad \text{for } 0 < \theta < \pi$$

6. Find the bounded solution of the following boundary-value problem:

$$u_{xx}(x, y) + u_{yy}(x, y) = 0, \quad x > 0, \quad 0 < y < 1$$
$$u(0, y) = f(y), \quad 0 < y < 1$$
$$u_y(x, 0) = u_y(x, 1) = 0, \quad x > 0$$

3.2 EVOLUTION EQUATIONS: INITIAL-BOUNDARY-VALUE PROBLEMS FOR HEAT EQUATION

We now consider some further examples where separation of variables can be used to solve problems in partial differential equations. These examples will involve physical systems that are not in a state of equilibrium but rather are evolving with time. For this reason, we refer to the partial differential equations involved as "evolution equations"; the heat and wave equations will serve as the typical examples of evolution equations just as the Laplace equation was the prototype for the (nonevolutionary) elliptic equations.

While the heat and wave equations both describe systems that evolve with time, the two equations describe different kinds of evolutionary behavior. We refer to these types of behavior as "diffusionlike" and "wavelike" evolution, respectively. It will be easier to discuss the two kinds of behavior and the differences between them after we have solved several example problems. We begin by considering the diffusionlike evolution, that is, the behavior that is characteristic of solutions to the heat equation.

This first example appears again in Chapter 8 as Example 8.3.3, where the problem is treated by discrete methods. You may find it interesting to read that example after reading this one to see the parallels in the two treatments.

EXAMPLE 3.2.1 _____

Mixed Initial-Boundary-Value Problem

We begin by considering the following initial-boundary-value problem for the heat equation:

$$
\begin{aligned}
u_t(x, t) &= Du_{xx}(x, t), \quad 0 < x < 1, \quad t > 0 \\
u(x, 0) &= f(x), \quad 0 < x < 1 \\
u(0, t) &= u(1, t) = 0, \quad t > 0
\end{aligned}
\tag{3.2.1}
$$

Here D denotes a given, positive constant and $f(x)$ represents a given function of x. Then $u(x, t)$ could represent the temperature at position x and time $t > 0$ in a thin, heat-conducting rod where the physical constants K, δ, and C (thermal conductivity, density, and heat capacity) are such that $K/\delta C = D$. The lateral surface of the rod is insulated so that the heat is forced to move only in the axial direction, which we label the x direction. The initial distribution of temperature is described by $f(x)$, and the ends of the rod are maintained at temperature zero for all $t > 0$. Then solving this problem will allow us to predict the temperature distribution inside the rod for all times greater than zero. Since the auxiliary conditions in this problem are a mixture of initial and boundary conditions, this is sometimes referred to as a "mixed" initial-boundary-value problem. We are about to prove, by construction, that a solution to this problem exists. In Chapter 6 we show that this solution is unique.

Lateral surface insulated

$u(0, t) = 0$ $u(x, 0) = f(x)$ $u(1, t) = 0$

Separation of Variables   If we suppose that $u(x, t) = X(x)T(t)$, then substituting this into the Equation (3.2.1) leads to

$$X''(x)/X(x) = T'(t)/DT(t) = -\mu$$

This separates in the usual way into the following two problems for the factors $X(x)$ and $T(t)$:

$$-X''(x) = \mu X(x), \qquad X(0) = X(1) = 0 \tag{3.2.2a}$$
$$T'(t) = -\mu DT(t) \tag{3.2.2b}$$

Evidently, (3.2.2a) is a Sturm–Liouville problem with the following eigenvalues and eigenfunctions:

$$\mu_n = (n\pi)^2, \qquad X_n(x) = \sin n\pi x, \qquad n = 1, 2, \ldots$$

(this is Example 2.3.1 for the case $L = 1$).

Setting $\mu = \mu_n$ in (3.2.2b), we solve this equation to obtain the family of solutions

$$T_n(t) = C_n \exp[-(n\pi)^2 Dt], \qquad n = 1, 2, \ldots$$

where C_n denotes an arbitrary constant. Then for each n

$$u_n(x, t) = C_n \exp[-(n\pi)^2 Dt]\sin n\pi x$$

satisfies the partial differential equation and the two homogeneous boundary conditions, and by the principle of superposition, the same is true for the formal solution

$$u(x, t) = \sum_{n=1}^{\infty} C_n \exp[-(n\pi)^2 Dt]\sin n\pi x$$

At $t = 0$, this reduces to

$$u(x, 0) = \sum_{n=1}^{\infty} C_n \sin n\pi x$$

and hence the initial condition is satisfied if the constants C_n are chosen such that

$$\sum_{n=1}^{\infty} C_n \sin n\pi x = f(x), \qquad 0 < x < 1$$

The orthogonality of the eigenfunctions implies that in order for this equality to hold, it is necessary for the C_n to satisfy

$$C_n = 2 \int_0^1 f(x)\sin n\pi x \, dx = f^n, \qquad n = 1, 2, \ldots$$

The completeness of the family of eigenfunctions implies that this condition is also sufficient for the initial condition to be satisfied. Then we have

$$u(x, t) = \sum_{n=1}^{\infty} f^n \exp[-(n\pi)^2 Dt]\sin n\pi x \tag{3.2.3}$$

as the solution to problem (3.2.1).

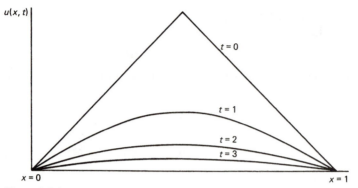

Figure 3.2.1

Discussion of Solution   If we take for the initial function $f(x)$,

$$f(x) = \begin{cases} x & \text{for } 0 < x < \frac{1}{2} \\ 1 - x & \text{for } \frac{1}{2} < x < 1 \end{cases}$$

then we can compute

$$f^n = 2(-1)^{n+1}/[\pi^2(2n - 1)^2], \qquad n = 1, 2, \ldots$$

We can plot $u(x, t)$ versus x for $t = 0, 1, 2, 3$ to obtain Figure 3.2.1.

The profiles in Figure 3.2.1 represent the evolution with time of the piecewise linear initial state $f(x)$. The profiles decrease uniformly with increasing time just as the profiles in Figure 3.1.1 decrease uniformly with increasing distance from the side $x = 1$ of the square. The striking similarity between Figures 3.1.1 and 3.2.1 suggests that the influence of the boundary temperature distribution $u(1, y)$ in Example 3.1.1 propagates through *space* in a manner that is much like the propagation in *time* of the influence of the initial state $u(x, 0)$ in Example 3.2.1. One way in particular that the solution in Example 3.1.1 resembles that of Example 3.2.1 is in the "smoothing action" that is common to them both. That is, the solutions to these examples are generally very smooth, independent of the smoothness of the auxiliary data in the problem. Saying this another way, it can be shown that when the auxiliary data (boundary or initial data) have some minimal smoothness, say, these functions are square-integrable functions of their arguments, the corresponding solution are infinitely differentiable functions of their arguments in the interior of the domain where the partial differential equation is to hold.

An additional similarity between the solution of this problem and the solution of the problem in Example 3.1.1 becomes evident if we write

$$u^n(t) = f^n \exp[-(n\pi)^2 Dt], \qquad n = 1, 2, \ldots \qquad (3.2.4)$$

Then the $u^n(t)$ represent the time-dependent generalized Fourier coefficients for the function $u(x, t)$ relative to the eigenfunctions $X_n(x)$ associated with this problem, and we are motivated to think of the solution procedure as consisting of steps similar to those appearing in Figure 3.1.2. That is:

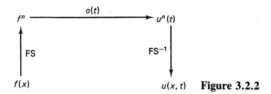

$f(x)$ $u(x, t)$ **Figure 3.2.2**

1. The initial state $f(x)$ is transformed to a generalized Fourier series in terms of the eigenfunctions $X_n(x)$: $f(x) \to f^n$.
2. The generalized Fourier coefficients u^n [as given in (3.2.4)] for the solution are obtained from the separation-of-variables procedure: $f^n \to u^n(t)$.
3. The solution $u(x, t)$ can be evaluated by summing the series:

$$\sum_{n=1}^{\infty} u^n(t)X_n(x)$$

The solution can be *approximated* by summing finitely many terms of this series: $u^n(t) \to u(x, t)$.

If we denote the operation indicated in step 1 by $\mathscr{F}_n$, then step 3 consists of the inverse operation $\mathscr{F}_n^{-1}$. Let us also denote the operation indicated in step 2 by $\sigma(t)$. Then we can represent the solution procedure for this problem schematically as in Figure 3.2.2. If we denote the composition of these three steps by $S(t)$, then

$$S(t)[f(x)] = \mathscr{F}_n^{-1}[\sigma(t)\{\mathscr{F}_n[f(x)]\}]$$
$$= \sum_{n=1}^{\infty} f^n \exp[-(n\pi)^2 Dt]\sin n\pi x$$

and we can think of the solution operator for this problem for the heat equation as the mapping that carries the initial state $f(x)$ into the time-t-state $u(x, t)$. In viewing the solution operator in this way, the smoothing action becomes very evident. The rate at which the Fourier coefficients of a function decrease with increasing n reflects the smoothness of the (periodic extension of that) function (see Exercise 15 in Section 2.1). In particular, this is true of the initial state $f(x)$ and the solution $u(x, t)$. Once the function $f(x)$ is specified, the Fourier coefficients can be computed, and the smoother the function $f(x)$, the more rapidly will the coefficients f^n decrease with increasing n. However, the coefficients $u^n(t)$ all contain the convergence factor $\exp[-(n\pi)^2 Dt]$, which means that for any $t > 0$, the Fourier coefficient $u^n(t)$ decreases more rapidly with increasing n than any power of $1/n$. From this it can be shown that $u(x, t)$ is, for each $t > 0$, infinitely differentiable with respect to both x and t. All this is true independent of the properties of $f(x)$. Thus the mathematical model evolves from an initial state $f(x)$, which need not be especially smooth, into subsequent states

$$u(x, t) = S(t)[f(x)]$$

which are extremely smooth. On reflection, it is clear that this type of behavior is consistent with the physical process of heat conduction. Suppose the initial temperature distribution in our heat-conducting rod is the following discontinuous function:

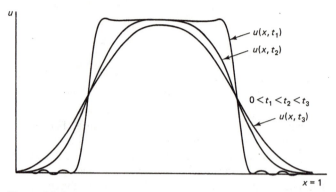

Figure 3.2.3

$$f(x) = \begin{cases} 1 & \text{if } \frac{1}{3} < x < \frac{2}{3} \\ 0 & \text{if } x < \frac{1}{3} \text{ or } x > \frac{2}{3} \end{cases}$$

It is easy to imagine that the sharp jump in the temperature that occurs at $x = \frac{1}{3}$ and $x = \frac{2}{3}$ will be immediately "smoothed out" by the conduction process and that the temperatures for $t > 0$ will vary continuously with x. This is illustrated in Figure 3.2.3, which shows several time profiles of the solution $u(x, t)$ corresponding to this initial state.

The profile $u(x, t_1)$ corresponds to a temperature distribution at a time t_1 very close to $t = 0$. At that point, the exponential convergence factors are not yet very effective, and so the profile, which was computed by truncating the infinite series after just 20 terms, still shows a little of the oscillation characteristic of a Fourier series trying to converge to a discontinuous function. Nevertheless, the sharp jump in the temperature occurring at $x = \frac{1}{3}$ and $x = \frac{2}{3}$ appears to be smoothed off. The later profiles, at t_2 and t_3, exhibit none of this oscillation since by now the exponential convergence factors are causing the series to converge very rapidly; only a few terms are needed to give a very accurate approximation to $u(x, t)$. Also it is clear that the temperature is varying very smoothly with x for these values of t. Plotting later profiles would show that the temperatures decrease uniformly to an equilibrium state in which the temperature is zero over the entire length of the rod.

Comparison of Continuous and Semidiscrete Solutions   It is interesting to compare the solution (3.2.3) with the solution (1.2.10) of Section 1.2. There we had

$$\mathbf{U}(t) = \sum_{n=1}^{N-1} \Gamma_n \exp[\alpha_n t] \mathbf{W}_n \tag{3.2.5}$$

as the solution of the semidiscrete initial-boundary-value problem

$$\mathbf{U}'(t) = [A]\mathbf{U}(t), \qquad \mathbf{U}(0) = \mathbf{C} \tag{3.2.6}$$

The matrix $[A]$ in (3.2.6) is the symmetric $(N - 1) \times (N - 1)$ negative definite matrix of Section 1.2. Then the eigenvectors $\mathbf{W}_n$ form an orthogonal basis for R^{N-1}, and it is not hard to show that the constants Γ_n are then related to the initial state vector **C** as follows:

$$\Gamma_n = (\mathbf{C}, \mathbf{W}_n)/\|\mathbf{W}_n\|^2 \quad \text{for } n = 1, 2, \ldots N - 1 \qquad (3.2.7)$$

This is analogous to

$$f^n = (f, \sin n\pi x)/\|\sin n\pi x\|^2, \qquad n = 1, 2, \ldots \qquad (3.2.8)$$

in (3.2.3), where

$$\|\sin n\pi x\|^2 = \int_0^1 |\sin n\pi x|^2 \, dx = \tfrac{1}{2}, \qquad n = 1, 2, \ldots$$

Since $[A]$ is negative definite, the eigenvalues $\alpha_1, \ldots, \alpha_{N-1}$ are all negative, and it follows that the solution $\mathbf{U}(t)$ in (3.2.5) decreases exponentially with increasing t. We can be more precise about this statement. Suppose that the eigenvalues of $[A]$ are arranged in decreasing order,

$$0 > \alpha_1 > \alpha_2 > \cdots > \alpha_{N-1}$$

Then for every $t > 0$, it follows that

$$\exp[\alpha_1 t] > \exp[\alpha_2 t] > \cdots > \exp[\alpha_{N-1} t] > 0$$

If we compute the inner product of $\mathbf{U}(t)$ with itself and remember that the vectors $\mathbf{W}_n$ are mutually orthogonal, then we find

$$\|\mathbf{U}(t)\|^2 \leq \exp[2\alpha_1 t] \sum_{n=1}^{N-1} \Gamma_n^2 \|\mathbf{W}_n\|^2$$

In view of (3.2.7),

$$\|\mathbf{U}(t)\|^2 \leq \exp[2\alpha_1 t] \sum_{n=1}^{N-1} (\mathbf{C}, \mathbf{W}_n)^2 = \exp[2\alpha_1 t]\|\mathbf{C}\|^2$$

that is,

$$\|\mathbf{U}(t)\| \leq \exp[\alpha_1 t]\|\mathbf{C}\| \quad \text{for all } t \geq 0 \qquad (3.2.9)$$

We can derive an analogous estimate for the solution (3.2.3) using the fact that the eigenfunctions $X_n(x) = \sin n\pi x$ form a complete orthogonal family in $L^2(0, 1)$. We note first from (3.2.3) that for $m = 1, 2, \ldots$

$$(u, \sin m\pi x) = \sum_{n=1}^{\infty} f^n \exp[-(n\pi)^2 Dt](\sin n\pi x, \sin m\pi x)$$
$$= f^m \exp[-(m\pi)^2 Dt]$$
$$\leq \exp[-\pi^2 Dt](f, \sin m\pi x)\|\sin n\pi x\|^2$$

where (f, g) denotes the inner product of f and g in $L^2(0, 1)$ and we have made use of (3.2.8) in replacing f^m. Now we use the completeness of the eigenfunctions once more by appealing to part 3 of Theorem 2.5.2. This tells us that

$$\|u\|^2 = \sum_{n=1}^{\infty} \left[\frac{(u, \sin n\pi x)}{\|\sin n\pi x\|} \right]^2$$

$$\leq \exp[-2\pi^2 Dt] \sum_{n=1}^{\infty} [(f, \sin n\pi x)]^2$$

$$\leq \exp[-2\pi^2 Dt]\|f\|^2$$

that is,

$$\|u\| \leq \exp[-\pi^2 Dt]\|f\| \quad \text{for } t \geq 0 \tag{3.2.10}$$

The estimate (3.2.10) will be used in Section 8.5 for analyzing the stability of certain numerical methods for solving the heat equation. ■ ■

Duhamel's Principle for Inhomogeneous Problems

EXAMPLE 3.2.2 _____

An Inhomogeneous Equation

In the previous example we discussed the fact that the solution procedure can be viewed as consisting of three distinct steps. In the first step, the data in the problem (the initial state) is expressed as a generalized Fourier series in terms of the eigenfunctions associated with the problem. In the second step we find the (time-dependent) Fourier coefficients for the solution, expressing the solution coefficients in terms of the data coefficients. Finally, in step 3, we reverse the procedure of step 1, converting the Fourier coefficients of the solution into function values by summing the series.

In the discrete and semidiscrete models of heat conduction that we discussed in Chapter 1, the data in the problem was related to the solution by an algebraic equation in the first (discrete) case and by a system of ordinary differential equations in the second (semidiscrete) case. What we are trying to illustrate in these examples is that analogous problems for the continuous model may be reduced to one or the other of these simpler situations, and the means for carrying out this reduction is provided by the eigenfunctions in the problem. Our purpose in this example is to focus on step 2 of the procedure in order to see precisely how the eigenfunctions expose the relationship between the problem input (data) and the output (solution).

Consider an initial-boundary-value problem for an inhomogeneous heat equation,

$$u_t(x, t) = Du_{xx}(x, t) + F(x, t), \quad 0 < x < 1, \quad t > 0$$
$$u(x, 0) = 0, \quad 0 < x < 1$$
$$u_x(0, t) = u_x(1, t) = 0, \quad t > 0 \tag{3.2.11}$$

Here, D denotes a positive constant and $F(x, t)$ represents a given function of x and t, which (for convenience) we suppose is continuous with respect to x, $0 < x < 1$, and t, $t \geq 0$.

The model (3.2.11) describes the conduction of heat in a thin rod of length 1 whose ends have been insulated so that no heat escapes from either end. Initially the rod is at temperature zero throughout the interior, and heat is being produced or removed in the interior at a rate described by the function $F(x, t)$. Here $F(x, t)$ positive corresponds to production of heat and $F(x, t)$ negative corresponds to heat being removed at the point x at time t. Physically, $F(x, t)$ represents some internal heat source or sink. Such a term could be the result of an internal chemical reaction in the conductor. Also, $F(x, t)$ could result from an embedded heating device such as an electrical resistance or it could reflect the fact that the conductor is undergoing radioactive decay.

The first step in solving the problem (3.2.11) is to identify the eigenfunctions associated with the problem. In order to do this, we consider, temporarily, the homogeneous version of the partial differential equation we are solving. In this case that is the heat equation. We assume a separated solution, that is, a solution of the form $X(x)T(t)$, and substitute this into the equation and the boundary conditions. This leads then to the following Sturm–Liouville problem:

$$-X''(x) = \mu X(x), \qquad X'(0) = X'(1) = 0 \qquad (3.2.12)$$

We consider only the $X(x)$ problem at this point as we are only interested now in identifying the eigenfunctions associated with this problem. The Sturm–Liouville problem (3.2.12) is identical with Example 2.3.2 (for $L = 1$). There we found the eigenfunctions to be

$$X_n(x) = \cos n\pi x, \qquad n = 0, 1, \ldots$$

Of course, since they are generated by a Sturm–Liouville problem, these eigenfunctions form a complete orthogonal family in $L^2(0, 1)$. Having identified the eigenfunctions for this problem, we now suppose that the solution $u(x, t)$ can be expressed in the form

$$u(x, t) = \tfrac{1}{2}u_0(t) + \sum_{n=1}^{\infty} u_n(t)\cos n\pi x \qquad (3.2.13)$$

Then (formally) we can compute

$$u_t(x, t) = \tfrac{1}{2}u_0'(t) + \sum_{n=1}^{\infty} u_n'(t)\cos n\pi x$$

$$u_{xx}(x, t) = \sum_{n=1}^{\infty} -(n\pi)^2 u_n(t)\cos n\pi x$$

and if we substitute these expressions into Equation (3.2.11), we obtain

$$\tfrac{1}{2}u_0'(t) + \sum_{n=1}^{\infty} [u_n'(t) + D(n\pi)^2 u_n(t)]\cos n\pi x = F(x, t)$$

For each $m = 0, 1, \ldots$, we can multiply both sides of this expression by $\cos m\pi x$ and integrate from 0 to 1. This leads to

$$u_m'(t) + D(m\pi)^2 u_m(t) = F_m(t), \qquad t > 0 \qquad (3.2.14)$$

where

$$F_m(t) = 2 \int_0^1 F(x, t) \cos m\pi x \, dx, \qquad m = 0, 1, \ldots \qquad (3.2.15)$$

In addition, (3.2.13) implies that

$$\tfrac{1}{2}u_0(0) + \sum_{n=1}^{\infty} u_n(0)\cos n\pi x = 0$$

Multiplying the two sides of this expression by cos $m\pi x$ and integrating as before, we obtain

$$u_m(0) = 0 \quad \text{for } m = 0, 1, \dots . \tag{3.2.16}$$

Together, (3.2.14) and (3.2.16) are a system of infinitely many ordinary differential equations and initial conditions for the unknown functions $u_m(t)$. This system is analogous to the semidiscrete model described by Equation (1.2.9). In Chapter 5 we show how this system can be further reduced by means of the Laplace transform to a system of algebraic equations analogous to the discrete model for the heat equation. For the present we confine our attention to the system of ordinary differential equations. Although (3.2.14) and (3.2.16) are, in fact, a system of equations, it is completely uncoupled, which means that each equation and initial condition contains just *one* of the unknown functions. This permits us to solve each equation independently of the others. Using the method of variation of parameters for linear ordinary differential equations, we find the solution of (3.2.14) and (3.2.16) to be, for $m = 0, 1, \dots$,

$$u_m(t) = \int_0^t \exp[-(m\pi)^2 D(t - \tau)]F_m(\tau)\, d\tau, \qquad t \geq 0 \tag{3.2.17}$$

At this point we have completed step 2 of the solution procedure. The final step is accomplished by substituting (3.2.17) into (3.2.13),

$$u(x, t) = \tfrac{1}{2}u_0(t) + \sum_{m=0}^{\infty} \int_0^t \exp[-(m\pi)^2 D(t - \tau)]F_m(\tau)\, d\tau \, \cos m\pi x$$

$$= \tfrac{1}{2}u_0(t) + \int_0^t \sum_{m=0}^{\infty} \exp[-(m\pi)^2 D(t - \tau)]F_m(\tau)\, d\tau \, \cos m\pi x \tag{3.2.18}$$

An interesting feature emerges if we observe from (3.2.15) that, in the notation introduced in the previous example,

$$F_m(t) = \mathscr{F}_n[F(x, t)]$$

Then, letting σ continue to be the operator that expresses the Fourier coefficients of the solution in terms of the Fourier coefficients of the data as in the previous example, we can write (3.2.18) in the form

$$u(x, t) = \int_0^t \mathscr{F}_n^{-1}[\sigma(t - \tau)\{\mathscr{F}_n[F(x, \tau)]\}]\, d\tau$$

or, letting $S(t)$ denote the composition, $\mathscr{F}_n^{-1}[\sigma(t)\{\mathscr{F}_n\}]$,

$$u(x, t) = \int_0^t S(t - \tau)[F(x, \tau)]\, d\tau \tag{3.2.19}$$

The relationship displayed in Equation (3.2.19) is sometimes called *Duhamel's principle*. What it really says is that the solution of the inhomogeneous equation can be expressed in terms of the solution operator for the homogeneous version of the same equation.

Application of Duhamel's Principle   As an application of Duhamel's principle for solving an inhomogeneous equation, we consider the following problem:

$$u_t(x, t) = Du_{xx}(x, t), \qquad 0 < x < 1, \qquad t > 0$$
$$u(x, 0) = 0, \qquad 0 < x < 1,$$
$$u_x(0, t) = p(t), \qquad u_x(1, t) = 0, \qquad t > 0 \tag{3.2.20}$$

Here $p(t)$ denotes a given function of t, which for convenience we suppose satisfies the condition $p(0) = 0$. This is a condition of compatibility between the boundary condition and the initial condition at $x = 0$. Physically, (3.2.20) describes one-dimensional heat conduction in a conductor that is insulated at the end $x = 1$. At the end $x = 0$, however, the heat flux is varying as a function of time. This problem could also be modeling one-dimensional diffusion in a tube that is sealed at the end $x = 1$ with the diffusing substance being injected at the end $x = 0$ at a time-dependent rate related to $p(t)$.

In the previous examples involving the heat equation, we have considered inhomogeneous terms in the initial condition and in the equation, but in this problem the inhomogeneity appears in the boundary condition. As it stands, then the problem cannot be solved by the method of separation of variables since the $X(x)$ problem arising in the separation procedure will not have homogeneous boundary conditions, as it must if it is to be a Sturm–Liouville problem. However, there is a change of dependent variable that will reduce this problem to one that can be solved by separation of variables. Let

$$v(x, t) = u(x, t) - p(t)\phi(x) \tag{3.2.21}$$

where $\phi(x)$ is any function of x satisfying the conditions

$$\phi'(0) = 1 \quad \text{and} \quad \phi'(1) = 0 \tag{3.2.22}$$

Although different choices for $\phi(x)$ lead to distinct functions $v(x, t)$, the combination $v(x, t) + p(t)\phi(x)$ is independent of the choice of $\phi(x)$ so long as (3.2.22) is satisfied. We illustrate this later with examples, but for now let us suppose that we have chosen a function $\phi(x)$ that satisfies (3.2.22). Two possible choices for $\phi(x)$ are

$$\phi_1(x) = x - x^2/2 \quad \text{and} \quad \phi_2(x) = 2/\pi \sin \pi x/2$$

For $\phi(x)$ satisfying (3.2.22) and $v(x, t)$ given by (3.2.21), it is easy to compute

$$v_t(x, t) = u_t(x, t) - p'(t)\phi(x)$$
$$v_{xx}(x, t) = u_{xx}(x, t) - p(t)\phi''(x)$$

so that if $u(x, t)$ satisfies the heat equation, then $v(x, t)$ satisfies

$$v_t(x, t) - Dv_{xx}(x, t) = Dp(t)\phi''(x) - p'(t)\phi(x)$$

In addition, if $u(x, t)$ satisfies the auxiliary conditions of problem (3.2.20), then

$$v(x, 0) = u(x, 0) - p(0)\phi(x) = 0$$
$$v_x(0, t) = u_x(0, t) - p(t)\phi'(0) = p(t) - p(t) = 0$$
$$v_x(1, t) = u_x(1, t) - p(t)\phi'(1) = 0 - 0 = 0$$

Here we have made use of the fact that $p(0) = 0$ and that $\phi(x)$ satisfies (3.2.22). Then $v(x, t)$ satisfies *homogeneous* boundary conditions and an *inhomogeneous* equation. That is,

$$v_t(x, t) - Dv_{xx}(x, t) = F(x, t), \qquad 0 < x < 1, \qquad t > 0$$
$$v(x, 0) = 0, \qquad 0 < x < 1$$
$$v_x(0, t) = 0, \qquad v_x(1, t) = 0, \qquad t > 0 \tag{3.2.23}$$

where we have let

$$F(x, t) = Dp(t)\phi''(x) - p'(t)\phi(x) \tag{3.2.24}$$

Since $p(t)$ and $\phi(x)$ are known, $F(x, t)$ is a *known* function. Then (3.2.23) is a problem of the form (3.2.11), and the solution is given by (3.2.18). That is,

$$v(x, t) = \tfrac{1}{2}v_0(t) + \sum_{m=1}^{\infty} v_m(t)\cos m\pi x \tag{3.2.25}$$

where for $m = 0, 1, \ldots$.

$$v_m(t) = \int_0^t \exp[-D(m\pi)^2(t - \tau)]F_m(\tau)\, d\tau, \qquad t \geq 0$$

and

$$F_m(t) = 2\int_0^1 [Dp(t)\phi''(x) - p'(t)\phi(x)]\cos m\pi x\, dx$$

Now we compute $v(x, t)$ and $u(x, t)$ for the two choices ϕ_1 and ϕ_2 for $\phi(x)$. We see that while the two choices for ϕ lead to different results for $v(x, t)$, the solution $u(x, t)$ is independent of the choice of $\phi(x)$ (as it should be).

First, for $\phi(x) = \phi_1(x) = x - x^2/2$, we have

$$F(x, t) = -Dp(t) - p'(t)(x - x^2/2)$$

Then it is straightforward to compute

$$F_0(t) = -2Dp(t) - \tfrac{2}{3}p'(t)$$
$$F_n(t) = -p'(t)\phi^n, \qquad n = 1, 2, \ldots$$

and from these we then compute

$$v_0(t) = -2D\int_0^t p(\tau)\, d\tau - \tfrac{2}{3}p(t)$$

$$v_n(t) = -\phi^n p(t) + (n\pi)^2 D\, \phi^n \int_0^t \exp[-(n\pi)^2 D(t - \tau)]p(\tau)\, d\tau \tag{3.2.26}$$

Then $v(x, t)$ equals

$$v(x, t) = \left[-\frac{1}{3} - \sum_{n=1}^{\infty} \phi^n \cos n\pi x\right]p(t)$$

$$- D\int_0^t \left[1 - \sum_{n=1}^{\infty} (n\pi)^2 \phi^n\right.$$

$$\left. \times \exp[-D(n\pi)^2(t - \tau)]\cos n\pi x\right]p(\tau)\, d\tau \tag{3.2.27}$$

But the first expression in brackets in (3.2.27) is nothing more than the eigenfunction expansion of the function $\phi_1(x)$, and it follows that (3.2.27) reduces to

$$v(x, t) = -\phi_1(x)p(t)$$

$$-D \int_0^t \left[1 - \sum_{n=1}^{\infty} (n\pi)^2 \phi^n \right.$$

$$\left. \times \exp[-D(n\pi)^2(t - \tau)]\cos n\pi x \right] p(\tau) \, d\tau \qquad (3.2.28)$$

Finally, since $u(x, t) = v(x, t) + \phi(x)p(t)$, we conclude that

$$u(x, t) = -D \int_0^t \left[1 - \sum_{n=1}^{\infty} (n\pi)^2 \phi^n \exp[-D(n\pi)^2(t - \tau)]\cos n\pi x \right] p(\tau) \, d\tau$$

If we now go through this same procedure for the choice

$$\phi(x) = \phi_2(x) = 2/\pi \sin \pi x/2$$

then we find

$$F(x, t) = [-D(\pi/2)^2 p(t) - p'(t)]\phi_2(x)$$

Just as we did before, we can calculate

$$v_0(t) = \frac{-8}{\pi^2}p(t) - 2D \int_0^t p(\tau) \, d\tau$$

$$v_n(t) = -\phi''p(t) + \phi''D\left\{ (n\pi)^2 - \left(\frac{\pi}{2}\right)^2 \right\} \int_0^t \exp[-(n\pi)^2 D(t - \tau)]p(\tau) \, d\tau$$

Then,

$$v(x, t)$$

$$= \left[\frac{-4}{\pi^2} - \sum_{n=1}^{\infty} \phi'' \cos n\pi x \right] p(t)$$

$$- D \int_0^t \left[1 - \sum_{n=1}^{\infty} \left\{ (n\pi)^2 - \left(\frac{\pi}{2}\right)^2 \right\} \phi'' \exp[-(n\pi)^2 D(t - \tau)]\cos n\pi x \right] p(\tau) \, d\tau$$

As before, the first expression in brackets here is just the eigenfunction expansion for the function $\phi_2(x)$. Then,

$$v(x, t) = -\phi_2(x)p(t)$$

$$-D \int_0^t \left[1 - \sum_{n=1}^{\infty} \left\{ (n\pi)^2 - \left(\frac{\pi}{2}\right)^2 \right\} \phi^n \right.$$

$$\left. \times \exp[-(n\pi)^2 D(t - \tau)]\cos n\pi x \right] p(\tau) \, d\tau \qquad (3.2.29)$$

and since $u(x, t) = v(x, t) + \phi(x)p(t)$, we can see that

$$u(x, t) = -D \int_0^t \left[1 - \sum_{n=1}^{\infty} \left\{ (n\pi)^2 - \left(\frac{\pi}{2}\right)^2 \right\} \phi^n \right.$$

$$\left. \times \exp[-(n\pi)^2 D(t - \tau)]\cos n\pi x]p(\tau) \right] d\tau \qquad (3.2.30)$$

Obviously the expression (3.2.28) for $v(x, t)$ differs from (3.2.29). However, the two expressions for $u(x, t)$ are identical if we can show that, for $n = 1, 2, \ldots$,

$$\{(n\pi)^2 - (\pi/2)^2]\phi_2^n = (n\pi)^2\phi_1^n \tag{3.2.31}$$

In Exercises 16 and 18 in Section 2.1 we computed

$$\phi_1^n = -2/(n\pi)^2, \qquad n = 1, 2, \ldots$$
$$\phi_2^n = (8/\pi^2)(1 - 4n^2)^{-1}, \qquad n = 1, 2, \ldots$$

from which (3.2.31) easily follows.

Steady-State Solutions   It will be convenient to use this example to illustrate the notion of a *steady-state* or *equilibrium solution*. The steady-state solution is what remains of the solution $u(x, t)$ to the initial-boundary-value problem after all the transient behavior has died out. It is important to realize that the steady-state solution *need not* be independent of time. To illustrate this point, consider the following examples.

   (a) Steady State Not Dependent on Time   Consider the problem (3.2.20) in the case that the boundary forcing term $p(t)$ is given by

$$p(t) = Ate^{-bt}, \qquad t > 0 \tag{3.2.32}$$

where A and b denote fixed, positive constants. Then (3.2.30) reduces to

$$u(x, t) = \Phi_0(t) + \sum_{n=1}^{\infty} \left\{ (n\pi)^2 - \left(\frac{\pi}{2}\right)^2 \right\} \phi_2^n \Phi_n(t) \cos n\pi x$$

where

$$\Phi_0(t) = -D \int_0^t p(\tau) \, d\tau = \frac{-AD}{b^2[1 - e^{-bt}(1 + bt)]}$$

$$\Phi_n(t) = D \int_0^t \exp[-(n\pi)^2 D(t - \tau)]p(\tau) \, d\tau$$

$$= \frac{At \exp[-bt]}{D(n\pi)^2 - b} + \frac{A \exp[-D(n\pi)^2 t] - \exp[-bt]}{[D(n\pi)^2 - b]^2}$$

Then as $t \to +\infty$, $\Phi_n(t) \to 0$ for $n = 1, 2, \ldots$ and $\Phi_0(t)$ tends to the constant $-AD/b^2$. It follows that $u(x, t)$ tends to the steady-state solution $u(x, \infty) = -AD/b^2$. This result is consistent with the fact, expressed in Chapter 1, that

$$\textbf{flux}(x, t) = -K \, \textbf{grad} \, u(x, \textbf{t})$$
$$= -Ku_x(x, t)\textbf{i}$$

Then if $p(t)$ is positive, the flux of heat at $x = 0$ is in the direction of $-\textbf{i}$; that is, *out of* the interval $(0, 1)$. Thus a positive $p(t)$ is consistent with a *loss of* heat and a corresponding decrease in the temperature. Evidently, the problem (3.2.20) in the case that $p(t)$ is given by (3.2.32) is a model for the situation in which a heat-conducting rod is initially at temperature zero and insulated at both ends [recall $p(0) = 0$]. As t increases

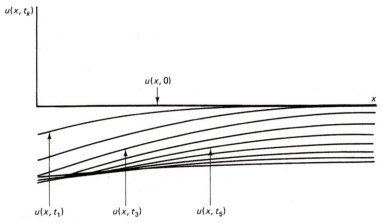

Figure 3.2.4

from zero, heat is extracted from the end $x = 0$ at a rate that at first increases and then decreases gradually to zero. The temperature decreases in response to the extraction of heat and eventually tends to a constant limiting value as the rate of extraction decreases to zero. Some profiles showing the temperature distribution as a function of x for various times for this example are shown in Figure 3.2.4. Since the solution $u(x, t)$ in this problem tends to a steady state that is independent of time, it follows that the steady-state solution $u_e(x)$ then satisfies

$$0 = u''(x), \qquad 0 < x < 1$$
$$u'(0) = p(\infty) = 0, \qquad u'(1) = 0$$

This is what the original problem (3.2.20) reduces to in the limit since $u_t(x, t)$ and $p(t)$ $\rightarrow 0$ as $t \rightarrow \infty$. The solution of this reduced problem is easily seen to be $u_S(x) = $ const. However, the value of this constant can only be found by solving the full problem and then finding the limiting value of the solution as $t \rightarrow \infty$. In some cases, a time-independent steady state can be found more directly (see Exercise 12).

(b) Periodic Steady State   In the previous example, the choice of the boundary forcing function $p(t)$ was such that the solution $u(x, t)$ tended to a steady state that was independent of time. Now consider the choice

$$p(t) = A \sin \Omega t, \qquad t > 0$$

where A and Ω denote fixed, positive constants. Then (3.2.20) can be thought of as a heat conduction problem that is driven by periodic forcing at the boundary. This could arise, for example, in treating one-dimensional heat conduction in a slab, one of whose faces is heated and cooled, say, by the rising and setting of the sun.

Proceeding as in part (a), we find

$$\Phi_0(t) = \frac{AD}{\Omega} [\cos \Omega t - 1]$$

and for $n = 1, 2, \ldots,$

$$\Phi_n(t) = \frac{AD(n\pi)^2\sin \Omega t - A\Omega \cos \Omega t + A\Omega \exp[-(n\pi)^2Dt]}{\Omega^2 + D^2(n\pi)^4}$$

Then, as $t \to +\infty$, the only transient term in the solution is the decaying exponential appearing in $\Phi_n(t)$, and it follows that $u(x, t)$ tends to the periodic steady state given by

$$u_S(x, t) = \frac{AD}{\Omega[\cos \Omega t - 1]} + D \sum_{n=1}^{\infty} \left\{(n\pi)^2 - \left(\frac{\pi}{2}\right)^2\right\} \phi_2^n T_n(t)\cos n\pi x$$

where

$$T_n(t) = \frac{AD(n\pi)^2\sin \Omega t - A\Omega \cos \Omega t}{\Omega^2 + D^2(n\pi)^4}$$

Evidently, as the end at $x = 0$ of the heat-conducting rod is periodically heated and cooled, the influence of the initial state dies out in time, and the temperature in the rod begins to vary periodically with time. Note that $T_n(t)$ can be written in the form

$$T_n(t) = \frac{A \sin[\Omega t - \theta_n]}{[\Omega^2 + D^2(n\pi)^4]^{1/2}}$$

where

$$\cos \theta_n \equiv D(n\pi)^2/[\Omega^2 + D^2(n\pi)^4]^{1/2}$$
$$\sin \theta_n \equiv \Omega/[\Omega^2 + D^2(n\pi)^4]^{1/2}$$

Thus the steady-state response to the periodic forcing function $p(t)$ is periodic with the same period as the forcing, but there is a shift in the phase of the response that varies with x. Profiles showing $u(x, t)$ plotted against x for several values of t are shown in Figure 3.2.5.   ■ ■

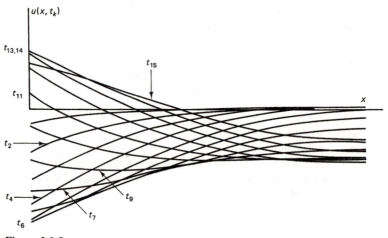

Figure 3.2.5

Influence of Lower Order Terms: Convection–Diffusion

We consider next an example that illustrates the effect on the solution of the lower order terms in the heat equation.

EXAMPLE 3.2.3 _____

Diffusion with Convection on Bounded Interval

Consider a diffusing substance occupying a region Ω in space and let $u(x, y, z, t)$ denote the concentration of this substance at the point (x, y, z) in Ω at time t. Let $\Phi(x, y, z, t)$ denote the flux field associated with $u(x, y, z, t)$. That is, the *direction* of diffusion at (x, y, z, t) is the direction of $\Phi(x, y, z, t)$, and the *rate* of diffusion at (x, y, z, t) equals the magnitude of $\Phi(x, y, z, t)$. Then, using arguments nearly identical to those used to derive the equation of heat conduction, we can derive the following *continuity equation* expressing the conservation of mass:

$$u_t + \mathbf{div}[\Phi] = S \quad \text{in } \Omega \tag{3.2.33}$$

where $S(x, y, z, t)$ is the source or sink term, S reflecting the creation or destruction of the diffusing substance, say, by chemical reaction. If S is positive, then substance is being created at (x, y, z, t) at a rate equal to $S(x, y, z, t)$. In this case we say S is a *source* term. If S is negative, then substance is being destroyed at (x, y, z, t), and we say S is a *sink* term. Here we suppose that

$$S(x, y, z, t) = cu(x, y, z, t) \tag{3.2.34}$$

for some constant c. Then whether S is a source or a sink depends on the sign of c, but in either case, the rate at which substance is created or destroyed is proportional to the present concentration.

In the event that the diffusing substance is not only diffusing through Ω but is being transported along with a "carrier substance" that is moving through Ω with velocity $\mathbf{V}(x, y, z, t)$, (3.2.33) must be modified as follows to include the convective transport:

$$\frac{du}{dt} + \mathbf{div}[\Phi] = S$$

where du/dt denotes the *total derivative* of u with respect to t,

$$\frac{du}{dt} = u_t + u_x x'(t) + u_y y'(t) + u_z z'(t)$$

But $x'(t)$, $y'(t)$, $z'(t)$ are the components of the vector $\mathbf{V}(t)$, and hence the continuity equation can be rewritten in the following notation:

$$u_t + \mathbf{V} \, \mathbf{grad} \, u + \mathbf{div}[\Phi] = S \tag{3.2.35}$$

A situation of this sort arises, for example, when a fluid contaminant diffuses in water that is moving or a gaseous contaminant diffuses in moving, uncontaminated air. Then $\mathbf{V}$ would represent the velocity field for the moving fluid or moving air. Of course when $\mathbf{V} = \mathbf{0}$, (3.2.35) reduces to (3.2.33), the equation for conservation of matter for diffusion in a stationary medium.

In the diffusion context, the law relating $\boldsymbol{\Phi}$ to u is known as *Fick's law of diffusion*. This law states

$$\boldsymbol{\Phi}(x, y, z, t) = -D \textbf{ grad } u(x, y, z, t) \tag{3.2.36}$$

where the positive parameter D is known as the diffusivity. As Fourier's law of heat conduction, Fick's law is just a quantification of the observation that a diffusing substance moves from areas of high concentration to areas of lower concentration at a rate that is proportional to the difference in the levels of concentration.

Using (3.2.36) and (3.2.34) in (3.2.35) leads to the following equation, governing the convection–diffusion of the substance through the region Ω:

$$u_t + V_1 u_x + V_2 u_y + V_3 u_z = \textbf{div}[D \textbf{ grad } u] + cu \tag{3.2.37}$$

where V_1, V_2, V_3 denote the x, y, and z components of the velocity $\mathbf{V}$. We assume here that each of these components is constant.

One-dimensional Convection–Diffusion Now consider the special case that Ω is a thin pipe whose lateral surface is impervious to the passage of the diffusing substance. Then we are dealing with one-dimensional convection–diffusion, and Equation (3.2.37) reduces to

$$u_t(x, t) = D u_{xx}(x, t) - V_1 u_x(x, t) + cu(x, t) \tag{3.2.38}$$

where we have labeled the axial direction along the pipe as the x direction. Suppose that the pipe is of finite length, occupying the interval $(0, 1)$ on the x axis, and that the initial concentration distribution of the substance is described by the known function $f(x)$; that is,

$$u(x, 0) = f(x), \qquad 0 < x < 1 \tag{3.2.39}$$

Finally, suppose that the end $x = 0$ of the pipe is closed to the flow of the diffusing substance while the end at $x = 1$ is open but the concentration there is maintained at the level zero. Then,

$$u_x(0, t) = 0, \qquad u(1, t) = 0, \qquad t > 0 \tag{3.2.40}$$

We now use the method of eigenfunction expansion to solve the initial-boundary-value problem consisting of Equation (3.2.38) together with the conditions (3.2.39) and (3.2.40).

Transformation to Eliminate Lower Order Terms Although it is not essential to do so, we shall remove the lower order terms in Equation (3.2.38) by a change of dependent variable. That is, we let

$$u(x, t) = \exp[\alpha x + \beta t]v(x, t) \tag{3.2.41}$$

Then

$$u_t = (v_t + \beta v)\exp[\alpha x + \beta t]$$
$$u_x = (v_x + \alpha v)\exp[\alpha x + \beta t]$$
$$u_{xx} = (v_{xx} + 2\alpha v_x + \alpha^2 v)\exp[\alpha x + \beta t]$$

and it follows from (3.2.38) that

$$v_t = Dv_{xx} + (2\alpha D - V_1)v_x + (D\alpha^2 - V_1\alpha + c - \beta)v$$

The parameters α and β denote constants whose value up to this point has not been specified. However, if we choose

$$\alpha = V_1/2D \quad \text{and} \quad \beta = c - V_1^2/4D \tag{3.2.42}$$

then (3.2.38) reduces to the heat equation. It is easy to see how the conditions (3.2.39) and (3.2.40) are correspondingly modified and to determine that $v(x, t)$ satisfies

$$v_t(x, t) = Dv_{xx}(x, t), \quad 0 < x < 1, \quad t > 0$$
$$v(x, 0) = f(x)\exp[-(V_1/2D)x], \quad 0 < x < 1$$
$$v_x(0, t) + (V_1/2D)v(0, t) = 0$$
$$v(1, t) = 0 \tag{3.2.43}$$

Separation of Variables   The assumption that $v(x, t) = X(x)T(t)$ causes (3.2.43) to separate into the following two problems:

$$-X''(x) = \mu X(x), \quad X'(0) + \alpha X(0) = X(1) = 0 \tag{3.2.44a}$$
$$T'(t) = -D\mu T(t) \tag{3.2.44b}$$

Clearly, (3.2.44a) is the Sturm–Liouville problem, and it is similar to Example 2.3.3. Proceeding as we did in that example, we consider the three possibilities $\mu = 0$, $\mu < 0$, and $\mu > 0$.

In the case $\mu = 0$, the general solution to (3.2.44a) is given by $X(x) = Ax + B$ for constants A and B. Then the boundary conditions imply

$$A + B = 0 \quad \text{and} \quad A + \alpha B = 0$$

This pair of equations for A and B has a nontrivial solution if and only if $\alpha = 1$ (i.e., $V_1 = 2D$), in which case $X_0(x) = A(x - 1)$ is the eigenfunction corresponding to the eigenvalue zero. For purposes of this example, we suppose that the parameters V_1 and D are such that α is *not* equal to 1 and that zero is therefore not an eigenvalue for this problem.

For μ negative, say, $\mu = -\sigma^2$ for σ real and not zero, the general solution to (3.2.44a) can be expressed in the form

$$X(x) = A \sinh \sigma x + B \sinh \sigma(1 - x)$$

Then the boundary conditions become

$$\sigma A + (\alpha \sinh \sigma - \sigma \cosh \sigma)B = 0$$
$$\sinh \sigma A = 0$$

If σ is not zero, then this pair of equations has a nontrivial solution if and only if σ satisfies

$$\alpha \sinh \sigma = \sigma \cosh \sigma, \quad \text{i.e.,} \quad \tanh \sigma = \sigma/\alpha \tag{3.2.45}$$

In Exercise 10 you are asked to show that nonzero solutions to (3.2.45) exist if and only if $\alpha > 1$, that is, if and only if $V_1 > 2D$. For the purposes of this example we suppose that V_1 and D are such that $\alpha > 1$. In this case there are two nonzero solutions to (3.2.45), one positive and one negative, having the same absolute value. These lead then to a *single* negative eigenvalue, $\mu_0 = -\sigma_0^2$, where σ_0 satisfies (3.2.45). The eigenfunction corresponding to this negative eigenvalue is then

$$X_0(x) = B \sinh \sigma_0(1 - x) \qquad (3.2.46)$$

The remaining eigenvalues must be positive, say, $\mu = \sigma^2$ for σ real and not zero. For $\mu = \sigma^2$ the general solution to Equation (3.2.45a) can be written in the form

$$X(x) = A \sin \sigma x + B \sin \sigma(1 - x)$$

The functions $\sin \sigma x$ and $\sin \sigma(1-x)$ are independent if $\sin \sigma$ is not zero. We assume then that $\sin \sigma$ is not zero.

The boundary condition at $x = 1$ implies

$$X(1) = A \sin \sigma = 0$$

and since $\sin \sigma$ is not zero, then A must be zero. Then the boundary condition at $x = 0$ becomes

$$-B[\sigma \cos \sigma - \alpha \sin \sigma] = 0$$

and in order for B to be different from zero, it is necessary that σ satisfy

$$\sigma \cos \sigma = \alpha \sin \sigma, \quad \text{i.e.,} \quad \tan \sigma = \sigma/\alpha \qquad (3.2.47)$$

Figure 3.2.6 shows $F(\sigma) = \tan \sigma$ and $G(\sigma) = \sigma/\alpha$ plotted versus σ on the same set of axes, and we see from this figure that (3.2.47) has infinitely many positive solutions σ_n. It is easy to see that the roots of (3.2.47) occur in positive–negative pairs. That is, for each positive root σ, $-\sigma$ is also a root. It is also evident from Figure 3.2.6 that if α is *positive*, then the positive roots σ_n satisfy

$$\sigma_1 < \sigma_2 < \sigma_3 < \cdots \quad \text{with} \quad n\pi < \sigma_n < (n + \tfrac{1}{2})\pi$$

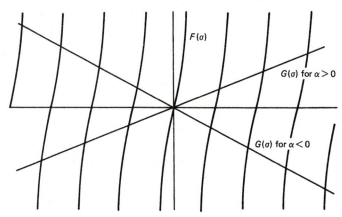

Figure 3.2.6

Likewise, when α is *negative*, the positive roots σ_n satisfy

$$\sigma_1 < \sigma_2 < \sigma_3 < \cdots \quad \text{with} \quad (n - \tfrac{1}{2})\pi < \sigma_n < n\pi$$

In either case, α positive or α negative, there are infinitely many roots to (3.2.47) and, correspondingly, infinitely many eigenvalues $\mu_n = \sigma_n^2$ and eigenfunctions

$$X_n(x) = B \sin \sigma_n(1 - x)$$

Note that when $\alpha = 0$, the eigenvalue equation (3.2.47) reduces to $\cos \sigma = 0$, with roots at zero and $(n - \tfrac{1}{2})\pi$ for $n = 1, 2, \ldots$. Evidently when α is positive, the eigenvalues are shifted to the left, and when α is negative, they are shifted to the right. Since $\alpha = V_1/2D$, it is apparent that one effect of the lower order term $[V_1 u_x]$ is to shift the eigenvalues to the left or the right depending on the sign of V_1.

Equation (3.2.44b) has the solution

$$T_n(t) = C_n \exp[-\mu_n Dt], \qquad n = 0, 1, \ldots$$

and then

$$v(x, t) = C_0 \exp[-\mu_0 Dt]\sinh \sigma_0(1 - x)$$
$$+ \sum_{n=1}^{\infty} C_n \exp[-\mu_n Dt]\sin \sigma_n(1 - x) \qquad (3.2.48)$$

satisfies, formally, all of the conditions of the problem except the inhomogeneous initial condition. This condition will be satisfied as well if we choose the constants C_n as follows:

$$C_n = \frac{1}{M_n} \int_0^1 f(x)e^{-\alpha x} X_n(x)\, dx, \qquad n = 0, 1, \ldots \qquad (3.2.49)$$

where

$$M_n = \int_0^1 X_n(x)^2\, dx, \qquad n = 0, 1, \ldots$$

Here we have made use of the fact, ensured by Theorem 2.3.1, that the family of eigenfunctions $X_n(x)$ is complete and orthogonal.

Discussion of Solution   In view of (3.2.41) and (3.2.42), $u(x, t)$, the solution of the original problem, is given by

$$u(x, t) = e^{\alpha x} e^{\beta t} v(x, t)$$
$$= \sum_{n=0}^{\infty} C_n \exp\left[\frac{V_1 x}{2D}\right]\exp\left[-\left\{\left(\frac{V_1}{2D}\right)^2 - \frac{c}{D} + \mu_n\right\} Dt\right]X_n(x) \qquad (3.2.50)$$

Since the eigenvalue μ_0 is negative, it is evident from (3.2.50) that there is the possibility that the first term (the $n = 0$ term) in the series for $u(x, t)$ grows exponentially with t. To see whether this, in fact, happens, we recall that

$$\mu_0 = -\sigma_0^2 = -\alpha^2 \tanh^2 \sigma_0 = -(V_1/2D)^2 \tanh^2 \sigma_0$$

Then, since $|\tanh \sigma| \leq 1$, it follows that

$$\mu_0 + (V_1/2D)^2 = (V_1/2D)^2[1 - \tanh^2\sigma_0] \geq 0$$

Then if $c < 0$

$$\mu_0 + (V_1/2D)^2 - c/D \geq 0$$

and the $n = 0$ term in the series (3.2.50) decays exponentially with increasing t. On the other hand, if c satisfies the condition

$$c > c^* \equiv D\mu_0 + V_1^2/4D$$

then the $n = 0$ term in the series for $u(x, t)$ grows exponentially with t.

Recall from (3.2.34) that c positive corresponds to an internal source of diffusing material. That is, c positive implies that material is being created inside the pipe. It is then apparent that when c is greater than c^*, the diffusing material is being created sufficiently rapidly inside the pipe that the concentration $u(x, t)$ increases exponentially with t. When c is nonnegative but less than or equal to c^*, then $u(x, t)$ is either stationary or decreasing with increasing t. We interpret this to mean that the diffusion together with the convection are able to disperse the material more rapidly than the source can produce it. Finally, if c is negative, then S represents a sink, and material is being destroyed inside the pipe. In this case the concentration $u(x, t)$ decreases steadily with increasing t.

We observe that when $c > c^*$, there is *no steady-state solution for this problem.* As t increases to infinity, the solution $u(x, t)$ increases without bound. If $c < c^*$, then the solution decreases to zero as t increases. In this case $u(x, t)$ identically zero is the steady state. Finally, in the case that $c = c^*$, you can show that the steady-state solution is (see Exercise 14)

$$u(x, \infty) \equiv \lim_{t \to \infty} u(x, t) = C_0\exp[V_1x/2D]X_0(x)$$

We interpret this to be the equilibrium distribution of concentration that occurs when source production exactly balances the dispersive effect of the convection and diffusion. Increasing the rate of production beyond this value causes the concentration to increase without bound, while a production rate less than c^* cannot keep up with the convection–diffusion, and the concentration decreases eventually to zero. ■ ■

Diffusion in More Than One Dimension

All of the examples for the heat equation that we have considered have involved the independent variables t (time) and x, a spatial variable. We refer to such examples as examples of the one-dimensional heat equation, referring to the fact that only one spatial dimension is involved. We can consider problems in more spatial dimensions, but the manipulations become rapidly unwieldy as the number of variables increases. Consider the following example.

EXAMPLE 3.2.4 ———————————————————————————

Two-dimensional Diffusion Problem

Let Ω denote the unit square, $0 < x, y < 1$, and let $u(x, y, t)$ denote the concentration at (x, y, t) of a substance that is moving in Ω by pure diffusion. If the initial distribution of material is given by the known function $f(x, y)$, and if the edges of the unit square are impervious to the diffusing substance, then $u(x, y, t)$ must satisfy

$$u_t(x, y, t) = D[u_{xx}(x, y, t) + u_{yy}(x, y, t)] \quad \text{in } \Omega \quad \text{for } t > 0$$
$$u(x, y, 0) = f(x, y) \quad \text{in } \Omega$$
$$u_y(x, 0) = u_y(x, 1) = 0, \quad 0 < x < 1$$
$$u_x(0, y) = u_x(1, y) = 0, \quad 0 < y < 1 \tag{3.2.51}$$

The initial distribution $f(x, y)$ evolves via pure diffusion into the state $u(x, y, t)$ at time t. In this example the absence of lower order terms indicates that there is no convection and no sources or sinks. Treating this as a two-dimensional problem is tantamount to the assumption that the distribution of material is uniform in the direction normal to the plane of Ω or else the diffusing substance consists of a thin layer contained between two impenetrable plates. This example is treated by numerical means in Section 8.6.

Separation of Variables in Several Variables If we suppose that $u(x, y, t) = \Phi(x, y)T(t)$ then (3.2.51) separates into

$$-\nabla^2\Phi(x, y) = \mu\Phi(x, y), \qquad \Phi_y(x, 0) = \Phi_y(x, 1) = 0, \tag{3.2.52a}$$
$$\Phi_x(0, y) = \Phi_x(1, y) = 0$$
$$T'(t) = -D\mu\, T(t), \qquad t > 0 \tag{3.2.52b}$$

Problem (3.2.52a) is an eigenvalue problem similar to Example 3.1.4. It separates further, under the assumption that $\Phi(x, y) = X(x)Y(y)$, into

$$X''(x)/X(x) = \mu - Y''(y)/Y(y) \equiv -\sigma$$

That is,

$$-X''(x) = \sigma X(x), \qquad X'(0) = X'(1) = 0 \tag{3.2.53a}$$
$$-Y''(y) = (\sigma + \mu)Y(y), \qquad Y'(0) = Y'(1) = 0 \tag{3.2.53b}$$

These are each Sturm–Liouville problems with eigenvalues and eigenfunctions as follows:

$$\sigma_n = (n\pi)^2, \qquad X_n(x) = \cos n\pi x, \qquad n = 0, 1, \ldots \tag{3.2.54a}$$
$$\sigma_n + \mu_{mn} = (m\pi)^2, \qquad Y_m(y) = \cos m\pi y, \qquad m = 0, 1, \ldots \tag{3.2.54b}$$

Then

$$\mu_{mn} = (m^2 + n^2)\pi^2, \qquad \Phi_{mn}(x, y) = \cos n\pi x \cos m\pi y, \qquad m, n = 0, 1, \ldots$$

are the two-dimensional eigenvalues and eigenfunctions for this problem. The family $\{\Phi_{mn} : m, n = 0, 1, \ldots\}$ is a complete orthogonal family in $L^2(\Omega)$.

The solution to (3.2.52b) is then

$$T_{mn}(t) = A_{mn}\exp[-(m^2 + n^2)\pi^2 Dt], \qquad m, n = 0, 1, \ldots$$

for arbitrary constants A_{mn}. Then superposition implies that

$$u(x, y, t) = \sum_{m=0}^{\infty} \sum_{n=0}^{\infty} A_{mn}\exp[-(m^2 + n^2)\pi^2 Dt]\cos n\pi x \cos m\pi y$$

satisfies all the conditions of the problem except the inhomogeneous initial condition. This condition will be satisfied as well if we choose the constants A_{mn} so that

$$A_{mn} = 4 \int_0^1 \int_0^1 f(x, y)\cos n\pi x \cos m\pi y \, dx \, dy = f^{m,n}$$

for $m, n = 1, 2, \ldots$, and

$$A_{00} = \int_0^1 \int_0^1 f(x, y) \, dx \, dy \equiv f^* = \text{average value of } f \text{ on } \Omega$$

Then we can write

$$u(x, y, t) = f^* + \sum_{m=1}^{\infty} \sum_{n=1}^{\infty} f^{m,n}\exp[-\mu_{mn} Dt]\Phi_{mn}(x, y) \qquad (3.2.55)$$

and since $\mu_{mn} = (m^2 + n^2)\pi^2$, it is apparent that as t tends to plus infinity, $u(x, y, t)$ tends to the constant steady state f^*. In other words, the equilibrium distribution of the diffusing substance is achieved when the concentration is uniformly equal to the average value f^* of $f(x, y)$.

Associated Inhomogeneous Problem   If we adopt the notation

$$u(x, y, t) = \mathcal{F}_n^{-1}[\sigma(t)\mathcal{F}_n\{f(x, y)\}] = S(t)[f(x, y)]$$

to represent (3.2.55), then the Duhamel principle implies that

$$v(x, y, t) = S(t)[f(x, y)] + \int_0^t S(t - \tau)[F(x, y, \tau)] \, d\tau$$

solves the *inhomogeneous* problem

$$v_t(x, y, t) = D[v_{xx}(x, y, t) + v_{yy}(x, y, t)] + F(x, y, t)$$
$$v(x, y, 0) = f(x, y) \quad \text{in } \Omega$$
$$v_y(x, 0) = v_y(x, 1) = 0, \qquad 0 < x < 1$$
$$v_x(0, y) = v_x(1, y) = 0, \qquad 0 < y < 1 \qquad (3.2.56)$$

Evidently, the fact that this problem involves diffusion in two space dimensions leads to a two-parameter family of eigenfunctions that, in turn, leads to a double Fourier series representation for the solution. Similarly, diffusion in three space dimensions could be expected to lead to a three-parameter family of eigenfunctions and a triple Fourier series representation for the solution. It is also evident that as the number of independent variables in the problem increases, the computational difficulty in solving the problem increases very rapidly. It is safe to say that this increase in difficulty is not associated with exact solutions only. The computational difficulty of computing approximate solutions by numerical methods also increases rapidly with the number of independent variables (see Section 8.6).   ■ ■

Exercises: Initial-Boundary-Value Problems for the Heat Equation

1. Find $u(x, t)$ satisfying

$$u_t(x, t) = Du_{xx}(x, t), \qquad 0 < x < 1, \qquad t > 0$$

$$u(x, 0) = \begin{cases} 2x & \text{if } 0 < x < \frac{1}{2} \\ 1 & \text{if } \frac{1}{2} < x < 1 \end{cases}$$

$$u(0, t) = 0, \qquad u_x(1, t) = 0, \qquad t > 0$$

2. Solve Exercise 1 in the case that the initial condition is changed to

$$u(x, 0) = \begin{cases} 1 & \text{if } \frac{1}{3} < x < \frac{2}{3} \\ 0 & \text{if } 0 < x < \frac{1}{3} \text{ or } \frac{2}{3} < x < 1 \end{cases}$$

3. Find $u(x, t)$ satisfying

$$u_t(x, t) = Du_{xx}(x, t), \qquad 0 < x < 2, \qquad t > 0$$

$$u(x, 0) = \begin{cases} 2x & \text{if } 0 < x < \frac{1}{2} \\ 1 & \text{if } \frac{1}{2} < x < \frac{3}{2} \\ 4 - 2x & \text{if } \frac{3}{2} < x < 2 \end{cases}$$

$$u(0, t) = 0, \qquad u(2, t) = 0, \qquad t > 0$$

Compute $u_x(1, t)$ for this solution and show that this derivative equals zero for $t > 0$. In view of this, are the solutions to Exercises 1 and 3 equal for all x and t? Are they equal for $0 < x < 1$ and $t > 0$?

4. Consider the problem

$$u_t(x, t) = Du_{xx}(x, t), \qquad 0 < x < 2, \qquad t > 0$$

$$u(x, 0) = f(x), \qquad 0 < x < 2$$

$$u(0, t) = 0, \qquad u(2, t) = 0, \qquad t > 0$$

How can the function $f(x)$ be chosen so that the solution to this problem is equal, for $0 < x < 1$ and $t > 0$, to the solution of Exercise 2?

5. Suppose that $u(x, t)$ satisfies

$$u_t(x, t) = Du_{xx}(x, t), \qquad 0 < x < 1, \qquad t > 0$$

$$u(x, 0) = 0, \qquad 0 < x < 1$$

$$u(0, t) = 1, \qquad u(1, t) = 0, \qquad t > 0$$

This problem cannot be solved directly by eigenfunction expansion since the boundary conditions are not homogeneous. However, we can reduce this problem to one we can solve as follows. Let $v(x, t) = u(x, t) - (1 - x)$ for $0 < x < 1, t > 0$. Then show that

$$v_t(x, t) = Dv_{xx}(x, t), \qquad 0 < x < 1, \qquad t > 0$$

$$v(x, 0) = x - 1, \qquad 0 < x < 1$$

$$v(0, t) = 0, \qquad v(1, t) = 0, \qquad t > 0$$

6. Solve Exercise 5 by finding $v(x, t)$ and then noting that

$$u(x, t) = v(x, t) + 1 - x$$

Note that the initial condition for $u(x, t)$ is not consistent with the boundary condition at $x = 0$. Which if either of the two conditions does your solution satisfy? Inconsistent auxiliary conditions occur frequently in applications. Can you give a physical interpretation for this problem?

7. Find $u(x, t)$ satisfying

$$u_t(x, t) = Du_{xx}(x, t), \qquad 0 < x < 1, \qquad t > 0$$
$$u(x, 0) = 0, \qquad 0 < x < 1$$
$$u(0, t) = 1, \qquad u_x(1, t) = 0, \qquad t > 0$$

8. Solve for $u(x, t)$ if

$$u_t(x, t) = Du_{xx}(x, t) + \alpha u(x, t), \qquad 0 < x < 1, \qquad t > 0$$
$$u(x, 0) = x(1 - x), \qquad 0 < x < 1$$
$$u(0, t) = 0, \qquad u(1, t) = 0, \qquad t > 0$$

where α, D are given positive constants.

9. Find $u(x, t)$ such that

$$u_t(x, t) = Du_{xx}(x, t) + \alpha u(x, t), \qquad 0 < x < 1, \qquad t > 0$$
$$u(x, 0) = x, \qquad 0 < x < 1$$
$$u(0, t) = 0, \qquad u_x(1, t) = 1, \qquad t > 0$$

where α, D are given positive constants.

10. Consider the eigenvalue equation

$$\tanh \sigma = \sigma/\alpha$$

where α denotes a real parameter. By plotting the monotone function $F(\sigma) = \tanh \sigma$ and the linear function $G(\sigma) = \sigma/\alpha$ against σ on the same set of axes, observe that the graphs of F and G must intersect either once (at $\sigma = 0$) or three times (one of the three intersections being $\sigma = 0$). By comparing the linear function $G(\sigma)$ with the tangent line at zero to $F(\sigma)$, show that the three intersections occur when α is greater than 1. Finally, show that the two nonzero intersections of F and G are the negatives of one another; that is, show that if σ satisfies $F(\sigma) = G(\sigma)$, then also $F(-\sigma) = G(-\sigma)$.

11. Find $u(x, t)$ such that

$$u_t(x, t) = Du_{xx}(x, t) - \beta u_x(x, t), \qquad 0 < x < 1, \qquad t > 0$$
$$u(x, 0) = f(x), \qquad 0 < x < 1$$
$$u(0, t) = 0, \qquad u_x(1, t) = 0, \qquad t > 0$$

Does the solution tend to a steady state?

12. Show that the solution of the problem

$$v_t(x, t) = v_{xx}(x, t), \qquad 0 < x < 1, \qquad t > 0$$
$$v(x, 0) = 0, \qquad 0 < x < 1$$
$$v(0, t) = 1, \qquad v(1, t) = 0, \qquad t > 0$$

tends to a time-independent steady-state solution $v_e(x)$ as $t \to \infty$. Show (by direct computation) that $v_e(x)$ satisfies

$$0 = v''(x), \qquad 0 < x < 1$$
$$v(0) = 1, \qquad v(1) = 0$$

Observe that this is what the original problem reduces to when all mention of the t variable is suppressed.

13. Find the steady-state solutions for Exercises 7 and 9.

14. Consider the solution (3.2.50) for Example 3.2.3. Show that in the case that the source strength c is equal to the value c^*, $u(x, t)$ tends to the steady state,

$$u(x, \infty) = C_0 \exp[V_1 x/2k]\sinh \sigma_0(1 - x)$$

15. Solve for $u(x, t)$ if

$$u_t(x, t) = Du_{xx}(x, t) + F(x, t), \qquad 0 < x < 1, \qquad t > 0$$
$$u(x, 0) = 0, \qquad 0 < x < 1$$
$$u(0, t) = 0, \qquad u_x(1, t) = 0, \qquad t > 0$$

where $F(x, t)$ is a given function of x, t. Find the steady-state solution for this problem if

(a) $F(x, t) = 1 - x$

(b) $F(x, t) = 1 - \exp[-t]$

16. Solve for $u(x, y, t)$ if

$$u_t(x, y, t) = D \nabla^2 u(x, y, t), \qquad 0 < x, y < 1, \qquad t > 0$$
$$u(x, y, 0) = f(x, y), \qquad 0 < x, y < 1$$
$$u(0, y, t) = u(1, y, t) = 0, \qquad 0 < y < 1, \qquad t > 0$$
$$u_y(x, 0, t) = u_y(x, 1, t) = 0, \qquad 0 < x < 1, \qquad t > 0$$

where $f(x, y)$ denotes a given function of x, y. Does the solution of this problem tend to a steady state? If so, find it.

17. Solve for $u(x, y, t)$ if

$$u_t(x, y, t) = D \nabla^2 u(x, y, t), \qquad 0 < x, y < 1, \qquad t > 0$$
$$u(x, y, 0) = f(x, y), \qquad 0 < x, y < 1$$
$$u_x(0, y, t) = u_x(1, y, t) = 0, \qquad 0 < y < 1, \qquad t > 0$$
$$u_y(x, 0, t) = u_y(x, 1, t) = 0, \qquad 0 < x < 1, \qquad t > 0$$

where $f(x, y)$ denotes a given function of x, y. Does the solution of this problem tend to a steady state? If so, find it.

18. An experiment is designed to measure the diffusion of radioactive radon. A radon source is placed at one end of a thin, air-filled tube of length L, and a detector is placed at the opposite end that is insulated against the passage of radon. Initially the concentration of radon in the tube is zero, and we want to measure the concentration as a function of time at the end $x = L$ of the pipe. We can model this situation mathematically as follows. Let $C(x, t)$ denote the concentration of radon at position x and time t. Then we can show that $C(x, t)$ satisfies

$$C_t(x, t) = DC_{xx}(x, t) - \sigma C(x, t), \qquad 0 < x < L, \qquad t > 0$$

where D denotes the diffusivity of the radon in air and σ is a positive parameter whose magnitude reflects the loss of radon through the sides of the tube. The initial condition must be

$$C(x, 0) = 0 \quad \text{for } 0 < x < L$$

and the boundary conditions are

$$C(0, t) = C_0 > 0 \quad \text{and} \quad C_x(L, t) = 0 \quad \text{for } t > 0$$

Find $C(L, t)$ for $t > 0$. Find the limit as $t \to \infty$ of $C(L, t)$ in the case $\sigma = 0$ *and* in the case $\sigma > 0$.

19. Use the Weierstrass M-test to show that the solution (3.2.3) for the initial boundary-value problem of Example 3.2.1 converges uniformly on any region of the form

$$Q[t_0] = \{0 < x < 1, t \geq t_0 \text{ for } t_0 > 0\}$$

provided that the initial state $f(x)$ is in $L^2(0, 1)$. Show also that the series in (3.2.3) may be differentiated any number of times with respect to either x or t and the differentiated series will also converge uniformly on $Q[t_0]$.

3.3 EVOLUTION EQUATIONS: INITIAL-BOUNDARY-VALUE PROBLEMS FOR WAVE EQUATION

In Section 3.2 we considered examples of evolution equations that describe what we call diffusionlike evolution. We defer until later a precise description of the characteristics of diffusionlike evolution. Instead, we now consider several examples of evolution equations where the evolutionary behavior is *not* diffusionlike. These examples illustrate a type of evolutionary behavior we refer to as "wavelike." When we have collected a sufficient number of examples of each type, we shall compare and contrast these two kinds of behavior.

EXAMPLE 3.3.1 ───

Mixed Initial-Boundary-Value Problem
    In Example 3.2.1 we began our study of the heat equation by considering a simple initial-boundary-value problem. The problem analogous to that example but posed for the wave equation reads as follows:

$$u_{tt}(x, t) = a^2 u_{xx}(x, t) \quad \text{for } 0 < x < 1, \quad t > 0$$
$$u(x, 0) = f(x), \quad u_t(x, 0) = g(x), \quad 0 < x < 1$$
$$u(0, t) = u(1, t) = 0, \quad t > 0 \tag{3.3.1}$$

Here a denotes a given, real constant and $f(x)$, $g(x)$ denote known functions of x. Note that (3.3.1) involves two initial conditions for the solution $u(x, t)$ at $t = 0$ while in Example 3.2.1 there is only one. This is a consequence of the fact that the wave equation contains a second-order derivative with respect to the variable t and the heat equation is only first order in t. Then, in a manner of speaking, solving the wave equation involves two integrations with respect to t and hence two auxiliary conditions involving the t variable. The heat equation, requiring only one integration in t, needs only one initial condition to determine a unique solution. While this explanation is an oversimplification, we shall see that problem (3.3.1), as (3.2.1), is well posed. We discuss some physical interpretations for this initial-boundary-value problem at the conclusion of the example.

Separation of Variables   Assuming that $u(x, t) = X(x)T(t)$ leads to the following separated problems:

$$-X''(x) = \mu X(x), \quad X(0) = X(1) = 0 \tag{3.3.2a}$$
$$-T''(t) = a^2 \mu T(t) \tag{3.3.2b}$$

The Sturm–Liouville problem (3.3.2a) is the same one that occurs in Example 3.2.1. The eigenvalues and eigenfunctions are

$$\mu_n = (n\pi)^2, \quad X_n(x) = \sin n\pi x, \quad n = 1, 2, \ldots$$

The general solution of (3.3.2b) (with $\mu = \mu_n$) can be written in the form

$$T_n(t) = A_n \cos n\pi a t + B_n \sin n\pi a t, \quad n = 1, 2, \ldots$$

where A_n, B_n denote a family of pairs of arbitrary constants. Then superposition implies that

$$u(x, t) = \sum_{n=1}^{\infty} [A_n \cos n\pi a t + B_n \sin n\pi a t] \sin n\pi x \qquad (3.3.3)$$

satisfies (at least formally) the partial differential equation and the two homogeneous boundary conditions. Note that (3.3.3) implies

$$u(x, 0) = \sum_{n=1}^{\infty} A_n \sin n\pi x$$

$$u_t(x, 0) = \sum_{n=1}^{\infty} n\pi a B_n \sin n\pi x$$

Then we use the fact that the eigenfunctions $\{\sin n\pi x: n = 1, 2, \ldots\}$ are a complete orthogonal family in $L^2(0, 1)$ in order to conclude in the usual way that we have

$$u(x, 0) = f(x), \qquad 0 < x < 1$$

provided the constants A_n satisfy

$$A_n = f^n = 2 \int_0^1 f(x) \sin n\pi x \, dx, \qquad n = 1, 2, \ldots$$

In addition, the other initial condition,

$$u_t(x, 0) = g(x), \qquad 0 < x < 1$$

will be satisfied if the constants B_n satisfy

$$B_n = \frac{1}{n\pi a} g^n = \frac{2}{n\pi a} \int_0^1 g(x) \sin n\pi x \, dx, \qquad n = 1, 2, \ldots$$

Then (3.3.3) becomes

$$u(x, t) = \sum_{n=1}^{\infty} f^n \cos n\pi a t \sin n\pi x$$
$$+ \sum_{n=1}^{\infty} \frac{g^n}{n\pi a} \sin n\pi a t \sin n\pi x \qquad (3.3.4)$$

where, as usual, f^n and g^n denote the generalized Fourier coefficients for the functions $f(x)$ and $g(x)$ in terms of the eigenfunctions $\sin n\pi x$.

D'Alembert's Solution   Simple trigonometric identities can be used to show that

$$\cos n\pi a t \sin n\pi x = \tfrac{1}{2}[\sin n\pi(x + at) + \sin n\pi(x - at)]$$
$$\sin n\pi a t \sin n\pi x = \tfrac{1}{2}[\cos n\pi(x - at) - \cos n\pi(x + at)]$$
$$= \tfrac{1}{2} n\pi \int_{x-at}^{x+at} \sin n\pi z \, dz$$

If we substitute these two results into (3.3.4), then we obtain

$$u(x, t) = \sum_{n=1}^{\infty} \frac{1}{2} f^n \{\sin n\pi(x + at) + \sin n\pi(x - at)\} \qquad (3.3.5)$$

$$+ \frac{1}{2a} \sum_{n=1}^{\infty} g^n \int_{x-at}^{x+at} \sin n\pi z \, dz$$

Now we know from the discussions of Chapter 2 that

$$\sum_{n=1}^{\infty} f^n \sin n\pi x \quad \text{and} \quad \sum_{n=1}^{\infty} g^n \sin n\pi z$$

converge to $\tilde{f}(x)$ and $\tilde{g}(z)$, the *odd* periodic extensions of $f(x)$ and $g(z)$ of period 2, provided $\tilde{f}(x)$ and $\tilde{g}(z)$ and their first derivatives are sectionally continuous. Then (3.3.5) can be rewritten as

$$u(x, t) = \tfrac{1}{2}[\tilde{f}(x + at) + \tilde{f}(x - at)]$$
$$+ \frac{1}{2a} \int_{x-at}^{x+at} \tilde{g}(z) \, dz \qquad (3.3.6)$$

The representations (3.3.4), (3.3.5), and (3.3.6) are all equivalent ways of writing the solution $u(x, t)$ for Example 3.3.1. However, (3.3.6) has the important advantage that it is in *"closed form"*; this means that (3.3.6) does not require us to sum an infinite series in order to evaluate $u(x, t)$. The form (3.3.6) is called the *D'Alembert form of the solution* for Example 3.3.1.

If it seems hard to accept that (3.3.6) can, in fact, satisfy all the conditions of problem (3.3.1) without the use of an infinite series, then we can verify each of the conditions directly. That is, if (3.3.6) holds, then

$$u_t(x, t) = \tfrac{1}{2}[\tilde{f}'(x + at)(a) + \tilde{f}'(x - at)(-a)]$$
$$+ \tfrac{1}{2}a[\tilde{g}(x + at)(a) - \tilde{g}(x - at)(-a)]$$
$$u_x(x, t) = \tfrac{1}{2}[\tilde{f}'(x + at)(1) + \tilde{f}'(x - at)(1)]$$
$$+ \tfrac{1}{2}a[\tilde{g}(x + at)(1) - \tilde{g}(x - at)(1)]$$
$$u_{tt}(x, t) = \tfrac{1}{2}[\tilde{f}''(x + at)(a)^2 + \tilde{f}''(x - at)(-a)^2]$$
$$+ \tfrac{1}{2}a[\tilde{g}'(x + at)(a)^2 - \tilde{g}'(x - at)(-a)^2]$$
$$u_{xx}(x, t) = \tfrac{1}{2}[\tilde{f}''(x + at)(1)^2 + \tilde{f}''(x - at)(1)^2]$$
$$+ \tfrac{1}{2}a[\tilde{g}'(x + at)(1)^2 - \tilde{g}'(x - at)(1)^2]$$

Then at each point where $\tilde{f}''$ and $\tilde{g}'$ exist, the partial differential equation is satisfied. It is also clear from (3.3.6) and the preceding expression for $u_t(x, t)$ that the initial conditions are satisfied. Finally, since $\tilde{f}(x)$ and $\tilde{g}(x)$ denote the *odd* extensions of $f(x)$ and $g(x)$,

$$\int_{-at}^{at} \tilde{g}(z) \, dz = 0 \quad \text{for all } t > 0$$

and

$$[\tilde{f}(at) + \tilde{f}(-at)] = 0 \quad \text{for all } t > 0$$

Then $u(0, t) = 0$ for all $t > 0$. Finally, since $\tilde{f}$ and $\tilde{g}$ are periodic with period 2, it follows that $u(1, t) = 0$ for $t > 0$. This establishes (3.3.6) as a solution for the problem (3.3.1). In Chapter 6 we prove that the solution to the problem (3.3.1) is unique. Since

(3.3.4) and (3.3.5) can also be shown to satisfy all the conditions of the problem, these must be *equivalent* ways of writing the *unique* solution $u(x, t)$.

Since the convergence properties of the infinite series appearing in (3.3.4) and (3.3.5) are determined by the smoothness properties of the functions $\tilde{f}(x)$ and $\tilde{g}(x)$, unless these functions are very smooth, the series are liable to converge very slowly. This makes (3.3.4) and (3.3.5) computationally inconvenient for evaluating the solution $u(x, t)$. Fortunately, we have the closed-form expression (3.3.6) at our disposal. The only inconvenient feature of (3.3.6) is that it requires us to compute with the extensions $\tilde{f}$ and $\tilde{g}$ instead of the functions f and g themselves. We show now how this difficulty is surmounted.

Method of Characteristics   In order to evaluate $u(x, t)$ for fixed values $x = x_0$ and $t = t_0$, note first that there exist unique positive integers M and N (depending on x_0 and t_0) such that

$$-1 < x_0 + at_0 - 2N \equiv P_N < 1$$
$$-1 < x_0 - at_0 + 2M \equiv Q_M < 1 \tag{3.3.7}$$

There are six cases to consider, and there is a formula for $u(x_0, t_0)$ in each case.

Case 1. $Q_M < P_N < 0$:

$$u(x_0, t_0) = -\tfrac{1}{2}[f(-Q_M) + f(-P_N)] - \frac{1}{2a}\int_{-P_N}^{-Q_M} g(s)\,ds$$

Case 2. $Q_M < 0 < P_N$:

$$u(x_0, t_0) = \tfrac{1}{2}[f(P_N) - f(-Q_M)] + \frac{1}{2a}\int_{-Q_M}^{P_N} g(s)\,ds$$

Case 3. $0 < Q_M < P_N$:

$$u(x_0, t_0) = \tfrac{1}{2}[f(Q_M) + f(P_N)] + \frac{1}{2a}\int_{Q_M}^{P_N} g(s)\,ds$$

Case 4. $P_N < Q_M < 0$:

$$u(x_0, t_0) = -\tfrac{1}{2}[f(-Q_M) + f(-P_N)] + \frac{1}{2a}\int_{-Q_M}^{-P_N} g(s)\,ds$$

Case 5. $P_N < 0 < Q_M$:

$$u(x_0, t_0) = \tfrac{1}{2}[f(Q_M) - f(-P_N)] + \frac{1}{2a}\int_{Q_M}^{-P_N} g(s)\,ds$$

Case 6. $0 < P_N < Q_M$:

$$u(x_0, t_0) = \tfrac{1}{2}[f(Q_M) + f(P_N)] - \frac{1}{2a}\int_{P_N}^{Q_M} g(s)\,ds$$

We show how these formulas are derived in Cases 1 and 5 and leave for the exercises the other four cases.

In case 1 we have $Q_M < P_N < 0$, and so

$$\tilde{f}(Q_M) = -f(-Q_M) \quad \text{and} \quad \tilde{f}(P_N) = -f(-P_N)$$

In addition,

$$\int_{x_0 - at_0}^{x_0 + at_0} \tilde{g}(s)\, ds = \int_{x_0 - at_0}^{Q_M} \tilde{g}(s)\, ds + \int_{Q_M}^{P_N} \tilde{g}(s)\, ds + \int_{P_N}^{x_0 + at_0} \tilde{g}(s)\, ds$$

According to (3.3.7), the first and third integrals on the right side of this last expression are integrals of a periodic function $\tilde{g}(x)$ over a fundamental period. Therefore these two integrals are zero. Moreover, since $\tilde{g}(x)$ is odd, and we have $Q_M < P_N < 0$, then it follows that

$$\int_{Q_M}^{P_N} \tilde{g}(s)\, ds = -\int_{-P_N}^{-Q_M} g(s)\, ds$$

Then the Case 1 result for $u(x_0, t_0)$ follows.

To derive the Case 5 result, recall that in Case 5 we have $P_N < 0 < Q_M$, and hence,

$$\tilde{f}(P_N) = -f(-P_N), \qquad \tilde{f}(Q_M) = f(Q_M).$$

By exactly the same argument used in Case 1, we have

$$\int_{x_0 - at_0}^{x_0 + at_0} \tilde{g}(s)\, ds = \int_{x_0 - at_0}^{Q_M} \tilde{g}(s)\, ds + \int_{Q_M}^{P_N} \tilde{g}(s)\, ds + \int_{P_N}^{x_0 + at_0} \tilde{g}(s)\, ds$$

$$= \int_{Q_M}^{P_N} \tilde{g}(s)\, ds$$

Now if $-1 < P_N < 0$, then $1 < P_N + 2 < 2$, and by the periodicity of $\tilde{g}(x)$,

$$\int_{Q_M}^{P_N} \tilde{g}(s)\, ds = \int_{Q_M}^{P_N + 2} \tilde{g}(s)\, ds$$

Now for any α such that $1 < \alpha < 2$, it is clear that $1 - (\alpha - 1) = 2 - \alpha$ and the interval $2 - \alpha < x < \alpha$ is a symmetric interval about $x = 1$. Then since $\tilde{g}$ is the odd two-periodic extension of g, it follows that the integral of $g(x)$ from $2 - \alpha$ to α is zero. Applying this result in the case $\alpha = P_N + 2$ leads to the result

$$\int_{Q_M}^{P_N} \tilde{g}(s)\, ds = \int_{Q_M}^{P_N + 2} \tilde{g}(s)\, ds$$

$$= \int_{Q_M}^{\alpha} \tilde{g}(s)\, ds = \int_{Q_M}^{2 - \alpha} \tilde{g}(s)\, ds$$

$$= \int_{Q_M}^{-P_N} \tilde{g}(s)\, ds$$

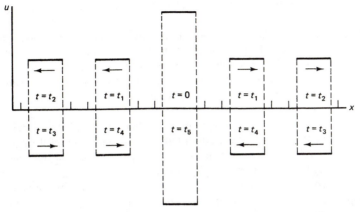

Figure 3.3.1

This proves that the formula for $u(x_0, t_0)$ in Case 5 is correct. The other cases are proved in a similar fashion, taking advantage of the definitions of the extensions, $\tilde{f}$ and $\tilde{g}$. The procedure we have described is a variation on what is called the *method of characteristics*. This method will be illustrated again in example 3.3.4 and once again in more detail in Chapter 7.

We apply this method to (3.3.1) for data $f(x)$ and $g(x)$ as follows;

$$f(x) = \begin{cases} 1 & \text{if } \frac{4}{9} < x < \frac{5}{9} \\ 0 & \text{otherwise on } (0, 1) \end{cases} \quad \text{and} \quad g(x) = 0$$

Then we can plot the resulting $u(x, t)$ versus x for several values of t as shown in Figure 3.3.1.

Discussion of Solution   Looking at Figure 3.3.1, one sees that at $t = 0$, the solution $u(x, 0)$ is the single square pulse located near the center of the interval $(0, 1)$. At $t = t_1$, this single pulse has split into two pieces, $\frac{1}{2}\tilde{f}(x + at)$ and $\frac{1}{2}\tilde{f}(x - at)$, located to the left and right (respectively) of the initial pulse. At $t = t_2$ these two pieces have moved further from the middle of the interval $(0, 1)$, and at $t = t_3$ they have been reflected from the ends of the interval and are now moving back toward the center. At $t = t_4$ they are closer to the center of the interval, but the solution still consists of two separated pieces. At $t = t_5$ the two pieces have come together to form a single pulse identical to the initial pulse but having the opposite sign. Continuing to plot $u(x, t)$ versus x for larger values of t will show that the solution at any instant consists of the sum of the two pieces $\frac{1}{2}\tilde{f}(x + at)$ and $\frac{1}{2}\tilde{f}(x - at)$, that appear to move back and forth across the interval $(0, 1)$, reflecting off the ends of the interval at $x = 0$ and $x = 1$ by changing direction (and sign) to move back toward the center of the interval. Note that unlike the solution to the heat equation, the solution of Example 3.3.1 does not decay with increasing t. That is,

$$\int_0^1 u(x, t)^2 \, dx = \frac{1}{9} \quad \text{for all } t > 0$$

We describe this by saying that the solution does not decrease in intensity with time.

Physical Interpretation of Problem   Physically, the equation and conditions of Example 3.3.1 could be describing the propagation of small-amplitude (acoustic) pressure waves in a thin closed pipe of length 1. Then the unknown u represents the longitudinal displacement of the gas molecules from their equilibrium position. In particular, $u(x, t)$ equals the displacement away from the equilibrium position that occurs at position x at time t.

The initial condition on u for the example pictured in Figure 3.3.1 corresponds to the occurrence at $t = 0$ of a small disturbance near the center of the pipe. The fact that u_t is initially zero means that the air in the pipe is initially at rest, and the boundary conditions are consistent with a pipe with closed ends.

According to Figure 3.3.1, the disturbance represented by the initial function $f(x)$ propagates up and down the pipe, reflecting off the ends. An observer at a fixed position in the pipe, seeing these reflections over and over, would refer to them as "echoes." According to the figure, not only does the intensity of the solution not decrease with time but the shape of the waveform remains the same for all $t > 0$. We shall see in Example 3.3.3 that both of these conditions can be altered by the presence of lower order terms in the equation.

Method Applied to Other Boundary Conditions   This version of the method of characteristics can be applied with boundary conditions other than the ones considered in Example 3.3.1. For example, if the boundary conditions are

$$u_x(0, t) = u_x(1, t) = 0, \qquad t > 0$$

then we simply extend the data functions $f(x)$ and $g(x)$ to the interval $-1 < x < 1$ as even functions and extend these even functions to the whole real line $(-\infty, \infty)$ as periodic functions of period 2. This will result in slightly different formulas for $u(x, t)$ than those listed for Cases 1–6, but the formulas for these boundary conditions are derived by the same methods. For the boundary conditions

$$u(0, t) = u_x(1, t) = 0, \qquad t > 0$$

we extend the data functions $f(x)$ and $g(x)$, *first* as functions symmetric about the line $x = 1$; that is,

$$\tilde{f}(x) = \begin{cases} f(x) & \text{if } 0 < x < 1 \\ f(2 - x) & \text{if } 1 < x < 2 \end{cases}$$

This ensures that the boundary condition at $x = 1$ will be satisfied. *Next*, extend these functions (which are now defined on the interval $0 < x < 2$) to the interval $(-2, 0)$ as *odd* functions. This ensures that the boundary condition at $x = 0$ is satisfied. *Finally*, extend these functions from the interval $(-2, 2)$ to the whole real line $(-\infty, \infty)$ as periodic functions of period 4. That is, $f(x)$ and $g(x)$ are extended according to

$$\tilde{f}(x) = \begin{cases} f(x) & \text{if } 0 < x < 1 \\ f(2 - x) & \text{if } 1 < x < 2 \\ -f(-x) & \text{if } -1 < x < 0 \\ -f(x - 2) & \text{if } -2 < x < -1 \end{cases} \qquad (3.3.8)$$

and

$$\tilde{f}(x) = \tilde{f}(x + 4N) \quad \text{for all } x \text{ and all integers N}$$

Finally, if we wish to consider the boundary conditions

$$u(1, t) = u_x(0, t) = 0, \qquad t > 0$$

we simply observe that if $u(x, t) = v(1 - x, t), 0 < x < 1, t > 0$, then $v(x, t)$ satisfies

$$v(0, t) = v_x(1, t) = 0, \qquad t > 0$$

and this change of dependent variable reduces the new problem to one we solved before. ■ ■

Comparison of Continuous and Semidiscrete Models

It is interesting to compare the solution of the continuous wave equation with the solution of the semidiscrete wave equation that we considered in Chapter 1. The semidiscrete analogue of problem (3.3.1) is the problem

$$\mathbf{U}''(t) = \beta^2[A]\mathbf{U}(t), \qquad \mathbf{U}(0) = \mathbf{F}, \qquad \mathbf{U}'(0) = \mathbf{G} \qquad (3.3.9)$$

for the unknown $N - 1$ vector $\mathbf{U}(t)$. Here β denotes a real constant, $\mathbf{F}$ and $\mathbf{G}$ denote given initial vectors analogous to the data functions $f(x)$ and $g(x)$ in (3.3.1), and $[A]$ is the negative definite $(N - 1) \times (N - 1)$ symmetric matrix of Section 1.2, with the $N - 1$ negative eigenvalues $\alpha_1, \ldots, \alpha_{N-1}$ and the $N - 1$ orthogonal eigenvectors $\mathbf{W}_1, \ldots, \mathbf{W}_{N-1}$. These eigenvectors are then a basis for R_{N-1}.

If we let $\theta_n^2 = -\alpha_n$, then the solution to (3.3.9) can be written as

$$\mathbf{U}(t) = \sum_{n=1}^{N-1} [C_n \exp(i\theta_n \beta t) + D_n \exp(-i\theta_n \beta t)]\mathbf{W}_n$$

Note that since $\theta_n^2 = -\alpha_n$, the negative eigenvalues of $[A]$ lead to the imaginary exponents that produce a solution that oscillates with time. The solution may be written in the alternative form

$$\mathbf{U}(t) = \sum_{n=1}^{N-1} [A_n \cos \theta_n \beta t + B_n \sin \theta_n \beta t]\mathbf{W}_n \qquad (3.3.10)$$

which will be more convenient for finding choices of the constants A_n and B_n so that the initial conditions will be satisfied. The first initial condition requires that the A_n satisfy the condition

$$\mathbf{U}(0) = \sum_{n=1}^{N-1} A_n \mathbf{W}_n = \mathbf{F}$$

Since the eigenvectors here are an orthogonal basis, we easily find that in order for the initial condition to be satisfied, it is necessary and sufficient that

$$A_n = (\mathbf{F}, \mathbf{W}_n)/\|\mathbf{W}_n\|^2, \qquad n = 1, 2, \ldots, N - 1$$

The second initial condition requires that the constants B_n satisfy

$$\mathbf{U}'(0) = \sum_{n=1}^{N-1} B_n \theta_n \beta \mathbf{W}_n = \mathbf{G}$$

Once again, using the orthogonality of the eigenvectors, we find

$$B_n = \frac{1}{(\theta_n \beta)} \frac{(\mathbf{G}, \mathbf{W}_n)}{\|\mathbf{W}_n\|^2}$$

Now solution (3.3.10) becomes

$$\mathbf{U}(t) = \sum_{n=1}^{N-1} \left[\cos(\beta \theta_n t)(\mathbf{F}, \mathbf{w}_n) + \frac{1}{\beta \theta_n} \sin \beta \theta_n t (\mathbf{G}, \mathbf{w}_n) \right] \mathbf{w}_n \qquad (3.3.11)$$

where, for $n = 1, \ldots, N - 1$, $\mathbf{w}_n = \mathbf{W}_n / \|\mathbf{W}_n\|$; that is, these are eigenvectors of unit length. Compare (3.3.11) with (3.3.4), the solution to the continuous problem.

We can derive an estimate for the norm of the solution $\mathbf{U}(t)$ in terms of the norms of the data vectors $\mathbf{F}$ and $\mathbf{G}$ that is analogous to the estimate (3.2.9) of the previous section. We use the orthogonality of the eigenvectors to write

$$\|\mathbf{U}(t)\|^2 = (\mathbf{U}(t), \mathbf{U}(t))$$

$$= \sum_{n=1}^{N-1} \left[(\mathbf{F}, \mathbf{w}_n)\cos(\beta \theta_n t) + \frac{(\mathbf{G}, \mathbf{w}_n)\sin \beta \theta_n t}{\beta \theta_n} \right]^2$$

$$\leq \sum_{n=1}^{N-1} \left[|(\mathbf{F}, \mathbf{w}_n)| + \left| \frac{(\mathbf{G}, \mathbf{w}_n)}{\beta \theta_n} \right| \right]^2$$

$$\leq \sum_{n=1}^{N-1} 2 \left[|(\mathbf{F}, \mathbf{w}_n)|^2 + \left| \frac{(\mathbf{G}, \mathbf{w}_n)}{\beta \theta_n} \right|^2 \right]$$

In the last step, we have made use of the identity

$$(P - Q)^2 \geq 0, \quad \text{i.e.,} \quad P^2 + Q^2 \geq 2PQ$$

Finally, since $\theta_1 < \theta_2 < \cdots < \theta_{N-1}$, we arrive at

$$\|\mathbf{U}(t)\|^2 \leq 2 \|\mathbf{F}\|^2 + 2/(\beta \theta_1)^2 \|\mathbf{G}\|^2 \qquad (3.3.12)$$

This estimate is analogous to the estimate (3.2.9) for the solution to the semidiscrete heat equation. Note that (3.2.9) implies that the solution to the semidiscrete heat equation decreases in norm, exponentially with t. Estimate (3.3.12), on the other hand, does not imply any such decrease for the norm of the solution to the wave equation.

An estimate similar to (3.3.12) can be derived for the solution to the continuous problem. From (3.3.4) and the orthogonality [in $L^2(0, 1)$] of the eigenfunctions $X_n(x) = \sin n\pi x$, it follows that, for $m = 1, 2, \ldots$,

$$\int_0^1 u(x, t)\sin m\pi x \, dx = (u, X_m)$$

$$= \left\{ f^m \cos m\pi a t + \frac{g^m}{m\pi a} \sin m\pi a t \right\} \|X_m\|^2$$

Then, using the same identity we used to get (3.3.12),

$$(u, X_m)^2 \leq [2\{f^m\}^2 + 2\{g^m/\pi a\}^2] \|X_m\|^4$$

Finally, part 3 of Theorem 2.5.2 implies

$$\|u\|^2 \leq \|f\|^2 + (1/\pi a)^2 \|g\|^2 \qquad (3.3.13)$$

which is the analogue for the wave equation of the estimate (3.2.10) for the heat equation.

Duhamel's Principle

In Exercise 10 at the end of the chapter you are asked to solve an initial-boundary-value problem for the wave equation driven by a periodic forcing function at the boundary. A problem of this nature can arise in studying the propagation of small-amplitude fluid waves in experimental device called a "ripple tank." The equations for the ripple tank example are derived in Section 7.6, where the problem is treated by the method of characteristics. This same problem is solved in Chapter 5 by still another method.

When faced with an initial-boundary-value problem for the wave equation in which the boundary conditions are inhomogeneous, it is necessary to perform a change of dependent variable to transform the boundary conditions to homogeneous conditions before the method of separation of variables can be applied. This is the method suggested in Exercise 10 where you will find that the transformation moves the inhomogeneous term from the boundary condition into the wave equation. We describe now a method for dealing with an inhomogeneous wave equation.

EXAMPLE 3.3.2 _____

Inhomogeneous Wave Equation

In Section 3.2 we observed that our eigenfunction expansion technique for solving the heat equation consisted of the following steps: The data was transformed from a function into a sequence of Fourier coefficients, an auxiliary problem was solved in order to determine a sequence of time-dependent Fourier coefficients for the solution, and finally the solution was transformed from a sequence of (time-dependent) Fourier coefficients to a function by summing the Fourier series.

We can view our solution to problem (3.3.1) for the wave equation in a similar fashion. Consider the form (3.3.4) of the solution to this problem and adopt the following notation:

1. Transform the data $f(x)$ from a function on $(0, 1)$ to a sequence of Fourier coefficients in terms of the eigenfunctions $\{\sin n\pi x : n = 1, 2, \ldots\}$

$$f(x) \xrightarrow{\ \mathscr{F}_n\ } f^n = \mathscr{F}_n\{f\}$$

2. An auxiliary problem is solved to find a sequence of time-dependent Fourier coefficients for the solution

$$f^n \xrightarrow{\ \sigma(t)\ } f^n \cos n\pi a t = \sigma(t)\mathscr{F}_n\{f(x)\}$$

3. The solution is transformed from a sequence of time-dependent Fourier coefficients to a function by reversing step 1 and summing the series:

$$f''\cos n\pi at \xrightarrow{\mathscr{F}_n^{-1}} u(x, t) = \mathscr{F}_n^{-1}\{f''\cos n\pi at\}$$

$$= \sum_{n=1}^{\infty} f''\cos n\pi at \sin n\pi x$$

If we denote the composition of these three operations by $S(t)$,

$$S(t)[f(x)] = \mathscr{F}_n^{-1}\{\sigma(t)[\mathscr{F}_n\{f(x)\}]\}$$

Then (3.3.4) can be expressed as

$$u(x, t) = S(t)[f(x)] + \int_0^t S(\tau)[g(x)] \, d\tau \tag{3.3.14}$$

That is, the part of the solution $u(x, t)$ that denotes the response to the data $g(x)$ is not just $S(t)$ acting on $g(x)$. Instead, we have

$$g'' \longrightarrow g''\left(\frac{1}{n\pi a}\right)\sin n\pi at = \int_0^t \sigma(\tau)g'' \, d\tau$$

We can now use these results to solve an associated *inhomogeneous* problem. Consider

$$u_t(x, t) = a^2 u_{xx}(x, t) + F(x, t) \qquad 0 < x < 1, \qquad t > 0$$
$$u(x, 0) = 0, \qquad u_t(x, 0) = 0, \qquad 0 < x < 1$$
$$u(0, t) = u(1, t) = 0, \qquad t > 0 \tag{3.3.15}$$

Here $F(x, t)$ denotes a given forcing term in the equation. The eigenfunctions associated with this problem have already been identified in Example 3.3.1. They are the functions

$$X_n(x) = \sin n\pi x, \qquad n = 1, 2, \dots$$

and hence we assume a solution for (3.3.15) of the form

$$u(x, t) = \sum_{n=1}^{\infty} u_n(t)\sin n\pi x \tag{3.3.16}$$

Substituting this into (3.3.15) leads to

$$\sum_{n=1}^{\infty} u_n''(t)X_n(x) = -\sum_{n=1}^{\infty} a^2(n\pi)^2 u_n(t)X_n(x) + F(x, t)$$

$$\sum_{n=1}^{\infty} u_n(0)X_n(x) = 0, \qquad \sum_{n=1}^{\infty} u_n'(0)X_n(x) = 0$$

Then the fact that the eigenfunctions are a complete orthogonal family implies that, for $n = 1, 2, \dots$,

$$u_n''(t) + (n\pi a)^2 u_n(t) = F_n(t), \qquad t > 0$$
$$u_n(0) = 0, \qquad u_n'(0) = 0 \tag{3.3.17}$$

where

$$F_n(t) = 2 \int_0^1 F(x, t)\sin n\pi x \, dx, \qquad n = 1, 2, \ldots$$

$$= \mathscr{F}_n[F(x, t)] \tag{3.3.18}$$

Problem (3.3.17) can be solved by the method of variation of parameters or (as we shall see in the next chapter) by using the Laplace transform. The solution is given by

$$u_n(t) = \int_0^t \cos n\pi a(t - \tau)F_n(\tau) \, d\tau, \qquad n = 1, 2, \ldots$$

$$= \int_0^t \sigma(t - \tau)\mathscr{F}_n[F(x, \tau)] \, d\tau \tag{3.3.19}$$

Then the solution $u(x, t)$ to problem (3.3.15) is given by

$$u(x, t) = \mathscr{F}_n^{-1}\{u_n(t)\}$$

$$= \int_0^t \sum_{n=1}^{\infty} \sigma(t - \tau)F_n(\tau) \, d\tau \, \sin n\pi x$$

$$= \int_0^t S(t - \tau)[F(x, \tau)] \, d\tau \tag{3.3.20}$$

This is the Duhamel principle as it applies to the wave equation. As was the case for the heat equation, Duhamel's principle allows us to write the solution to an inhomogeneous wave equation in terms of the solution operator for the homogeneous equation. ∎ ∎

Influence of Lower Order Terms: Dispersion

We consider now an example that illustrates the effect of lower order terms on the solution of the wave equation.

EXAMPLE 3.3.3 ───

Dispersive Wave Motion
   In Section 1.4 we derived the so-called telegraph equation

$$u_{tt}(x, t) = a^2 u_{xx}(x, t) + cu_t(x, t) + du(x, t)$$

which reduces to the wave equation when $c = d = 0$. More generally, the equation

$$u_{tt}(x, t) = a^2 u_{xx}(x, t) + bu_x(x, t) + cu_t(x, t) + du(x, t) \tag{3.3.21}$$

contains all possible lower order terms and reduces to the one-dimensional wave equation when the coefficients of the lower order terms equal zero. It is not difficult to show that a variable change of the form

$$u(x, t) = \exp[\alpha x + \beta t]v(x, t)$$

when substituted into (3.3.21) leads to the equation

$$v_{tt}(x, t) = a^2 v_{xx}(x, t) + [b + 2\alpha a^2] v_x(x, t) + [c - 2\beta] v_t(x, t) \quad (3.3.22)$$
$$+ [d + c\beta + b\alpha + a^2\alpha^2 - \beta^2] v(x, t)$$

The constants α and β are, so far, arbitrary, but we see that by choosing them correctly, we can cause two of the three lower order terms in (3.3.22) to vanish. Note that even if one or even two of the coefficients b, c, and d equals zero in (3.3.21), there is in general no choice of α and β that will eliminate *all three* of the lower order terms in (3.3.22). Therefore, the best we can do for a wave equation with lower order terms is to reduce it to a wave equation with *just one* lower order term, and we have a free choice as to which of the three possible lower order terms that can be. We consider then the example

$$u_{tt}(x, t) = a^2 u_{xx}(x, t) - b^2 u(x, t), \quad 0 < x < 1, \quad t > 0$$
$$u(x, 0) = f(x), \quad u_t(x, 0) = 0, \quad 0 < x < 1$$
$$u(0, t) = u(1, t) = 0, \quad t > 0 \quad (3.3.23)$$

Here a and b denote known real constants and $f(x)$ is a given function defined on $0 < x < 1$. In view of the remarks we have just made, any equation of the form (3.3.21) could be reduced to the equation in (3.3.23) by a transformation of the type described. Thus if we can solve (3.3.23), then we can solve *any* wave equation containing lower order terms with constant coefficients. The equation in (3.3.23) is called the Klein–Gordon equation.

Separation of Variables   If we make the usual assumption that the solution is separable, $u(x, t) = X(x)T(t)$, then the equation reduces to

$$T''(t)/a^2 T(t) = X''(x)/X(x) - (b/a)^2$$

This leads to the following two separated problems:

$$-X''(x) = \mu X(x), \qquad X(0) = X(1) = 0 \qquad (3.3.24a)$$
$$-T''(t) = [\mu a^2 + b^2]T(t) \qquad T'(0) = 0 \qquad (3.3.24b)$$

We are by now familiar with the fact that problem (3.3.24a) is a Sturm–Liouville problem with the following eigenvalues and eigenfunctions:

$$\mu_n = (n\pi)^2, \qquad X_n(x) = \sin n\pi x, \qquad n = 1, 2, \ldots$$

Moreover, we can solve problem (3.3.24b) to obtain

$$T_n(t) = A_n \cos n\pi c_n t, \qquad n = 1, 2, \ldots$$

where

$$(n\pi c_n)^2 = \mu_n a^2 + b^2 = (n\pi a)^2 + b^2$$

that is,

$$c_n = [a^2 + (b/n\pi)^2]^{1/2}$$

Then, proceeding as we did in Example 3.3.1, we find

$$u(x, t) = \sum_{n=1}^{\infty} f^n \cos n\pi c_n t \sin n\pi x \tag{3.3.25}$$

This is analogous to solution (3.3.4) found in Example 3.3.1. Using the trigonometric identities as we did there, we can convert this representation for the solution to the equivalent form

$$u(x, t) = \sum_{n=1}^{\infty} \frac{1}{2} f^n \{\sin n\pi(x + c_n t) + \sin n\pi(x - c_n t)\} \tag{3.3.26}$$

This representation is analogous to solution (3.3.5) found in Example 3.3.1. In that example we went further, reasoning that since

$$\sum_{n=1}^{\infty} f^n \sin n\pi x \quad \text{converges to } \tilde{f}(x) \tag{3.3.27}$$

it follows that

$$\sum_{n=1}^{\infty} f^n \sin n\pi(x \pm at) \quad \text{must converge to } \tilde{f}(x \pm at) \tag{3.3.28}$$

This led to solution (3.3.6). We used there the observation that the argument x on both sides of Equation (3.3.27) can be replaced by *any* algebraic expression, and when that expression is $x \pm at$, the result is (3.3.28). Note, however, for the substitution to be valid, it is essential that the algebraic expression that is substituted does not depend on n. On a superficial level, it is clear that if the argument depended on n, then the series, summed over the index n, would converge to $\tilde{f}$ evaluated at an argument that still depends on the index n. Of course, this makes no sense, and we conclude that the argument must be independent of n for the substitution to be valid.

In order to understand what goes on here on a somewhat deeper level, consider for a moment the relationship between the functions $\sin \pi x$ and $\sin \pi(x - at)$. The function $\sin \pi x$ is everywhere defined, and in particular at the point $x = \frac{1}{2}$ it assumes its maximum value of 1; it assumes that value at infinitely many other points too, but we are focusing on just this one point. The function $\sin \pi(x - at)$ is also everywhere defined, and for the fixed value $t = 0$ it assumes the value 1 at $x = \frac{1}{2}$. We suppose here that the parameter a is a positive real number.

For $t = t_0 > 0$, $\sin \pi(x - at_0)$ assumes the value 1 at the point $x_0 = \frac{1}{2} + at_0$, which is somewhere to the right of $x = \frac{1}{2}$. At the time $t = t_1 > t_0$ the function $\sin \pi(x - at_1)$ assumes the value 1 at the point $x_1 = \frac{1}{2} + at_1$, which is still further to the right of $x = \frac{1}{2}$. In fact, the distance between the points x_1 and x_0 is exactly $a(t_1 - t_0)$, and if we think of $t_1 - t_0$ as the time required for the maximum point on the graph of $\sin \pi(x - at)$ to travel from x_0 to x_1, then the parameter a must represent the *speed* at which the points on the graph are moving to the right. For this reason, we refer to $\sin \pi(x - at)$ as a "*traveling wave*," and we say that the speed of the wave is a. More precisely, the function $\sin \pi(x - at)$ is a traveling wave moving from left to right with wave speed a. Then $\sin \pi x$ is a traveling wave for which the wave speed is zero; that is, it is a "stationary wave." In just the same way, the function $\sin \pi(x + at)$ can be viewed as a traveling wave that is moving from right to left with wave speed a.

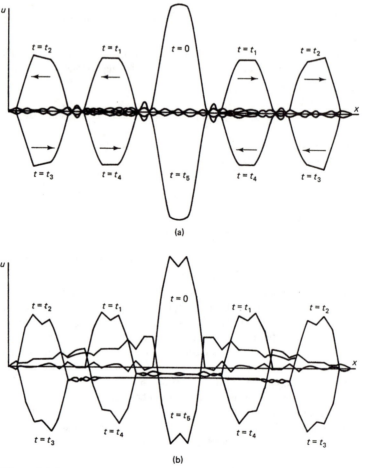

Figure 3.3.2

In (3.3.28), each of the functions $\sin n\pi(x \pm at)$ can be viewed as a traveling wave, moving to the left or to the right with speed equal to a. For $t = 0$ the series sums to $\tilde{f}(x)$. For $t = t_0$ each of the functions has translated its graph a distance at_0 to the left/right, and since each graph translates the same amount, the series sums to the same waveform $\tilde{f}$ but translated to the left/right by the amount at_0. That is, the series sums to $\tilde{f}(x \pm at_0)$. The significant point here is that at any time $t > 0$, the individual waves $\sin n\pi(x \pm at)$ have *all moved the same distance and in the same direction,* and so the series sums to the function $\tilde{f}$ translated through that same distance and direction.

In view of these remarks, one can clearly see that in the series (3.3.26) each of the functions $\sin n\pi(x \pm c_n t)$ is a traveling wave, but each travels at a different speed. The wave $\sin n\pi(x \pm c_n t)$ travels at the speed $c_n = [a^2 + (b/n\pi)^2]^{1/2}$, and unless $b = 0$ (and the lower order term in the equation vanishes), there is a different speed for each n. We say that the wave speed is a function of the *wave number n*.

The series (3.3.26) sums to $\bar{f}(x)$ at $t = 0$, but for $t > 0$, since each of the graphs $\sin n\pi(x \pm c_n t)$ has translated by a different amount, the series no longer sums to a translation of $\bar{f}$. In other words, waves of different frequencies travel at different speeds, so that as t increases from zero, the Fourier components of $\bar{f}(x)$ become dispersed and the waveform $\bar{f}$ becomes increasingly distorted. We refer to this phenomenon where the wave speed depends on the wave number as *dispersion*. Evidently the effect of low order terms in the wave equation is to introduce dispersion. The related notion of *numerical dispersion* is discussed in Section 9.4.

In Figure 3.2.2a we plot $u(x, t)$ versus x for several values of t beginning with $t = 0$. In Figure 3.3.2a the profiles are computed with $b = 0$, and in Figure 3.3.2b we plot the same profiles for $b > 0$. In both cases, the solution $u(x, t)$ is computed from the same initial functions $f(x)$ and $g(x)$ that were used to generate Figure 3.3.1. Note that the solution shown in Figure 3.3.2a is only an approximation to the exact solution because the evaluation was carried out by summing finitely many terms of the Fourier series. The waveforms in Figure 3.3.1 are exact since they were generated by the closed-form expression (3.3.6).

In Figure 3.3.2a, the waveform of the solution is not distorted by propagation. That is, the shape of the waves at $t = t_k$ for $k = 1, 2, \ldots$ is the same as the shape of the wave at $t = 0$. At least this appears to be true as far as we can tell from the sum of finitely many terms of the series. The slow convergence of the Fourier series representation for solutions of the wave equation is one reason for using the method of characteristics as illustrated by (3.3.6) whenever possible.

These profiles correspond to solutions $u(x, t)$ computed from (3.3.26) with $b > 0$. Here the waveforms are somewhat distorted, and there is "noise" clearly present in the picture. (The profiles of the apparently random waveform in Figure 3.3.2b are the noise.) Increasing the dispersion (i.e., increasing the magnitude of the parameter in Figure 3.3.2b) will increase both the distortion and the noise.

The presence of the lower order term $b^2 u(x, t)$ in Equation (3.3.18) introduces dispersion into the solution. Lower order terms $u_t(x, t)$ and/or $u_x(x, t)$ will introduce not only dispersion but exponential damping as well (see Exercises 7–9).

EXAMPLE 3.3.4

Method of Characteristics: Longitudinal Vibration of a Thin Rod

We consider in this example a problem for which separating the variables leads to an eigenvalue problem that is *not* a Sturm–Liouville problem. Even though we are able to find eigenvalues and eigenfunctions for this problem, the eigenfunctions will not be an orthogonal family, and we do not know whether they form a basis for the space L^2. For this reason, separation of variables will be abandoned as a method for solving this problem, and the solution will be carried out instead by the method of characteristics.

We consider a homogeneous rod of length 1 with uniform cross sections. Assume the cross-sectional dimensions are small compared to 1 and that as the rod is stretched or compressed along its longitudinal axis, these cross sections remain parallel to one another. Let

$u(x, t)$ = position at time t of plane cross section whose equilibrium position is x

Then the plane cross section whose position is x when the bar is in a state of equilibrium displaces to the position $u(x, t)$ at time t. Similarly, the cross section whose equilibrium position is $x + dx$ displaces to $u(x + dx, t)$ at time t. If we write

$$u(x + dx, t) \simeq u(x, t) + u_x(x, t) \, dx$$

then the segment of rod that occupies the interval $(x, x + dx)$ in the equilibrium condition experiences the following change in length at time t:

$$dL = u_x(x, t) \, dx$$

This change is the result of a tension or compression force F, which according to Hooke's law is related to dL by

$$F(x, t) = \frac{-EA \, dL}{dx} = -EAu_x(x, t)$$

where A denotes the area of the cross section and E denotes Young's modulus of the rod material. Viewing the segment as a free body, the net force acting on the body is

$$F(x, t) - F(x + dx, t) = -EA[u_x(x, t) - u_x(x + dx, t)]$$
$$\simeq EAu_{xx}(x, t) \, dx$$

Then Newton's law requires that this force equal the mass times the acceleration of the segment, which is to say

$$\sigma A \, dx \, u_{tt}(x, t) = EAu_{xx}(x, t) \, dx$$

that is,

$$u_{tt}(x, t) = (E/\sigma)u_{xx}(x, t)$$

where σ denotes the material density of the rod. Thus we see that the displacement $u(x, t)$ satisfies the wave equation.

Suppose that the end of the rod at $x = 0$ is fixed and the other end of the rod is free. We assume the rod to be initially at rest, but at $t = 0$ the free end is struck by a mass M moving with velocity v in the direction of the axis of the rod. The mass M is then joined with the rod, which oscillates in the longitudinal direction. Then the initial-boundary-value problem satisfied by the displacement function is

$$u_{tt}(x, t) = a^2 u_{xx}(x, t), \qquad 0 < x < 1, \qquad t > 0 \qquad (3.3.29)$$
$$u(x, 0) = 0,$$
$$u_t(x, 0) = \begin{cases} 0 & \text{for } 0 < x < 1 \\ -v & \text{for } x = 1^+ \end{cases}$$
$$u(0, t) = 0 \quad \text{and} \quad Mu_{tt}(1, t) = -EAu_x(1, t), \qquad t > 0$$

Here $a^2 \equiv E/\sigma$. The inhomogeneous initial condition means that for h tending to zero through positive values,

$$\lim_{h \to 0} u_t(1 - h, 0) = 0 \quad \text{and} \quad \lim_{h \to 0} u_t(1 + h, 0) = -v$$

that is, $u_t(x, 0)$ is discontinuous at $x = 1$.

Separation of Variables If we suppose that $u(x, t) = X(x)T(t)$, then we are led in the usual way to the separated equations

$$-X''(x) = \mu^2 X(x) \quad \text{and} \quad -T''(t) = \mu^2 a^2 T(t)$$

It follows easily that $X(x)$ and $T(t)$ must satisfy

$$X(0) = 0 \quad \text{and} \quad T(0) = 0$$

In addition, the boundary condition at $x = 1$ implies

$$MX(1)T''(t) = -EAX'(1)T(t) \quad \text{for } t > 0$$

That is,

$$\frac{-EA}{M}\frac{X'(1)}{X(1)} = \frac{T''(t)}{T(t)} = -\mu^2\left(\frac{E}{\sigma}\right)$$

Then the boundary condition satisfied by $X(x)$ at $x = 1$ is

$$X'(1) = \mu^2 \beta X(1)$$

where $\beta \equiv M/(\sigma A)$, the ratio of the mass of the weight to the mass of the rod (which equals σA times the length, which is 1). Then the two separated problems coming from (3.3.29) are

$$
\begin{align}
-X''(x) = \mu^2 X(x), && X(0) = X'(1) - \mu^2 \beta X(1) = 0 && \text{(3.3.30a)} \\
-T''(t) = T(t), && T(0) = 0 && \text{(3.3.30b)}
\end{align}
$$

Then (3.3.30) is clearly an eigenvalue problem, but it is *not* a Sturm–Liouville problem because the eigenvalue parameter μ^2 appears not just in the equation but in the boundary condition at $x = 1$. Then we are not entitled to use any of the conclusions of the Sturm–Liouville theorem (Theorem 2.3.1). In particular we are not entitled to conclude that the eigenfunctions of problem (3.3.30a) form an orthogonal basis for the vector space $L^2(0, 1)$, nor can we assume that the eigenvalues are (necessarily) all real.

Before we drop this approach to the problem, we point out that problem (3.3.30a) *does have* an infinite number of real, positive eigenvalues μ_n^2 where the numbers μ_n denote the roots of the eigenvalue equation,

$$\mu \tan \mu = 1/\beta, \quad n = 1, 2, \ldots$$

The corresponding eigenfunctions are then

$$X_n(x) = \sin \mu_n x, \quad n = 1, 2, \ldots$$

These eigenfunctions satisfy

$$(X_m, X_n) = -2\beta X_m(1)X_n(1)$$

and since this is not zero, the eigenfunctions are not orthogonal. Even if the eigenfunctions can be shown to form a basis for $L^2(0, 1)$, the fact that they are not orthogonal presents a problem in trying to solve for the coefficients in an eigenfunction expansion of an arbitrary function. For all of these reasons we abandon this approach to solving problem (3.3.29) and pursue now an alternative approach.

Method of Characteristics   We have seen previously in Equation (3.3.6) that a function of the form

$$u(x, t) = F(x + at) + G(x - at) \tag{3.3.31}$$

is a solution to the wave equation. In fact, it can be immediately verified that for arbitrary smooth functions F and G of one variable, $u(x, t)$ given by (3.3.31) solves the wave equation.

The boundary condition at $x = 0$ implies that

$$u(0, t) = F(at) + G(-at) = 0 \quad \text{for } t > 0$$

and this condition is satisfied if

$$G(z) = -F(-z) \quad \text{for } z > 0$$

Then,

$$u(x, t) = F(at + x) - F(at - x), \quad 0 < x < 1, \quad t > 0 \tag{3.3.32}$$

Next, the initial conditions imply, for $0 < x < 1$,

$$u(x, 0) = F(x) - F(-x) = 0$$
$$u_t(x, 0) = a[F'(x) - F'(-x)] = 0$$

But this requires that $F(x)$ and $F'(x)$ are *both* even functions of x for $0 < x < 1$. The only function $F(x)$ with this property is a constant function, and without affecting the solution $u(x, t)$, we can take this constant to be zero. Then,

$$F(x) = 0 \quad \text{for } -1 < x < 1 \tag{3.3.33}$$

We can extend our knowledge of the function $F(x)$ outside the interval $(-1, 1)$ by using the boundary condition at $x = 1$. This implies

$$Ma[F''(at + 1) - F''(at - 1)] = -EA[F'(at + 1) + F'(at - 1)]$$

and if we let $z = at + 1$, this can be rewritten

$$F''(z) + 1/\beta F'(z) = F''(z - 2) - 1/\beta F'(z - 2)$$

Now (3.3.33) implies that for $1 < z < 3$, the right side of this equation is identically zero. Then,

$$F''(z) = -1/\beta F'(z) \quad \text{for } 1 < z < 3$$

and it follows that

$$F'(z) = C \exp[-z/\beta], \quad \text{for } 1 < z < 3$$

In order to evaluate the constant of integration, we recall the discontinuity in $u_t(x, 0)$ at $x = 1$. In order to have

$$\lim_{x \to 1^+} u_t(x, 0) = -v$$

we must require $a[F'(1^+) - F'(-1^+)] = -v$. But $F'(-1^+) = 0$ by (3.3.28) and $F'(1^+) = -v/a$ if $C = -v/a \exp[1/\beta]$; that is,

$$F'(z) = -v/a \exp[-(z - 1)/\beta], \quad 1 < z < 3 \tag{3.3.34}$$

It follows from this that

$$F(z) = \beta v/a \, \exp[-(z - 1)/\beta] + C', \qquad 1 < z < 3$$

where the constant C' must be chosen so that $u(x, t)$ is continuous at $x = 1$ for all $t \geq 0$. Since $u(1^-, 0) = 0$ in particular, this amounts to the condition

$$0 = F(1^+) - F(-1^+) = F(1^+)$$

that is,

$$C' = -v\beta/a$$

Then, finally

$$F(z) = -v\beta/a(1 - \exp[-(z - 1)/\beta]), \qquad 1 < z < 3 \qquad (3.3.35)$$

Before discussing the solution $u(x, t)$, we go one more step, extending $F(z)$ to the interval $(3, 5)$. On this interval, we use (3.3.34) in the boundary condition at $x = 1$ in order to write

$$F''(z) + 1/\beta F'(z) = F''(z - 2) - 1/\beta F'(z - 2)$$
$$= 2v/(a\beta) \, \exp[-(z - 3)/\beta]$$

Then,

$$F'(z) = C \exp[-z/\beta] + 2v/(a\beta)(z - 3) \exp[-(z - 3)/\beta]$$

where the constant C must here be chosen so that $u_t(1, t)$ is continuous for t strictly greater than zero (this derivative is discontinuous at $t = 0$, but for $t > 0$ it is continuous). Now,

$$u_t(1, t) = a[F'(at + 1) - F'(at - 1)]$$

and this derivative is continuous at $t = 2/a$, in particular, if

$$F'(3^-) - F'(1^-) = F'(3^+) - F'(1^+)$$

Since $F'(1^-) = 0$, this reduces to

$$F'(3^+) = F'(3^-) + F'(1^+)$$

that is,

$$C \exp[-3/\beta] = -v/a \exp[-2/\beta] - v/a$$

or

$$C = -v/a(\exp[1/\beta] + \exp[3/\beta])$$

Then, for $3 < z < 5$

$$F'(z) = 2v/(a\beta) \, (z - 3)\exp[-(z - 3)/\beta]$$
$$- v/a(\exp[-(z - 1)/\beta] - \exp[-(z - 3)/\beta]) \qquad (3.3.36)$$

We integrate this to get $F(z)$, choosing the constant of integration so that $u(1, t)$ is continuous for $t > 0$; applying this in particular at $t = 2/a$ leads to

$$F(z) = \beta v/a \, \exp[-(z - 1)/\beta]$$
$$- \beta v/a[1 + (2/\beta)(z - 3)]\exp[-(z - 3)/\beta] \qquad (3.3.37)$$

for $3 < z < 5$. In this way, applying the continuity of $u_t(1, t)$ and $u(1, t)$ successively at $t = 2/a, 4/a, \ldots$, we can extend our information about $F(z)$ to the intervals $(5, 7)$, $(7, 9), \ldots$. Now we can discuss the solution $u(x, t)$.

Discussion of Solution   For $0 < t < 1/a$:

$$F(at - x) = 0 \quad \text{for } 0 < x < 1$$

and

$$u(x, t) = F(at + x) = -v\beta/a(1 - \exp[-(at + x - 1)/\beta])$$

This is a wave (the result of the impact of the weight at the end $x = 1$) traveling from the end $x = 1$ toward the end $x = 0$. At $t = 1/a$ it reaches the end $x = 0$, where it is reflected, and the solution then consists of the sum of two waves.

For $1/a < t < 2/a$:

$$F(at - x) = \begin{cases} -v\beta/a(1 - \exp[-(at - x - 1)/\beta] & \text{for } x < at - 1 \\ 0 & \text{for } x > at - 1 \end{cases}$$

$$F(at + x) = -v\beta/a(1 - \exp[-(at + x - 1)/\beta]) \quad \text{for } 0 < x < 1$$

$$u(x, t) = F(at + x) - F(at - x)$$

At $t = 2/a$, the reflected wave $F(at - x)$, which is traveling from $x = 0$ toward $x = 1$, reaches the end $x = 1$ and is reflected a second time. It will then be traveling *from* $x = 1$ *toward* $x = 0$ and so is represented by $F(at + x)$ but given now by (3.3.37) instead of (3.3.35). Then we have:

For $2/a < t < 3/a$:

$$F(at - x) = -v\beta/a(1 - \exp[-(at - x - 1)/\beta]) \quad \text{for } 0 < x < 1$$

$$F(at + x) = \begin{cases} \text{given by (3.3.30)} & \text{for } x < 3 - at \\ \text{given by (3.3.32)} & \text{for } x > 3 - at \end{cases}$$

$$u(x, t) = F(at + x) - F(at - x)$$

We can continue in this way, extending our computation of $u(x, t)$ to intervals of the form $(N/a, (N + 1)/a)$.

Physical Interpretations   We point out that an ideal gas enclosed in a tube of uniform cross section behaves in essentially the same way as a rod executing longitudinal vibrations. In particular, let

$$u(x, t) = \text{location at time } t \text{ of thin cross-sectional}$$
$$\text{slab of particles whose equilibrium location is } x$$
$$\sigma(x, t) = \text{deviation at position } x \text{ and time } t \text{ from an}$$
$$\text{``equilibrium density''}$$
$$p(x, t) = \text{deviation at position } x \text{ and time } t \text{ from an}$$
$$\text{``equilibrium pressure''}$$
$$\phi(x, t) = \text{velocity potential at position } x \text{ and time } t$$
$$v(x, t) = \text{longitudinal velocity of gas particles at}$$
$$\text{position } x \text{ and time } t$$

Then each of these functions can be shown to satisfy the wave equation with the same constant a^2, depending on the parameters in the ideal gas equation. At a *closed end* of the tube, the boundary conditions that apply to each of the quantities are

$$u = 0, \qquad v = 0, \qquad \phi_x = 0, \qquad p_x = 0, \qquad \sigma_x = 0$$

At an *open end* of the tube the boundary conditions satisfied by each of the unknowns are

$$u_x = 0, \qquad v_x = 0, \qquad \phi = 0, \qquad p = 0, \qquad \sigma = 0$$

Other types of boundary conditions arise from ends that are partially closed in some specific way, or they may also arise from Neumann (derivative) boundary conditions under a transformation of the dependent variable.

Using this interpretation, Example 3.3.1 could describe the displacement from the equilibrium location or the longitudinal velocity of gas particles in a tube with both ends closed, or it could equally well describe the deviation of the gas pressure or density from the equilibrium state in a tube with both ends open. In any case, the influence of the initial state propagates in the tube in a wavelike fashion. The solutions to each of the example problems in this section illustrates what we call *wavelike evolution*. ■ ■

Wave Equation in Two Dimensions

Each of the examples we have considered for the wave equation have involved the one-dimensional wave equation. When the number of spatial dimensions increases beyond 1, the solution techniques do not change significantly but the computations become more complex (see Exercise 12).

Exercises: Initial-Boundary-Value Problems for Wave Equation

1. Suppose $u(x, t)$ satisfies

$$\begin{aligned}
u_{tt}(x, t) &= a^2 u_{xx}(x, t), & 0 < x < 1, & \quad t > 0 \\
u(x, 0) &= x(1 - x), & 0 < x < 1 \\
u_t(x, 0) &= 0, & 0 < x < 1 \\
u(0, t) &= u(1, t) = 0, & t > 0
\end{aligned}$$

Evaluate $u(0.3, t)$ at $t = 5/a$, $t = 7/a$, and $t = 10/a$.

2. Repeat Exercise 1 if the boundary conditions

$$u(0, t) = u(1, t) = 0, \qquad t > 0$$

are replaced by the conditions

$$u_x(0, t) = u_x(1, t) = 0, \qquad t > 0$$

3. Repeat Exercise 1 if the boundary conditions

$$u(0, t) = u(1, t) = 0, \qquad t > 0$$

are replaced by the conditions

$$u(0, t) = u_x(1, t) = 0, \qquad t > 0$$

4. Repeat Exercise 1 if the initial conditions are replaced by

$$u(x, 0) = 0 \quad \text{for } 0 < x < 1$$

$$u_t(x, 0) = \begin{cases} 1 & \text{if } \frac{1}{4} < x < \frac{3}{4} \\ 0 & \text{otherwise} \end{cases}$$

5. Repeat Exercise 4 if the boundary conditions

$$u(0, t) = u(1, t) = 0, \qquad t > 0$$

are replaced by the conditions

$$u_x(0, t) = u_x(1, t) = 0, \qquad t > 0$$

6. Repeat Exercise 4 if the boundary conditions

$$u(0, t) = u(1, t) = 0, \qquad t > 0$$

are replaced by the conditions

$$u(0, t) = u_x(1, t) = 0, \qquad t > 0$$

7. Obtain a solution of the form (3.3.25) or (3.3.26) for the problem

$$u_{tt}(x, t) = a^2 u_{xx}(x, t) - \beta^2 u_t(x, t)$$
$$u(x, 0) = 0, \qquad u_t(x, 0) = g(x), \qquad 0 < x < 1$$
$$u(0, t) = u(1, t) = 0, \qquad t > 0$$

8. Obtain a solution of the form (3.3.25) or (3.3.26) for the problem

$$u_{tt}(x, t) = a^2 u_{xx}(x, t) - \beta^2 u_x(x, t)$$
$$u(x, 0) = 0, \qquad u_t(x, 0) = g(x), \qquad 0 < x < 1$$
$$u(0, t) = u(1, t) = 0, \qquad t > 0$$

9. Find $u(x, t)$ satisfying

$$u_{tt}(x, t) = a^2 u_{xx}(x, t) - \beta^2 u_x(x, t) - \sigma u_t(x, t)$$
$$u(x, 0) = f(x), \qquad u_t(x, 0) = 0, \qquad 0 < x < 1$$
$$u(0, t) = u(1, t) = 0, \qquad t > 0$$

10. Solve for $u(x, t)$ if

$$u_{tt}(x, t) = a^2 u_{xx}(x, t)$$
$$u(x, 0) = 0, \qquad u_t(x, 0) = 0, \qquad 0 < x < 1$$
$$u(0, t) = A \sin \Omega t, \qquad u(1, t) = 0, \qquad t > 0$$

where a, A, and Ω denote given constants. *Hint:* Let

$$v(x, t) = u(x, t) - A(1 - x)\sin \Omega t$$

Then show that $v(x, t)$ satisfies an inhomogeneous equation with homogeneous boundary conditions and solve for $v(x, t)$.

11. Repeat Exercise 10 if the boundary conditions

$$u(0, t) = A \sin \Omega t, \qquad u(1, t) = 0, \qquad t > 0$$

are replaced by

$$u(0, t) = A \sin \Omega t, \qquad u_x(1, t) = 0, \qquad t > 0$$

Here try the change of variables,

$$v(x, t) = u(x, t) - A\phi(x)\sin \Omega t$$

where $\phi(x)$ is such that $\phi(0) = 1$ and $\phi'(1) = 0$.

12. Find $u(x, y, t)$ satisfying

$$
\begin{aligned}
u_{tt}(x, t) &= a^2\nabla^2 u(x, y, t), & 0 < x, y < 1, & \quad t > 0 \\
u(x, y, 0) &= f(x, y), & u_t(x, y, 0) = 0 & \\
u(x, 0, t) &= u(x, 1, t) = 0, & 0 < x < 1, & \quad t > 0 \\
u(0, y, t) &= u(1, y, t) = 0, & 0 < y < 1, & \quad t > 0
\end{aligned}
$$

where $f(x, y)$ is given. Solve this problem first by the method of characteristics and then by the method of eigenfunction expansion.

13. Consider the longitudinal vibration of a thin rod with a weight attached to one end. If the length of the rod is 1 and the weight is attached at the end $x = 1$, then the longitudinal displacement $u(x, t)$ satisfies

$$Mu_{tt}(1, t) = -ESu_x(1, t), \qquad t > 0$$

at the weighted end. If the end $x = 0$ is held fixed and if the rod is initially at rest and with initial displacement equal to a given function $f(x)$, then $u(x, t)$ satisfies

$$
\begin{aligned}
u_{tt}(x, t) &= (ES/m)u_{xx}(x, t), & 0 < x < 1, & \quad t > 0 \\
u(x, 0) &= f(x), & u_t(x, 0) = 0, & \quad 0 < x < 1 \\
u(0, t) &= 0, & Mu_{tt}(1, t) = -ESu_x(1, t)
\end{aligned}
$$

Here M, m denote (respectively) the mass of the attached weight and of the rod, E denotes the Young modulus of the rod material, and S denotes the area of the cross sections of the rod. Show that this problem leads to a nonorthogonal family of eigenfunctions. Solve the problem by the method of characteristics as in Example 3.3.4.

Chapter 4

Integral Transforms

In Chapters 2 and 3 we saw that the method of eigenfunction expansion is a suitable tool for solving problems in partial differential equations in which the spatial domain is a bounded set. In Chapter 5 we consider problems for which the spatial region is not bounded. For these problems we need a new tool, the integral transform. In Chapter 5 we shall see that the integral transform plays the same role in problems on unbounded domains that the eigenfunction expansions played in problems where the spatial domain was bounded.

In this chapter we develop the basic properties of two integral transforms, the Fourier transform and the Laplace transform. These are by no means the only integral transforms of importance in applied mathematics, but they are the most generally useful. We concentrate on those properties of the transforms that have application in solving problems in partial differential equations, although integral transforms are useful for other purposes as well.

In Chapter 2 we introduced the class of square-integrable functions on a *bounded* set as a convenient setting in which to study the Fourier series. Here we find it convenient to develop the Fourier transform in the class of functions that are square integrable on the *unbounded* set $(-\infty, \infty)$ or $(0, \infty)$.

4.1 FUNCTION SPACE $L^2(a, b)$ WHEN (a, b) IS UNBOUNDED

In Chapter 2 we introduced the notion of the vector space $L^2(a, b)$ for (a, b) an interval of finite length. The vectors in this vector space are the functions that are defined on the interval (a, b) and satisfy the condition

$$\int_a^b f(x)^2 \, dx < \infty \qquad (4.1.1)$$

Note that although condition (4.1.1) is satisfied by every function that is continuous on (a, b), it is not necessary that $f(x)$ be continuous or even sectionally continuous. We saw examples in Section 2.5 of functions that became infinite at points in the interval (a, b) and yet satisfied the condition (4.1.1). Evidently, discontinuities, even infinite jump discontinuities, are allowable in the functions of $L^2(a, b)$, provided the infinite discontinuities are such that the (improper) integral (4.1.1) is convergent. The class of square-integrable functions [i.e., the class $L^2(a, b)$] is large enough to accommodate most functions encountered in applied mathematics. In particular, for most naturally occurring problems in partial differential equations the data and the solutions are functions that are at least square integrable.

In this chapter we wish to consider the space $L^2(a, b)$ in the case that the interval (a, b) is not of finite length. Of course, in this case, the integral in condition (4.1.1) becomes an improper integral of the second kind. Thus in addition to being square integrable on every bounded subinterval of the infinite interval (a, b), a function $f(x)$ in $L^2(a, b)$ must also be such that the improper integral over the infinite interval is convergent.

EXAMPLE 4.1.1 ——

Functions in $L^2(1, \infty)$

 (a) Consider $(a, b) = (1, \infty)$ and $f(x) = x^p$ for p real. Then,

$$\int_1^\infty f(x)^2 \, dx = \begin{cases} (2p + 1)^{-1} & \text{if } (2p + 1) < 0 \\ \infty & \text{if } (2p + 1) \geq 0 \end{cases}$$

Evidently, in order for x^p to be square integrable on $(1, \infty)$, it is necessary for the function to "die off at infinity" sufficiently fast. In this case sufficiently fast means p must be strictly less than $-\frac{1}{2}$.

 (b) On the same interval $(1, \infty)$, consider the sectionally continuous functions

$$g(x) = \begin{cases} 1 & \text{if } n < x < n + (1/n^2), \, n = 1, 2, \ldots \\ 0 & \text{otherwise} \end{cases}$$

and

$$h(x) = \begin{cases} n^{1/2} & \text{if } n < x < n + (1/n^3), \, n = 1, 2, \ldots \\ 0 & \text{otherwise} \end{cases}$$

Then,

$$\int_1^\infty g(x)^2 \, dx = \sum_{n=1}^\infty 1 \frac{1}{n^2} < \infty$$

and

$$\int_1^\infty h(x)^2 \, dx = \sum_{n=1}^\infty n \frac{1}{n^3} < \infty$$

Each of the functions $g(x)$ and $h(x)$ are sectionally continuous but *not* continuous. The function $g(x)$ is everywhere bounded but does *not* go to zero as x goes to infinity. The function $h(x)$ not only does not go to zero at infinity, it is not even *bounded*. The purpose of showing these two functions as examples of functions in $L^2(1, \infty)$ is to offset the common misconception created by examples such as 4.1.1(a). It is often (incorrectly) believed that in order for a function to be square integrable on an infinite interval, it is necessary for the function to approach zero as the magnitude of its argument becomes infinite. The functions $g(x)$ and $h(x)$ are counterexamples.   ■ ■

Example 4.1.1(b) notwithstanding, there are some inferences that *can* be drawn regarding the "behavior at infinity" for functions in $L^2(a, b)$ when (a, b) is not of finite length.

Proposition 4.1.1.   Suppose that $f(x)$ and $f'(x)$ both belong to $L^2(a, b)$ where $b - a = +\infty$. Then $f(x) \rightarrow 0$ as $|x| \rightarrow \infty$.

Proof.   Suppose that $b = +\infty$. Observe that

$$[f(x) \pm f'(x)]^2 \geq 0 \quad \text{for all } x$$

and hence, for any real numbers s, t,

$$\int_s^t f(x)^2 \, dx + \int_s^t f'(x)^2 \, dx \geq 2 \left| \int_s^t f(x)f'(x) \, dx \right|$$

But,

$$2 \int_s^t f(x)f'(x) \, dx = f(t)^2 - f(s)^2$$

and if $f(x)$ and $f'(x)$ are both in $L^2(a, b)$, then

$$\lim_{\substack{t \to \infty \\ s \to \infty}} \int_s^t f(x)^2 \, dx = \lim_{\substack{t \to \infty \\ s \to \infty}} \int_s^t f'(x)^2 \, dx = 0$$

where s and t tend independently to infinity. Together, these results imply that $f(x)^2$ tends to a constant value as x tends to $+\infty$, and since $f(x)$ is in $L^2(a, b)$, this constant must be zero. A similar argument applies when $a = -\infty$.

We should point out that in order for $f'(x)$ to be square integrable it is necessary that $f(x)$ be at least continuous. If $f(x)$ is less smooth than this, $f'(x)$ is not defined (in the usual sense) at points where $f(x)$ is discontinuous. Even when the notion of derivative is generalized so as to permit discontinuous functions to have (generalized) derivatives, the generalized derivative of a function with a jump discontinuity is not square integrable.

One of the principal reasons for introducing the vector space $L^2(a, b)$ when (a, b) is of *finite* length is to gain access to the notion of an orthogonal basis of vectors (functions). We saw in Chapter 2 that for (a, b) of finite length, every regular Sturm–Liouville problem generates an orthogonal basis of eigenfunctions in $L^2(a, b)$. For example, in $L^2(-\pi, \pi)$ the Sturm–Liouville problem,

$$-u''(x) = \mu u(x), \qquad -\pi < x < \pi$$
$$u(-\pi) = u(\pi), \qquad u'(-\pi) = u'(\pi)$$

had eigenvalues

$$\mu_n = n^2, \qquad n = 0, 1, 2, \ldots$$

and for each eigenvalue the pair of eigenfunctions

$$\exp[inx], \exp[-inx]$$

These eigenfunctions are orthogonal on $(-\pi, \pi)$ and complete in the sense that any function $f(x)$ in $L^2(-\pi, \pi)$ can be uniquely expressed in the form

$$f(x) = \sum_{n=-\infty}^{\infty} f^n \exp[inx] \tag{4.1.2}$$

The meaning of (4.1.2) is the following:

$$\|f - f_N\|^2 = \int_{-\pi}^{\pi} [f(x) - f_N(x)]^2 \, dx \to 0 \quad \text{as } N \to +\infty \tag{4.1.3}$$

where

$$f_N(x) = \sum_{n=-N}^{N} f^n \exp[inx] \tag{4.1.4}$$

and

$$f^n = \frac{1}{2\pi} \int_{-\pi}^{\pi} f(x) \exp[-inx] \, dx = \frac{(f, e^{-inx})}{2\pi} \tag{4.1.5}$$

Note from (4.1.5) that altering the function $f(x)$ at a discrete set of points x does not affect the Fourier coefficients f^n. As a result, to say that f and g in $L^2(-\pi, \pi)$ are equal to one another does not necessarily mean that $f(x) = g(x)$ at each x in $(-\pi, \pi)$. Equality in $L^2(-\pi, \pi)$ means, precisely, that $\|f - g\| = 0$, and hence functions in $L^2(-\pi, \pi)$ must be thought of in this sense, which is more general than the customary pointwise sense of interpreting functions. That is, the functions belonging to $L^2(-\pi, \pi)$ are in fact equivalence classes of pointwise functions where two pointwise functions are said to be equivalent if they differ at most on a discrete set of points in $(-\pi, \pi)$. It is because of this equivalence class interpretation of the functions in $L^2(-\pi, \pi)$ that we are able to say that there is a one-to-one correspondence between functions f in $L^2(-\pi, \pi)$ and their Fourier coefficients given by (4.1.5). In addition, it follows from part 3 of Theorem 2.5.2 that if the function $f(x)$ is square integrable on $(-\pi, \pi)$, then the sequence of Fourier coefficients is "square summable"; that is,

$$\frac{1}{2\pi} \sum_{n=-\infty}^{\infty} |f^n|^2 = \|f\|^2 = \int_{-\pi}^{\pi} |f(x)|^2 \, dx \tag{4.1.6}$$

Because for each integer N, we have

$$f_N(x) = f_N(x + 2\pi) \quad \text{for all } x$$

it follows that the eigenfunction expansion (4.1.2) converges to the 2π-periodic extension of the function $f(x)$. On an *unbounded* interval (a, b), however, one cannot expect to expand arbitrary functions in $L^2(a, b)$ in terms of periodic eigenfunctions on (a, b). This is because the eigenfunctions cannot be simultaneously periodic and square integrable on an unbounded interval.

Then in place of the periodic Sturm–Liouville problem, consider the following singular Sturm–Liouville problem:

$$-u''(x) = \mu u(x), \qquad -\infty < x < \infty \qquad (4.1.7)$$
$$u(x) \quad \text{and} \quad u'(x) \quad \text{bounded for all } x$$

For $\mu = \alpha^2$, each of the functions

$$\exp[i\alpha x] \quad \text{and} \quad \exp[-i\alpha x]$$

is a solution of (4.1.7) for *all* real values α (not just integer values). For functions in $L^2(-\infty, \infty)$ this suggests the following analogue of (4.1.2):

$$f(x) = \int_{-\infty}^{\infty} F(\alpha)\exp[i\alpha x] \, d\alpha \qquad (4.1.8)$$

for $F(\alpha)$ given by (4.11). We can show that for f in $L^2(-\infty, \infty)$,

$$\|f - f_N\|^2 = \int_{-\infty}^{\infty} [f(x) - f_N(x)]^2 \, dx \to 0 \quad \text{as } N \to +\infty \qquad (4.1.9)$$

where

$$f_N(x) = \int_{-N}^{N} F(\alpha)\exp[i\alpha x] \, d\alpha \qquad (4.1.10)$$

and

$$F(\alpha) = \frac{1}{2\pi} \int_{-\infty}^{\infty} f(x)\exp[-i\alpha x] \, dx = \frac{(f, e^{-i\alpha x})}{2\pi} \qquad (4.1.11)$$

In addition, it is true that for $f(x)$ in $L^2(-\infty, \infty)$ the function $F(\alpha)$ is a square-integrable function of α and that

$$\frac{1}{2\pi} \int_{-\infty}^{\infty} |F(\alpha)|^2 \, d\alpha = \|f\|^2 = \int_{-\infty}^{\infty} |f(x)|^2 \, dx \qquad (4.1.12)$$

Establishing the validity of Equations (4.1.9)–(4.1.12) is beyond the scope of this text. We simply make note of the analogy between these results and the formulas (4.1.3)–(4.1.6).

The function $F(\alpha)$ defined in (4.1.11) is called the *Fourier integral transform* of the function $f(x)$. It is clear from (4.1.9) and (4.1.10) that there is a one-to-one correspondence between $f(x)$ in $L^2(-\infty, \infty)$ and $F(\alpha)$, which is also in $L^2(-\infty, \infty)$ (with respect to the variable α). Here both $f(x)$ and $F(\alpha)$ denote equivalence classes of pointwise functions defined on $(-\infty, \infty)$. Then the functions in $L^2(-\infty, \infty)$, as those in $L^2(a, b)$ for (a, b) bounded, are "generalized functions" to the extent that we have had to weaken slightly the way in which we think of pointwise values for a function in $L^2(-\infty, \infty)$. For a function $f(x)$ in $L^2(-\infty, \infty)$, we cannot, in general, speak of "the value" of f at the single point x. Instead we speak of the integral average of the values of f^2 on intervals containing x. In many physical situations we may talk about the value at x of a quantity (say, temperature) as if we mean the pointwise value. However, all

TABLE 4.1.1

Fourier series, $L^2(-\pi, \pi)$, n = integer	Fourier transform, $L^2(-\infty, \infty)$, $\alpha \in (-\infty, \infty)$
$\mu_n = n^2$	$\mu(\alpha) = \alpha^2$
$\exp[inx]$, $\exp[-inx]$	$\exp[i\alpha x]$, $\exp[-i\alpha x]$
$f(x) = \displaystyle\sum_{n=-\infty}^{\infty} f^n e^{inx}$	$f(x) = \displaystyle\int_{-\infty}^{\infty} F(\alpha) e^{i\alpha x}\, dx$
$f^n = (f, e^{-inx})/2\pi$	$F(\alpha) = (f, e^{-i\alpha x})/2\pi$
$\displaystyle\int_{-\pi}^{\pi} |f(x)|^2\, dx = \sum_{-\infty}^{\infty} |f^n|^2$	$\displaystyle\int_{-\infty}^{\infty} |f(x)|^2\, dx = \int_{-\infty}^{\infty} |F(\alpha)|^2\, d\alpha$

we are really able to measure is an integral average of the quantity over some small region surrounding x. Then this value is not pointwise but rather an integral average value, which is to say that the $L^2(-\infty, \infty)$ setting is physically meaningful.

A summary of the comparison of Fourier series in $L^2(-\pi, \pi)$ and Fourier transforms in $L^2(-\infty, \infty)$ can be displayed as in Table 4.1.1.

The first two lines of Table 4.1.1 indicate that in $L^2(-\pi, \pi)$ the eigenvalues and eigenfunctions form countably infinite collections while in $L^2(-\infty, \infty)$ they are continuously distributed; that is, the eigenvalues and eigenfunctions in $L^2(-\pi, \pi)$ can be put in one-to-one correspondence with the integers while the eigenvalues and eigenfunctions in $L^2(-\infty, \infty)$ are in one-to-one correspondence with the real numbers.

Lines 3 and 4 show that in $L^2(-\pi, \pi)$ the correspondence between $f(x)$ and its sequence of Fourier coefficients f^n is one-to-one whereas in $L^2(-\infty, \infty)$ the one-to-one correspondence is between $f(x)$ and its Fourier transform $F(\alpha)$.

Line 5 of the table contains the discrete and continuous versions of *Parseval's relation*. The discrete version says that if $f(x)$ is square integrable, then f^n is square summable. The continuous version says that if $f(x)$ is a square-integrable function of x, then $F(\alpha)$ is a square-integrable function of α.

While our primary concern in this book is with *real-valued* functions, it is more general to consider the space $L^2(a, b)$ for *complex-valued* functions. For complex valued functions, it is the integral

$$\int_a^b f(x)\bar{f}(x)\, dx = \int_a^b |f(x)|^2\, dx$$

that must be finite in order for $f(x)$ to belong to $L^2(a, b)$. Here $\bar{f}(x)$ denotes the complex conjugate of $f(x)$. Since $\bar{f}(x) = f(x)$ for real-valued $f(x)$, every function that is square integrable in the real sense is also square integrable in the complex sense. The spaces $L^2(-\pi, \pi)$ and $L^2(-\infty, \infty)$ of Table 4.1.1 should be interpreted in sense of complex square integrable functions since it is possible for a real-valued $f(x)$ to have a complex-valued Fourier transform $F(\alpha)$. Similarly a real-valued $f(x)$ in $L^2(-\pi, \pi)$ can have complex-valued (exponential) Fourier coefficients.

We present here the $L^2(-\infty, \infty)$ theory of the Fourier transform. There are other settings in which the Fourier transform can be developed, but this one is the most convenient for our purposes.

4.2 THE FOURIER TRANSFORM

In the last section, we introduced the notion of the Fourier integral transform and indicated (without rigorously justifying anything) that the transform would play the same role in the vector space $L^2(-\infty, \infty)$ that the Fourier series plays in $L^2(a, b)$ when (a, b) is of finite length.

In $L^2(a, b)$ we can compute the Fourier coefficients f^n from $f(x)$ by integrating, and we can recover $f(x)$ from the Fourier coefficients by summing the appropriate series of eigenfunctions. In $L^2(-\infty, \infty)$ we can compute the Fourier transform $F(\alpha)$ from $f(x)$ by integration, and according to (4.1.8), we can recover $f(x)$ from the Fourier transform by another integration. However, in most cases, *carrying out* this integration requires us to perform contour integration in the complex plane and to apply the theory of residues. An alternative approach is to compute many examples of Fourier transforms and thereby generate a "dictionary" of transform pairs f and F. Then in order to recover f from its transform, we look in this dictionary instead of doing the integration. We can solve a large number of problems in partial differential equations with only a small dictionary of transforms.

In Section 4.1 we used the notation $F(\alpha)$ for the Fourier transform of the function $f(x)$, and for the remainder of this text we use that notation to indicate the Fourier transform. The function itself will be denoted by a lowercase letter such as $f(x)$ or $g(x)$ and the Fourier transform of the function will be denoted by the corresponding uppercase letter $F(\alpha)$, $G(\alpha)$. Also, we use English letters, x, y, z, . . . to denote the variable in the space $L^2(-\infty, \infty)$ (not transformed) and Greek letters α, β, . . . to denote the variables in the transform space. Finally, in a few instances we use the notation $\mathscr{F}[f](\alpha)$ in place of $F(\alpha)$.

EXAMPLE 4.2.1 _____

Fourier Transforms

   (a)   For A a positive real number, consider the function

$$f(x) = \begin{cases} 1 & \text{if } |x| < A \\ 0 & \text{if } |x| > A \end{cases}$$

Then it is easy to check that $f(x)$ belongs to $L^2(-\infty, \infty)$, and we proceed to compute the Fourier transform $F(\alpha)$:

$$F(\alpha) = \frac{1}{2\pi} \int_{-A}^{A} 1 \, \exp[-i\alpha x] \, dx$$

$$= -1/(2\pi i \alpha)\exp[-i\alpha x]|_{-A}^{A} = \sin(\alpha A)/(\alpha \pi)$$

Frequently in the text we shall use the notation $I_A(x)$ for the function $f(x)$ in this example.

   (b)   For b a positive constant, consider $f(x) = \exp[-b|x|]$. Then, $f(x)$ is square integrable on $(-\infty, \infty)$, and we can compute

$$F(\alpha) = \frac{1}{2\pi} \left[\int_{-\infty}^{0} e^{bx} e^{-i\alpha x} \, dx + \int_{0}^{\infty} e^{-bx} e^{-i\alpha x} \, dx \right]$$

$$= \frac{1}{2\pi} \left(\frac{\exp[(b - i\alpha)x]}{b - i\alpha} \Big|_{-\infty}^{0} - \frac{\exp[-(b + i\alpha)x]}{b + i\alpha} \Big|_{0}^{\infty} \right)$$

Now,

$$\lim_{x \to -\infty} |\exp[(b - i\alpha)x]| = \lim_{x \to -\infty} \exp[bx] = 0$$

and

$$\lim_{x \to \infty} |\exp[-(b + i\alpha)x]| = \lim_{x \to \infty} \exp[-bx] = 0$$

so that we obtain

$$F(\alpha) = \frac{1}{2\pi} \left[\frac{1}{b - i\alpha} + \frac{1}{b + i\alpha} \right]$$

$$= \frac{b}{\pi(b^2 + \alpha^2)}$$

Note that $F(\alpha)$ is square integrable with respect to α on $(-\infty, \infty)$.

(c) For positive constant c, we consider $f(x) = \exp[-cx^2]$, which, as for the previous examples, is in $L^2(-\infty, \infty)$. Then,

$$F(\alpha) = \frac{1}{2\pi} \int_{-\infty}^{\infty} \exp[-cx^2]\exp[-i\alpha x] \, dx$$

$$= \frac{1}{2\pi} \int_{-\infty}^{\infty} \exp\left[-c\left(x^2 + \frac{i\alpha x}{c} - \frac{\alpha^2}{4c^2}\right) - \frac{\alpha^2}{4c}\right] dx$$

$$= \exp\left[\frac{-\alpha^2}{4c}\right] \frac{1}{2\pi} \int_{-\infty}^{\infty} \exp[-c(x + i\alpha)^2] \, dx$$

where, in the second step, we completed the square in the exponent. Now it is recorded in most tables of definite integrals that

$$\int_{-\infty}^{\infty} \exp[-c(x + i\alpha)^2] \, dx = \int_{-\infty}^{\infty} \exp[-cz^2] \, dz = \left[\frac{\pi}{c}\right]^{1/2}$$

and it follows that

$$F(\alpha) = [4\pi c]^{-1/2}\exp[-\alpha^2/4c] \qquad \blacksquare \blacksquare$$

Operational Properties of the Fourier Transform

In Example 4.2.1, we computed the Fourier transform for three specific functions. We derive now several general properties of the Fourier transform, properties that apply to *any* function in $L^2(-\infty, \infty)$. We refer to these as the *operational properties* of the transform.

Proposition 4.2.1   (Linearity, Homogeneity).   For arbitrary f and g in $L^2(-\infty, \infty)$ and arbitrary constants a, b

   (i) $\mathscr{F}[af(x) + bg(x)] = aF(\alpha) + bG(\alpha)$
   (ii) $\mathscr{F}[f(ax)] = (1/|a|)F(\alpha/a)$ $(a \neq 0)$

The proof of this proposition is based on trivial observations about the properties of the integral.

Proposition 4.2.2   (Differentiation and Transformation)

   (i) For arbitrary $f(x)$ and $f'(x)$ both in $L^2(-\infty, \infty)$,

$$\mathscr{F}[f'(x)] = i\alpha F(\alpha)$$

   If $f''(x)$ is also in $L^2(-\infty, \infty)$, then

$$\mathscr{F}[f''(x)] = -\alpha^2 F(\alpha)$$

   (ii) For $f(x)$ and $xf(x)$ both in $L^2(-\infty, \infty)$,

$$\mathscr{F}[xf(x)] = iF'(\alpha)$$

Proof.   For $f'(x)$ in $L^2(-\infty, \infty)$ we can compute

$$\mathscr{F}[f'(x)] = \frac{1}{2\pi} \int_{-\infty}^{\infty} f'(x)\exp[-i\alpha x]\, dx$$

$$= \frac{1}{2\pi} \left[f(x)e^{-i\alpha x} \Big|_{-\infty}^{\infty} - \int_{-\infty}^{\infty} f(x)(-i\alpha)e^{-i\alpha x}\, dx \right]$$

where we have integrated by parts. Now if $f(x)$ and $f'(x)$ both are in $L^2(-\infty, \infty)$, then $f(x) \to 0$ as $|x| \to \infty$. Then the boundary term in the preceding integral vanishes, and this leads to

$$\mathscr{F}[f'(x)] = (i\alpha) \frac{1}{2\pi} \int_{-\infty}^{\infty} f(x)e^{-i\alpha x}\, dx = i\alpha\, F(\alpha)$$

If $f''(x)$ is in $L^2(-\infty, \infty)$ as well, then apply the result just proved to the functions $f'(x)$ and $f''(x)$.

Iterating this result leads to

$$\mathscr{F}[f^{(n)}(x)] = (i\alpha)^n F(\alpha) \quad \text{for } n = 0, 1, \ldots, N$$

provided that $f^{(n)}(x)$ belongs to $L^2(-\infty, \infty)$ for $n = 0, 1, \ldots, N$.
   To prove part (ii) write

$$F(\alpha) = \frac{1}{2\pi} \int_{-\infty}^{\infty} f(x)\exp[-ix\alpha]\, dx$$

and differentiate both sides with respect to α.

Proposition 4.2.2 expresses the Fourier transform duality between differentiation and multiplication; that is, differentiating $f(x)$ corresponds to multiplying $F(\alpha)$ by α, and conversely, multiplying $f(x)$ by x corresponds to differentiating $F(\alpha)$.

Proposition 4.2.3 (Shifting Properties). For $f(x)$ in $L^2(-\infty, \infty)$ and c a real constant,

(i) $\mathcal{F}[f(x - c)] = e^{-i\alpha c}F(\alpha)$
(ii) $\mathcal{F}[e^{icx}f(x)] = F(\alpha - c)$

Proof. For f in $L^2(-\infty, \infty)$

$$2\pi \, \mathcal{F}[f(x - c)] = \int_{-\infty}^{\infty} f(x - c)\exp[-i\alpha x] \, dx$$

$$= \int_{-\infty}^{\infty} f(z)\exp[-i\alpha(z + c)] \, dz$$

$$= 2\pi e^{-i\alpha c}F(\alpha)$$

Similarly,

$$2\pi \, \mathcal{F}[e^{icx} f(x)] = \int_{-\infty}^{\infty} f(x)e^{icx}e^{-i\alpha x} \, dx$$

$$= \int_{-\infty}^{\infty} f(x)\exp[-ix(\alpha - c)] \, dx = 2\pi F(\alpha - c)$$

Proposition 4.2.4 (Repeated Transformation). For $f(x)$ in $L^2(-\infty, \infty)$, $F(\alpha)$ is also in $L^2(-\infty, \infty)$, and

$$\mathcal{F}[2\pi F(x)] = f(-\alpha)$$

that is, treating $F(\alpha)$ as a function of x and computing the Fourier transform yields $f(-\alpha)$.

Proof. Write the Fourier transform formula

$$F(\alpha) = \frac{1}{2\pi} \int_{-\infty}^{\infty} f(x)\exp[-i\alpha x] \, dx = \mathcal{F}[f(x)]$$

and the inversion formula

$$f(z) = \int_{-\infty}^{\infty} F(\beta)\exp[i\beta z] \, d\beta$$

In the second integral, replace z by $-\alpha$ and replace β by x and compare with the integral above.

We can apply Proposition 4.2.4 in order to double the number of entries in our Fourier transform dictionary:

From Example 4.2.1(a)

$$\mathcal{F}[I_A(x)] = \sin(\alpha A)/(\pi\alpha)$$

Then,

$$\mathcal{F}[\sin(Ax)/x] = \tfrac{1}{2}I_A(\alpha)$$

From Example 4.2.1(b)

$$\mathcal{F}[\exp[-b|x|]] = (b/\pi)(b^2 + \alpha^2)^{-1}$$

and by the proposition,

$$\mathcal{F}[2b/(b^2 + x^2)] = \exp[-b|\alpha|]$$

Finally,

$$\mathcal{F}[\exp(-cx^2)] = [4\pi c]^{-1/2}\exp[-\alpha^2/4c]$$

and by the proposition,

$$\mathcal{F}[(\pi/c)^{1/2}\exp(-x^2/4c)] = \exp[-c\alpha^2]$$

We collect all these Fourier transform pairs in a table in the endpapers.

For f and g in $L^2(-\infty, \infty)$, define

$$f*g(x) = \int_{-\infty}^{\infty} f(x - y)g(y)\, dy$$

Then $f*g$ is called the *convolution product* of f and g. It can be shown that for f and g in $L^2(-\infty, \infty)$ it follows that $f*g$ belongs to $L^1(-\infty, \infty)$. In addition, the convolution product has the following properties. For arbitrary f and g in $L^2(-\infty, \infty)$,

    (i) $f*g(x) = g*f(x)$
    (ii) $f*(g + h)(x) = f*g(x) + f*h(x)$

The importance of the convolution product relative to the Fourier transform lies in the following.

Proposition 4.2.5   (Product of Transforms). For f and g in $L^2(-\infty, \infty)$,

$$\mathcal{F}[f*g] = 2\pi F(\alpha)G(\alpha)$$

Proof. For f and g in $L^2(-\infty, \infty)$ we have, by definition,

$$2\pi\,\mathcal{F}[f*g] = \int_{-\infty}^{\infty} \exp[-i\alpha x]f*g(x)\, dx$$

$$= \int_{-\infty}^{\infty} \exp[-i\alpha x]\int_{-\infty}^{\infty} f(x - y)g(y)\, dy\, dx$$

$$= \int_{-\infty}^{\infty} \exp[-i\alpha y]g(y)\int_{-\infty}^{\infty} \exp[-i(x - y)\alpha]f(x - y)\, dx\, dy$$

In this last step, we (formally) interchanged the order of integration. Justifying this interchange requires techniques that are beyond the scope of this text. However, if we accept this interchange as valid, then letting $z = x - y$ in the last integral leads to

$$2\pi \ \mathcal{F}[f*g] = \int_{-\infty}^{\infty} \exp[-i\alpha y]g(y) \ dy \int_{-\infty}^{\infty} \exp[-i\alpha z]f(z) \ dz$$
$$= 2\pi G(\alpha)2\pi F(\alpha)$$

We conclude this section with some examples that illustrate, at an elementary level, how the Fourier transform can be used in solving differential equations. In order to keep the examples simple, the equations in these examples are all ordinary differential equations.

EXAMPLES 4.2.2 _____

Applications of the Transform
   (a) *Linear Equation with Constant Coefficients*
   Consider the problem of finding a function $y(x)$ in $L^2(-\infty, \infty)$ satisfying

$$y''(x) - b^2y(x) = f(x) \quad \text{for } -\infty < x < \infty$$

Here, $f(x)$ denotes a given function in $L^2(-\infty, \infty)$. Note that there are no boundary conditions in this problem, or to be more precise, the condition that the solution is to belong to $L^2(-\infty, \infty)$ plays the role of the boundary conditions; that is, if we were to look for a solution to this equation in a class of functions smaller than $L^2(-\infty, \infty)$, the problem might have *no* solution, while if we looked in a larger class, it is possible that there may be *more* than one solution.
   If we let $Y(\alpha) = \mathcal{F}[y(x)]$, then

$$\mathcal{F}[y''(x)] = -\alpha^2 Y(\alpha)$$

and

$$-(\alpha^2 + b^2)Y(\alpha) = F(\alpha)$$

where $F(\alpha) = \mathcal{F}[f(x)]$. The effect of the Fourier transform is to cause the *differential* equation in the unknown function $y(x)$ to become an *algebraic* equation in the transform $Y(\alpha)$. We solve this algebraic equation to get

$$Y(\alpha) = -F(\alpha)/(\alpha^2 + b^2)$$
$$= -(\pi/b)[b/(\pi(b^2 + \alpha^2))]F(\alpha)$$

We have written the expression for $Y(\alpha)$ in this way because

$$\mathcal{F}^{-1}[b/(\pi(b^2 + \alpha^2))] = \exp[-b|x|]$$

[see Example 4.2.1(b)]. Then using Proposition 4.2.5, the convolution formula, we obtain

$$y(x) = -\frac{1}{2b}\int_{-\infty}^{\infty} \exp[-b|x - z|]f(z) \ dz$$

This function $y(x)$ is a solution for the differential equation and can be shown to be the unique solution in $L^2(-\infty, \infty)$. Note that the solution process consists of the following steps:

1. Apply the Fourier transform to convert the differential equation for $y(x)$ to an algebraic equation for $Y(\alpha)$.
2. Solve the algebraic equation to obtain $Y(\alpha)$.
3. Invert the Fourier transform using the convolution formula if necessary.

Schematically, we represent this as

$$
\begin{array}{ccc}
& \text{solve} & \\
F(\alpha) - & \text{algebraic} & \to Y(\alpha) \\
& \text{equation} & \\
\mathscr{F} \uparrow & & \downarrow \mathscr{F}^{-1} \\
& \text{solve} & \\
f(x) - & \text{differential} & \to y(x) \\
& \text{equation} &
\end{array}
$$

If we let S denote the operation of solving the differential equation and σ denotes the operation of solving the algebraic equation, then

$$
y(x) = S[f(x)] = \mathscr{F}^{-1} \cdot \sigma \cdot \mathscr{F}[f(x)]
$$

that is, S consists of the composition of $\mathscr{F}$ with σ followed by $\mathscr{F}^{-1}$.

(b) *Linear Equation with Variable Coefficients*
Consider the problem of finding a function $y(x)$ that satisfies

$$
y''(x) - xy(x) = 0 \quad \text{for } -\infty < x < \infty
$$

We will proceed formally in this example, letting $Y(\alpha)$ denote the Fourier transform of $y(x)$. Then, according to Proposition 4.2.2(ii),

$$
\mathscr{F}[xy(x)] = iY'(\alpha)
$$

and the second-order differential equation becomes

$$
-\alpha^2 Y(\alpha) = iY'(\alpha)
$$

In this example, the differential equation for $y(x)$ is transformed into a differential, not algebraic, equation in $Y(\alpha)$. This is a consequence of the fact that the original equation had variable rather than constant coefficients. In general, we can only expect that the transformed equation is a linear algebraic equation in the transform of the unknown function if the original equation is a linear differential equation having constant coefficients.

As sometimes happens, the differential equation for $Y(\alpha)$ in this example can be solved. The solution is this case is

$$
Y(\alpha) = C \exp[-i\alpha^3/3]
$$

where C denotes an arbitrary constant. Note that since no auxiliary conditions were specified for this problem [not even the condition that the solution belong to $L^2(-\infty, \infty)$], it is to be expected that the solution will contain some degree of arbitrariness. The function $Y(\alpha)$ is not one that appears in our dictionary of Fourier transform pairs, and we cannot easily find the inverse in order to complete the solution of our problem. The difficulty inherent in the original problem resulting from the variable coefficient shows itself in the third step of our solution process in the form of a $Y(\alpha)$ for which no inverse is readly available.

Although we are not able to complete this example by finding an inverse for $Y(\alpha)$, we can perhaps recognize the original second-order equation as one that is well known in applied mathematics. It is called *Airy's equation*, and the solution is one of the so-called special functions of mathematical physics called the *Airy function*. This function is denoted Ai(x) and is called a special function because it cannot be expressed in terms of finitely many elementary functions. It is usually expressed in the form of an infinite series or as a combination of other special functions such as Bessel functions. We encounter this function again in connection with a partial differential equation called the *dispersion equation*.

Let us note finally that

$$Y(\alpha) = C \exp[i\alpha^3/3] = C[\cos(\alpha^3/3) + i \sin(\alpha^3/3)]$$

which is an evidently oscillatory function of α defined on $-\infty < \alpha < \infty$. Such a function is not an element of $L^2(-\infty, \infty)$, and it follows that $y(x)$ is not then in $L^2(-\infty, \infty)$ as a function of x. This raises the question then of how were we able to apply the Fourier transform to this problem if neither $y(x)$ nor $Y(\alpha)$ are in $L^2(-\infty, \infty)$? The answer is that we cannot apply the $L^2(-\infty, \infty)$ theory of the Fourier transform, but there *is* a setting, more general than the $L^2(-\infty, \infty)$ setting, in which the Fourier transform can be developed. In this setting, called the distributional setting, the "functions" are treated in a way still more general than the way they are treated in $L^2(-\infty, \infty)$. In the distributional setting, the notion of pointwise values for the generalized functions is weakened even more than it is in the $L^2(-\infty, \infty)$ setting, but in return operations of analysis such as the Fourier transform or differentiation can be applied in much more generality. With respect to the present example, the functions $y(x)$ and $Y(\alpha)$ can be placed into the framework of distribution theory, and the differential equation can be solved in that setting by means of the Fourier transform.

For more information about the Fourier transform in the framework of distribution theory, the reader is referred to the following:

F. G. Friedlander, *Introduction to the Theory of Distributions*, Cambridge University Press, Cambridge, 1982

H. Bremmermann, *Distributions, Complex Variables, and the Fourier Transform*, Addison-Wesley, Reading, Mass., 1965

G. Lighthill, *Fourier Analysis and Generalized Functions*, Cambridge University Press, Cambridge, 1960.

(c) *The Sampling Theorem*

As our final example of an application of the Fourier transform, we consider, not a differential equation, but a result from the area of signal processing called the *sampling theorem*.

Suppose $f(x)$ in $L^2(-\infty, \infty)$ is such that its Fourier transform $F(\alpha)$ satisfies

$$F(\alpha) = 0 \quad \text{for } |\alpha| > A$$

for some positive real number A; for example, $f(x) = \sin(Ax)/x$ is such a function. Functions with this property are said to be *band limited* with *cutoff frequency* equal to A. Then

$$f(x) = \int_{-A}^{A} F(\alpha)\exp[ix\alpha]\,d\alpha \tag{4.2.1}$$

Now the fact that $F(\alpha)$ vanishes for $|\alpha| > A$ means $F(\alpha)$ belongs to $L^2(-A, A)$, and hence we can write

$$F(\alpha) = \sum_{n=-\infty}^{\infty} F^n \exp\left(\frac{in\pi\alpha}{A}\right), \qquad |\alpha| < A \tag{4.2.2}$$

and expect this (exponential) Fourier series to converge to $F(\alpha)$ on $(-A, A)$. Of course the series converges to the $2A$-periodic extension of F outside $(-A, A)$, but that does not concern us since we only use (4.2.2) on the interval $(-A, A)$. Here we use the notation

$$F^n = \frac{1}{2A}\int_{-A}^{A} F(\alpha)\exp\left[\frac{-in\pi\alpha}{A}\right]\,d\alpha \tag{4.2.3}$$

Now comparing (4.2.3) with (4.2.1) leads to the observation that

$$2AF^n = f(-n\pi/A) \quad \text{for } n = \text{integer}$$

That is,

$$F^n = [2A]^{-1}f(-n\pi/A) \tag{4.2.4}$$

If we use (4.2.4) in (4.2.2), we obtain

$$F(\alpha) = \frac{1}{2A}\sum_{m=-\infty}^{\infty} f\left(\frac{m\pi}{A}\right)\exp\left[-\frac{im\pi\alpha}{A}\right] \tag{4.2.5}$$

where we have made the substitution $m = -n$. Finally, if we use (4.2.5) in (4.2.1) and carry out the integration with respect to α, we obtain

$$f(x) = \sum_{m=-\infty}^{\infty} f\left(\frac{m\pi}{A}\right)\frac{\sin[Ax - m\pi]}{Ax - m\pi} \tag{4.2.6}$$

This result says that it is possible to reconstruct the function $f(x)$ on the entire line $(-\infty, \infty)$ by "sampling" its values at the discrete points $x_m = m\pi/A$, $m = $ integer. This provides a way of transforming a band-limited time signal $f(x)$, defined on $(-\infty, \infty)$ but not necessarily periodic, from a continuous signal into a discrete signal $f_n = f(n\pi/A)$ and then returning to the continuous signal via (4.2.6). A function that is not band limited but is periodic can be transformed from a continuous signal to a discrete signal by computing its Fourier coefficients. We can then return to the continuous signal by summing the Fourier series. Functions that are neither band limited nor periodic generally require some approximation procedure either in transforming to a discrete signal or returning to the continuous signal.

Applying this same principle in the reverse direction, suppose that

$$f(x) = 0 \quad \text{for } |x| > A$$

Then for any $L > A$, we can write

$$f(x) = \sum_{n=-\infty}^{\infty} f^n \exp\left[\frac{in\pi x}{L}\right]$$

where

$$f^n = \frac{1}{2L} \int_{-A}^{A} f(x)\exp\left[\frac{-in\pi x}{L}\right] dx$$

and the series converges to the $2L$-periodic extension of $f(x)$. But if we recall that $f(x)$ must also belong to $L^2(-\infty, \infty)$ and that the Fourier transform of $f(x)$ is given by

$$F(\alpha) = \frac{1}{2\pi} \int_{-A}^{A} f(x)\exp[-i\alpha x] dx$$

then we see at once that

$$f^n = (\pi/L)F(n\pi/L) \tag{4.2.7}$$

For example, if $f(x) = I_A(x)$, then from Example 2.2.1(a),

$$F(\alpha) = \sin A\alpha/\pi\alpha$$

But then (4.2.7) implies that for any $L > A$, we have

$$f^n = 1/n\pi \sin(An\pi/L)$$

Not only does this give us another way of obtaining the Fourier coefficients for certain functions, it also illustrates the connection between the Fourier transform and Fourier series for those functions that have both.

In this same vein, suppose $f(x)$ belongs to $L^2(-\infty, \infty)$ and that $F(\alpha)$ in $L^2(-\infty, \infty)$ denotes its Fourier transform. Then for $W > 0$ a fixed constant, let

$$f_W(x) = \sum_{n=-\infty}^{\infty} f(nW)I_{W/2}(x - nW) \tag{4.2.8}$$

That is, $f_W(x)$ is a "staircase function" approximation to $f(x)$. The parameter W is the step width of the staircase. Now let

$$F_W(\alpha) = \frac{1}{2\pi} \int_{-\infty}^{\infty} f_W(x)\exp[-ix\alpha] dx$$

$$= \frac{1}{2\pi} \int_{-\infty}^{\infty} \sum_{n=-\infty}^{\infty} f(nW)I_W(x - nW)\exp[-ix\alpha] dx$$

$$F_W(\alpha) = \sum_{n=-\infty}^{\infty} f(nW) \frac{1}{2\pi} \int_{-\infty}^{\infty} I_W(x - nW)\exp[-ix\alpha] dx$$

$$= \sum_{n=-\infty}^{\infty} f(nW)\mathscr{F}\{I_W(x - nW)\}$$

But transform pair 1 and operational property 5 of the Fourier transform tables (see the endpapers) give the result

$$\mathscr{F}\{I_W(x - nW)\} = \exp[-inW\alpha](\sin W\alpha)/\pi\alpha$$

and then,

$$F_W(\alpha) = \sum_{n=-\infty}^{\infty} f(nW)\exp[-inW\alpha](\sin W\alpha)/\pi\alpha \qquad (4.2.9)$$

Formula (4.2.9) allows us to compute an approximate Fourier transform for the $f(x)$ transform based on just a discrete set of sample values for $f(x)$. ■ ■

Exercises: Fourier Transform

1. Compute the Fourier transform of the following functions:

(a) $f(x) = \begin{cases} 4 & \text{if } 3 < x < 7 \\ 0 & \text{otherwise} \end{cases}$

(b) $g(x) = xI_2(x) = \begin{cases} x & |x| < 2 \\ 0 & |x| > 2 \end{cases}$

(c) $h(x) = \begin{cases} x - 1 & \text{if } 0 < x < 2 \\ 0 & \text{otherwise} \end{cases}$

Compute the transform in two ways: first, using the definition and, second, using the result of Example 4.2.1(a) together with the shifting properties of Proposition 4.2.3. For f, g, and h compute

$$\|f\|^2 = \int_{-\infty}^{\infty} |f(x)|^2 \, dx$$

2. Compute the Fourier transform of the following functions:

(a) $f(x) = \begin{cases} -1 & \text{if } -2 < x < -1 \\ 2 & \text{if } 0 < x < 2 \\ 0 & \text{otherwise} \end{cases}$

(b) $g(x) = \begin{cases} \sin x & \text{if } |x| < \pi \\ 0 & \text{if } |x| > \pi \end{cases}$

(c) $h(x) = \begin{cases} \cos x & \text{if } |x| < \frac{1}{2}\pi \\ 0 & \text{otherwise} \end{cases}$

3. Determine which of the functions in Exercises 1 and 2 belong to $L^2(-\infty, \infty)$? For which do the Fourier transforms also belong to $L^2(-\infty, \infty)$?

4. Compute the Fourier transforms of the following functions:

(a) $f(x) = \begin{cases} e^{-ax}\sin bx & \text{if } x > 0 \\ 0 & \text{if } x < 0 \end{cases}$

(b) $g(x) = \begin{cases} e^{-ax}\cos bx & \text{if } x > 0 \\ 0 & \text{if } x < 0 \end{cases}$

where a and b denote positive constants. Are these functions in $L^2(-\infty, \infty)$?

5. Compute the Fourier transform of:
   (a) $x \exp[-b|x|]$ (b) $x^2\exp[-b|x|]$
   If b denotes a positive constant, are these functions in $L^2(-\infty, \infty)$?

6. If $f(x)$ in $L^2(-\infty, \infty)$ has Fourier transform $F(\alpha)$, then what is the function of x whose transform is the function:
   (a) $H(\alpha) = F(\alpha - 1)$

(b) $H(\alpha) = \cos(b\alpha)F(\alpha)$, b = real constant

(c) $H(\alpha) = \exp[-t\alpha^2]F(\alpha)$, t = positive real constant

(d) $H(\alpha) = F(\alpha - 1)F(\alpha + 1)$

(e) $H(\alpha) = \alpha^2\exp[-\alpha^2]$

7. Give an example of an ordinary differential equation for the unknown function $u(x)$, $-\infty < x < \infty$, with the property that the Fourier transform transforms the equation in $u(x)$ into the identical equation in $U(\alpha)$.

8. Is it possible to find two functions in $L^2(-\infty, \infty)$ neither one of which is identically zero but whose convolution product is zero?

9. Consider a function $f(x)$ in $L^2(-\infty, \infty)$ with Fourier transform $F(\alpha)$ and let

$$g(x) = \mathcal{F}^{-1}[F(\alpha)I_A(\alpha)] \quad \text{for some } A > 0$$

Then $g(x)$ is a band-limited function with cutoff frequency equal to A. In what way, if any, is $g(x)$ related to $f(x)$?

10. Let k denote a positive integer and let

$$f_k(x) = \begin{cases} 1 & \text{if } |x - k| < 1 \\ 0 & \text{if } |x - k| > 1 \end{cases}$$

Compute the Fourier transform $F_k(\alpha)$ for $k = 1, 2, \ldots$ and sketch the graphs of $f_k(x)$ and $F_k(\alpha)$ for $k = 1, 2, 3$. Relate the changes in $f_k(x)$ as k varies to the corresponding changes in $F_k(\alpha)$.

11. Repeat Exercise 10 for

$$f_k(x) = \begin{cases} k & \text{if } |x| < 1/k \\ 0 & \text{if } |x| > 1/k \end{cases}$$

As the graph of $f_k(x)$ becomes "tall and thin," what happens to the graph of the transform?

12. Repeat Exercise 10 for

$$f_k(x) = \begin{cases} 1 & \text{if } |x| < k \\ 0 & \text{if } |x| > k \end{cases}$$

As the graph of $f_k(x)$ becomes wider, what corresponding change occurs in the graph of the transform?

13. Let the functions $f_k(x)$ be as in Exercise 10 and define

$$g_N(x) = \sum_{k=-N}^{N} f_k(x), \quad N = 1, 2, \ldots$$

Compute the Fourier transform of $g_N(x)$.

14. Let k denote a positive integer and let

$$F_k(\alpha) = \begin{cases} 1 & \text{if } |\alpha - k| < 1 \\ 0 & \text{if } |\alpha - k| > 1 \end{cases}$$

For $k = 1, 2, \ldots$ find the functions $f_k(x)$ with Fourier transforms equal to $F_k(\alpha)$.

15. Let k denote a positive integer and let

$$F_k(\alpha) = \begin{cases} 1 & \text{if } |\alpha - k| < \frac{1}{2} \\ -1 & \text{if } |\alpha + k| < \frac{1}{2} \\ 0 & \text{otherwise} \end{cases}$$

For $k = 1, 2, \ldots$ find the functions $f_k(x)$ with Fourier transforms equal to $F_k(\alpha)$.

16. Let k denote a positive integer and let

$$F_k(\alpha) = \begin{cases} 1 & \text{if } -2k < \alpha < -k \text{ or } k < \alpha < 2k \\ 0 & \text{if } |\alpha| > 2k \\ -1 & \text{if } |\alpha| < k \end{cases}$$

For $k = 1, 2, \ldots$ find the functions $f_k(x)$ with Fourier transforms equal to $F_k(\alpha)$.

17. $F(\alpha) = 0$ for $|\alpha| > A$, implies that F belongs to $L^2(-B, B)$ for any $B \geq A$. Then how does the sampling formula (4.2.6) change if we expand $F(\alpha)$ as follows:

$$F(\alpha) = \sum_{n=-\infty}^{\infty} F^n \exp\left[\frac{in\pi\alpha}{B}\right], \qquad B > A$$

instead of as in (4.2.2)? Does this change require more frequent sampling of $f(x)$ or less frequent?

18. We have introduced the Fourier transform by viewing it as the analogue of the Fourier series for functions that are everywhere defined but are not periodic. In the language of signal processing, each is a frequency domain representation for the functions $f(x)$, which we refer to as the "frequency spectrum"; when $f(x)$ is periodic, the frequency spectrum is discrete, and when $f(x)$ is not periodic, it is continuous. Consider the reverse situation in which the "time domain" function is discrete; that is, suppose

$$f_n = f(x_n), \qquad n = \text{integer}$$

and let

$$F(\alpha) = \frac{1}{2L} \sum_{n=-\infty}^{\infty} f_n \exp\left[\frac{in\pi\alpha}{L}\right]$$

Then it follows from the results of Chapter 2 that

$$f_n = \int_{-L}^{L} F(\alpha) \exp\left[\frac{-in\pi\alpha}{L}\right] d\alpha$$

Verify this last remark and then discuss the following implications:

Periodic time signal $\Leftrightarrow$ discrete frequency spectrum
Nonperiodic time signal $\Leftrightarrow$ continuous frequency spectrum
Discrete time signal $\Leftrightarrow$ periodic frequency spectrum

19. In connection with the previous problem, consider the following discrete time signal:

$$f_n = \begin{cases} 1 & \text{if } n = -2, -1, 0, 1, 2 \\ 0 & \text{otherwise} \end{cases}$$

This is a discrete square pulse. Compute $F(\alpha)$ (with $L = \pi$) for this discrete time signal and show that

$$F(\alpha) = \frac{1}{2\pi} \frac{\sin[5\alpha/2]}{\sin[\alpha/2]}$$

20. Compute $F(\alpha)$ for the following discrete time signals:

(a) $f_n = \begin{cases} 1 & \text{if } n = 2, 3, 4, 5, 6 \\ 0 & \text{otherwise} \end{cases}$

(b) $g_n = \begin{cases} 1 & \text{if } n = 2, 3, 4 \\ -1 & \text{if } n = -2, -3, -4 \\ 0 & \text{otherwise} \end{cases}$

(c) $h_n = \begin{cases} 1 & \text{if } n = -1, 0, 1, n = 3, 4, 5, 6 \\ 0 & \text{otherwise} \end{cases}$

4.3 THE LAPLACE TRANSFORM

The Fourier transform is an integral transform that is defined for functions in $L^2(-\infty, \infty)$. Functions that are defined on $(-\infty, \infty)$ but *do not* belong to $L^2(-\infty, \infty)$ include all constant functions, $\sin x$, $\cos x$, all polynomials in x, and $\exp[x]$. The fact that so many commonly occurring functions are excluded from $L^2(-\infty, \infty)$ is inconvenient. For problems in which these functions occur as data or as the solution, the L^2 theory of the Fourier transform does not apply.

For problems in which the natural domain is $(0, \infty)$ rather than $(-\infty, \infty)$, this inconvenience can be avoided by using a related integral transform called the Laplace transform. This transform is defined for functions $f(t)$ whose domain is $(0, \infty)$ *or* for functions whose domain is $(-\infty, \infty)$ and satisfy $f(t) = 0$ for $t < 0$. For such a function $f(t)$, the Laplace transform of f is the function $\hat{f}(s)$ defined as

$$\hat{f}(s) = \int_0^\infty f(t)e^{-st}\, dt \tag{4.3.1}$$

Note that it is not necessary that $f(t)$ belong to $L^2(-\infty, \infty)$ in order for the Laplace transform of f to exist. In fact, the Laplace transform of $f(t)$ exists if there exist constants $M > 0$ and b such that

$$|f(t)| \leq Me^{bt} \quad \text{for } t \geq 0 \tag{4.3.2}$$

When we say the Laplace transform of $f(t)$ exists, what we mean is that the integral (4.3.1), which is an improper integral, converges for that particular $f(t)$. We collect these facts formally in the next propostion.

Proposition 4.3.1. Suppose that $f(t)$ is such that

(i) dom $f = (0, \infty)$ or dom $f = (-\infty, \infty)$ and $f(t) = 0$ for $t < 0$.
(ii) (4.3.2) holds for constants $M > 0$ and b.

Then $\hat{f}(s)$ given by (4.3.1) exists for $s > b$ and satisfies

$$|\hat{f}(s)| \leq M/(s - b) \tag{4.3.3}$$

Proof. If $f(t)$ satisfies hypotheses (i) and (ii), then

$$|\hat{f}(s)| = \left| \int_0^\infty f(t)e^{-st}\, dt \right|$$

$$|\hat{f}(s)| \leq \int_0^\infty |f(t)|e^{-st}\, dt \leq M \int_0^\infty \exp[-(s-b)t]\, dt$$

$$\leq M/(s-b)$$

Note that we have used t to denote the independent variable in this section. In many of the applications we consider, the condition that $f(t) = 0$ for $t < 0$ is consistent with the interpretation of the variable t as time.

We use the notations $\hat{f}(s)$ and $\mathcal{L}[f(t)]$ to indicate the Laplace transform of the function $f(t)$. The transform variable is denoted by s, which we assume to be a real, positive number.

Functions that satisfy condition (4.3.2) are said to be functions of *exponential type*. These include all constants, $\sin t$, $\cos t$, all polynomials in t, and $\exp[t]$. We compute the Laplace transform for some functions of exponential type.

EXAMPLE 4.3.1

Laplace Transforms

(a)   Consider the constant function $f(t) = 1$ for $t > 0$,

$$\mathcal{L}[1] = \int_0^\infty 1e^{-st}\, dt = -\frac{1}{s} e^{-st} \bigg|_0^\infty$$

$$= 1/s \quad \text{for } s > 0$$

(b)   The function $f(t) = e^{kt}$ for k a real constant is of exponential type (with $M = 1$ and $b = k$). Then

$$\mathcal{L}[e^{kt}] = \int_0^\infty e^{kt}e^{-st}\, dt$$

$$= \frac{-1}{s-k} \exp[-(s-k)t]\big|_0^\infty$$

$$= 1/(s-k) \quad \text{for } s > k$$

Additional formulas can be generated in this way directly from definition (4.3.1). However, we find it easier to generate Laplace transform pairs using the so-called operational properties. ■ ■

Operational Properties of Laplace Transform

Just as for the Fourier transform, the Laplace transform has a number of properties we refer to as "operational properties." These are not specific transform pairs such as those derived in Example 4.3.1. Rather they are general properties that apply to *all* functions of exponential type and their transforms.

Proposition 4.3.2 (Linearity, Homogeneity). For arbitrary functions f and g of exponential type and arbitrary constants a, b

(i) $\mathscr{L}[af(t) + bg(t)] = a\hat{f}(s) + b\hat{g}(s)$

(ii) $\mathscr{L}[f(at)] = (1/a)\hat{f}(s/a), \, a \neq 0$

Proposition 4.3.3 (Differentiation and Laplace Transformation). For arbitrary $f(t)$ of exponential type,

(i) if $f'(t)$ is also of exponential type,

$$\mathscr{L}[f'(t)] = s\hat{f}(s) - f(0)$$

If $f''(t)$ is also of exponential type,

$$\mathscr{L}[f''(t)] = s^2\hat{f}(s) - sf(0) - f'(0)$$

(ii) $\mathscr{L}[tf(t)] = -d/ds[\hat{f}(s)]$

$\mathscr{L}[t^2f(t)] = (-d/ds)^2[\hat{f}(s)]$

Proof. Under the hypotheses of (i),

$$\mathscr{L}[f'(t)] = \int_0^\infty f'(t)e^{-st} \, dt$$

$$= f(t)e^{-st}\Big|_0^\infty - \int_0^\infty f(t)(-se^{-st}) \, dt$$

$$= -f(0) + s\int_0^\infty f(t)e^{-st} \, dt$$

$$= s\hat{f}(s) - f(0)$$

Here, we integrated by parts and used the fact that if $f(t)$ is of exponential type, then for $s > b$

$$\lim_{t \to \infty} f(t)e^{-st} = 0$$

Since $f''(t)$ is the derivative of $f'(t)$, we can apply this result once again to obtain

$$\mathscr{L}[f''(t)] = s\mathscr{L}[f'(t)] - f'(0)$$

$$= s^2\hat{f}(s) - sf(0) - f'(0)$$

Further iterations leads to formulas for the transform of derivatives of higher order.

Note that the formula for the transform of the derivative of $f(t)$ of order N involves the values at $t = 0$ of $f(t)$ and all of its derivatives up to the order $N - 1$. These initial values are introduced by the boundary terms in the integration by parts. The boundary values do not appear in the corresponding formulas for the Fourier transform of the derivatives of $f(x)$ because if $f(x)$ and $f'(x)$ both belong to $L^2(-\infty, \infty)$, then it follows from Proposition 4.1.1 that $f(x)$ vanishes as $|x| \to \infty$.

To prove (ii), note that (formally)

$$\frac{d}{ds} \int_0^\infty f(t)\exp[-st]\,dt = \int_0^\infty -tf(t)\exp[-st]\,dt$$

that is,

$$\frac{d}{ds}\hat{f}(s) = \mathcal{L}[-tf(t)]$$

But if $f(t)$ is of exponential type, then it is easy to show that $tf(t)$ must also be of exponential type, and then the formal result is valid.

Proposition 4.3.3 has many applications including the generation of additional transform pairs:

EXAMPLE 4.3.2 _____

Applications of Proposition 4.3.3

    (a) From Example 4.3.1(a) we have $\mathcal{L}[1] = 1/s$. Then by (ii),

$$\mathcal{L}[t] = -\frac{d}{ds}\frac{1}{s} = \frac{1}{s^2}$$

$$\mathcal{L}[t^2] = -\frac{d}{ds}\frac{1}{s^2} = \frac{2}{s^3}$$

$$\mathcal{L}[t^3] = -\frac{d}{ds}\frac{2}{s^3} = \frac{6}{s^4}$$

$$\vdots \qquad\qquad \vdots \qquad\qquad \vdots$$

$$\mathcal{L}[t^n] = -\frac{d}{ds}\frac{(n-1)!}{s^n} = \frac{n!}{s^{n+1}}$$

    (b) For $f(t) = \sin t$ we have $f'(t) = \cos t$ and $f''(t) = -f(t)$, and hence by (i),

$$\mathcal{L}[-\sin t] = s^2\hat{f}(s) - sf(0) - f'(0)$$

Then,

$$-\hat{f}(s) = s^2\hat{f}(s) - 0 - 1$$

and

$$\hat{f}(s) = 1/[s^2 + 1]$$

In addition,

$$\mathcal{L}[\cos t] = s\hat{f}(s) - f(0) = s/[s^2 + 1]$$

    (c) Using (b) together with part (ii) of the proposition leads to

$$\mathcal{L}[t \sin t] = -\frac{d}{ds}\frac{1}{s^2 + 1}$$

$$= \frac{2s}{(s^2 + 1)^2}$$

$$\mathcal{L}[t \cos t] = -\frac{(s^2 + 1) - 2s^2}{(s^2 + 1)^2}$$

$$= \frac{s^2 - 1}{(s^2 + 1)^2} \qquad \blacksquare \blacksquare$$

Proposition 4.3.4 (Shifting Properties of Laplace Transform). For arbitrary $f(t)$ of exponential type and real constant $b, b > 0$

(i) $\mathcal{L}[e^{bt}f(t)] = \hat{f}(s - b)$
(ii) $\mathcal{L}[H(t - b)f(t - b)] = e^{-bs}\hat{f}(s)$

where $H(t)$ denotes the Heaviside step function,

$$H(t) = \begin{cases} 0 & \text{if } t < 0 \\ 1 & \text{if } t > 0 \end{cases}$$

Proof. By definition,

$$\mathcal{L}[e^{bt}f(t)] = \int_0^\infty f(t)e^{bt}e^{-st}\,dt$$

$$= \int_0^\infty f(t)e^{-(s-b)t}\,dt = \hat{f}(s - b)$$

Similarly,

$$\mathcal{L}[H(t - b)f(t - b)] = \int_b^\infty f(t - b)e^{-st}\,dt$$

The change of variable $\tau = t - b$ reduces this integral to exactly $e^{-bs}\hat{f}(s)$.

The function $f_b(t) \equiv H(t - b)f(t - b)$ in part (ii) of the proposition is nothing more than the function $f(t)$ "delayed" by the amount b. The graph of $f_b(t)$ is just the graph of $f(t)$ translated b units to the right (if b is positive).

EXAMPLE 4.3.3

Applications of Proposition 4.3.4

(a) Part (i) of the proposition implies

$$\mathcal{L}[e^{kt}t] = 1/(s - k)^2$$
$$\mathcal{L}[e^{kt}\sin t] = 1/[(s - k)^2 + 1]$$
$$\mathcal{L}[e^{kt}\cos t] = (s - k)/[(s - k)^2 + 1]$$

(b)   Part (ii) can be used to find $\hat{f}(s)$ for

$$f(t) = \begin{cases} 3 & 2 < t < 4 \\ 0 & \text{otherwise} \end{cases}$$
$$= 3H(t - 2) - 3H(t - 4)$$

We have

$$\hat{f}(s) = (3/s)e^{-2s} - (3/s)e^{-4s}$$ ■ ■

Proposition 4.3.5   (Integration and Laplace transformation).   For arbitratry $f(t)$ of exponential type,

(i) $\mathscr{L}\left[\int_0^t f(\tau)\,d\tau \right] = \frac{1}{s}\hat{f}(s)$

(ii) $\mathscr{L}\left[\frac{1}{t} f(t) \right] = \int_s^\infty \hat{f}(z)\,dz$

provided that $\lim_{t \to 0}(1/t)f(t)$ exists.

Proof.   Let $F(t) = \int_0^t f(\tau)\,d\tau$. Then $F'(t) = f(t)$ and $F(0) = 0$. Then

$$\hat{f}(s) = \mathscr{L}[F'(t)] = s\,\mathscr{L}[F(t)] - 0$$

and result (i) follows.

If we let $\hat{F}(s) = \int_s^\infty \hat{f}(z)\,dz$, then

$$\frac{d}{ds}\hat{F}(s) = -\hat{f}(s) = -\mathscr{L}[f(t)].$$

But if we denote the inverse Laplace transform of $\bar{F}(s)$ by $G(t)$ then

$$\frac{d}{ds}\hat{F}(s) = \mathscr{L}[-tG(t)]$$

and it follows that $tG(t) = f(t)$; that is, $G(t) = (1/t)f(t)$. These formal steps are valid provided $G(t)$ has a Laplace transform, which it does if the limit of $(1/t)f(t)$ as $t \to 0$ exists.

Results (i) and (ii) of Proposition 4.3.5 are duals of one another. That is, (i) says that integration with respect to the variable t corresponds to division by the parameter s while (ii) states that this correspondence is reciprocal in the sense that integration with respect to s corresponds to division by t. Proposition 4.3.3 states that the same duality applies for the *inverses* of these operations; that is, differentiation with respect to t corresponds to multiplication by s [proposition 4.3.3(i)] while differentiation with respect to s corresponds to division by t [Proposition 4.3.3(ii)]. The operations of multiplication and division are inverses of one another, just as differentiation and integration are inverse operations of each other. Then Proposition 4.3.3(i) implies correspondence between the operations of t-differentiation and s-multiplication, while Proposition 4.3.5(i) implies that the correspondence holds for the inverse operations as well. Part (ii) of each of these propositions implies that the correspondences are maintained for the dual statements.

EXAMPLE 4.3.4 ⎯⎯⎯⎯⎯⎯⎯⎯⎯⎯⎯⎯⎯⎯⎯⎯⎯⎯⎯⎯⎯⎯⎯⎯⎯⎯⎯⎯

Special Transform Formulas

(a)   We derive a special transform formula that will be of use in solving problems involving the heat equation. Then we derive two related formulas using Proposition 4.3.5.

Let

$$f(t) = (\pi t)^{-1/2} \exp[-k^2/4t], \qquad k = \text{positive constant} \qquad (4.3.4)$$

Then

$$\hat{f}(s) = \int_0^\infty f(t) e^{-st} \, dt$$

$$= \pi^{-1/2} \exp[-ks^{1/2}] \int_0^\infty \exp\left[-\left(\frac{k^2}{4t} - ks^{1/2} + st\right)\right] t^{-1/2} \, dt$$

$$= \pi^{-1/2} \exp[-ks^{1/2}] \int_0^\infty \exp\left[-\left(\frac{k}{2t^{1/2}} - (st)^{1/2}\right)^2\right] t^{-1/2} \, dt$$

Now the substitution $t = (\mu/2)^2$ produces the result

$$\hat{f}(s) = \pi^{-1/2} \exp[-ks^{1/2}] \int_0^\infty \exp\left[-\left(\frac{k}{\mu} - \frac{\mu s^{1/2}}{2}\right)^2\right] d\mu \qquad (4.3.5)$$

For $M = \text{const}$, putting $\tau = M/\mu$ in (4.3.5) leads to

$$\hat{f}(s) = \pi^{-1/2} \exp[-ks^{1/2}] \int_0^\infty \exp\left[-\left(\frac{k\tau}{M} - \frac{Ms^{1/2}}{2\tau}\right)^2\right] M\tau^{-2} \, d\tau \qquad (4.3.6)$$

If we choose $M = 2ks^{-1/2}$ in (4.3.6), then

$$\hat{f}(s) = \pi^{-1/2} \exp[-ks^{1/2}] \int_0^\infty \exp\left[-\left(\frac{\tau s^{1/2}}{2} - \frac{k}{\tau}\right)^2\right] 2ks^{-1/2}\tau^{-2} \, d\tau$$

$$= 2k(\pi s)^{-1/2} \exp[-ks^{1/2}] \int_0^\infty \exp\left[-\left(\frac{k}{\mu} - \frac{\mu s^{1/2}}{2}\right)^2\right] \mu^{-2} \, d\mu \qquad (4.3.7)$$

Now adding the two expressions for $\hat{f}(s)$, (4.3.5) and (4.3.7), we get

$$2\hat{f}(s) = 2(\pi s)^{-1/2} \exp[-ks^{1/2}] \int_0^\infty \exp\left[-\left(\frac{k}{\mu} - \frac{\mu s^{1/2}}{2}\right)^2\right]\left[\frac{s^{1/2}}{2} + \frac{k}{\mu^2}\right] d\mu$$

$$= 2(\pi s)^{-1/2} \exp[-ks^{1/2}] \int_{-\infty}^\infty \exp[-z^2] \, dz$$

This last definite integral is well known to have the value $\pi^{1/2}$, which leaves us with

$$\hat{f}(s) = s^{-1/2} \exp[-ks^{1/2}], \qquad s > 0$$
$$f(t) = (\pi t)^{-1/2} \exp[-k^2/4t], \qquad k = \text{positive constant} \qquad (4.3.8)$$

(b) Applying (ii) of Proposition 4.3.5 to $f(t)$ given by 4.3.4, we have

$$\mathcal{L}\left[(\pi t^3)^{-1/2}\exp\left[\frac{-k^2}{4t}\right]\right] = \int_s^\infty z^{-1/2}\exp[-kz^{1/2}]\,dz$$
$$= 2/k\,\exp[-ks^{1/2}]$$

That is,

$$f(t) = k/(4\pi t^3)^{1/2}\exp[-k^2/4t]$$
$$\hat{f}(s) = \exp[-ks^{1/2}] \tag{4.3.9}$$

(c) Apply (i) of Proposition 4.3.5 with

$$f(t) = k/(4\pi t^3)^{1/2}\exp[-k^2/4t]$$

Then,

$$\mathcal{L}\left[\int_0^t f(\tau)\,d\tau\right] = \frac{1}{s}\hat{f}(s) = \frac{1}{s}\exp[-ks^{1/2}]$$

Now note that

$$\int_0^t f(\tau)\,dt = \int_0^t \frac{k}{(4\pi\tau^3)^{1/2}}\exp\left[\frac{-k^2}{4\tau}\right]d\tau$$
$$= 2\pi^{-1/2}\int_{k/(4t)^{1/2}}^\infty \exp[-\mu^2]\,d\mu$$
$$\equiv \operatorname{erfc}[k/(4t)^{1/2}]$$

That is, the antiderivative of $f(t)$ is $F(t) = \operatorname{erfc}[k/(4t)^{1/2}]$, and then, according to (i) of the proposition, the Laplace transform pair for $F(t)$ is given by

$$F(t) = \operatorname{erfc}[k/(4t)^{1/2}]$$
$$\hat{f}(s) = 1/s\,\exp[-ks^{1/2}] \tag{4.3.10}$$

The complementary error function, $\operatorname{erfc}(z)$, is defined as follows:

$$\operatorname{erfc}(z) = \frac{2}{\pi^{1/2}}\int_z^\infty \exp[-s^2]\,ds$$

$$= 1 - \operatorname{erf}(z) \equiv 1 - \frac{2}{\pi^{1/2}}\int_0^z \exp[-s^2]\,ds$$

The error function $\operatorname{erf}(z)$ is a tabulated function and is even included as a library function in some computer subroutine libraries. Laplace transform solutions to the heat equation are often expressed in terms of the functions $\operatorname{erf}(z)$ or $\operatorname{erfc}(z)$. ■ ■

Convolution and Laplace Transform

In the previous section we defined the convolution product of two functions in the following way:

$$f*g(t) = \int_{-\infty}^\infty f(\tau)g(t-\tau)\,d\tau$$

Now if $f(t)$ and $g(t)$ each vanish for negative values of their argument, then

$$f(\tau) = 0 \quad \text{for } \tau < 0 \quad \text{and} \quad g(t - \tau) = 0 \quad \text{for } \tau > t$$

Then the convolution integral reduces to

$$f*g(t) = \int_0^t f(\tau)g(t - \tau) \, d\tau \tag{4.3.11}$$

Then we have:

Proposition 4.3.6 (Convolution and Laplace Transform). For $f(t)$ and $g(t)$ arbitrary functions of exponential type, vanishing for $t < 0$,

$$\mathcal{L}[f*g(t)] = \hat{f}(s)\hat{g}(s)$$

for $f*g$ defined by (4.3.11).

The proof of this proposition is nearly identical to the proof of Proposition 4.2.5 and is omitted.

EXAMPLE 4.3.5 ───

Additional Transform Formulas
    Suppose $\hat{f}(s)$ denotes the Laplace transform of the function $f(t)$. We shall find the function whose transform is

$$\hat{h}(s) = \hat{f}(s) \frac{\sinh[as^{1/2}]}{\sinh[bs^{1/2}]}, \quad a, b = \text{real constants}$$

In order to find the inverse transform, we first write

$$\frac{\sinh[as^{1/2}]}{\sinh[bs^{1/2}]} = \frac{\exp[as^{1/2}] - \exp[-as^{1/2}]}{\exp[bs^{1/2}] - \exp[-bs^{1/2}]}$$

$$= \frac{\exp[(a - b)s^{1/2}] - \exp[-(a + b)s^{1/2}]}{1 - \exp[-2bs^{1/2}]}$$

where we have expressed the hyperbolic sine in terms of exponentials and then multiplied the quotient top and bottom by $\exp[-bs^{1/2}]$. Now we use the fact that

$$\sum_{n=0}^{\infty} x^n = \frac{1}{1 - x} \quad \text{for } |x| < 1$$

to write

$$\frac{\sinh[as^{1/2}]}{\sinh[bs^{1/2}]} = \sum_{n=0}^{\infty} \exp[-\alpha_n s^{1/2}] - \sum_{n=0}^{\infty} \exp[-\beta_n s^{1/2}]$$

where

$$\alpha_n = (2n + 1)b - a \quad \text{and} \quad \beta_n = (2n + 1)b + a$$

Then we have

$$\hat{h}(s) = \sum_{n=0}^{\infty} \hat{f}(s) \exp[-\alpha_n s^{1/2}] - \sum_{n=0}^{\infty} \hat{f}(s) \exp[-\beta_n s^{1/2}]$$

and it follows from (4.3.9) and Proposition 4.3.6 that

$$h(t) = \sum_{n=0}^{\infty} \int_0^t \frac{\alpha_n}{(4\pi\tau^3)^{1/2}} \exp\left[\frac{-\alpha_n^2}{4\tau}\right] f(t-\tau)\,d\tau$$
$$- \sum_{n=0}^{\infty} \int_0^t \frac{\beta_n}{(4\pi\tau^3)^{1/2}} \exp\left[\frac{-\beta_n^2}{4\tau}\right] f(t-\tau)\,d\tau$$

The substitution

$$z = \alpha_n/(4\tau)^{1/2}$$

in the first series of integrals and

$$z = \beta_n/(4\tau)^{1/2}$$

in the second series of integrals reduces this to

$$h(t) = \frac{2}{\pi^{1/2}} \sum_{n=0}^{\infty} \int_{\alpha_n/(4t)^{1/2}}^{\infty} \exp[-z^2] f\left(t - \left(\frac{\alpha_n}{2z}\right)^2\right) dz$$
$$- \frac{2}{\pi^{1/2}} \sum_{n=0}^{\infty} \int_{\beta_n/(4t)^{1/2}}^{\infty} \exp[-z^2] f\left(t - \left(\frac{\beta_n}{2z}\right)^2\right) dz$$

Computations such as the one in this example will be useful in solving problems for the heat equation on a bounded interval by means of the Laplace transform. ■ ■

With just this very brief introduction to the Fourier and Laplace transforms, we are now in a position to begin applying the transforms to solve problems in partial differential equations. This is the subject of the next chapter.

Exercises: Laplace Transform

1. Find the Laplace transform of the following functions:
   (a) $\cosh at$
   (b) $t \sinh at$
   (c) $t^{1/2}$
   (d) $(t + 1)^2$
   (e) $(t - 2)\exp[t + 2]$
   (f) $(4t + 1)\exp[2t - 2]$
   (g) $1/t \sin 3t$
   (h) $f(t) = \begin{cases} 1 & \text{if } 1 < t < 2 \\ -2 & \text{if } 3 < t < 5 \\ 0 & \text{otherwise} \end{cases}$

2. Find the functions $f(t)$ whose Laplace transforms are the following functions:
   (a) $1/(s - 2)$ (b) $2/(s^2 - 3s + 2)$
   (c) $s/(s^2 - 3s + 2)$ (d) $1/[s(s^2 - 3s + 2)]$
   (e) $(s^2 - 4)/(s^2 + 4)^2$ (f) $1/(s^4 - 81)$
   (g) $\exp[-3s]/s^3$ (h) $(1/s)1/(s^2 + 1)$

3. For $k = 0, 1, 2, \ldots,$ let

$$f_k(t) = \begin{cases} 1 & \text{if } 2k < t < 2k + 1 \\ 0 & \text{otherwise} \end{cases}$$

Find $\hat{f}_k(s)$ for $k = 0, 1, 2.$

4. Let $f_k(t)$ be as in Exercise 3. Find the Laplace transform of

$$g_N(t) = \sum_{k=0}^{N} f_k(t)$$

5. If a and b denote distinct real constants, show that

(a) $\dfrac{\cosh[as]}{\cosh[bs]} = \displaystyle\sum_{n=0}^{\infty} (-1)^n \exp[-\alpha_n s] + \sum_{n=0}^{\infty} (-1)^n \exp[-\beta_n s]$

(b) $\dfrac{\cosh[as]}{\sinh[bs]} = \displaystyle\sum_{n=0}^{\infty} \exp[-\alpha_n s] + \sum_{n=0}^{\infty} \exp[-\beta_n s]$

(c) $\dfrac{\sinh[as]}{\cosh[bs]} = \displaystyle\sum_{n=0}^{\infty} (-1)^n \exp[-\alpha_n s] - \sum_{n=0}^{\infty} (-1)^n \exp[-\beta_n s]$

(d) $\dfrac{\sinh[as]}{\sinh[bs]} = \displaystyle\sum_{n=0}^{\infty} \exp[-\alpha_n s] - \sum_{n=0}^{\infty} \exp[-\beta_n s]$

where

$$\alpha_n = (2n + 1)b - a \qquad \beta_n = (2n + 1)b + a$$

6. Let $f(t)$ have as its Laplace transform $\hat{f}(s)$. If a and b denote distinct real constants, then derive the following inversion formulas:

(a) $\mathcal{L}^{-1}\left[\dfrac{\hat{f}(s)\cosh[as]}{\cosh[bs]}\right] = \displaystyle\sum_{n=0}^{\infty} (-1)^n H(t - \alpha_n)f(t - \alpha_n) + \sum_{n=0}^{\infty} (-1)^n H(t - \beta_n)f(t - \beta_n)$

(b) $\mathcal{L}^{-1}\left[\dfrac{\hat{f}(s)\cosh[as]}{\sinh[bs]}\right] = \displaystyle\sum_{n=0}^{\infty} H(t - \alpha_n)f(t - \alpha_n) + \sum_{n=0}^{\infty} H(t - \beta_n)f(t - \beta_n)$

(c) $\mathcal{L}^{-1}\left[\dfrac{\hat{f}(s)\sinh[as]}{\cosh[bs]}\right] = \displaystyle\sum_{n=0}^{\infty} (-1)^n H(t - \alpha_n)f(t - \alpha_n) - \sum_{n=0}^{\infty} (-1)^n H(t - \beta_n)f(t - \beta_n)$

(d) $\mathcal{L}^{-1}\left[\dfrac{\hat{f}(s)\sinh[as]}{\sinh[bs]}\right] = \displaystyle\sum_{n=0}^{\infty} H(t - \alpha_n)f(t - \alpha_n) - \sum_{n=0}^{\infty} H(t - \beta_n)f(t - \beta_n)$

where

$$\alpha_n = (2n + 1)b - a, \qquad \beta_n = (2n + 1)b + a$$

7. Let

$$f(t) = \begin{cases} t & \text{if } 0 < t < 1 \\ 2 - t & \text{if } 1 < t < 2 \\ 0 & \text{otherwise} \end{cases}$$

Then find:

(a) $\mathcal{L}^{-1}[s\hat{f}(s)]$

(b) $\mathcal{L}^{-1}[s^2\hat{f}(s)]$

(c) $\mathcal{L}^{-1}[\exp(-2s)s\hat{f}(s)]$

(d) $\mathcal{L}^{-1}[s\hat{f}(s - 1)]$

8. Find the inverse transforms of each of the following in terms of $f(t) = \mathcal{L}^{-1}[\hat{f}(s)]$:

(a) $\hat{f}(s)1/s$

(b) $\hat{f}(s)1/(s + h), h > 0$

(c) $\hat{f}(s)\exp[-ks^{1/2}]$

(d) $\hat{f}(s)\exp[-k(s + h)^{1/2}], h > 0$

(e) $\hat{f}(s)s^{-1}\exp[-ks^{1/2}]$

(f) $\hat{f}(s)s^{-1}\exp[-k(s + h)^{1/2}], h > 0$

(g) $\hat{f}(s)s^{-1/2}\exp[-ks^{1/2}]$

Chapter 5

Problems on Unbounded Domains

In Chapter 3 we solved boundary-value and initial-boundary-value problems in which the spatial domain was bounded in at least one direction, and in many of the examples it was bounded in all directions. Problems of this sort very often can be solved by the method of eigenfunction expansion, and the examples in Chapter 3 were intended to show how this method is applied. In this chapter we consider problems in which at least one of the independent variables in the problem ranges over an unbounded interval. Then one or the other (or in some cases, both) of the integral transforms we discussed in Chapter 4 can be brought to bear in order to solve the problem.

In the first section of this chapter we consider elementary examples of the Laplace, heat, and wave equations for which a spatial variable ranges over the whole real line $(-\infty, \infty)$ and the Fourier transform is the appropriate tool. In the second section, the spatial domain in each of the examples will be the half-line $(0, \infty)$. As we shall see, this allows the solution to be affected by influences from the spatial boundary [when the domain is $(-\infty, \infty)$, no influences enter through the spatial boundary]. We investigate these cases and show how *either* the Fourier or the Laplace transform might be applied to solve the problem. Finally, in Section 5.3 we consider some problems involving inhomogeneous equations.

5.1 ELEMENTARY EXAMPLES ON $(-\infty, \infty)$

We begin, as we did in Chapter 3, by considering some simple boundary-value problems for the Laplace equation.

EXAMPLE 5.1.1 _____

Boundary-Value Problems for Laplace Equation

Let H denote the *half-space* consisting of the points (x, y) such that $-\infty < x < \infty$, $y > 0$, and consider the problem of finding a function $u(x, y)$ satisfying

$$u_{xx}(x, y) + u_{yy}(x, y) = 0 \quad \text{in } H$$
$$u(x, 0) = f(x), \qquad -\infty < x < \infty$$
$$u(x, y) \text{ bounded for } (x, y) \text{ in } H$$

In Section 1.3 we showed how a problem of this sort could arise in the analysis of steady-state heat conduction. This (Dirichlet) problem could also arise in connection with the following physical situation. Imagine a field in which water of variable depth stands on top of the soil. This could be the result of water confined between impervious walls. We suppose that the soil surface is approximately planar and that the soil itself is completely homogeneous and fully saturated. We arrange our coordinates so that the positive y axis is directed *into* the soil and the x and z axes are parallel to the plane surface of the field. In general, u, the hydraulic head in the soil, will be a function of x, y, and z. However, if the soil is completely homogeneous and if the water is distributed on the surface in such a way that the depth of water varies with x but not with z, then $u = u(x, y)$ will vary only with x and y. In this case, we are entitled to suppress the z variable in the problem and to view this as a two-dimensional problem with x and y the independent variables.

In a saturated situation, the continuity equation reduces to the statement that the divergence of the fluid flux vanishes, and when we add to this the constitutive relation that the flux is proportional to the gradient of the hydraulic head, then we are led to the conclusion that $u = u(x, y)$ satisfies Laplace's equation in H. The boundary condition $u(x, 0) = f(x)$ is then the statement that the depth of the ponded water on the surface of the field varies with x according to the function $f(x)$.

The assumption that the field extends to $\pm\infty$ in the x direction (and tacitly the z direction as well) amounts to deciding to neglect the effects of the boundaries of the field. Consequently we could expect the solution to the problem to be most accurate in that part of the field away from the boundaries and to be of diminishing accuracy as we evaluate it at points nearer the boundary.

Solution by Integral Transform   Proceeding now with the solution, we note that if we were to suppose that $u(x, y) = X(x)Y(y)$, then in the usual way, we would come to the conclusion that $X(x)$ must satisfy

$$-X''(x) = \mu X(x), \qquad -\infty < x < \infty$$
$$X(x) \text{ and } X'(x) \text{ bounded for all } x$$

This is exactly the singular Sturm–Liouville problem (4.1.7) in Section 4.1. This problem was the one we used to motivate the development of the Fourier transform, and for this reason, we try to solve this problem by means of the Fourier transform. Later in the chapter when we have accumulated several examples, we isolate those characteristics of a problem that indicate when the Fourier transform ought to be used.

Let $U(\alpha, y)$ denote the Fourier transform (with respect to the variable x) of the solution $u(x, y)$,

$$U(\alpha, y) = \frac{1}{2\pi} \int_{-\infty}^{\infty} u(x, y)\exp[-i\alpha x] \, dx$$

Then it follows that

$$\mathcal{F}[u_{xx}(x, y)] = -\alpha^2 U(\alpha, y)$$

and

$$\mathcal{F}[u_{yy}(x, y)] = \frac{d^2}{dy^2} U(\alpha, y)$$

That is, since we applied the Fourier transform in the x variable, the derivative $u_{xx}(x, y)$ interacts with the transform according to Proposition 4.2.2. On the other hand, the variable y simply plays the role of a parameter as far as the transform with respect to x is concerned, and hence the derivative $u_{yy}(x, y)$ commutes with the Fourier transform in x. In view of these remarks, the Laplace equation in the unknown function $u(x, y)$ is transformed to the following equation in the unknown transform $U(\alpha, y)$:

$$-\alpha^2 U(\alpha, y) + U''(\alpha, y) = 0 \quad \text{for } y > 0$$
$$U(\alpha, 0) = F(\alpha)$$
$$U(\alpha, y) \text{ bounded for } y > 0$$

Here $F(\alpha)$ denotes the transform of the boundary function $f(x)$.

The general solution to this linear ordinary differential equation in the variable y can be written in the form

$$U(\alpha, y) = A \exp[-|\alpha|y] + B \exp[|\alpha|y], \quad y > 0$$

where we have expressed the solution in terms of the absolute value of α in order to make it easier to distinguish the bounded from the unbounded component of the solution (recall that α is real and can assume *both* positive *and* negative values). The component $\exp[|\alpha|y]$ grows without bound as $y \to +\infty$, and hence the condition that $U(\alpha, y)$ remain bounded implies that $B = 0$. The condition at $y = 0$ implies that $A = F(\alpha)$, and thus the solution to the transformed problem is

$$U(\alpha, y) = F(\alpha)\exp[-|\alpha|y] \tag{5.1.1}$$

It only remains to invert the transform to complete the solution. The convolution formula states that

$$u(x, y) = (1/2\pi)f * \mathcal{F}^{-1}[\exp[-|\alpha|y]]$$

and then transform pair 5 in the Fourier transform table (see endpapers) provides the final step, leading to the solution

$$u(x, y) = \frac{1}{\pi} \int_{-\infty}^{\infty} \frac{y}{[(x-z)^2 + y^2]} f(z) \, dz \tag{5.1.2}$$

We can also write this as

$$u(x, y) = \frac{1}{\pi} \int_{-\infty}^{\infty} \frac{y}{[z^2 + y^2]} f(x-z) \, dz$$

and then the change of variable

$$z = y \tan \theta, \quad \text{i.e.,} \quad \theta = \tan^{-1}[z/y]$$

reduces this to

$$u(x, y) = \frac{1}{\pi} \int_{-\pi/2}^{\pi/2} f(x - y \tan \theta) \, d\theta \qquad (5.1.3)$$

From this last representation it is clear that if $f(x)$ is continuous, then $u(x, y)$ tends to $f(x)$ as y decreases to zero so that the boundary condition is satisfied. That $u(x, y)$ satisfies the equation can be verified directly by differentiating either (5.1.2) or (5.1.3) under the integral sign.   ■ ■

EXAMPLE 5.1.1 (Part 2) _____

Neumann Problem

We can also solve the Neumann boundary-value problem for Laplace's equation on the half-space H; that is, consider the following problem for the unknown function $v(x, y)$;

$$v_{xx}(x, y) + v_{yy}(x, y) = 0 \quad \text{in } H$$
$$v_y(x, 0) = g(x), \qquad -\infty < x < \infty$$
$$v(x, y) \text{ bounded in } H$$

To solve this problem, we note that if we differentiate the equation for $v(x, y)$ with respect to y, we have

$$[v_{xx}(x, y)]_y + [v_{yy}(x, y)]_y = 0$$
$$[v_y(x, y)]_{xx} + [v_y(x, y)]_{yy} = 0$$

Evidently, the function $w(x, y) = v_y(x, y)$ satisfies

$$w_{xx}(x, y) + w_{yy}(x, y) = 0 \quad \text{in } H$$
$$w(x, 0) = g(x), \qquad -\infty < x < \infty$$
$$w(x, y) \text{ bounded in } H$$

This is a special case of the following more general principle: *If the function u satisfies a linear differential equation with constant coefficients, then every derivative of u satisfies (formally) the same equation.* Each derivative of u satisfies the differential equation only formally until it is shown that the derivative in fact exists and is sufficiently smooth to be able to satisfy the equation. However, these are technical points that we avoid for the present.

In this case, the problem for $w(x, y)$ is one we have already solved. The solution is most conveniently expressed in the form

$$w(x, y) = \frac{1}{\pi} \int_{-\infty}^{\infty} \frac{y}{y^2 + z^2} g(x - z) \, dx$$

that is,

$$v_y(x, y) = \frac{1}{2\pi} \int_{-\infty}^{\infty} g(x - z) \frac{2y}{y^2 + z^2} \, dz$$

Integrating this with respect to y leads to

$$v(x, y) = \frac{1}{2\pi} \int_{-\infty}^{\infty} g(x - z) \log[y^2 + z^2] \, dz + C(x)$$

Here $C(x)$ denotes a function of x that must satisfy

$$C''(x) = 0 \quad \text{and} \quad C(x) \text{ bounded for all } x$$

Then $C(x) = $ const, and the solution for $v(x, y)$ is

$$v(x, y) = \frac{1}{2\pi} \int_{-\infty}^{\infty} g(x - z) \log[y^2 + z^2] \, dz + C \qquad (5.1.4)$$

The constant C is indeterminate, as is usual in Neumann-type boundary-value problems; given any solution $v(x, y)$, *any constant* can be added to $v(x, y)$ without altering the fact that $v(x, y)$ satisfies all the conditions of the problem.

    Of course we could also have solved this Neumann problem by directly applying the Fourier transform as we did to solve the Dirichlet problem before it. The only difference between the two problems would have been in the solving of the transformed problem, which would have had a derivative boundary condition at $y = 0$. The execution of this approach is left to the exercises.

Discussion of Solution   It can be seen from (5.1.2) and (5.1.4) that the solutions to these two boundary-value problems each have the "organic behavior" illustrated by the solution to Example 3.1.5. That is, at every point (x, y) the solution value depends on *all* of the data values. Thus if we were to increase the data functions $f(x)$ or $g(x)$ on *any* interval of positive length, no matter how small, the solution value $u(x, y)$ or $v(x, y)$ would be correspondingly changed at *every* point in (x, y) in H. As we pointed out in Chapter 3, this behavior is consistent with the fact that Laplace's equation and elliptic partial differential equations in general very often are describing physical systems that are in a state of equilibrium. Then increasing the state of the solution on the boundary requires a complete rearrangement of solution values in the interior of the region in order to bring the system into a new equilibrium state.   ■ ■

Physical Interpretations of Laplace Equation

In addition to the physical example of saturated flow in a porous medium, the following are physical situations that are governed by Laplace's equation or its inhomogeneous version, the Poisson equation.

Steady-State Conduction of Heat   Here u denotes the temperature as a function of position in a heat-conducting solid after all the transient temperature behavior has died out. Then the distribution of temperature in the interior of the region is completely determined by the temperature distribution on the boundary of the region (Dirichlet problem) or it is determined up to an additive constant by the prescription of the heat flux on the boundary (Neumann problem). Other boundary-value problems, such as the Robin problem, arise when modeling heat loss at the boundary through the mechanism of convection according to Newton's law of cooling.

Steady-State Diffusion   In this case u denotes the concentration of the diffusing substance as a function of position. The interpretation of the boundary conditions is very similar to the heat conduction interpretation.

Static Electric Field   Maxwell's equations imply that the electric potential function u for a static electric field in a vacuum satisfies Poisson's equation where the inhomogeneous term on the right side of the equation is proportional to the volume density of charges in the region. In this application, the Dirichlet problem amounts to specifying the electric potential on the boundary of the region while the Neumann problem corresponds to specifying the charge density on the boundary; that is, the charge density on the boundary is proportional to the normal derivative of u on the boundary. At an insulated surface this normal derivative vanishes. Then the electric force field is obtained by computing the gradient of u.

Potential Flow of an Incompressible Fluid   For a steady flow of an incompressible fluid, the continuity equation reduces to the condition that the divergence of the velocity field vanishes. Then the assumption that the flow is irrotational implies that the velocity field is the gradient of a potential function u (which is a function of position only). Combining these equations shows that u satisfies Laplace's equation in the region occupied by the fluid. At a solid, stationary boundary the normal derivative of the potential function must vanish. In an unbounded region, the potential u must also satisfy some "condition at infinity."

We can summarize these last remarks by saying that in each of the preceding applications, the governing equation is the equation

$$\mathbf{div}[\sigma \ \mathbf{grad} \ u] = 0$$

where the interpretation of the equation, of the unknown u, of the parameter σ, and of the quantity $\sigma \ \mathbf{grad} \ u$, is as follows for the various applications (note that after each interpretation for the parameter σ we have listed in parentheses the letter most commonly used for σ in that application):

Heat conduction:

$$\text{Equation} = \text{conservation of thermal energy}$$
$$u = \text{temperature}$$
$$\sigma = \text{thermal conductivity } (K)$$
$$-\sigma \ \mathbf{grad} \ u = \text{heat flux}$$

Diffusion:

$$\text{Equation} = \text{conservation of mass}$$
$$u = \text{concentration of diffusing material}$$
$$\sigma = \text{diffusivity } (D)$$
$$-\sigma \ \mathbf{grad} \ u = \text{mass flux}$$

Electrostatics:

$$\text{Equation} = \text{Gauss's law for electric fields}$$
$$u = \text{electric potential}$$
$$\sigma = \text{dielectric constant } (\epsilon)$$
$$-\sigma \, \mathbf{grad} \, u = \sigma \mathbf{E}, \mathbf{D} \text{ (electric induction field)}$$

Potential flow:

$$\text{Equation} = \text{conservation of mass}$$
$$u = \text{velocity potential}$$
$$\sigma = 1$$
$$-\sigma \, \mathbf{grad} \, u = \text{velocity field}$$

Exercises: Laplace Equation on Unbounded Sets

1. Solve the Neumann problem

$$v_{xx}(x, y) + v_{yy}(x, y) = 0 \quad \text{in } H$$
$$v_y(x, 0) = g(x), \qquad -\infty < x < \infty$$
$$v(x, y) \text{ bounded in } H$$

by direct application of the Fourier transform. What is the solution in the case that $g(x) = I_1(x)$?

2. Solve the Dirichlet problem of Example 5.1.1 in the special case that $f(x) = hI_1(x)$ for h a constant. Then $u(x, y)$ can be interpreted as the hydraulic head in a saturated field underneath an irrigation canal of rectangular cross section having depth h. Show that in this case the integral (5.1.2) can be explicitly evaluated to give

$$u(x, y) = (h/\pi)[\arctan[(1 - x)/y] + \arctan[(1 + x)/y]]$$

The fluid flux at the soil–water interface at the bottom of the canal is proportional to $u_y(x, 0)$. Find $u_y(x, 0)$. Does $u_y(x, 0)$ change sign at any point as x varies from $-\infty$ to $+\infty$? What would be the physical meaning of such a sign change?

3. How is the solution to Exercise 2 altered if $f(x) = hI_1(x)$ is replaced by

$$f(x) = \begin{cases} -1 & \text{if } -1 < x < 0 \\ 1 & \text{if } 0 < x < 1 \\ 0 & \text{otherwise} \end{cases}$$

4. Verify that the function $u(x, y) = xy$ satisfies

$$\nabla^2 u(x, y) = 0 \quad \text{for } -\infty < x < \infty, \quad y > 0$$
$$u(x, 0) = 0 \quad \text{for } -\infty < x < \infty$$

Explain why we cannot simply add this function to the solution to the problem of Example 5.1.1 in order to get a second solution for that problem.

5. Find the bounded solution of

$$2u_{xx}(x, y) + 3u_{yy}(x, y) = 0, \qquad -\infty < x < \infty, \qquad y > 0$$
$$u(x, 0) = \begin{cases} 1 & \text{for } x > 0 \\ 0 & \text{for } x < 0 \end{cases}$$

Describe the behavior of $u(x, y)$ and of $u_x(x, y)$ as (x, y) approaches the origin along lines $y = kx$ for k equal to various constant values.

6. Find the steady-state distribution of temperature in the half-space $H = \{y > 0\}$ if the boundary temperature is

(a) $u(x, 0) = \begin{cases} 1 & \text{if } -1 < x < 0 \\ 1 - x & \text{if } 0 < x < 1 \\ 0 & \text{otherwise} \end{cases}$

(b) $u(x, 0) = 1$, $0 < x < 10$ and $= 0$ otherwise
(c) $u(x, 0) = I_A(x)$
Does the solution for (c) tend to the solution for (b) as $A \to \infty$?

7. Consider the problem in the infinite strip,

$$\nabla^2 u(x, y) = 0, \qquad -\infty < x < \infty, \qquad 0 < y < 1$$
$$u(x, 0) = f(x), \qquad u(x, 1) = g(x), \qquad -\infty < x < \infty$$
$$u(x, y) \text{ bounded for } |x| \to \infty$$

(a) Show that the solution to the transformed problem is given by

$$U(\alpha, y) = G(\alpha) \frac{\sinh \alpha y}{\sinh \alpha} + F(\alpha) \frac{\sinh \alpha(1 - y)}{\sinh \alpha}$$

where $F(\alpha)$ and $G(\alpha)$ denote the Fourier transforms of the data functions.
(b) Show that for $0 < z < 1$ and all real α,

$$\frac{\sinh \alpha z}{\sinh \alpha} = \frac{\sinh z|\alpha|}{\sinh |\alpha|}$$

$$= \sum_{n=0}^{\infty} \exp[-(2n + 1) - z]|\alpha| - \sum_{n=0}^{\infty} \exp[-(2n + 1) + z]|\alpha|$$

(c) Use the result of (b) in (a) in order to carry out the inversion of the Fourier transform to obtain the solution $u(x, y)$.

8. Solve Exercise 7 in the special cases:
(a) $f(x) = I_2(x)$ $g(x) = 0$
(b) $f(x) = 0$ $g(x) = I_2(x)$

Note: To solve (b) just modify the solution for (a). Do not solve a second problem.

9. Find the bounded function $v(x, y)$ that satisfies

$$\nabla^2 v(x, y) = 0, \qquad -\infty < x < \infty, \qquad 0 < y < 1$$
$$v_y(x, 0) = f(x), \qquad v_y(x, 1) = g(x), \qquad -\infty < x < \infty$$

[*Hint:* Let $w(x, y) = v_y(x, y)$ and use the results of Exercise 7 to find $w(x, y)$. Then find a y antiderivative of $w(x, y)$ to get $v(x, y)$.]

10. Find the bounded function $u(x, y)$ such that

$$\nabla^2 u(x, y) = 0, \qquad -\infty < x < \infty, \qquad 0 < y < 1$$
$$u(x, 0) = f(x), \qquad u_y(x, 1) = 0, \qquad -\infty < x < \infty$$

How is the solution changed if the boundary conditions are changed to

$$u_y(x, 0) = 0, \qquad u(x, 1) = f(x), \qquad -\infty < x < \infty$$

EXAMPLE 5.1.2 ───

Initial-Value Problems for Heat Equation
        Consider the following problem for the unknown function $u = u(x, t)$:

$$u_t(x, t) = Du_{xx}(x, t), \qquad -\infty < x < \infty, \qquad t > 0$$
$$u(x, 0) = f(x), \qquad -\infty < x < \infty$$
$$u(x, t) \text{ bounded for all } x \quad \text{and} \quad t > 0$$

We could think of this as representing one-dimensional heat conduction in an infinitely long rod in which the initial distribution of temperature is given by the known function $f(x)$. Of course infinitely long rods, heat conducting or otherwise, do not exist in the real world. However, if we are interested in studying one-dimensional conduction without having to consider the effect on the solution of whatever conditions apply at the ends of the rod, then we assume the rod is infinitely long. The solution is only valid then on an interval that is some positive distance from the ends of the rod and for only a finite period of time. The length of time the solution is valid depends on the distance from the interval of interest to the ends of the rod and the value of D.

Solution by Integral Transform   We solve this problem by means of the Fourier transform in the variable x. That is, we let

$$U(\alpha, t) = \frac{1}{2\pi} \int_{-\infty}^{\infty} u(x, t) \exp[-i\alpha x] \, dx$$

Then, as in the previous example,

$$\mathscr{F}[u_{xx}(x, t)] = -\alpha^2 U(\alpha, t)$$
$$\mathscr{F}[u_t(x, t)] = \frac{d}{dt} U(\alpha, t)$$

so that the original initial-value problem is transformed to

$$U'(\alpha, t) = -D\alpha^2 U(\alpha, t), \qquad t > 0$$
$$U(\alpha, 0) = F(\alpha)$$

Here the prime denotes differentiation with respect to t and α is only playing the role of a parameter. The solution of this ordinary differential equation is easily found to be

$$U(\alpha, t) = C \exp[-D\alpha^2 t]$$

and the initial condition implies then

$$U(\alpha, t) = F(\alpha) \exp[-D\alpha^2 t]$$

Entry 4 of the table of Fourier transform pairs gives

$$\mathscr{F}^{-1}[\exp(-D\alpha^2 t)] = [\pi/(Dt)]^{1/2} \exp[-x^2/(4Dt)]$$

and then the convolution formula leads to

$$u(x, t) = (4\pi Dt)^{-1/2} \int_{-\infty}^{\infty} \exp\left[-\frac{(x - z)^2}{4Dt}\right] f(z) \, dz \qquad (5.1.5)$$

If we introduce the change of variable,

$$\sigma = (x - z)/(4Dt)^{1/2}$$

in the integral, then (5.1.5) reduces to

$$u(x, t) = \frac{1}{\pi^{1/2}} \int_{-\infty}^{\infty} \exp[-\sigma^2] f(x - \sigma(4Dt)^{1/2}) \, d\sigma \qquad (5.1.6)$$

From representation (5.1.6), it is easy to check that if $f(x)$ is continuous, then $u(x, t) \rightarrow f(x)$ as $t \rightarrow 0$ for each x.

Note that we have not used the "condition at infinity" that requires $u(x, t)$ to be bounded for all x and $t > 0$. Yet we show in Chapter 6 that this condition is needed if the problem is to have a unique solution. In fact, the heat equation can be shown to have a solution that vanishes for $t = 0$ and grows without bound as $|x| \rightarrow \infty$ for $t > 0$. Requiring that the solution to our initial-value problem remain bounded for all x and $t > 0$ *rules out* this unbounded solution and causes the solution to the initial-value problem to be unique. This unbounded solution is, of course, not in $L^2(-\infty, \infty)$, but in addition, it is not even of exponential growth so that it does not fall into *any* of the settings for the Fourier transform. In other words, by selecting the Fourier transform as our solution method, we have tacitly imposed an "auxiliary condition" that rules out the unbounded component of the solution to our initial-value problem.

EXAMPLE 5.1.2 (Part 2) ──

Conduction with Convection

Consider next the following variation on the original initial-value problem:

$$v_t(x, t) = Dv_{xx}(x, t) + Vv_x(x, t), \qquad -\infty < x < \infty, \qquad t > 0$$
$$v(x, 0) = f(x), \qquad -\infty < x < \infty$$
$$v(x, t) \text{ bounded for all } x \quad \text{and} \quad t > 0$$

Here V denotes a constant, and as we shall see, this partial differential equation describes the conduction of heat in a medium that is moving with velocity V parallel to the x axis.

If we apply the Fourier transform in the same way as in the previous problem, we obtain the following ordinary differential equation for the unknown transform $V(\alpha, t)$:

$$V'(\alpha, t) = (-D\alpha^2 + i\alpha V)V(\alpha, t), \qquad t > 0$$
$$V(\alpha, 0) = F(\alpha)$$

Then, solving, we get

$$V(\alpha, t) = F(\alpha)\exp[-\alpha^2 Dt]\exp[i\alpha Vt]$$

This is the same as the solution for $U(\alpha, t)$ in the previous problem except for the "shift factor" $\exp[i\alpha Vt]$. Then entry 5 of the table of Fourier transform operational properties (also in the endpapers) together with (5.1.5) lead to the solution

$$v(x, t) = (4\pi Dt)^{-1/2} \int_{-\infty}^{\infty} \exp\left[-\frac{(x + Vt - z)^2}{4Dt}\right] f(z) \, dz \qquad (5.1.7)$$

Or, using (5.1.6) instead of (5.1.5),

$$v(x, t) = \frac{1}{\pi^{1/2}} \int_{-\infty}^{\infty} \exp[-\sigma^2] f(x + Vt - \sigma(4Dt)^{1/2}) \, d\sigma \qquad (5.1.8)$$

Evidently,

$$v(x, t) = u(x + Vt, t) \qquad (5.1.9)$$

where $u(x, t)$ is the solution of the original "pure conduction" problem. Before discussing the interpretation of this result further, let us consider one more variation on the original initial-value problem. ■ ■

EXAMPLE 5.1.2 (Part 3) ————————————————————————————————

Conduction with Dissipation
Consider

$$w_t(x, t) = Dw_{xx}(x, t) + Sw(x, t), \qquad -\infty < x < \infty, \qquad t > 0$$
$$w(x, 0) = f(x), \qquad -\infty < x < \infty$$
$$w(x, t) \text{ bounded for all } x \quad \text{and} \quad t > 0$$

where S denotes a constant. We shall see that this term acts as a source or a sink in the equation (depending on whether S is positive or negative) and causes the solution to increase or decrease exponentially with increasing t.

Proceeding as we did before, we come to the following initial-value problem for an ordinary differential equation in the unknown transform $W(\alpha, t) = \mathcal{F}_x[w(x, t)]$:

$$W'(\alpha, t) = (-D\alpha^2 + S)W(\alpha, t), \qquad t > 0$$
$$W(\alpha, 0) = F(\alpha)$$

As in the previous two cases, the prime here denotes differentiation with respect to t, and α plays the role of a parameter in the equation. In this case the solution for $W(\alpha, t)$ is

$$W(\alpha, t) = F(\alpha)\exp[-\alpha^2 Dt]\exp[St]$$

The factor $\exp[St]$ does not interact with the Fourier transform or its inverse; it behaves as a constant multiplier, and hence we obtain, for $w(x, t)$,

$$w(x, t) = \exp[St]u(x, t) \qquad (5.1.10)$$

where $u(x, t)$ is given by (5.1.5) or (5.1.6).

Discussion of Solution: Diffusionlike Evolution We now examine some of the properties of the solutions of these three initial-value problems. Each is an example of a particular type of evolution process that we refer to as "diffusionlike" evolution; the first example illustrates what we call "pure diffusion," and the second and third examples illustrate the effect of lower order terms on the pure diffusion process. Although the processes of heat conduction and gaseous diffusion are physically different, they are both modeled by the equation that is referred to frequently as the "heat equation" and less frequently as the "conduction–diffusion equation." When discussing these equations, we sometimes use the terminology of heat conduction; other times we use the language associated with the diffusion interpretation. Although the mathematics is the same in either case, sometimes one interpretation leads more easily through physical intuition to mathematical insight than the other.

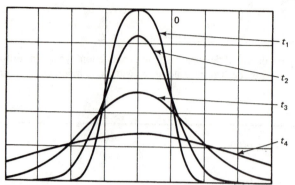

Figure 5.1.1

To see what is meant by diffusionlike evolution, consider Example 5.1.2 (part 1) in the case that the initial state is given by

$$f(x) = I_a(x)$$

where a denotes a small, positive constant. Then (5.1.6) reduces to

$$u(x, t) = (\pi)^{-1/2} \int_{(x-a)/(4Dt)^{1/2}}^{(x+a)/(4Dt)^{1/2}} \exp[-z^2] \, dz \qquad (5.1.11)$$

Now, $\exp[-z^2]$ is positive for all z, and so it follows from (5.1.11) that for each $t > 0$, $u(x, t) > 0$ *for all x*. This means that even though the initial state is zero everywhere except in the arbitrarily small interval $(-a, a)$, subsequent states $u(x, t)$ for each t positive (no matter how small) are positive *everywhere*. It is as if the influence of the initial state propagates with infinite speed, and we often say that solutions of the conduction–diffusion equation have the property of "infinite speed of propagation." A more meaningful way of describing this phenomenon would be to refer to it as "infinite smoothing of the data." The motivation for this terminology comes from the fact that an initial state that is only piecewise smooth and whose graph has sharp "corners" will be immediately smoothed out so that $u(x, t)$ is, for each $t > 0$ (no matter how small), an infinitely differentiable function of both x and t. This phenomenon is illustrated in Figure 5.1.1 showing $u(x, 0)$ and $u(x, t_k)$ for $0 < t_1 < t_2 < t_3 < t_4$ plotted on the same axis. Evidently $u(x, t_1)$ is a smoothed-out version of the discontinuous initial state $u(x, 0)$. It is clear from (5.1.11) that for any $t > 0$ $u(x, t)$ is infinitely differentiable with respect to both x and t. It also follows from (5.1.11) and the mean-value theorem for integrals that for fixed x

$$0 \le u(x, t) \le a/(\pi t)^{1/2} \exp[-\sigma^2/4t]$$

for some σ, $x - a < \sigma < x + a$. Therefore, even though the influence of the initial state propagates with infinite speed, the strength of this influence dies out as $\exp[-\sigma^2/4t]$.

A more realistic interpretation for the speed of propagation associated with the conduction–diffusion equation comes from considering the following initial state:

$$f(x) = \begin{cases} 1 & \text{for } x > 0 \\ 0 & \text{for } x < 0 \end{cases} \qquad (5.1.12)$$

The solution corresponding to this initial state is then

$$u(x, t) = \frac{1}{\pi^{1/2}} \int_{-\infty}^{x/(4Dt)^{1/2}} \exp[-z^2] \, dz$$

and since

$$\int_{-\infty}^{\infty} \exp[-z^2] \, dz = \pi^{1/2}$$

it is clear that $u(x, t)$ assumes for each x and $t > 0$ a value between 0 and 1. Moreover, along parabolic curves of the form

$$x = M(4Dt)^{1/2}, \qquad M = \text{const} > 0$$

$u(x, t)$ is constant. In particular, for a fixed value of the constant M, let u_M denote the corresponding constant value of u along the parabolic curve. Then for each $t = t^* > 0$, we have $u = u_M$ at just one point $x = X(t)$, and this point moves with the speed

$$X'(t) = MD^{1/2}t^{-1/2} \quad \text{for } t > 0$$

For each positive value of t this speed is finite and is a more meaningful "speed of propagation" to associate with the conduction or diffusion process modeled by this partial differential equation.

For the example in part 2 we found the solution to be

$$v(x, t) = u(x + Vt, t)$$

where $u(x, t)$ denotes the solution just discussed. In particular, the solution $v(x, t)$ corresponding to the initial state (5.1.12) must then be constant along curves of the form

$$x = -Vt + M(4Dt)^{1/2}$$

It follows that the speed of propagation associated with this solution is

$$X'(t) = -V + MD^{1/2}t^{-1/2}$$

This differs from the previously found diffusion speed by the constant $-V$. Evidently the term $Vv_x(x, t)$ appearing in the partial differential equation for $v(x, t)$ has the effect of superimposing on the diffusion solution a constant-speed convection. We refer to the partial differential equation satisfied by $v(x, t)$ as a convection–conduction or convection–diffusion equation.

Looking now at part 3 of the example, we see from (5.1.10) that the effect of the term $Sw(x, t)$ in the partial differential equation for $w(x, t)$ is to cause the pure diffusion solution to be multiplied by the "modifier" $\exp[St]$. When S is negative, this multiplier is a moderator (for $t > 0$, it makes the solution smaller), and we say that the equation models diffusion with "dissipation." That is, in the parlance of heat conduction, the term $Sw(x, t)$ represents a heat sink that dissipates thermal energy at a rate proportional to the present value of the temperature. ■ ■

The three examples considered here have in common the feature that the solution of the initial-value problem is, for each $t > 0$, infinitely smooth, and this smoothness is independent of the smoothness of the initial state $f(x)$. The solutions are generally expressed in terms of an integral that is improper but can be shown to be convergent. These integrals depend parametrically on the variables x and t and are not only convergent but differentiable infinitely often with respect to both x and t. This smoothing action in which an initial state that need not be particularly smooth is carried immediately into a state that is infinitely smooth is one of the characteristic features of diffusionlike evolution.

Exercises: Initial-Value Problems for Heat Equation

1. Find the bounded solution of

$$u_t(x, t) = Du_{xx}(x, t), \quad -\infty < x < \infty, \quad t > 0$$

$$u(x, 0) = \begin{cases} 1 & \text{if } |x - 1| < 0.1 \\ -1 & \text{if } |x + 1| < 0.1 \\ 0 & \text{otherwise} \end{cases}$$

Express the solution in terms of the error function.

2. Solve Exercise 2 in the case that the equation is replaced by

$$u_t(x, t) = Du_{xx}(x, t) + Vu_x(x, t), \quad -\infty < x < \infty, \quad t > 0$$

3. Solve Exercise 1 in the case that the equation is replaced by

$$u_t(x, t) = Du_{xx}(x, t) + Bu(x, t), \quad -\infty < x < \infty, \quad t > 0$$

4. Suppose $u(x, t)$ is a bounded function satisfying

$$u_t(x, t) = Du_{xx}(x, t), \quad -\infty < x < \infty, \quad t > 0$$
$$u(x, 0) = f(x), \quad -\infty < x < \infty$$

where the given function $f(x)$ and its derivative $f'(x)$ both belong to $L^2(-\infty, \infty)$. Show that:
(a) $\mathcal{F}[u(x, t)] = F(\alpha)\exp[-D\alpha^2 t] = \mathcal{F}[f*K]$ where $K(x, t) = \mathcal{F}^{-1}[\exp(-D\alpha^2 t)]$
(b) $\mathcal{F}[\partial_x u(x, t)] = i\alpha F(\alpha)\exp[-D\alpha^2 t] = \mathcal{F}[f'*K]$
(c) Based on (a) and (b), show that

$$u_x(0, t) = \pi^{-1/2} \int_{-\infty}^{\infty} \exp[-z^2]f'(-z(4Dt)^{1/2}) \, dz$$

5. Deduce from the results of Exercise 4 that if $f(x)$ is an even function of x, then $u_x(0, t) = 0$ for all $t > 0$.

6. Suppose $u(x, t)$ is a bounded function satisfying

$$u_t(x, t) = Du_{xx}(x, t) + Vu_x(x, t), \quad -\infty < x < \infty, \quad t > 0$$
$$u(x, 0) = f(x), \quad -\infty < x < \infty$$

where the given function $f(x)$ and its derivative $f'(x)$ both belong to $L^2(-\infty, \infty)$. Derive a formula analogous to the one from Exercise 4(c) for $u_x(0, t)$ in this case. Is it still true that $u_x(0, t)$ is zero for all $t > 0$ if $f(x)$ is even? How would you answer if the equation were replaced by

$$u_t(x, t) = Du_{xx}(x, t) + Bu(x, t), \quad -\infty < x < \infty, \quad t > 0$$

7. For $u(x, t)$ the solution to the initial-value problem in Exercise 4, show that:
   (a) If $f(x)$ is continuous, then
   $$\lim_{t \to 0^+} u(x, t) = f(x)$$
   (b) $u_t(x, t) = \left(\dfrac{D}{\pi t}\right)^{1/2} \displaystyle\int_{-\infty}^{\infty} -z \exp[-z^2] f'(x - z(4Dt)^{1/2}]\, dz$
   Deduce from this that $\lim_{t \to 0^+} u_t(x, t) = \infty$.

8. Use the results of Exercise 7 to give conditions on $f(x)$ sufficient to imply that
   (a) $u_t(0, t) = 0$
   (b) $u_t(0, t) > 0$
   (c) $u_t(0, t) < 0$

EXAMPLE 5.1.3

Initial-Value Problems with Wavelike Solutions
   We look now at some examples of initial-value problems for evolution equations
in which the solutions exhibit behavior that we describe as being typical of "wavelike"
evolution.
   Consider the first-order initial-value problem for the unknown function $u(x, t)$
where

$$u_t(x, t) = Cu_x(x, t), \qquad -\infty < x < \infty, \qquad t > 0$$
$$u(x, 0) = f(x), \qquad -\infty < x < \infty$$

where C denotes a constant.
   We should point out that this example and the next few to follow (as well as the
problems in Example 5.1.2) could be solved by applying the Laplace transform in the t
variable. However, when the initial condition is inhomogeneous, this will lead to an
inhomogeneous ordinary differential equation in which the unknown is the Laplace transform of the function $u(x, t)$. In general, this is less desirable than the situation in which
the transformed differential equation is homogeneous with an inhomogeneous side condition. For that reason we have decided to use the Fourier transform. Later we present
examples in which the Laplace transform is the technique to be preferred. We can also
treat this problem by the *method of characteristics*. This method is applied to this example
in Section 7.1.

Solution by Integral Transform   If we apply the Fourier transform to our first-order
initial-value problem, then we are led to the following problem in the unknown transform
$U(\alpha, t) = \mathcal{F}_x[u(x, t)]$:

$$U'(\alpha, t) = i\alpha C U(\alpha, t), \qquad t > 0$$
$$U(\alpha, 0) = F(\alpha)$$

Here the prime denotes differentiation with respect to the independent variable t; the
transform variables α plays the role of a parameter in this equation.
   This problem is easily solved. The solution is

$$U(\alpha, t) = F(\alpha)\exp[i\alpha Ct] \tag{5.1.13}$$

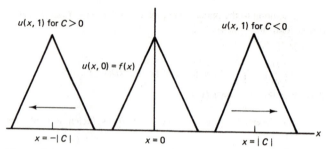

Figure 5.1.2

and then, according to entry 5 in the table of Fourier transform operational properties, we have

$$u(x, t) = f(x + Ct) \qquad -\infty < x < \infty, \qquad t > 0 \qquad (5.1.14)$$

It is evident from Figure 5.1.2 that the solution consists of the waveform $f(x)$ moving to the left or right according to whether the constant C is positive or negative, respectively; that is, at $t = 0$ the peak of the initial state $f(x)$ is located at $x = 0$, while at $t = 1$ it is at the point $x = -C$.

A solution of the form (5.1.14) is called a *traveling wave*, and C is often referred to as the *group or phase velocity* for the traveling wave. The fact that the waveform $f(x)$ is transmitted without distortion is described by saying that this is nondispersive wave propagation.

Note also that the solution $u(x, t)$ is constant along lines

$$x + Ct = \text{const}$$

These are the characteristic curves associated with the original first-order equation, and it is evident from (5.1.14) that the initial values of $u(x, t)$ propagate along these lines. See Chapter 7 for a more complete discussion of characteristics. ■ ■

Exercises: Initial-Value Problems with Wavelike Solutions

1. Solve for $u(x, t)$ such that

$$u_t(x, t) = Cu_x(x, t) - u(x, t), \qquad -\infty < x < \infty, \qquad t > 0$$

$$u(x, 0) = \begin{cases} 1 - |x|/a & \text{for } |x| < a \\ 0 & \text{for } |x| > a \end{cases}$$

Sketch the graph of $u(x, t)$ for all x where $u(x, t)$ is different from zero for $t = 0, t = 1/C$, $t = 4/C$.

2. Consider the problem

$$u_t(x, t) = Cu_x(x, t), \qquad -\infty < x < \infty, \qquad t > 0$$

$$u(x, 0) = \begin{cases} 1 & \text{if } x > 0 \\ 0 & \text{if } x < 0 \end{cases}$$

For each $t > 0$, let $\Sigma(t) = \{(x, t) : u(x, t) > 0\}$. Describe $\Sigma(t)$ in the case $C = 1$ and in the case $C = -1$ for $t > 0$.

EXAMPLE 5.1.3 (Part 2) ———————————————————————————

Second-Order Wave Equation

Next, we consider a pure initial-value problem for the (second-order) wave equation,

$$u_{tt}(x, t) = a^2 u_{xx}(x, t), \qquad -\infty < x < \infty, \qquad t > 0$$
$$u(x, 0) = f(x), \qquad u_t(x, 0) = g(x), \qquad -\infty < x < \infty$$

Here $f(x)$ and $g(x)$ denote given functions that are everywhere defined. This initial-value problem could describe the propagation of acoustic waves in a thin infinite pipe.

Solution by Fourier Transform Applying the Fourier transform in the x variable, we obtain the transformed problem

$$U''(\alpha, t) = -\alpha^2 a^2 U(\alpha, t), \qquad t > 0$$
$$U(\alpha, 0) = F(\alpha), \qquad U'(\alpha, 0) = G(\alpha)$$

Here, the unknown is $U(\alpha, t) = \mathscr{F}_x[u(x, t)]$, the Fourier transform (in x) of the unknown function $u(x, t)$. The general solution to this problem can be written as follows:

$$U(\alpha, t) = C_1 \exp[ia\alpha t] + C_2 \exp[-ia\alpha t]$$

and then the initial conditions imply that

$$U(\alpha, t) = \tfrac{1}{2}[F(\alpha) + G(\alpha)/(ia\alpha)]\exp[ia\alpha t] \qquad (5.1.15)$$
$$+ \tfrac{1}{2}[F(\alpha) - G(\alpha)/(ia\alpha)]\exp[-ia\alpha t]$$

In order to find the inverse transform here, we note that if $h(x)$ is a function satisfying $h'(x) = g(x)$, then the Fourier transforms of $h(x)$ and $g(x)$ are related as follows:

$$H(\alpha) = G(\alpha)/(i\alpha)$$

Then in terms of $H(\alpha)$, we have

$$U(\alpha, t) = \tfrac{1}{2}[F(\alpha) + H(\alpha)/a]\exp[ia\alpha t] + \tfrac{1}{2}[F(\alpha) - H(\alpha)/a]\exp[-ia\alpha t] \quad (5.1.16)$$

Just as inverting (5.1.13) leads to (5.1.14), inverting (5.1.16) leads to

$$u(x, t) = \tfrac{1}{2}[f(x + at) + h(x + at)/a] + \tfrac{1}{2}[f(x - at) - h(x - at)/a] \quad (5.1.17)$$

If we make use of the fact that $h' = g$ implies

$$h(x) = \int_c^x g(z)\, dz \qquad c = \text{arbitrary const}$$

then (5.1.17) can be written in the form

$$u(x, t) = \tfrac{1}{2}[f(x + at) + f(x - at)] + \frac{1}{2a}\int_{x-at}^{x+at} g(z)\, dz \qquad (5.1.18)$$

The solution (5.1.18) is the *D'Alembert* form of the solution to the initial-value problem for the wave equation. When the solution is written in this form, it is easy to check that the initial conditions are satisfied provided the data $f(x)$ and $g(x)$ are continuous functions and that the wave equation is satisfied in the pointwise sense provided f, f', and f'' and g and g' are all continuous. When f and g are not this smooth, the equation is satisfied

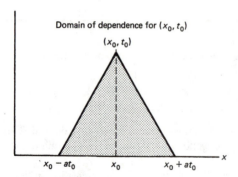

Figure 5.1.3

in some more general sense but not necessarily in the pointwise sense. We discuss this point in more detail later (see Section 7.4).

Several features of the solution to the initial-value problem for the wave equation can be discovered by examining solutions (5.1.17) and (5.1.18).

First, from (5.1.18) it is apparent that the solution to the initial-value problem for the one-dimensional wave equation is not more regular than the data functions $f(x)$ and $g(x)$. This is in contrast to what we found for the heat equation and Laplace's equation where the solutions were infinitely differentiable, independent of the smoothness of the initial and boundary data. This lack of the smoothing action of the solution operator is typical of the wave equation independent of the number of space dimensions.

Second, from (5.1.17), we can see that the solution of the wave equation consists of two traveling waves; one is moving to the left and one is moving to the right, each with velocity equal to the (finite) constant $|a|$. In addition, this wave propagation is nondispersive since the waveforms $f(x)$ and $h(x)$ are simply translated along the x axis.

Note further that at the fixed point (x_*, t_*) the solution value $u(x_*, t_*) = u_*$ depends on the values of $f(x)$ at the two points $x_* + at_*$ and $x_* - at_*$ and on the values of $g(x)$ over the entire interval $(x_* - at_*, x_* + at_*)$. The values of $f(x)$ and $g(x)$ outside of this set of points have *no effect* on the solution value at (x_*, t_*), and we say then that the interval $(x_* - at_*, x_* + at_*)$ is the *domain of dependence* for the point (x_*, t_*) (see Figure 5.1.3). Clearly, the domain of dependence depends on the point (x_*, t_*); in fact, it is just the interval on the x axis that is cut off by the two characteristics through (x_*, t_*),

$$x - at = C_1 = x_* - at_* \quad \text{and} \quad x + at = C_2 = x_* + at_*$$

Through each point in the plane there passes one characteristic of the form $x - at = C_1$ and one of the form $x + at = C_2$; choosing $C_1 = x_* - at_*$ and $C_2 = x_* + at_*$ selects precisely those two characteristics that pass through (x_*, t_*). The domain of dependence can be thought of as containing just that data that, starting at $t = 0$ and traveling with speed equal to $|a|$, has had time to reach the point $x = x_*$ by the time $t = t_*$.

It is similarly evident that if $I(0) = (x_1, x_2)$ denotes a fixed interval and if $f(x)$ and $g(x)$ are zero for x outside of $I(0)$, then for each fixed value $t = t_0 > 0$, the solution

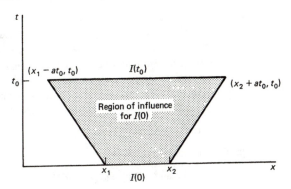

Figure 5.1.4

value $u(x, t_0)$ must also equal zero for all x outside of a corresponding interval $I(t_0) = (x_1 - at_0, x_2 + at_0)$. The trapezoidal area bounded on the bottom by $I(0)$, on the top by $I(t_0)$, and on the left and right, respectively, by the characteristics

$$x + at = x_2 \quad \text{and} \quad x - at = x_1$$

is referred to as the *domain of influence* for $I(0)$ (Figure 5.1.4). For each $t_0 > 0$, the domain of influence for $I(0)$ is the set of points (x, t), $t \le t_0$, that contain at least one point of $I(0)$ in their domain of dependence; it is the set of points (x, t), $t \le t_0$, where the solution value $u(x, t)$ is influenced by data values from $I(0)$. The fact that the size of $I(t)$ grows with constant speed equal to a is another manifestation of the finite speed of propagation associated with wavelike evolution. ∎∎

EXAMPLE 5.1.3 (Part 3) ─────────────────────────────────────

Dispersive Wave Motion
     We recall from Chapter 3, Example 3.3.3, that the effect of lower order terms in the wave equation is to introduce dispersion into the solution of the equation. Consider then the so-called Klein–Gordon equation considered in (3.3.18) of Chapter 3,

$$u_{tt}(x, t) = a^2 u_{xx}(x, t) - d^2 u(x, t)$$

Without considering initial conditions, note that if we apply the Fourier transform in the x variable to this equation, we obtain

$$U''(\alpha, t) = -(a^2\alpha^2 + d^2)U(\alpha, t)$$

as the ordinary differential equation in the unknown transform $U(\alpha, t)$. The general solution of this equation may be written in the form

$$U(\alpha, t) = C_1\exp[i\alpha Q(\alpha, d)t] + C_2\exp[-i\alpha Q(\alpha, d)t] \qquad (5.1.19)$$

where

$$Q(\alpha, d) = [a^2 + (d/\alpha)^2]^{1/2}$$

When initial conditions are prescribed, it will be possible to evaluate the arbitrary constants C_1 and C_2 in terms of the Fourier transforms of the initial data. However, when the parameter $d \neq 0$, we are unable to carry out the inversion of the exponential terms appearing in (5.1.19) without resorting to techniques that are not elementary.

Even without being able to complete the inversion, we shall see that the solution to this equation will exhibit dispersive, wavelike evolution. Compare the solution (5.1.19) and (5.1.16). In (5.1.16) there appear the exponential "shifting factors"

$$\exp[i\alpha at] \quad \text{and} \quad \exp[-i\alpha at]$$

while in (5.1.19) there appear the factors

$$\exp[i\alpha Q(\alpha, d)t] \quad \text{and} \quad \exp[-i\alpha Q(\alpha, d)t]$$

According to entry 5 in the table of Fourier transform operational properties, multiplying a Fourier transform by a factor of the form $\exp[\pm i\alpha at]$ introduces a translation by $\pm at$ in the inverse transform. Then the solution operator associated with (5.1.16) operates on waveforms determined by the initial data by propagating them along the x axis at the constant speed a.

Reasoning by analogy, it would apppear that the factors $\exp[\pm i\alpha Q(\alpha, d)t]$ lead to solutions in which the waveforms are translated with the speed of propagation equal to $Q(\alpha, d)$. When $d = 0$, we have $Q(\alpha, 0)$ equal to the constant a, but for $d \neq 0$, the factor $Q(\alpha, d)$ depends on α; the translation speed of a component of frequency α depends on α. The waveform $f(x)$, which is composed of components of every frequency according to the Fourier inversion formula,

$$f(x) = \int_{-\infty}^{\infty} F(\alpha)\exp[i\alpha x] \, d\alpha$$

will distort as it is propagated by the operators

$$\exp[i\alpha Q(\alpha, d)t] \quad \text{and} \quad \exp[-i\alpha Q(\alpha, d)t]$$

because the various frequency components do not all translate together. Distortion produced by components of different frequencies traveling at different speeds is called *dispersion*. We point out that any constant-coefficient equation of the form

$$u_{tt}(x, t) = Au_{xx}(x, t) + Bu_x(x, t) + Cu_t(x, t) + Du(x, t)$$

can be reduced to the Klein–Gordon equation by a change of dependent variable of the form

$$u(x, t) = \exp[\sigma x + \beta t]v(x, t)$$

It follows that *any* of the lower order terms appearing in the equation with the same principal part as the wave equation can introduce dispersion into the solution. ■ ■

EXAMPLE 5.1.3 (Part 4)

Dispersion Equation

As our final example of an initial-value problem with a wavelike solution, consider the problem

$$u_t(x, t) = u_{xxx}(x, t), \quad -\infty < x < \infty, \quad t > 0$$
$$u(x, 0) = f(x), \quad -\infty < x < \infty$$

This equation is not any of the types we have previously considered; it is sometimes referred to as the *dispersion equation*.

If we apply the Fourier transform in the x variable to this problem, we obtain the following initial-value problem for an ordinary differential equation in $U(\alpha, t)$, the transform of the unknown function $u(x, t)$:

$$U'(\alpha, t) = (i\alpha)^3 U(\alpha, t), \quad t > 0$$
$$U(\alpha, 0) = F(\alpha)$$

This initial-value problem is easily solved to obtain

$$U(\alpha, t) = F(\alpha)\exp[-i\alpha^3 t] \tag{5.1.20}$$

Now we compare this solution with the following previously obtained transform solutions:

$$U(\alpha, y) = F(\alpha)\exp[-y|\alpha|] \qquad \text{(Laplace equation)}$$
$$U(\alpha, t) = F(\alpha)\exp[-t\alpha^2] \qquad \text{(heat equation)}$$
$$U(\alpha, t) = C_1\exp[iat\alpha] + C_2\exp[-iat\alpha] \qquad \text{(wave equation)}$$
$$U(\alpha, t) = C_1\exp[itQ\alpha] + C_2\exp[-itQ\alpha] \qquad \text{(Klein–Gordon equation)}$$

Of these four, the solution (5.1.20) is most similar to the Klein–Gordon solution. That is, in each of these examples the solution operator for the problem may be thought of as being composed of the following pieces:

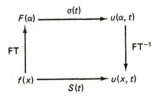

In every case, the operation $\sigma(t)$ amounts to multiplying by a function of the form $\exp[t\phi(\alpha)]$. In the case of the Laplace and heat equations, $\phi(\alpha)$ is *real* valued, while for the wave and Klein–Gordon equations it is *complex* valued (purely imaginary to be precise). This observation leads us to guess that for solution operators of this form, a real-valued $\phi(\alpha)$ is associated with the smoothing action, while a complex- or imaginary-valued $\phi(\alpha)$ may be linked to wavelike evolution (which according to our few examples involves translation without smoothing).

Looking just at the functions $\phi(\alpha)$ for the wave and Klein–Gordon equations, we see that $\phi(\alpha)$ for the wave equation is *linear* in α, whereas the $\phi(\alpha)$ for the Klein–Gordon equation is a *nonlinear* function of α. This suggests that when a complex-valued $\phi(\alpha)$ is linear in α, the solution involves nondispersive, wavelike evolution. But when a complex-valued $\phi(\alpha)$ is *not* linear in α, components of different frequencies will propagate at different speeds, and dispersive wave motion will result. See also Section 9.4 for related remarks.

Then, based on these remarks and the fact that in (5.1.20) we have $\phi(\alpha) = -i\alpha^3$, we would expect the solution of the dispersion equation to exhibit the properties of dispersive, wavelike evolution. This much is known without even inverting the solution (5.1.20). Inverting this function involves techniques we prefer to avoid in this text. However, we have the following result from Example 4.2.2(b):

$$\mathcal{F}[\text{Ai}(x)] = C \exp[-i\alpha^3/3]$$

where Ai(x) denotes the so-called Airy function of mathematical physics. Then formula 2 in the table of Fourier transform operational properties leads to

$$\mathcal{F}^{-1}[\exp[-i\alpha^3 t]] = (3t)^{-1/3} \text{Ai}[(x^3/3t)^{1/3}]$$

Using this in connection with the convolution formula, we find the following inverse for the $U(\alpha, t)$ in (5.1.20):

$$u(x, t) = \frac{1}{2\pi(3t)^{1/3}} \int_{-\infty}^{\infty} \text{Ai}\left[\frac{(x - y)}{(3t)^{1/3}}\right] f(y) \, dy$$

$$= \frac{1}{2\pi} \int_{-\infty}^{\infty} \text{Ai}(z) f(x - z(3t)^{1/3}) \, dz \qquad (5.1.21)$$

■ ■

Without more knowledge about the Airy function, solution (5.1.21) is not particularly useful. However, based on our observations about the transform of the solution, we already know some of the qualitative behavior of the solution $u(x, t)$.

Exercises: Initial-Value Problems with Wavelike Solutions

Solve for $u(x, t)$:

1. $u_{tt}(x, t) = a^2 u_{xx}(x, t)$, $-\infty < x < \infty$, $t > 0$
   $u(x, 0) = 0$, $u_t(x, 0) = \sin \pi x$, $-\infty < x < \infty$

2. $u_{tt}(x, t) = a^2 u_{xx}(x, t)$, $-\infty < x < \infty$, $t > 0$
   $u(x, 0) = \sin 3x$, $u_t(x, 0) = 0$, $-\infty < x < \infty$

3. $u_{tt}(x, t) = a^2 u_{xx}(x, t)$, $-\infty < x < \infty$, $t > 0$
   $u(x, 0) = 0$, $u_t(x, 0) = f(x)$, $-\infty < x < \infty$
   where $f(x)$ denotes the 2π-periodic function satisfying

   $$f(x) = \begin{cases} -1, & -\pi < x < 0 \\ 1, & 0 < x < \pi \end{cases}$$

4. Suppose $u(x, t)$ satisfies

   $$u_{tt}(x, t) = a^2 u_{xx}(x, t) \qquad -\infty < x < \infty, \qquad t > 0$$
   $$u(x, 0) = f(x), \qquad -\infty < x < \infty$$
   $$u_t(x, 0) = g(x), \qquad -\infty < x < \infty$$

   Show that:
   (a) If $f(0) = f(L) = 0$ and $f(x)$ is $2L$-periodic and if $g(x) = 0$ for all x, then

   $$u(0, t) = u(L, t) = 0 \quad \text{for all } t$$

   (b) If

   $$f(-x) = f(x) \quad \text{for } 0 < x < L$$
   $$f(x + 2L) = f(x) \quad \text{for all } x$$

and if $g(x) = 0$ for all x, then

$$u_x(0, t) = u_x(L, t) = 0$$

5. For $u(x, t)$ as in Exercise 4, let $f(x) = 0$ for all x. Then find conditions on $g(x)$ similar to those in (a) or (b) of Exercise 4 sufficient to imply $u(0, t) = u(L, t) = 0$ for all t.

6. Suppose $u(x, t)$ satisfies

$$u_{tt}(x, t) = 16u_{xx}(x, t), \qquad -\infty < x < \infty, \qquad t > 0$$

$$u(x, 0) = \begin{cases} 4 - x^2, & |x| < 2 \\ 0, & |x| > 2 \end{cases}$$

$$u_t(x, 0) = \begin{cases} 1, & |x| < 1 \\ 0, & |x| > 1 \end{cases}$$

Then for each $t > 0$, find the set of points x where $u(x, t) \neq 0$.

7. Consider the problem

$$u_{tt}(x, t) = a^2 u_{xx}(x, t) \qquad 0 < x < L, \qquad t > 0$$
$$u(x, 0) = f(x), \qquad 0 < x < L$$
$$u_t(x, 0) = g(x), \qquad 0 < x < L$$
$$u(0, t) = u(L, t) = 0$$

Replace this problem by an equivalent initial-value problem on the interval $(-\infty, \infty)$ by extending the data functions $f(x)$ and $g(x)$ in such a way that the boundary conditions will be automatically satisfied. *Hint:* See Exercise 4. Solve the equivalent problem.

8. Repeat Exercise 7 if the boundary conditions

$$u(0, t) = u(L, t) = 0$$

are replaced by the conditions

$$u_x(0, t) = u_x(L, t) = 0$$

9. Repeat Exercise 7 if the boundary conditions

$$u(0, t) = u(L, t) = 0$$

are replaced by the conditions

$$u(0, t) = u_x(L, t) = 0$$

EXAMPLE 5.1.4 ————————————————————————————

Although the techniques of this first section are elementary, they provide us with the tools to treat some fairly sophisticated problems. The following example illustrates this point.

Random Advection
We now consider a partial differential equation in which one of the coefficients is not known in a deterministic sense but instead is known only in a probabilistic sense. Therefore we begin this example by recalling a few notions from probability theory. Let

z denote a real-valued random variable and let $P[a < z < b]$ denote the probability that z assumes a value between the real numbers a and b. If

$$P[t < z < t + dt] = p(t)\, dt$$

for some real-valued, nonnegative function $p(t)$, then we say that $p(t)$ is the probability density function associated with the random variable z. Clearly if $p(t)$ is a probability density function, then it must satisfy

$$\int_{-\infty}^{\infty} p(t)\, dt = 1$$

Now we can define the following notions in connection with z:

$$\langle\!\langle z \rangle\!\rangle = \text{expected value or average value of } z$$

$$= \int_{-\infty}^{\infty} tp(t)\, dt$$

$$\langle\!\langle f(z) \rangle\!\rangle = \text{expected value of real-valued function } f(z)$$

$$= \int_{-\infty}^{\infty} f(t)p(t)\, dt$$

For example, for positive constants t_* and D, let

$$p(t) = (4\pi D)^{-1/2}\exp[-(t - t_*)^2/4D] \qquad (5.1.22)$$

Then

$$\int_{-\infty}^{\infty} p(t)\, dt = 1$$

and $p(t)$ is said to be the probability density function associated with a *normally distributed random variable* with mean value t_* and variance $\sigma^2 = (2D)$.

Now suppose that $u(x, t)$ satisfies

$$u_t(x, t) + Cu_x(x, t) = 0 \qquad -\infty < x < \infty, \qquad t > 0$$
$$u(x, 0) = u_0(x), \qquad -\infty < x < \infty \qquad (5.1.23)$$

where C denotes a given constant and $u_0(x)$ is a given initial function. Then as we saw in Example 5.1.3, the solution of this problem is given by

$$U(x, t) = u_0(x - Ct)$$

This solution is just the waveform $u_0(x)$, translating from left to right along the x axis with speed equal to C. Now in very many physical applications, the parameter C, which we may think of as a *speed of advection*, is not precisely known. Instead, C may be known only as a normally distributed random variable with mean value C_* and variance equal to σ^2. If C is not known precisely, then it is unrealistic to expect to be able to compute the solution $u(x, t)$ precisely. However, we can compute the expected value of $u(x, t)$. That is, we apply our formula for the expected value of a real-valued function of a random variable,

$$«u(x, t)» = \int_{-\infty}^{\infty} u_0(x - Ct)(4\pi D)^{-1/2}\exp\left[\frac{-(C - C_*)^2}{4D}\right] dC$$

$$= (4\pi Dt^2)^{-1/2} \int_{-\infty}^{\infty} u_0(r)\exp\left[\frac{-(x - C_*t - r)^2}{4Dt^2}\right] dr$$

$$= (4\pi Ds)^{-1/2} \int_{-\infty}^{\infty} u_0(r)\exp\left[\frac{-(y - r)^2}{4Ds}\right] dr \qquad (5.1.24)$$

Here we have adopted the notation

$$y = x - C_*t, \qquad s = t^2$$

If we compare the last form of (5.1.24) with Equation (5.1.5), it is evident that the function

$$v(y, s) = «u(x, t)»$$

solves the problem

$$v_s(y, s) = Dv_{yy}(y, s), \qquad -\infty < y < \infty, \qquad t > 0$$
$$v(y, 0) = u_0(y), \qquad -\infty < y < \infty \qquad (5.1.25)$$

But

$$|t| = s^{1/2} \quad \text{and} \quad x = y + C_*s^{1/2}$$

and it follows then from the chain rule that

$$\frac{\partial}{\partial s} = \tfrac{1}{2}C_*s^{-1/2}\frac{\partial}{\partial x} + \tfrac{1}{2}s^{-1/2}\frac{\partial}{\partial t} \quad \text{and} \quad \frac{\partial}{\partial y} = \frac{\partial}{\partial x}$$

Substituting these expressions in (5.1.25) then leads to the following problem for «u(x, t)»:

$$\frac{\partial}{\partial t}«u(x, t)» + C_*\frac{\partial}{\partial x}«u(x, t)» = 2D|t|\frac{\partial^2}{\partial x^2}«u(x, t)»$$
$$«u(x, 0)» = u_0(x) \qquad (5.1.26)$$

The interesting conclusion here is the following: When $u(x, t)$ satisfies an advection equation such as (5.1.23) in which the parameter C is known only as a normally distributed random variable with mean C_* and variance $2D$, then «u(x, t)», the expected value of $u(x, t)$, satisfies a convection–diffusion equation with time-dependent diffusivity. ■ ■

5.2 EXAMPLES ON SEMIBOUNDED REGIONS

Example 5.1.1 involved boundary-value problems on the half-plane $\{-\infty < x < \infty, 0 < y < \infty\}$. The half-plane is the simplest example of a set that has a boundary. In this section we begin by considering some boundary-value problems for Laplace's equation on unbounded sets with slightly more complex boundariès. This will give us a chance to see other ways in which the boundary can influence the solution to a boundary-value problem.

EXAMPLE 5.2.1

Boundary-Value Problem on Quarter-Plane
  Let Q denote the *quarter-plane* $\{x > 0, y > 0\}$ and consider the following boundary-value problem for the unknown function $u(x, y)$:

$$u_{xx}(x, y) + u_{yy}(x, y) = 0 \quad \text{in } Q$$
$$u(x, 0) = f(x), \qquad x > 0$$
$$u_x(0, y) = g(y), \qquad y > 0$$

Here $f(x)$ and $g(y)$ denote given data, and we add the *condition at infinity* that we look only for the solution that is bounded on Q.
  We could think of this physically as a problem in the steady-state conduction of heat in a thin heat-conducting plate in which two of the edges are so remote that they can be treated as being infinitely distant. Along the edge $y = 0$, the temperature is specified, and along the edge $x = 0$, the flux of heat is controlled, and from this information we are to construct the temperature distribution in the interior of the plate.
  Although the range of both variables in this problem is the half-line $(0, \infty)$, the Laplace transform is *not* the correct integral transform to apply here. The equation is second order in both variables, and therefore applying the Laplace transform in either variable would require us to know *both* the function $u(x, y)$ *and* its first derivative along one edge of Q. Since we do not have this information, it would be inconvenient to use the Laplace transform on this problem.
  This problem could be solved using so-called half-range Fourier transforms. These are variations on the Fourier transform that are suited to treating problems in which one or more of the spatial variables ranges over the half-line $(0, \infty)$. However, we would be forced to develop modified transform formulas and properties analogous to the results summarized in the Fourier transform table in the endpapers. We can avoid doing this and get by with just the results we have already developed by using some simple techniques of extension.
  We split the original problem into the following two subproblems:

Subproblem A: Find a function that is bounded on Q and satisfies

$$u_{xx}^A(x, y) + u_{yy}^A(x, y) = 0 \quad \text{in } Q$$
$$u^A(x, 0) = f(x), \qquad x > 0$$
$$u_x^A(0, y) = 0, \qquad y > 0$$

Subproblem B: Find a function that is bounded on Q and satisfies

$$u_{xx}^B(x, y) + u_{yy}^B(x, y) = 0 \quad \text{in } Q$$
$$u^B(x, 0) = 0, \qquad x > 0$$
$$u_x^B(0, y) = g(y), \qquad y > 0$$

If we can solve these two subproblems, then by superposition, the solution of the original problem is given by

$$u(x, y) = u^A(x, y) + u^B(x, y)$$

In order to solve subproblem A, let $f_E(x)$ denote the even extension of $f(x)$ to the whole real line $(-\infty, \infty)$; that is,

$$f_E(x) = \begin{cases} f(x) & \text{if } x > 0 \\ f(-x) & \text{if } x < 0 \end{cases}$$

Now suppose that $w(x, y)$ satisfies

$$w_{xx}(x, y) + w_{yy}(x, y) = 0 \quad \text{in } H$$
$$w(x, 0) = f_E(x), \qquad -\infty < x < \infty$$
$$w(x, y) \text{ bounded on } H$$

This problem has already been solved in Example 5.1.1 where we found

$$w(x, y) = \frac{1}{\pi} \int_{-\infty}^{\infty} \frac{y}{[(x - z)^2 + y^2]} f_E(z) \, dz$$

and we claim now that $w(x, y) = u^A(x, y)$ for (x, y) in Q.

To see this, note first that $w(x, y)$ satisfies all of the conditions of subproblem A with the possible exception of the homogeneous boundary condition along the edge $x = 0$. To see that this boundary condition is satisfied as well, write

$$w_x(x, y) = \frac{y}{\pi} \int_{-\infty}^{\infty} \frac{f_E'(x - z)}{y^2 + z^2} \, dz$$

At $x = 0$ this becomes

$$w_x(0, y) = \frac{y}{\pi} \int_{-\infty}^{\infty} \frac{f_E'(-z)}{[y^2 + z^2]} \, dz$$

Since $f_E(x)$ is even, its derivative is an odd function of its argument, and it follows that the integrand in the preceding integral is then an odd function of z. But the integral of an odd function over a symmetric interval is zero, and hence $w_x(0, y) = 0$ for $y > 0$. Then $w(x, y)$ satisfies *all* of the conditions of subproblem A, and since the solution of this problem can be shown to be unique, it follows that $w(x, y) = u^A(x, y)$ for (x, y) in Q.

Note that if we let $G(x - z, y) = y/\pi[(x - z)^2 + y^2]$, then

$$w(x, y) = \int_{-\infty}^{\infty} G(x - z, y) f_E(x) \, dz$$

$$= \int_{-\infty}^{0} G(x - z, y) f_E(z) \, dz + \int_{0}^{\infty} G(x - z, y) f_E(z) \, dz$$

$$= -\int_{\infty}^{0} G(x + z, y) f_E(-z) \, dz + \int_{0}^{\infty} G(x - z, y) f_E(z) \, dz$$

Here we made the change of variable $-z \to z$ in the integral over the negative axis.

Then, since $f_E(-z) = f_E(z)$ for all z and $f_E(z) = f(z)$ for $z > 0$, this reduces to

$$u^A(x, y) = \int_0^\infty [G(x + z, y) + G(x - z, y)]f(z)\, dz \qquad (5.2.1)$$

and subproblem A is solved.

To solve subproblem B, we extend the function $g(y)$ to the entire y axis as an *odd* function of y. That is,

$$g_0(y) = \begin{cases} g(y) & \text{for } y > 0 \\ -g(-y) & \text{for } y < 0 \end{cases}$$

Then suppose that $v(x, y)$ solves the problem

$$v_{xx}(x, y) + v_{yy}(x, y) = 0 \quad \text{in } H$$
$$v_x(0, y) = g_0(y), \qquad y > 0$$
$$v(x, y) \text{ bounded in } H$$

This problem has also been solved in Example 5.1.1 (part 2). We have from (5.1.4) that

$$v(x, y) = \frac{1}{2\pi} \int_{-\infty}^\infty \log[x^2 + (y - z)^2]g_0(z)\, dz + C$$

for C an arbitrary constant of integration. If we let

$$N(x, y - z) = 1/2\pi \log[x^2 + (y - z)^2]$$

then

$$v(x, y) = \int_{-\infty}^\infty N(x, y - z)g_0(z)\, dz$$

Proceeding as we did with $w(x, y)$ in the solution of subproblem A, we can write $v(x, y)$ in the form of an integral over $(0, \infty)$ with $g(y)$ rather than the extension $g_0(y)$ in the integrand; that is,

$$v(x, y) = \int_0^\infty [N(x, y - z) - N(x, y + z)]g(z)\, dz + C$$

From this integral it is evident that if $C = 0$, then $v(x, 0) = 0$. Then $v(x, y)$ satisfies all of the conditions of subproblem B (including the homogeneous boundary condition at $y = 0$), and we conclude by uniqueness that $v(x, y) = u^B(x, y)$ in Q; that is,

$$u^B(x, y) = \int_0^\infty [N(x, y - z) - N(x, y + z)]g(z)\, dz \quad \text{for } (x, y) \text{ in } Q \qquad (5.2.2)$$

Adding $u^A(x, y)$ from (5.2.1) and $u^B(x, y)$ from (5.2.2) yields the solution $u(x, y)$ of the quarter-plane boundary-value problem. The solution $u^A(x, y)$ represents the response of the plate to the temperature specification $f(x)$, while the solution $u^B(x, y)$ represents the temperature response inside the plate to the flux specification $g(y)$. It is because the problem is linear that superposition applies, and we can simply add these two responses in order to obtain the compound response to the combined specifications.

Note that extending $f(x)$ as an *even* function ensured that the solution $w(x, y)$ of the half-plane problem would *automatically* satisfy the homogeneous condition

$w_x(0, y) = 0$. Likewise, extending the data function $g(y)$ as an *odd* function of y ensured that the solution $v(x, y)$ of the other half-plane problem would automatically satisfy the homogenous condition $v(x, 0) = 0$. This technique of extending the data functions in such a way as to ensure that certain homogeneous boundary conditions are automatically satisfied is often referred to as the *method of images*.   ■ ■

Exercises: Boundary-Value Problems on Semibounded Regions

1. Solve the problem of Example 5.2.1 in the case that

$$g(y) = 0 \quad \text{for all } y, \qquad f(x) = \begin{cases} 1 & \text{for } 0 < x < 1 \\ 0 & \text{for } x > 1 \end{cases}$$

2. Find the bounded solution of the problem

$$u_{xx}(x, y) + u_{yy}(x, y) = 0, \qquad x > 0, \qquad y > 0$$
$$u_x(0, y) = g(y), \qquad y > 0$$
$$u_y(x, 0) = f(x), \qquad x > 0$$

3. Solve the previous exercise in the special case that

$$g(y) = 0 \quad \text{for all } y, \qquad f(x) = \begin{cases} 1 & \text{for } 0 < x < 1 \\ 0 & \text{for } x > 1 \end{cases}$$

4. Find the bounded solution of the problem

$$u_{xx}(x, y) + u_{yy}(x, y) = 0, \qquad x > 0, \qquad y > 0$$
$$u(0, y) = g(y), \qquad y > 0$$
$$u(x, 0) = f(x), \qquad x > 0$$

5. Solve the previous exercise in the special case that

$$g(y) = \begin{cases} 1 & 1 < y < 2 \\ 0 & \text{otherwise} \end{cases}, \qquad f(x) = \begin{cases} -1 & 2 < x < 3 \\ 0 & \text{otherwise} \end{cases}$$

6. Find the bounded solution of the following problem:

$$u_{xx}(x, y) + u_{yy}(x, y) = 0, \qquad x > 0, \qquad 0 < y < 1$$
$$u(0, y) = 0, \qquad 0 < y < 1$$
$$u(x, 0) = f(x), \qquad u(x, 1) = g(x), \qquad x > 0$$

Hint: Extend the data functions f and g such that the boundary condition $u(0, y) = 0$ is automatically satisfied. Note that in the case of an inhomogeneous condition at $x = 0$ we can split the problem into two subproblems one of which is the preceding problem and the other is Example 3.1.7.

7. Find the bounded solution of the following problem:

$$u_{xx}(x, y) + u_{yy}(x, y) = 0, \qquad x > 0, \qquad 0 < y < 1$$
$$u_x(0, y) = 0, \qquad 0 < y < 1$$
$$u(x, 0) = f(x), \qquad u(x, 1) = g(x), \qquad x > 0$$

We could interpret this problem as one in which we are trying to determine the hydraulic head at each point in a semi-infinite slab of porous medium. The face $x = 0$ of the slab is an impervious boundary while the head is specified along the faces $y = 0$ and $y = 1$.

8. Find the bounded solution of the following problem:

$$u_{xx}(x, y) + u_{yy}(x, y) = 0, \qquad x > 0, \qquad 0 < y < 1$$
$$u_x(0, y) = 0, \qquad 0 < y < 1$$
$$u_y(x, 0) = 0, \qquad u(x, 1) = g(x), \qquad x > 0$$

In each of the examples of the previous section that involved an evolution equation, the spatial domain was the entire real line $(-\infty, \infty)$. Since this is a set without a boundary, there was no opportunity in these examples for the solution to exhibit a response to any boundary influence. The following examples on the half-line $(0, \infty)$ will illustrate how the integral transform techniques represent the influence on the solution of conditions at the boundary $x = 0$.

EXAMPLE 5.2.2

Diffusionlike Evolution on Semi-infinite Interval

Consider the problem of finding an unknown function $u(x, t)$ that is bounded for $x > 0$, $t > 0$, and satisfies, for positive constant D and given functions $f(x)$ and $g(t)$,

$$u_t(x, t) = Du_{xx}(x, t), \qquad x > 0, \qquad t > 0$$
$$u(x, 0) = f(x), \qquad x > 0$$
$$u(0, t) = g(t), \qquad t > 0$$

Here, $u(x, t)$ could represent the concentration at time $t > 0$ and position x of a substance diffusing into a thin pipe lying along the x axis. One end of the pipe is at $x = 0$ and the other end is sufficiently far away that we may treat it as being "at infinity." The concentration in the pipe is given initially by the data function $f(x)$, and material is being injected into the pipe at the end $x = 0$ according to the schedule prescribed in the data function $g(t)$. Then $u(x, t)$ represents the response of the concentration to this initial state and boundary concentration.

As in the previous example we split this problem into two subproblems. While this is not strictly necessary, it will be more convenient to proceed this way.

Subproblem A: Find a function $u^A(x, t)$ bounded for $x > 0$, $t > 0$, and satisfying

$$u_t^A(x, t) = Du_{xx}^A(x, t), \qquad x > 0, \qquad t > 0$$
$$u^A(x, 0) = f(x), \qquad x > 0$$
$$u^A(0, t) = 0, \qquad t > 0$$

Subproblem B: Find a function $u^B(x, t)$, bounded for $x > 0$, $t > 0$, and satisfying

$$u_t^B(x, t) = Du_{xx}^B(x, t), \qquad x > 0, \qquad t > 0$$
$$u^B(x, 0) = 0, \qquad x > 0$$
$$u^B(0, t) = g(t), \qquad t > 0$$

If we can solve these two subproblems, then the solution $u(x, t)$ of the original problem is (by superposition) just the sum of $u^A(x, t)$ and $u^B(x, t)$. Clearly, $u^A(x, t)$ will represent that part of the solution that is the response to the initial state, while $u^B(x, t)$ represents the response to the "forcing at the boundary." We shall see that $u^A(x, t)$ is a transient response in the sense that it decreases in magnitude as t increases. On the other hand, $u^B(x, t)$ will contain that part of the response that describes the behavior of the concentration after all transient behavior has died out. This does not (necessarily) mean that this part of the solution is independent of time.

To solve subproblem A, we extend $f(x)$ to the whole real axis $(-\infty, \infty)$ as an odd function of x. If we denote this odd extension by $f_0(x)$, then we can consider the initial-value problem for the unknown function $w(x, t)$,

$$w_t(x, t) = Dw_{xx}(x, t), \qquad -\infty < x < \infty, \qquad t > 0$$
$$w(x, 0) = f_0(x), \qquad -\infty < x < \infty$$
$$w(x, t) \text{ bounded for all } x \quad \text{and} \quad t > 0$$

This problem has been solved previously in Example 5.1.2 where we found

$$w(x, t) = \int_{-\infty}^{\infty} K(x - z, t)f_0(z) \, dz$$

where

$$K(x - z, t) = (4\pi Dt)^{-1/2}\exp[-(x - z)^2/(4Dt)]$$

Then we can proceed as we did in the quarter-plane example to show that

$$w(x, t) = \int_{0}^{\infty} [K(x - z, t) - K(x + z, t)]f(z) \, dz$$

It is clear from this last expression that $w(0, t) = 0$, which means that $w(x, t)$ satisfies all of the conditions of subproblem A. By uniqueness of the solution to that problem, it follows that $w(x, t) = u^A(x, t)$ for $x > 0$ and $t > 0$. Then,

$$u^A(x, t) = \int_{0}^{\infty} [K(x - z, t) - K(x + z, t)]f(z) \, dz \qquad (5.2.3)$$

A change of variables such as the one that reduced (5.1.5) to (5.1.6) allows this to be written in the form

$$u^A(x, t) = \pi^{-1/2}\left[\int_{x(4Dt)^{-1/2}}^{\infty} \exp[-z^2]f(z(4Dt)^{1/2} - x) \, dz \right.$$
$$\left. + \int_{-x(4Dt)^{-1/2}}^{\infty} \exp[-z^2]f(x + z (4Dt)^{1/2}) \, dz \right] \qquad (5.2.4)$$

From this last representation, it is easy to show that if $f(x)$ is continuous, then as t decreases to zero, $u^A(x, t) \to f(x)$ at each $x > 0$.

In order to solve subproblem B, we apply the Laplace transform in the t variable. The Laplace transform applies in this case because

  (i) the t variable ranges over the interval $(0, \infty)$ and
  (ii) the equation is first order in the t variable with a corresponding single initial condition $u^B(x, 0) = 0$.

If we let $\hat{u}(x, s)$ denote the Laplace transform (with respect to t) of the unknown function $u^B(x, t)$, then

$$\mathcal{L}[u_t^B(x, t)] = s\hat{u}(x, s) - 0$$

$$\mathcal{L}[u_{xx}^B(x, t)] = \frac{d^2}{dx^2} \hat{u}(x, s)$$

and subproblem B transforms to

$$\frac{d^2}{dx^2} \hat{u}(x, s) = \frac{s}{D} \hat{u}(x, s), \quad x > 0$$

$$\hat{u}(0, s) = \hat{g}(s) = \mathcal{L}[g(t)]$$
$$\hat{u}(x, s) \text{ bounded for } x > 0$$

The general solution to this equation can be written in the form

$$\hat{u}(x, s) = C_1 \exp[x(s/D)^{1/2}] + C_2 \exp[-x(s/D)^{1/2}]$$

In order that the solution remain bounded for $x > 0$, it is necessary that $C_1 = 0$. Then the other boundary condition implies that $C_2 = \hat{g}(s)$, and we have

$$\hat{u}(x, s) = \hat{g}(s) \exp[-x(s/D)^{1/2}]$$

Entry 7 in the Laplace transform pairs table in the endpapers together with the convolution formula for the Laplace transform then leads to

$$u^B(x, t) = \int_0^t x[4\pi D(t - \tau)^3]^{-1/2} \exp\left[\frac{-x^2}{4D(t - \tau)}\right] g(\tau) \, d\tau \qquad (5.2.5)$$

The change of independent variable

$$z = x/[4D(t - \tau)]^{1/2}$$

reduces (5.2.5) to

$$u^B(x, t) = 2\pi^{-1/2} \int_{x(4Dt)^{-1/2}}^{\infty} \exp[-z^2] g\left(t - \frac{x^2}{4Dz^2}\right) dz \qquad (5.2.6)$$

If we rewrite (5.2.6) in the form

$$u^B(x, t) = 2\pi^{-1/2} \int_0^{\infty} \exp[-z^2] g\left(t - \frac{x^2}{4Dz^2}\right) dz \qquad (5.2.7)$$

$$- 2\pi^{-1/2} \int_0^{x(4Dt)^{-1/2}} \exp[-z^2] g\left(t - \frac{x^2}{4Dz^2}\right) dz$$

then it is clear that if $g(t)$ is bounded for all $t > 0$, the second integral in the preceding approaches zero as $t \to \infty$; that is, the upper limit in the integral tends to zero. Then this part of the solution may be viewed as a "transient" portion of the solution $u^B(x, t)$, and the other integral then represents the equilibrium solution. If we consider the special case

$$g(t) = A \cos \Omega t, \qquad A = \text{real constant}$$

then we can show (see Exercise 3 following Example 5.2.4) that

$$2\pi^{-1/2} \int_0^\infty \exp[-z^2] \cos\left[\Omega\left(t - \frac{x^2}{4Dz^2}\right)\right] dz$$

$$= \exp\left[-x\left(\frac{\Omega}{2D}\right)^{1/2}\right] \cos\left[\Omega t - x\left(\frac{\Omega}{2D}\right)^{1/2}\right]$$

Then we have, for $g(t) = A \cos \Omega t$,

$$u^B(x, t) \to A \exp[-x(\Omega/2D)^{1/2}] \cos [\Omega(t - x)/(2D\Omega)^{1/2}] \qquad (5.2.8)$$

as $t \to \infty$. We observe that a periodic input from the boundary produces a response with the following properties:

(a) The response tends to an equilibrium solution as $t \to \infty$.
(b) The amplitude of the equilibrium response is attenuated by the factor $\exp[-x(\Omega/2D)^{1/2}]$; that is, the attenuation increases exponentially with x and with $(\Omega/2D)^{1/2}$.
(c) The equilibrium solution translates similar to a traveling wave having frequency Ω and phase velocity equal to $(2D\Omega)^{1/2}$. In particular, it is significant that the phase velocity is frequency dependent. Then if the input from the boundary were composed of components of different frequencies, these components would not translate at the same speed, and dispersion would result

We shall compare this behavior with the behavior of the solution to the analogous problem for the wave equation in order to see diffusionlike and wavelike evolution differ. ■ ■

Exercises: Diffusion on Semi-infinite Interval

1. Solve the problem in Example 5.2.2 for

$$f(x) = 0, \quad x > 0, \qquad g(t) = 1, \quad t > 0$$

Note that these auxiliary conditions are inconsistent since $f(0) \neq g(0)$. Which if either of the two conditions does your solution satisfy?

2. Solve the problem in Example 5.2.2 for

$$f(x) = 0, \quad x > 0, \qquad g(t) = 1 - \exp[-bt], \quad t > 0$$

For b a large positive constant, this $g(t)$ rapidly approaches the $g(t)$ of Exercise 1 as t increases from zero and these data functions lead to consistent auxiliary conditions.

3. Find the bounded function $u(x, t)$ satisfying

$$u_t(x, t) = Du_{xx}(x, t), \qquad x > 0, \qquad t > 0$$
$$u(x, 0) = f(x), \qquad x > 0$$
$$u_x(0, t) = g(t), \qquad t > 0$$

4. Solve Exercise 3 in the special case,

(a) $f(x) = \begin{cases} 1, & 0 < x < 1 \\ 0, & x > 1 \end{cases}, \qquad g(t) = 0, \quad t > 0$

(b) $f(x) = 0, \quad x > 0, \qquad g(t) = \begin{cases} 1 & 0 < t < 1 \\ 0 & t > 1 \end{cases}$

5. For the solution $u^B(x, t)$ of subproblem B of Example 5.2.2, compute

$$\Phi(t) = -u_x^B(0, t) \quad \text{and} \quad Q(t) = \int_0^t \Phi(\tau)\, d\tau$$

If we interpret the solution in the context of diffusion, then $\Phi(t)$ is related to the flux of material through the end $x = 0$ and $Q(t)$ is related to the total amount of material that has passed through the end $x = 0$ (per unit area) in the time interval $(0, t)$. [*Hint*: Carry out the differentiation on the transform: that is,

$$\hat{u}_x(x, s) = -(s/D)^{1/2}\exp[-x(s/D)^{1/2}]\hat{g}(s)$$

so

$$\hat{u}_x(0, s) = -(s/D)^{1/2}\hat{g}(s) \quad \text{and} \quad \hat{\Phi}(s) = 1/(sD)^{1/2}s\hat{g}(s)$$

Then use the operational properties of the transform to invert this last expression. Similarly, $\hat{Q}(s) = 1/s\hat{\Phi}(s)$ can be inverted using operational properties. *Note*: $\mathcal{L}[(\pi t)^{-1/2}] = s^{-1/2}$.]

6. Consider Exercise 3 in the case $f(x) = u_0$ and $g(t) = -\Phi_0$ are both constant. Then the problem approximately models the cooling of the surface of the earth after sunset on a clear, windless night. If we let $\Phi(x, t) = -u_x(x, t)$, then it is easy to show that $\Phi(x, t)$ satisfies

$$\Phi_t(x, t) = D\Phi_{xx}(x, t), \qquad x > 0, \qquad t > 0$$
$$\Phi(x, 0) = 0, \qquad x > 0$$
$$\Phi(0, t) = \Phi_0$$

Then show

$$\Phi(x, t) = \Phi_0 \operatorname{erfc}[x/(4Dt)^{1/2}]$$

and

$$u(x, t) = \Phi_0 \int_x^\infty \operatorname{erfc}\left[\frac{z}{(4Dt)^{1/2}}\right] dz$$

7. To solve

$$u_t(x, t) = Du_{xx}(x, t), \qquad x > 0, \qquad t > 0$$
$$u(x, 0) = T_0$$
$$u(0, t) = T_1 > T_0$$

it is sufficient to solve

$$v_t(x, t) = Dv_{xx}(x, t), \qquad x > 0, \qquad t > 0$$
$$v(x, 0) = T_0 - T_1 = \delta T$$
$$v(0, t) = 0$$

Then show that

$$u(x, t) = T_1 + \delta T \operatorname{erf}[x/(4Dt)^{1/2}]$$

At what time (time expressed in terms of D) does $u(1, t) = T_1 + \delta T/2$? At what times does $u(1, t) = T_1$? At what time does $u(2, t) = T_1 + \delta T/2$? How is this time affected if the parameter D is doubled? Evaluate $-u_x(0, t)$. How is this quantity affected if D is doubled?

8. Consider the problem

$$u_t(x, t) = Du_{xx}(x, t), \qquad x > 0, \qquad t > 0$$
$$u(x, 0) = u_0 \qquad x > 0$$
$$-u_x(0, t) + Bu(0, t) = 0 \qquad t > 0$$

where B and u_0 denote constants. Note that

$$v(x, t) = u(x, t) - 1/B \, u_x(x, t)$$

solves the problem

$$v_t(x, t) = Dv_{xx}(x, t), \qquad x > 0, \qquad t > 0$$
$$v(x, 0) = u_0, \qquad x > 0$$
$$v(0, t) = 0, \qquad t > 0$$

Find $v(x, t)$ and note that $v(x, t) \to u_0$ as $x \to \infty$.

9. From the fact that $u(x, t)$ and $v(x, t)$ in the previous exercise are related by

$$u_x(x, t) - Bu(x, t) = Bv(x, t)$$
$$u(x, t) \text{ bounded for } x > 0$$

deduce that

$$u(x, t) = B \int_0^x v(x + z, t)\exp[-Bz] \, dz$$

Show that

$$u(0, t) = u_0\exp[B^2Dt]\text{erfc}[B(Dt)^{1/2}]$$

10. Think of the earth as a semi-infinite one-dimensional conductor of heat that experiences periodic heating and cooling at $x = 0$. The effect of daily temperature fluctuation (i.e., frequency equal to Ω) has been observed to penetrate to a depth of 3–4 feet beneath the surface while seasonal fluctuations (i.e., frequency equal to $\Omega/365$) produce effects that are detectable at depths of 60–70 feet. Show that these observations are roughly consistent with the amplitude attenuation effects predicted by (5.2.8).

11. Consider the problem of finding the bounded solution for

$$u_t(x, t) = Du_{xx}(x, t) + Vu_x(x, t), \qquad x > 0, \qquad t > 0$$
$$u(x, 0) = 0, \qquad x > 0$$
$$u(0, t) = g(t), \qquad t > 0$$

Solve this problem by a change of dependent variable designed to eliminate the lower order term in the equation.

12. Consider the same problem for

$$u_t(x, t) = Du_{xx}(x, t) + Vu_x(x, t), \qquad x > 0, \qquad t > 0$$
$$u(x, 0) = 0, \qquad x > 0$$
$$u_x(0, t) = g(t), \qquad t > 0$$

Possibly the solution of Exercise 8 can be of help in this problem.

13. Consider the problem of finding the bounded solution for

$$u_t(x, t) = Du_{xx}(x, t) - Pu(x, t), \qquad x > 0, \qquad t > 0$$
$$u(x, 0) = 0, \qquad x > 0$$
$$u(0, t) = g(t), \qquad t > 0$$

Solve this problem by a change of dependent variable designed to eliminate the lower order term in the equation.

In order to see how wavelike evolution is distinguished from diffusionlike evolution, we now consider some examples that exhibit the wavelike behavior.

EXAMPLE 5.2.3

Wavelike Evolution on Semi-infinite Interval

Consider the problem of finding an unknown function $u(x, t)$ satisfying

$$u_t(x, t) = Cu_x(x, t), \quad x > 0, \quad t > 0$$
$$u(x, 0) = f(x), \quad x > 0$$
$$u(0, t) = g(t), \quad t > 0$$

This first-order wave equation, or advection equation, as it is sometimes called, is discussed also in Chapter 7 from a different point of view. There the method of characteristics is used to solve the preceding initial-boundary-value problem.

In Example 5.1.3 we saw that if $U(x)$ is a smooth function of one variable, then

$$u(x, t) = U(x + Ct) \tag{5.2.9}$$

satisfies the first-order wave equation. With regard to the boundary and initial condition, there are two cases to consider:

Case 1: Sending Mode, $C < 0$

The solution $u(x, t)$ given by (5.2.9) is constant along characteristic lines $x + Ct = $ const, and for $C < 0$ these lines have *positive* slope. In this case, the quarter-plane $\{x > 0, t > 0\}$ is divided into two complementary regions by the characteristic $x + Ct = 0$ passing through the origin (see Figure 5.2.1).

Each point in the *upper* portion of the quarter-plane lies on a characteristic of the form $x + Ct = CT < 0$ where T denotes the point where the characteristic intercepts the t-axis.

Each point in the *lower* portion of the quarter-plane lies on a characteristic of the form $x + Ct = X > 0$ where X denotes the point where this characteristic crosses the x axis.

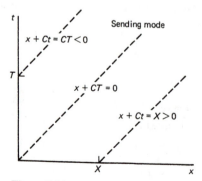

Figure 5.2.1

This means that at each point in the *upper* part of the quarter-plane Q, $u(x, t)$ is given by

$$u(x, t) = g(t + x/C)$$

that is, the solution is controlled by the *boundary* condition. At each point in the *lower* half of Q, we have

$$u(x, t) = f(x + Ct)$$

that is, the solution is determined by the *initial* condition. For $t > 0$, if we let

$$Q_B(t) = [x \text{ in } Q : x + Ct < 0]$$
$$Q_I(t) = [x \text{ in } Q : x + Ct > 0]$$

then for each $t > 0$

$$u(x, t) = \begin{cases} g(t + x/C) & \text{for } x \text{ in } Q_B(t) \\ f(x + Ct) & \text{for } x \text{ in } Q_I(t) \end{cases}$$

We can write this as

$$u(x, t) = g(t + x/C)H_{x/c}(t) + f(x + Ct)[1 - H_{x/c}(t)] \qquad (5.2.10)$$

where

$$H_{x/c}(t) = \begin{cases} 0 & \text{if } t < x/C \\ 1 & \text{if } t > x/C \end{cases}$$

It is not difficult to interpret the meaning of the solution (5.2.10). It is as if we were sending a signal into a one-dimensional transmission line at the end $x = 0$. The signal travels along the line with speed equal to C and at any time t, $Q_B(t)$ represents that part of the line that the signal has had time to reach. At any point x in $Q_B(t)$, $u(x, t) = g$, the signal at the boundary, evaluated at the earlier time $t + x/C$ (recall that x/C is negative). This accounts for the fact that the signal took x/C units of time to reach the position x.

For any $t > 0$, $Q_I(t)$ is the part of the line where the signal from the boundary has not yet arrived, and there the solution $u(x, t)$ is determined by the initial state in the line. However, unless the initial state was constant, $u(x, t)$ is not a constant in this part of the line. It is as if the initial state propagates down the line, "pushed ahead" of the oncoming signal from the boundary.

Case 2: Receiving Mode, $C > 0$

For $C > 0$, the characteristics $x + Ct = M$ are lines with negative slope (see Figure 5.2.2). This means that each point in Q is on one and only one of these characteristics.

Different choices for the constant M produce different straight lines in the family of characteristics. Evidently M can be taken to be X, the x intercept of the characteristic, *or* it can be taken equal to CT, where T denotes the t intercept of this same characteristic. Correspondingly, the solution $u(x, t)$ can be given by *either* of the expressions

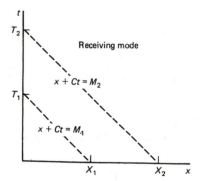

Figure 5.2.2

$$u(x, t) = f(x + Ct), \qquad (x, t) \text{ in } Q \qquad (5.2.11a)$$
$$u(x, t) = g(t + x/C), \qquad (x, t) \text{ in } Q \qquad (5.2.11b)$$

What is important to notice here is that these two expressions must be *equal* at every point in Q. This means that we cannot arbitrarily impose *both* the initial *and* the boundary condition on the solution when C is positive. We can impose *only one* of these conditions, and then the other is determined by (5.2.11). Another way to say this is to say that the initial-boundary-value problem for $u(x, t)$ is not well posed when C is positive.

If we assume that the initial condition has been imposed, then we can view this as a transmission line in a "receiving mode" where the signal is being received at the end $x = 0$ of the line. The signal there is the boundary function $g(t)$.

Application to Convection–Diffusion Equation   The fact that the sign of the parameter C in this problem has a bearing on how many auxiliary conditions must be imposed for the problem to be well posed has some interesting consequences when we consider an initial-boundary-value problem for a convection–diffusion equation:

$$u_t(x, t) = Du_{xx}(x, t) + Vu_x(x, t), \qquad x > 0, \qquad t > 0$$
$$u(x, 0) = f(x), \qquad x > 0$$
$$u(0, t) = g(t), \qquad t > 0$$

By rescaling the independent variables in the problem, we can bring the equation to the form

$$u_\tau(\sigma, \tau) = (1/P)u_{\sigma\sigma}(\sigma, \tau) + u_\sigma(\sigma, \tau)$$

where $P = VL/D$ is called the *Peclet number* for L equal to some *characteristic length* associated with the problem. Then for large values of P, the equation is convection dominated, while for small values of P it is diffusion dominated.

This problem can be solved exactly using the methods presented in the previous example. In that case the size of P is not an issue. However, when trying to solve such problems by some approximation technique, the size of P does make a difference. To see why, suppose that we are dealing with a convection-dominated case and that to the accuracy of our approximation, the equation behaves similar to the first-order wave equation we have been considering in this example. Then, as we have just seen, if V is positive, we have a problem with the boundary condition at $x = 0$. Although this condition is perfectly proper (and in fact, necessary) for the parabolic convection–diffusion equation, it makes the problem for the hyperbolic first-order wave equation overdetermined. That is, there will in general be no solution that satisfies all of the conditions of the first-order problem. Then applying an approximation scheme to a convection–diffusion problem in the convection-dominated case with V positive would very likely encounter some sort of computational difficulty near $x = 0$. This analysis explains the source of the difficulty, but resolving the difficulty would require the use of so-called boundary-layer techniques to match the *outer approximation* (which satisfies the equation) with the *inner approximation* (which satisfies the boundary condition at $x = 0$). ■ ■

EXAMPLE 5.2.3 (Part 2) ――――――――――――――――――――――――――――――――――――

Second-Order Wave Equation
     We now consider the initial-boundary-value problem on a semi-infinite interval for the second-order wave equation,

$$u_{tt}(x, t) = a^2 u_{xx}(x, t), \qquad x > 0, \qquad t > 0$$
$$u(x, 0) = f(x), \qquad u_t(x, 0) = g(x), \qquad x > 0$$
$$u(0, t) = h(t), \qquad t > 0$$

Here a denotes a real parameter. In order to illustrate more clearly how the boundary influences the solution, we split this problem into two subproblems. Then, by superposition, the solution to the initial-boundary-value problem will be the sum of the solutions to the following two subproblems.

     Subproblem A: Suppose $u^A(x, t)$ satisfies

$$u_{tt}^A(x, t) = a^2 u_{xx}^A(x, t), \qquad x > 0, \qquad t > 0$$
$$u^A(x, 0) = f(x), \qquad u_t^A(x, 0) = g(x), \qquad x > 0$$
$$u^A(0, t) = 0, \qquad t > 0$$

     Subproblem B: Suppose $u^B(x, t)$ satisfies

$$u_{tt}^B(x, t) = a^2 u_{xx}^B(x, t), \qquad x > 0, \qquad t > 0$$
$$u^B(x, 0) = 0, \qquad u_t^B(x, 0) = 0, \qquad x > 0$$
$$u^B(0, t) = h(t), \qquad t > 0$$

Laplace Transform Solution of Subproblem B   We solve subproblem B first, and for this purpose the Laplace transform in the variable t will be the convenient transform to use. That is, t ranges over $(0, \infty)$, the equation contains a second derivative with respect to t, and the auxiliary conditions specify both $u(x, 0)$ and $u_t(x, 0)$. Then we have

$$\mathcal{L}[u_{tt}^B(x, t)] = s^2 \hat{u}(x, s) - s \cdot 0 - 0$$

Although x also ranges over $(0, \infty)$ and the equation also contains a second-order derivative with respect to x, the single auxiliary condition at $x = 0$ is not sufficient for applying the Laplace transform in the x variable.

In its transformed state, subproblem B assumes the form

$$\frac{d^2}{dx^2} \hat{u}(x, s) = \left(\frac{s}{a}\right)^2 \hat{u}(x, s), \qquad x > 0$$

$$\hat{u}(0, s) = \hat{h}(s) = \mathcal{L}[h(t)]$$

The general solution of this differential equation may be written in the form

$$\hat{u}(x, s) = C_1 \exp[xs/a] + C_2 \exp[-xs/a]$$

In choosing the parameters C_1 and C_2 so as to satisfy the auxiliary conditions, we recall that for the wave equation there is no condition at infinity. Instead we note that in view of entry 6 in the table of Laplace transform operational properties in the endpapers, the exponentials

$$\exp[xs/a] \quad \text{and} \quad \exp[-xs/a]$$

act as "shifting operators" to the left and to the right, respectively. A shift to the left propagates a signal that originates at $x = 0$, *out of* the interval $(0, \infty)$, while a shift to the right has the effect of propagating the signal *into* the interval $(0, \infty)$. Then choosing $C_1 = 0$ causes the signal originating at $x = 0$ to propagate into the region $\{x > 0, t > 0\}$. Then choosing

$$C_2 = \hat{h}(s)$$

causes that signal to be the signal $h(t)$. Then we have

$$\hat{u}(x, s) = \hat{h}(s) \exp[-xs/a]$$

which then leads to

$$u^B(x, t) = h(t - x/a)H(t - x/a), \qquad x > 0, \qquad t > 0 \qquad (5.2.12)$$

At the fixed point x, $u^B(x, t)$ is equal to the signal $h(t)$ delayed by the amount of time it takes the signal to reach the point x, namely, x/a units of time.

Fourier Transform Solution of Subproblem A In order to solve subproblem A, we extend the initial functions $f(x)$ and $g(x)$ to the interval $(-\infty, \infty)$ as *odd* functions of x. If we denote these extensions, respectively, by $f_0(x)$ and $g_0(x)$ and if $w(x, t)$ satisfies

$$w_{tt}(x, t) = a^2 w_{xx}(x, t), \qquad -\infty < x < \infty, \qquad t > 0$$
$$w(x, 0) = f_0(x), \qquad w_t(x, 0) = g_0(x), \qquad -\infty < x < \infty$$

then it will automatically follow that $w(0, t) = 0$. Then, $w(x, t)$ satisfies all of the conditions of subproblem A, and since the solution of subproblem A is unique, we are entitled to conclude that

$$u^A(x, t) = w(x, t), \qquad x > 0, \qquad t > 0 \qquad (5.2.13)$$

From (5.1.18) we have

$$w(x, t) = \tfrac{1}{2}[f_0(x + at) + f_0(x - at)] + \frac{1}{2a} \int_{x-at}^{x+at} g_0(s) \, ds$$

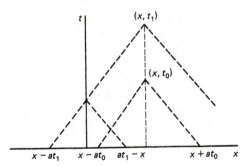

Figure 5.2.3

and so it follows from (5.2.13) and the definition of the odd extension of a function that, for $0 < t < x/a$,

$$u^A(x, t) = \tfrac{1}{2}[f(x + at) + f(x - at)] + \frac{1}{2a} \int_{x-at}^{x+at} g(s) \, ds \qquad (5.2.14)$$

and for $x/a < t$

$$u^A(x, t) = \tfrac{1}{2}[f(at + x) - f(at - x)] + \frac{1}{2a} \int_{at-x}^{at+x} g(s) \, ds$$

Discussion of Solution   The form of the solution in (5.2.14) can be most easily understood in terms of the characteristic lines in Figure 5.2.3.

We consider a fixed value for x and two fixed times t_0 and t_1 satisfying

$$t_0 < x/a < t_1 \qquad (5.2.15)$$

Then:

For t_0 the domain of dependence is $[x - at_0, x + at_0]$.
For t_1 the domain of dependence is $[at_1 - x, at_1 + x]$.

The result for t_1, is a consequence of the fact that the backward characteristics through the point (x, t_1) cut off the interval $[x - at_1, x + at_1]$ where $x - at_1$ is *negative* for t_1 satisfying (5.2.15). Recall that $w(x, t)$, and hence $u^A(x, t)$ as well, are based on the *odd* extensions of the data functions. These odd extensions satisfy

$$f_0(x - at_1) = -f(at_1 - x)$$

and

$$\int_{x-at_1}^{at_1-x} g_0(s) \, ds = 0$$

Using these two results in the D'Alembert solution for $w(x, t)$ leads at once to (5.2.14).

From the figure one can interpret this result physically in the following way. The data values at $at_1 - x$ propagate to the left along a characteristic line until they hit the boundary at $x = 0$. There they are reflected and propagate to the right along another characteristic line to the point (x, t_1).

Note that since the second-order equation permits propagation in both directions, there is no possibility of incompatibility (as there was in the first-order wave equation) between the initial and boundary conditions.

Consider the special case that

$$f(x) = g(x) = 0 \quad \text{and} \quad h(t) = A \cos \Omega t$$

This is the problem that we must solve in analyzing the so-called ripple tank example treated in Section 7.6. There the problem is expressed as a pair of first-order equations. Here we have the solution

$$u(x, t) = \begin{cases} 0 & \text{if } 0 < t < x/a \\ A \cos \Omega(t - x/a) & \text{if } t > x/a \end{cases}$$

For purposes of comparing this with the solution for Example 5.2.2 [with $f(x) = 0$], we note that:

(a) There is *no* transient response.

(b) The amplitude of the solution is *not* attenuated as a function of x or of Ω.

(c) The group velocity of the traveling-wave solution is equal to the constant a. In particular, it is not dependent on the frequency Ω so that there is *no* dispersion.

Comparing these properties with those found for the solution of Example 5.2.2 points up some of the ways in which diffusionlike evolution differs from wavelike evolution. It will be interesting to pursue this comparison in the situation where the spatial interval is bounded. ∎ ∎

Exercises: Wavelike Evolution on Semi-infinite Interval

1. Find $u(x, t)$ such that

$$u_t(x, t) + 4u_x(x, t) = -2u(x, t), \qquad x > 0, \qquad t > 0$$

$$u(x, 0) = 0, \qquad x > 0$$

$$u(0, t) = \begin{cases} \sin t, & 0 < t < \pi \\ 0, & t > \pi \end{cases}$$

Is this problem well posed? Find the smallest time T such that $u(x, T) = 0$ for all x satisfying $0 < x < 10\pi$. Is there any time T such that $u(x, t)$ is zero for all $x > 10\pi$ and $t > T$?

2. Find $u(x, t)$ such that

$$u_{tt}(x, t) = a^2 u_{xx}(x, t), \qquad x > 0, \qquad t > 0$$

$$u(x, 0) = u_t(x, 0) = 0, \qquad x > 0$$

$$u(0, t) = \begin{cases} t(1 - t), & 0 < t < 1 \\ 0, & t > 1 \end{cases}$$

For a fixed $L > 0$ can you find T depending on L such that $u(x, t)$ is zero for $0 < x < L$ for $t > T$? Can you find a T depending on L such that $u(x, t)$ is zero for all $x > L$ and all $t > T$?

3. Find $u(x, t)$ such that

$$u_{tt}(x, t) = a^2 u_{xx}(x, t), \qquad x > 0, \qquad t > 0$$

$$u(x, 0) = u_t(x, 0) = 0, \qquad x > 0$$

$$u_x(0, t) = \begin{cases} 1, & 0 < t < 1 \\ 0, & t > 1 \end{cases}$$

For a fixed $L > 0$ can you find T depending on L such that $u(x, t)$ is zero for $0 < x < L$ for $t > T$? Can you find a T depending on L such that $u(x, t)$ is zero for all $x > L$ and all $t > T$?

4. Find $u(x, t)$ such that

$$u_{tt}(x, t) = a^2 u_{xx}(x, t), \qquad x > 0, \qquad t > 0$$

$$u(x, 0) = u_t(x, 0) = 0, \qquad x > 0$$

$$u_x(0, t) - Bu(0, t) = g(t), \qquad t > 0$$

If $g(t) = 0$ for $t > T$ for some T, is there a corresponding τ such that $u(x, t) = 0$ for $0 < x < 1$ for $t > \tau$?

5. Find $u(x, t)$ such that

$$u_{tt}(x, t) = a^2 u_{xx}(x, t), \qquad x > 0, \qquad t > 0$$

$$u(x, 0) = u_t(x, 0) = 0, \qquad x > 0$$

$$-Ku_x(0, t) = Mu_{tt}(0, t) + g(t), \qquad t > 0$$

A problem of this form arises in connection with analyzing the torsional vibration of a semi-infinite shaft that carries a massive propellor at the end $x = 0$. This propellor experiences a time-dependent torsional loading $g(t)$ that induces then a time-dependent torsional deflection $u(x, t)$ at position x in the shaft.

6. Suppose $u(x, t)$ satisfies

$$u_{tt}(x, t) = a^2 u_{xx}(x, t), \qquad x > 0, \qquad t > 0$$

$$u(x, 0) = u_t(x, 0) = 0, \qquad x > 0$$

$$u(0, t) = f(t), \qquad t > 0$$

Can $f(t)$ be chosen such that $u(x, t)$ is not identically zero but $u(1, t)$ is zero for all $t > 10$?

7. Suppose $u(x, t)$ satisfies

$$u_{tt}(x, t) = a^2 u_{xx}(x, t), \qquad x > 0, \qquad t > 0$$

$$u(x, 0) = f(x), \qquad u_t(x, 0) = 0, \qquad x > 0$$

$$u(0, t) = 0, \qquad t > 0$$

Can $f(x)$ be chosen such that $u(x, t)$ is not identically zero but $u(x, 10) = 0$ for all $x < 1$?

EXAMPLE 5.2.4 ——————————————————————————————

Wavelike Versus Diffusionlike Evolution on Bounded Interval
We consider the following two problems:

$$u_t(x, t) = Du_{xx}(x, t), \qquad 0 < x < 1, \qquad t > 0$$
$$u(x, 0) = 0, \qquad 0 < x < 1$$
$$u(0, t) = f(t), \qquad u(1, t) = 0, \qquad t > 0 \tag{5.2.16}$$

for a positive real constant D and

$$
\begin{aligned}
v_{tt}(x, t) &= a^2 v_{xx}(x, t), &\quad 0 < x < 1, &\quad t > 0 \\
v(x, 0) &= 0, \quad v_t(x, 0) = 0, &\quad 0 < x < 1 \\
v(0, t) &= f(t), \quad v(1, t) = 0, &\quad t > 0
\end{aligned}
\tag{5.2.17}
$$

for a positive real constant a.

We have solved each of these problems previously, in Chapter 3, using the technique of eigenfunction expansion (after subtracting off an appropriate function to make the boundary conditions homogeneous). Here we intend to apply the Laplace transform in t to solve the problems, and the solutions we construct will be quite different from those constructed in Chapter 3. In spite of the different appearance, the solutions are (theoretically) equivalent in the sense that they will yield identical numerical results when precisely evaluated. Since either form of solution involves an infinite series, neither can be evaluated precisely but must be truncated after some finite number of terms. This leads to some practical difference between the two forms of solution; for a given number of terms in the series, the eigenfunction expansion is more accurate for large values of t while the Laplace transform solution produces more accurate results when t is small.

Solution of Diffusion Problem   Applying the Laplace transform in t to problem (5.2.16) leads to the problem

$$
\begin{aligned}
\frac{d^2}{dx^2} \hat{u}(x, s) - \frac{s}{D} \hat{u}(x, s) &= 0, \quad 0 < x < 1 \\
\hat{u}(0, s) = \hat{f}(s), \quad \hat{u}(1, s) &= 0
\end{aligned}
$$

Since this is a two-point boundary-value problem on the bounded interval $(0, 1)$, we choose to write the general solution in the form

$$
\hat{u}(x, s) = C_1 \sinh x\sigma + C_2 \sinh(1 - x)\sigma
$$

where we let $\sigma \equiv (s/D)^{1/2}$. Then the boundary conditions imply that

$$
\begin{aligned}
\hat{u}(x, s) &= \hat{f}(s) \frac{\sinh(1 - x)\sigma}{\sinh \sigma} \\
&= \hat{f}(s) \sum_{n=0}^{\infty} \exp[-(2n + x)\sigma] - \hat{f}(s) \sum_{n=0}^{\infty} \exp[-(2n + 2 - x)\sigma]
\end{aligned}
$$

Entry 7 in the table of Laplace transform pairs together with the convolution formula leads to

$$
\begin{aligned}
u(x, t) &= \int_0^t \sum_{n=0}^{\infty} \frac{2n + x}{[4\pi D(t - \tau)^3]^{1/2}} \exp\left[-\frac{(2n + x)^2}{4D(t - \tau)}\right] f(\tau)\, d\tau \\
&+ \int_0^t \sum_{n=0}^{\infty} \frac{2n + 2 - x}{[4\pi D(t - \tau)^3]^{1/2}} \exp\left[-\frac{(2n + 2 - x)^2}{4D(t - \tau)}\right] f(\tau)\, d\tau \quad (5.2.18)
\end{aligned}
$$

We now consider solution (5.2.18) in the special case

$$
f(t) = A \cos \Omega t
$$

and in addition, we introduce a change of variables in the integrals to bring (5.2.18) to the form

$$
u(x, t) = 2A\pi^{-1/2} \sum_{n=0}^{\infty} \int_{(2n+x)/(4Dt)^{1/2}}^{\infty} \exp[-z^2]
$$

$$
\times \cos \Omega\left[t - \frac{(2n + x)^2}{4Dz^2}\right] dz
$$

$$
- 2A\pi^{-1/2} \sum_{n=0}^{\infty} \int_{(2n+2-x)/(4Dt)^{1/2}}^{\infty} \exp[-z^2]
$$

$$
\times \cos \Omega\left[t - \frac{(2n + 2 - x)^2}{4Dz^2}\right] dz \qquad (5.2.19)
$$

We can also write (5.2.19) more compactly in the form

$$
u(x, t) = 2A\pi^{-1/2} \sum_{n=0}^{\infty} \int_{\alpha_n}^{\infty} \exp[-z^2]\cos \Omega[t - \mu_n(z)]\, dz
$$

$$
- 2A\pi^{-1/2} \sum_{n=0}^{\infty} \int_{\beta_n}^{\infty} \exp[-z^2]\cos \Omega[t - \phi_n(z)]\, dz
$$

where

$$
\alpha_n = (2n + x)/(4Dt)^{1/2}, \qquad \beta_n = (2n + 2 - x)/(4Dt)^{1/2}
$$

and

$$
\mu_n(z) = (2n + x)^2/(4Dz^2), \qquad \phi_n(z) = (2n + 2 - x)^2/(4Dz^2)
$$

Note that as $t \to \infty$, α_n, $\beta_n \to 0$. Then we express (5.2.19) finally in the form

$$
u(x, t) = 2A\pi^{-1/2} \sum_{n=0}^{\infty} \int_{0}^{\infty} \exp[-z^2]\cos \Omega[t - \mu_n(z)]\, dz
$$

$$
- 2A\pi^{-1/2} \sum_{n=0}^{\infty} \int_{0}^{\alpha_n} \exp[-z^2]\cos \Omega[t - \mu_n(z)]\, dz
$$

$$
- 2A\pi^{-1/2} \sum_{n=0}^{\infty} \int_{0}^{\infty} \exp[-z^2]\cos \Omega[t - \phi_n(z)]\, dz
$$

$$
+ 2A\pi^{-1/2} \sum_{n=0}^{\infty} \int_{0}^{\beta_n} \exp[-z^2]\cos \Omega[t - \phi_n(z)]\, dz
$$

In this form, it becomes clear that as $t \to \infty$ the solution $u(x, t)$ tends toward the following time-dependent equilibrium solution:

$$
u_s(x, t) = 2A\pi^{-1/2} \sum_{n=0}^{\infty} \int_{0}^{\infty} \exp[-z^2]\cos \Omega[t - \mu_n(z)]\, dz
$$

$$
- 2A\pi^{-1/2} \sum_{n=0}^{\infty} \int_{0}^{\infty} \exp[-z^2]\cos \Omega[t - \phi_n(z)]\, dz \qquad (5.2.20)
$$

We can show that (see Exercise 3)

$$2A\pi^{-1/2} \int_0^\infty \exp[-z^2]\cos \Omega[t - \mu_n(z)]\, dz$$

$$= A \exp\left[\frac{-\Omega(2n + x)}{(2\Omega D)^{1/2}}\right] \cos \Omega\left[t - \frac{(2n + x)}{(2\Omega D)^{1/2}}\right]$$

Then if we write out the first few terms of the series for $u_s(x, t)$, the steady-state solution to which $u(x, t)$ tends, we see

$$
\begin{aligned}
u_s(x, t) = \ & A \exp[-\Omega x/(2\Omega D)^{1/2}]\cos \Omega[t - x/(2\Omega D)^{1/2}] \\
& - A \exp[-\Omega(2 - x)/(2\Omega D)^{1/2}]\cos \Omega[t - (2 - x)/(2\Omega D)^{1/2}] \\
& + A \exp[-\Omega(2 + x)/(2\Omega D)^{1/2}]\cos \Omega[t - (2 + x)/(2\Omega D)^{1/2}] \\
& - A \exp[-\Omega(4 - x)/(2\Omega D)^{1/2}]\cos \Omega[t - (4 - x)/(2\Omega D)^{1/2}] \ldots
\end{aligned}
$$

Then, at a fixed point (x, t), the steady solution consists of the sum of a series of attenuated traveling waves, all traveling with speed equal to $(2\Omega D)^{1/2}$. Each term in the series experiences a phase lag consistent with a distance traveled:

$\cos \Omega[t - x/(2\Omega D)^{1/2}]$ has phase lag $\tau_0 = x/(2\Omega D)^{1/2}$ corresponding to the distance from the end $x = 0$ to the position x.

$\cos \Omega[t - (2 - x)/(2\Omega D)^{1/2}]$ has phase lag $\tau_1 = (2 - x)/(2\Omega D)^{1/2}$, which is consistent with a distance $2 - x = 1 + 1 - x$, which corresponds to the distance from the end $x = 0$ to the end $x = 1$ plus the distance $1 - x$ from $x = 1$ to position x. This is the distance traveled by a wave that originates at $x = 0$ and is reflected once from the end $x = 1$ before arriving at position x.

$\cos \Omega[t - (2 + x)/(2\Omega D)^{1/2}]$ has phase lag $\tau = (2 + x)/(2\Omega D)^{1/2}$, which is consistent with the distance $2 + x = 1 + 1 + x$. This is the distance traveled by a wave that originates at $x = 0$, reflects off the end $x = 1$, reflects again off the end $x = 0$, and finally arrives at position x.

The steady-state solution can evidently be viewed as the sum of infinitely many reflected attenuated traveling waves. But the effects of all these waves are felt *simultaneously*, with no delay, as if the waves were all traveling with infinite speed. This is in contrast with that we shall see with the solution to (5.2.17).

Solution of Wave Problem Applying the Laplace transform to the problem (5.2.17) leads to the ordinary differential equation

$$\frac{d^2}{dx^2} \hat{u}(x, s) = \left(\frac{s}{a}\right)^2 \hat{u}(x, s), \qquad 0 < x < 1$$

$$\hat{u}(0, s) = \hat{f}(s), \qquad \hat{u}(1, s) = 0$$

The general solution is expressed in the form

$$\hat{u}(x, s) = C_1 \sinh xs/a + C_2 \sinh(1 - x)s/a$$

and then the boundary conditions lead to

$$\hat{u}(x, s) = \hat{f}(s) \frac{\sinh(1 - x)s/a}{\sinh s/a}$$

$$= \hat{f}(s) \sum_{n=0}^{\infty} \exp\left[\frac{-(2n + x)s}{a}\right] - \hat{f}(s) \sum_{n=0}^{\infty} \exp\left[\frac{-(2n + 2 - x)s}{a}\right]$$

Recalling the shifting property of the Laplace transform, we invert this to find

$$u(x, t) = \sum_{n=0}^{\infty} f\left[t - \frac{(2n + x)}{a}\right] H\left[t - \frac{(2n + x)}{a}\right]$$

$$- \sum_{n=0}^{\infty} f\left[t - \frac{(2n + 2 - x)}{a}\right] H\left[t - \frac{(2n + 2 - x)}{a}\right]$$

Of course, in the special case $f(t) = A \cos \Omega t$, this becomes

$$u(x, t) = A \sum_{n=0}^{\infty} \cos \Omega\left[t - \frac{(2n + x)}{a}\right] H\left[t - \frac{(2n + x)}{a}\right]$$

$$- A \sum_{n=0}^{\infty} \cos \Omega\left[t - \frac{(2n + 2 - x)}{a}\right] H\left[t - \frac{(2n + 2 - x)}{a}\right] \qquad (5.2.21)$$

Writing out the first few terms of this series, we see

$$\begin{aligned} u(x, t) = \ & A \cos \Omega[t - x/a]H[t - x/a] \\ & - A \cos \Omega[t - (2 - x)/a]H[t - (2 - x)/a] \\ & + A \cos \Omega[t - (2 + x)/a]H[t - (2 + x)/a] \\ & - A \cos \Omega[t - (4 - x)/a]H[t - (4 - x)/a] + \cdots \end{aligned}$$

Comparing this with the steady-state component of the solution to (5.2.16), we find some similarities and some differences. The solution (5.2.21) consists of the sum of a series of traveling waves, all with amplitude equal to A (no attenuation) and all traveling with speed equal to the constant a (constant, not frequency dependent). But the most important difference between the wave equation solution and the diffusion equation solution lies in the following observation: For fixed, positive x and t, the solution $u(x, t)$ given by (5.2.21) is zero until time $t = x/a$, when it first feels the influence of the forcing function at $x = 0$. Then, for $x/a < t < (2 - x)/a$

$$u(x, t) = A \cos \Omega[t - x/a]$$

At time $t = (2 - x)/a$, the reflection of the input wave off the end $x = 1$ first reaches the position x, and we have, for $(2 - x)/a < t < (2 + x)/a$,

$$u(x, t) = A \cos \Omega[t - x/a] - A \cos \Omega[t - (2 - x)/a]$$

At $t = (2 + x)/a$ the effect of the next reflection is first felt, and so on, ad infinitum. But note that these various reflections are *not* all felt *simultaneously*, as was the case with the steady-state component of the solution to (5.2.16). Each term in the series (5.2.21) experiences a phase shift τ_n *and a delay* $H(t - \tau_n)$ because the influence of a signal that travels with finite speed cannot be felt before the signal has had time to travel from its point of origination to the point of detection. Thus we are motivated to associate with the solution to the diffusion problem (5.2.16) an infinite speed of propagation and with the solution to the wave problem (5.2.17) the finite speed of propagation a.

Part of our purpose in presenting this example is to illustrate how the solution to a problem may be expressed in various equivalent ways each intended to expose certain properties of the solution. For example, consider the solution to (5.2.16):

(a) In the form (5.2.18), it is clear that the solution to (5.2.16) is infinitely differentiable with respect to both x and t for all x, $0 < x < 1$, and all t, $t > 0$, and that this is true independent of the smoothness of $f(t)$.

(b) In the form (5.2.19) and its variants, it becomes clear that the solution tends to a steady-state solution. It is also evident that because of the frequency-dependent attenuation factors, the influence of low-frequency forcing will penetrate to a greater depth than high-frequency periodic forcing at the boundary.

Thus it is important not only to be able to solve the problem but also to express the solution in a form that will be convenient for extracting the particular information desired.
■ ■

Exercises: Initial-Boundary-Value Problems on Bounded Intervals

1. Solve problem (5.2.16) in the case

$$f(t) = \begin{cases} 1 & \text{for } 0 < t < 1 \\ 0 & \text{for } t > 1 \end{cases}$$

2. Solve problem (5.2.17) for $f(t)$ as in Exercise 1.

3. Let

$$I(\Omega) = \int_0^\infty \exp[-z^2]\cos\left[\Omega\frac{(t - x^2)}{4Dz^2}\right] dz$$

Write

$$2\cos[\Omega(t - x^2)/(4Dz^2)] = \exp[i\Omega t]\exp[-i\Omega x^2/(4Dz^2)] + \exp[-i\Omega t]\exp[i\Omega x^2/(4Dz^2)]$$

Then, we can show that

$$\int_0^\infty \exp[-z^2]\exp\left[\frac{a^2}{z^2}\right] dz = \tfrac{1}{2}\pi^{1/2}\exp[-2a]$$

For

$$a^2 = \pm i\Omega x^2/4D \quad (\text{with } i^2 = -1)$$

show

$$a = \tfrac{1}{2}(1 \pm i)(\Omega/2D)^{1/2}$$

and

$$2I(\Omega) = \pi^{1/2}\exp[-x(\Omega/2D)^{1/2}]\cos[\Omega t - x(\Omega/2D)^{1/2}]$$

4. Solve for $u(x, t)$ satisfying

$$u_t(x, t) = Du_{xx}(x, t), \qquad 0 < x < 1, \qquad t > 0$$

$$u(x, 0) = 0, \qquad 0 < x < 1$$
$$u(0, t) = f(t), \qquad u_x(1, t) = 0, \qquad t > 0$$

5. Find $u(x, t)$ in Exercise 4 for the following special cases:

   (a) $f(t) = A \cos \Omega t$ (b) $f(t) = \begin{cases} 1, & 0 < t < 1 \\ 0, & t > 1 \end{cases}$

6. Under certain assumptions about the properties of the soil, the volumetric water content $u(x, t)$ in a one-dimensional column of soil can be shown to satisfy

$$aCu_t(x, t) = u_{xx}(x, t) - au_x(x, t), \qquad 0 < x < 1, \qquad t > 0$$
$$u(x, 0) = u_0(x), \qquad 0 < x < 1$$
$$-Du_x(0, t) + K = f(t), \qquad t > 0$$
$$-Du_x(1, t) + K = 0, \qquad t > 0$$

where a, C, D, K denote positive constants related to the properties of the soil. The boundary conditions imply that the flux of moisture is respectively equal to $f(t)$ and zero at the ends $x = 0$ and $x = 1$ of the column of soil. This problem then models what might be referred to as "flux-controlled" infiltration of moisture into the column. Solve for $u(x, t)$ by introducing a change of dependent variable designed to eliminate the lower order term in the equation.

7. Under the same assumptions on the soil, one can study the loss of moisture from a column of soil by the mechanisms of surface evaporation and root extraction. The equation and conditions that approximately describe this situation are

$$aCu_t(x, t) = u_{xx}(x, t) - au_x(x, t) - aC(u - u^*), \qquad 0 < x < 1, \qquad t > 0$$
$$u(x, 0) = u_0(x), \qquad 0 < x < 1$$
$$-Du_x(0, t) + K = E(u - u^*), \qquad t > 0$$
$$-Du_x(1, t) + K = 0, \qquad t > 0$$

where a, C, D, K, u^* are constants relating to the soil properties and E is a constant related to the evaporation rate. Here, $u_0(x)$ denotes the initial distribution of moisture in the soil column. Introduce a change of variable to eliminate the lower order terms in the equation and solve.

8. Solve for $u(x, t)$,

$$u_t(x, t) = Du_{xx}(x, t), \qquad 0 < x < 1, \qquad t > 0$$
$$u(x, 0) = 0, \qquad 0 < x < 1$$
$$u_x(0, t) = f(t), \qquad u_x(1, t) = 0, \qquad t > 0$$

How does the solution change if the boundary condition at $x = 1$ is changed to

$$u(1, t) = 0$$

9. Solve for $u(x, t)$,

$$u_{tt}(x, t) = a^2 u_{xx}(x, t), \qquad 0 < x < 1, \qquad t > 0$$
$$u(x, 0) = u_t(x, 0) = 0, \qquad 0 < x < 1$$
$$u(0, t) = f(t), \qquad u_x(1, t) = 0, \qquad t > 0$$

10. Consider the problem

$$u_{tt}(x, t) = a^2 u_{xx}(x, t), \qquad 0 < x < 1, \qquad t > 0$$
$$u(x, 0) = u_t(x, 0) = 0, \qquad 0 < x < 1$$
$$u(0, t) = 0, \qquad -Ku_x(1, t) = Mu_{tt}(1, t) + A \sin \Omega t$$

Find $\hat{u}(x, s)$ in terms of a, K, M, and Ω. For what values of Ω does the denominator of $\hat{u}(x, s)$ have a zero of order 2? We say σ is a zero of order 2 for the function $P(s)$ if $P(s)$ can be written in the form

$$P(s) = (s - \sigma)F(s) \quad \text{for } F(\sigma) = 0$$

The values of Ω for which the denominator of $\hat{u}(x, s)$ has a zero of order 2 are the *resonant frequencies* of this problem. This problem models the torsional vibration of a shaft that is rigidly fixed at the end $x = 0$. At the end $x = 1$, a massive propellor experiences periodic torsional loading, as, for example, the propellor of a ship might. Note that we are not as interested in the solution $u(x, t)$ of the problem as we are in the resonant frequencies.

11. Suppose $u(x, t)$ satisfies

$$u_{tt}(x, t) = a^2 u_{xx}(x, t), \quad 0 < x < 2, \quad t > 0$$
$$u(x, 0) = u_t(x, 0) = 0, \quad 0 < x < 2$$
$$u(0, t) = p(t), \quad u_x(2, t) = 0, \quad t > 0$$

Can $p(t)$ be chosen such that $u(x, t)$ is not identically zero but $u(1, t)$ is zero for all $t > 10$?

12. Suppose $u(x, t)$ satisfies

$$u_{tt}(x, t) = a^2 u_{xx}(x, t), \quad 0 < x < 1, \quad t > 0$$
$$u(x, 0) = f(x), \quad u_t(x, 0) = 0, \quad 0 < x < 1$$
$$u(0, t) = 0, \quad u_x(1, t) = 0$$

Can $f(x)$ be chosen such that $u(x, t)$ is not identically zero but $u(\frac{1}{2}, t)$ is zero for all $t > 10$?

5.3 Inhomogeneous Equations

In all of the examples of this chapter, the partial differential equations we have considered have been homogeneous. We now consider several examples in which the equation is inhomogeneous.

EXAMPLE 5.3.1 _____

Boundary-Value Problem for Poisson's Equation

Let H denote the half-space of Example 5.1.1 and consider the problem of finding an unknown function $u(x, y)$ that is bounded in H and satisfies

$$u_{xx}(x, y) + u_{yy}(x, y) = f(x, y) \quad \text{in } H$$
$$u(x, 0) = 0, \quad -\infty < x < \infty$$

A problem of this sort arises in the following contexts:

(a) Find the steady-state distribution $u = u(x, y)$ of temperature in H when the temperature at the boundary is zero and the region H contains internal heat sources or sinks distributed according to the function $f(x, y)$. These heat sources or sinks could be the result of chemical reactions or radioactivity occurring in H.

(b) Find the hydraulic head $u = u(x, y)$ in the saturated region H when the head at the boundary of H is zero and H contains pumping wells or injection wells distributed according to the function $f(x, y)$.

Solution by Integral Transform We assume that $f(x, y)$ is a known function belonging to $L^2(H)$ and we let $f_0(x, y)$ denote the extension of $f(x, y)$ to all of R^2 as an odd function of y. This choice of extension is dictated by the homogeneous Dirichlet boundary condition. That is, if we let $w(x, y)$ denote the solution of

$$w_{xx}(x, y) + w_{yy}(x, y) = f_0(x, y) \quad \text{in } R^2$$

then we see that $w(x, y)$ automatically satisfies $w(x, 0) = 0$.

Let $W(\alpha, y)$ and $F(\alpha, y)$ denote (respectively) Fourier transforms with respect to x of $w(x, y)$ and $f(x, y)$. Then,

$$-\alpha^2 W(\alpha, y) + \frac{d^2}{dy^2} W(\alpha, y) = F(\alpha, y), \quad -\infty < y < \infty$$

We now apply a second Fourier transformation, this time in the variable y. This leads to the following algebraic equation for the unknown (double) Fourier transform $W(\alpha, \beta)$ of $w(x, y)$ in terms of the known (double) Fourier transform $F(\alpha, \beta)$ of $f(x, y)$,

$$-(\alpha^2 + \beta^2)W(\alpha, \beta) = F(\alpha, \beta)$$

This is the analogue of the equation

$$-(m^2 + n^2)\pi^2 u_{mn} = F_{mn}$$

which arose in solving subproblem E in Example 3.1.3. It is interesting to compare these two inhomogeneous problems.

Solving for $W(\alpha, \beta)$,

$$W(\alpha, \beta) = \frac{-\pi}{\alpha} \frac{\alpha F(\alpha, \beta)}{\pi[\alpha^2 + \beta^2]}$$

Here we have written $W(\alpha, \beta)$ in this way to facilitate the application of formula 2 of the table of Fourier transform pairs in the endpapers. Using this formula, in which β is the transform variable and α is only a parameter, together with the convolution formula, we obtain

$$W(\alpha, y) = \frac{-1}{2\alpha} \int_{-\infty}^{\infty} \exp[-|\alpha(y - z)|]F(\alpha, z) \, dz$$

Here we have written $\sqrt{\alpha^2} = |\alpha|$. Now $F(\alpha, z)$ is an odd function of z because of the way in which we extended $f(x, y)$. This allows us to write

$$W(\alpha, y) = \frac{-1}{2\alpha} \int_0^{\infty} \{\exp[-|\alpha(y - z)|] + \exp[-|\alpha(y + z)|]\}F(\alpha, z) \, dz$$

Now the function $(1/\alpha)\exp[-k|\alpha|]$ is not one of the entries in the table of transform pairs, and so in order to complete the inversion, we need a trick. We suppose that

$$F(\alpha, z) = \frac{\partial}{\partial z} G(\alpha, z) \tag{5.3.1}$$

for some functions $G(\alpha, z)$ that satisfies $G(\alpha, 0) = 0$. Then,

$$W(\alpha, y) = \frac{-1}{2\alpha} \int_0^\infty \{\exp[-|a(y - z)|] + \exp[-|\alpha(y + z)|]\} G_z(\alpha, z) \, dz$$

$$= \frac{1}{2\alpha} \int_0^\infty \frac{\partial}{\partial z} \{\exp[-|\alpha(y - z)|] + \exp[-|\alpha(y + z)|]\} G(\alpha, z) \, dz$$

$$= \frac{-1}{2} \int_0^\infty \{\exp[-|\alpha(y - z)|] + \exp[-|\alpha(y + z)|]\} G(\alpha, z) \, dz$$

We used integration by parts to move the z-differentiation from the G onto the exponentials [here we needed the fact that $G(\alpha, 0)$ is zero]. Then differentiating the exponentials with respect to z causes the parameter α to come down and cancel the $1/\alpha$, which is preventing us from inverting the transform. Now we can apply formula 4 in the table of Fourier transform pairs to obtain

$$w(x, y) = \frac{-1}{4\pi} \int_{-\infty}^\infty \int_0^\infty \left\{ \frac{|y - z|}{(x - s)^2 + (y - z)^2} + \frac{|y + z|}{(x - s)^2 + (y + z)^2} \right\} g(s, z) \, ds \, dz$$

In order to express this in terms of the data function $f(x, y)$ instead of the function $g(x, y)$, note that

$$\frac{2r}{(x - s)^2 + r^2} = \frac{\partial}{\partial r} \log[(x - s)^2 + r^2]$$

Then,

$$w(x, y) = \frac{1}{8\pi} \int_{-\infty}^\infty \int_0^\infty \frac{\partial}{\partial z} \{\log[(x - s)^2 + (y - z)^2]$$

$$- \log[(x - s)^2 + (y + z)^2]\} g(s, z) \, ds \, dz$$

and if we integrate by parts once more, this time moving the differentiation back onto the $g(s, z)$, we get [in view of (5.3.1)]

$$w(x, y) = \frac{-1}{8\pi} \int_{-\infty}^\infty \int_0^\infty \log \frac{(x - s)^2 + (y - z)^2}{(x - s)^2 + (y + z)^2} f(s, z) \, ds \, dz \tag{5.3.2}$$

It is apparent from (5.3.2) that $w(x, 0) = 0$, and since the solution of the original problem can be shown to be unique, it follows that

$$u(x, y) = w(x, y) \quad \text{for } (x, y) \text{ in } H$$

Note that if the boundary condition were inhomogenous in this boundary-value problem for Poisson's equation, we could split the problem into two subproblems. One of the problems would be this example, and Example 5.1.1 would be the other. ■ ■

We now solve some inhomogeneous evolution equations and then explain the relationship between the solutions to the homogeneous and inhomogeneous problems.

EXAMPLE 5.3.2 _____

Inhomogeneous Diffusion Equation

In this example we consider the problem of finding a bounded function $u = u(x, t)$ that satisfies

$$u_t(x, t) = Du_{xx}(x, t) + f(x, t), \qquad -\infty < x < \infty, \qquad t > 0$$
$$u(x, 0) = 0, \qquad -\infty < x < \infty$$

for $f(x, t)$ a given function of x and t.

This problem could arise in the following situations:

(a) $u(x, t)$ denotes the concentration of a substance that is diffusing in the x direction while material is being created or destroyed at position x and time t at a rate equal to $f(x, t)$. This production or destruction of material could be the result of chemical reaction. The initial condition is consistent with a situation in which the initial concentration is everywhere zero.

(b) In the presence of certain simplifying assumptions $u(x, t)$ could denote the moisture content in a partially saturated porous medium. Conditions must be such that the flow of moisture is one-dimensional and $f(x, t)$ could model the loss of moisture through root uptake.

Solution by Integral Transform If we apply to this problem the Laplace transform in the t variable, then the transformed problem reads

$$s\hat{u}(x, s) - 0 = D\frac{d^2}{dx^2}\hat{u}(x, s) + \hat{f}(x, s)$$

where $\hat{u}(x, s)$ and $\hat{f}(x, s)$ denote the Laplace transforms of $u(x, t)$ and $f(x, t)$, respectively.

Next, we apply the Fourier transform in the x variable to this transformed problem. This leads to

$$s\hat{U}(\alpha, s) = -D\alpha^2\hat{U}(\alpha, s) + F(\alpha, s)$$

We can easily solve this algebraic equation for the unknown (double) transform $\hat{U}(\alpha, s)$ of $u(x, t)$,

$$\hat{U}(\alpha, s) = F(\alpha, s)/(s + D\alpha^2)$$

We begin the inversion procedure by removing the Laplace transform in t. We use entry number 3 in the table of Laplace transform pairs together with convolution, and we find

$$U(\alpha, t) = \int_0^t \exp[-\alpha^2 D(t - \tau)]F(\alpha, \tau)\, d\tau$$

Then entry 4 in the table of Fourier transform pairs provides the inverse of $\exp[-\alpha^2 D(t - \tau)]$, so that we have

$$u(x, t) = \int_0^t \int_{-\infty}^{\infty} K(x - z, t - \tau)f(z, \tau)\, dz\, d\tau \qquad (5.3.3)$$

where

$$K(x - z, t - \tau) = [4\pi D(t - \tau)]^{-1/2}\exp[-(x - z)^2/4D(t - \tau)]$$

that is,

$$K(x, t) = \mathscr{F}^{-1}[\exp[-Dt\alpha^2]]$$ ■ ■

EXAMPLE 5.3.3 ──

Inhomogeneous Wave Equations

First-Order Example   For C a given constant and $f(x, t)$ a given function, consider the following inhomogeneous first-order wave equation:

$$u_t(x, t) = Cu_x(x, t) + f(x, t), \qquad -\infty < x < \infty, \qquad t > 0$$
$$u(x, 0) = 0, \qquad -\infty < x < \infty$$

We apply first the Laplace transform in t followed by the Fourier transform in x,

$$s\hat{u}(x, s) - 0 = C\frac{d}{dx}\hat{u}(x, s) + \hat{f}(x, s)$$
$$s\hat{U}(\alpha, s) = iC\alpha\hat{U}(\alpha, s) + \hat{F}(\alpha, s)$$

Here we are using the same notation as in the previous example. Then,

$$\hat{U}(\alpha, s) = \hat{F}(\alpha, s)/(s - iC\alpha)$$

and if we invert the Laplace transform first, we get

$$U(\alpha, t) = \int_0^t \exp[i\alpha C(t - \tau)]F(\alpha, \tau)\,d\tau$$

The exponential in the integrand acts as a shift, according to formula 5 in the table of Fourier transform operational properties, so that inverting the Fourier transform leads to the result

$$u(x, t) = \int_0^t f(x + C(t - \tau), \tau)\,d\tau \tag{5.3.4}$$

Second-Order Inhomogeneous Wave Equation   The analogous problem for the second-order wave equation is the following:

$$u_{tt}(x, t) = a^2 u_{xx}(x, t) + f(x, t), \qquad -\infty < x < \infty, \qquad t > 0$$
$$u(x, 0) = 0, \qquad u_t(x, 0) = 0, \qquad -\infty < x < \infty$$

Proceeding in the same way as in the first-order example, we apply first the Laplace transform in t,

$$s^2\hat{u}(x, s) - 0 = a^2\frac{d^2}{dx^2}\hat{u}(x, s) + \hat{f}(x, s)$$

followed by the Fourier transform in x,

$$s^2\hat{U}(\alpha, s) = -a^2\alpha^2\hat{U}(\alpha, s) + \hat{F}(\alpha, s)$$

Then,

$$\hat{U}(\alpha, s) = \hat{F}(\alpha, s)/(s^2 + a^2\alpha^2)$$

Now, we remove the Laplace transform in t using formula 4 of the table of Laplace transform pairs,

$$U(\alpha, t) = \frac{1}{a\alpha} \int_0^t \sin[a\alpha(t - \tau)]F(\alpha, \tau) \, d\tau$$

$$= \frac{\pi}{a} \int_0^t \frac{1}{(\pi\alpha)} \sin[a(t - \tau)\alpha]F(\alpha, \tau) \, d\tau$$

Finally, using formula 1 in the table of Fourier transform operational properties, we can invert the Fourier transform

$$u(x, t) = \frac{1}{2a} \int_0^t \int_{-\infty}^{\infty} I_{a(t-\tau)} (x - z)f(z, \tau) \, dz \, d\tau$$

$$= \frac{1}{2a} \int_0^t \int_{x-a(t-\tau)}^{x+a(t-\tau)} f(z, \tau) \, dz \, d\tau \qquad (5.3.5)$$

The second form of the integral in (5.3.5) follows from the fact that

$$I_{a(t-\tau)}(x - z) = \begin{cases} 1 & \text{if } |x - z| < a(t - \tau) \\ 0 & \text{if } |x - z| > a(t - \tau) \end{cases}$$

$$= \begin{cases} 1 & \text{if } x - a(t - \tau) < z < x + a(t - \tau) \\ 0 & \text{otherwise} \end{cases}$$

We can now describe how the solution to an inhomogeneous equation is related to the solution to the corresponding homogeneous equation with an inhomogeneous initial condition. This observation is sometimes referred to as Duhamel's principle. We discussed Duhamel's principle previously in Section 3.3. ■ ■

5.4 DUHAMEL'S PRINCIPLE

The initial-value problem for the ordinary differential equation

$$u'(t) = Au(t), \qquad u(0) = g$$

has as its solution

$$u(t) = e^{At}g, \qquad t \geq 0$$

Here A and g denote real constants. The corresponding inhomogeneous problem

$$u'(t) = Au(t) + f(t), \qquad u(0) = g$$

has the solution

$$u(t) = e^{At}g + \int_0^t e^{A(t-\tau)}f(\tau) \, d\tau$$

Evidently, the solution to the homogeneous problem provides the "seed" for the solution to the inhomogeneous problem. In this case, the seed is the function exp[At].

A similar state of affairs exists with respect to partial differential equations. Consider the initial-value problem

$$u_t(x, t) = Du_{xx}(x, t), \qquad u(x, 0) = g(x)$$

The solution is given by (see Example 5.1.2)

$$u(x, t) = \int_{-\infty}^{\infty} K(x - z, t)g(z)\, dz$$

where

$$K(x, t) = \mathcal{F}^{-1}[\exp[-\alpha^2 Dt]]$$

If we adopt the notation

$$S(t)[g(x)] = \int_{-\infty}^{\infty} K(x - z, t)g(z)\, dz \tag{5.4.1}$$

then we will see that $S(t)$ plays the same role for the initial-value problem for the heat equation that exp[At] plays in the initial-value problem for the ordinary differential equation. Consider the inhomogeneous problem

$$u_t(x, t) = Du_{xx}(x, t) + f(x, t), \qquad u(x, 0) = g(x) \tag{5.4.2}$$

As we can see from Examples 5.1.2 and 5.3.2, the solution to this problem is, in terms of the notation of (5.4.1),

$$u(x, t) = S(t)[g(x)] + \int_0^t S(t - \tau)[f(x, \tau)]\, d\tau \tag{5.4.3}$$

For the first-order wave equation

$$u_t(x, t) = Cu_x(x, t) + f(x, t), \qquad u(x, 0) = g(x) \tag{5.4.4}$$

if we adopt the notation

$$S(t)[g(x)] = g(x + Ct) \tag{5.4.5}$$

then the solution of (5.4.4) can be expressed as

$$u(x, t) = S(t)[g(x)] + \int_0^t S(t - \tau)[f(x, \tau)]\, d\tau \tag{5.4.6}$$

This is verified by comparing (5.4.6) with (5.3.4).

Finally for the initial-value problem for the second-order homogeneous wave equation

$$u_{tt}(x, t) = a^2 u_{xx}(x, t)$$
$$u(x, 0) = 0, \qquad u_t(x, 0) = g(x) \tag{5.4.7}$$

the solution has been found in Example 5.1.3 to be

$$u(x, t) = \frac{1}{2a} \int_{x-at}^{x+at} g(z)\, dz$$

In this case, we let

$$S(t)[g(x)] = \frac{1}{2a} \int_{x-at}^{x+at} g(z)\, dz \qquad (5.4.8)$$

Then the inhomogeneous problem

$$u_{tt}(x, t) = a^2 u_{xx}(x, t) + f(x, t)$$
$$u(x, 0) = h(x), \qquad u_t(x, 0) = g(x) \qquad (5.4.9)$$

has as its solution

$$u(x, t) = \frac{\partial}{\partial t} S(t)[h(x)] + S(t)[g(x)] + \int_0^t S(t - \tau)[f(x, \tau)]\, d\tau \qquad (5.4.10)$$

That (5.4.10) is, in fact, correct can be verified by examining the solutions from Examples 5.1.3 and 5.3.3. To understand the form of (5.4.10), consider the following initial-value problem for a second-order ordinary differential equation:

$$u''(t) + A^2 u(t) = f(t), \qquad u(0) = h, \qquad u'(0) = g$$

Here, A, h, and g denote real constants.

The solution of this problem is

$$u(t) = \cos(At)h + \frac{1}{A}\sin(At)g + \int_0^t \frac{1}{A}\sin[A(t - \tau)]f(\tau)\, d\tau$$
$$= \frac{d}{dt}\left\{\frac{1}{A}\sin(At)\right\}h + \frac{1}{A}\sin(At)g + \int_0^t \frac{1}{A}\sin[A(t - \tau)]f(\tau)\, d\tau$$

We conclude that while $S(t)$ given by (5.4.1) and (5.4.5) (corresponding to initial-value problems for first-order equations) are analogous to $\exp[At]$, the $S(t)$ defined in (5.4.8) (for the solution to an initial-value problem for a second-order equation) is analogous to $1/A \sin(At)$. These analogies go further.

For $S(t)$ given by (5.4.1) or by (5.4.5),

$$S(0) = \text{identity}$$

in the sense that for both of these operators,

$$S(0)[g(x)] = g(x) \quad \text{for all functions } g(x)$$

This is analogous to the fact that $\exp[At] = 1$ at $t = 0$.

Similarly, for $S(t)$ given by (5.4.8)

$$S(0) = 0 \quad \text{and} \quad \frac{d}{dt} S(0) = \text{identity}$$

in the sense that

$$S(0)[g(x)] = 0$$

and

$$\frac{d}{dt} S(t)[g(x)] = \frac{1}{2a} [g(x + at)(a) - g(x - at)(-a)]$$

so

$$\frac{d}{dt} S(0)[g(x)] = \tfrac{1}{2}[g(x) + g(x)] = g(x)$$

for all functions $g(x)$. This corresponds to the behavior of the function $1/A \sin(At)$ and its derivative $\cos(At)$ at $t = 0$.

The observation that the solution to an inhomogeneous evolution equation can be deduced from the solution to the corresponding homogeneous initial-value problem is generally referred to as Duhamel's principle, although it is usually developed in a different way. While Duhamel's principle does not add any information that we did not already have from the previous two sections, it does bring to light the structure of the solution to the inhomogeneous problem.

Exercises: Inhomogeneous Problems and Duhamel's Principle

1. Solve for $u(x, y)$,

$$\nabla^2 u(x, y) = f(x, y), \qquad x > 0, \qquad y > 0$$
$$u_x(0, y) = 0, \qquad y > 0$$
$$u(x, 0) = 0, \qquad x > 0$$

2. Solve for $u(x, y)$,

$$\nabla^2 u(x, y) = f(x, y), \qquad x > 0, \qquad 0 < y < 1$$
$$u_x(0, y) = 0, \qquad 0 < y < 1$$
$$u(x, 0) = 0, u(x, 1) = 0, \qquad x > 0$$

How does the solution change if the boundary condition at $x = 0$ is changed to $u(0, y) = 0$, $0 < y < 1$?

In the following inhomogeneous initial-boundary-value problems identify the solution operator $S(t)$ for the homogeneous problem and then use Duhamel's principle to solve the inhomogeneous problem.

3. Find a bounded function $u(x, t)$ such that

$$u_t(x, t) = Du_{xx}(x, t) - u(x, t) + f(x, t), \qquad -\infty < x < \infty, \qquad t > 0$$
$$u(x, 0) = 0, \qquad -\infty < x < \infty$$

4. Find a bounded function $u(x, t)$ such that

$$u_t(x, t) = Du_{xx}(x, t) + f(x, t), \qquad x > 0, \qquad t > 0$$
$$u(x, 0) = 0, \qquad x > 0$$
$$u(0, t) = 0, \qquad t > 0$$

Solve again if the boundary condition at $x = 0$ is changed to,
(a) $u_x(0, t) = 0$
(b) $u_x(0, t) - hu(0, t) = 0$ [Hint: See Exercise 8 following Example 5.2.2.]

5. Solve for $u(x, t)$ if

$$u_{tt}(x, t) = a^2u_{xx}(x, t) + f(x, t), \qquad x > 0, \qquad t > 0$$
$$u(x, 0) = u_t(x, 0) = 0, \qquad x > 0$$
$$u(0, t) = 0, \qquad\qquad\qquad t > 0$$

6. Solve Exercise 5 if the boundary condition at $x = 0$ is changed to

$$u_x(0, t) = 0, t > 0$$

7. In Exercises 5 and 6, show how you would evaluate $u(x, t)$ at a point (x, t) for which $x - at < 0$.

Chapter 6

Uniqueness and Continuous Dependence on Data

In Chapter 1 we briefly discussed the notion of a well-posed problem for a partial differential equation. A well-posed problem is one for which a solution not only exists but in addition the solution is unique and depends continuously on the data. In Chapters 3 and 5 we solved a number of problems in partial differential equations, thereby establishing the existence of a solution for each of those problems. In this chapter we prove for some of these examples that the solution is also unique and depends continuously on the data. This will prove that these examples are well-posed problems. The proofs of uniqueness and continuous dependence will be based on one or the other of the following techniques:

- **(a)** Energy integral arguments adapted from integral identities
- **(b)** Maximum–minimum principles

The energy integrals are the subject of Section 6.2, and in Section 6.3 the maximum–minimum principles are discussed. We begin, however, with a brief additional discussion of well-posedness for problems in partial differential equations.

6.1 WELL-POSED PROBLEMS IN PARTIAL DIFFERENTIAL EQUATIONS

As we have said before, a problem in partial differential equations is said to be well posed if and only if (1) a solution exists, (2) the solution is unique, and (3) the solution depends continuously on the data. If any one of these three conditions is violated, then the problem is said to be ill-posed. While ill-posed problems are not without interest, the analysis of such a problem is usually a complicated matter. We do not attempt to solve any ill-posed problems in this text.

The existence and uniqueness of a solution to a partial differential equation is controlled to a large extent by the *number* of the auxiliary conditions imposed. If the number of conditions is too few, the solution will not be unique. If this number is too many, no solution will exist. This is similar to the situation that exists with respect to ordinary differential equation problems. Continuous dependence on the data, however, is controlled by the *type* of auxiliary conditions that are imposed, and in this respect partial differential equations problems are not at all similar to problems in ordinary differential equations. For a given equation, it is difficult to characterize the type of conditions that will lead to a well-posed problem for that equation. Instead, then, we list several examples of problems that are well posed. In Sections 6.2 and 6.3 we prove the well-posedness for some of them. In the exercises at the end of this section we give a few examples of ill-posed problems where the solutions do not depend continuously on the data.

Boundary-Value Problems on Bounded Domains

Let D denote a bounded region in the plane with "smooth" boundary S. Most of the examples we present carry over with little or no change to the case where D is a region in R^3 or even R^N. However, to simplify things as much as possible, we consider only examples involving partial differential equations in two independent variables. We are not specific about what is meant by a smooth boundary for a region D other than to say that included in this class are "reasonable" regions such as disks or sectors of disks, rectangles or unions of finitely many rectangles, and other nonpathological sets.

In each of the following examples, F and g denote arbitrary functions defined and, say, continuous on D and S, respectively.

EXAMPLE 6.1.1

Dirichlet Problem

$$\nabla^2 u(x, y) = F(x, y) \quad \text{in } D \quad \text{and} \quad u = g \quad \text{on } S$$

This problem has one and only one solution and that solution depends continuously on the data F and g. ■ ■

EXAMPLE 6.1.2

Neumann Problem

$$\nabla^2 u(x, y) = F(x, y) \quad \text{in } D \quad \text{and} \quad \partial_N u = g \quad \text{on } S$$

Here we are using the notation $\partial_N u = \nabla \mathbf{u} \cdot \mathbf{N}$, where $\mathbf{N}$ denotes the unit outward normal to S.

This problem has a solution (provided F and g satisfy a compatibility condition to be specified later). This solution is unique up to an additive constant and depends continuously on the data. ■ ■

EXAMPLE 6.1.3 _____

Robin Problem

$$\nabla^2 u(x, y) = F(x, y) \quad \text{in } D \quad \text{and} \quad a\, \partial_N u + bu = g \quad \text{on } S$$

Here, $a(x, y)$ and $b(x, y)$ appearing in the boundary condition denote (continuous) functions defined on S and satisfying the condition $ab > 0$ on S. Then this problem has a unique solution that depends continuously on the data.　■ ■

When the domain D is not bounded, the solution to a boundary-value problem may be required to satisfy "conditions at infinity." There may be some freedom in choosing the precise form of these conditions, and their form may also vary with the number of independent variables and with the form of the condition that applies on the accessible part of the boundary.

For example, let D denote the "half-space"

$$\{(x, y) : x \in R, y > 0\}$$

The "accessible part" of the boundary is then the real axis

$$S = \{(x, 0) : x \in R\}$$

The region

$$D = \{(x, y) : x^2 + y^2 > 1\}$$

is another example of an unbounded region. For this D, the accessible part of the boundary is the set

$$S = \{x^2 + y^2 = 1\}$$

Then for *either* of these two examples of an unbounded region D, the following are examples of well-posed problems.

EXAMPLE 6.1.4 _____

Dirichlet Problem

$$\nabla^2 u(x, y) = F(x, y) \quad \text{in } D$$
$$u = g \quad \text{on } S$$
$$u(x, y) \text{ is bounded for all } (x, y) \text{ in } D$$　■ ■

EXAMPLE 6.1.5 _____

Neumann Problem

$$\nabla^2 u(x, y) = F(x, y) \quad \text{in } D$$
$$\partial_N u = g \quad \text{on } S$$
$$u(x, y) \to 0 \quad \text{as} \quad [x^2 + y^2] \to \infty$$

When the region D is unbounded, there is no compatibility condition imposed on F and g, as there is for the Neumann problem when the region D is bounded.　■ ■

EXAMPLE 6.1.6 ───

Robin Problem

$$\nabla^2 u(x, y) = F(x, y) \quad \text{in } D$$
$$a\,\partial_N u + bu = g \quad \text{on } S$$
$$u(x, y) \text{ bounded for all } (x, y) \text{ in } D$$

■ ■

Initial-Value Problems

We list now some examples of well-posed problems for some pure initial-value problems. Pure initial-value problems are sometimes referred to as *Cauchy problems*.

EXAMPLE 6.1.7 ───

Heat Equation

$$u_t(x, t) = u_{xx}(x, t) + F(x, t), \quad -\infty < x < \infty, \quad t > 0$$
$$u(x, 0) = g(x), \quad -\infty < x < \infty$$
$$u(x, t) \text{ bounded for all } x \text{ and } t > 0$$

■ ■

EXAMPLE 6.1.8 ───

Wave Equation

$$u_{tt}(x, t) = u_{xx}(x, t) + F(x, t), \quad -\infty < x < \infty, \quad t > 0$$
$$u(x, 0) = g_1(x), \quad u_t(x, 0) = g_2(x), \quad -\infty < x < \infty$$

■ ■

Note that the wave equation needs two initial conditions, while just one is sufficient for the heat equation. Note also that Example 6.1.7 for the heat equation includes a condition at infinity for well-posedness, while Example 6.1.8 for the wave equation requires none. The reason for this is related to the fact (developed in Chapter 5) that the wave equation describes processes that propagate with finite speed, while processes that are described by the heat equation can, in a sense, be said to propagate with infinite speed.

Mixed Initial-Boundary-Value Problems

We list here four examples of well-posed problems for evolution equations in which both initial conditions and boundary conditions are involved. We suppose through all four examples that not both of the coefficient functions $a(t)$, $b(t)$ appearing in any one boundary condition are identically zero. If both of these coefficients were to vanish identically, the boundary condition would disappear.

EXAMPLE 6.1.9 _____

Heat Equation on Semi-infinite Interval

$$u_t(x, t) = u_{xx}(x, t) + F(x, t), \qquad x > 0, \qquad t > 0$$
$$u(x, 0) = g(x), \qquad x > 0$$
$$a(t)u_x(0, t) + b(t)u(0, t) = f(t), \qquad t > 0$$
$$u(x, t) \text{ bounded for } x > 0, \quad t > 0 \qquad \blacksquare\blacksquare$$

EXAMPLE 6.1.10 _____

Wave Equation on Semi-infinite Interval

$$u_{tt}(x, t) = u_{xx}(x, t) + F(x, t), \qquad x > 0, \qquad t > 0$$
$$u(x, 0) = g_1(x), \qquad u_t(x, 0) = g_2(x), \qquad x > 0$$
$$a(t)u_x(0, t) + b(t)u(0, t) = f(t), \qquad t > 0 \qquad \blacksquare\blacksquare$$

These two examples involve an unbounded (spatial) interval so that the problem for the heat equation requires a condition at infinity. The next two examples are for a bounded (spatial) interval, which we denote by $I = (x_1, x_2)$, $-\infty < x_1 < x_2 < \infty$.

EXAMPLE 6.1.11 _____

Heat Equation on Bounded Interval

$$u_t(x, t) = u_{xx}(x, t) + F(x, t), \qquad x \in I, \qquad t > 0$$
$$u(x, 0) = g(x), \qquad x \in I$$
$$a_1(t)u_x(x_1, t) + b_1(t)u(x_1, t) = f_1(t), \qquad t > 0$$
$$a_2(t)u_x(x_2, t) + b_2(t)u(x_2, t) = f_2(t), \qquad t > 0 \qquad \blacksquare\blacksquare$$

EXAMPLE 6.1.12 _____

Wave Equation on Bounded Interval

$$u_{tt}(x, t) = u_{xx}(x, t) + F(x, t), \qquad x \in I, \qquad t > 0$$
$$u(x, 0) = g_1(x), \qquad u_t(x, 0) = g_2(x), \qquad x \in I$$
$$a_1(t)u_x(x_1, t) + b_1(t)u(x_1, t) = f_1(t), \qquad t > 0$$
$$a_2(t)u_x(x_2, t) + b_2(t)u(x_2, t) = f_2(t), \qquad t > 0 \qquad \blacksquare\blacksquare$$

These examples are not the only examples of well-posed problems for these equations, but they are representative examples that illustrate the number and type of auxiliary conditions that combine with the equations to form well-posed problems.

Exercises: Well-Posed Problems

1. Let H denote the half-plane $\{(x, y) : -\infty < x < \infty, y > 0\}$. Then verify that for N a positive integer $u(x, y) = (1/N^2)\sin(Nx)\sinh(Ny)$ satisfies

$$\nabla^2 u(x, y) = 0 \quad \text{in } H$$
$$u(x, 0) = 0, \qquad u_y(x, 0) = 1/N \sin(Nx), \qquad -\infty < x < \infty$$

Also verify that the trivial solution $v(x, y) = 0$ satisfies

$$\nabla^2 v(x, y) = 0 \quad \text{in } H$$
$$v(x, 0) = 0, \qquad v_y(x, 0) = 0, \qquad -\infty < x < \infty$$

Note that by choosing N sufficiently large, the data in the u problem can be made arbitrarily close to the data in the v problem (uniformly in x). On the other hand, the difference between $u(x, y)$ and $v(x, y)$ becomes arbitrarily large as N increases. Show this by selecting a point (x, y) in H such that

$$|u(x, y) - v(x, y)| \to \infty \quad \text{as} \quad N \to \infty$$

Then this is an example of a problem in which the solution does not depend continuously on the data; that is, it is an-ill posed problem for Laplace's equation.

2. Let D denote the region $(0, 1) \times (0, L)$ for $L > 0$. Then verify that for positive integers M and N,

$$u(x, y) = \sin(M\pi x)\sin(N\pi y)$$

satisfies

$$u_{xx}(x, y) - u_{yy}(x, y) = 0 \quad \text{in } D$$
$$u(x, 0) = u(x, L) = 0, \qquad 0 < x < 1$$
$$u(0, y) = u(1, y) = 0, \qquad 0 < y < L$$

provided M, N satisfy the condition $L^2 = (M/N)^2$; that is, L must be a rational number.

Verify that $u(x, y) = 0$ is a solution for any value of L. Then this problem has infinitely many solutions if L is a rational number but only the trivial solution if L is an irrational number. This is an example of an ill-posed problem for the wave equation. The solution does not depend continuously on the data (the dimensions of the domain are considered part of the data in the problem).

Following is a list of partial differential equations and for each a domain where the equation is to be satisfied is specified. Add to each a sufficient number of auxiliary conditions to form a well-posed problem.

3. $u_t(x, t) = Du_{xx}(x, t)$ for $0 < x < 1$, $t > 0$

4. $u_t(x, t) = Du_{xx}(x, t)$ for $-\infty < x < \infty$, $t > 0$

5. $u_t(x, t) = Du_{xx}(x, t) - u(x, t)$ for $x > 0$, $t > 0$

6. $u_{xx}(x, y) + u_{yy}(x, y) = 0$ for $-\infty < x < \infty$, $y > 0$

7. $u_{xx}(x, y) + u_{yy}(x, y) + 2u(x, y) = 0$ for $0 < x, y < 1$

8. $u_{xx}(x, y) + u_{yy}(x, y) - u_x(x, y) = 0$ for $x > 0$, $0 < y < 1$

9. $\nabla^2 u(r, \theta) = 0$ for $r < 1$

10. $u_{tt}(x, t) = a^2 u_{xx}(x, t)$ for $-\infty < x < \infty$, $t > 0$

11. $u_{tt}(x, t) = a^2 u_{xx}(x, t) - u_t(x, t)$ for $x > 0$, $t > 0$

12. $u_{tt}(x, t) = a^2 u_{xx}(x, t) - u(x, t)$ for $0 < x < L$, $t > 0$

6.2 GREEN'S IDENTITIES AND ENERGY INEQUALITIES

The so-called Green's identities all stem from a single integral identity known as the divergence theorem.

Theorem 6.2.1 (Divergence Theorem). Let Ω denote a bounded region in R^N having a smooth boundary Σ with $\mathbf{N} = \mathbf{N}(x_1, \ldots, x_N)$ denoting the unit outward-pointing normal vector to Σ. Then for any smooth vector-valued function

$$\mathbf{W}(x_1, \ldots, x_N) = \sum_{j=1}^{N} w_j(x_1, \ldots, x_N)\mathbf{e}_j$$

we have

$$\int_\Omega \text{div } \mathbf{W} \, d\Omega = \int_\Sigma \mathbf{W} \cdot \mathbf{N} \, dS, \qquad d\Omega = dx_1 \cdots dx_N \tag{6.2.1}$$

When $N = 3$,

$$\text{div } \mathbf{W} = w_{1,x}(x, y, z) + w_{2,y}(x, y, z) + w_{3,z}(x, y, z)$$

and Equation (1.2.1) can be seen to be a discrete version of Equation (6.2.1). In the one-dimensional case, Ω is an interval (a, b), and the boundary of Ω is just the set of points $\{a\}$ and $\{b\}$. Then $\mathbf{N}(a) = -\mathbf{i}$, $\mathbf{N}(b) = \mathbf{i}$, $\text{div } \mathbf{W} = w'(x)$, and (6.2.1) reduces to

$$\int_a^b w'(x) \, dx = w(b) - w(a)$$

which is just the fundamental theorem of calculus.

If we apply the divergence theorem in the special case that

$$\mathbf{W} = U \text{ grad } V$$

for smooth, scalar-valued functions U and V, then

$$\text{div}[U \text{ grad } V] = U \nabla^2 V + \text{grad } U \cdot \text{grad } V \tag{6.2.2}$$

The identity (6.2.2) follows from many applications of the usual product rule for derivatives. Then (6.2.1) becomes

$$\int_\Omega [U \nabla^2 V + \text{grad } U \cdot \text{grad } V] \, d\Omega = \int_\Sigma U \, \partial_N V \, dS \tag{6.2.3}$$

where

$$\partial_N V = \text{grad } V \cdot \mathbf{N} = \text{outward normal derivative}$$
$$= \text{directional derivative of } V \text{ in direction of } \mathbf{N}$$

The identity (6.2.3) is known as *Green's first identity*. In the one-dimensional case (6.2.3) reduces to

$$\int_a^b [U(x)V''(x) + U'(x)V'(x)] \, dx = U(x)V'(x)\big|_a^b \tag{6.2.4}$$

which is just the integration-by-parts formula.

From Green's first identity, it is easy to derive *Green's second identity*,

$$\int_\Omega [U \nabla^2 V - V \nabla^2 U] \, d\Omega = \int_\Sigma [U \, \partial_N V - V \, \partial_N U] \, dS \tag{6.2.5}$$

Note that the divergence theorem and Green's identities call for Ω to have a "piecewise smooth" boundary. Specifying precisely what is meant by smooth can get complicated. We shall only say that examples of regions with piecewise smooth boundaries include disks or segments of disks and rectangles or finite unions of rectangles in two dimensions and spheres or segments of spheres and rectangular boxes or finite unions of boxes in three dimensions.

In addition, these results call for a certain degree of smoothness for the functions in the integrands of the integrals that occur. For the divergence theorem, it is sufficient for $\mathbf{W}$ to be continuously differentiable in Ω and continuous on the set $\Omega + \Sigma$ consisting of Ω together with its boundary Σ. For the Green identities, then, it is sufficient for U and V to be two times continuously differentiable in Ω and continuous on $\Omega + \Sigma$. In the lemmas to follow, this is what we mean by a "smooth solution."

We now give several applications of Green's identities.

Uniqueness for Mixed Boundary-Value Problem

Lemma 6.2.1. Suppose Ω is a region in R^N ($N = 2, 3$) whose boundary consist of two disjoint, complementary parts Σ_A and Σ_B. Let F, f, and g denote given data functions and consider the problem

$$\nabla^2 u = F \quad \text{in } \Omega$$
$$u = f \quad \text{on } \Sigma_A$$
$$\partial_N u = g \quad \text{on } \Sigma_B$$

for the unknown function u. Then this problem has at most one smooth solution.

Proof. In proving uniqueness, we always begin by supposing that the problem has two distinct solutions. We then proceed to show that the difference of the two solutions is identically zero and hence that the solutions are not distinct.

Suppose then the preceding problem has two smooth solutions denoted by u and v. Then let $w = u - v$, and note that w must satisfy

$$\nabla^2 w = 0 \quad \text{in } \Omega$$
$$w = 0 \quad \text{on } \Sigma_A$$
$$\partial_N w = 0 \quad \text{on } \Sigma_B$$

Then by Green's first identity, choosing $U = V = w$, we have

$$\int_\Omega w \, \nabla^2 w \, d\Omega = \int_\Sigma w \, \partial_N w \, dS - \int_\Omega |\mathbf{grad}\, w|^2 \, d\Omega$$

$$0 = 0 - \int_\Omega |\mathbf{grad}\, w|^2 \, d\Omega$$

Note that

$$\int_\Sigma w \, \partial_N w \, dS = \int_{\Sigma_A} w \, \partial_N w \, dS + \int_{\Sigma_B} w \, \partial_N w \, dS = 0$$

since $w = 0$ on Σ_A and $\partial_N w = 0$ on Σ_B.

Now it follows from the preceding that $|\textbf{grad } w|$ vanishes on $\Omega + \Sigma$, which is to say w is a constant in $\Omega + \Sigma$. But w is smooth on $\Omega + \Sigma$ and is zero on Σ_A. Then w must be identically zero on $\Omega + \Sigma$, which means u and v are identically equal on $\Omega + \Sigma$. This proves the lemma.

It follows from Lemma 6.2.1 that the following examples from Chapter 3 have at most one smooth solution:

> **Example 3.1.1:** Here $F = 0$, Ω is the unit square, and $\Sigma = \Sigma_A$ (with Σ_B empty).

> **Example 3.1.3:** Here $F \neq 0$, Ω is the unit square, and Σ_A consists of the vertical sides of Ω while the horizontal sides make up Σ_B

Since we constructed smooth solutions for each of these problems in Chapter 3, it follows that those are the *unique* solutions.

Compatibility of Data in Neumann Problem

Lemma 6.2.2. Let Ω denote a bounded region in R^N ($N = 2, 3$) having smooth boundary Σ. The Neumann problem

$$\nabla^2 u = F \quad \text{in } \Omega, \qquad \partial_N u = g \quad \text{on } \Sigma \qquad (6.2.6)$$

has *no* solution unless the data functions F and g satisfy the compatibility condition

$$\int_\Omega F \, d\Omega = \int_\Sigma g \, dS \qquad (6.2.7)$$

Proof. We apply Green's first identity with $U = 1$ and $V = u$. Then **grad** U is zero, and

$$\int_\Omega \nabla^2 u \, d\Omega = \int_\Sigma \partial_N u \, dS$$

Then if u satisfies (6.2.6), (6.2.7) follows.

Note that when the compatibility condition is satisfied, the solution exists but is unique only up to an additive constant. This was already observed in Example 3.1.2.

Uniqueness in Mixed Problem for Heat Equation

Lemma 6.2.3. Let Ω denote a bounded region in R^N ($N = 1, 2, 3$) having smooth boundary Σ. Then the initial-boundary-value problem for the heat equation,

$$\begin{aligned}
u_t &= D \, \nabla^2 u + F && \text{for } x \text{ in } \Omega \text{ and } t > 0 \\
u(x, 0) &= f(x) && \text{for } x \text{ in } \Omega \\
u(x, t) &= g(x, t) && \text{for } x \text{ on } \Sigma, t > 0
\end{aligned} \qquad (6.2.8)$$

has at most one smooth solution.

Proof. We begin in the usual way by supposing there are *two* solutions u and v and $w = u - v$. Then w satisfies a homogeneous version of (6.2.8):

$$w_t = D \, \nabla^2 w \quad \text{for } x \text{ in } \Omega \text{ and } t > 0$$
$$w(x, 0) = 0 \qquad \text{for } x \text{ in } \Omega$$
$$w(x, t) = 0 \qquad \text{for } x \text{ on } \Sigma, \, t > 0 \qquad (6.2.9)$$

Now define

$$J(t) = \int_\Omega w(x, t)^2 \, d\Omega \quad \text{for } t \geq 0$$

Then $J(t) \geq 0$, $J(0) = 0$, and

$$J'(t) = 2 \int_\Omega w(x, t) w_t(x, t) \, d\Omega$$

$$= 2D \int_\Omega w(x, t) \, \nabla^2 w(x, t) \, d\Omega$$

Now we apply Green's first identity with $U = V = w$ to get

$$J'(t) = -2D \int_\Omega |\mathbf{grad} \, w|^2 \, d\Omega \leq 0$$

Here we have made use of the fact that w vanishes on Σ. Since $J'(t) \leq 0$, it follows that $J(t)$ is a nonincreasing, nonnegative function of t. But $J(0) = 0$, and it follows that $J(t)$ is identically zero. But then $w(x, t)$ must be zero for all x in Ω and all $t > 0$. This proves that $u = v$ in Ω for all $t > 0$.

The function $J(t)$ is called the "energy integral" associated with (6.2.9) since $J(t)$ is related to the total energy in the system modeled by (6.2.9). The fact that $J(t)$ is nonincreasing is related to the fact that this system "dissipates" energy. A system for which $J'(t) = 0$ would be called "conservative" since in such a system the energy must remain constant. We see an example of such a system in the next lemma.

Lemma 6.2.3 implies that the problem solved in Example 3.2.1 has at most one solution. There $N = 1$ and Ω is the interval $(0, 1)$ and F and g are zero. The dissipative nature of this system is reflected in the profiles plotted in Figure 3.2.1. There the temperature profiles can be seen to decay uniformly from the initial state toward a zero steady state.

Uniqueness in Mixed Problem for Wave Equation

Lemma 6.2.4. The problem,

$$u_{tt}(x, t) = a^2 u_{xx}(x, t) + F(x, t), \qquad 0 < x < L, \qquad t > 0$$
$$u(x, 0) = f(x), \qquad u_t(x, 0) = g(x), \qquad 0 < x < L$$
$$u(0, t) = p(t), \qquad u(L, t) = q(t), \qquad t > 0$$

has at most one smooth solution.

Proof. Suppose there are two solutions u and v and let w denote their difference. Then $w(x, t)$ satisfies

$$
\begin{aligned}
w_{tt}(x, t) &= a^2 w_{xx}(x, t), && 0 < x < L, && t > 0 \\
w(x, 0) &= 0, & w_t(x, 0) &= 0, && 0 < x < L \\
w(0, t) &= 0, & w(L, t) &= 0, && t > 0
\end{aligned}
$$

Define

$$
J(t) = \int_0^L [w_t(x, t)^2 + a^2 w_x(x, t)^2] \, dx
$$

Then,

$$
J'(t) = 2 \int_0^L [w_t w_{tt} + a^2 w_x w_{xt}] \, dx
$$

Now we apply the one-dimensional version of Green's first identity, (6.2.4), with the choices $U = w_t$ and $V = w$. This implies

$$
\int_0^L w_x w_{xt} \, dx = w_t w_x \Big|_0^L - \int_0^L w_t w_{xx} \, dx
$$

and if we use this in the expression for $J'(t)$, we get

$$
J'(t) = 2 \int_0^L w_t[w_{tt} - a^2 w_{xx}] \, dx + a w_x w_t \Big|_0^L
$$
$$
= 0
$$

Here we have made use of the fact that $w(x, t)$ satisfies a homogeneous wave equation and homogeneous boundary conditions. In particular, $w(0, t) = w(L, t) = 0$ for all t implies that $w_t(0, t) = w_t(L, t) = 0$ for all t. It follows that $J(t) = \text{const}$, and then the initial conditions imply that this constant is zero. Since $J(t) = 0$, it follows that $w(x, t)$ is a constant, and then the initial conditions are used once more to infer that $w(x, t) = 0$. Then u and v are identically equal, and the problem has at most one solution.

Here $J(t)$ is, as it was in the previous lemma, related to the total energy in the system being described by this initial-boundary-value problem. More specifically,

$\int_0^L u_x(x, t)^2 \, dx$ is related to the potential energy produced by displacing the system away from its equilibrium state and

$\int_0^L u_t(x, t)^2 \, dx$ is related to the kinetic energy in the system as a result of movement in the system.

In the special case that $F = p = q = 0$, there is no energy coming into the system from external sources, and

$$
J(t) = \int_0^L [u_t(x, t)^2 + a^2 u_x(x, t)^2] \, dx
$$

can be shown to be a constant. This implies that in the absence of externally supplied energy, the energy in the system is constant; that is, the system is conservative. This lemma and discussion apply to Example 3.3.1 and problem (5.2.17).

Exercises: Green's Identities

1. Suppose Ω is a two-dimensional region with smooth boundary Σ. Let α and β be constants such that $\alpha\beta > 0$. Then show that there can be at most one smooth solution for

$$\nabla^2 u(x, y) = F(x, y) \quad \text{in } \Omega$$
$$\alpha u + \beta \, \partial_N u = g \quad \text{on } \Sigma$$

Does the conclusion change if (a) $\alpha\beta \geq 0$ or (b) $\alpha\beta < 0$?

2. Given $F(x, y)$, suppose that $u(x, y)$ is a smooth solution for

$$\nabla^2 u(x, y) + u = F(x, y) \quad \text{in } \Omega$$
$$\alpha u + \beta \, \partial_N u = 0 \quad \text{on } \Sigma$$

Show that if $v(x, y)$ satisfies

$$\nabla^2 v(x, y) + v(x, y) = 0 \quad \text{in } \Omega$$
$$\alpha v + \beta \, \partial_N v = 0 \quad \text{on } \Sigma$$

then, necessarily,

$$\int_\Omega F(x, y)v(x, y) \, dx \, dy = 0$$

that is, the solution $u(x, y)$ does not exist for every F.

3. Suppose $u(x, y)$ satisfies

$$\nabla^2 u(x, y) + ku(x, y) = F(x, y) \quad \text{in } \Omega$$
$$\partial_N u = g \quad \text{on } \Sigma$$

Show:
(a) If $k < 0$, then this problem has at most one smooth solution.
(b) If $k = 0$, then any two smooooth solutions differ by a constant.
(c) What conclusions regarding uniqueness can be drawn if k is positive?

4. Suppose $u(x, y)$ satisfies

$$u_{xx}(x, y) + u_{yy}(x, y) = 0, \quad x > 0, \quad y > 0$$
$$u(0, y) = f(y), \quad y > 0$$
$$u(x, 0) = g(x), \quad x > 0$$
$$u(x, y) \to 0 \quad \text{as} \quad x \to \infty \quad \text{or} \quad y \to \infty$$

Then show that this problem has at most one smooth solution. On the other hand, if $u(x, y)$ satisfies

$$u_{xx}(x, y) + u_{yy}(x, y) = 0 \quad \text{for all } x, \quad 0 < y < \pi$$
$$u(x, 0) = f(x), \quad u(x, \pi) = g(x), \quad \text{for all } x$$

then $u(x, y) + e^x \sin y$ is also a solution. Explain.

5. Suppose Ω is as in Exercise 1 and that $u(x, y, t)$ satisfies

$$
\begin{aligned}
u_t(x, y, t) &= D \nabla^2 u(x, y, t) + F(x, y, t) && \text{in } \Omega \text{ for } t > 0 \\
u(x, y, 0) &= f(x, y) && \text{in } \Omega \\
\partial_N u(x, y, t) &= g(x, y, t) && \text{on } \Sigma \text{ for } t > 0
\end{aligned}
$$

If D denotes a positive constant, show that the problem has at most one smooth solution. Does the conclusion change if D is a negative constant?

6. Repeat Exercise 5 if the boundary condition is replaced by

$$
\alpha u(x, y, t) + \beta \, \partial_N u(x, y, t) = g(x, y, t) \quad \text{on } \Sigma \text{ for } t > 0
$$

where α and β are constants that satisfy $\alpha\beta > 0$.

7. Suppose that $u(x, t)$ satisfies

$$
\begin{aligned}
u_t(x, t) &= D u_{xx}(x, t) + F(x, t), && x > 0, && t > 0 \\
u(x, 0) &= f(x), && x > 0 \\
u(0, t) &= g(t), && t > 0 \\
& u(x, t) \to 0 \quad \text{as} \quad x \to \infty
\end{aligned}
$$

Show that this problem has at most one smooth solution.

8. Repeat Exercise 7 in the case that the boundary condition

$$
u(0, t) = g(t) \quad \text{for } t > 0
$$

is replaced by the condition

$$
\alpha u(0, t) + \beta u_x(0, t) = g(t) \quad \text{for } t > 0
$$

Is it necessary for the constants α, β to satisfy $\alpha\beta > 0$?

9. Prove that the problem

$$
\begin{aligned}
u_t(x, t) &= D u_{xx}(x, t) + F(x, t), && 0 < x < L, && t > 0 \\
u(x, 0) &= f(x), && 0 < x < L \\
u(0, t) &= g(t), && u_x(L, t) = h(t), && t > 0
\end{aligned}
$$

has at most one smooth solution.

10. Suppose $u(x, t)$ satisfies

$$
\begin{aligned}
u_t(x, t) &= u_{xx}(x, t), && 0 < x < L, && t > 0 \\
u(x, 0) &= f(x), && 0 < x < L \\
u(0, t) &= 0, && u_x(L, t) = 0, && t > 0
\end{aligned}
$$

and let

$$
J(t) = \int_0^L u(x, t)^2 \, dx
$$

What can you say about $J(0)$? Is $J(t)$ increasing, decreasing, or constant?

11. Show that the problem

$$
\begin{aligned}
u_{tt}(x, t) &= a^2 u_{xx}(x, t) + F(x, t), && 0 < x < L, && t > 0 \\
u(x, 0) &= f(x), && u_t(x, 0) = g(x), && 0 < x < L \\
u(0, t) &= 0, && u_x(L, t) = 0, && t > 0
\end{aligned}
$$

has at most one smooth solution.

12. Suppose $u(x, t)$ is the solution of the initial-boundary-value problem in Exercise 11 in the case $F(x, t) = 0$, and let

$$J(t) = \int_0^L [u_t(x, t)^2 + a^2 u_x(x, t)^2] \, dx$$

Compute $J(0)$ in terms of the data in the problem. Is $J(t)$ increasing, decreasing, or constant?

13. Suppose $u(x, t)$ satisfies

$$u_{tt}(x, t) = a^2 u_{xx}(x, t) - u_t(x, t), \qquad 0 < x < L, \qquad t > 0$$
$$u(x, 0) = f(x), \qquad u_t(x, 0) = g(x), \qquad 0 < x < L$$
$$u(0, t) = 0, \qquad u(L, t) = 0, \qquad t > 0$$

If $J(t)$ is defined as in the previous exercise, then is $J(t)$ increasing, decreasing, or constant?

14. Prove that Example 6.1.10 has at most one solution.

15. Prove that Example 6.1.12 has at most one solution.

6.3 MAXIMUM–MINIMUM PRINCIPLES

In Section 1.3 we observed that solutions to the discrete Laplace equation satisfied the so-called discrete max–min principle. You may wish to reread that section now since we are about to show that this behavior carries over to the continuous Laplace equation.

Solutions to elliptic and parabolic partial differential equations exhibit certain typical behavior that we can collectively describe by saying that the solutions satisfy max–min principles. There is *no* such principle that applies to solutions of partial differential equations of hyperbolic type.

Max–Min Principles for Laplace Equation

Suppose that $f(x)$ denotes a function of one variable that is continuous on the *closed* interval $[a, b]$. Suppose further that $f''(x)$ exists and is continuous on the *open* interval (a, b). If we let

$$m = \min\{f(a), f(b)\} \quad \text{and} \quad M = \max\{f(a), f(b)\}$$

then we can draw the following conclusions,

    (a) If $f''(x) \geq 0$ for $a < x < b$
        then $f(x) \leq M$ for $a \leq x \leq b$
    (b) If $f''(x) = 0$ for $a < x < b$
        then $m \leq f(x) \leq M$ for $a \leq x \leq b$
    (c) If $f''(x) \leq 0$ for $a < x < b$
        then $f(x) \geq m$ for $a \leq x \leq b$

The sketches in Figure 6.3.1 indicate why these statements must be true.

The conditions that $f(x)$ be continuous on the *closed* interval $[a, b]$ and $f''(x)$ be continuous on the open interval (a, b) are necessary in order that these max–min principles hold. This is evident from Figure 6.3.2.

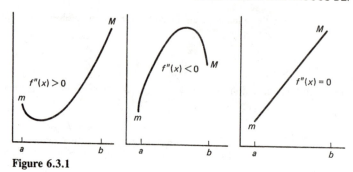

Figure 6.3.1

In the first two sketches in Figure 6.3.2 $f''(x)$ is not continuous on the open interval, and in the third sketch $f(x)$ is not continuous on the closed interval.

Max–min principles analogous to these are true when the one-dimensional Laplacian d^2/dx^2 is replaced by the Laplacian in two or more variables.

Theorem 6.3.1. Let Ω denote a bounded region in R^N having smooth boundary Σ, and suppose u is continuous on the closed set $\Omega + \Sigma$ and that $\nabla^2 u$ is continuous on Ω. Let

$$m = \min_{\Sigma} u \quad \text{and} \quad M = \max_{\Sigma} u$$

Then

    (a) $\nabla^2 u \geq 0$ in Ω implies that $u \leq M$ in $\Omega + \Sigma$
    (b) $\nabla^2 u = 0$ in Ω implies that $m \leq u \leq M$ in $\Omega + \Sigma$
    (c) $\nabla^2 u \leq 0$ in Ω implies that $u \geq m$ in $\Omega + \Sigma$

Note: Result (c) follows from applying result (a) to $-u$.

A proof of Theorem 6.3.1 can be found in P. DuChateau and D. Zachmann, *Partial Differential Equations,* Schaum's Outline Series, McGraw-Hill, New York, 1986. Theorem 6.3.1 is referred to as the "weak" maximum principle since it does not preclude the possibility of an interior maximum or minimum. A stronger result is true, however.

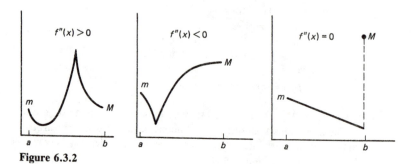

Figure 6.3.2

Theorem 6.3.2. Let Ω denote a bounded region in R^N, and suppose u is continuous on the closed set $\Omega + \Sigma$ and that $\nabla^2 u$ is continuous on Ω. Let

$$m = \min_\Sigma u \quad \text{and} \quad M = \max_\Sigma u$$

If $\nabla^2 u \geq 0$ in Ω, then $u = M$ at *some* point inside Ω implies that $u = M$ at *all* points in $\Omega + \Sigma$. If $\nabla^2 u \leq 0$ in Ω, then $u = m$ at *some* point inside Ω implies that $u = m$ at *all* points in $\Omega + \Sigma$.

Theorem 6.3.2 is called the "strong" maximum principle. It adds to the conclusion of the weak maximum principle the remark that unless u is a constant, the maximum and minimum values occur on the boundary of Ω. A proof of Theorem 6.3.2 can be found in M.H. Protter and H. F. Weinberger, *Maximum Principles in Differential Equations,* Prentice-Hall, Englewood Cliffs, N.J., 1967.

The hypothesis that Ω be bounded is necessary in order for the max–min principles to hold as stated. For example,

$$u(x, y) = e^x \sin y$$

satisfies Laplace's equation in the strip $\{-\infty < x < \infty, \, 0 < y < \pi\}$ and u is zero on the boundary of the strip so that $m = M = 0$. However, u assumes both positive and negative values *inside* the strip in violation of the conclusions of both Theorems 6.3.1 and 6.3.2.

The max–min principle has many applications. The following theorems will indicate some of the ways in which it can be used.

Continuous Dependence Result for Dirichlet Problem

Theorem 6.3.4. Let Ω denote a bounded region in R^N having smooth boundary denoted by Σ, and suppose

$$\nabla^2 u = 0 \quad \text{in } \Omega \qquad u = g \quad \text{on } \Sigma \tag{6.3.1}$$

where we assume that u is continuous on the closed set $\Omega + \Sigma$ and $\nabla^2 u$ is continuous on Ω. Then u depends continuously on the data in the problem.

Note: A solution u of (6.3.1) such that u is continuous on the closed set $\Omega + \Sigma$ and $\nabla^2 u$ is continuous on Ω is said to be a *classical solution* of the boundary-value problem. This means that the partial differential equation and the boundary conditions are satisfied in the classical pointwise sense. It is not uncommon to encounter problems in partial differential equations where no classical solution exists. For instance, several of the solutions constructed in Chapters 3 and 5 are not classical solutions. In such cases it is often possible to find a function that satisfies the conditions of the problem in a sense that is weaker than the classical pointwise sense. One such notion of "weak solution" is discussed in Chapter 7.

Proof. Let u_1, u_2 denote solutions of the boundary-value problem (6.3.1) corresponding, respectively, to data functions g_1 and g_2. Then the principle of superposition implies that the difference $w = u_1 - u_2$ satisfies

$$\nabla^2 w = 0 \quad \text{in } \Omega, \qquad w = g_1 - g_2 \quad \text{on } \Sigma$$

If we let

$$M = \max_{\Sigma} (g_1 - g_2) \quad \text{and} \quad m = \min_{\Sigma} (g_1 - g_2)$$

then Theorem 6.3.1 implies

$$m \leq u_1 - u_2 \leq M \quad \text{in } \Omega + \Sigma$$

This statement means that if two functions each satisfy Laplace's equation in the region Ω but assume different values on the boundary Σ, then the functions must be different in Ω. However, their difference in Ω nowhere exceeds the maximum of their difference on Σ; that is, the solution of (6.3.1) depends continuously on the data.

Monotonicity Result for Poisson's Equation

Theorem 6.3.5. Let Ω be a bounded region in R^N having smooth boundary Σ. Consider the problem

$$\nabla^2 u = F \quad \text{in } \Omega, \qquad u = g \quad \text{on } \Sigma \tag{6.3.2}$$

Let u_1 and u_2 denote classical solutions to this problem corresponding to forcing functions F_1 and F_2, respectively. Then,

$$F_1 \geq F_2 \quad \text{in } \Omega \quad \text{implies that} \quad u_1 \leq u_2 \quad \text{in } \Omega + \Sigma \tag{6.3.3}$$

Proof. The difference $w = u_1 - u_2$ satisfies

$$\nabla^2 w = F_1 - F_2 \geq 0 \quad \text{in } \Omega$$
$$w = 0 \qquad\qquad\qquad \text{on } \Sigma$$

Then

$$M = \max_{\Sigma} w = 0$$

and Theorem 6.3.1 implies that $w = u_1 - u_2$ is nonpositive in $\Omega + \Sigma$. Theorem 6.3.2 implies further that $u_1 - u_2$ is strictly negative in Ω if $F_1 \geq F_2$.

Max–Min Principles for Heat Equation

There are analogues of the max–min principle that apply to the heat equation. If $u(x, t)$ satisfies

$$u_t(x, t) = u_{xx}(x, t), \qquad 0 < x < L, \qquad t > 0$$

then at each point (x, t) where $u_{xx} > 0$, we have $u_t > 0$, so that $u(x, t)$ is *increasing* with increasing t (see Figure 6.3.3). Similarly, where $u_{xx} < 0$, $u(x, t)$ is a *decreasing* function of t.

Evidently, a function $u(x, t)$ that satisfies the heat equation on an interval with constant boundary conditions will tend to a linear function of x as $t \to \infty$. For a linear function of x, the maximum and minimum values occur at the endpoints of the interval on which it is defined. Then for the function $u(x, t)$, the extreme values must occur at

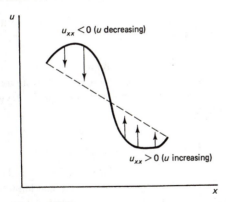

u

$u_{xx} < 0$ (*u* decreasing)

$u_{xx} > 0$ (*u* increasing)

x

Figure 6.3.3

the endpoints of the interval $(0, L)$ or else they must occur when $t = 0$ (in which case they may be located at interior points of the interval). This observation generalizes to:

Theorem 6.3.6. Let Ω denote a bounded region in R^N with smooth boundary Σ and let T denote a positive constant. Suppose u is continuous for x in $\Omega + \Sigma$ and $0 \leq t \leq T$ and that $\nabla^2 u$ and u_t are continuous for x in Ω and $0 < t < T$ and let

$$M_0 = \max\{u : x \text{ in } \Omega + \Sigma, t = 0\}$$
$$M_\Sigma = \max(u : x \text{ on } \Sigma, 0 < t < T\}$$
$$M = \max\{M_0, M_\Sigma\}$$

Let m_0, m_Σ, m denote similarly defined minimum values for u. Then:

(a) If $u_t - \nabla^2 u \leq 0$ in Q_T, then $u \leq M$ in $Q_T + \Sigma$
(b) If $u_t - \nabla^2 u \geq 0$ in Q_T, then $u \geq m$ in $Q_T + \Sigma$
(c) If $u_t - \nabla^2 u = 0$ in Q_T, then $m \leq u \leq M$ in $Q_T + \Sigma$

where

$$Q_T = \{(x, t) : x \in \Omega \text{ and } 0 < t < T)$$
$$Q_T + \Sigma = ((x, t) : x \in \Omega + \Sigma \text{ and } 0 \leq t \leq T)$$

Theorem 6.3.6 is referred to as the "weak" maximum principle. A "strong" version of this theorem, stated for the one-dimensional heat equation follows.

Theorem 6.3.7. Suppose $u(x, t)$ is continuous for $0 \leq x \leq 1$ and $0 \leq t$ and that u and u_{xx} are continuous for $0 < x < 1$, $t > 0$. Suppose further that $u(x, t)$ satisfies

$$u_t(x, t) - u_{xx}(x, t) \leq 0, \qquad 0 < x < 1, \qquad t > 0$$

and let

$$M_0 = \max(u : 0 \leq x \leq 1, t = 0)$$
$$M_\Sigma = \max(u(0, t), u(1, t) : t \geq 0)$$
$$M = \max(M_0, M_\Sigma)$$

If $u(x_0, t_1) = M$ for *some* (x_0, t_1) such that $0 < x_0 < 1$ and $t_1 > 0$, then it follows that $u(x, t) = M$ for *all* (x, t), $0 \le x \le 1$, $0 \le t \le t_1$.

Proofs of Theorems 6.3.6 and 6.3.7 can be found in the book by Protter and Weinberger mentioned in the paragraph following Theorem 6.3.2. Note that in the case of a heat-conducting rod of length 1 with initial temperature equal everywhere to 100 and ends maintained at temperature 0, we have $M = 100$ and $m = 0$. Then Theorem 6.3.6 states that

$$0 \le u(x, t) \le 100 \quad \text{for } 0 \le x \le 1 \quad \text{and} \quad t \ge 0$$

Theorem 6.3.7 implies that for $t > 0$, $u(x, t)$ must lie *strictly* between the values of 0 and 100 for all x, $0 < x < 1$. That is, the influence of the zero end temperatures is felt *instantly* throughout the interior of the rod. This is another manifestation of the infinite speed of propagation associated with the heat equation.

A frequently useful extension of the max–min principle for the heat equation is the following.

Theorem 6.3.8. Suppose $u(x, t)$ is continuous for $0 \le x \le 1$ and $0 \le t$ and that u_t and u_{xx} are continuous for $0 < x < 1$, $t > 0$. Suppose further that $u(x, t)$ satisfies

$$u_t(x, t) - u_{xx}(x, t) = 0, \qquad 0 < x < 1, \qquad t > 0$$

and let M and m be as defined in Theorem 6.3.7. Then either $u(x, t)$ is constant for $0 \le x \le 1$, $t \ge 0$, or else;

(a) At boundary point $x_0 = 0, 1$, where $u(x_0, t) = M$, we have

$$\partial_N u(x_0, t) > 0$$

(b) At boundary point $x_0 = 0, 1$, where $u(x_0, t) = m$, we have

$$\partial_N u(x_0, t) < 0$$

Note that $\partial_N u(0, t) = -u_x(0, t)$, $\partial_N u(1, t) = u_x(1, t)$.

A proof of this theorem can be found in the book by Protter and Weinberger (see the paragraph following Theorem 6.3.2) and in J. R. Cannon, "The One Dimensional Heat Equation," *Encyclopedia of Mathematics and Its Applications*, vol. 23, Addison-Wesley, Reading, Mass., 1984.

Monotonicity Result for the Heat Equation

Theorem 6.3.9. Let Ω denote a bounded set in R^N with smooth boundary Σ and let U_1 and U_2 denote smooth solutions of

$$\begin{aligned}
U_t - \nabla^2 U &= F(x, t) && \text{for } x \text{ in } \Omega \text{ and } t \text{ in } (0, T) \\
U(x, 0) &= g(x) && \text{for } x \text{ in } \Omega \\
U(x, t) &= h(x, t) && \text{for } x \text{ on } \Sigma \text{ and } t \text{ in } (0, T)
\end{aligned}$$

corresponding to data sets $[F_1, g_1, h_1]$ and $[F_2, g_2, h_2]$, respectively. If the data sets satisfy

$$F_1(x, t) \geq F_2(x, t) \quad \text{for } x \text{ in } \Omega \text{ and } t \text{ in } (0, T)$$
$$g_1(x) \geq g_2(x) \quad \text{for } x \text{ in } \Omega$$
$$h_1(x, t) \geq h_2(x, t) \quad \text{for } x \text{ on } \Sigma \text{ and } t \text{ in } (0, T) \qquad (6.3.4)$$

then the solutions satisfy

$$U_1(x, t) \geq U_2(x, t) \quad \text{for } x \text{ in } \Omega \text{ and } t \text{ in } (0, T) \qquad (6.3.5)$$

Proof. Let $W(x, t) = U_1(x, t) - U_2(x, t)$. Then superposition implies that W satisfies

$$W_t - \nabla^2 W = F_1(x, t) - F_2(x, t) \geq 0 \quad \text{in } Q_T$$

Using Theorem 6.3.6(ii), we conclude that $W \geq m$ in $Q_T + \Sigma$. From (6.3.4), we know that

$$W(x, 0) = g_1(x) - g_2(x) \geq 0 \quad \text{for } x \text{ in } \Omega$$

so that $m_0 \geq 0$. In addition,

$$W(x, t) = h_1(x, t) - h_2(x, t) \geq 0 \quad \text{for } x \text{ on } \Sigma \text{ and } t \text{ in } (0, T)$$

and hence $m_\Sigma \geq 0$. Then $m = \min[m_0, m_\Sigma] \geq 0$, and it follows that $W(x, t) = U_1(x, t) - U_2(x, t) \geq 0$ in $Q_T + \Sigma$.

The result in Theorem 6.3.9 makes is possible to "order" solutions to initial-boundary-value problems when the data is ordered. In this sense, the theorem is a sort of monotonicity result. The next theorem is more along the lines of a continuous dependence on the data statement.

Continuous Dependence Result for Heat Equation

Theorem 6.3.10. Let $u_1(x, t)$ and $u_2(x, t)$ denote smooth solutions of

$$u_t(x, t) = u_{xx}(x, t), \quad 0 < x < 1, \quad t > 0$$
$$u(x, 0) = f(x), \quad 0 < x < 1$$
$$u(0, t) = u(1, t) = 0, \quad t > 0$$

corresponding to data functions $f = f_1$ and $f = f_2$, respectively. Then for all x, t, $0 \leq x \leq 1, t \geq 0$,

$$m \leq u_1(x, t) - u_2(x, t) \leq M \qquad (6.3.6)$$

where

$$m = \min [f_1(x) - f_2(x)], \quad M = \max [f_1(x) - f_2(x)]$$

and the max and min are taken over $0 \leq x \leq 1$.

Proof. Let $w(x, t)$ denote the difference $u_1(x, t) - u_2(x, t)$ and note that $w(x, t)$ satisfies

$$w_t(x, t) = w_{xx}(x, t), \qquad 0 < x < 1, \qquad t > 0$$
$$w(x, 0) = f_1(x) - f_2(x), \qquad 0 < x < 1$$
$$w(0, t) = w(1, t) = 0, \qquad t > 0$$

We now apply Theorem 6.3.6 to the initial-boundary-value problem for $w(x, t)$. Clearly, the initial and boundary conditions imply

$$M_\Sigma = m_\Sigma = 0, \qquad M_0 = M, \qquad m_0 = m$$

Then Theorem 6.3.6 implies that the max and min values for $w(x, t)$ occur, not on the lateral boundary, but initially. Then result (6.3.6) follows.

This theorem is thought of as a "continuous-dependence-on-the-data" result since it asserts that the difference in solutions corresponding to different data can nowhere exceed the difference in the data.

In this section we stated max–min principles for the Laplace and heat equations. In a more advanced course in partial differential equations it could be shown that these principles extend to elliptic and parabolic partial differential equations in general. We repeat that there is no analogous principle that applies to the wave equation in particular nor to hyperbolic equations in general.

Exercises: Max–Min Principles

1. Suppose Ω denotes the region $\{x, y : 0 < x < 2, 0 < y < 4 - x^2\}$ and that $u(x, y)$ satisfies

$$u_{xx} + u_{yy} = 0 \text{ in } \Omega$$
$$u(0, y) = 0, \qquad 0 < y < 4$$
$$u(x, 0) = x(4 - x^2), \qquad 0 < x < 2$$
$$u(x, 4 - x^2) = 0, \qquad 0 < x < 2$$

Then show that

$$0 < u(x, y) < x(4 - x^2 - y) \text{ in } \Omega$$

2. Suppose $u(x, y)$ satisfies

$$u_{xx}(x, y) + u_{yy}(x, y) = (\sin xy)^2 \text{ in } x^2 + y^2 < 1$$
$$u(x, y) = 0 \text{ on } x^2 + y^2 = 1$$

Does $u(x, y)$ change sign at any point in $x^2 + y^2 \le 1$?

3. The function

$$u(x, y) = \frac{1 - (x^2 + y^2)}{(1 - x)^2 + y^2}$$

satisfies Laplace's equation in $x^2 + y^2 < 1$. In addition, $u(x, y)$ is positive there and vanishes on $x^2 + y^2 = 1$ except at $(1, 0)$, where $u = 1$. Does this violate either Theorem 6.3.1 or 6.3.2?

4. Suppose $u_1(x, y)$ and $u_2(x, y)$ satisfy

$$u_{xx}(x, y) + u_{yy}(x, y) = 0 \qquad \text{in } \Omega$$
$$u(x, y) = g(x, y) \quad \text{on } \Sigma \text{ the boundary of } \Omega$$

for data $g = g_1, g_2$, respectively. If $g_1 \ge g_2$ on Σ, then show that $u_1 \ge u_2$ in $\Omega + \Sigma$.

5. Show that the solution of

$$-\nabla^2 u(x, y) = xy(x - \pi)(y - \pi) \quad \text{on } \Omega = \{0 < x, y < \pi\}$$
$$u = 0 \qquad\qquad\qquad \text{on boundary of } \Omega$$

satisfies

$$u(x, y) \geq \tfrac{1}{2} \sin x \sin y \quad \text{on } \Omega + \Sigma$$

and

$$u_x(0, y) \geq \tfrac{1}{2} \sin y, \qquad 0 < y < \pi$$

[*Hint:* $\phi(x, y) = \tfrac{1}{2} \sin x \sin y$ satisfies

$$-\nabla^2 \phi(x, y) = \sin x \sin y \quad \text{in } \Omega$$
$$\phi(x, y) = 0 \qquad\qquad \text{on boundary of } \Omega]$$

6. Suppose that $u(x, y)$ satisfies

$$\nabla^2 u(x, y) = f(x, y) \quad \text{on } \Omega$$
$$u = g \qquad\qquad \text{on boundary } \Sigma \text{ of } \Omega$$

where Ω is a bounded set contained in the square $(0, A) \times (0, B)$. Show that for some constant $C > 0$ depending on Ω,

$$|u(x, y)| \leq \max_\Sigma |g| + C \max_\Omega |f|$$

[*Hint:* $v(x, y) = \max_\Sigma |g| + (e^A - e^x) \max_\Omega |f|$ satisfies

$$v \geq \max_\Sigma |g| \quad \text{on } \Sigma \quad \text{and} \quad \nabla^2 v \leq f \quad \text{in } \Omega]$$

7. Use Theorem 6.3.10 to conclude that there is at most one solution to

$$u_t(x, t) = u_{xx}(x, t), \qquad 0 < x < 1, \qquad t > 0$$
$$u(x, 0) = f(x), \qquad 0 < x < 1$$
$$u(0, t) = u(1, t) = 0, \qquad t > 0 \tag{A}$$

Suppose then that $U(x, t)$ satisfies

$$U_t(x, t) = U_{xx}(x, t), \qquad -\infty < x < \infty, \qquad t > 0$$
$$U(x, 0) = F(x), \qquad -\infty < x < \infty$$

where

$$F(x) = f(x), \qquad 0 < x < 1$$
$$\quad = -f(-x) \qquad -1 < x < 0$$

and

$$F(x) = F(x + 2) \quad \text{for all } x$$

Then show that $U(x, t)$ satisfies all the conditions of problem (A) and that, necessarily, $U(x, t) = u(x, t)$ for $0 \leq x \leq 1, t \geq 0$.

8. Use Theorem 6.3.8 to conclude that there is at most one solution to

$$v_t(x, t) = v_{xx}(x, t), \qquad 0 < x < 1, \qquad t > 0$$
$$v(x, 0) = g(x), \qquad 0 < x < 1$$
$$v_x(0, t) = v_x(1, t) = 0, \qquad t > 0 \tag{B}$$

Derive a result that is analogous to Theorem 6.3.10 asserting continuous dependence on the data for this problem.

9. Suppose that $V(x, t)$ satisfies

$$V_t(x, t) = V_{xx}(x, t), \qquad -\infty < x < \infty, \qquad t > 0$$
$$V(x, 0) = G(x), \qquad -\infty < x < \infty$$

where

$$G(x) = \begin{cases} g(x), & 0 < x < 1 \\ g(-x), & -1 < x < 0 \end{cases}$$

and

$$G(x) = G(x + 2) \quad \text{for all } x$$

Then show that $V(x, t)$ satisfies all of the conditions of problem (B) and that necessarily $V(x, t) = v(x, t)$ for $0 \le x \le 1$, $t \ge 0$.

10. Show that if the following problem has any solution, it is unique and depends continuously on the data:

$$u_t(x, t) = u_{xx}(x, t), \qquad 0 < x < 1, \qquad t > 0$$
$$u(x, 0) = f(x), \qquad 0 < x < 1$$
$$u(0, t) = u_x(1, t) = 0, \qquad t > 0 \tag{C}$$

Show that $u(x, t) = W(x, t)$ for $0 \le x \le 1$ and $t \ge 0$, where $W(x, t)$ solves

$$W_t(x, t) = W_{xx}(x, t), \qquad -\infty < x < \infty, \qquad t > 0$$
$$W(x, 0) = F(x), \qquad -\infty < x < \infty$$

where

$$F(x) = \begin{cases} f(x), & 0 < x < 1 \\ -f(2 - x), & 1 < x < 2 \\ -f(-x), & -2 < x < 0 \end{cases}$$

and

$$F(x) = F(x + 4) \quad \text{for all } x$$

11. Consider the problem,

$$u_t(x, t) - D(x, t)u_{xx}(x, t) = -\sin \pi x, \qquad 0 < x < 1, \qquad t > 0$$
$$u(x, 0) = 0, \qquad 0 < x < 1$$
$$u(0, t) = u(1, t) = 0, \qquad t > 0$$

where for some constant $C > 0$, the variable coefficient $D(x, t)$ satisfies

$$0 < D(x, t) \le C, \qquad 0 \le x \le 1, \qquad t \ge 0$$

Conclude that for $0 \le x \le 1$, $t \ge 0$,
(a) $u(x, t) \le 0$
(b) $u(x, t) \le [\exp(-\pi^2 Ct) - 1]\sin \pi x \, (1/C\pi^2) \equiv \Phi(x, t)$
(c) $u_x(0, t) \le [\exp(-\pi^2 Ct) - 1]/(\pi C)$ for $t > 0$
[*Hint:* Show that the function $\Phi(x, t)$ appearing in (b) satisfies

$$\Phi_t - C\Phi_{xx} = -\sin \pi x, \qquad 0 < x < 1, \qquad t > 0$$
$$\Phi(x, 0) = 0, \qquad 0 < x < 1$$
$$\Phi(0, t) = \Phi(1, t) = 0, \qquad t > 0$$

and

$$\Phi_{xx}(x, t) > 0, \qquad 0 < x < 1, \qquad t > 0$$

Then show that $w(x, t) = u(x, t) - \Phi(x, t)$ satisfies

$$w_t - D(x, t)w_{xx}(x, t) = [D(x, t) - C]\Phi_{xx}(x, t) \leq 0$$
$$w(x, 0) = w(0, t) = w(1, t) = 0$$

Finally, use Theorem 6.3.6.]

Chapter 7

First-Order Equations

Our discussion to this point has, for the most part, dealt with partial differential equations of second order. Specifically, we have seen how to use Fourier series, the Fourier transform, and the Laplace transform to construct solutions to boundary and initial-boundary-value problems for the Laplace, heat, and wave equations.

In this chapter we examine first-order partial differential equations. We shall find that certain curves, called *characteristics*, provide crucial information about the nature of the solution. Thus, our emphasis will switch from solution techniques based on separation of variables and series transform methods to solution techniques based on characteristic curves.

7.1 CONSTANT-COEFFICIENT ADVECTION EQUATION

Characteristics

The equation

$$au_x + u_t = 0, \quad a = \text{const} \tag{7.1.1}$$

for $u = u(x, t)$ is called a *first-order wave equation* or an *advection equation*. Equation (7.1.1) arises in mathematical models that involve the movement of a wave in one direction with no change in the shape of the wave. The constant a will be shown to represent the *signal speed*, or *celerity*, of the wave.

We solve the partial differential equation (7.1.1) by finding curves in the xt plane along which it reduces to an ordinary differential equation. Let C be a curve defined by $x = x(t)$. Then, on C, we have $u(x, t) = u(x(t), t)$. Differentiation of u along C yields

$$\frac{du}{dt} = \frac{\partial u}{\partial x}\frac{dx}{dt} + \frac{\partial u}{\partial t} \tag{7.1.2}$$

Comparing the right side of (7.1.2) with the left side of (7.1.1), we see that if we require

$$\frac{dx}{dt} = a \quad \text{on } C \tag{7.1.3}$$

then

$$\frac{du}{dt} = 0 \quad \text{on } C \tag{7.1.4}$$

A curve along which (7.1.3) holds is a *characteristic curve*, or simply a *characteristic* of Equation (7.1.1). Equation (7.1.4) implies that the solution u is constant along a characteristic.

Solving the characteristic equation (7.1.3) yields

$$x = at + x_0, \qquad x_0 = \text{const} \tag{7.1.5}$$

Thus, the characteristics of the constant-coefficient equation (7.1.1) are seen to be straight lines in the xt plane. Note that at time $t = 0$ the characteristic defined by (7.1.5) intersects the x axis at position x_0. Also, for every unit increase of t, a point on the characteristic moves a units in the positive x direction if a is positive and a units in the negative x direction if a is negative.

Initial-Value Problem

Next consider the *initial-value problem*

$$au_x + u_t = 0, \qquad -\infty < x < \infty, t > 0 \tag{7.1.6a}$$
$$u(x, 0) = f(x) \tag{7.1.6b}$$

An initial-value problem of the form (7.1.6a, b) is also referred to as a *Cauchy problem*. From Equation (7.1.5) we see that the points (x, t) and $(x - at, 0)$ lie on the same characteristic (see Figures 7.1.1 and 7.1.2). Since u is constant along a characteristic,

$$u(x, t) = u(x - at, 0) = f(x - at)$$

This establishes the following result.

Theorem 7.1.1. If $f(x)$ is continuously differentiable, then

$$u(x, t) = f(x - at)$$

solves the initial-value problem defined by Equations (7.1.6a, b).

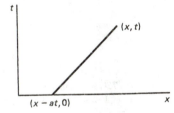

Figure 7.1.1 A characteristic of Equation (7.1.1) for the case $a > 0$.

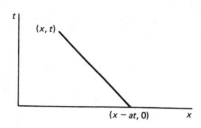

Figure 7.1.2 A characteristic of Equation (7.1.1) for the case $a < 0$.

Recall that in Section 5.1 the Fourier transform was used to obtain the solution given in Theorem 7.1.1.

The hypothesis that f be continuously differentiable ensures that

$$u_x = f'(x - at) \quad \text{and} \quad u_t = f'(x - at) \cdot (-a)$$

are continuous. Thus, $u(x, t)$ has the required differentiability to be called a solution of Equation (7.1.1), or more specifically a *classical solution* of (7.1.1).

In many applications f is allowed to be only piecewise continuously differentiable or even piecewise continuous. In these cases $u(x, t) = f(x - at)$ is still referred to as a solution of the initial-value problem (7.6a, b), but it should be called a *weak*, or *generalized, solution*. Generalized solutions are discussed in Section 7.4.

EXAMPLE 7.1.1 _____

Consider the initial-value problem

$$2u_x + u_t = 0, \quad -\infty < x < \infty, t > 0$$
$$u(x, 0) = e^{-x^2}, \quad -\infty < x < \infty$$

Since $a = 2$ and $f(x) = e^{-x^2}$, Theorem 7.1.1 gives the solution $u(x, t) = e^{-(x - 2t)^2}$ (Figure 7.1.3). ■ ■

EXAMPLE 7.1.2 _____

The initial-value problem

$$-u_x + u_t = 0$$
$$u(x, 0) = \begin{cases} 0, & |x| > 1 \\ 1 - |x|, & |x| \leq 1 \end{cases}$$

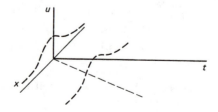

Figure 7.1.3 Solution of Example 7.1.1 with u = 1 along the characteristic x = 2t.

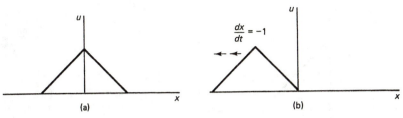

Figure 7.1.4   (a) Solution of Example 7.1.2 at t = 0. (b) Solution of Example 7.1.2 at $t = 1$.

has (generalized) solution

$$u(x, t) = \begin{cases} 0, & x > 1 - t \\ 1 - |x + t|, & -1 - t < x < 1 - t \\ 0, & x < -1 - t \end{cases}$$

(See Figures 7.1.4a, b.)   ■ ■

Initial-Boundary-Value Problem

An *initial-boundary-value problem* for the constant-coefficient, first-order wave equation has the form

$$au_x + u_t = 0, \qquad x > 0, t > 0 \tag{7.1.7a}$$
$$u(x, 0) = f(x), \qquad x > 0 \tag{7.1.7b}$$
$$u(0, t) = g(t), \qquad t > 0 \tag{7.1.7c}$$

The initial condition is given by (7.1.7b) and the boundary condition by (7.1.7c).

    We first consider the case that a is positive. Then the characteristic through the origin divides the first quadrant into two regions (Figure 7.1.5):

$$R_1 : x > at \quad \text{and} \quad R_2 : x < at$$

In R_1 the solution u is determined by the initial condition to be $u(x, t) = f(x - at)$. Given a point $(\bar{x}, \bar{t})$ in R_2 we can trace back along the characteristic $x = a(t - \bar{t}) + \bar{x}$ to the point $(0, \bar{t} - \bar{x}/a)$. This shows that in R_2

$$u(x, t) = u(0, t - x/a) = g(t - x/a)$$

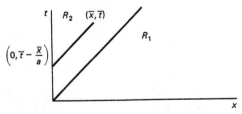

Figure 7.1.5 R_1, region of influence of initial condition $f(x)$; R_2, region of influence of boundary condition $g(t)$.

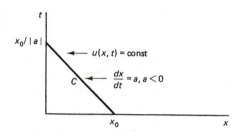

Figure 7.1.6 Information about solution cannot be specified at two points on same characteristic.

and we see that the solution to (7.1.7a)–(7.1.7c) is given by

$$
u(x, t) = \begin{cases} f(x - at), & x > at, \, a > 0 \\ g\left(\dfrac{at - x}{a}\right), & x < at, \, a > 0 \end{cases} \tag{7.1.8}
$$

The condition for $u(x, t)$ defined by Equation (7.1.8) to be continuous across the line $x = at$ is that $f(0) = g(0)$. The condition for the differentiability of u across $x = at$ is left as an exercise.

If $a < 0$, the initial-boundary-value problem (7.1.7) is not well posed. To see this, note that for $x_0 > 0$, the points $(x_0, 0)$ and $(0, x_0/|a|)$ lie on the same characteristic. The initial condition (7.1.7b) states $u(x_0, 0) = f(x_0)$, while the boundary condition (7.1.7c) gives $u(0, x_0/|a|) = g(x_0/|a|)$. The fact that u is constant along a characteristic requires that

$$
f(x_0) = g(x_0/|a|) \tag{7.1.9}
$$

for arbitrary $x_0 > 0$. Thus, the boundary condition g and the initial condition f must satisfy the compatibility condition (7.1.9), and we see that for general f and g the initial-boundary-value problem (7.1.7a–c) will not have a solution (Figure 7.1.6).

In many applications that involve a first-order differential equation of the form (7.1.1), we find that x plays the role of a space variable and t a time variable. As the next example illustrates, this is not always the case.

EXAMPLE 7.1.3 _____

Let $u(x, t)$ denote the population density of individuals of age x at time t. Consider this population during a time when no individuals leave the population by, say, migration or death. Then, applying the argument that age increases with time, we find that $u(x, t)$ satisfies the balance law

$$
u_x + u_t = 0
$$

If we assume that at time $t = 0$ the age structure in the population is given by $u(x, 0) = f(x)$ and that the number of age zero (newborn) individuals is given at any time t by $u(0, t) = g(t)$, then we find that the population density in this age-structured population satisfies the initial-boundary-value problem

$$u_x + u_t = 0, \qquad x > 0, t > 0$$
$$u(x, 0) = f(x), \qquad x > 0$$
$$u(0, t) = g(t), \qquad t > 0$$

■ ■

Exercises

1. In each of the following initial-value problems, solve for $u(x, t)$ and sketch the graph of $u(x, t)$ versus x for $t = 0$, $t = 1$, and $t = 2$.
   (a) $u_x + u_t = 0$, $u(x, 0) = 1/(1 + x^2)$
   (b) $u_x - u_t = 0$, $u(x, 0) = 1/(1 + x^2)$
   (c) $2u_x + u_t = 0$, $u(x, 0) = \begin{cases} 1 - |x|, & |x| < 1 \\ 0, & |x| > 1 \end{cases}$

   (d) $2u_x - u_t = 0$, $u(x, 0) = \begin{cases} 1 - |x|, & |x| < 1 \\ 0, & |x| > 1 \end{cases}$

   (e) $2u_x + u_t = 0$, $u(x, 0) = e^{-x^2}$
   (f) $3u_x + 2u_t = 0$, $u(x, 0) = \sin x$

2. In each of the following initial-boundary-value problems, solve for $u(x, t)$ and sketch the graph of $u(x, t)$ versus x at time $t = 0$, $t = 1$, and $t = 2$.
   (a) $u_x + u_t = 0$, $x > 0$, $t > 0$
   $u(x, 0) = e^{-x^2}$, $x > 0$
   $u(0, t) = 1$, $t > 0$
   (b) $u_x + u_t = 0$, $x > 0$, $t > 0$
   $u(x, 0) = e^{-x^2}$, $x > 0$
   $u(0, t) = e^{-t^2}$, $t > 0$
   (c) $2u_x + u_t = 0$, $x > 0$, $t > 0$
   $u(x, 0) = \begin{cases} 1 - x, & 0 < x < 1 \\ 0, & x \geq 1 \end{cases}$

   $u(0, t) = 1$, $t > 0$
   (d) $2u_x + u_t = 0$, $x > 0$, $t > 0$
   $u(x, 0) = \begin{cases} x - 1, & 0 < x < 1 \\ 1, & x \geq 1 \end{cases}$

   $u(0, t) = 0$, $t > 0$
   (e) $3u_x + u_t = 0$, $x > 0$, $t > 0$
   $u(x, 0) = 0$, $x > 0$
   $u(0, t) = \begin{cases} t, & 0 < t < 1 \\ 1, & t \geq 1 \end{cases}$

   (f) $u_x + u_t = 0$, $x > 0$, $t > 0$
   $u(x, 0) = 0$, $x > 0$
   $u(0, t) = \sin t$, $t > 0$

3. Given the initial-value problem

$$3u_x + u_t = 5, \qquad u(x, 0) = e^x$$

use the change of dependent variable $u(x, t) = v(x, t) + 5t$ to obtain an advection equation (7.1.1) for v. Then determine $u(x, t)$.

4. Given the initial-value problem

$$4u_x + u_t = u, \qquad u(x, 0) = \cos x$$

use the change of dependent variable $v(x, t) = e^{-t}u$ to obtain a constant-coefficient advection equation (7.1.1) for $v(x, t)$. Then find $u(x, t)$.

5. Find a change of dependent variable that reduces the constant-coefficient equation $au_x + u_t = cu$ to the advection equation

$$av_x + v_t = 0$$

6. Determine the conditions on the initial condition $f(x)$ and the boundary condition $g(t)$ in (7.1.7) that ensure that the solution (7.1.8) is continuously differentiable.

7.2 LINEAR AND QUASI-LINEAR EQUATIONS

Classification

The general *first-order, quasi-linear equation* for $u(x, t)$ has the form

$$a(x, t, u) u_x + b(x, t, u)u_t = c(x, t, u) \tag{7.2.1}$$

An *initial-value*, or *Cauchy, problem* consists of (7.2.1) together with an initial condition

$$u(x, 0) = f(x) \tag{7.2.2}$$

If c is identically zero, (7.2.1) is said to be *homogeneous*. If a and b are independent of u, (7.2.1) is called *almost linear*. If a and b are independent of u and c has the form

$$c(x, t, u) = d(x, t) u + S(x, t)$$

then (7.2.1) is *linear*.

Initial-Value Problem

Before discussing the solution of the initial-value problem (7.2.1) and (7.2.2), let us review some results from the last section. For the constant-coefficient, homogeneous equation $au_x + u_t = 0$, the characteristic curves are the straight lines defined by $dx/dt = a$. These lines can be described by

$$x = ar + s, \quad t = r \tag{7.2.3}$$

where r is a parameter that fixes the position on the characteristic and s specifies the point where the characteristic intersects the x axis (see Figure 7.2.1).

Let u be a solution of $au_x + u_t = 0$ restricted to a characteristic of the form (7.2.3). Differentiation of u along the characteristic yields

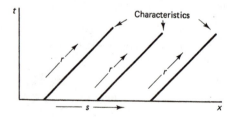

Figure 7.2.1 Parametric representation of characteristics with r running along characteristics and s running along the x axis.

$$\frac{du}{dr} = \frac{\partial u}{\partial x}\frac{dx}{dr} + \frac{\partial u}{\partial t}\frac{dt}{dr} = au_x + u_t$$

which shows that along a characteristic the (homogeneous) partial differential equation $au_x + u_t = 0$ reduces to the (homogeneous) ordinary differerential equation $du/dr = 0$. This shows that the solution to $au_x + u_t = 0$ is constant along characteristics. The reduction of the partial differential equation to an ordinary differential equation is the key to a solution technique known as the *method of characteristics*.

We now turn our attention to the initial-value problem defined by (7.2.1) and (7.2.2). As a first step, let us introduce a curve C defined by

$$C: \quad \begin{cases} x = x(r), & x(0) = s \\ t = t(r), & t(0) = 0 \end{cases}$$

As in the preceding, we shall use r to fix a position along C and s to specify where the curve C intersects the x axis. Consider the restriction of $u(x, t)$ to the curve C, $u = u(x(r), t(r))$. Using the chain rule to differentiate with respect to r, we find

$$\frac{du}{dr} = \frac{\partial u}{\partial x}\frac{dx}{dr} + \frac{\partial u}{\partial t}\frac{dt}{dr}$$

If we require that

$$\frac{dx}{dr} = a \quad \text{and} \quad \frac{dt}{dr} = b$$

then, along C, the partial differential equation (7.2.1) reduces to the ordinary differential equation

$$\frac{du}{dr} = c$$

The equations

$$\frac{dx}{dr} = a, \quad \frac{dt}{dr} = b; \quad \frac{du}{dr} = c \qquad (7.2.4)$$

are called the *characteristic equations* of Equation (7.2.1).

If we eliminate the parameter r, the characteristic equations can be expressed in the alternate form

$$\frac{dx}{a} = \frac{dt}{b} = \frac{du}{c} \qquad (7.2.4a)$$

In this form the term du/c requires some interpretation when $c = 0$. When $c = 0$ we must conclude that $du = 0$ along the characteristic C. This is just a restatement of the fact that the solution of the homogeneous equation $au_x + bu_t = 0$ is constant along characteristics.

A curve $(x(r), t(r))$ in the xt plane along which $x'(r) = a$ and $t'(r) = b$ holds is called a *characteristic ground curve* for Equation (7.2.1). A space curve $(x(r), t(r), u(r))$ along which (7.2.4) holds is called a *characteristic space curve* for Equation (7.2.1). In most discussions both characteristic ground curves and characteristic space curves are referred to as simply *characteristics*. The context generally determines whether the curve lies in the *xt* plane or in *xtu* space.

EXAMPLE 7.2.1

For the constant-coefficient, homogeneous, linear equation

$$3u_x + 2u_t = 0$$

the characteristic (ground) curves are defined by

$$\frac{dx}{dr} = 3, \quad x(0) = s; \quad \frac{dt}{dr} = 2, \quad t(0) = 0$$

Solving for $x(t)$ and $t(r)$, we find $x(r) = 3r + s$, $t(r) = 2r$. If we eliminate the parameter r, the characteristics are seen to be straight lines in the *xt* plane defined by

$$x - s = \tfrac{3}{2}t$$

We are viewing x as a function of the parameter r that measures a position along the characteristic. However, the expression for $x(t)$ contains the additional parameter s that fixes the point where a characteristic intersects the x axis. Thus, it would be proper to view x as a function of both r and s and write $x = x(r, s)$. This point will be pursued shortly.

■ ■

EXAMPLE 7.2.2

Given the linear, variable-coefficient equation

$$xtu_x + u_t = u$$

the characteristic equations (7.2.4a) are

$$\frac{dx}{xt} = \frac{dt}{1} = \frac{du}{u}$$

From the first equality it is easy to see that

$$\ln(x) = \tfrac{1}{2}t^2 + \text{const}$$

If we again let s denote the point where a characteristic intersects the x axis, then the constant in the last equation is seen to be $\ln(s)$, and we conclude that $x = s \cdot e^{t^2/2}$ along a characteristic (ground or space) curve. To examine the behavior of u along a characteristic, we use the characteristic equation $dt = du/u$ to conclude that

$$\ln(u) = t + \text{const} \quad \text{or} \quad u = \text{const} \cdot e^t$$

along a characteristic. As we show in what follows, the initial condition for u at time $t = 0$ determines the constant in the expression for u along a characteristic.

■ ■

Associated with each point in *xtu* space is a vector

$$\mathbf{v}(x, t, u) = (a(x, t, u), b(x, t, u), c(x, t, u))$$

The vector field $\mathbf{v}(x, t, u)$ is called the *characteristic vector field* associated with Equation

(7.2.1). From (7.2.4a) it follows that the characteristics of (7.2.1) are the field lines of the vector field $\mathbf{v}$. That is, at each point of a characteristic C, $\mathbf{v}$ is tangent to C.

Theorem 7.2.1. Let S be a smooth surface in xtu space defined by $u = z(x, t)$. The function $z(x, t)$ solves Equation (7.2.1) if and only if at every point of S the characteristic vector field is tangent to S.

Proof. At an arbitrary point on S, a vector normal to S is given by

$$\mathbf{n} = (z_x, z_t, -1)$$

If the characteristic vector field is tangent to S, then

$$(z_x, z_t, -1) \cdot (a, b, c) = 0$$

which shows $az_x + bz_t - c = 0$ so $z(x, t)$ solves Equation (7.2.1).

To establish the converse, assume $u = z(x, t)$ solves (7.2.1) and let P be an arbitrary point on S. Since z satisfies (7.2.1),

$$az_x + bz_t - c = 0 \quad \text{at } P$$

This shows the vector (a, b, c) to be orthogonal to the normal vector $(z_x, z_t, -1)$ to S at P. Therefore, $\mathbf{v} = (a, b, c)$ lies in the tangent plane to S at P.

A uniqueness theorem from ordinary differential equations shows that a given characteristic of (7.2.1) either lies entirely in the solution surface S or does not intersect S provided a, b, and c are continuously differentiable with $ab \neq 0$. In the following discussion let us assume that a, b, and c satisfy these conditions.

The geometric content of Theorem 7.2.1 is that a smooth solution of the first-order equation (7.2.1) can be viewed as a surface made up of characteristic curves. We now use this point of view to construct a solution to the initial-value problem (7.2.1) and (7.2.2). There is no added difficulty, and some new insights can be gained if we consider an initial condition slightly more general than (7.2.2).

Let Γ be a curve defined by

$$\Gamma = \{(x, t, u) : x = p(s), t = q(s), u = u(s), s_1 < s < s_2\}$$

and assume that

$$a \cdot p'(s) - b \cdot q'(s) \neq 0 \quad \text{along } \Gamma$$

This ensures that $(p(s), q(s))$ is not a characteristic (ground) curve of Equation (7.2.1). We shall refer to Γ as the *initial curve* and consider the solution of the partial differential equation

$$a(x, t, u)u_x + b(x, t, u)u_t = c(x, t, u) \tag{7.2.5}$$

subject to the initial condition

$$u(x, t) = u(p(s), q(s)) = f(s) \quad \text{along } \Gamma \tag{7.2.6}$$

If the initial curve Γ is defined by $p(s) = s$ and $q(s) = 0$ with $s_1 = -\infty$, $s_2 = \infty$, then Equations (7.2.5) and (7.2.6) reduce to Equations (7.2.1) and (7.2.2).

The solution of the initial-value problem (7.2.5) and (7.2.6) can be viewed as a surface S with parametric representation

$$x = x(r, s), \qquad t = t(r, s), \qquad u = u(r, s)$$

where s is the parameter used to fix position along the curve Γ. The parameter r varies along characteristics of Equation (7.2.5) and is assigned the value $r = 0$ on Γ. Thus, we require that the characteristic equation (7.2.4) hold as r varies. This leads us to the following system for the determination of the surface S:

$$\frac{\partial x}{\partial r} = a, \qquad x(0, s) = p(s)$$

$$\frac{\partial t}{\partial r} = b, \qquad t(0, s) = q(s) \qquad\qquad (7.2.7)$$

$$\frac{\partial u}{\partial r} = c, \qquad u(0, s) = f(s)$$

In the sense that only differentiation with respect to r is involved, the system (7.2.7) can be viewed as three first-order ordinary differential equations with initial values for x, t, and u provided by p, q, and f.

Suppose that the system (7.2.7) has been solved for $x = x(r, s)$, $t = t(r, s)$, and $u = u(r, s)$. If r and s can be written in terms of x and t as $r = r(x, t)$, $s = s(x, t)$, then

$$u = u(r, s) = u(r(x, t), s(x, t))$$

provides the solution of Equations (7.2.5) and (7.2.6).

The inverse-function theorem from calculus states that a sufficient condition for the solvability of $x = x(r, s)$, $t = t(r, s)$ in a neighborhood of Γ to obtain $r = r(x, t)$, $s = s(x, t)$ is that the Jacobian

$$\frac{\partial(x, t)}{\partial(r, s)} = \begin{vmatrix} x_r & x_s \\ t_r & t_s \end{vmatrix} = x_r t_s - x_s t_r$$

be nonzero along Γ. Since $x_r = a$, $t_r = b$, $x_s = p'$, and $t_s = q'$, it follows that

$$\frac{\partial(x, t)}{\partial(r, s)} = aq' - bp' \quad \text{along } \Gamma$$

In the description of Γ it was assumed that $aq' - bp' \neq 0$ along Γ. Therefore, at least near Γ, r and s can be solved for in terms of x and t.

This method of passing a characteristic of the partial differential equation (7.2.5) through each point of the initial curve to generate a solution surface is known as the *method of characteristics or Lagrange's method*. It settles the question of the solvability of the initial-value problem in a neighborhood of a noncharacteristic initial curve.

If Γ is such that $(x, t) = (p(s), q(s))$ is a characteristic (ground) curve of (7.2.5), then either the problem (7.2.5) and (7.2.6) has no solution or it has infinitely many solutions. See Exercises 6 and 7.

Unless Equation (7.2.5) has some special structure, constant coefficients, for example, the local solution obtained by the method of characteristics may not exist over all of the xt plane. The following examples illustrate the constructive aspects of the method.

EXAMPLE 7.2.3 _____

Given the linear initial-value problem

$$u_x + u_t = u$$
$$u(x, 0) = \cos x$$

we look for a solution in the form $x = x(r, s)$, $t = t(r, s)$, $u = u(r, s)$ subject to the conditions (7.2.7), which, for this problem, read

$$\frac{\partial x}{\partial r} = 1, \qquad x(0, s) = s$$

$$\frac{\partial t}{\partial r} = 1, \qquad t(0, s) = 0$$

$$\frac{\partial u}{\partial r} = u, \qquad u(0, s) = \cos s$$

Solving for x, t, and u, we find

$$x = r + s, \qquad t = r, \qquad u = \cos(s) \cdot e^r$$

Solving for r and s, $r = t$ and $s = x - t$, yields the solution

$$u = e^t \cos(x - t)$$

which is valid for all x and t. ■ ■

EXAMPLE 7.2.4 _____

For the linear initial-value problem

$$2xtu_x + u_t = u, \qquad u(x, 0) = x$$

the characteristic system is

$$\frac{\partial x}{\partial r} = 2xt, \qquad x(0, s) = s$$

$$\frac{\partial t}{\partial r} = 1, \qquad t(0, s) = 0$$

$$\frac{\partial u}{\partial r} = u, \qquad u(0, s) = s$$

Solving first for $t = r$, we then see

$$\frac{\partial x}{\partial r} = 2xr \quad \text{or} \quad \frac{\partial x}{x} = 2r \, \partial r$$

which shows that $x = se^{t^2}$. As in the last example, we have $u = se^t$. Now, from $r = t$ and $u/x = e^{r-r^2}$, it follows that the solution is

$$u = xe^{t-t^2}$$

To illustrate the use of the alternative characteristic equations (7.2.4a) in the solution of this problem, consider the system

$$\frac{dx}{2xt} = \frac{dt}{1} = \frac{du}{u}, \qquad x = s, u = s \text{ when } t = 0$$

Using the first equality, we see $x = se^{t^2}$, while the second equality gives $u = se^t$. Eliminating s again produces the solution $u = xe^{t-t^2}$. ∎ ∎

EXAMPLE 7.2.5

Consider the quasi-linear initial-value problem

$$xuu_x + tuu_t = -xt$$
$$u = 5 \quad \text{on } (x, t) = (p(s), q(s)) = (s, 1/s), \qquad s > 0$$

For this problem, the characteristic equations (7.2.4a) read

$$\frac{\partial x}{\partial r} = xu, \qquad \frac{\partial t}{\partial r} = tu, \qquad \frac{\partial u}{\partial r} = -xt$$

From the first two equations we conclude

$$\frac{1}{x}\frac{\partial x}{\partial r} = \frac{1}{t}\frac{\partial t}{\partial r} \quad \text{so } \ln x = \ln t + \ln \phi(s)$$

where $\ln \phi(s)$ denotes an arbitrary, r-independent "constant" of integration. From the condition $x = s$ and $t = 1/s$ on Γ, we can fix $\phi(s)$ and find that $x = s^2t$. Now in the equation for u_r replace x by s^2t and use the last two of the characteristic equations to conclude that

$$u\frac{\partial u}{\partial r} = -s^2t\frac{\partial t}{\partial r} \quad \text{so } u^2 + s^2t^2 = \psi(s)$$

where $\psi(s)$ represents the constant of integration. Finally we fix $\psi(s)$ by using the equations $u = 5$ and $t = 1/s$ on Γ and find $\psi(s) = 26$. Thus the solution is given by $u^2 + s^2t^2 = 26$. Since $s^2 = x/t$, $u^2 + xt = 26$, or

$$u = \sqrt{26 - xt}$$

Notice that the solution constructed here is only valid for $xt \le 26$. ∎ ∎

Exercises

1. Use the method of characteristics to solve the following initial-value problems. State the region on which the solution is valid.
   (a) $u_x + u_t = u$, $u(x, 0) = \cos x$
   (b) $u_x + xu_t = u^2$, $u(x, 0) = 1$, $x > 0$

(c) $uu_x - tu_t = x$, $u = 0$ on Γ: $x = 2s$, $t = s$
(d) $2xtu_x + u_t - u = 0$, $u(x, 0) = x^2$
(e) $u_x - tu_t = 1 + u$, $u(0, t) = t$
(f) $xu_x + tu_t = 1$, $u = x^2 + t$ on Γ: $x = 1 - s$, $t = s$, $0 < s < 1$
(g) $tu_x + u_t = x$, $u(x, 0) = x^2$
(h) $u_x + xu_t = u^2$, $u(x, 0) = 1$
(i) $x^2 u_x + t^2 u_t = u^2$, $u(x, 4x) = 1$
(j) $xu_x - tu_t = 0$, $u(x, x) = x^2$
(k) $u_x + u_t = -u$, $u(x, 0) = 1 + \cos x$
(l) $xu_x + u_t = -tu$, $u(x, 0) = f(x)$

2. (a) Given the initial-value problem

$$au_x + u_t = S(x, t), \qquad -\infty < x < \infty, t > 0$$
$$u(x, 0) = f(x), \qquad -\infty < x < \infty$$

where a is a constant, use the method of characteristics to develop the solution

$$u(x, t) = \int_0^t S(ar + x - at, r)\, dr + f(x - at)$$

(b) Obtain a solution in the case $S(x, t) = xt$, $f(x) = \cos x$.

3. Given the initial-value problem

$$q(t)u_x + u_t = cu, \qquad -\infty < x < \infty, t > 0, c = \text{const}$$
$$u(x, 0) = f(x), \qquad -\infty < x < \infty, t > 0$$

use the method of characteristics to derive the solution

$$u(x, t) = e^{ct} f\left(x - \int_0^t q(\tau)\, d\tau\right)$$

4. An age-structured population in which older individuals are removed at a faster rate than younger ones and no individual survives past age $x = L$ can be modeled by the initial-value problem

$$u_x + u_t = \frac{-c}{L - x} u, \qquad 0 < x < L, t > 0$$
$$u(0, t) = b(t), \qquad t > 0$$
$$u(x, 0) = f(x), \qquad 0 < x < L$$

Use the method of characteristics to find the age-structured population density $u(x, t)$ and give physical interpretations of the positive constant c and the positive functions $b(t)$ and $f(x)$.

5. If u satisfies $(2t + 1)u_x + u_t = u$, show that $v = e^{-t}u$ satisfies $(2t + 1)v_x + v_t = 0$. Then find the solution to the original equation subject to $u(x, 0) = x$.

6. Show that the initial-value problem $u_x + u_t = 1$, $u(x, x) = x^2$ does not have a solution. *Hint*: The line $x = t$ along which the initial condition is given is a characteristic.

7. Show that the initial-value problem $tu_x + xu_t = cu$, $c = \text{const}$, $u(x, x) = f(x)$ can have a solution only if f has the form $f(x) = bx^c$, $b = \text{const}$. If f has the required form, show that $u = g(x^2 - t^2)[\frac{1}{2}(x + t)]^c$ is a solution for any function g satisfying $g(0) = b$.

8. Given the initial-value problem

$$au_x + u_t = d(x, t)u, \qquad -\infty < x < \infty, t > 0$$
$$u(x, 0) = f(x), \qquad -\infty < x < \infty$$

where a is a constant, use the method of characteristics to develop the solution

$$u(x, t) = f(x - at)\exp\left[\int_0^t d(a\rho + x - at, \rho)\, d\rho\right]$$

9. Use the results of Exercises 2 and 8 together with the principle of superposition to obtain the solution to

$$au_x + u_t = d(x, t)u + S(x, t), \qquad -\infty < x < \infty,\, t > 0$$
$$u(x, 0) = f(x), \qquad\qquad -\infty < x < \infty$$

where a is a constant.

7.3 CONSERVATION LAW EQUATIONS

Integral and Differential Formulations of Conservation Laws

A *conservation law* states that the change in the total amount of a substance or material within some fixed region is equal to the flux of that substance across the the boundary of the region. In this section, we consider conservation laws that involve the flow of material in only one direction.

Let $u = u(x, t)$ denote the linear density at position x at time t of some material M. For example, u might represent fluid mass per unit length of pipe in a one-dimensional fluid flow. Here, the units associated with u would be fluid mass per unit length. To express these units we write $\{u\} = ML^{-1}$. Alternatively, u could represent the number of vehicles per unit length of highway in a one-dimensional traffic flow. In this case we would assign u the units of vehicles per unit length of highway.

Let F denote the one-dimensional flux (vector) of the quantity whose density is u. The units associated with F are given by the units of material M per unit area per unit time. For example, in fluid flow in a pipe, F would have units of fluid mass per unit cross section of flow per unit time. That is, $\{F\} = M \cdot L^{-2}T^{-1}$. In general, the flux of a substance u will depend on the value of u, the value of the derivative, or gradient, of u, the position x, and the time t. We assume that F depends explicitly on only u and write $F = F(u)$. Of course, since $u = u(x, t)$, F is x and t dependent through its dependence on u.

To formulate the *integral form of a conservation law*, let I denote an interval $a < x < b$ and consider the time interval $c < t < d$. A conservation law states that the change in M content of I from time $t = c$ to time $t = d$ is equal to the flow of substance M into I across $x = a$ minus the flow out of I across $x = b$ during the time period $c < t < d$ (see Figure 7.3.1). In equation form this conservation principle is

$$\int_a^b \left[u(x, d) - u(x, c)\right] dx = \int_c^d \left[F(u(a, t)) - F(u(b, t))\right] dt \qquad (7.3.1)$$

On the right side of Equation (7.3.1), the term $F(u(a, t))$ represents the flow (per unit of cross section) into the interval I while $F(u(b, t))$ accounts for flow out of I across the right boundary. We say that the integral form of the conservation law holds on some region R of the xt plane provided (7.3.1) holds for every rectangle $a < x < b$,

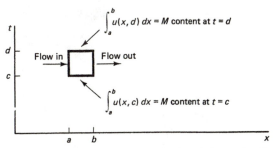

Figure 7.3.1   Change in M content of (a, b) is equal to flow in at $x = a$ minus flow out at $x = b$.

$c < t < d$ contained in R. Notice that for u to satisfy Equation (7.3.1), u need not be differentiable, but only well enough behaved for the integrals to exist.

Let us now assume that $F = F(u)$ is a continuously differentiable function and that u is continuously differentiable with respect to x and t. By the fundamental theorem of calculus, we have

$$\int_c^d \frac{\partial}{\partial t} u(x, t) \, dt = u(x, d) - u(x, c)$$

and

$$-\int_a^b \frac{\partial}{\partial x} F(u(x, t)) \, dx = F(u(a, t)) - F(u(b, t))$$

These last two equations combined with Equation (7.3.1) allows us to conclude that

$$\int_c^d \int_a^b \left[\frac{\partial}{\partial t} u(x, t) + \frac{\partial}{\partial x} F(u(x, t)) \right] dx \, dt = 0$$

Since the intervals (a, b) and (c, d) are arbitrary, it follows that if the quantity in square brackets in the last equation is continuous, then that quantity must be identically zero. Thus, provided u and F are continuously differentiable, we obtain a *one-dimensional conservation law* for u. Namely,

$$\frac{\partial u}{\partial t} + \frac{\partial F(u)}{\partial x} = 0 \quad \text{or} \quad u_t + F(u)_x = 0 \tag{7.3.2}$$

EXAMPLE 7.3.1 ——

Let u denote car traffic density on a single-lane road with u measured in cars per kilometer. The flux of traffic is given as the product of the local traffic speed and the local traffic density. If we observe that traffic speed v decreases exponentially as traffic density increases, then

$$v = Ve^{-ku}$$

where V and k denote positive constants. Now the traffic flux can be written in terms of u as

$$F(u) = Ve^{-ku} \cdot u$$

and a conservation law of the form (7.3.2) reads

$$u_t + (Ve^{-ku} \cdot u)_x = 0 \qquad\qquad \blacksquare\ \blacksquare$$

Initial-Value Problems for Conservation Laws

The solution of the conservation law (7.3.2) is sometimes simpler if (7.3.2) is rewritten in a slightly different form. By the chain rule, we have

$$F(u)_x = \frac{dF}{du}\frac{\partial u}{\partial x} = F'(u) \cdot u_x$$

Thus, if we set $a(u) = F'(u)$, then Equation (7.3.2) can be expressed as

$$a(u)u_x + u_t = 0, \qquad t > 0 \tag{7.3.3}$$

Let us now consider the initial-value problem consisting of (7.3.3) subject to the initial condition

$$u(x, 0) = f(x), \qquad -\infty < x < \infty \tag{7.3.4}$$

This problem can be solved, at least for small t, using the method of characteristics described in Section 7.2. Formulating the characteristic equations (7.2.7) for the problem (7.3.3) and (7.3.4) shows the solution to be defined by

$$\frac{\partial x}{\partial r} = a(u), \qquad \frac{\partial t}{\partial r} = 1, \qquad \frac{\partial u}{\partial r} = 0 \tag{7.3.5}$$

$$x(0, s) = s, \qquad t(0, s) = 0, \qquad u(0, s) = f(s) \tag{7.3.6}$$

Recall that r is a parameter that varies along the characteristics of (7.3.3) and s is a parameter that varies along the initial curve, the x axis in this case. The condition $u_r = 0$ implies u is constant along characteristics. Therefore, $a(u)$ is constant with respect to r and Equations (7.3.5) and (7.3.6) can be solved to obtain

$$x = a(u)r + s, \qquad t = r, \qquad u = f(s) \tag{7.3.7}$$

Eliminating r and s from (7.3.7) indicates that a solution of the initial-value problem (7.3.3) and (7.3.4) is defined implicitly by the equation

$$u = f(x - a(u)t) \tag{7.3.8}$$

Let us now examine the conditions under which Equation (7.3.8) does provide a solution of the initial-value problem. By setting $t = 0$ in Equation (7.3.5), it is clear that the initial condition $u(x, 0) = f(x)$ is satisfied. If we assume the functions f and a to be continuously differentiable, then a careful application of the chain rule yields

$$u_x = f'(x - a(u)t)[1 - a'(u)u_x t]$$

which can be solved for u_x to obtain

$$u_x = \frac{f'(x - a(u)t)}{1 + f'(x - a(u)t) \cdot a'(u) \cdot t}$$

(7.3.9)

Similarly,

$$u_t = \frac{-a(u)f'(x - a(u)t)}{1 + f'(x - a(u)t) \cdot a'(u) \cdot t}$$

(7.3.10)

Equations (7.3.9) and (7.3.10) show that (7.3.3) is satisfied provided the denominator expression in (7.3.9) and (7.3.10) does not vanish. This establishes the following result.

Theorem 7.3.1. Assume the functions f and a are continuously differentiable. Then the solution $u = u(x, t)$ to the initial-value problem (7.3.3) and (7.3.4) is defined implicitly by

$$u = f(x - a(u)t)$$

provided $1 + f'(x - a(u)t) \cdot a'(u) \cdot t$ remains positive.

EXAMPLE 7.3.2 _____

Consider the intitial-value problem

$$(u^2)_x + u_t = 0, \qquad t > 0$$
$$u(x, 0) = \tan^{-1}x, \qquad -\infty < x < \infty$$

Since $(u^2)_x = 2uu_x$, $a(u) = 2u$. By Theorem 7.3.1, u is defined implicitly by

$$u = \tan^{-1}(x - 2ut)$$

Even though this implicit solution cannot be solved for u as an explicit function of x and t, it can be used to display the solution behavior. Taking the tangent of both sides of the last equation yields

$$x = 2ut + \tan u$$

Now, curves of constant u, $u = \bar{u}$, can be plotted as straight lines in the xt plane. For example, the points for which $u = \frac{1}{4}\pi$ lie on the line $x = \frac{1}{2}\pi t + 1$. Alternatively, we can obtain a plot of u versus x for fixed $t = \bar{t}$ by using $x = 2u\bar{t} + \tan u$ to plot x against t. ■ ■

EXAMPLE 7.3.3 _____

Consider the same equation as in the last example but with the initial condition $u(x, 0) = -x$. Now the solution is given by

$$u = -(x - 2ut)$$

which can be solved to give $u = x/(2t - 1)$. Note that the solution in this example fails to hold for $t \geq \frac{1}{2}$. ■ ■

EXAMPLE 7.3.4

Consider the initial-boundary-value problem

$$(\tfrac{1}{3}u^3)_x + u_t = 0, \qquad x > 0, t > 0$$
$$u(x, 0) = \sqrt{x}, \qquad x > 0$$
$$u(0, t) = 0, \qquad t > 0$$

Since $a(u) = u^2$, the solution is defined implicitly by

$$u = \sqrt{x - u^2 t} \quad \text{for } x > 0$$

so $u^2 = x - u^2 t$, which shows

$$u = \sqrt{x/(1 + t)} \quad \text{for } x > 0$$

It is easy to verify that $u(0, t) = 0$. ■ ■

Exercises

1. Solve the conservation law equation initial-value problem $uu_x + u_t = 0$, $u(x, 0) = x$.

2. Solve the conservation law equation initial-value problem $(u^2)_x + u_t = 0$, $u(x, 0) = x + 1$.

3. (a) Obtain an implicit solution for $(e^u)_x + u_t = 0$, $u(x, 0) = x$.
   (b) Use (a) to solve for x in terms of u and t.
   (c) Use (b) to obtain a graph of u versus x at time $t = 1$.

4. (a) Obtain an implicit solution for $uu_x + u_t = 0$, $u(x, 0) = e^x$.
   (b) Use (a) to solve for x in terms of u and t.
   (c) Use (b) to obtain a graph of u versus x at time $t = 1$.

5. (a) Given the conservation law initial-value problem $(ue^{-u})_x + u_t = 0$, $u(x, 0) = x$, draw characteristics from $x = -2$, $x = 0$, and $x = 2$ forward in time and assign to each characteristic a u value.
   (b) Obtain an implicit solution to the problem in (a).

6. Use Theorem 7.2.1 to solve

$$uu_x + u_t = 0$$

$$u(x, 0) = \begin{cases} 1, & x \le 0 \\ 1 - x, & 0 < x < 1 \\ 0, & x \ge 1 \end{cases}$$

and show that the solution obtained from (7.3.8) is valid only for $t < 1$.

7. Consider the constant-coefficient conservation law

$$(au)_x + u_t = 0, \qquad -\infty < x < \infty, t > 0, a = \text{const}$$

subject to the initial condition

$$u(x, 0) = f(x), \qquad -\infty < x < \infty$$

Use Theorem 7.2.1 to obtain the solution $u(x, t) = f(x - at)$.

8. Let v denote traffic velocity in miles per hour and let u denote traffic density in cars per mile. Assume $v = 60 \exp(-u/20)$, and traffic flux $F(u) = uv(u)$. Use Theorem 7.3.1 to obtain an implicit solution to the conservation law initial-value problem

$$F(u)_x + u_t = 0$$

$$u(x, 0) = \begin{cases} 200, & x < 0 \\ 200(1 - x), & 0 \le x \le 1 \\ 0, & x > 1 \end{cases}$$

Give a physical interpretation of the initial condition and the solution.

9. In traffic flow assume the velocity has the form

$$v(u) = \alpha u + \beta$$

Let U and V denote the maximum traffic density and the maximum traffic velocity, respectively.
   (a) Determine values for the constants α and β so that $v(0) = V$ and $v(U) = 0$.
   (b) Determine the density at which the flux $F(u) = u \cdot v(u)$ is a maximum.

10. In traffic flow assume the velocity has the form

$$v(u) = \alpha u^2 + \beta$$

Let U and V denote the maximum traffic density and the maximum traffic velocity, respectively.
   (a) Determine values for the constants α and β so that $v(0) = V$ and $v(U) = 0$.
   (b) Determine the density at which the flux $F(u) = u \cdot v(u)$ is a maximum.

11. Rework Exercise 8 using the velocity density relationship of Exercise 9.

12. Rework Exercise 8 using the velocity density relationship of Exercise 10.

7.4 GENERALIZED SOLUTIONS

In the last section we used the method of characteristics to solve an initial-value conservation law problem of the form

$$F(u)_x + u_t = a(u)u_x + u_t = 0, \qquad -\infty < x < \infty, t > 0 \qquad (7.4.1)$$

$$u(x, 0) = f(x), \qquad -\infty < x < \infty \qquad (7.4.2)$$

Under the hypotheses of Theorem 7.3.1, the method of characteristics provides a continuously differentiable solution. In many applications the initial condition $u(x, 0) = f(x)$ is not continuously differentiable and, therefore, Theorem 7.3.1 does not apply. In this section we discuss the construction of solutions to problems with only piecewise smooth or even just piecewise continuous, initial conditions.

One condition required for Theorem 7.3.1 to provide a continuously differentiable solution is that the expression

$$1 + f'(x - a(u)t)a'(u)t$$

remain positive. However, if the functions f' and a' are of opposite sign, then for some positive t the implicitly defined solution of Theorem 7.3.1 will fail to have well-defined derivatives u_x and u_t. That is, even if the initial condition $f(x)$ is continuously differentiable, the solution may develop a discontinuity for some positive value of t. These kinds of discontinuities will be examined in this section.

In terms of the geometry of the characteristics, there are generally two ways in which Theorem 7.3.1 can fail to apply to the initial-value problem. First, there may be some region of the xt plane that is not covered by any characteristic of (7.4.1). In this

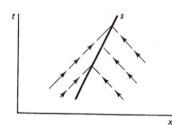

Figure 7.4.1 Essential discontinuity $x = s(t)$ with characteristics converging on s.

case, the solution in the characteristic free region is defined by what is called an *expansion fan*.

A second situation to which Theorem 7.3.1 does not apply occurs when some region of the xt plane is covered by two distinct characteristics carrying contradictory information about the value of the solution. In this case, the double-valued nature of the method of characteristics solution is reconciled by introducing in the solution a discontinuity called a *shock*.

To discuss expansion fans and shocks, we must extend the notion of solution from a classical, continuously differentiable setting of Theorem 7.3.1. We say that $u(x, t)$ is a *weak solution*, or a *generalized solution*, of (7.4.1) in some region R of the xt plane provided $u(x, t)$ satisfies the integral form of the conservation law (7.3.1) in R. Moreover, we are interested in generalized solutions whose only discontinuities are those that are essential to prevent multivaluedness. In terms of the characteristics of Equation (7.4.1) these "essential" discontinuities can be characterized as follows. Suppose $u(x, t)$ is a generalized solution that is discontinuous across a curve $x = s(t)$ (see Figure 7.4.1). Geometrically, we view the discontinuity curve as essential if the characteristics drawn from either side of $s(t)$ intersect along $s(t)$ as t increases. As we shall see in what follows, these kinds of discontinuities are shocks.

Solution Fans

Consider the conservation law (7.4.1) subject to the initial condition

$$u(x, 0) = \begin{cases} u_1, & x < 0, \ u_1 = \text{const} \\ u_2, & x > 0, \ u_2 = \text{const} \end{cases} \tag{7.4.3}$$

and assume that $u_2 > u_1$. Also, assume that $a(u)$ is increasing, so that $a(u_2) > a(u_1)$. The method of characteristics shows that

$$u(x, t) = \begin{cases} u_1, & x < a(u_1)t, \ t > 0 \\ u_2, & x > a(u_2)t, \ t > 0 \end{cases}$$

but does not provide a solution in the wedge

$$W: \qquad a(u_1)t < x < a(u_2)t$$

because no characteristics lie in this region (see Figure 7.4.2).

In the region W we look for a solution of the form

$$u(x, t) = g\left(\frac{x}{t}\right).$$

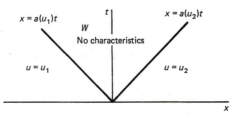

Figure 7.4.2 Wedge-shaped region that is not covered by characteristics.

Exercise 4 provides some motivation for looking for a solution in this form. If g is assumed to be continuously differentiable, then

$$u_x = g'\left(\frac{x}{t}\right) \cdot \left(\frac{1}{t}\right) \quad \text{and} \quad u_t = g'\left(\frac{x}{t}\right) \cdot \left(-\frac{x}{t^2}\right)$$

Putting these derivatives into the conservation law (7.4.1) yields

$$a\left(g\left(\frac{x}{t}\right)\right) \cdot g'\left(\frac{x}{t}\right) + g'\left(\frac{x}{t}\right) \cdot \left(-\frac{x}{t}\right) = 0$$

Assuming for now that $g' \neq 0$, the last equation shows

$$a\left(g\left(\frac{x}{t}\right)\right) = \frac{x}{t} \quad \text{or} \quad a(g(\cdot)) = (\cdot)$$

Thus, we see that the functions a and g are inverses of each other, $g(\cdot) = a^{-1}(\cdot)$.

Finally, consider the values of $g(x/t)$ on the lines $x = a(u_i)t$, $i = 1, 2$. Setting $x = a(u_i)t$ in $u(x, t) = g(x/t)$ gives

$$u(a(u_i)t, t) = g(a(u_i)) = u_i, \qquad i = 1, 2$$

where the last equality follows from $g = a^{-1}$. The preceding discussion shows that a continuous generalized solution to the conservation law (7.4.2) subject to the intital conditon (7.4.3) is given by

$$u(x, t) = \begin{cases} u_1, & x < a(u_1)t \\ g(x/t), & a(u_1)t < x < a(u_2)t \\ u_2, & x > a(u_2)t \end{cases} \tag{7.4.4}$$

The part of the solution defined by $g(x/t)$ is called an *expansion wave* or an *expansion fan*. The verification that the piecewise defined solution (7.4.4) actually solves the integral equation form of the conservation law equation is left as an exercise.

EXAMPLE 7.4.1 _____

Consider the initial-value problem

$$u^2 u_x + u_t = 0, \qquad -\infty < x < \infty$$

$$u(x, 0) = \begin{cases} 1, & x < 0 \\ 2, & x > 0 \end{cases}$$

Here, $u_1 = 1$, $u_2 = 2$, and $a(u) = u^2$. From $a(\cdot) = (\cdot)^2$, it follows that $g(\cdot) = a^{-1}(\cdot)$ $= \sqrt{(\cdot)}$. Now the solution is seen to be

$$u(x, t) = \begin{cases} 1, & x < t \\ \sqrt{x/t}, & t < x < 4t \\ 2, & x > 4t \end{cases}$$

■ ■

A slight modification of the method of solution of Example 7.4.1 allows us to deal with a discontinuity in the initial condition at a value of x other than zero. As the next example shows, all that is required is a shift of the origin of the x axis to the point of discontinuity.

EXAMPLE 7.4.2 _____

Consider the initial-value problem

$$u^2 u_x + u_t = 0$$

$$u(x, 0) = \begin{cases} 1, & x < 3 \\ 2, & x > 3 \end{cases}$$

This problem is identical to the one considered in Example 7.4.1 except that the discontinuity in the initial condition is at $x = 3$ rather than $x = 0$. A translation of $x = 3$ to $x = 0$ is accomplished by replacing x by $x - 3$. Thus, the solution of this problem is

$$u(x, t) = \begin{cases} 1, & x - 3 < t \\ \sqrt{\dfrac{x - 3}{t}}, & t < x - 3 < 4t \\ 2, & x - 3 > 4t \end{cases}$$

■ ■

As a final example, we illustrate how an expansion fan can be used to "fill a gap" left between two nonconstant parts of a solution.

EXAMPLE 7.4.3 _____

Consider the initial-value problem

$$uu_x + u_t = 0$$

$$u(x, 0) = \begin{cases} x + 1, & x < 0 \\ x + 2, & x > 0 \end{cases}$$

Here, $a(u) = u$. The initial condition

$$f(x) = x + 1 \quad \text{for } x < 0$$

provides an implicit solution

$$u = f(x - a(u)t)$$

or

$$u = (x - ut) + 1$$

in the region covered by characteristics that begin on the half-line $x < 0$. This argument shows that in the region $x < t$ the solution is given by

$$u(x, t) = \frac{x + 1}{t + 1}$$

A similar argument shows that on the region covered by characteristics drawn from the positive x axis, namely, $x > 2t$, the solution is given by

$$u(x, t) = \frac{x + 2}{t + 1}$$

Now, an expansion fan can be used to define the solution in the region $t < x < 2t$, which is not covered by any characteristic drawn from the x axis. Since $a(u) = u$ and $g(\cdot) = a^{-1}(\cdot)$, it follows that $g(u) = u$. On the left edge of the fan, $x = t$ and u has the value $u = 1$, and on the right edge of the fan, $x = 2t$ and $u = 2$. Therefore, in the region $t < x < 2t$, the solution is given by $u(x, t) = x/t$. Putting the pieces of the solution together yields

$$u(x, t) = \begin{cases} \dfrac{x + 1}{t + 1}, & x < t \\[2ex] \dfrac{x}{t}, & t < x < 2t \\[2ex] \dfrac{x + 2}{t + 1}, & x > 2t \end{cases}$$
■ ■

Shocks

In our discussion of expansion fans, we have seen how a discontinuity in the initial condition can result in a region of the xt plane covered by no characteristics of the conservation law. As the next example shows, it may be the case that a discontinuity in the initial condition produces a region of the xt plane that is doubly covered by characteristics.

EXAMPLE 7.4.4 ───

For the initial-value problem

$$uu_x + u_t = 0$$

$$u(x, 0) = \begin{cases} 2, & x < 0 \\ 1, & x > 0 \end{cases}$$

we see that the characteristics drawn from the negative x axis satisfy $dx/dt = a(u(x, 0)) = 2$ and along these characteristics $u = 2$. Similarly, the characteristics from the positive x axis are defined by $dx/dt = 1$ and carry the value $u = 1$. Thus, the region

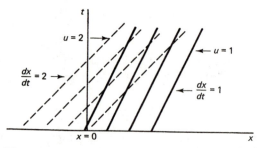

Figure 7.4.3 Characteristics doubly cover part of xt plane and carry contradictory information about solution.

$t < x < 2t$ is doubly covered by characteristics (Figure 7.4.3). Moreover, these two families of characteristics carry contradictory information about the value of the solution. ∎ ∎

The initial condition in Example 7.4.4 is only piecewise continuous. In view of this example, we might be tempted to conclude that the failure of the solution to be single valued can be attributed to the presence of a discontinuity in $u(x, 0)$. The following example shows that even a smooth initial condition can give rise to contradictory information about u being transmitted along intersecting characteristics.

EXAMPLE 7.4.5 _____

For the problem

$$uu_x + u_t = 0$$

$$u(x, 0) = \frac{1}{x^2 + 1}$$

$a(u) = u$. Therefore, the characteristic through $(x, t) = (0, 0)$, defined by $x'(t) = a(u(0, 0)) = 1$, is $x = t$. Along this characteristic $u(x, t)$ has the constant value $u(0, 0) = 1$. Similarly, the characteristic through $(x, t) = (1, 0)$ is $x = \frac{1}{2}t + 1$; along this characteristic $u(x, t) = u(1, 0) = \frac{1}{2}$. Since the characteristics $x = t$ and $x = \frac{1}{2}t + 1$ intersect at $(x, t) = (2, 2)$ we are led to the impossible conclusion that u is simultaneously equal to 1 and equal to $\frac{1}{2}$ at the point $(x, t) = (2, 2)$ (Figure 7.4.4). Exercise 13 outlines an approach for showing that the first intersection of characteristics carrying contradictory information occurs at

$$(x_B, t_B) = (\sqrt{3}, \tfrac{8}{9}\sqrt{3})$$

In analogy with breaking water waves, t_B is called the *break time*. ∎ ∎

The last two examples illustrate how a conservation law can have an apparently double-valued solution as a result of intersecting characteristics that carry contradictory information about the solution. From a physical point of view, we expect a conservation law to have a single-valued solution. The apparent double valuedness of the solutions in

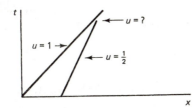

Figure 7.4.4 Characteristics carrying contradictory information intersect at (2, 2).

the last two examples can be reconciled by introducing a discontinuity called a *shock* in the solution.

Let $S : x = s(t)$ be a curve in the xt plane across which a solution of (7.4.1) is discontinuous. Assume the conservation law equation (7.4.1) has generalized solution $u(x, t)$ defined by

$$u(x, t) = \begin{cases} u_1(x, t), & x < s(t) \\ u_2(x, t), & x > s(t) \end{cases}$$

with $u_1(s(t), t) \neq u_2(s(t), t)$. Also, assume that in a neighborhod of S, $u_1(x, t)$ and $u_2(x, t)$ are continuously differentiable (Figure 7.4.5). Given the conservation law

$$F(u)_x + u_t = 0 \quad \text{or} \quad a(u)u_x + u_t = 0 \tag{7.4.5}$$

we say that the curve S defined by $x = s(t)$ is a *shock* provided the following conditions hold along S:

$$\text{S1:} \qquad \frac{ds}{dt} = \frac{F(u_1) - F(u_2)}{u_1 - u_2} \tag{7.4.6}$$

$$\text{S2:} \qquad a(u_1) > \frac{ds}{dt} > a(u_2) \tag{7.4.7}$$

The *shock condition* S1 is a statement of conservation of material M across the shock. Let $x_1 < s(t) < x_2$. Since $u(x, t)$ represents the linear density of material, or substance, M, we can write the quantity of that substance in the interval (x_1, x_2) at time t as

$$Q(t) = \int_{x_1}^{x_2} u(x, t)\, dx = \int_{x_1}^{s(t)} u_1(x, t)\, dx + \int_{s(t)}^{x_2} u_2(x, t)\, dx \tag{7.4.8}$$

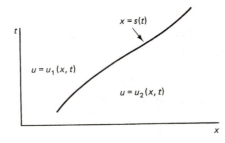

Figure 7.4.5 Curve $S: x = s(t)$ is shock; $u_1(x, t)$ is solution behind shock; $u_2(x, t)$ is solution in front of shock.

For the material M, whose density is u, to be conserved, the time rate of change of Q must equal the flow of u across $x = x_1$ minus the flow of u across $x = x_2$. That is,

$$\frac{dQ}{dt} = F(u_1(x_1, t)) - F(u_2(x_2, t)) \tag{7.4.9}$$

To obtain another expression for dQ/dt, recall from calculus that if

$$G(t) = \int_{x_0}^{s(t)} g(x)\, dx$$

with x_0 constant, $g(x)$ continuous, and $s(t)$ differentiable, then

$$G'(t) = g(s(t)) \cdot s'(t) \tag{7.4.10}$$

Using (7.4.10) and the fact that

$$\int_{s(t)}^{x_2} u_2(x, t)\, dx = -\int_{x_2}^{s(t)} u_2(x, t)\, dx$$

to differentiate Equation (7.4.8), we can also write

$$\frac{dQ}{dt} = u_1(s(t), t)s'(t) - u_2(s(t), t)s'(t) \tag{7.4.11}$$

Since Equation (7.4.9) holds in the limit as $x_1 \to s(t)^-$ and $x_2 \to s(t)^+$, (7.4.9) and (7.4.11) imply

$$[u_1(s(t), t) - u_2(s(t), t)]\frac{ds}{dt} = F(u_1(s(t), t)) - F(u_2(s(t), t))$$

which is equivalent to the shock condition S1. See Exercise 12 for an alternate derivation of the shock condition S1.

The geometric interpretation of the shock condition S2 is that characteristics starting on either side of the shock S, when continued forward in time, intersect S. When the characteristics to the left of S carry different u values than those to the right of S, a discontinuity along S is unavoidable if the solution is to be single valued. In this sense, a shock is an essential discontinuity in the solution.

In the context of compressible fluid flow, the condition S2 asserts that fluid that crosses a shock S must experience an increase in entropy. For this reason S2 is often referred to as the *entropy shock condition*. To illustrate the application of the shock conditions S1 and S2, we conclude this section with two examples.

EXAMPLE 7.4.6 _____

Consider the initial-value problem

$$(e^u)_x + u_t = 0 \quad \text{or} \quad e^u u_x + u_t = 0$$

$$u(x, 0) = \begin{cases} 2, & x < 0 \\ 1, & x > 0 \end{cases}$$

Here $F(u) = e^u$ and $a(u) = F'(u) = e^u$. The characteristics drawn from the negative x axis satisfy $x'(t) = e^2$, while those drawn from the positive x axis satisfy $x'(t) = e$. Thus the region $et < x < e^2t$ is doubly covered by characteristics. The characteristics from the negative x axis carry the value $u = 2$ and those drawn from the positive x axis carry the value $u = 1$. To reconcile this double valuedness of u, we introduce a shock $x = s(t)$ behind which $u = 2$ and in front of which $u = 1$. According to the shock condition S1,

$$s'(t) = \frac{F(2) - F(1)}{2 - 1} = \frac{e^2 - e}{2 - 1} = e(e - 1)$$

Sketching some characteristics into the xt plane, we see the shock begins at $(x, t) = (0, 0)$. Thus S is defined by

$$x = s(t) = e(e - 1)t$$

and we have

$$u(x, t) = \begin{cases} 2, & x < e(e - 1)t \\ 1, & x > e(e - 1)t \end{cases}$$

It is left as an exercise to show that the entropy condition is satisfied across the shock. ■ ■

EXAMPLE 7.4.7 _____

For the initial-value problem

$$2uu_x + u_t = 0$$

$$u(x, 0) = \begin{cases} \sqrt{|x|}, & x < 0 \\ 0, & x > 0 \end{cases}$$

the characteristics from the positive x axis are the vertical lines $x = const$ and along them u is zero. The characteristics from the negative x axis satisfy

$$\frac{dx}{dt} = a(u(x, 0)) = 2\sqrt{|x|}$$

and, therefore, point forward in x with increasing t. Along these characteristics u is positive. Thus, points in the region $x > 0$, $t > 0$ will be assigned two u values unless a shock is present. Clearly, the solution in front of the shock is given by $u_2 = 0$. To find the solution behind the shock, we use the implicit solution $u = f(x - a(u)t)$ with $f(x) = \sqrt{|x|}$ and $a(u) = 2u$ to conclude, from Theorem 7.3.1, that the solution from the negative x axis satisfies

$$u = \sqrt{|x - 2ut|}$$

or $u^2 = |x - 2ut| = -x + 2ut$. Using the quadratic formula to solve for u, we find

$$u_1(x, t) = t + \sqrt{t^2 - x}$$

To apply the shock condition S1, $F(u)$ is required. From $F'(u) = 2u$, we conclude $F(u) = u^2 + \text{const}$. With no loss of generality we can choose the constant of integration to be zero and take $F(u) = u^2$. (Why?) Now S1 implies

$$\frac{ds}{dt} = \frac{u_1^2 - u_2^2}{u_1 - u_2} = \frac{u_1^2 - 0}{u_1 - 0} = u_1(s(t), t)$$

or

$$\frac{ds}{dt} = t + \sqrt{t^2 - s(t)}$$

The form of the last equation suggests we look for a solution that is quadratic in t. If we suppose that $s(t) = ct^2$, $c = \text{const}$, then we have

$$2c = 1 + \sqrt{1 - c}$$

from which we see $(2c - 1)^2 = 1 - c$ or $4c^2 - 3c = 0$. The solution $c = 0$ is seen to be extraneous, which leaves us with $c = \frac{3}{4}$. Thus, we find the equation of the shock to be $x = s(t) = 3t^2/4$. Now the solution of the initial-value problem is given by

$$u(x, t) = \begin{cases} t + \sqrt{t^2 - x}, & x < 3t^2/4 \\ 0, & x > 3t^2/4 \end{cases}$$

Notice that $a(u_1(s(t), t)) = 2(t + t/2) = 3t$, $a(u_2(s(t), t)) = 0$, and $s'(t) = 3t/2$ so the entropy condition S2

$$a(u_1) > s'(t) > a(u_2) \quad \text{or} \quad 3t > 3t/2 > 0$$

is satisfied. ∎ ∎

Exercises

1. For each of the following initial-value problems, sketch a few characteristics from the x axis forward in time and identify a region of the xt plane that is not covered by any characteristics. Then use an expansion fan to obtain a continuous generalized solution valid for $t > 0$.

   (a) $uu_x + u_t = 0$, $u(x, 0) = \begin{cases} 1, x < 0 \\ 3, x > 0 \end{cases}$

   (b) $uu_x + u_t = 0$, $u(x, 0) = \begin{cases} 2, x < 1 \\ 4, x > 1 \end{cases}$

   (c) $uu_x + u_t = 0$, $u(x, 0) = \begin{cases} -3, x < 0 \\ -2, x > 0 \end{cases}$

   (d) $(u^2)_x + u_t = 0$, $u(x, 0) = \begin{cases} 1, x < 0 \\ 3, x > 0 \end{cases}$

   (e) $(e^u)_x + u_t = 0$, $u(x, 0) = \begin{cases} 0, x < 0 \\ 1, x > 0 \end{cases}$

   (f) $(u^4)_x + u_t = 0$, $u(x, 0) = \begin{cases} 0, x < 0 \\ 1, x > 0 \end{cases}$

2. Find the continuous generalized solution to the initial-value problem

$$uu_x + u_t = 0, \qquad u(x, 0) = \begin{cases} x - 1, & x < 0 \\ x + 1, & x > 0 \end{cases}$$

3. Obtain a continuous generalized solution to the initial-value problem

$$uu_x + u_t = 0, \qquad u(x, 0) = \begin{cases} 0, & x < 0 \\ x, & 0 < x < 1, t > 0 \\ 1, & x > 1 \end{cases}$$

Then follow the solution backward in time to $t = -1$ and plot $u(x, -1)$.

4. Obtain a continuous generalized solution to the initial-value problem

$$uu_x + u_t = 0, \qquad u(x, 0) = \begin{cases} 0, & x < 0 \\ x/\epsilon, & 0 < x < \epsilon, t > 0 \\ 1, & x > \epsilon \end{cases}$$

Then allow the parameter ϵ to approach zero in your solution.

5. (a) Given the conservation law equation $(u^2/2)_x + u_t = 0$, show that

$$u(x, t) = \begin{cases} 1, & x < 2t \\ 3, & x > 2t \end{cases}$$

is a generalized solution satisfying the initial condition

$$u(x, 0) = \begin{cases} 1, & x < 0 \\ 3, & x > 0 \end{cases}$$

Hint: Any rectangle through which the line $x = 2t$ passes can be written as the union of three rectangles one of which has $x = 2t$ as a diagonal.
   (b) From the points $(x, t) = (1, 1)$ and $(x, t) = (3, 1)$ draw characteristics forward in time. Do these characteristics intersect the line $x = 2t$?
   (c) It can be shown that the initial-value problem $(u^2/2)_x + u_t = 0$, $u(x, 0) = 1$, $x < 0$, $u(x, 0) = 3$, $x > 0$, has only one continuous generalized solution. Find it.

6. For each of the following initial-value problems find a generalized solution and give the equation of the shock.

(a) $uu_x + u_t = 0$, $u(x, 0) = \begin{cases} 1, x < 0 \\ 0, x > 0 \end{cases}$

(b) $uu_x + u_t = 0$, $u(x, 0) = \begin{cases} 3, x < 0 \\ 1, x > 0 \end{cases}$

(c) $(e^u)_x + u_t = 0$, $u(x, 0) = \begin{cases} 1, x < 0 \\ 0, x > 0 \end{cases}$

(d) $(e^{-u})_x + u_t = 0$, $u(x, 0) = \begin{cases} 0, x < 0 \\ 1, x > 0 \end{cases}$

(e) $uu_x + u_t = 0$, $u(x, 0) = \begin{cases} 5, x < 2 \\ 3, x > 2 \end{cases}$

7. Find a generalized solution to the initial-value problem

$$2uu_x + u_t = 0$$

$$u(x, 0) = \begin{cases} 1, & x < 0 \\ 1 - x, & 0 < x < 1 \\ 0, & x > 1 \end{cases}$$

8. Find a generalized solution to the initial-value problem

$$2uu_x + u_t = 0$$

$$u(x, 0) = \begin{cases} 1, & x < 0 \\ 1 - x/\epsilon, & 0 < x < \epsilon \\ 0, & x > \epsilon \end{cases}$$

and express the shock in terms of x, t, and ϵ.

9. Obtain a generalized solution to the initial-value problem

$$2uu_x + u_t = 0$$

$$u(x, 0) = \begin{cases} 1 - |x|, & |x| < 1 \\ 0, & |x| > 1 \end{cases}$$

10. Show that the solution (7.4.4) satisfies the integral form of the conservation law (7.3.1).

11. Show that the solution of Example 7.4.6 satisfies the entropy condition S2.

12. Suppose that $x = s(t)$ is a shock and that the solution to

$$F(u)_x + u_t = 0$$

is given by $u = u_1$ for $x < s$ and $u = u_2$ for $x > s$. From the following figure:

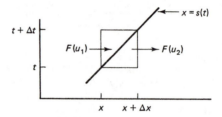

argue that to first order in Δx and Δt, the following material balance equation holds:

$$(u_1 - u_2) \, \Delta x = [F(u_1) - F(u_2)] \, \Delta t.$$

Use this material balance equation and

$$s'(t) = \frac{\Delta x}{\Delta t} \quad \text{as } \Delta x \text{ and } \Delta t \to 0$$

to derive the shock condition S1.

13. (a) Use Theorem 7.3.1 to show that a smooth solution of Example 7.4.5 satisfies

$$x - ut = \pm \left[\frac{1}{u} - 1 \right]^{1/2}$$

(b) Show that the characteristics of $uu_x + u_t = 0$ are given by

$$x - ut = \xi, \qquad \xi = \text{const}$$

(c) Use (a) and (b) to show that

$$\xi = \pm \left[\frac{1}{u} - 1 \right]^{1/2}$$

is the position where the characteristic carrying the value u intersects the x axis.

(d) Show that along the characteristic $x - ut = \xi$,

$$u_x = \frac{f'(\xi)}{1 - [2\xi/(\xi^2 + 1)^2]t}$$

(e) Show that the break time

$$t_B = \min \left\{ t : t(\xi) = \frac{(\xi^2 + 1)^2}{2\xi} \right\}$$

is $t_B = \frac{8}{9}\sqrt{3}$.

7.5 APPLICATIONS OF SCALAR CONSERVATION LAWS

Traffic Flow

Suppose that cars are traveling along a one-lane road on which no passing is allowed. Assume that no traffic enters or leaves this stretch of road. Let x denote the distance measured along the road in the direction of traffic flow. Take d to be a typical distance between consecutive cars. Choose D to be a representative distance over which significant variations in traffic density occur. If d is much smaller than D, $d << D$, then we can idealize the traffic distribution to be a *continuum* rather than a discrete collection of individual cars.

To use a *continuum model* to describe the flow of traffic along the road, we first associate with each point x and each time t a traffic density function denoted by $u(x, t)$. Let h be a distance intermediate to d and D in the sense that $d << h << D$. Let $N(x, t)$ be the number of cars in the interval $(x - h, x + h)$ at time t and define a traffic density by

$$u(x, t) = N(x, t)/2h$$

EXAMPLE 7.5.1 ───

After a football game, cars leave the stadium along a single-lane road. Near the stadium, the typical distance separating cars is 10 ft, while 3 miles down the road away from the stadium most cars are about 100 ft apart. Here it would be appropriate to choose

$$d = (10 \text{ ft} + 100 \text{ ft})/2 \approx 50 \text{ ft}$$
$$D = 3 \text{ miles} \approx 16,000 \text{ ft}$$
$$h = 0.25 \text{ miles} \approx 1000 \text{ ft}$$

Then h is an order of magnitude larger that d and an order of magnitude smaller that D. With $h = 0.25$ miles, the units associated with $u(x, t)$ are cars per quarter mile. A simple rescaling, namely, multiplication of $u(x, t)$ by 4, would allow the local traffic density to be expressed in the units of cars per mile. ■ ▪

It would be more correct to say that the traffic density is a function of not just x and t but rather of x, t, and the choice of h and write $u = u(x, t; h)$ instead of $u = u(x, t)$. This would emphasize the dependence of the traffic density function on the somewhat arbitrary choice of h. However, in an analysis based on a continuum approach, it is assumed that small variations in the choice of h do not produce significantly different density functions.

Paralleling our development of a continuum traffic density, we can also identify with each point x and each time t a continuum velocity. Let $v(x, t)$ denote the average of the velocities of the cars in the interval $(x - h, x + h)$ at time t. The units associated with v are length per time (e.g., miles per hour).

At this point in our model development, we have two independent variables, x and t, and two dependent variables, u and v. If our aim is to develop a mathematical model in the form of a scalar conservation law, we must introduce a *constitutive relationship* betwen u and v that permits us to eliminate one of these dependent variables in terms of the other. To this end, we assume that v is a function of x and t only through its dependence on u. That is, we assume $v = v(u)$. Observations support this assumption. In fact, we expect v to be a decreasing function of u.

At any point x on the road and at any time t, a measure of traffic flow is obtained by forming the product of u and v to obtain a traffic flux F representing the number of cars per unit time passing the point x. If u has units of cars per mile and v has units of miles per hour, then the traffic flux

$$F(u) = v(u) \cdot u$$

is expressed in units of cars per hour.

Now, the argument leading to the integral form of a conservation law given by Equation (7.3.1) applies to the one-dimensional traffic flow described in the preceding. If we further assume that the traffic density u and the traffic flux $F(u)$ are continuously differentiable, then we can conclude that the traffic flow is governed by the conservation law

$$F(u)_x + u_t = 0$$

We now present three examples that illustrate the use of the method of characteristics in dealing with this continuum model of traffic flow.

EXAMPLE 7.5.2 _____

Solution Defined Implicitly

Let V denote the maximum traffic velocity and let U represent some reference traffic density at which the traffic velocity is less than maximal. For this example, we assume a constitutive relation between density and velocity of the form

$$v = v(u) = Ve^{-u/U}$$

Notice that as u approaches zero, the traffic velocity approaches V, so low traffic density corresponds to high velocity. Also, at $u = U$ the traffic velocity is given by $v = V/e$, which is less than the maximal value V. According to our continuum model of traffic flow, the traffic flux is given by $F(u) = v(u)u$, and we have the conservation law

$$F(u)_x + u_t = 0 \quad \text{or} \quad (uVe^{-u/U})_x + u_t = 0$$

to consider. To complete the formulation of this example problem, let us assume an initial condition of the form

$$u(x, 0) = \begin{cases} U, & x < 0 \\ U(1 - x), & 0 < x < 1 \\ 0, & x > 1 \end{cases}$$

Expanding the x derivative in the conservation law gives

$$a(u)u_x + u_t = 0, \qquad a(u) = V(1 - u/U)e^{-u/U}$$

Now, we see that the characteristics drawn from the negative x axis are the vertical lines $x = $ const. Along these characteristics u has the value $u = U$. Thus, we see that $u(x, t) = U$ for $x < 0, t > 0$. The characteristics drawn from the part of the x axis where $x > 1$ satisfy $dx/dt = a(u(x, 0)) = a(0) = V$. Therefore, in the region $x > Vt + 1, t > 0$, the solution is given by $u(x, t) = 0$. In the remainder of the upper half of the xt plane the solution is defined implicitly by $u = f(x - a(u)t)$ to be $u = U(1 - (x - a(u)t))$ or

$$u = U(1 - (x - V(1 - u/U)e^{-u/U} \cdot t))$$

If we want to plot u as a function of x for some fixed time t, we see immediately that $u = U$ for $x < 0$ and that $u = 0$ for $x > Vt + 1$. However, the plotting of the implicitly defined solution in the region $0 < x < Vt + 1$ is no easy job if we insist on plotting u as a function of x. Another way to obtain a graph of u versus x is to plot x as a function of u or of u/U and then rotate and reflect this graph to see u plotted in terms of x. Solving the implicit solution for x, we find

$$x = (1 - u/U) + V(1 - u/U)e^{-u/U} \cdot t$$

For fixed t, it is not difficult to plot x against u/U for $0 < u/U < 1$. ■■

EXAMPLE 7.5.3 ——————————————————————

Solution with an Expansion Fan

Let V again denote the maximum traffic velocity, but in this example let U denote the traffic density at which the traffic stalls and $v = 0$. Assume that v is related to u by $v(u) = V(1 - u/U)$. Suppose that at time $t = 0$ a roadblock that has been in place at $x = 0$ for some time is removed. This situation could be modeled by the following conservation law initial-value problem:

$$F(u)_x + u_t = 0, \qquad F(u) = v(u)u = V(u - u^2/U)$$

$$u(x, 0) = \begin{cases} U, & x < 0 \\ 0, & x > 0 \end{cases}$$

Expanding the x derivative in the conservation law gives

$$a(u)u_x + u_t = 0, \qquad a(u) = V(1 - 2u/U)$$

The characteristics from the negative x axis satisfy $x'(t) = a(U) = -V$, which shows the solution in the region $x < -Vt, t > 0$, to be given by $u(x, t) = U$. The characteristics from the positive x axis satisfy $x'(t) = a(0) = V$, so the solution in the region $x > Vt$, $t > 0$, is $u(x, t) = 0$. In the region $-Vt < x < Vt, t > 0$, there are no characteristics.

Here, we obtain a solution in the form of an expansion fan according to $u(x, t) = g(x/t)$, where g is the inverse of the function a. Setting $w = a(u)$ and solving for u, we find

$$g(w) = \tfrac{1}{2}U(1 - w/V)$$

Thus, the expansion fan part of the solution is given by

$$u(x, t) = \tfrac{1}{2}U(1 - x/Vt), \qquad -Vt < x < Vt$$

and the solution is complete.   ■ ■

EXAMPLE 7.5.4

Solution with a Shock

As in the last example, let V denote the maximum traffic velocity and U the density at which the traffic stalls and assume that $v(u) = V(1 - u/U)$. Consider the initial-value problem

$$F(u)_x + u_t = 0, \qquad F(u) = u \cdot v(u) = V(u - u^2/U)$$

$$u(x, 0) = \begin{cases} U/2, & x < 0 \\ U(1 + x)/2, & 0 < x < 1 \\ U, & x > 1 \end{cases}$$

Since $a(u) = V(1 - 2u/U)$, we see that the characteristics from the negative x axis satisfy $x'(t) = a(U/2) = 0$. Along these vertical characteristics u has the value $U/2$, so the method of characteristics predicts that $u(x, t) = U/2$ for $x < 0, t > 0$. However, a characteristic from a typical point $\bar{x}$ on $x > 1$ defined by $x'(t) = a(U) = -V$, $x(0) = \bar{x}$ is $x - \bar{x} = -Vt$. Along such a characteristic u has the value U. Now, the vertical characteristics, $x = $ const, carry the value $u = U/2$. The characteristics $x - \bar{x} = -Vt, \bar{x} > 1$, carry the value $u = U$. Since the region $-Vt < x < 0, t > 1$, is covered by characteristics along which $u = U/2$ and characteristics along which $u = U$, the traffic density is predicted to be double valued there unless a shock is introduced. Since the earliest intersection of characteristics that carry contradictory u values is at $(x, t) = (0, 1/V)$, we expect this to be the point of shock formation. Before we proceed with the determination of the shock, we find the solution in the triangular region $0 < x < 1 - Vt, 0 < t < 1/V$. From the implicit solution $u = f(x - a(u)t)$, we find that in that region

$$u = U(1 + (x - a(u)t))/2, \qquad a(u) = V(1 - 2u/U)$$

from which we determine that

$$u(x, t) = \frac{U}{2}\left\{\frac{1 + x - Vt}{1 - Vt}\right\}, \qquad 0 < x < 1 - Vt, 0 < t < 1/V \qquad (7.5.1)$$

This last piece of the solution supports our expectation of a shock forming at position $(x, t) = (0, 1/V)$. To determine the path of the shock, we use the shock condition S1 with $u_1 = U/2$ and $u_2 = U$ to write

$$\frac{ds}{dt} = \frac{F(u_1) - F(u_2)}{u_1 - u_2} = -\frac{V}{2}$$

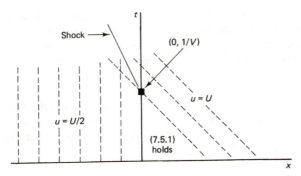

Figure 7.5.1 Characteristic diagram for Example 7.5.4.

Thus, we see the shock $x = s(t)$ is the straight half-line

$$x = -\tfrac{1}{2}V(t - 1/V), \qquad t > 1/V$$

and to the left of the shock $u = U/2$ while on the right of the shock $u = U$. The solution is displayed graphically in Figure 7.5.1. ■ ■

Porous Media Flow

Consider the vertical flow of a fluid, say, water, in a porous medium, a soil, for example. Assume the x axis points downward in the direction of gravity and the fluid can be treated as incompressible. The assumption of incompressibility implies that conservation of fluid volume is equivalent to conservation of fluid mass. Let $u(x, t)$ denote the volumetric water content at a depth x at time t measured as the ratio of water-filled pore space to total pore space in the porous medium. The value of u can range from $u = 0$, the case of a perfectly dry porous medium, up to the upper bound $u = 1$, which corresponds to a fully saturated porous medium.

Let F denote the vertical flux of fluid through the porous medium. In general, F depends on both the values of u and u_x as well as the character of the porous medium. The dependence of F on u can be roughly explained by noting that when a soil is very dry, the only means of water transport is through a thin film of water that covers the solid grains that make up the porous medium. On the other hand, when the soil is saturated, all of the pore space is available for fluid transport. Thus, under just the driving force of gravity, more water will move through a wet soil than through a dry soil.

In partially saturated flow, the dependence of F on u_x is connected to the phenomenon of capillarity that causes water to move from a wet region to a dry region. In many situations the fluid transport by gravity effects dominates the fluid transport due to capillary effects. In these situations, we can model a vertical, partially saturated flow with a scalar conservation law.

To formulate a conservation law for porous flow, we need an expression relating the flux F to the water content u. Experimental results support the assumption that $F(u)$ has the form

$$F(u) = Ku^n$$

where K and n are parameters that are soil dependent. Usually n turns out to be larger than 1. Here, K is the saturated hydraulic conductivity. It represents the flux value under conditions of fully saturated gravity flow. If we assume that K and n are constants, we can model vertical infiltration using the conservation law

$$K(u^n)_x + u_t = 0$$

The natural problem to pose for this conservation law has the form of an initial-boundary-value problem. The initial condition is used to specify the moisture distribution in the soil at time $t = 0$ while the boundary condition is used to provide information about the water content at the surface.

EXAMPLE 7.5.5

Solution with a Shock and an Expansion Wave
    Using the notation presented in the preceding, consider a gravity-dominated flow in a porous medium modeled by

$$
\begin{aligned}
(u^2)_x + u_t &= 0, \qquad x > 0, t > 0 \\
u(x, 0) &= \tfrac{1}{2}, \qquad x > 0
\end{aligned}
$$

$$
u(0, t) = \begin{cases} 1, & 0 < t < 1 \\ \tfrac{1}{2}, & t > 1 \end{cases}
$$

For this example, we have chosen $K = 1$ and $n = 2$. This choice gives $F(u) = u^2$ from which it follows that $a(u) = 2u$. Referring to the characteristic diagram of Figure 7.5.2, we find a straight-line shock defined by $x = 3t/2$ separates a region where $u = 1$ from a region where $u = \tfrac{1}{2}$. Applying the shock condition S1 with $u_1 = 1$ and $u_2 = \tfrac{1}{2}$, we

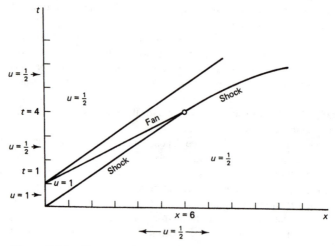

Figure 7.5.2 Characteristic diagram for Example 7.5.5.

find the shock connects the points $(x, t) = (0, 0)$ and $(x, t) = (6, 4)$. The region $t - 1 < x < 2(t - 1)$, $0 < t < 4$, is not covered by any characteristics. In this region the solution is given by the expansion fan

$$u(x, t) = \frac{x}{2(t - 1)}$$

To track the shock beyond the position $(x, t) = (6, 4)$, we must use the shock condition S1 in the form

$$\frac{ds}{dt} = \frac{F(u_1(s, t)) - F(\tfrac{1}{2})}{u_1(s, t) - \tfrac{1}{2}}$$

where $u_1(x, t)$ is the expansion fan solution

$$u_1(x, t) = \frac{x}{2(t - 1)}$$

Thus, condition S1 provides a differential equation for $s(t)$; namely,

$$s'(t) = \frac{1}{2(t - 1)} s(t) + \tfrac{1}{2}, \qquad s(4) = 6$$

Using the integrating factor $(t - 1)^{-1/2}$, we find the solution to be

$$s(t) = (t - 1) + \sqrt{3(t - 1)}$$

This shows the solution of the initial-boundary value problem to be given by

$$u(x, t) = \begin{cases} \tfrac{1}{2}, & 0 < x < t - 1, 0 < t < 4 \\[2mm] \dfrac{x}{2(t - 1)}, & t - 1 < x < 2(t - 1), 0 < t < 4 \\[2mm] 1, & 2(t - 1) < x < 3t/2, 0 < t < 4 \\[1mm] \tfrac{1}{2}, & x > 3t/2, 0 < t < 4 \end{cases}$$

$$u(x, t) = \begin{cases} \tfrac{1}{2}, & 0 < x < t - 1, t > 4 \\[2mm] \dfrac{x}{2(t - 1)}, & t - 1 < x < s(t), t > 4 \\[2mm] \tfrac{1}{2}, & x > s(t), t > 4 \end{cases}$$

■ ■

Exercises

1. Calculate the jump across the shock in Example 7.5.5 and plot it as a function of time.

2. Let V denote the maximum traffic velocity and U the maximum density at which traffic will move. Assume the traffic velocity is related to the traffic density by $v = V(1 - u/U)$ and that the traffic flux is $F(u) = v(u) \cdot u$. Solve the traffic flow equation $F(u)_x + u_t = 0$ subject to each of the following initial conditions and give a physical interpretation of your solution.

(a) $u(x, 0) = \begin{cases} U/4, x < 0 \\ U/6, x > 0 \end{cases}$

(b) $u(x, 0) = \begin{cases} U/6, x < 0 \\ U/4, x > 0 \end{cases}$

(c) $u(x, 0) = \begin{cases} U/4, \ x < 0 \\ 3U/4, x > 0 \end{cases}$

(d) $u(x, 0) = \begin{cases} U, & x < 0 \\ U(1 - x/L), 0 < x < L \\ 0, & x > L \end{cases}$

(e) $u(x, 0) = \begin{cases} 0, & x < 0 \\ Ux/L, 0 < x < L \\ U, & x > 0 \end{cases}$

(f) $u(x, 0) = \begin{cases} 0, & x < 0 \\ 3Ux/4L, 0 < x < L \\ 3U/4, & x > L \end{cases}$

3. In the traffic flow equation $(vu)_x + u_t = 0$, assuming that velocity v is a function of only traffic density u, show that the local traffic density appears unchanged to an observer at $x = x(t)$ who moves so that $dx/dt = d(uv)/du$. *Hint:* Consider the value of u along the characteristics of the traffic flow equation.

4. In the traffic flow equation, show that if $v = v(u)$ satisfies $v'(u) < 0$, then the rate of propagation of small variations in traffic density cannot exceed the speed of an individual vehicle at that density.

5. In the traffic flow equation $(vu)_x + u_t = 0$ assume the speed density relationship $v = V(1 - u/U)$, where V and U denote, respectively, the maximum speed and maximum density. Suppose cars are traveling along the road with uniform density $u = U/3$ and uniform speed $v = 2V/3$. At time $t = 0$ a truck enters the road at $x = 0$ and travels along it at a speed $V/3$ until $x = L$, where it exits the road.
   (a) In the xt plane sketch the path of the truck and the shock that is present while the truck is on the road.
   (b) Find the time at which the car immediately behind the truck catches up with the undisturbed traffic ahead of the truck.
   (c) Repeat 4(a) and 4(b) assuming the truck travels at speed $V/2$.

6. Using the equation $(u^2)_x + u_t = 0$ to represent the volumetric water content in a vertical porous medium, obtain the solution to each of the following initial-boundary-value problems and give a physical explanation of each solution:
   (a) $u(x, 0) = \frac{1}{2}, u(0, t) = \frac{1}{4}$
   (b) $u(x, 0) = \frac{1}{2}, u(0, t) = \frac{1}{2}e^{-t}$
   (c) $u(x, 0) = \frac{1}{2}, u(0, t) = \begin{cases} 0, 0 < t < 1 \\ \frac{1}{2}, t > 1 \end{cases}$
   (d) $u(x, 0) = \frac{1}{2}, u(0, t) = \begin{cases} 1, 0 < t < 1 \\ \frac{1}{3}, t > 1 \end{cases}$

7.6 SYSTEMS OF FIRST-ORDER EQUATIONS

Classification

In the previous sections, we have seen how some physical problems can be described by a single first-order hyperbolic partial differential equation. For example, we have considered a continuum model of traffic flow, evolution of an age-structured population, and flow in a porous media. In each of these applications we characterized the physical problem with a single function of the two variables, $u = u(x, t)$. In many situations, more than one dependent variable is required to model the physical system.

EXAMPLE 7.6.1 ───

As we saw in Section 1.4, the current $i = i(x, t)$ and voltage $v = v(x, t)$ at position x and time t in a transmission line satisfy the first-order equations

$$\frac{\partial i}{\partial x} + C\frac{\partial v}{\partial t} = -Gv, \qquad \frac{\partial v}{\partial x} + L\frac{\partial i}{\partial t} = -Ri$$

where R, L, C, and G denote, respectively, resistance, inductance, capacitance, and leakage conductance per unit length of transmission line. ■ ■

EXAMPLE 7.6.2 ───

The first-order system

$$(\rho v)_x + \rho_t = 0$$

$$vv_x + v_t = -\frac{1}{\rho}p_x$$

$$vp_x + p_t = -\gamma p v_x$$

governs the one-dimensional flow of an ideal gas with velocity $v = v(x, t)$, density $\rho = \rho(x, t)$, and pressure $p = p(x, t)$. Here, γ is a physical constant determined by the specific heat of the gas. ■ ■

Problems such as these present us with systems of first-order partial differential equations. The general *quasi-linear system* of n first-order partial differential equations in two independent variables has the form

$$\sum_{j=1}^{n} a_{ij}\frac{\partial u_j}{\partial x} + \sum_{j=1}^{n} b_{ij}\frac{\partial u_j}{\partial t} = c_i, \qquad i = 1, 2, \ldots, n \qquad (7.6.1)$$

where a_{ij}, b_{ij}, and c_i may depend on x, t, and $u_1, u_2, \ldots, u_n$. If each a_{ij} and b_{ij} is at most a function of x and t, the system (7.6.1) is called *almost linear*. If, in addition, each c_i depends linearly on $u_1, u_2, \ldots, u_n$,

$$c_i = \sum_{j=1}^{n} r_{ij}u_j + S_i$$

with r_{ij} and S_i functions of at most x and t, the system is said to be *linear*. If $c_i = 0$ for $i = 1, 2, \ldots, n$, the system is called *homogeneous*. If C, G, R, and L depend at most on x and t, the transmission line equations are linear. The ideal-gas equations are quasi-linear.

In terms of the $n \times n$ matrices $\mathbf{A} = [a_{ij}]$ and $\mathbf{B} = [b_{ij}]$ and the column vectors $\mathbf{u} = [u_1, u_2, \ldots, u_n]^T$ and $\mathbf{c} = [c_1, c_2, \ldots, c_n]^T$, the system of equations (7.6.1) can be written as

$$\mathbf{A}\mathbf{u}_x + \mathbf{B}\mathbf{u}_t = \mathbf{c} \tag{7.6.2}$$

EXAMPLE 7.6.3 ───

To express the transmission line equations in the matrix notation of (7.6.2), introduce the notation

$$\mathbf{u} = \begin{bmatrix} i \\ v \end{bmatrix}, \qquad \mathbf{A} = \begin{bmatrix} 1 & 0 \\ 0 & 1 \end{bmatrix}, \qquad \mathbf{B} = \begin{bmatrix} 0 & C \\ L & 0 \end{bmatrix}, \qquad \mathbf{c} = \begin{bmatrix} -Gv \\ -Ri \end{bmatrix} \qquad \blacksquare\ \blacksquare$$

In most applications the matrix $\mathbf{B}$ is nonsingular, and in all that follows we assume this to be the case and, therefore, take $\det(\mathbf{B}) \neq 0$. Associated with the system (7.6.3) is a *characteristic polynomial* defined by

$$F(\lambda) = \det(\mathbf{A} - \lambda\mathbf{B})$$

Since $\mathbf{A}$ and $\mathbf{B}$ are $n \times n$ matrices and $\det(\mathbf{B}) \neq 0$, the polynomial F has degree n.

If $F(\lambda)$ has n distinct real zeros, we classify the first-order system (7.6.3) as *hyperbolic*. The system is also called hyperbolic if $F(\lambda)$ has n real zeros and the generalized eigenvalue problem $(\mathbf{A} - \lambda\mathbf{B})\mathbf{u} = 0$ has n linearly independent solutions. If $F(\lambda)$ has no real zeros, then (7.6.3) is called *elliptic*. If $F(\lambda)$ has n real zeros but $(\mathbf{A} - \lambda\mathbf{B})\mathbf{u} = 0$ does not have n linearly independent solutions, then the system (7.6.3) may be classified as *parabolic*. An exhaustive classification cannot be carried out when $F(\lambda)$ has both real and complex zeros.

Characteristics

The systems of first-order equations that arise in physical problems are often of hyperbolic type. Therefore, we confine the remainder of our discussion of systems of first-order equations to those of hyperbolic type.

In our analysis of a single first-order equation, the characteristics played a central role. The notion of characteristics is also important in the analysis of systems of equations. Given the hyperbolic system (7.6.2), let $\lambda_1, \lambda_2, \ldots, \lambda_n$ be the n real zeros of the characteristic polynomial

$$F(\lambda) = \det(\mathbf{A} - \lambda\mathbf{B})$$

Then the *characteristics* for the system are the curves in the xt plane satisfying

$$\frac{dx}{dt} = \lambda_i, \qquad i = 1, 2, \ldots, n$$

If the system (7.6.2) is linear, then the characteristic directions λ_i depend at most on x and t. If (7.6.2) is quasi-linear, the characteristic directions may be solution dependent.

EXAMPLE 7.6.4

To classify the transmission line equations, we define the characteristic polynomial

$$F(\lambda) = \det(\mathbf{A} - \lambda\mathbf{B}) = \det\left\{\begin{bmatrix} 1 & 0 \\ 0 & 1 \end{bmatrix} - \lambda\begin{bmatrix} 0 & C \\ L & 0 \end{bmatrix}\right\} = 1 - CL\lambda^2$$

Since the parameters C and L are positive, $F(\lambda)$ has the distinct real roots $\lambda_1 = \sqrt{1/CL}$, $\lambda_2 = -\sqrt{1/CL}$. This shows that the transmission line equations are hyperbolic. If we assume C and L to be constants, then the characteristics are straight lines defined by

$$\frac{dx}{dt} = \lambda_i, \qquad i = 1, 2 \qquad \blacksquare\ \blacksquare$$

Given a hyperbolic system of equations, it is possible to transform it to an equivalent system in which each equation of the transformed system involves differentiation in only one direction, namely, along one of the characteristics. As in the case of a single equation, we shall see that being able to view the partial differential equations as ordinary differential equations along the characteristics leads to a powerful method for solving the hyperbolic system (7.6.2). This method is known as the *method of characteristics*.

Constant-Coefficient, Homogeneous Systems

If the system of hyperbolic equations (7.6.2) has constant coefficients and is homogeneous, then the method of characteristics can be used to obtain a complete solution of the system subject to specified initial conditions for each of the unknowns. Consider the system of n first-order partial differential equations

$$\sum_{j=1}^{n} a_{ij}\frac{\partial u_j}{\partial x} + \sum_{j=1}^{n} b_{ij}\frac{\partial u_j}{\partial t} = 0, \qquad i = 1, 2, \ldots, n \tag{7.6.3}$$

for the n unknown functions $u_j(x, t), j = 1, 2, \ldots, n$. We shall look for a solution of (7.6.3) subject to initial conditions of the form

$$u_j(x, 0) = f_j(x), \qquad j = 1, 2, \ldots, n \tag{7.6.4}$$

Each of the coefficients a_{ij}, b_{ij} is assumed to be a constant.

In terms of the $n \times n$ matrices $\mathbf{A} = [a_{ij}]$ and $\mathbf{B} = [b_{ij}]$ and the column vectors $\mathbf{u} = [u_1, u_2, \ldots, u_n]^T$ and $\mathbf{f} = [f_1, f_2, \ldots, f_n]^T$, the initial-value problem (7.6.3) and (7.6.4) can be written as

$$\mathbf{A}\mathbf{u}_x + \mathbf{B}\mathbf{u}_t = \mathbf{0}, \qquad -\infty < x < \infty, t > 0 \tag{7.6.5}$$

$$\mathbf{u}(x, 0) = \mathbf{f}(x) \qquad -\infty < x < \infty \tag{7.6.6}$$

We now describe the method of characteristics for solving the hyperbolic initial-value problem (7.6.3) and (7.6.4) or the matrix version (7.6.5) and (7.6.6). The major difficulty presented by the system of equations (7.6.3) is the coupling of the unknowns $u_j(x, t)$ in the sense that in general each equation involves all unknowns. The method of solution requires that we form certain linear combinations of the components of the unknown vector $\mathbf{u}$,

$$z_i = \sum_{j=1}^{n} q_{ij} u_j, \qquad i = 1, 2, \ldots, n$$

or

$$\mathbf{z} = \mathbf{Qu}, \qquad \mathbf{Q} = [q_{ij}]$$

Here the q_{ij} represent constants to be determined. Also, we shall form certain linear combinations of the n equations (7.6.3) so that the groupings of equations and the groupings of the components of $\mathbf{u}$ dictated by these linear combinations will allow us to uncouple the unknowns and obtain n first-order equations of the form

$$\lambda_i \frac{\partial z_i}{\partial x} + \frac{\partial z_i}{\partial t} = 0, \qquad i = 1, 2, \ldots, n$$

Since each equation for each z_i is a constant-coefficient advection equation, the method of solution of Section 7.1 allows us to determine $z_i(x, t)$. Finally we recover the original unknowns from $\mathbf{u} = Q^{-1}\mathbf{z}$. With this outline of the solution method in mind, let us now proceed to determine linear combinations of equations and unknowns that will uncouple the system (7.6.3).

Consider the system of equations in the matrix form (7.6.3). Let $\mathbf{P} = [p_{ij}]$ be an invertible $n \times n$ matrix whose construction will be discussed in what follows. The form of $\mathbf{P}$ is fixed by the requirement that

$$\mathbf{PA} = \mathbf{DPB}$$

where $\mathbf{D}$ represents a diagonal matrix

$$\mathbf{D} = \begin{bmatrix} \lambda_1 & & & \\ & \lambda_2 & & \\ & & \ddots & \\ & & & \lambda_n \end{bmatrix}$$

Now, if we multiply each term of Equation (7.6.3) by the matrix $\mathbf{P}$, we find $\mathbf{PAu}_x + \mathbf{PBu}_t = 0$ or, using $\mathbf{PA} = \mathbf{DPB}$, that $\mathbf{DPBu}_x + \mathbf{PBu}_t = 0$. The entries in the matrix $\mathbf{P}$ will fix the proper linear combination of equations required to uncouple the unknowns.

To determine a linear combination of unknowns that leads to an uncoupled system, define a matrix $\mathbf{Q} = \mathbf{PB}$ and set $\mathbf{z} = \mathbf{Qu}$. Then the equalities

$$\mathbf{DPBu}_x + \mathbf{PBu}_t = \mathbf{DQu}_x + \mathbf{Qu}_t = \mathbf{D(Qu)}_x + \mathbf{(Qu)}_t = \mathbf{Dz}_x + \mathbf{z}_t$$

imply that the components of the vector $\mathbf{z}$ satisfy the uncoupled system

$$\lambda_i \frac{\partial z_i}{\partial x} + \frac{\partial z_i}{\partial t} = 0, \qquad i = 1, 2, \ldots, n$$

in which the ith equation involves differentiation only along the ith characteristic.

Having grouped the components of the unknown vector $\mathbf{u}$ in Equations (7.6.3), we must form the same groupings in the initial conditon. We obtain initial conditions for $\mathbf{z}$ from

$$\mathbf{z}(x, 0) = \mathbf{Q}\mathbf{u}(x, 0) \quad \text{or} \quad z_i(x, 0) = \sum_{j=1}^{n} q_{ij} f_j(x) = g_i(x)$$

Recall from Section 7.1 that the constant-coefficient advection equation $\lambda z_x + z_t = 0$, $z(x, 0) = g(x)$, has the solution $z(x, t) = g(x - \lambda t)$. Thus, we see that the solution to the initial-value problem for $\mathbf{z}$ is given by

$$z_i(x, t) = g_i(x - \lambda_i t), \quad i = 1, 2, \ldots, n$$

As we show next, the matrix $\mathbf{P}$ is invertible. Since $\mathbf{B}$ has been assumed to be invertible, the product $\mathbf{Q} = \mathbf{PB}$ is invertible. Thus we can recover the original unknown $\mathbf{u}(x, t)$ from $\mathbf{u} = \mathbf{Q}^{-1}\mathbf{z}$. All that remains to complete the description of the solution method for the initial-value problem (7.6.3) and (7.6.4) is a means of determining the matrix $\mathbf{P}$. To make the determination of $\mathbf{P}$ easy to follow, we show how to construct $\mathbf{P}$ for the case of a system of two equations, $n = 2$. The case of general n involves no new ideas.

We require a matrix $\mathbf{P}$ with the property that $\mathbf{PA} = \mathbf{DPB}$, where $\mathbf{D}$ is diagonal. For the case $n = 2$ we have

$$\begin{bmatrix} p_{11} & p_{12} \\ p_{21} & p_{22} \end{bmatrix} \begin{bmatrix} a_{11} & a_{12} \\ a_{21} & a_{22} \end{bmatrix} = \begin{bmatrix} \lambda_1 & 0 \\ 0 & \lambda_2 \end{bmatrix} \begin{bmatrix} p_{11} & p_{12} \\ p_{21} & p_{22} \end{bmatrix} \begin{bmatrix} b_{11} & b_{12} \\ b_{21} & b_{22} \end{bmatrix}$$

or

$$\begin{bmatrix} p_{11} & p_{12} \\ p_{21} & p_{22} \end{bmatrix} \begin{bmatrix} a_{11} & a_{12} \\ a_{21} & a_{22} \end{bmatrix} = \begin{bmatrix} \lambda_1 p_{11} & \lambda_1 p_{12} \\ \lambda_2 p_{21} & \lambda_2 p_{22} \end{bmatrix} \begin{bmatrix} b_{11} & b_{12} \\ b_{21} & b_{22} \end{bmatrix}$$

The transpose of the product of square matrices is the product of the transposes in reverse order, $(\mathbf{PA})^T = \mathbf{A}^T \mathbf{P}^T$. Applying this result to form the transpose of both sides of the last matrix equation gives

$$\begin{bmatrix} a_{11} & a_{21} \\ a_{12} & a_{22} \end{bmatrix} \begin{bmatrix} p_{11} & p_{21} \\ p_{12} & p_{22} \end{bmatrix} = \begin{bmatrix} b_{11} & b_{21} \\ b_{12} & b_{22} \end{bmatrix} \begin{bmatrix} \lambda_1 p_{11} & \lambda_2 p_{21} \\ \lambda_1 p_{12} & \lambda_2 p_{22} \end{bmatrix}$$

which is equivalent to

$$\begin{bmatrix} a_{11} & a_{21} \\ a_{12} & a_{22} \end{bmatrix} \begin{bmatrix} p_{11} \\ p_{12} \end{bmatrix} = \lambda_1 \begin{bmatrix} b_{11} & b_{21} \\ b_{12} & b_{22} \end{bmatrix} \begin{bmatrix} p_{11} \\ p_{12} \end{bmatrix} \tag{7.6.7}$$

and

$$\begin{bmatrix} a_{11} & a_{21} \\ a_{12} & a_{22} \end{bmatrix} \begin{bmatrix} p_{21} \\ p_{22} \end{bmatrix} = \lambda_2 \begin{bmatrix} b_{11} & b_{21} \\ b_{12} & b_{22} \end{bmatrix} \begin{bmatrix} p_{21} \\ p_{22} \end{bmatrix} \tag{7.6.8}$$

To be able to obtain nonzero vectors

$$\mathbf{p}_1 = [p_{11}, p_{12}]^T \quad \text{and} \quad \mathbf{p}_2 = [p_{21}, p_{22}]^T$$

it must be the case that λ_1 and λ_2 are zeros of the characteristic polynomial

$$F(\lambda) = \det(\mathbf{A} - \lambda\mathbf{B}) = \det(\mathbf{A}^T - \lambda\mathbf{B}^T)$$

Otherwise, we know from linear algebra that the only solutions of (7.6.7) and (7.6.8) are the zero solutions $\mathbf{p}_1 = \mathbf{0}$ and $\mathbf{p}_2 = \mathbf{0}$. For the matrix $\mathbf{P}$ to be invertible, we require both $\mathbf{p}_1$ and $\mathbf{p}_2$ to be nonzero. We are now in a position to state an algorithm for determining the matrix $\mathbf{P}$.

ALGORITHM 7.1

INPUT:

The $n \times n$ coefficient matrices $\mathbf{A}$ and $\mathbf{B}$ from (7.6.3).

OUTPUT:

An $n \times n$ matrix $\mathbf{P}$ and $\lambda_1, \lambda_2, \ldots, \lambda_n$.

Step 1.

Form the characteristic polynomial $F(\lambda) = \det(\mathbf{A} - \lambda\mathbf{B})$.

Step 2.

Find the n real zeros of $F(\lambda)$, $\lambda_1, \lambda_2, \ldots, \lambda_n$.

Step 3.

Determine the entries of the ith row of the matrix $\mathbf{P}$ by finding a nonzero vector $\mathbf{p}_i = [p_{i1}, p_{i2}, \ldots, p_{in}]^T$ satisfying

$$\mathbf{A}^T\mathbf{p}_i = \lambda_i\mathbf{B}^T\mathbf{p}_i$$

Now the method of characteristics for the hyperbolic system (7.6.3) and (7.6.4) can be summarized as follows. After using Algorithm 7.1 to construct a nonsingular matrix $\mathbf{P}$, we form linear combinations of the original partial differential equations (7.6.3). These linear combinations of equations are defined by $\mathbf{PAu}_x + \mathbf{PBu}_t = 0$. The matrix $\mathbf{P}$ has been constructed so that $\mathbf{PA} = \mathbf{DPB}$ where $\mathbf{D}$ is a diagonal matrix with the λ_i on the diagonal. Thus in terms of these linear combinations of equations we have $\mathbf{DPBu}_x + \mathbf{PBu}_t = 0$. Next, we form linear combinations of the original unknowns according to the calculation $\mathbf{z} = \mathbf{Qu}$ with $\mathbf{Q} = \mathbf{PB}$. Since both $\mathbf{P}$ and $\mathbf{B}$ are invertible, $\mathbf{Q}$ also has an inverse. In terms of $\mathbf{z}$ the problem now reads $\mathbf{Dz}_x + \mathbf{z}_t = 0$. Since $\mathbf{D}$ is a diagonal matrix, the z system is uncoupled and the ith equation involves only differentiation of z_i in the ith characteristic direction λ_i. Initial conditions for $\mathbf{z}$ are provided by

$$\mathbf{z}(x, 0) = \mathbf{Qu}(x, 0) = \mathbf{Qf}(x) = \mathbf{g}(x)$$

The uncoupled initial-value problem for $\mathbf{z}$ is easily solved to give $z_i(x, t) = g_i(x - \lambda_i t)$. Finally $\mathbf{u}(x, t)$ is obtained from $\mathbf{u} = \mathbf{Q}^{-1}\mathbf{z}$. The essential steps in the solution process are given next in the form of an algorithm.

ALGORITHM 7.2 Method of Characteristics for Initial-Value Problem (7.6.3) and (7.6.4)

INPUT:

The $n \times n$ coefficient matrices $\mathbf{A}$ and $\mathbf{B}$ from (7.6.3).
The n initial-value functions $f_i(x)$ from (7.6.4).
The n zeros of $F(\lambda)$, $\lambda_1, \lambda_2, \ldots, \lambda_n$ from Algorithm 7.1.
The $n \times n$ matrix $\mathbf{P}$ from Algorithm 7.1.

OUTPUT:

The solution $\mathbf{u}(x, t)$ of the initial-value problem (7.6.3) and (7.6.4).

Step 1.

Calculate $\mathbf{Q} = \mathbf{PB}$.

Step 2.

Set $\mathbf{g}(x) = \mathbf{Qf}(x)$.

Step 3.

Set $z_i(x, t) = g_i(x - \lambda_i t)$.

Step 4.

Solve the system $\mathbf{Qu} = \mathbf{z}$.

EXAMPLE 7.6.5 ──

Consider the system of first-order equations

$$3\,\frac{\partial u_1}{\partial x} + 2\,\frac{\partial u_2}{\partial x} + \frac{\partial u_1}{\partial t} + \frac{\partial u_2}{\partial t} = 0$$

$$5\,\frac{\partial u_1}{\partial x} + 2\,\frac{\partial u_2}{\partial x} - \frac{\partial u_1}{\partial t} + \frac{\partial u_2}{\partial t} = 0$$

subject to the initial conditions

$$u_1(x, 0) = \sin x, \qquad u_2(x, 0) = e^x$$

In the matrix from (7.6.3) and (7.6.4) this initial-value problem has the form

$$\begin{bmatrix} 3 & 2 \\ 5 & 2 \end{bmatrix}\begin{bmatrix} u_1 \\ u_2 \end{bmatrix}_x + \begin{bmatrix} 1 & 1 \\ -1 & 1 \end{bmatrix}\begin{bmatrix} u_1 \\ u_2 \end{bmatrix}_t = 0$$

$$\begin{bmatrix} u_1(x, 0) \\ u_2(x, 0) \end{bmatrix} = \begin{bmatrix} \sin x \\ e^x \end{bmatrix}$$

or $\mathbf{A}u_x + \mathbf{B}u_t = 0$, $\mathbf{u}(x, 0) = \mathbf{f}(x)$, where

$$\mathbf{A} = \begin{bmatrix} 3 & 2 \\ 5 & 2 \end{bmatrix}, \qquad \mathbf{B} = \begin{bmatrix} 1 & 1 \\ -1 & 1 \end{bmatrix}, \qquad \mathbf{f}(x) = \begin{bmatrix} \sin x \\ e^x \end{bmatrix}$$

Setting $\det(\mathbf{A} - \lambda \mathbf{B}) = 0$, we find $\lambda_1 = -1$, $\lambda_2 = 2$. To find the first row of matrix $\mathbf{P}$, we need a nonzero solution to $\mathbf{A}^T\mathbf{p}_1 = \lambda_1\mathbf{B}^T\mathbf{p}_1$, or

$$\begin{bmatrix} 3 & 5 \\ 2 & 2 \end{bmatrix}\begin{bmatrix} p_{11} \\ p_{12} \end{bmatrix} = -1 \begin{bmatrix} 1 & -1 \\ 1 & 1 \end{bmatrix}\begin{bmatrix} p_{11} \\ p_{12} \end{bmatrix} \Rightarrow \begin{bmatrix} 4 & 4 \\ 3 & 3 \end{bmatrix}\begin{bmatrix} p_{11} \\ p_{12} \end{bmatrix} = 0$$

Now we see that p_{11} and p_{12} need only satisfy $p_{11} + p_{12} = 0$ and $p_{11}^2 + p_{12}^2 \neq 0$. A convenient choice is $p_{11} = 1$ and $p_{12} = -1$. Similarly, $\mathbf{p}_2$ satisfies

$$\begin{bmatrix} 3 & 5 \\ 2 & 2 \end{bmatrix}\begin{bmatrix} p_{21} \\ p_{22} \end{bmatrix} = 2 \begin{bmatrix} 1 & -1 \\ 1 & 1 \end{bmatrix}\begin{bmatrix} p_{21} \\ p_{22} \end{bmatrix} \Rightarrow \begin{bmatrix} 1 & 7 \\ 0 & 0 \end{bmatrix}\begin{bmatrix} p_{21} \\ p_{22} \end{bmatrix} = 0$$

so $p_{21} + 7p_{22} = 0$. We choose $p_{21} = 7$ and $p_{22} = -1$. Now, according to Algorithm 7.1, the matrix $\mathbf{P}$ is given by

$$\mathbf{P} = \begin{bmatrix} 1 & -1 \\ 7 & -1 \end{bmatrix}$$

To check our calculations, we compute

$$\mathbf{PA} = \begin{bmatrix} 1 & -1 \\ 7 & -1 \end{bmatrix}\begin{bmatrix} 3 & 2 \\ 5 & 2 \end{bmatrix} = \begin{bmatrix} -2 & 0 \\ 16 & 12 \end{bmatrix}$$

and

$$\mathbf{PB} = \begin{bmatrix} 1 & -1 \\ 7 & -1 \end{bmatrix}\begin{bmatrix} 1 & 1 \\ -1 & 1 \end{bmatrix} = \begin{bmatrix} 2 & 0 \\ 8 & 6 \end{bmatrix}$$

to verify that

$$\mathbf{PA} = \begin{bmatrix} \lambda_1 & 0 \\ 0 & \lambda_2 \end{bmatrix}\mathbf{PB}, \qquad \lambda_1 = -1, \ \lambda_2 = 2$$

Now, the method of characteristics tells us that if we set $\mathbf{z} = \mathbf{PB}\mathbf{u}$,

$$\begin{bmatrix} z_1 \\ z_2 \end{bmatrix} = \begin{bmatrix} 2 & 0 \\ 8 & 6 \end{bmatrix}\begin{bmatrix} u_1 \\ u_2 \end{bmatrix} = \begin{bmatrix} 2u_1 \\ 8u_1 + 6u_2 \end{bmatrix}$$

then $\mathbf{z}$ satisfies

$$\lambda_1 \frac{\partial z_1}{\partial x} + \frac{\partial z_1}{\partial t} = 0, \qquad \lambda_2 \frac{\partial z_2}{\partial x} + \frac{\partial z_2}{\partial t} = 0$$

$$z_1(x, 0) = 2 \sin x, \qquad z_2(x, 0) = 8 \sin x + 6e^x$$

Solving these constant-coefficient advection equations yields

$$z_1(x, 0) = 2 \sin(x - \lambda_1 t) = 2 \sin(x + t)$$

and

$$z_2(x, 0) = 8 \sin(x - \lambda_2 t) + 6e^{x - \lambda_2 t}$$
$$= 8 \sin(x - 2t) + 6e^{x - 2t}$$

Finally, the equations $z_1 = 2u_1$ and $z_2 = 8u_1 + 6u_2$ show that

$$u_1(x, t) = \sin(x + t)$$

and

$$u_2(x, t) = \tfrac{1}{6}[8 \sin(x - 2t) + 6e^{x - 2t} - 8 \sin(x + t)]$$

Exercise 6 illustrates an alternative method for solving this example problem. ■ ■

Physical Application

To illustrate how the method of characteristics applies to an initial-boundary-value problem, we present a final example. The following example shows how the method of characteristics can be applied in a graphical setting. It also illustrates an important point regarding the specification of boundary conditions. Namely, on a boundary, exactly one condition on the solution is specified for each family of characteristics that point forward in time from that boundary. If we specify less boundary conditions, the problem will be underdetermined and generally have several solutions. More boundary data would lead to an overspecified problem with no solution in general.

EXAMPLE 7.6.6 ───

Consider a long tank with rectangular cross section and let x denote distance measured along the tank from one end. Assume that water in the tank has depth u and velocity v and that water moves only in the x direction. Then we can take $u = u(x, t)$ and $v = v(x, t)$ to characterize the water depth and water velocity along the tank at position x at time t. If we assume the tank is horizontal, ignore the viscosity of the water, and assume the pressure in the water is hydrostatic, then conservation of fluid mass and conservation of fluid momentum imply the following pair of quasi-linear first-order hyperbolic differential equations:

$$v u_x + u v_x + u_t = 0, \qquad g u_x + v v_x + v_t = 0$$

Here g denotes the gravitational acceleration constant. To obtain a constant-coefficient system of equations related to these quasi-linear tank equations, we consider the following linearization. Assume that $u(x, t) = U + h(x, t)$, where $h(x, t)$ represents a long, shallow wave imposed on the constant mean depth U. The assumption that the waves are shallow means $h(x, t)$ is small while the assumption that the waves are long means $h_x(x, t)$ is small. Also assume that the fluid velocity $v(x, t)$ is small. A standard linear-

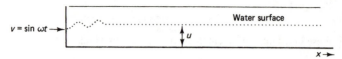

Figure 7.6.1   Diagram of ripple tank.

ization argument is that products of the small quantities h and v are negligible compared to v and h. If we put $u = U + h(x, t)$ and v into the quasi-linear tank equations and drop the terms vh_x, hv_x, and vv_x, then the constant-coefficient linearized tank equations result, namely,

$$Uv_x + h_t = 0, \qquad gh_x + v_t = 0$$

These constant-coefficient equations could be used to model the fluid motion in an apparatus called a ripple tank, a long tank with a movable wall on one end (Figure 7.6.1). This wall can be used to generate different wave modes, which may be of interest in coastal construction design, for example. Let us consider a situation in which the water is initially at rest, $h(x, 0) = 0$ and $v(x, 0) = 0$. Beginning at time $t = 0$, assume the wall $x = 0$ is set into sinusoidal motion. Since the fluid adjacent to the wall moves with the wall, the resulting boundary condition is $v(0, t) = \sin \omega t$. If we further assume that the other end of the tank is sufficiently far away that no waves reach it during the simulation, then we can consider the following initial-boundary-value problem:

$$\begin{array}{lll} Uv_x + h_t = 0, & x > 0, t > 0 & (7.6.9) \\ gh_x + v_t = 0, & x > 0, t > 0 & (7.6.10) \\ h(x, 0) = 0 = v(x, 0), & x > 0 & (7.6.11) \\ v(0, t) = \sin \omega t, & t > 0 & (7.6.12) \end{array}$$

In matrix form, $\mathbf{A}\mathbf{u}_x + \mathbf{B}\mathbf{u}_t = 0$, the linearized equations read

$$\begin{bmatrix} 0 & U \\ g & 0 \end{bmatrix} \begin{bmatrix} h \\ v \end{bmatrix}_x + \begin{bmatrix} 1 & 0 \\ 0 & 1 \end{bmatrix} \begin{bmatrix} h \\ v \end{bmatrix}_t = 0$$

where $\mathbf{u} = [h, v]^T$. The characteristic polynomial

$$F(\lambda) = \det(\mathbf{A} - \lambda\mathbf{B}) = \lambda^2 - gU$$

for this system is easily seen to have roots $\lambda_1 = \sqrt{gU}$ and $\lambda_2 = -\sqrt{gU}$. From Algorithm 7.1, it follows that the matrix $\mathbf{P}$ can be taken to be

$$\mathbf{P} = \begin{bmatrix} \sqrt{gU} & U \\ \sqrt{gU} & -U \end{bmatrix}$$

Since $\mathbf{B}$ is the identity matrix, $\mathbf{Q} = \mathbf{P}\mathbf{B} = \mathbf{P}$ and Algorithm 7.2 states that new dependent variables should be defined by

$$\mathbf{z} = \mathbf{Q}\mathbf{u} = \mathbf{P}\mathbf{u}$$

This gives

$$z_1 = \sqrt{gU} \, h + Uv \qquad z_2 = \sqrt{gU} \, h - Uv$$

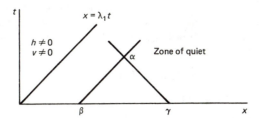

Figure 7.6.2 Characteristic diagram for ripple tank initial-boundary-value problem.

and it follows that z_1 and z_2 satisfy the uncoupled equations

$$\lambda_1 \frac{\partial z_1}{\partial x} + \frac{\partial z_1}{\partial t} = 0, \qquad \lambda_2 \frac{\partial z_2}{\partial x} + \frac{\partial z_2}{\partial t} = 0$$

The last pair of equations assert that z_1 is constant along the forward-facing characteristics $dx/dt = \lambda_1$ and that z_2 is constant along the backward-facing characteristics $dx/dt = \lambda_2$. Thus, z_1 and z_2 represent groupings of the physical variables that remain constant along characteristics. Such groupings are called *Riemann invariants*. As we see next these invariants can be used to solve our initial-boundary-value problem for h and v.

We first argue that h and v are both zero in the region $x > \lambda_1 t$. Let the points α and β lie on a forward-facing characteristic drawn from the positive x axis and α and γ lie on a backward-facing characteristic with $\gamma > \beta$. See Figure 7.6.2. The initial conditions $h(x, 0) = 0 = v(x, 0)$ imply that z_1 is zero at β and that z_2 is zero at γ. The invariance of z_1 along the characteristic joining α and β and the invariance of z_2 along the characteristic joining α and γ imply that both z_1 and z_2 are zero at α. This establishes the result that $h = v = 0$ in the region $x > \lambda_1 t$. This region is called the *zone of quiet*.

We now use the method of characteristics to solve for h and v in the region $0 < x < \lambda_1 t$. Let a be any point in that region and let b denote the point on $x = 0$ that is joined to a by a forward-facing characteristic $x'(t) = \lambda_1$. See Figure 7.6.3. Let c and d denote points on the x axis that are joined to a and b, respectively, by backward-facing characteristics, $x'(t) = \lambda_2$.

The invariance of z_2 along the characteristic joining b and d and the fact that $z_2 = 0$ at d implies

$$\sqrt{gU}\, h(b) = Uv(b).$$

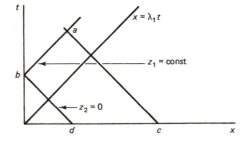

Figure 7.6.3 Riemann invariant z_2 is zero on backward characteristics and z_1 is a on each forward-facing characteristic.

This shows that v and h cannot be independently prescribed along the boundary $x = 0$. From the line $x = 0$ there is only one forward-facing characteristic, and therefore only one boundary condition can be prescribed there. Since z_2 is zero along the characteristic ac, we also conclude that

$$\sqrt{gU}\, h(a) = Uv(a)$$

Finally, the invariance of z_1 along ab shows

$$\sqrt{gU}\, h(a) + Uv(a) = \sqrt{gU}\, h(b) + Uv(b)$$

which together with the last two equations give $v(a) = v(b)$.

If the coordinates of a are (x, t), then the coordinates of b are $(0, t - x/\sqrt{gU})$. Therefore,

$$v(x, t) = \sin \omega(t - x/\sqrt{gu})$$

and, since $\sqrt{gU}\, h(a) = Uv(a)$,

$$h(x, t) = \sqrt{U/g}\, \sin \omega(t - x/\sqrt{gU})$$

for $0 < x < \lambda_1 t$. See also Example 5.2.3 (Part 2). ■ ■

Examples 7.6.5 and 7.6.6 illustrate the application of Algorithms 7.1 and 7.2 in the solution of a constant-coefficient, homogeneous hyperbolic system of first-order partial differential equations. If we drop either of the assumptions of constant coefficients or the homogeneous assumption, the application of the method of characteristics generally fails to provide an exact solution. However, as we show in Chapter 9, it is possible to use numerical methods to implement the method of characteristics.

Exercises

1. Let $\mathbf{u} = (u_1, u_2)^T = (u, v)^T$. Describe each of the systems of equations using the terms linear, almost linear, quasilinear, and homogeneous. Then classify as either elliptic, parabolic, or hyperbolic

   (a) $4u_x + v_x - 6u_t + 3v_t = 0$
       $2u_x + v_x + 5u_t + 3v_t = 0$
   (b) $4u_x + v_x - 6u_t + 3vt = 2u^2 + 7v$
       $2u_x + v_x + 5u_t + 3v_t = u - 9v^3$
   (c) $4u_x + v_x - 6u_t + 3v_t = 2u + 7v + 5$
       $2u_x + v_x + 5u_t + 3v_t = u - 9v + 12$
   (d) $2u_x - 2v_x + u_t - 3v_t = v$
       $u_x - 4v_x \qquad + v_t = u$
   (e) $e^u u_x - 2v_x + u_t - 3v_t = v$
       $u_x - 4v_x \qquad + v_t = u$
   (f) $u_x = v_t$
       $u_t = -v_x$
   (g) $u_x = v_t$
       $u_t = v$

2. Let $\mathbf{u} = (u_1, u_2, u_3) = (u, v, w)^T$. Classify the system

$$3u_x + u_t = 0$$
$$v_x + w_x + v_t = 0$$
$$2v_x + 4w_x + w_t = 0$$

3. Show that the one-dimensional second-order wave equation $w_{tt} - c^2 w_{xx} = 0$ can be expressed as a system of first-order equations of the form
   (a) $u_t - cu_x = v$, $v_t + cv_x = 0$
   (b) $v_x = u_t$, $v_t = c^2 u_x$

4. Show that if u and v are twice continuously differentiable and satisfy the (Cauchy–Riemann) system

$$u_x = v_y, \qquad u_y = -v_x$$

then u and v each solve Laplace's equation,

$$u_{xx} + u_{yy} = 0, \qquad v_{xx} + v_{yy} = 0$$

5. Use the method of characteristics to solve the initial-value problem
   (a) $4u_x - 6v_x + u_t = 0$, $u(x, 0) = \sin x$
       $u_x - 3v_x + v_t = 0$, $v(x, 0) = \cos x$
   (b) $3u_x + 2v_x + u_t + v_t = 0$, $u(x, 0) = \sin x$
       $5u_x + 2v_x - u_t + v_t = 0$, $v(x, 0) = e^x$
   (c) $u_x + v_x + u_t = 0$, $u(x, 0) = f(x)$
       $4u_x + v_x + v_t = 0$, $v(x, 0) = g(x)$
   (d) $2u_x - 2v_x + u_t - 3v_t = 0$, $u(x, 0) = f(x)$
       $u_x - 4v_x + v_t = 0$, $v(x, 0) = g(x)$
   (e) $2u_x + v_x + u_t = 0$, $u(x, 0) = f(x)$
       $2u_x + 3u_x + v_t = 0$, $v(x, 0) = g(x)$
   (f) $4u_x + v_x + u_t = 0$, $u(x, 0) = f(x)$
       $u_x + 4v_x + v_t = 0$, $v(x, 0) = g(x)$

6. Reconsider the system of equations of Example 7.6.5:

$$3\frac{\partial u_1}{\partial x} + 2\frac{\partial u_2}{\partial x} + \frac{\partial u_1}{\partial t} + \frac{\partial u_2}{\partial t} = 0; \qquad u_1(x, 0) = f(x) = \sin x$$

$$5\frac{\partial u_1}{\partial x} + 2\frac{\partial u_2}{\partial x} - \frac{\partial u_1}{\partial t} + \frac{\partial u_2}{\partial t} = 0; \qquad u_2(x, 0) = g(x) = e^x$$

Develop the solution by the *method of undetermined coefficients* as follows.
   (a) Show that for a solution of the form

$$u_1(x, t) = af(x - \lambda_1 t) + bg(x - \lambda_1 t) + cf(x - \lambda_2 t) + dg(x - \lambda_2 t)$$
$$u_2(x, t) = \alpha f(x - \lambda_1 t) + \beta g(x - \lambda_1 t) + \gamma f(x - \lambda_2 t) + \delta g(x - \lambda_2 t)$$

   the initial conditions require

$$a + c = 1; \qquad \alpha + \gamma = 0$$
$$b + d = 0; \qquad \beta + \delta = 1$$

   (b) Show that for the first-order equations $\mathbf{A}u_x + \mathbf{B}u_t = 0$ to hold, it must be the case that

$$(\mathbf{A} - \lambda_1 \mathbf{B})\begin{bmatrix} a \\ \alpha \end{bmatrix} = 0; \qquad (\mathbf{A} - \lambda_1 \mathbf{B})\begin{bmatrix} b \\ \beta \end{bmatrix} = 0$$

$$(\mathbf{A} - \lambda_2\mathbf{B})\begin{bmatrix} c \\ \gamma \end{bmatrix} = \mathbf{0}; \qquad (\mathbf{A} - \lambda_2\mathbf{B})\begin{bmatrix} d \\ \delta \end{bmatrix} = \mathbf{0}$$

where $\lambda_1 = -1$ and $\lambda_2 = 2$.

(c) Use part (b) to conclude that

$$4a - 3\alpha = 0; \qquad 4b - 3\beta = 0; \qquad c = 0; \qquad d = 0$$

(d) Use the equations from parts (a) and (c) to obtain the solution to the system of Example 7.6.5.

7. Rework Exercise 5 using the method of undetermined coefficients developed in Exercise 6.

8. Beginning with the system

$$Uv_x + h_t = 0$$
$$gh_x + v_t = 0$$

show that $v_{tt} - c^2v_{xx} = 0$ and $h_{tt} - c^2h_{xx} = 0$ and identify the constant c. (In this connection, see Section 5.2.)

9. Given the ripple-tank model equations of Example 7.6.6 in the form

$$Uv_x + h_t = 0, \qquad x > 0, \qquad t > 0$$
$$gh_x + v_t = 0, \qquad x > 0, \qquad t > 0$$
$$h(x, 0) = 0 = v(x, 0), \qquad x > 0$$
$$v(0, t) = \begin{cases} \sin \omega t, & 0 < t < \pi/\omega \\ 0, & t > \pi/\omega \end{cases}$$

use the method of characteristic to obtain the solution. Give a physical interpretation of your results.

PART TWO

NUMERICAL METHODS

Chapter 8

Finite-Difference Methods for Parabolic Equations

In Chapters 1–7 we presented a number of methods for obtaining the solution to a partial differential equation subject to boundary and/or initial conditions. The solution techniques that we have focused on include separation of variables, superposition, Fourier series, Fourier transform, Laplace transform, and the method of characteristics. The main strength of these solution methods is that when they can be applied, they produce the exact solution to the problem at hand. An attractive feature of an exact solution is that it often displays the roles and relative importances of the physical parameters that appear in the problem. Thus, from an exact solution, one can often gain not only quantitative, but also qualitative, information about the solution.

The exact-solution techniques also have some serious weaknesses. First, each of the exact-solution methods we have considered can only be applied to a restricted class of problems. For example, the separation of variables–superposition–Fourier series approach is generally limited to linear, constant-coefficient equations. If the equation has variable coefficients or is nonlinear, another method for dealing with the problem must usually be found. A second weakness of exact solutions is that the expressions for them are often very complicated (e.g., Examples 3.2.3 and 5.2.4). The computational effort required to evaluate some exact solutions greatly exceeds the effort required to obtain the solution. Even though the exact solution may be known, from a quantitative standpoint all that is often available is an approximate solution. For example, if a solution is given as a Fourier series, all that can be computed is the partial sum of some finite number of terms. Of course, this approximation can be made arbitrarily close to the exact solution by summing enough terms.

In Chapters 8–10 we introduce a numerical method, called the finite-difference method, for approximating the solution to a partial differential equations problem. To apply the finite-difference method, all derivatives in a differential equation are replaced by approximating difference quotients. This procedure reduces the differential equation problem to a system of algebraic equations.

The finite-difference method of dealing with a partial differential equations problem has a number of advantages. The finite-difference method applies to problems with variable coefficients and even to nonlinear problems. For many problems, the finite-difference method produces a quantitative approximation to the solution with much less computational effort than would be required with an exact solution. A finite-difference approximation to a solution can generally be made arbitrarily accurate.

The major disadvantage of finite-difference methods is that they seldom display qualitative information about the solution. Also, in many cases it is difficult to estimate the discrepancy between a finite-difference approximation and the corresponding exact solution.

8.1 DIFFERENCE FORMULAS

Suppose that $u = u(x, t)$ is a function of two independent variables. The first partial derivatives of u are defined as limits of difference quotients:

$$u_x(x, t) = \frac{\partial u}{\partial x}(x, t) = \lim_{h \to 0} \frac{u(x + h, t) - u(x, t)}{h}$$

and

$$u_t(x, t) = \frac{\partial u}{\partial t}(x, t) = \lim_{k \to 0} \frac{u(x, t + k) - u(x, t)}{k}$$

Instead of allowing h and k to approach zero in the definitions of u_x and u_t, we can use the respective difference quotients to approximate these partial derivatives:

$$u_x(x, t) \simeq \frac{u(x + h, t) - u(x, t)}{h} \tag{8.1.1}$$

$$u_t(x, t) \simeq \frac{u(x, t + k) - u(x, t)}{k} \tag{8.1.2}$$

The amount by which the difference quotient differs from the derivative it approximates is called a *truncation error* (TE). The truncation errors for finite-difference approximations such as (8.1.1) and (8.1.2) depend on the behavior of the function u at the point (x, t) as well as the choices of h and k. To analyze the truncation errors associated with the approximations (8.1.1) and (8.1.2), we require Taylor's theorem in two variables.

Theorem 8.1.1 Taylor's Theorem. Suppose that $u(x, t)$ and all of its partial derivatives of order less than or equal to $n + 1$ are continuous on

$$\Omega = \{(x, t) : a \le x \le b, c \le t \le d\}$$

Let (x_0, t_0) be a point in Ω. For every (x, t) in Ω there exist ξ between x and x_0 and τ between t and t_0 such that

$$u(x, t) = P_n(x, t) + R_n(x, t)$$

where

$$P_n(x, t) = u(x_0, t_0) + \left[\frac{\partial u}{\partial x}(x_0, t_0) \cdot (x - x_0) + \frac{\partial u}{\partial t}(x_0, t_0) \cdot (t - t_0) \right]$$

$$+ \left[\frac{\partial^2 u}{\partial x^2}(x_0, t_0) \frac{(x - x_0)^2}{2} + \frac{\partial^2 u}{\partial x \, \partial t}(x_0, t_0) \cdot (x - x_0)(t - t_0) \right.$$

$$+ \left. \frac{\partial^2 u}{\partial t^2}(x_0, t_0) \frac{(t - t_0)^2}{2} \right] + \cdots$$

$$+ \left[\frac{1}{n!} \sum_{j=0}^{n} \frac{n!}{(n - j)! j!} \frac{\partial^n u}{\partial x^{n-j} \, \partial t^j}(x_0, t_0)(x - x_0)^{n-j}(t - t_0)^j \right]$$

and

$$R_n(x, t) = \frac{1}{(n + 1)!} \sum_{j=0}^{n+1} \frac{(n + 1)!}{(n + 1 - j)! j!} \frac{\partial^{n+1} u(\xi, \tau)}{\partial x^{n+1-j} \, \partial t^j} \cdot (x - x_0)^{n+1-j}(t - t_0)^j$$

Here, P_n is the *Taylor polynomial of degree n in two variables* for the function u about (x_0, t_0) and $R_n(x, t)$ is the *remainder* associated with $P_n(x, t)$.

The proof of Theorem 8.1.1 can be found in any standard advanced calculus book. See, for example, W. Fulks, *Advanced Calculus*, Wiley, New York, 1978, p. 331.

EXAMPLE 8.1.1

The Taylor polynomial of degree 3 for $u(x, t) = \sin(2x + t)$ about $(\pi, 0)$ is determined from

$$P_3(x, t) = u(\pi, 0) + \frac{\partial u}{\partial x}(\pi, 0)(x - \pi) + \frac{\partial u}{\partial t}(\pi, 0)(t - 0)$$

$$+ \frac{1}{2} \left[\frac{\partial^2 u}{\partial x^2}(\pi, 0)(x - \pi)^2 + 2\frac{\partial^2 u}{\partial x \, \partial t}(\pi, 0)(x - \pi)t + \frac{\partial^2 u}{\partial t^2}(\pi, 0)t^2 \right]$$

$$+ \frac{1}{6} \left[\frac{\partial^3 u}{\partial x^3}(\pi, 0)(x - \pi)^3 + 3\frac{\partial^3 u}{\partial x^2 \, \partial t}(\pi, 0)(x - \pi)^2 t \right.$$

$$+ \left. 3\frac{\partial^3 u}{\partial x \, \partial t^2}(\pi, 0)(x - \pi)t^2 + \frac{\partial^3 u}{\partial t^3}(\pi, 0)t^3 \right]$$

Evaluating each of these partial derivatives at $(x_0, t_0) = (\pi, 0)$ gives

$$P_3(x, t) = 2(x - \pi) + t + \tfrac{1}{6}[-8(x - \pi)^3 - 3 \cdot 4(x - \pi)^2 t - 3 \cdot 2(x - \pi)t^2 - t^3]$$

This polynomial will give a close approximation to $\sin(2x + t)$ provided x is close to π and t is close to zero. For example,

$$P_3(\pi + 0.05, 0.03) = 0.1296338$$

while

$$\sin[2(\pi + 0.05) + 0.03] = 0.1296341$$

■ ■

Derivations of Difference Formulas

Taylor's theorem can be used to derive the difference formula (8.1.1) and the associated truncation error. Assume that u is twice continuously differentiable and that h is positive. From Taylor's theorem, we have

$$u(x + h, t) = u(x, t) + u_x(x, t)h + u_{xx}(\xi, t)\frac{h^2}{2}, \qquad x < \xi < x + h$$

which can be rearranged to read

$$u_x(x, t) = \frac{u(x + h, t) - u(x, t)}{h} - u_{xx}(\xi, t)\frac{h}{2} \tag{8.1.3}$$

The finite-difference formula (8.1.3) for $u_x(x, t)$ is known as the *forward difference formula* for u_x. In (8.1.3) we find the partial derivative $u_x(x, t)$ expressed as a difference quotient,

$$\frac{u(x + h, t) - u(x, t)}{h}$$

plus a truncation error,

$$-u_{xx}(\xi, t)\frac{h}{2}$$

If u is twice continuously differentiable, so that u_{xx} is bounded on the interval $[x, x + h]$, then the truncation error can be made small by choosing h small.

By just changing the sign of h in the preceding derivation of the forward difference formula, we obtain the *backward difference formula* for $u_x(x, t)$. Namely,

$$u_x(x, t) = \frac{u(x, t) - u(x - h, t)}{h} + \text{TE}$$
$$\text{TE} = u_{xx}(\xi, t)h/2, \qquad x - h < \xi < x \tag{8.1.4}$$

Based on the definition of a partial derivative, we expect a difference quotient to become a better approximation to the derivative as the *step size* h approaches zero. This expectation is supported by the forward and backward difference formulas (8.1.3) and (8.1.4) in which the truncation error is proportional to h.

As $h \to 0$, $h^2 < h$. Therefore, it would be desirable to have a finite-difference formula with a truncation error proportional to h^2 rather than h. Taylor's theorem can be used to derive a centered difference formula for $u_x(x, t)$ with a truncation error proportional to h^2. By Taylor's Theorem 8.1.1,

$$u(x + h, t) = u(x, t) + u_x(x, t)h + u_{xx}(x, t)\frac{h^2}{2} + u_{xxx}(\xi_1, t)\frac{h^3}{6} \tag{8.1.5}$$

and

$$u(x - h, t) = u(x, t) - u_x(x, t)h + u_{xx}(x, t)\frac{h^2}{2} - u_{xxx}(\xi_2, t)\frac{h^3}{6} \tag{8.1.6}$$

where $x < \xi_1 < x + h$ and $x - h < \xi_2 < x$. Subtracting (8.1.6) from (8.1.5) and solving for u_x yields

$$u_x(x, t) = \frac{u(x + h, t) - u(x - h, t)}{2h} - [u_{xxx}(\xi_1, t) + u_{xxx}(\xi_2, t)]\frac{h^2}{12} \quad (8.1.7)$$

The truncation error in the difference formula (8.1.7) can be simplified. If u_{xxx} is continuous, the intermediate-value theorem implies that

$$\tfrac{1}{2}[u_{xxx}(\xi_1, t) + u_{xxx}(\xi_2, t)] = u_{xxx}(\xi, t), \qquad \xi_1 < \xi < \xi_2 \quad (8.1.8)$$

Using (8.1.8) in (8.1.7) together with $x - h < \xi_2 < x$ and $x < \xi_1 < x + h$, we obtain the *centered difference formula* for $u_x(x, t)$:

$$u_x(x, t) = \frac{u(x + h, t) - u(x - h, t)}{2h} + \text{TE}$$

$$\text{TE} = -\tfrac{1}{6}h^2 u_{xxx}(\xi, t), \qquad x - h < \xi < x + h$$

We also require finite-difference formulas for the higher order derivatives of u. To obtain a difference formula for $u_{xx}(x, t)$, we use Taylor's theorem to write

$$u(x + h, t) = u + u_x h + u_{xx}\frac{h^2}{2} + u_{xxx}\frac{h^3}{6} + u_{xxxx}(\xi_1, t)\frac{h^4}{24} \quad (8.1.9)$$

and

$$u(x - h, t) = u - u_x h + u_{xx}\frac{h^2}{2} - u_{xxx}\frac{h^3}{6} + u_{xxxx}(\xi_2, t)\frac{h^4}{24} \quad (8.1.10)$$

where u, u_x, u_{xx}, and u_{xxx} are evaluated at (x, t) and $x < \xi_1 < x + h$ and $x - h < \xi_2 < x$. Adding Equations (8.1.9) and (8.1.10) and then solving for u_{xx} gives a *centered difference formula* for $u_{xx}(x, t)$:

$$u_{xx}(x, t) = \frac{u(x - h, t) - 2u(x, t) + u(x + h, t)}{h^2} + \text{TE}$$

$$\text{TE} = -\frac{h^2}{24}[u_{xxxx}(\xi_1, t) + u_{xxxx}(\xi_2, t)]$$

$$= -\frac{h^2}{12}u_{xxxx}(\xi, t), \qquad x - h < \xi < x + h \quad (8.1.11)$$

We have illustrated the derivation of difference formulas using x partial derivatives. Analogous formulas hold for the t partial derivatives of u. For example, the *forward difference formula* for $u_t(x, t)$ is

$$u_t(x, t) = \frac{u(x, t + k) - u(x, t)}{k} + \text{TE}$$

$$\text{TE} = -\frac{k}{2}u_{tt}(x, \tau), \qquad t < \tau < t + k \quad (8.1.12)$$

Taylor's theorem can also be used to derive difference formulas for the mixed partials of u. For example,

$$u_{xt}(x, t) = \frac{u(x + h, t + k) - u(x + h, t - k) - u(x - h, t + k) + u(x - h, t - k)}{4hk}$$

$$+ \text{ TE} \quad (8.1.13)$$

$$\text{TE} = -\frac{h^2}{6} u_{xxxt}(\xi_1, \tau_1) - \frac{k^2}{6} u_{xttt}(\xi_2, \tau_2), \qquad \begin{array}{l} x - h < \xi_1, \xi_2 < x + h, \\ t - k < \tau_1, \tau_2 < t + k \end{array}$$

In practice, we usually are not able to provide values for the derivatives that appear in the various truncation error expressions. Therefore, we describe a truncation error such as that in (8.1.11) as being of order h^2, which means the truncation error is proportional to h^2. A function $f(h)$ is said to be of the *order of magnitude* $g(h)$ as $h \to 0$, where $g(h)$ is a nonnegative function, if

$$\lim_{h \to 0} \frac{f(h)}{g(h)} = \text{const}$$

In the order-of-magnitude notation, or *O-notation*, we write

$$f(h) = O(g(h)) \quad \text{as } h \to 0$$

In the context of finite differences, the qualifier as $h \to 0$ is usually omitted.

EXAMPLE 8.1.2 _____

We say $\sin 2h$ is $O(h)$ as $h \to 0$ since

$$\lim_{h \to 0} \frac{\sin 2h}{h} = 2$$

We say $\cos h$ is $O(1)$ as $h \to 0$ since

$$\lim_{h \to 0} \frac{\cos h}{1} = 1$$

For the difference formula (8.1.11), we say the truncation error is of order ("is big O of") h^2 provided u_{xxxx} is continuous since

$$\lim_{h \to 0} \frac{\text{TE}}{h^2} = \lim_{h \to 0} -\tfrac{1}{12}u_{xxxx}(\xi, t) = -\tfrac{1}{12}u_{xxxx}(x, t) \qquad \blacksquare\blacksquare$$

Grid Notation

The statements of the various finite-difference formulas can be greatly simplified if we adopt the following subscript–superscript notation. A *grid*, or *mesh*, in the xt plane is a set of points

$$(x_n, t_j) = (x_0 + nh, t_0 + jk)$$

where n and j are integers and (x_0, t_0) is a reference point. The (x_n, t_j) are called *grid points*, *mesh points*, or *nodes*. The positive numbers h and k are, respectively, called the x and t *grid spacings*, *mesh spacings*, or *step sizes*. In general, we allow the x grid

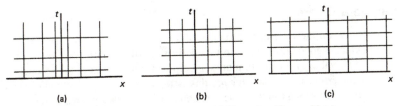

Figure 8.1.1   Nonuniform grid (a), square grid (b), and uniform grid (c).

spacing h to vary with n and the t grid spacing k to vary with j to allow for a *nonuniform grid*. However, if h and k are constants, the grid is called *uniform*. If $h = k = \text{const}$, the grid is said to be *square*. See Figure 8.1.1.

We denote the value of $u(x, t)$ at the grid point (x_n, t_j) by

$$u_n^j = u(x_n, t_j)$$

Table 8.1.1 displays some commonly used difference formulas and their associated truncation errors using the compact subscript–superscript and order-of-magnitude notations.

TABLE 8.1.1   FINITE-DIFFERENCE EXPRESSIONS WITH ORDER
OF MAGNITUDE OF TRUNCATION ERRORS

Forward difference for u_x:

$$u_x(x_n, t_j) = \frac{u_{n+1}^j - u_n^j}{h} + O(h) \tag{8.1.14}$$

Backward difference for u_x:

$$u_x(x_n, t_j) = \frac{u_n^j - u_{n-1}^j}{h} + O(h) \tag{8.1.15}$$

Centered difference for u_x:

$$u_x(x_n, t_j) = \frac{u_{n+1}^j - u_{n-1}^j}{2h} + O(h^2) \tag{8.1.16}$$

Centered difference for u_{xx}:

$$u_{xx}(x_n, t_j) = \frac{u_{n-1}^j - 2u_n^j + u_{n+1}^j}{h^2} + O(h^2) \equiv \frac{\delta_x^2 u_n^j}{h^2} + O(h)^2 \tag{8.1.17}$$

Forward difference for u_t:

$$u_t(x_n, t_j) = \frac{u_n^{j+1} - u_n^j}{k} + O(k) \tag{8.1.18}$$

Backward difference for u_t:

$$u_t(x_n, t_j) = \frac{u_n^j - u_n^{j-1}}{k} + O(k) \tag{8.1.19}$$

Centered difference for u_t:

$$u_t(x_n, t_j) = \frac{u_n^{j+1} - u_n^{j-1}}{2k} + O(k^2) \tag{8.1.20}$$

Centered difference for u_{tt}:

$$u_{tt}(x_n, t_j) = \frac{u_n^{j-1} - 2u_n^j + u_n^{j+1}}{k^2} + O(k^2) \equiv \frac{\delta_t^2 u_n^j}{k^2} + O(k^2) \tag{8.1.21}$$

Rounding/Data Errors

Based on the difference formulas of Table 8.1.1, it would appear that the best way to obtain a good approximation to a derivative using a difference quotient is to use a very small value of h or k. In infinite precison arithmetic and in the absence of any measurement errors in the data for u_n^j, this is the case. However, if roundoff or measurement errors are present in the determination of the u_n^j values, then a different strategy is called for.

To illustrate the interplay between roundoff or data measurement error and truncation error, we consider the calculation of u_t by the forward difference formula. Suppose that v_n^j represents a measured or computed value of u_n^j that differs from the exact u_n^j value by an amount e_n^j, so

$$u_n^j = v_n^j + e_n^j$$

Also, suppose that E is an upper bound for the errors e_n^j,

$$|e_n^j| \leq E \quad \text{for all } n, j$$

Finally, assume that M is an upper bound for $u_{tt}(x, t)$,

$$|u_{tt}(x, t)| \leq M \quad \text{for all } x, t$$

Now, if we apply the forward difference formula to estimate $u_t(x_n, t_j)$, we find

$$u_t(x_n, t_j) = \frac{u_n^{j+1} - u_n^j}{k} - \frac{k}{2} u_{tt}(x_n, \tau_j), \qquad t_j < \tau_j < t_{j+1}$$

or

$$u_t(x_n, t_j) = \frac{v_n^{j+1} - v_n^j}{k} + \frac{e_n^{j+1} - e_n^j}{k} - \frac{k}{2} u_{tt}(x_n, \tau_j)$$

Thus, we see that the estimate of u_t consists of three parts: a computed difference quotient plus a rounding-measurement error plus a truncation error. The total error incurred in using the difference quotient $(v_n^{j+1} - v_n^j)/k$ to approximate $u_t(x_n, t_j)$ can be bounded as follows.

$$\left| \frac{e_n^{j+1} - e_n^j}{k} - \frac{k}{2} u_{tt}(x_n, \tau_j) \right| \leq \left| \frac{e_n^{j+1} - e_n^j}{k} \right| + \left| \frac{k}{2} u_{tt}(x_n, \tau_j) \right| \leq \frac{2E}{k} + \frac{Mk}{2}$$

The upper bound to the total error, $2E/k + Mk/2$, can be viewed as a worst-case situation in which $e_n^{j+1} = E = -e_n^j$ and $u_{tt}(x_n, \tau_j) = -M$. In general, the computed difference quotient will not be in error by this much. Notice that the upper bound on the total error approaches infinity as $k \to 0$ and as $k \to \infty$. To maintain control of the total error, we should choose k to minimize the maximum total error. In this example that would require choosing $k = 2\sqrt{E/M}$.

EXAMPLE 8.1.3 ─────────────────────────────

Suppose that $u(x, t)$ represents the displacement at position x and time t of a vibrating elastic string. Assume that a high-speed camera is used to track a particle on the string as it vibrates and the position of the particle, measured in millimeters from equilibrium, is recorded at intervals of 10^{-4} sec. Given that our recording device is only accurate to two decimal places, we take $E = 5 \times 10^{-3}$ mm. Suppose that our observations support an estimate of the maximum acceleration of the particle of $u_{tt} = 10$ mm/sec^2. Then we can set $M = 10$ mm/sec^2. Now, if we were to use the forward difference method to estimate the velocity of the particle, we see from the preceding discussion that we should take

$$k = 2\sqrt{5 \times 10^{-3}/10} = 2\sqrt{5} \times 10^{-1} \text{ sec}$$

to minimize the maximum of the sum of the measurement error and the truncation error. This value of k is much larger than the smallest value available, namely, $k = 10^{-4}$. ■ ■

Exercises

1. Given the function $u(x, t) = e^{-t}\sin x$:
   (a) Find $u_x(\pi/4, 0)$, $u_t(\pi/4, 0)$, $u_{xx}(\pi/4, 0)$, and $u_{tt}(\pi/4, 0)$.
   (b) Use the forward difference formula (8.1.14) and the values $h = 0.5$, $h = 0.1$, and $h = 0.01$ to approximate $u_x(\pi/4, 0)$. Compare with the exact value from (a).
   (c) Use the centered difference formula (8.1.16) and the values $h = 0.5$, $h = 0.1$, and $h = 0.01$ to approximate $u_x(\pi/4, 0)$. Compare with the exact value from (a).
   (d) Use the forward difference formula (8.1.18) and the values $k = 0.5$, $k = 0.1$, and $k = 0.01$ to approximate $u_t(\pi/4, 0)$. Compare with the exact value from (a).
   (e) Use the centered difference formula (8.1.20) and the values $k = 0.5$, $k = 0.1$, and $k = 0.01$ to approximate $u_t(\pi/4, 0)$. Compare with the exact value from (a).
   (f) Use the centered difference formula (8.1.17) and the values $h = 0.5$, $h = 0.1$, and $h = 0.01$ to approximate $u_{xx}(\pi/4, 0)$. Compare with the exact value from (a).
   (g) Use the centered difference formula (8.1.21) and the values $k = 0.5$, $k = 0.1$, and $k = 0.01$ to approximate $u_{tt}(\pi/4, 0)$. Compare with the exact value from (a).

2. Given $u(x, t) = e^{2x+3t}$, determine the Taylor polynomial $P_3(x, t)$ of degree 3 about $(0, 0)$ and the associated remainder. Compare $P_3(0.2, 0.1)$ with $u(0.2, 0.1)$.

3. Given $u(x, t) = 1 + x - t + x^2 + xt + t^2$, determine the Taylor polynomials $P_1(x, t)$, $P_2(x, t)$, and $P_3(x, t)$ about $(0, 0)$ and their associated remainders.

4. For positive h and k, write the Taylor polynomial of degree 3 for $u(x + h, t + k)$ about the point (x, t) and provide an expression for the remainder.

5. Use Taylor polynomials of degree 3 for $u(x \pm h, t \pm k)$ to derive the centered difference formula (8.1.13) for u_{xt}.

6. Let $\{h_i\}$ be a sequence of positive numbers and define the nonuniform grid $x_n = x_0 + \sum_{i=1}^{n} h_i$ so that $x_n - x_{n-1} = h_n$.
   (a) Derive the difference formula

   $$u_x(x_n, t_j) = \frac{u_{n+1}^j - u_{n-1}^j}{h_{n+1} + h_n} + \text{TE}$$

   (b) Show that the truncation error in (a) has the form

   $$\text{TE} = \tfrac{1}{2}u_{xx}(x_n, t_j)(h_{n+1} - h_n) + \frac{O(h_{n+1}^3) + O(h_n^3)}{h_{n+1} + h_n}$$

This shows that better accuracy can be obtained with the difference formula (a) if the grid spacings h_n are changed gradually with increasing n.

(c) Derive the difference formula

$$u_{xx}(x_n, t_j) = \frac{h_{n+1}u_{n-1}^j - (h_n + h_{n+1})u_n^j + h_nu_{n+1}^j}{0.5(h_nh_{n+1}^2 + h_{n+1}h_n^2)} + \text{TE}$$

7. Derive the difference formula

$$u_x(x, t) = \frac{-3u(x, t) + 4u(x + h, t) - u(x + 2h, t)}{2h} + \text{TE}$$

and show $\text{TE} = O(h^2)$.

8. Replace h by $-h$ in the difference formula of Exercise 7 to obtain a difference formula for $u_x(x, t)$ in terms of u at (x, t), $(x - h, t)$, and $(x - 2h, t)$.

9. Given a uniform grid $x_1, x_2, \ldots, x_n, \ldots$, with $x_{n+1} - x_n = h$:

(a) Show that the quadratic function of x defined by

$$q(x) = u_{n-1}^j \frac{(x - x_n)(x - x_{n+1})}{(x_{n-1} - x_n)(x_{n-1} - x_{n+1})} + u_n^j \frac{(x - x_{n-1})(x - x_{n+1})}{(x_n - x_{n-1})(x_n - x_{n+1})}$$

$$+ u_{n+1}^j \frac{(x - x_{n-1})(x - x_n)}{(x_{n+1} - x_{n-1})(x_{n+1} - x_n)}$$

interpolates the data points (x_{n-1}, u_{n-1}^j), (x_n, u_n^j), (x_{n+1}, u_{n+1}^j).

(b) From $q'(x_n)$, obtain the difference formula (8.1.16).

(c) From $q''(x_n)$, obtain the difference formula (8.1.17)

10. Show that $e^h - 1 = O(\sin h)$ as $h \to 0$.

11. Show that $\sin(e^{h^2} - 1) = O(h^2)$ as $\to 0$.

8.2 FINITE-DIFFERENCE EQUATIONS FOR $u_t - a^2u_{xx} = S$

Discretization

The finite-difference formulas of the last section can be used to develop efficient methods for approximating the solution to a partial differential equation. Suppose that a partial differential equation (PDE), expressed symbolically as $L[u] = S$, holds on some region Ω of the xt plane. Let us cover the region with a finite-difference grid $(x_n, t_j) = (nh, jk)$. If all the derivatives in the PDE are replaced by difference quotients, the result is a *finite-difference equation* (FDE). When we use difference quotients to approximate derivatives in this manner, we say that we have *differenced*, or *discretized*, the continuous PDE problem to obtain a discrete FDE problem. This procedure is illustrated in Figure 8.2.1. Compare Figure 8.2.1 with Figure 1.2.1.

The solution of the FDE U_n^j approximates $u_n^j = u(x_n, t_j)$, the solution of the PDE at the grid points. If the discretization is to provide a useful approximation, the solution of the PDE should very nearly satisfy the FDE when the grid spacings h and k are taken sufficiently small. The amount by which the solution to the PDE fails to satisfy the FDE

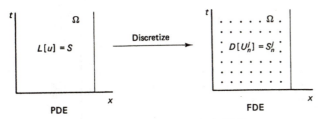

Figure 8.2.1   Discretization of PDE to obtain FDE.

is called the *local truncation error*; it may be expressed as

$$T_n^j = D[u_n^j] - S_n^j$$

The FDE is said to be *consistent with* the PDE if

$$\lim_{h,k \to 0} T_n^j = 0$$

To say that a FDE is consistent with a PDE means that the FDE converges to the PDE as the grid spacings approach zero. We also want the solution of the FDE to converge to the solution of the PDE.

    We say that a FDE is *convergent* if

$$\lim_{h,k \to 0} |U_n^j - u_n^j| = 0 \quad \text{for all } (x_n, t_j) \text{ in } \Omega$$

It is possible for a FDE to be consistent but not convergent. It is also possible for the U_n^j to converge to the solution of a PDE other than the one that was used to derive the FDE. These points will be discussed in more detail in Section 8.5.

    In addition to the consistency and convergence properties of a FDE, we must be concerned with the manner in which the FDE propagates roundoff errors. For example, suppose we use a FDE to approximate a PDE whose solution is known to be bounded. If, in the course of solving the FDE, the rounding errors grow without bound, then the unbounded solution of the FDE cannot be expected to provide a good approximation to the bounded solution of the PDE. In this context, we can describe a FDE as *stable* if, in the solution of the FDE, rounding errors remain bounded. In Section 8.5, we give a more thorough treatment of the important topics of consistency, convergence, and stability.

Dirichlet Initial-Boundary-Value Problem

We use the one-dimensional diffusion equation to illustrate the discretization of a continuous PDE problem to obtain a discrete FDE problem. Specifically, we consider the initial-boundary-value problem

$$u_t - a^2 u_{xx} = S(x, t), \qquad 0 < x < L, \, t > 0 \tag{8.2.1}$$

$$u(x, 0) = f(x), \qquad 0 < x < L \tag{8.2.2}$$

$$u(0, t) = p(t), \qquad u(L, t) = q(t), \qquad t > 0 \tag{8.2.3}$$

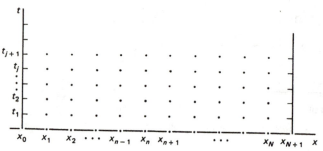

Figure 8.2.2 Finite-difference grid for initial-boundary-value problem (8.2.1)–(8.2.3) with time step k and space step $h = L/(N + 1)$.

Let us introduce a uniform x grid as indicated in Figure 8.2.2 with $x_0 = 0$, $x_{N+1} = L$, and $h = L/(N + 1)$. Similarly, we use a uniform t grid with $t_0 = 0$ and $t_{j+1} - t_j = k$.

The initial condition (8.2.2) provides values for U_n^0. Namely,

$$U_n^0 = f_n = f(x_n), \qquad n = 1, 2, \ldots, N \tag{8.2.4}$$

Similarly, the boundary conditions (8.2.3) give

$$U_0^j = p^j = p(t_j), \qquad U_{N+1}^j = q^j = q(t_j), \qquad j = 1, 2, \ldots \tag{8.2.5}$$

It remains for us to obtain from the PDE (8.2.1) a FDE that will allow us to calculate values for U_n^j, $n = 1, 2, \ldots, N$, $j = 1, 2, \ldots$.

TABLE 8.2.1 FINITE-DIFFERENCE METHODS FOR $u_t - a^2 u_{xx} = S$

Forward difference method:

$$\frac{U_n^{j+1} - U_n^j}{k} - a^2 \frac{U_{n-1}^j - 2U_n^j + U_{n+1}^j}{h^2} = S_n^j \tag{8.2.6}$$

or

$$U_n^{j+1} = (1 + r\delta_x^2)U_n^j + kS_n^j, \qquad r = a^2 k/h^2$$

Backward difference method:

$$\frac{U_n^{j+1} - U_n^j}{k} - a^2 \frac{U_{n-1}^{j+1} - 2U_n^{j+1} + U_{n+1}^{j+1}}{h^2} = S_n^{j+1} \tag{8.2.7}$$

or

$$(1 - r\delta_x^2)U_n^{j+1} = U_n^j + kS_n^j, \qquad r = a^2 k/h^2$$

Crank–Nicolson method:

$$\frac{U_n^{j+1} - U_n^j}{k} - \frac{a^2}{2} \frac{\delta_x^2 U_n^j + \delta_x^2 U_n^{j+1}}{h^2} = \frac{S_n^j + S_n^{j+1}}{2} \tag{8.2.8}$$

or

$$\left(1 - \frac{r}{2}\delta_x^2\right) U_n^{j+1} = \left(1 + \frac{r}{2}\delta_x^2\right) U_n^j + \frac{k}{2} [S_n^j + S_n^{j+1}], \qquad r = \frac{a^2 k}{h^2}$$

Table 8.2.1 displays three commonly used FDEs for (8.2.1). In each finite-difference method of Table 8.2.1, $u_t(x_n, t_j)$ is approximated by $(U_n^{j+1} - U_n^j)/k$. For the forward difference method (8.2.6) we use the centered difference formula (8.1.17) to approximate u_{xx} at time level j. In the backward difference method u_{xx} is discretized at time t_{j+1}. The Crank–Nicolson method uses the average of the u_{xx} differences at t_j and t_{j+1}.

We postpone until the next section any discussion of the computation of the solutions of the finite-difference methods given in Table 8.2.1. Notice, however, that the forward difference method (8.2.6) is *explicit* in the sense that only one U_n^{j+1} appears in each difference equation, and it can be expressed explicitly in terms of U values that have already been calculated at time t_j. In contrast, the backward and Crank–Nicolson difference methods each involve more than one unknown U value at time t_{j+1} and are therefore called *implicit* methods. With an implicit method, we are required to solve a system of linear equations for the U^{j+1} values.

In each of the finite-difference methods of Table 8.2.1, there is a truncation error associated with the approximation of u_t by a difference quotient and a truncation error associated with the replacement of u_{xx} by a difference quotient. Together these truncation errors define the local truncation error for each finite-difference method. The following theorem shows that each of the methods in Table 8.2.1 is consistent with the PDE (8.2.1).

Theorem 8.2.1. Suppose the initial-boundary-value problem has solution $u(x, t)$.

(a) The forward difference method (8.2.6) has local truncation error $O(k + h^2)$ provided u_{tt} and u_{xxxx} are bounded.

(b) The backward difference method (8.2.7) has local truncation error $O(k + h^2)$ provided u_{tt} and u_{xxxx} are bounded.

(c) The Crank–Nicolson method (8.2.8) has local truncation error $O(k^2 + h^2)$ provided u_{tttt} and u_{xxxx} are bounded.

Proof of (a). Let $u(x, t)$ be the solution of $u_t - a^2 u_{xx} = S(x, t)$. By the difference formulas (8.1.11) and (8.1.12),

$$[u_t(x_n, t_j) - a^2 u_{xx}(x_n, t_j)] = \frac{u_n^{j+1} - u_n^j}{k} - a^2 \frac{\delta_x^2 U_n^j}{h^2}$$
$$- \frac{k}{2} u_{tt}(x_n, \tau_j) + a^2 \frac{h^2}{12} u_{xxxx}(\xi_n, t_j)$$

where $t_j < \tau_j < t_{j+1}$ and $x_{n-1} < \xi_n < x_{n+1}$. The amount by which the solution of $u_t - a^2 u_{xx} = S(x, t)$ fails to satisfy the difference equation (8.2.6) is

$$T_n^j = \frac{k}{2} u_{tt}(x_n, \tau_j) - a^2 \frac{h^2}{12} u_{xxxx}(\xi_n, t_j) = O(k + h^2)$$

provided u_{tt} and u_{xxxx} are bounded. The proofs of parts (b) and (c) follow from similar arguments.

Matrix Formulation of Difference Methods

If one of the methods of Table 8.2.1 is applied to the initial-boundary-value problem, the initial condition (8.2.4) and the boundary conditions must be incorporated into the finite-difference formulation of the problem. A matrix formulation of the finite-difference method is a convenient way to display how the initial and boundary data are accounted for in a system of finite-difference equations. For example, consider the application of the forward difference method to the problem (8.2.1)–(8.2.3). Let $\mathbf{U}^j$ denote a column vector containing entries U_n^j, $n = 1, 2, \ldots, N$. That is,

$$\mathbf{U}^j = [U_1^j, U_2^j, \ldots, U_N^j]^T$$

Using (8.2.4) to initialize $\mathbf{U}^j$ and (8.2.5) to "close" the system of difference equations (8.2.6) by eliminating the U_0^j and U_{N+1}^j values we obtain

$$\begin{bmatrix} U_1^{j+1} \\ U_2^{j+1} \\ \vdots \\ U_n^{j+1} \\ \vdots \\ U_N^{j+1} \end{bmatrix} = \begin{bmatrix} (1-2r) & r & & & \\ r & (1-2r) & r & & \\ & \ddots & \ddots & \ddots & \\ & & r & (1-2r) & r \\ & & & \ddots & \ddots \\ & & & & r & (1-2r) \end{bmatrix} \begin{bmatrix} U_1^j \\ U_2^j \\ \vdots \\ U_n^j \\ \vdots \\ U_N^j \end{bmatrix} + \begin{bmatrix} kS_1^j + rp^j \\ kS_2^j \\ \vdots \\ kS_n^j \\ \vdots \\ kS_N^j + rq^j \end{bmatrix}$$

$$(8.2.9)$$

or

$$\mathbf{U}^{j+1} = B\mathbf{U}^j + \mathbf{c}, \qquad j = 1, 2, \ldots$$

subject to the initial condition

$$\mathbf{U}^0 = \mathbf{f} \qquad (8.2.10)$$

Similarly, the backward difference method can be formulated in matrix notation as follows:

$$\begin{bmatrix} (1+2r) & -r & & & \\ -r & (1+2r) & -r & & \\ & \ddots & \ddots & \ddots & \\ & & -r & (1+2r) & -r \\ & & & \ddots & \ddots \\ & & & & -r & (1+2r) \end{bmatrix} \begin{bmatrix} U_1^{j+1} \\ U_2^{j+1} \\ \vdots \\ U_n^{j+1} \\ \vdots \\ U_N^{j+1} \end{bmatrix} = \begin{bmatrix} U_1^j \\ U_2^j \\ \vdots \\ U_n^j \\ \vdots \\ U_N^j \end{bmatrix} + \begin{bmatrix} kS_1^{j+1} + rp^{j+1} \\ kS_2^{j+1} \\ \vdots \\ kS_n^{j+1} \\ \vdots \\ kS_N^{j+1} + rq^{j+1} \end{bmatrix}$$

$$(8.2.11)$$

or

$$A\mathbf{U}^{j+1} = \mathbf{U}^j + \mathbf{c}, \qquad j = 1, 2, \ldots$$

subject to the initial condition

$$\mathbf{U}^0 = \mathbf{f} \qquad (8.2.12)$$

Derivative Boundary Conditions

Now that we have seen how to formulate FDEs for the initial Dirichlet boundary value problem (8.2.1)–(8.2.3), let us consider a problem in which derivative boundary conditions are present. For example, consider the initial-boundary-value problem

$$u_t - a^2 u_{xx} = S(x, t), \qquad 0 < x < L, t > 0 \tag{8.2.13}$$

$$u(x, 0) = f(x), \qquad 0 < x < L \tag{8.2.14}$$

$$u_x(0, t) = p(t), \qquad u_x(L, t) = q(t), \qquad t > 0 \tag{8.2.15}$$

The derivative boundary conditions (8.2.15) are called Neumann boundary conditions. Any of the discretization methods of Table 8.2.1 can be used to approximate the PDE (8.2.13). In each of these methods the replacement of the u_{xx} term is correct to order h^2.

If we were to use one-sided differences to approximate $u_x(0, t)$ and $u_x(L, t)$, then we would introduce an order h error into our finite-difference approximation of (8.2.13)–(8.2.15). The simplest way to maintain order h^2 accuracy in a set of difference equations for (8.2.13)–(8.2.15) is to use centered differences for the boundary derivatives. The calculation of a centered difference at a grid point requires a neighbor grid point on both the right and left. To permit the centered difference calculations of the boundary derivatives, we define a uniform grid $x_0, x_1, x_2, \ldots, x_N, x_{N+1}$ with $x_1 = 0$, $x_N = L$, and $x_{n+1} - x_n = h$. See Figure 8.2.3. Other indexing schemes could be used if we must avoid using the zero subscript. One reason we have labeled the grid in this way is that the number of unknowns in the x direction is N, as was the case when Dirichlet boundary conditions were imposed.

Using the centered difference formula (8.1.16) to approximate the left boundary condition $u_x(x_1, t_j) = p(t_j)$ gives

$$\frac{U_2^j - U_0^j}{2h} = p^j \quad \text{or} \quad U_0^j = U_2^j - 2hp^j \tag{8.2.16}$$

Similarly, as a centered difference approximation to $u_x(x_N, t_j) = q(t_j)$, we have

$$\frac{U_{N+1}^j - U_{N-1}^j}{2h} = q^j \quad \text{or} \quad U_{N+1}^j = U_{N-1}^j + 2hq^j \tag{8.2.17}$$

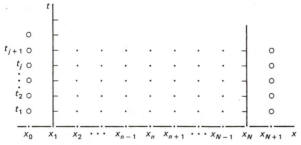

Figure 8.2.3 Finite-difference grid for initial-boundary-value problem (8.2.13)–(8.2.15), $x_1 = 0$, $x_N = L$.

Consider the application of the forward difference method to the initial-boundary-value problem (8.2.13)–(8.2.15). We have

$$U_1^{j+1} = rU_0^j + (1 - 2r)U_1^j + rU_2^j + kS_1^j$$
$$U_n^{j+1} = rU_{n-1}^j + (1 - 2r)U_n^j + rU_{n+1}^j + kS_n^j, \qquad n = 2, 3, \ldots, N - 1$$
$$U_N^{j+1} = rU_{N-1}^j + (1 - 2r)U_N^j + rU_{N+1}^j + kS_N^j$$

If (8.2.16) is used to eliminate U_0^j from the equation for U_1^{j+1} and (8.2.17) is used to eliminate U_{N+1}^j from the equation for U_N^{j+1}, then we obtain the following matrix representation of the forward difference method for the Neumann boundary value problem (8.2.13)–(8.2.15):

$$
\begin{bmatrix} U_1^{j+1} \\ U_2^{j+1} \\ \vdots \\ U_n^{j+1} \\ \vdots \\ U_N^{j+1} \end{bmatrix}
=
\begin{bmatrix}
(1-2r) & 2r & & & & \\
r & (1-2r) & r & & & \\
 & \ddots & \ddots & \ddots & & \\
 & & r & (1-2r) & r & \\
 & & & \ddots & \ddots & \ddots \\
 & & & & 2r & (1-2r)
\end{bmatrix}
\begin{bmatrix} U_1^j \\ U_2^j \\ \vdots \\ U_n^j \\ \vdots \\ U_N^j \end{bmatrix}
+
\begin{bmatrix} kS_1^j - 2hrp^j \\ kS_2^j \\ \vdots \\ kS_n^j \\ \vdots \\ kS_N^j + 2hrq^j \end{bmatrix}
$$

$$(8.2.18)$$

Summary of Methods

The matrix formulations of a number of finite-difference equations can be conveniently expressed if we use the following notation. Define column vectors of length N by

$$
\mathbf{U}^j = \begin{bmatrix} U_1^j \\ U_2^j \\ \vdots \\ U_n^j \\ \vdots \\ U_N^j \end{bmatrix}, \quad
\mathbf{s}^j = \begin{bmatrix} S_1^j \\ S_2^j \\ \vdots \\ S_n^j \\ \vdots \\ S_N^j \end{bmatrix}, \quad
\mathbf{f} = \begin{bmatrix} f_1 \\ f_2 \\ \vdots \\ f_n \\ \vdots \\ f_N \end{bmatrix}, \quad
\mathbf{e}_1 = \begin{bmatrix} 1 \\ 0 \\ \vdots \\ 0 \\ \vdots \\ 0 \end{bmatrix}, \quad
\mathbf{e}_N = \begin{bmatrix} 0 \\ 0 \\ \vdots \\ 0 \\ \vdots \\ 1 \end{bmatrix}
$$

and define $N \times N$ tridiagonal matrices $\mathbf{F}_1$, $\mathbf{F}_2$, $\mathbf{F}_3$, and $\mathbf{F}_4$ by

$$
\mathbf{F}_1 = \begin{bmatrix}
2 & -1 & & & & \\
-1 & 2 & -1 & & & \\
 & -1 & 2 & -1 & & \\
 & & & \ddots & & \\
 & & & -1 & 2 & -1 \\
 & & & & -1 & 2
\end{bmatrix}, \quad
\mathbf{F}_2 = \begin{bmatrix}
2 & -2 & & & & \\
-1 & 2 & -1 & & & \\
 & -1 & 2 & -1 & & \\
 & & & \ddots & & \\
 & & & -1 & 2 & -1 \\
 & & & & -2 & 2
\end{bmatrix}
$$

$$\mathbf{F}_3 = \begin{bmatrix} 2 & -2 & & & & \\ -1 & 2 & -1 & & & \\ & -1 & 2 & -1 & & \\ & & & \ddots & & \\ & & & -1 & 2 & -1 \\ & & & & -1 & 2 \end{bmatrix}, \qquad \mathbf{F}_4 = \begin{bmatrix} 1 & -1 & & & & \\ -1 & 2 & -1 & & & \\ & -1 & 2 & -1 & & \\ & & & \ddots & & \\ & & & -1 & 2 & -1 \\ & & & & -1 & 1 \end{bmatrix}$$

Using this notation, we display a number of finite-difference methods in Table 8.2.2

TABLE 8.2.2   FINITE-DIFFERENCE METHODS FOR $u_t - a^2 u_{xx} = S(x, t)$, $0 < x < L, t > 0$, $u(x, 0) = f(x)$ VARIOUS BOUNDARY CONDITIONS

Dirichlet–Dirichlet
$u(0, t) = p(t)$
$u(L, t) = q(t)$

$$h = L/(N + 1)$$

$$x_0 \quad x_1 \cdots \quad x_{n-1} x_n \quad x_{n+1} \cdots x_{N+1}$$

Forward difference method:

$$\mathbf{U}^{j+1} = (I - r\mathbf{F}_1)\mathbf{U}^j + k\mathbf{s}^j + rp^j\mathbf{e}_1 + rq^j\mathbf{e}_N, \qquad \mathbf{U}^0 = \mathbf{f}$$

Backward difference method:

$$(I + r\mathbf{F}_1)\mathbf{U}^{j+1} = \mathbf{U}^j + k\mathbf{s}^{j+1} + rp^{j+1}\mathbf{e}_1 + rq^{j+1}\mathbf{e}_N, \qquad \mathbf{U}^0 = \mathbf{f}$$

Crank–Nicolson method:

$$\left(I + \frac{r}{2}\mathbf{F}_1\right)\mathbf{U}^{j+1} = \left(I - \frac{r}{2}\mathbf{F}_1\right)\mathbf{U}^j + \frac{k}{2}(\mathbf{s}^j + \mathbf{s}^{j+1}) + \frac{r}{2}(p^j + p^{j+1})\mathbf{e}_1 + \frac{r}{2}(q^j + q^{j+1})\mathbf{e}_N$$

Neumann–Neumann
$u_x(0, t) = p(t)$
$u_x(L, t) = q(t)$

$$h = L/(N - 1)$$

$$x_0 \quad x_1 \quad x_2 \cdots \quad x_{n-1} \; x_n x_{n+1} \cdots \quad x_N \quad x_{N+1}$$

Forward difference method:

$$\mathbf{U}^{j+1} = (I - r\mathbf{F}_2)\mathbf{U}^j + k\mathbf{s}^j - 2hrp^j\mathbf{e}_1 + 2hrq^j\mathbf{e}_N, \qquad \mathbf{U}^0 = \mathbf{f}$$

Backward difference method:

$$(I + r\mathbf{F}_2)\mathbf{U}^{j+1} = \mathbf{U}^j + k\mathbf{s}^{j+1} - 2hrp^{j+1}\mathbf{e}_1 + 2hrq^{j+1}\mathbf{e}_N, \qquad \mathbf{U}^0 = \mathbf{f}$$

Crank–Nicolson method:

$$\left(I + \frac{r}{2}\mathbf{F}_2\right)\mathbf{U}^{j+1} = \left(I - \frac{r}{2}\mathbf{F}_2\right)\mathbf{U}^j + \frac{k}{2}\left(\mathbf{s}^j + \mathbf{s}^{j+1}\right) - rh(p^j + p^{j+1})\mathbf{e}_1 + rh(q^j + q^{j+1})\mathbf{e}_N$$

Neumann–Dirichlet
$u_x(0, t) = p(t)$
$u(L, t) = q(t)$

$$h = L/N$$

$$x_0 \quad x_1 \quad x_2 \cdots \quad x_{n-1} \; x_n \; x_{n+1} \cdots x_{N+1}$$

Forward difference method:

$$\mathbf{U}^{j+1} = (I - r\mathbf{F}_3)\mathbf{U}^j + k\mathbf{s}^j - 2hrp^j\mathbf{e}_1 + rq^j\mathbf{e}_N, \qquad \mathbf{U}^0 = \mathbf{f}$$

Backward difference method:

$$(I + r\mathbf{F}_3)\mathbf{U}^{j+1} = \mathbf{U}^j + k\mathbf{s}^{j+1} - 2hrp^{j+1}\mathbf{e}_1 + rq^{j+1}\mathbf{e}_N, \qquad \mathbf{U}^0 = \mathbf{f}$$

Crank–Nicolson method:

$$\left(I + \frac{r}{2}\mathbf{F}_3\right)\mathbf{U}^{j+1} = \left(I - \frac{r}{2}\mathbf{F}_3\right)\mathbf{U}^j + \frac{k}{2}(\mathbf{s}^j + \mathbf{s}^{j+1}) - rh(p^j + p^{j+1})\mathbf{e}_1 + \frac{r}{2}(q^j + q^{j+1})\mathbf{e}_N$$

The finite-difference methods of Table 8.2.2 are intended to illustrate an approach and do not exhaust the boundary conditions that may be imposed on the one-dimensional diffusion equation. A number of other initial-boundary-value problems are considered in the exercises.

Exercises

1. Given the initial-boundary-value problem

$$u_t - u_{xx} = 0, \qquad 0 < x < 1, t > 0$$
$$u(x, 0) = 100 \sin \pi x, \qquad 0 < x < 1$$
$$u(0, t) = 0, \qquad u(1, t) = 0, \qquad t > 0$$

and the uniform grid $0 = x_0 < x_1 < x_2 < x_3 < x_4 = 1$, formulate a system of finite-difference equations using
(a) the forward difference method (8.2.6),
(b) the backward difference method (8.2.7), and
(c) the Crank–Nicolson method (8.2.8).
(d) Express each of the methods (a)–(c) in matrix notation.

2. Given the initial-boundary-value problem

$$u_t - 4u_{xx} = xe^{-t}, \qquad 0 < x < 1, t > 0$$
$$u(x, 0) = x^2, \qquad 0 < x < 1$$
$$u(0, t) = 0, \qquad u(1, t) = 1, \qquad t > 0$$

and the uniform grid $0 = x_0 < x_1 < x_2 < x_3 < x_4 < x_5 = 1$, formulate a system of finite-difference equations using
(a) the forward difference method (8.2.6),
(b) the backward difference method (8.2.7), and
(c) the Crank–Nicolson method (8.2.8).
(d) Express each of the methods (a)–(c) in matrix notation.

3. Given the initial-boundary-value problem

$$u_t - u_{xx} = 0, \qquad 0 < x < 4, t > 0$$
$$u(x, 0) = 0, \qquad 0 < x < 4$$
$$u_x(0, t) = 1 - e^{-t}, \qquad u_x(4, t) = \sin t$$

and the uniform grid $0 = x_1 < x_2 < x_3 < x_4 = 4$, formulate a system of finite-difference equations using
(a) the forward difference method (8.2.6),
(b) the backward difference method (8.2.7), and
(c) the Crank–Nicolson method (8.2.8).
(d) Express each of the finite-difference systems (a)–(c) in matrix notation.

4. Given the initial-boundary-value problem

$$u_t - u_{xx} = xt^2, \qquad 0 < x < \pi, t > 0$$
$$u(x, 0) = \sin x, \qquad u_x(0, t) = 1, \qquad u(\pi, t) = \sin t$$

and the uniform grid $0 = x_1 < x_2 < x_3 < x_4 < x_5 = \pi$, formulate a system of finite-difference equations using

(a) the forward difference method (8.2.6),

(b) the backward difference method (8.2.7), and

(c) the Crank–Nicolson method (8.2.8).

(d) Express each of the finite-difference systems (a)–(c) in matrix notation.

5. Given the initial-boundary-value problem

$$u_t - u_{xx} - 3u_x - 5u = 0, \qquad 0 < x < 1, t > 0$$
$$u(x, 0) = f(x), \qquad 0 < x < 1$$
$$u(0, t) = p(t), \qquad u(1, t) = q(t), \qquad t > 0$$

formulate a forward difference method that has local truncation error $O(k + h^2)$.

6. Repeat Exercise 5 using a backward difference method.

7. In the forward difference method (8.2.6) for the PDE $u_t - a^2 u_{xx} = 0$:

   (a) Show that if $r = \frac{1}{2}$, then U_n^{j+1} is the arithmetic average of U_{n-1}^j and U_{n+1}^j.

   (b) Show that if $r = \frac{1}{3}$, then U_n^{j+1} is the arithmetic average of U_{n-1}^j, U_n^j, and U_{n+1}^j. This is called Schmidt's method.

   (c) What does the stability restriction $r \leq \frac{1}{2}$ imply about the weights assigned to U_{n-1}^j, U_n^j, and U_{n+1}^j when expressing U_n^{j+1} as a weighted average defined by (8.2.6)?

8. Show that the choice $r = \frac{1}{6}$ reduces the local truncation error of the forward difference method (8.2.6) from $O(k + h^2)$ to $O(k^2 + h^4)$. (*Hint:* $u_{tt} = a^2 u_{txx} = a^4 u_{xxxx}$.)

8.3 COMPUTATIONAL METHODS

In the last section, a number of finite-difference methods for the one-dimensional diffusion equation were introduced. In this section, we consider some computational aspects of these methods.

The notion of a *finite-difference stencil* or a *computational molecule* is useful when developing a computational algorithm for a difference method. A finite-difference stencil, or simply a stencil, is a schematic that displays the nodes that are involved in a given finite-difference method. Also, for difference methods that are derived by Taylor series methods, it is sometimes helpful to indicate on the stencil the point about which Taylor series are expanded. Figure 8.3.1 shows the difference stencils for the three methods discussed in Section 8.2.

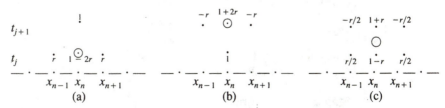

Figure 8.3.1 Finite difference stencils for (a) forward difference, (b) backward difference, and (c) Crank-Nicolson methods. A • denotes a grid point in FDE; a ○ denotes the expansion point in Taylor series development of the method.

To serve as an example problem in our discussion of computational methods, consider the following initial-boundary-value problem:

$$u_t - a^2 u_{xx} = S(x, t), \qquad 0 < x < L, \, t < 0 \qquad (8.3.1)$$
$$u(x, 0) = f(x), \qquad 0 < x < L \qquad (8.3.2)$$
$$u(0, t) = p(t), \qquad u(L, t) = q(t), \qquad t > 0 \qquad (8.3.3)$$

The forward difference method is the simplest finite-difference scheme for approximating the solution to (8.3.1)–(8.3.3). From the stencil displayed in Figure 8.3.1(a) or from the matrix formulation (8.2.9), we see that the value for U_n^{j+1} is a linear combination of the values of U_{n-1}^j, U_n^j, U_{n+1}^j, and S_n^j. The boundary conditions are accounted for in the difference equations whose stencils are centered at x_1 and x_N. The forward difference method for (8.3.1)–(8.3.3) is given next in algorithm form.

ALGORITHM 8.1   Forward Difference Method: Dirichlet Initial-Boundary-Value Problem

Step 1. Document

This algorithm uses the forward difference method (8.2.9) to approximate the solution to the initial-boundary-value problem

$$u_t - a^2 u_{xx} = S(x, t), \, 0 < x < L, \, t > 0$$
$$u(x, 0) = f(x), \, 0 < x < L$$
$$u(0, t) = p(t); \, u(L, t) = q(t), \, t > 0$$

INPUT

Real a^2, diffusivity
Real L, endpoint
Real k, time step
Integer *jmax*, number of time steps
Integer *nmax*, number of computational nodes
Function $S(x, t)$, right side of diffusion equation
Function $f(x)$, initial condition
Functions $p(t)$, $q(t)$, boundary conditions at $x = 0$ and $x = L$

OUTPUT

t, time for $j = 0, 1, \ldots, jmax$ and
U_n^j, approximate solution for $n = 0, 1, \ldots, nmax + 1$

Step 2. Define a grid

Set $h = L/(nmax + 1)$
    $r = a^2 k / h^2$
If $r > 1/2$, output message that method is unstable.

Step 3. Initialize numerical solution

Set $t = 0$
$$x_0 = 0$$
$$V_0 = (p(0) + f(0))/2$$
For $n = 1, 2, \ldots, nmax$ set
$$x_n = x_{n-1} + h$$
$$V_n = f(x_n)$$
Set $V_{nmax+1} = (q(0) + f(L))/2$
$$x_{nmax+1} = L$$
Output t
For $n = 0, 1, \ldots, nmax + 1$
    Output x_n, V_n

Step 4. Begin time stepping

For $j = 1, 2, \ldots, jmax$
    Do steps 5–7

Step 5. Advance solution one time step

For $n = 1, 2, \ldots, nmax$ set
$$U_n = rV_{n-1} + (1 - 2r)V_n + rV_{n+1} + kS(x_n, t)$$
Set $t = t + k$
$$U_0 = p(t)$$
$$U_{nmax+1} = q(t)$$

Step 6. Output numerical solution

Output t
For $n = 0, 1, \ldots, nmax + 1$
    Output x_n, U_n

Step 7. Prepare for next time step

For $n = 0, 1, \ldots, nmax + 1$ set
$$V_n = U_n$$

EXAMPLE 8.3.1 _____

Consider the initial-boundary-value problem

$$u_t - u_{xx} = 0, \qquad 0 < x < 1, t > 0 \tag{8.3.4}$$
$$u(x, 0) = 100 \sin \pi x, \qquad 0 < x < 1 \tag{8.3.5}$$
$$u(0, t) = 0 = u(1, t), \qquad t > 0 \tag{8.3.6}$$

Using separation of variables as in Section 3.2, the exact solution is found to be

TABLE 8.3.1 COMPARISON OF FORWARD
DIFFERENCE NUMERICAL
SOLUTION AND EXACT
SOLUTION WITH $k = 0.005$,
$h = 0.1$

$T = 0.50$	Numerical	Exact
$X = 0.0$	0.000000	0.000000
$X = 0.1$	0.204463	0.222241
$X = 0.2$	0.388912	0.422728
$X = 0.3$	0.535291	0.581836
$X = 0.4$	0.629273	0.683989
$X = 0.5$	0.661656	0.719188
$X = 0.6$	0.629273	0.683989
$X = 0.7$	0.535291	0.581836
$X = 0.8$	0.388912	0.422728
$X = 0.9$	0.204463	0.222241
$X = 1.0$	0.000000	0.000000

$u(x, t) = 100e^{-\pi^2 t}\sin \pi x$. Choosing $k = 0.005$, $nmax = 9$, and $jmax = 100$ gives the numerical solution shown in Table 8.3.1. Generally the truncation error for the forward difference method is $O(k + h^2)$. When $r = a^2 k/h^2$ is chosen to be $r = \frac{1}{6}$, the truncation error can be shown to be $O(k^2 + h^4)$. See Exercise 6 of Section 8.2. The numerical solution with $r = \frac{1}{6}$ is displayed in Table 8.3.2.   ■ ■

EXAMPLE 8.3.2 _____

In Figure 8.3.2 we show the result of using Algorithm 8.1 on the example problem (8.3.4)–(8.3.6) with a choice of $k = 0.025$ and $nmax = 9$. This choice produces an h

TABLE 8.3.2 COMPARISON OF FORWARD
DIFFERENCE NUMERICAL
SOLUTION AND EXACT
SOLUTION WITH $k = 0.001667$,
$h = 0.1$

$T = 0.50$	Numerical	Exact
$X = 0.0$	0.000000	0.000000
$X = 0.1$	0.222040	0.222022
$X = 0.2$	0.422346	0.422311
$X = 0.3$	0.581309	0.581262
$X = 0.4$	0.683370	0.683314
$X = 0.5$	0.718538	0.718479
$X = 0.6$	0.683370	0.683314
$X = 0.7$	0.581309	0.581262
$X = 0.8$	0.422346	0.422311
$X = 0.9$	0.222040	0.222022
$X = 1.0$	0.000000	0.000000

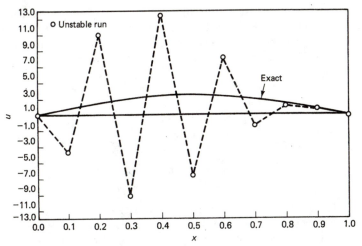

$h = 0.1 \quad k = 0.025 \quad t = 0.375$

Figure 8.3.2 Solid line is exact solution; dot is numerical solution.

value of 0.1 and thus an $r = a^2 k/h^2$ of 2.5. The oscillations exhibited by the numerical solution are typical of the output obtained from an unstable finite-difference method. As we show in Section 8.5, the forward difference method is stable if and only if $r \leq \frac{1}{2}$. ∎ ∎

From Algorithm 8.1 and the preceding examples, we see the forward difference method can be easily implemented to approximate the solution of the initial-boundary-value problem (8.3.1)–(8.3.3). However, the stability restriction that $r \leq \frac{1}{2}$ makes the forward difference method impractical under some conditions. For example, if $a^2 = 10$ and if we choose $h = 0.01$, then, to make $r \leq \frac{1}{2}$, we must restrict the time step to satisfy $k \leq 5 \times 10^{-6}$. To approximate the solution at time $t = 5$ with the forward difference method would require one million time steps in this case. In situations such as this, one of the implicit methods is usually selected. As we show in Section 8.5, the backward difference and Crank–Nicolson methods are stable for all choices of r.

Tridiagonal Linear Equations

If we use either the backward difference or the Crank–Nicolson method, then we must solve a system of linear algebraic equations to advance the solution from time t_j to time t_{j+1}. In each method these linear equations can be expressed in the form

$$\mathbf{AU}^{j+1} = \mathbf{BU}^j + \mathbf{s} \equiv \mathbf{d}^j$$

Thus, the problem of advancing the backward difference or Crank–Nicolson numerical solution reduces to the problem of solving a linear system of the form

$$\mathbf{AU} = \mathbf{d} \tag{8.3.7}$$

The stencils for these two difference methods show that the unknown U_n^{j+1} is computationally linked to only two other unknowns, U_{n-1}^{j+1} and U_{n+1}^{j+1}. The implication of this linkage of the unknowns is that the coefficient matrix $\mathbf{A}$ in (8.3.7) is *tridiagonal*. That is, all of the nonzero entries in $\mathbf{A}$ occur on the diagonal, the subdiagonal, and the superdiagonal:

$$
\mathbf{A} = \begin{bmatrix}
b_1 & c_1 & & & & \\
a_2 & b_2 & c_2 & & & \\
 & a_3 & b_3 & c_3 & & \\
 & & \ddots & \ddots & \ddots & \\
 & & & a_{N-1} & b_{N-1} & c_{N-1} \\
 & & & & a_N & b_N
\end{bmatrix} \equiv [\vec{\mathbf{a}}; \vec{\mathbf{b}}; \vec{\mathbf{c}}] \tag{8.3.8}
$$

Tridiagonal matrices are frequently encountered in systems of linear algebraic equations that are derived from finite-difference methods. We now describe how such finite-difference, tridiagonal systems can be solved using elementary Gaussian elimination.

The tridiagonal matrix (8.3.8) is said to be *strictly diagonally dominant* if

$$
|b_1| > |c_1|, \qquad |b_n| > |a_n| + |c_n|, \qquad |b_N| > |a_N|
$$

The tridiagonal matrices that appear in finite-difference equations are often diagonally dominant.

EXAMPLE 8.3.3 _____

The tridiagonal matrix associated with the backward difference method (8.2.11) has

$$
a_2 = a_3 = \cdots = a_N = -r
$$
$$
b_1 = b_2 = \cdots = b_N = 1 + 2r
$$
$$
c_2 = c_3 = \cdots = c_{N-1} = -r
$$

so it is diagonally dominant.   ■ ■

From numerical analysis we have the following result.

Theorem 8.3.1.   If $\mathbf{A}$ is a strictly diagonally dominant $N \times N$ matrix, then $\mathbf{A}$ is nonsingular. Moreover, Gaussian elimination can be performed on any linear system of the form $\mathbf{AU} = \mathbf{d}$ to obtain its unique solution without row or column interchanges and the computations are stable with respect to the growth of rounding errors.

Theorem 8.3.1 allows us to develop an efficient algorithm for the solution of a diagonally dominant tridiagonal linear system. Since we are guaranteed that no pivoting (row or column interchanges) is necessary, we do not have to test for a zero divisor in the computation of a multiplier. Also, in the forward substitution phase of Gaussian elimination we need only eliminate the subdiagonal.

Since the algorithm for solving a tridiagonal system plays such an important role in the solution of finite-difference systems, we provide sample computer code for the implementation of Algorithm 8.2. See Figure 8.3.3.

ALGORITHM 8.2 Solution of a Tridiagonal Linear System

Step 1. Document

This algorithm uses Gaussian elimination to solve a system of N linear equations $\mathbf{A}\mathbf{U} = \mathbf{d}$ in the case that $\mathbf{A}$ is a diagonally dominant, tridiagonal matrix.

INPUT

Array $a_2, a_3, \ldots, a_N$, subdiagonal entries of $\mathbf{A}$
Array $b_1, b_2, \ldots, b_N$, diagonal entries of $\mathbf{A}$
Array $c_1, c_2, \ldots, c_{N-1}$, superdiagonal entries of $\mathbf{A}$
Array $d_1, d_2, \ldots, d_N$, right-hand side of system

OUTPUT

$\mathbf{U}$, solution to the linear system stored in the array $\mathbf{d}$.

Step 2. Forward substitute to eliminate subdiagonal

For $n = 2, 3, \ldots, N$ set
$$\text{ratio} = a_n/b_{n-1}$$
$$b_n = b_n - \text{ratio} \cdot c_{n-1}$$
$$d_n = d_n - \text{ratio} \cdot d_{n-1}$$

Step 3. Back substitute and store solution in array d

Set $d_N = d_N/b_N$
For $n = N - 1, N - 2, \ldots, 1$ set
$$d_n = (d_n - c_n \cdot d_{n+1})/b_n$$

```
SUBROUTINE TRIDI(N,A,B,C,D)
DIMENSION A(N), B(N), C(N), D(N)
DO 1 I = 2,N
   RATIO = A(I)/B(I-1)
   B(I) = B(I) - RATIO*C(I-1)
   D(I) = D(I) - RATIO*D(I-1)
1  CONTINUE
D(N) = D(N)/B(N)
DO 2 I = N-1,1,-1
   D(I) = (D(I) - C(I)*D(I+1))/B(I)
2  CONTINUE
RETURN
END
```

Figure 8.3.3 Sample FORTRAN computer code for the solution tridiagonal system $\mathbf{A}\mathbf{U} = \mathbf{d}$, where $\mathbf{A} = [\mathbf{a}; \mathbf{b}; \mathbf{c}]$.

Implicit Difference Methods

Now that we have an efficient method for solving a tridiagonal system of linear equations, the backward difference and the Crank–Nicolson methods can be implemented as easily as the forward difference method. Algorithm 8.3 outlines how the backward difference method applies to the initial-boundary-value problem (8.3.1)–(8.3.3).

A quick glance at Algorithm 8.3 reveals a number of computational inefficiencies. For example, the arrays **a**, **b**, and **c** that define the tridiagonal coefficient matrix are redefined for each time step even though they are the same for all time steps. Also, in the solution of the tridiagonal system the same set of ratios, or multipliers, are computed at each time step. These multipliers could be computed just once and then stored. Ex-

---

ALGORITHM 8.3   Backward Difference Method: Dirichlet Initial-Boundary-Value Problem

Step 1. Document

This algorithm uses the backward difference method (8.2.11) to approximate the solution to the initial-boundary-value problem

$$u_t - a^2 u_{xx} = S(x, t),\ 0 < x < L,\ t > 0$$
$$u(x, 0) = f(x),\ 0 < x < L$$
$$u(0, t) = p(t);\ u(L, t) = q(t),\ t > 0$$

INPUT

Real a^2, diffusivity
Real L, endpoint functions $S(x, t)$
Real k, time step
Integer $jmax$, number of time steps
Integer $nmax$, number of computational nodes
Function $S(x, t)$, right side of diffusion equation
Function $f(x)$, initial condition
Functions $p(t), q(t)$, boundary conditions at $x = 0$ and $x = L$

OUTPUT

t, time for $n = 0, 1, \ldots, jmax$ and
U_n^j, approximate solution for $n = 0, 1, \ldots, nmax + 1$

Step 2. Define a grid

Set $h = L/(nmax + 1)$
    $r = a^2 k/h^2$

Step 3. Initialize numerical solution

Set $t = 0$
  $x_0 = 0$
  $U_0 = (p(0) + f(0))/2$
For $n = 1, 2, \ldots, nmax$ set
  $x_n = x_{n-1} + h$
  $U_n = f(x_n)$
Set $U_{nmax+1} = (q(0) + f(L))/2$
  $x_{nmax} = L$
Output t
For $n = 0, 1, \ldots, nmax + 1$
  Output x_n, U_n

Step 4. Begin time stepping

For $j = 1, 2, \ldots, jmax$
  Do Steps 5–7

Step 5. Define tridiagonal system

Set $t = t + k$
For $n = 1, 2, \ldots, nmax$ set
  $a_n = -r$
  $b_n = (1 + 2r)$
  $c_n = -r$
  $d_n = U_n + k \cdot S(x_n, t)$
Set $d_1 = d_1 + r \cdot p(t)$
  $d_{nmax} = d_{nmax} + r \cdot q(t)$

Step 6. Advance solution one time step

CALL TRIDI $(nmax, a, b, c, d)$
For $n = 1, 2, \ldots, nmax$ set
  $U_n = d_n$
Set $U_0 = p(t)$
  $U_{nmax+1} = q(t)$

Step 7. Output numerical solution

Output t
For $n = 0, 1, \ldots, nmax + 1$
  Output x_n, U_n

ercises 7–10 show how these operations in Algorithm 8.3 can be made more computationally efficient.

In choosing the form of Algorithm 8.3, our intention was to help the reader see how to implement the backward difference method. Once a method is well understood, refining it with computational efficiency in mind is the natural next step. By beginning with a straightforward but somewhat inefficient algorithm, one can often avoid the common mistake of designing a very computationally efficient code that does not implement the desired method. Also, the structure of Algorithm 8.3 generalizes easily to accommodate problems with a diffusivity a^2 that depends on x and t or even on the solution $u(x, t)$ itself.

Algorithm 8.3 can be easily modified so that it implements the Crank–Nicolson method.

ALGORITHM 8.4   Crank–Nicolson Method: Dirichlet Initial-Boundary-Value Problem

Step 1. Document

This algorithm uses the Crank–Nicolson method to approximate the solution to the IBVP (8.3.1)–(8.3.3).

INPUT

diffusivity a^2; endpoint L; functions $S(x, t)$, $f(x)$, $p(t)$, $q(t)$; time step k; number of time steps *jmax;* number of computational nodes *nmax*

OUTPUT

approximation U_n^j to $u(x_n, t_j)$ for $n = 0, 1, \ldots, nmax + 1$ and $j = 0, 1, \ldots,$ *jmax*

The only step in this algorithm that differs from those of Algorithm 8.3 is:

Step 5. Define tridiagonal system

For $n = 1, 2, \ldots, nmax$
Set
$$a_n = -r/2$$
$$b_n = (1 + r)$$
$$c_n = -r/2$$
$$d_n = \frac{r}{2} U_{n-1} + (1 - r)U_n + \frac{r}{2} U_{n+1} + \frac{k}{2} (S(x_n, t) + S(x_n, t + k))$$
Set $d_1 = d_1 + r \cdot p(t + k)/2$
$$d_{nmax} = d_{nmax} + r \cdot q(t + k)/2$$
$$t = t + k$$

TABLE 8.3.3 COMPARISON OF CRANK–NICOLSON[a] AND BACKWARD IN TIME[b] WITH EXACT SOLUTION OF EXAMPLE 8.3.4

$T = 0.50$	Algorithm 8.4	Exact	Algorithm 8.3
$X = 0.0$	0.000000	0.000000	0.000000
$X = 0.1$	0.231190	0.222241	0.259880
$X = 0.2$	0.439750	0.422728	0.494322
$X = 0.3$	0.605264	0.581836	0.680376
$X = 0.4$	0.711530	0.683989	0.799829
$X = 0.5$	0.748147	0.719188	0.840990
$X = 0.6$	0.711530	0.683989	0.799829
$X = 0.7$	0.605264	0.581836	0.680376
$X = 0.8$	0.439750	0.422728	0.494322
$X = 0.9$	0.231190	0.222241	0.259880
$X = 1.0$	0.000000	0.000000	0.000000

[a]Algorithm 8.4.
[b]Algorithm 8.3.

EXAMPLE 8.3.4

Consider again the initial-boundary-value problem (8.3.4) and (8.3.5) whose exact solution is

$$u(x, t) = 100e^{-\pi^2 t}\sin \pi x$$

Choosing $k = 0.005$, $nmax = 9$, and $jmax = 100$ gives the numerical solutions shown in Table 8.3.3.

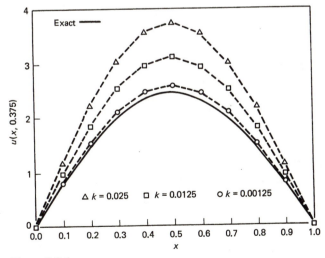

Figure 8.3.4

Figure 8.3.4 shows the result of using Algorithm 8.3 on the example problem (8.3.4)–(8.3.6) with various choices of k. With the choice $h = 0.1$ and $k = 0.025$, the forward difference method was seen to be unstable. Compare the $h = 0.1$, $k = 0.025$ curve of Figure 8.3.4 with Figure 8.3.2. The convergent property of the backward-in-time difference method can be observed in Figure 8.3.4   ■ ■

Algorithms 8.1–8.4 apply to the initial-boundary-value problem (8.3.1)–(8.3.3), which has Dirichlet boundary conditions specified at $x = 0$ and $x = L$. To illustrate how to arrange the computations when derivative boundary conditions are present, consider the following problem with Neumann boundary conditions at $x = 0$ and $x = L$.

$$u_t - a^2 u_{xx} = S(x, t), \qquad 0 < x < L, t > 0 \tag{8.3.9}$$

$$u(x, 0) = f(x), \qquad 0 < x < L \tag{8.3.10}$$

$$u_x(0, t) = p(t), \qquad u_x(L, t) = q(t), \qquad t > 0 \tag{8.3.11}$$

Algorithm 8.5 applies the backward difference method to the initial-boundary-value problem (8.3.9)–(8.3.11).

ALGORITHM 8.5   Backward Difference Method: Neumann Initial-Boundary Value Problem

Step 1. Document

This algorithm uses the backward difference method (8.2.11) to approximate the solution to the initial-boundary-value problem

$$u_t - a^2 u_{xx} = S(x, t), \qquad 0 < x < L, t > 0$$

$$u(x, 0) = f(x), \qquad 0 < x < L$$

$$u_x(0, t) = p(t), \, u_x(L, t) = q(t), \, t > 0$$

INPUT

Real a^2, diffusivity
Real L, endpoint functions $S(x, t)$
Real k, time step
Integer $jmax$, number of time steps
Integer $nmax$, number of computational nodes
Function $S(x, t)$, right side of diffusion equation
Function $f(x)$, initial condition
Functions $p(t)$, $q(t)$, boundary conditions at $x = 0$ and $x = L$

OUTPUT

t, time for $n = 0, 1, \ldots, jmax$ and
U_n^j, approximate solution for $n = 0, 1, \ldots, nmax + 1$

Step 2. Define a grid

Set $h = L/(nmax - 1)$
$r = a^2 k/h^2$

Step 3. Initialize numerical solution

Set $t = 0$
$x_0 = -h$
For $n = 1, 2, \ldots, nmax$ set
$x_n = x_{n-1} + h$
$U_n = f(x)$
Output t
For $n = 1, 2, \ldots, nmax$
Output x_n, U_n

Step 4. Begin time stepping

For $j = 1, 2, \ldots, jmax$
Do steps 5–7

Step 5. Define tridiagonal system

Set $t = t + k$
For $n = 1, 2, \ldots, nmax$ set
$a_n = -r$
$b_n = (1 + 2r)$
$c_n = -r$
$d_n = U_n + k \cdot S(x_n, t)$
Set $d_1 = d_1 - 2hr \cdot p(t)$
$d_{nmax} = d_{nmax} + 2hr \cdot q(t)$
$c_1 = -2r$
$a_{nmax} = -2r$

Step 6. Advance solution one time step

CALL TRIDI($nmax, a, b, c, d$)
For $n = 1, 2, \ldots, nmax$ set
$U_n = d_n$

Step 7. Output numerical solution

Output t
For $n = 1, 2, \ldots, nmax$
Output x_n, U_n

Algorithms 8.1 and 8.3–8.5 can be used as a basis for deriving computational finite-difference schemes for a variety of initial-boundary-value problems. See Exercises 1–3.

Finally, we should note that for a pure initial-value problem

$$u_t - a^2 u_{xx} = S(x, t), \qquad -\infty < x < \infty, \, t > 0$$
$$u(x, 0) = f(x), \qquad -\infty < x < \infty$$

or for an initial-boundary-value problem of the form

$$u_t - a^2 u_{xx} = S(x, t), \qquad x > 0, \, t > 0$$
$$u(x, 0) = f(x), \qquad x > 0$$
$$u(0, t) = p(t), \qquad t > 0$$

only the explicit, forward-in-time method is directly applicable. To apply one of the implicit methods without modification would result in an infinite system of linear equations involving an infinite number of unknowns.

Exercises

1. Modify Algorithm 8.1 so that it applies to the Neumann initial-boundary-value problem

$$u_t - a^2 u_{xx} = S(x, t), \qquad 0 < x < L, \, t > 0$$
$$u(x, 0) = f(x), \qquad 0 < x < L$$
$$u_x(0, t) = p(t), \qquad u_x(L, t) = q(t), \qquad t > 0$$

2. Modify Algorithm 8.1 so that it applies to the mixed initial-boundary-value problem

$$u_t - a^2 u_{xx} = S(x, t), \qquad 0 < x < L, \, t > 0$$
$$u(x, 0) = f(x), \qquad 0 < x < L$$
$$u_x(0, t) + u(0, t) = p(t), \qquad u(L, t) = q(t), \qquad t > 0$$

3. Modify Algorithm 8.3 so that it applies to the initial-boundary-value problem

$$u_t - a^2 u_{xx} + \beta u_x + \gamma u = S(x, t), \qquad 0 < x < L, \, t > 0$$
$$u(x, 0) = f(x), \qquad 0 < x < L$$
$$u(0, t) = p(t), \qquad u(L, t) = q(t), \qquad t > 0$$

and has local truncation error $O(k + h^2)$.

4. Use Algorithm 8.1 to approximate the solution of the initial-boundary-value problem

$$u_t - u_{xx} = -2e^{x-t}, \qquad 0 < x < 1, \, t > 0$$
$$u(x, 0) = e^x, \qquad 0 < x < 1$$
$$u(0, t) = e^{-t}, \qquad u(1, t) = e^{1-t}, \qquad t > 0$$

(a) Choose $k = 0.0025$ and $nmax = 9$ (so $h = 0.1$) and compare the numerical and exact solutions, $u(x, t) = e^{x-t}$, at time $t = 0.5$.

(b) Choose $k = 0.01$ and $nmax = 9$ and explain the numerical results.

5. Use Algorithm 8.3 or 8.4 to approximate the solution of the initial-boundary-value problem

$$u_t - u_{xx} = -2e^{x-t}, \qquad 0 < x < 1, \, t > 0$$
$$u(x, 0) = e^x, \qquad 0 < x < 1$$
$$u(0, t) = e^{-t}, \qquad u(1, t) = e^{1-t}, \qquad t > 0$$

(a) Choose $k = 0.0025$ and *nmax* $= 9$ (so $h = 0.1$) and compare the numerical and exact solutions, $u(x, t) = e^{x-t}$, at time $t = 0.5$.

(b) Choose $k = 0.01$ and *nmax* $= 9$ and compare the numerical and exact solutions at time $t = 0.5$.

(c) Choose $k = 0.01$ and *nmax* $= 99$ (so $h = 0.01$) and compare the numerical and exact solutions at time $t = 0.5$ at the positions $x = 0.1, 0.2, \ldots, 0.9$.

6. Use Algorithm 8.5 to approximate the solution of the initial-boundary-value problem

$$u_t - u_{xx} = -2e^{x-t}, \qquad 0 < x < 1, t > 0$$
$$u(x, 0) = e^x, \qquad\qquad 0 < x < 1$$
$$u_x(0, t) = e^{-t}, \qquad u_x(1, t) = e^{1-t}, \qquad t > 0$$

(a) Choose $k = 0.0025$ and *nmax* $= 9$ and compare the numerical and exact solutions, $u(x, t) = e^{x-t}$, at time $t = 0.5$.

(b) Choose $k = 0.01$ and *nmax* $= 9$ and compare the numerical and exact solutions at time $t = 0.5$.

(c) Choose $k = 0.01$ and *nmax* $= 99$ and compare the numerical and exact solutions at time $t = 0.5$.

7. Let $\mathbf{A} = [\mathbf{a}; \mathbf{b}; \mathbf{c}]$ be a strictly diagonally dominant tridiagonal $N \times N$ matrix $\mathcal{L} = [\boldsymbol{\alpha}; \mathbf{1}; \mathbf{0}]$ be a lower triangular matrix, and $\mathcal{U} = [\mathbf{0}; \boldsymbol{\beta}; \boldsymbol{\gamma}]$.

(a) Show that $\mathbf{A} = \mathcal{L}\mathcal{U}$ implies $\boldsymbol{\gamma} = \mathbf{c}$ and $\beta_1 = b_1$ and

$$\alpha_n = a_n / \beta_{n-1}, \qquad \beta_n = b_n - \alpha_n c_{n-1}, \qquad n = 2, 3, \ldots, N$$

(b) Show that by overwriting the arrays $\mathbf{a}$ and $\mathbf{b}$ with $\boldsymbol{\alpha}$ and $\boldsymbol{\beta}$ the $\mathcal{L}\mathcal{U}$ factorization of $\mathbf{A}$ can be stored using only three arrays.

8. Use the result of Exercise 7 to develop the following algorithm for factoring a diagonally dominant, tridiagonal matrix into a product of a lower triangular matrix $\mathcal{L}$ times an upper triangular matrix $\mathcal{U}$.

ALGORITHM 8.6 $\mathcal{L}\mathcal{U}$ Factorization of Tridiagonal Matrix

Step 1. Document

Given a diagonally dominant, tridiagonal matrix $\mathbf{A}$, this algorithm produces the factorization $\mathbf{A} = \mathcal{L}\mathcal{U}$, where $\mathcal{L}$ is lower triangular and $\mathcal{U}$ is upper triangular.

INPUT

Order of matrix $\mathbf{A}$ *nmax;* subdiagonal $\mathbf{a}$; diagonal $\mathbf{b}$; and superdiagonal $\mathbf{c}$ of $\mathbf{A} = [\mathbf{a}; \mathbf{b}; \mathbf{c}]$

OUTPUT

$\mathcal{L} = [\mathbf{a}; \mathbf{1}; \mathbf{0}]$ and $\mathcal{U} = [\mathbf{0}; \mathbf{b}; \mathbf{c}]$

Step 2. Overwrite arrays a and b

For $n = 2, 3, \ldots,$ *nmax* set

$$a_n = a_n/b_{n-1}$$
$$b_n = b_n - a_n \cdot c_{n-1}$$

Step 3. Output $\mathcal{L} = [a; 1; 0]$ and $\mathcal{U} = [0; b; c]$

---

9. Suppose that an $N \times N$ tridiagonal matrix $\mathbf{A}$ has the factorization

$$\mathbf{A} = [a; 1; 0][0; b; c] = \mathcal{L}\mathcal{U}$$

and consider the linear system $\mathbf{AV} = \mathbf{d}$, or $\mathcal{L}\mathcal{U}\mathbf{V} = \mathbf{d}$.
   (a) If we define $\mathbf{W} = \mathcal{U}\mathbf{V}$, show that the components of $\mathbf{W}$ can be determined recursively by $W_1 = d_1$, $W_n = d_n - a_n W_{n-1}$, $n = 2, 3, \ldots, N$.
   (b) Show that the components of $\mathbf{V}$ can be determined recursively by $V_N = W_N/b_N$, $V_n = (W_n - c_n V_{n+1})/b_n$, $n = N - 1, N - 2, \ldots, 1$.
   (c) Write an algorithm that accepts as input the $\mathcal{L}\mathcal{U}$ factorization of $\mathbf{A}$ and the array $\mathbf{d}$ and, on output, produces the solution $\mathbf{V}$ to the system $\mathbf{AV} = \mathbf{d}$.

10. Use the results of Exercises 8 and 9 to modify steps 4–6 of Algorithm 8.3 so that there is no need to redefine the tridiagonal system for each time step.

11. Consider the initial-boundary-value problem

$$u_t - u_{xx} = -2, \qquad 0 < x < 1, t > 0$$
$$u(x, 0) = x^2, \qquad 0 < x < 1$$
$$u(0, t) = t, \qquad u(1, t) = 1 + t, \qquad t > 0$$

whose exact solution is $u(x, t) = x^2 + t$.
   (a) Show that if the forward difference method, Algorithm 8.1, is used to approximate the solution of this problem, then the local truncation error is zero.
   (b) Use a stable value of $r = k/h^2$, $r \le \frac{1}{2}$, in Algorithm 8.1 to approximate the solution of this problem. If the algorithm is working properly, then the only discrepancies should be due to rounding. Such a test problem, with zero local truncation error, is a good way to test the validity of computer code.
   (c) Use an unstable value of r, $r > \frac{1}{2}$, and Algorithm 8.1 to approximate the solution of this problem and show that if the computation is continued long enough, then rounding errors grow to dominate the numerical solution.

8.4 Fourier's Method for Difference Equations

In Section 3.2 we saw how to use the method of separation of variables and Fourier series to solve the initial-value problem

$$u_t - u_{xx} = 0, \qquad 0 < x < 1, t > 0 \qquad (8.4.1)$$
$$u(x, 0) = f(x), \qquad 0 < x < 1 \qquad (8.4.2)$$
$$u(0, t) = 0 = u(1, t), \qquad t > 0 \qquad (8.4.3)$$

In Sections 8.2 and 8.3 we used finite-difference methods to approximate the solution of this problem. At first, it might appear that the two approaches, the Fourier series approach and the numerical finite-difference approach, are quite different.

Our aim in this section is to illustrate that the Fourier series method and the finite-difference method, when applied to problem (8.4.1)–(8.4.3), have a number of common features. By examining the similarities in the Fourier series and the finite-difference methods, we shall gain a better understanding of each method. As an added bonus, we shall discover an eigenvector expansion method that will allow us to determine the exact solution to a number of finite-difference equations.

Exact Solution by Fourier Series

For convenient reference we briefly review the solution of the continuous (nondiscretized) initial-boundary-value problem (8.4.1)–(8.4.3) by the Fourier series method. See also Example 3.2.1.

Step 1C Separate Variables.   We begin by looking for nontrivial solutions to (8.4.1) in the separated form $u(x, t) = v(x)w(t)$. Then $u_t = vw'$ and $u_{xx} = v''w$, so $vw' - v''w = 0$, or

$$\frac{w'}{w} = \frac{v''}{v} = \text{const} \tag{8.4.4}$$

Step 2C Impose Boundary Conditions.   From $v \neq 0$ and $v(0) = 0 = v(1)$ it follows that the separation constant in (8.4.4) must be negative, say, $-\gamma^2$, for $v(x)$ to oscillate. This argument leads to the eigenvalue problem

$$v'' = -\gamma^2 v, \qquad 0 < x < 1; \qquad v(0) = 0 = v(1) \tag{8.4.5}$$

which is seen to have eigenvalues γ_n^2 and eigenfunctions $v_n(x)$ given by

$$\gamma_n^2 = (n\pi)^2, \qquad v_n(x) = \sin n\pi x, \qquad n = 1, 2, \ldots \tag{8.4.6}$$

Step 3C Determine Transient Behavior.   From steps 1C and 2C, it follows that $w'/w = -(n\pi)^2$, $n = 1, 2, \ldots$. Thus, corresponding to each n value we have a solution w given by

$$w_n(t) = b_n e^{-(n\pi)^2 t}, \qquad n = 1, 2, \ldots \tag{8.4.7}$$

where b_n is a constant.

Step 4C Superposition.   We form a linear combination of the solutions $v_n(x)w_n(t)$ of (8.4.1) to express u in the form

$$u(x, t) = \sum_{n=1}^{\infty} b_n e^{-(n\pi)^2 t} \sin n\pi x \tag{8.4.8}$$

Step 5C Impose Initial Condition.   Since $u(x, 0) = f(x)$, it follows from (8.4.8) that

$$f(x) = \sum_{n=1}^{\infty} b_n \sin n\pi x, \qquad 0 < x < 1 \tag{8.4.9}$$

In the inner product

$$(\phi, \psi) = \int_0^1 \phi(x)\psi(x) \, dx$$

the family of eigenfunctions $v_n(x) = \sin n\pi x$, $n = 1, 2, \ldots$, is orthogonal. Using this orthogonality, the b_n in (8.4.9) are easily seen to be given by

$$b_n = \frac{(f(x), \sin n\pi x)}{(\sin n\pi x, \sin n\pi x)}$$

or

$$b_n = 2 \int_0^1 f(x)\sin n\pi x \, dx \tag{8.4.10}$$

This completes the solution of the initial-value problem (8.4.1)–(8.4.3) by the method of separation of variables and Fourier series. Collecting the results of steps 1C–5C, we find the solution is

$$u(x, t) = \sum_{n=1}^{\infty} b_n e^{-(n\pi)^2 t}\sin n\pi x, \qquad b_n = 2 \int_0^1 f(x)\sin n\pi x \, dx$$

Approximate Solution by Finite Differences

We now consider a finite-difference approximation to the continuous initial-boundary-value problem (8.4.1)–(8.4.3). We could discretize this initial-boundary-value problem using any of the methods of Section 8.2. For purposes of discussion we select the forward difference method (8.2.9).

Let a uniform x grid be defined by

$$0 = x_0 < x_1 < x_2 \cdots < x_N < x_{N+1} = 1, \qquad x_{n+1} - x_n = h = 1/(N + 1)$$

Also, let

$$0 = t_0 < t_1 < t_2 < \cdots, \qquad t_{j+1} - t_j = k$$

denote the uniformly spaced, discrete time values at which the solution of (8.4.1)–(8.4.3) is to be approximated. Using the forward difference method to form finite-difference equations for (8.4.1)–(8.4.3) yields

$$\frac{U_i^{j+1} - U_i^j}{k} - \frac{U_{i-1}^j - 2U_i^j + U_{i+1}^j}{h^2} = 0, \qquad \begin{array}{l} i = 1, 2, \ldots, N, \\ j = 0, 1, 2, \ldots \end{array} \tag{8.4.11}$$

$$U_i^0 = f_i, \qquad i = 1, 2, \ldots, N \tag{8.4.12}$$

$$U_0^j = 0 = U_{N+1}^j, \qquad j = 1, 2, \ldots \tag{8.4.13}$$

We present a five-step method for obtaining the exact solution of the finite difference system (8.4.11)–(8.4.13). The solution to the discrete problem, given in what follows in steps 1D–5D, is based on separation of indices, superposition, and orthogonality. Steps 1D–5D are direct analogues of the corresponding steps 1C–5C used in solving the continuous problem (8.4.1)–(8.4.3). Before we begin the solution of the discrete problem, we state a result that will be required in step 2D.

Theorem 8.4.1. The continuous eigenvalue problem

$$-v''(x) = \gamma v(x), \quad 0 < x < 1$$
$$v(0) = 0 = v(1)$$

has eigenvalues $\gamma_n = (n\pi)^2$ and corresponding eigenfunctions $v_n(x) = \sin n\pi x, n = 1, 2, \ldots$. The discretization of this problem,

$$-V_{n-1} + 2V_n - V_{n+1} = \lambda V_n, \quad n = 1, 2, \ldots, N$$
$$V_0 = 0 = V_{N+1}$$

on the uniform grid $x_i = ih$, $h = 1/(N + 1)$, can be written in matrix notation as

$$
\begin{bmatrix}
2 & -1 & & & & \\
-1 & 2 & -1 & & & \\
& -1 & 2 & -1 & & \\
& & \ddots & \ddots & \ddots & \\
& & & -1 & 2 & -1 \\
& & & & -1 & 2
\end{bmatrix}
\begin{bmatrix}
V_1 \\ V_2 \\ V_3 \\ \vdots \\ V_{N-1} \\ V_N
\end{bmatrix}
= \lambda
\begin{bmatrix}
V_1 \\ V_2 \\ V_3 \\ \vdots \\ V_{N-1} \\ V_N
\end{bmatrix}
\quad \text{or} \quad \mathbf{F}_1\mathbf{V} = \lambda\mathbf{V}
$$

The eigenvalues of the matrix $\mathbf{F}_1$ are

$$\lambda_n = 2 - 2\cos\frac{n\pi}{N+1}, \quad n = 1, 2, \ldots, N$$

and the corresponding eigenvectors are

$$
\mathbf{V}_n =
\begin{bmatrix}
\sin n\pi x_1 \\
\sin n\pi x_2 \\
\vdots \\
\sin n\pi x_N
\end{bmatrix}
\equiv \sin n\pi\mathbf{x}, \quad n = 1, 2, \ldots, N
$$

where $x_i = i/(N + 1)$, $i = 1, 2, \ldots, N$. Also, $\mathbf{V}_n \cdot \mathbf{V}_m = 0$ for $n \neq m$ and $\mathbf{V}_n \cdot \mathbf{V}_n = (N + 1)/2$.

The only part of Theorem 8.4.1 that requires proof are the expressions for the eigenvalues and eigenvectors of the matrix $\mathbf{F}_1$. See Exercise 8.

We now outline an eigenvector expansion method for solving the finite difference system (8.4.11)–(8.4.13).

Step 1D Separate Indices. We begin by looking for nontrivial solutions to (8.4.11) in the form $U_i^j = V_i W^j$. Then (8.4.11) yields

$$V_i \frac{W^{j+1} - W^j}{k} = \frac{V_{i-1} - 2V_i + V_{i+1}}{h^2} W^j$$

or after separating indices

$$\frac{h^2}{k} \frac{W^{j+1} - W^j}{W^j} = \frac{V_{i-1} - 2V_i + V_{i+1}}{V_i} = \text{const} \equiv -\lambda \qquad (8.4.14)$$

Compare (8.4.14) with (8.4.4).

Step 2D Impose Boundary Conditions. From $V_0 = 0 = V_{N+1}$ and (8.4.14), we obtain an eigenvalue problem of the form

$$-V_{i-1} + 2V_i - V_{i+1} = \lambda V_i, \qquad i = 1, 2, \ldots, N; \qquad V_0 = 0 = V_{N+1} \qquad (8.4.15)$$

Compare (8.4.15) with (8.4.5). Using the matrix $\mathbf{F}_1$, the eigenvalue problem (8.4.15) can be written as

$$\mathbf{F}_1 \mathbf{V} = \lambda \mathbf{V}$$

so, from Theorem 8.4.1 the eigenvalues and corresponding eigenvectors of (8.4.15) are

$$\lambda_n = 2 - 2 \cos \frac{n\pi}{N+1}, \qquad n = 1, 2, \ldots, N$$

and

$$\mathbf{V}_n = \sin n\pi x, \qquad n = 1, 2, \ldots, N \qquad (8.4.16)$$

Compare (8.4.16) with (8.4.6).

Step 3D Determine Transient Behavior. From (8.4.14) it follows that

$$\frac{h^2}{k} \frac{W_n^{j+1} - W_n^j}{W_n^j} = -\lambda_n$$

which yields the recursion formula

$$W_n^{j+1} = (1 - r\lambda_n)W_n^j, \qquad r = k/h^2, \qquad n = 1, 2, \ldots, N$$

where W_n^j denotes the j part, or time-dependent part, of the separated solution that is to be paired with λ_n and $\mathbf{V}_n$. If W_n^0 is known, then all W_n^j can be found from the preceding recursion formula to be

$$W_n^j = (1 - r\lambda_n)^j W_n^0, \qquad j = 1, 2, \ldots \qquad (8.4.17)$$

The correspondence between (8.4.17) and (8.4.7) will be examined after we complete the solution of the finite-difference system.

Step 4D Superposition. We form a linear combination of the solutions $W_n^j \mathbf{V}_n$ to express the solution of (8.4.11)–(8.4.13) in the form

$$U_i^j = \sum_{n=1}^N W_n^j \sin n\pi x_i, \qquad \begin{matrix} i = 1, 2, \ldots, N, \\ j = 1, 2, \ldots \end{matrix} \qquad (8.4.18)$$

Compare (8.4.18) with (8.4.8). If $\mathbf{U}^j$ denotes the column vector $[U_1^j, U_2^j, \ldots, U_N^j]^T$, then (8.4.18) can be expressed in the convenient vector form

$$\mathbf{U}^j = \sum_{n=1}^N W_n^j \sin n\pi\mathbf{x} \quad \text{or} \quad \mathbf{U}^j = \sum_{n=1}^N W_n^0 (1 - r\lambda_n)^j \sin n\pi\mathbf{x}$$

Step 5D Impose Initial Condition. With $\mathbf{f} = [f_1, f_2, \ldots, f_N]^T$, the initial condition (8.4.12) implies

$$\mathbf{f} = \sum_{n=1}^N W_n^0 \sin n\pi\mathbf{x} \qquad (8.4.19)$$

Compare (8.4.19) with (8.4.9). Since the matrix $\mathbf{F}_1$ is symmetric and the eigenvalues λ_n are distinct, it follows that the corresponding eigenvectors $\mathbf{V}_n$ are mutually orthogonal. If we form the inner (dot) product of each side of (8.4.19) with the vector $\sin m\pi\mathbf{x}$ and use the orthogonality relationship

$$\sin n\pi\mathbf{x} \cdot \sin m\pi\mathbf{x} = 0, \qquad m \neq n$$

we find

$$W_n^0 = \frac{\mathbf{f} \cdot \sin n\pi\mathbf{x}}{\sin n\pi\mathbf{x} \cdot \sin n\pi\mathbf{x}}, \qquad n = 1, 2, \ldots, N \qquad (8.4.20)$$

Compare (8.4.20) with (8.4.10). This completes the solution of the discrete problem (8.4.11)–(8.4.13).

The corresponding steps 1, 2, 4, and 5 of the continuous and discrete problems are very similar. The comparison of step 3C and step 3D requires some further study. From step 3C we see

$$\frac{W_n(t + k)}{W_n(t)} = e^{-(n\pi)^2 k} = 1 - (n\pi)^2 k + \cdots \qquad (8.4.21)$$

where the last equality follows by a Maclaurin expansion of the exponential. From step 3D

$$\frac{W_n^{j+1}}{W_n^j} = 1 - r\lambda_n = 1 - \frac{k}{h^2}\lambda_n$$

Since $\lambda_n = 2 - 2\cos(n\pi/N + 1) = (n\pi/N + 1)^2 + \cdots$ and $h = 1/(N + 1)$, it follows that

$$\frac{W_n^{j+1}}{W_n^j} = 1 - (n\pi)^2 k + \cdots \qquad (8.4.22)$$

Compare (8.4.22) with (8.4.21).

EXAMPLE 8.4.1 ———————————————————————————————————

To illustrate how to solve a system of difference equations using steps 1D–5D, consider the initial-boundary-value problem

$$u_t - u_{xx} = 0, \qquad 0 < x < 1, t > 0$$
$$u(x, 0) = x^2, \qquad 0 < x < 1$$
$$u(0, t) = 0 = u(1, t), \qquad t > 0$$

We choose the grid $\{x_0, x_1, x_2, x_3, x_4\} = \{0, \frac{1}{4}, \frac{1}{2}, \frac{3}{4}, 1\}$ so $h = \frac{1}{4}$. The choice of the time step k will be left open for now. The finite-difference system obtained from the forward difference method is given by (8.4.11)–(8.4.13) with $N = 3$ and $f_i = x_i^2$; $f_1 = \frac{1}{16}$, $f_2 = \frac{1}{4}$, and $f_3 = \frac{9}{16}$. For this example, with three interior grid points, we have, from (8.4.16),

$$\sin \pi x = \begin{bmatrix} \dfrac{\sqrt{2}}{2} \\ 1 \\ \dfrac{\sqrt{2}}{2} \end{bmatrix}, \qquad \sin 2\pi x = \begin{bmatrix} 1 \\ 0 \\ -1 \end{bmatrix}, \qquad \sin 3\pi x = \begin{bmatrix} \dfrac{\sqrt{2}}{2} \\ -1 \\ \dfrac{\sqrt{2}}{2} \end{bmatrix}$$

From (8.4.19) and (8.4.20) we find

$$W_1^0 = \frac{\sqrt{2}/32 + 1/4 + 9\sqrt{2}/32}{1/2 + 1 + 1/2} = \frac{4 + 5\sqrt{2}}{32}$$

$$W_2^0 = \frac{1/16 + 0 - 9/16}{1 + 0 + 1} = -\frac{1}{4}$$

$$W_3^0 = \frac{\sqrt{2}/32 - 1/4 + 9\sqrt{2}/32}{1/2 + 1 + 1/2} = \frac{-4 + 5\sqrt{2}}{32}$$

For $N = 3$ we have, from (8.4.16),

$$\lambda_1 = 2 - \sqrt{2}, \qquad \lambda_2 = 2, \qquad \lambda_3 = 2 + \sqrt{2}$$

We now have the ingredients necessary to write the solution of the difference system:

$$\mathbf{U}^j = W_1^0 (1 - r\lambda_1)^j \sin \pi x + W_2^0 (1 - r\lambda_2)^j \sin 2\pi x + W_3^0 (1 - r\lambda_3)^j \sin 3\pi x$$

where $r = k/h^2 = 16k$.

The choice of k has been left open until now. As we shall see, k cannot be chosen aribtrarily. Since the solution to the continuous problem is bounded on $0 < x < 1$, $t > 0$, we expect the solution to the discrete problem to be bounded for $i = 1, 2, 3$, $j = 1, 2, \dots$. A necessary and sufficient condition for the boundedness of the example solution is

$$|1 - r\lambda_n| \leq 1, \qquad n = 1, 2, 3$$

Since $\lambda_n > 0$, $n = 1, 2, 3$, and $r = 16k$, this boundedness condition amounts to requiring $r\lambda_n \leq 2$, or

$$k \le 2/(16\lambda_n)$$

which holds for $n = 1, 2, 3$ provided

$$k \le 2/(16 \cdot \lambda_3) \approx 0.035$$

This restriction on k is called a stability condition for the difference method. The notion of stability will be discussed in more detail in Section 8.5. ■ ■

Derivative Boundary Conditions

Example 8.4.1 shows how to use Fourier's method to solve the finite-difference equations for an initial-Dirichlet boundary-value problem. We now apply the Fourier method to difference equations that arise from an initial-boundary-value problem that involves a derivative boundary condition. The following result from linear algebra will be required.

Theorem 8.4.2. Let $\mathbf{F}$ be an $N \times N$ real matrix and let $\mathbf{F}^T$ denote the transpose of $\mathbf{F}$.

(i) $\mathbf{F}$ and $\mathbf{F}^T$ have the same eigenvalues.
(ii) If $\mathbf{F}$ and $\mathbf{F}^T$ have N distinct real eigenvalues $\lambda_1, \lambda_2, \ldots, \lambda_N$ with corresponding eigenvectors given by

$$\mathbf{F}\mathbf{V}_n = \lambda_n\mathbf{V}_n, \qquad \mathbf{F}^T\mathbf{W}_n = \lambda_n\mathbf{W}_n, \qquad n = 1, 2, \ldots, N$$

then

$$\mathbf{V}_n \cdot \mathbf{W}_m = 0 \quad \text{for } n \ne m \quad \text{and} \quad \mathbf{V}_n \cdot \mathbf{W}_n \ne 0$$

(iii) If $\mathbf{U}$ is any real N vector, then $\mathbf{U}$ has the eigenvector expansion

$$\mathbf{U} = \sum_{n=1}^{N} c_n\mathbf{V}_n, \qquad c_n = \frac{\mathbf{U} \cdot \mathbf{W}_n}{\mathbf{V}_n \cdot \mathbf{W}_n}$$

For the case of Dirichlet boundary conditions, Theorem 8.4.1 provides information about the eigenvectors of the finite-difference system. In Theorems 8.4.3–8.4.6, we present similar results for some eigenvalue problems involving derivative boundary conditions.

Theorem 8.4.3. The continuous eigenvalue problem

$$-v''(x) = \gamma v(x), \qquad 0 < x < 1$$
$$v'(0) = 0 = v'(1)$$

has eigenvalues $\gamma_n = [(n - 1)\pi]^2$ and corresponding eigenfunctions $v_n(x) = \cos(n - 1)\pi x, n = 1, 2, \ldots$ The discretization of this problem,

$$-V_{n-1} + 2V_n - V_{n+1} = \lambda V_n, \qquad n = 1, 2, \ldots, N$$
$$V_2 - V_0 = 0 = V_{N+1} - V_{N-1}$$

on the uniform grid $x_i = (i - 1)h$, $h = 1/(N - 1)$, can be written in matrix notation as

$$
\begin{bmatrix}
2 & -2 \\
-1 & 2 & -1 \\
 & -1 & 2 & -1 \\
 & & & \ddots \\
 & & & -1 & 2 & -1 \\
 & & & & -2 & 2
\end{bmatrix}
\begin{bmatrix}
V_1 \\ V_2 \\ V_3 \\ \vdots \\ V_{N-1} \\ V_N
\end{bmatrix}
= \lambda
\begin{bmatrix}
V_1 \\ V_2 \\ V_3 \\ \vdots \\ V_{N-1} \\ V_N
\end{bmatrix}
\quad \text{or} \quad \mathbf{F}_2 \mathbf{V} = \lambda \mathbf{V}
$$

The eigenvalues of the matrix $\mathbf{F}_2$ are

$$
\lambda_n = 2 - 2 \cos \frac{(n - 1)\pi}{N - 1}, \quad n = 1, 2, \ldots, N
$$

and the corresponding eigenvectors are

$$
\mathbf{V}_n =
\begin{bmatrix}
\cos(n - 1)\pi x_1 \\
\cos(n - 1)\pi x_2 \\
\vdots \\
\cos(n - 1)\pi x_N
\end{bmatrix}
\equiv \cos(\mathbf{n} - 1)\pi\mathbf{x}, \quad n = 1, 2, \ldots, N
$$

where $x_i = (i - 1)/(N - 1)$, $i = 1, 2, \ldots, N$. The eigenvectors of the matrix $\mathbf{F}_2^T$ are given by

$$
\mathbf{W}_n =
\begin{bmatrix}
\cos(n - 1)\pi x_1 \\
2 \cos(n - 1)\pi x_2 \\
2 \cos(n - 1)\pi x_3 \\
\vdots \\
2 \cos(n - 1)\pi x_i \\
\vdots \\
2 \cos(n - 1)\pi x_{N-1} \\
\cos(n - 1)\pi x_N
\end{bmatrix}, \quad n = 1, 2, \ldots, N
$$

Also, $\mathbf{V}_n \cdot \mathbf{W}_m = 0$ for $n \neq m$ and $\mathbf{V}_1 \cdot \mathbf{W}_1 = 2(N - 1) = \mathbf{V}_N \cdot \mathbf{W}_N$ and $\mathbf{V}_n \cdot \mathbf{W}_n = N - 1$ for $n = 2, 3, \ldots, N - 1$.

Theorem 8.4.4. The continuous eigenvalue problem

$$
-v''(x) = \gamma v(x), \quad 0 < x < 1
$$
$$
v'(0) = 0 = v(1)
$$

has eigenvalues

$$
\gamma_n = \left[\frac{(2n - 1)\pi}{2} \right]^2
$$

and corresponding eigenfunctions

$$
v_n(x) = \cos \frac{(2n - 1)\pi x}{2}, \quad n = 1, 2, \ldots
$$

The discretization of this problem,

$$-V_{n-1} + 2V_n - V_{n+1} = \lambda V_n, \qquad n = 1, 2, \ldots, N$$
$$V_2 - V_0 = 0 = V_{N+1}$$

on the uniform grid $x_i = (i - 1)h$, $h = 1/N$, can be written in matrix notation as

$$\begin{bmatrix} 2 & -2 & & & & \\ -1 & 2 & -1 & & & \\ & -1 & 2 & -1 & & \\ & & & \ddots & & \\ & & & -1 & 2 & -1 \\ & & & & -1 & 2 \end{bmatrix} \begin{bmatrix} V_1 \\ V_2 \\ V_3 \\ \vdots \\ V_{N-1} \\ V_N \end{bmatrix} = \lambda \begin{bmatrix} V_1 \\ V_2 \\ V_3 \\ \vdots \\ V_{N-1} \\ V_N \end{bmatrix} \qquad \text{or} \quad \mathbf{F}_3 \mathbf{V} = \lambda \mathbf{V}$$

The eigenvalues of the matrix $\mathbf{F}_3$ are

$$\lambda_n = 2 - 2\cos\frac{(2n-1)\pi}{2N}, \qquad n = 1, 2, \ldots, N$$

and the corresponding eigenvectors are

$$\mathbf{V}_n = \begin{bmatrix} \cos[(2n-1)\pi x_1/2] \\ \cos[(2n-1)\pi x_2/2] \\ \vdots \\ \cos[(2n-1)\pi x_N/2] \end{bmatrix} \equiv \cos\frac{(2n-1)\pi x}{2}, \qquad n = 1, 2, \ldots, N$$

where $x_i = (i - 1)/N$, $i = 1, 2, \ldots, N$. The eigenvectors of the matrix $\mathbf{F}_3^T$ are given by

$$\mathbf{W}_n = \begin{bmatrix} \cos[(2n-1)\pi x_1/2] \\ 2\cos[(2n-1)\pi x_2/2] \\ 2\cos[(2n-1)\pi x_3/2] \\ \vdots \\ 2\cos[(2n-1)\pi x_i/2] \\ \vdots \\ 2\cos[(2n-1)\pi x_{N-1}/2] \\ 2\cos[(2n-1)\pi x_N/2] \end{bmatrix}, \qquad n = 1, 2, \ldots, N$$

Also, $\mathbf{V}_n \cdot \mathbf{W}_m = 0$ for $n \neq m$ and $\mathbf{V}_n \cdot \mathbf{W}_n = N$.

The proofs of Theorems 8.4.4 and 8.4.5 require only elementary trigonometric identities. They are considered in the exercises.

Consider a two-level, homogeneous finite-difference system of the form

$$A(\mathbf{F})\mathbf{U}^{j+1} = B(\mathbf{F})\mathbf{U}^j \qquad (8.4.23)$$
$$\mathbf{U}^0 = \mathbf{f} \qquad (8.4.24)$$

where $A(\mathbf{F})$ and $B(\mathbf{F})$ are $N \times N$ matrices that are polynomial functions of the $N \times N$ matrix $\mathbf{F}$. We assume that $\mathbf{F}$ satisfies the hypotheses of Theorem 8.4.2.

EXAMPLE 8.4.2

Given the initial-boundary-value problem

$$u_t - a^2 u_{xx} = 0, \qquad 0 < x < 1, t > 0$$
$$u(x, 0) = f(x), \qquad 0 < x < 1$$
$$u_x(0, t) = 0 = u_x(1, t), \qquad t > 0$$

if the Crank–Nicolson method is used to discretize on the grid $x_i = ih$, $h = 1/(N - 1)$, then the resulting system of finite-difference equations can be written in the form

$$\left(I + \frac{r}{2} F_2\right) U^{j+1} = \left(I - \frac{r}{2} F_2\right) U^j, \qquad r = \frac{a^2 k}{h^2}$$

So we see that the $N \times N$ matrices $A(\mathbf{F}_2) = I + (r/2)\mathbf{F}_2$ and $B(\mathbf{F}_2) = I - (r/2)\mathbf{F}_2$ are (linear) polynomial functions of the $N \times N$ matrix $\mathbf{F}_2$. Here I denotes the $N \times N$ identity matrix. ∎∎

We look for a solution of the difference system (8.4.23)–(8.4.24) in the form of an eigenvector expansion,

$$\mathbf{U}^j = \sum_{n=0}^{N} c_n^j \mathbf{V}_n \tag{8.4.25}$$

where the $\mathbf{V}_n$ are eigenvectors of the matrix $\mathbf{F}$. The coefficients c_n^j are given by Theorem 8.4.2 as

$$c_n^j = \frac{\mathbf{U}^j \cdot \mathbf{W}_n}{\mathbf{V}_n \cdot \mathbf{W}_n}$$

where $\mathbf{W}_n$ is an eigenvector of $\mathbf{F}^T$. If we put the expression (8.4.25) into the difference equation (8.4.23), we have

$$A(\mathbf{F}) \sum_{n=0}^{N} c_n^{j+1} \mathbf{V}_n = B(\mathbf{F}) \sum_{n=0}^{N} c_n^j \mathbf{V}_n$$

Using the linearity property of matrix multiplication and the result that since A and B are polynomials in $\mathbf{F}$, $A(\mathbf{F})\mathbf{V}_n = A(\lambda_n)\mathbf{V}_n$ and $B(\mathbf{F})\mathbf{V}_n = B(\lambda_n)\mathbf{V}_n$, we have

$$\sum_{n=0}^{N} c_n^{j+1} A(\lambda_n)\mathbf{V}_n = \sum_{n=0}^{N} c_n^j B(\lambda_n)\mathbf{V}_n$$

Now, it follows from the linear independence of the eigenvectors $\mathbf{V}_n$ that

$$\frac{c_n^{j+1}}{c_n^j} = \frac{B(\lambda_n)}{A(\lambda_n)}$$

from which we conclude that

$$c_n^j = \left[\frac{B(\lambda_n)}{A(\lambda_n)}\right]^j c_n^0$$

Since $U^0 = f$ and f has the eigenvector expansion

$$f = \sum_{m=1}^{N} c_n^0 V_n, \qquad c_n^0 = \frac{f \cdot W_n}{V_n \cdot W_n}$$

it follows that the solution of the difference system (8.4.23) and (8.4.24) is given by

$$U^j = \sum_{n=0}^{N} \left[\frac{B(\lambda_n)}{A(\lambda_n)}\right]^j c_n^0 V_n, \qquad c_n^0 = \frac{f \cdot W_n}{V_n \cdot W_n}$$

EXAMPLE 8.4.3

Consider the initial-boundary-value problem

$$u_t - u_{xx} = 0, \qquad 0 < x < 1, t > 0$$
$$u(x, 0) = 1 - x^2, \qquad 0 < x < 1$$
$$u_x(0, t) = 0 = u(1, t)$$

If we use the backward difference method and a centered difference to approximate the derivative boundary condition, then on the grid

$$
\begin{array}{cccccc}
0 & \frac{1}{4} & \frac{1}{2} & \frac{3}{4} & 1 \\
\cdots \bullet & \!\!-\!\!\bullet\!\!-\!\! & \bullet & \!\!-\!\!\bullet\!\!-\!\! & \bullet & \cdots \\
x_0 & x_1 & x_2 & x_3 & x_4 & x_5
\end{array}
$$

the resulting difference system is

$$
\begin{bmatrix}
1 + 2r & -2r & 0 & 0 \\
-r & 1 + 2r & -r & 0 \\
0 & -r & 1 + 2r & -r \\
0 & 0 & -r & 1 + 2r
\end{bmatrix}
\begin{bmatrix}
U_1^{j+1} \\ U_2^{j+1} \\ U_3^{j+1} \\ U_4^{j+1}
\end{bmatrix}
=
\begin{bmatrix}
U_1^j \\ U_2^j \\ U_3^j \\ U_4^j
\end{bmatrix}
$$

$$
\begin{bmatrix}
U_1^0 \\ U_2^0 \\ U_3^0 \\ U_4^0
\end{bmatrix}
=
\begin{bmatrix}
f_1 \\ f_2 \\ f_3 \\ f_4
\end{bmatrix}
=
\begin{bmatrix}
1 \\ \frac{15}{16} \\ \frac{3}{4} \\ \frac{7}{16}
\end{bmatrix}
$$

which, with $N = 4$, can be written as

$$(I + rF_3)U^{j+1} = U^j, \qquad U^0 = f, \qquad r = k/h^2$$

where F_3 is the matrix of Theorem 8.4.4. Here we have

$$A(F_3) = I + rF_3 \text{ and } B(F_3) = I$$

According to Theorem 8.4.4,

$$\lambda_n = 2 - \cos\frac{(2n-1)\pi}{8}, \qquad n = 1, 2, 3, 4$$

$$\mathbf{V}_1 = \begin{bmatrix} 1 \\ \cos \ \pi/8 \\ \cos \ \pi/4 \\ \cos \ 3\pi/8 \end{bmatrix}, \qquad \mathbf{V}_2 = \begin{bmatrix} 1 \\ \cos \ 3\pi/8 \\ \cos \ 3\pi/4 \\ \cos \ 9\pi/8 \end{bmatrix}$$

$$\mathbf{V}_3 = \begin{bmatrix} 1 \\ \cos \ 5\pi/8 \\ \cos \ 5\pi/4 \\ \cos \ 15\pi/8 \end{bmatrix}, \qquad \mathbf{V}_4 = \begin{bmatrix} 1 \\ \cos \ 7\pi/8 \\ \cos \ 7\pi/4 \\ \cos \ 21\pi/8 \end{bmatrix}$$

and

$$\mathbf{W}_1 = \begin{bmatrix} 1 \\ 2 \cos \ \pi/8 \\ 2 \cos \ \pi/4 \\ 2 \cos \ 3\pi/8 \end{bmatrix}, \qquad \mathbf{W}_2 = \begin{bmatrix} 1 \\ 2 \cos \ 3\pi/8 \\ 2 \cos \ 3\pi/4 \\ 2 \cos \ 9\pi/8 \end{bmatrix}$$

$$\mathbf{W}_3 = \begin{bmatrix} 1 \\ 2 \cos \ 5\pi/8 \\ 2 \cos \ 5\pi/4 \\ 2 \cos \ 15\pi/8 \end{bmatrix}, \qquad \mathbf{W}_4 = \begin{bmatrix} 1 \\ 2 \cos \ 7\pi/8 \\ 2 \cos \ 7\pi/4 \\ 2 \cos \ 21\pi/8 \end{bmatrix}$$

Now, we have all the ingredients required to write the exact solution of the example difference system, and we find

$$\mathbf{U}^j = \sum_{n=0}^{N} (1 + r\lambda_n)^{-j} c_n^0 \mathbf{V}_n, \qquad c_n^0 = \frac{\mathbf{f} \cdot \mathbf{W}_n}{\mathbf{V}_n \cdot \mathbf{W}_n} \qquad \blacksquare \ \blacksquare$$

EXAMPLE 8.4.4 ───

Consider the Neumann initial-boundary-value problem

$$u_t - u_{xx} = 0, \qquad 0 < x < 1, t > 0$$
$$u(x, 0) = f(x), \qquad 0 < x < 1$$
$$u_x(0, t) = 0 = u_x(1, t), t > 0$$

A forward-in-time finite-difference system for this problem on the grid $x_1, x_2, \ldots, x_5,$

$$\begin{array}{ccccccc} 0 & \tfrac{1}{4} & \tfrac{1}{2} & \tfrac{3}{4} & 1 \\ \cdots \cdot \! - \! \cdot \! - \! \cdot \! - \! \cdot \! - \! \cdot \! \cdots \\ x_0 & x_1 & x_2 & x_3 & x_4 & x_5 & x_6 \end{array}$$

is given by

$$\mathbf{U}^{j+1} = (\mathbf{I} - r\mathbf{F}_2)\mathbf{U}^j, \qquad \mathbf{U}^0 = \mathbf{f} = [f_1, f_2, f_3, f_4, f_5]^T$$

where $\mathbf{I}$ is the identity matrix of order 5, $r = k/h^2$, and $\mathbf{F}_2$ is the matrix of Theorem 8.4.3 with $N = 5$. To solve this finite-difference system by Fourier's method, we need the eigenvalues λ_n of $\mathbf{F}_2$ and corresponding eigenvectors of $\mathbf{F}_2$ and $\mathbf{F}_2^T$. From Theorem 8.4.3 we know

$$\lambda_n = 2 - 2 \cos \frac{(n - 1)\pi}{4}, \qquad n = 1, \ldots, 5$$

so

$$\lambda_1 = 0, \qquad \lambda_2 = 2 - \sqrt{2}, \qquad \lambda_3 = 2, \qquad \lambda_4 = 2 + \sqrt{2}, \qquad \lambda_5 = 4$$

Also, from Theorem (8.4.3)

$$\mathbf{V}_1 = \begin{bmatrix} 1 \\ 1 \\ 1 \\ 1 \\ 1 \end{bmatrix}, \qquad \mathbf{V}_2 = \begin{bmatrix} 1 \\ \frac{1}{2}\sqrt{2} \\ 0 \\ -\frac{1}{2}\sqrt{2} \\ -1 \end{bmatrix}, \qquad \mathbf{V}_3 = \begin{bmatrix} 1 \\ 0 \\ -1 \\ 0 \\ 1 \end{bmatrix}$$

$$\mathbf{V}_4 = \begin{bmatrix} 1 \\ -\frac{1}{2}\sqrt{2} \\ 0 \\ \frac{1}{2}\sqrt{2} \\ -1 \end{bmatrix}, \qquad \mathbf{V}_5 = \begin{bmatrix} 1 \\ -1 \\ 1 \\ -1 \\ 1 \end{bmatrix}$$

with similar expressions for $\mathbf{W}_n$. Now, since $A(\lambda) = 1$ and $B(\lambda) = 1 - r\lambda$, it follows that the exact solution to the finite-difference system is

$$\mathbf{U}^j = \sum_{n=0}^{5} [1 - r\lambda_n]^j c_n^0 \mathbf{V}_n, \qquad c_n^0 = \frac{\mathbf{f} \cdot \mathbf{W}_n}{\mathbf{V}_n \cdot \mathbf{W}_n}$$

where the vectors $\mathbf{W}_n$, $n = 1, \ldots, 5$, are given by Theorem 8.4.3.

If we think of the example partial differential equation in the context of heat flow in a rod with insulated ends at $x = 0$ and $x = 1$, then we expect the solution of the differential equation to approach a constant in the limit as $t \to \infty$. Moreover, we would not expect the temperature to oscillate in time at any fixed x position. To ensure that the solution to the finite-difference system reflects this behavior requires some restriction of r. To illustrate, suppose the initial condition f is such that

$$c_1^0 = 1, \qquad c_2^0 = 0, \qquad c_3^0 = 0, \qquad c_4^0 = 0, \qquad c_5^0 = 1$$

Then, since $\lambda_1 = 0$ and $\lambda_5 = 4$, it follows that the solution to the difference equation is

$$\mathbf{U}^j = \mathbf{V}_1 + (1 - r \cdot 4)^j \mathbf{V}_5$$

or

$$\mathbf{U}^j = \begin{bmatrix} 1 \\ 1 \\ 1 \\ 1 \\ 1 \end{bmatrix} + (1 - 4r)^j \begin{bmatrix} 1 \\ -1 \\ 1 \\ -1 \\ 1 \end{bmatrix}$$

Now we see that unless we choose $r \leq \frac{1}{4}$, the numerical solution will exhibit oscillations in time at each node. Also, note that the choice $r = \frac{1}{4}$ leads to a numerical solution that is constant in space and time. Finally, if we choose $r = \frac{1}{2}$, then the numerical solution oscillates with no damping. ■ ■

By solving simple difference systems by Fourier methods, we can learn much about the behavior of the finite-difference method. In the exact solution of the difference equation, certain interdependencies between physical parameters, the grid parameters, and the various harmonics, or eigenvectors, are displayed. As Example 8.4.4 illustrates, we can use this information to guide us in choosing a finite-difference method to fit a particular physical problem.

Exact solutions to finite-difference equations also serve as a valuable tool in verifying computer code for the various difference methods. If a computer code is correctly written, it should produce the same results as the Fourier solution when both are applied to a test problem. In fact, for code verification, it is better to compare a computer code for solving finite-difference equations to the exact solution of the difference equations rather than the exact solution to the partial differential equation.

Exercises

1. For the case $N = 3$ write expressions for
   (a) the eigenvalues λ_n, $n = 1, 2, 3$, of the matrix $\mathbf{F}_2$;
   (b) the eigenvectors $\mathbf{V}_n$, $n = 1, 2, 3$, of the matrix $\mathbf{F}_2$;
   (c) the eigenvectors $\mathbf{W}_n$, $n = 1, 2, 3$, of the matrix $\mathbf{F}_2^T$.
   (d) Show directly that $\mathbf{V}_n \cdot \mathbf{W}_m = 0$ for $n \neq m$, $n, m = 1, 2, 3$.

2. For the case $N = 3$ write expressions for
   (a) the eigenvalues λ_n, $n = 1, 2, 3$, of the matrix $\mathbf{F}_3$;
   (b) the eigenvalues $\mathbf{V}_n$, $n = 1, 2, 3$, of the matrix $\mathbf{F}_3$;
   (c) the eigenvectors $\mathbf{W}_n$, $n = 1, 2, 3$, of the matrix $\mathbf{F}_3^T$.
   (d) Show directly that $\mathbf{V}_n \cdot \mathbf{W}_m = 0$ for $n \neq m$, $n, m = 1, 2, 3$.

3. Use Theorem 8.4.2 to express the vector $\mathbf{f} = [0, \frac{1}{4}, 1]^T$ as a linear combination of the eigenvectors of the matrix $\mathbf{F}_2$ with $N = 3$.

4. Use Theorem 8.4.2 to express the vector

$$\mathbf{f} = [0, \tfrac{1}{4}, \tfrac{1}{2}, \tfrac{3}{4}, 1]^T$$

   as a linear combination of the eigenvectors of the matrix $\mathbf{F}_2$ with $N = 5$.

5. Given the initial-boundary-value problem

$$u_t - u_{xx} = 0, \qquad 0 < x < 1, t > 0$$
$$u(x, 0) = x^2, \qquad 0 < x < 1$$
$$u_x(0, t) = 0 = u_x(1, t), \qquad t > 0$$

   and the grid $x_1 = 0$, $x_2 = \frac{1}{2}$, and $x_3 = 1$, so $N = 3$,
   (a) Use the matrix $\mathbf{F}_2$ to formulate a forward-in-time difference system.
   (b) Use the matrix $\mathbf{F}_2$ to formulate a backward-in-time difference system.
   (c) Use the matrix $\mathbf{F}_2$ to formulate a Crank–Nicolson difference system.
   (d) Use Fourier's method to solve the system (a).
   (e) Use Fourier's method to solve the system (b).
   (f) Use Fourier's method to solve the system (c).

6. Given the initial-boundary-value problem

$$u_t - u_{xx} = 0, \qquad 0 < x < 1, t > 0$$

$$u(x, 0) = x^2, \qquad 0 < x < 1$$
$$u_x(0, t) = 0 = u(1, t), \qquad t > 0$$

and the grid $x_1 = 0$, $x_2 = \frac{1}{3}$, and $x_3 = \frac{2}{3}$, so $N = 3$.

(a) Use the matrix F_3 to formulate a forward-in-time difference system.
(b) Use the matrix F_3 to formulate a backward-in-time difference system.
(c) Use the matrix F_3 to formulate a Crank–Nicolson difference system.
(d) Use Fourier's method to solve the system (a).
(e) Use Fourier's method to solve the system (b).
(f) Use Fourier's method to solve the system (c).

7. Given the initial-boundary-value problem

$$u_t - u_{xx} = 0, \qquad 0 < x < 1, \, t > 0$$
$$u(x, 0) = x, \qquad 0 < x < 1$$
$$u_x(0, t) = 0 = u_x(1, t), \qquad t > 0$$

and the grid $x_1 = 0$, $x_2 = \frac{1}{4}$, $x_3 = \frac{1}{2}$, $x_4 = \frac{3}{4}$, and $x_5 = 1$, so $N = 5$.

(a) Use the matrix F_2 to formulate a forward-in-time difference system.
(b) Use the matrix F_2 to formulate a backward-in-time difference system.
(c) Use the matrix F_2 to formulate a Crank–Nicolson difference system.
(d) Use Fourier's method to solve the system (a).
(e) Use Fourier's method to solve the system (b).
(f) Use Fourier's method to solve the system (c).

8. Given a tridiagonal eigenvalue problem of the form

$$\begin{bmatrix} b & c & & & & & \\ a & b & c & & & & \\ & a & b & c & & & \\ & & & \ddots & & & \\ & & & a & b & c & \\ & & & & a & b \end{bmatrix} \begin{bmatrix} V_1 \\ V_2 \\ V_3 \\ \vdots \\ V_{N-1} \\ V_N \end{bmatrix} = \lambda \begin{bmatrix} V_1 \\ V_2 \\ V_3 \\ \vdots \\ V_{N-1} \\ V_N \end{bmatrix}$$

it can be shown (*Schaum's Outline of Partial Differential Equations*, p. 179) that the eigenvalues and eigenvectors are given by

$$\lambda_n = b + 2c \sqrt{\frac{a}{c}} \sin \frac{n\pi}{N + 1}, \qquad n = 1, 2, \ldots, N$$

and

$$V_n = \begin{bmatrix} \beta \sin n\pi x_1 \\ \beta^2 \sin n\pi x_2 \\ \vdots \\ \beta^N \sin n\pi x_N \end{bmatrix}, \qquad n = 1, 2, \ldots, N, \, \beta = \sqrt{a/c}$$

where $x_i = i/(N + 1)$. Use this result to establish Theorem 8.4.1.

9. Given a uniform grid x_i, $i = 1, 2, \ldots, N$, with spacing h, show

$$-\cos m\pi x_{i-1} + 2 \cos m\pi x_i - \cos m\pi x_{i+1} = (2 - 2 \cos m\pi h)\cos m\pi x_i$$

10. In the notation of Theorem 8.4.3, show that for general N
(a) $F_2 V_n = \lambda_n V_n$, $n = 1, 2, \ldots, N$
(b) $F_2^T W_n = \lambda_n W_n$, $n = 1, 2, \ldots, N$
(c) $V_n \cdot W_m = 0$ if $n \neq m$, $m, n = 1, 2, \ldots, N$

11. In the notation of Theorem 8.4.4, show that for general N
   (a) $F_3 V_n = \lambda_n V_n, n = 1, 2, \ldots, N$
   (b) $F_3^T W_n = \lambda_n W_n, n = 1, 2, \ldots, N$
   (c) $V_n \cdot W_m = 0$ if $n \neq m, m, n = 1, 2, \ldots, N$

12. Given the eigenvalue problem

$$-v''(x) = \gamma v(x), \qquad 0 < x < 1$$
$$v'(0) = 0 = v'(1)$$

   (a) Show that if one-sided differences are used to approximate the derivative boundary conditions, then the discretization of this problem,

$$-V_{n-1} + 2V_n - V_{n+1} = \lambda V_n, \qquad n = 1, 2, \ldots, N$$
$$V_1 - V_0 = 0 = V_{N+1} - V_N$$

on the uniform grid $x_i = -h/2 + i \cdot h, i = 0, 1, 2, \ldots, N + 1, h = 1/N,$

```
      0                       ←h→                              1
    · —|— · — · · · · · — · — · — · — · · · · · — · — |— ·
    x₀      x₁    x           xᵢ₋₁ xᵢ   xᵢ₊₁        x_{N-1} x_N      x_{N+1}
```

can be written in matrix notation as

$$
\begin{bmatrix}
1 & -1 \\
-1 & 2 & -1 \\
 & -1 & 2 & -1 \\
 & & & \ddots \\
 & & & -1 & 2 & -1 \\
 & & & & -1 & 1
\end{bmatrix}
\begin{bmatrix}
V_1 \\ V_2 \\ V_3 \\ \vdots \\ V_{N-1} \\ V_N
\end{bmatrix}
= \lambda
\begin{bmatrix}
V_1 \\ V_2 \\ V_3 \\ \vdots \\ V_{N-1} \\ V_N
\end{bmatrix}
\quad \text{or} \quad F_4 V = \lambda V
$$

   (b) Verify that the eigenvalues of the matrix F_4 are given by

$$\lambda_n = 2 - 2 \cos \frac{(n-1)\pi}{N}, \qquad n = 1, 2, \ldots, N$$

and the corresponding eigenvectors are

$$
V_n =
\begin{bmatrix}
\cos(n-1)\pi x_1 \\
\cos(n-1)\pi x_2 \\
\vdots \\
\cos(n-1)\pi x_N
\end{bmatrix}
\equiv \cos(n-1)\pi x, \qquad n = 1, 2, \ldots, N
$$

   (c) Without doing a calculation, prove that $V_n \cdot V_m = 0$ for $n \neq m$. (*Hint:* $F_4 = F_4^T$.)
   (d) Show that $V_1 \cdot V_1 = N$ and $V_n \cdot V_n = N/2$ for $n = 2, 3, \ldots, N$

13. For the case $N = 3$ write expressions for
   (a) the eigenvalues $\lambda_n, n = 1, 2, 3,$ of the matrix F_4
   (b) the eigenvectors $V_n, n = 1, 2, 3,$ of the matrix F_4

14. Given the eigenvalue problem

$$-v''(x) = \gamma v(x), \qquad 0 < x < 1$$
$$v'(0) = 0 = v(1)$$

   (a) Show that if one-sided differences are used to approximate the derivative boundary conditions, then the discretization of this problem,

$$-V_{n-1} + 2V_n - V_{n+1} = \lambda V_n, \qquad n = 1, 2, \ldots, N$$
$$V_1 - V_0 = 0 = V_{N+1}$$

on the uniform grid $x_i = -h/2 + ih$, $h = 2/(2N + 1)$, $i = 0, 1, 2, \ldots, N + 1$,

can be written in matrix notation as

$$\begin{bmatrix} 1 & -1 & & & & \\ -1 & 2 & -1 & & & \\ & -1 & 2 & -1 & & \\ & & & \ddots & & \\ & & & -1 & 2 & -1 \\ & & & & -1 & 2 \end{bmatrix} \begin{bmatrix} V_1 \\ V_2 \\ V_3 \\ \vdots \\ V_{N-1} \\ V_N \end{bmatrix} = \lambda \begin{bmatrix} V_1 \\ V_2 \\ V_3 \\ \vdots \\ V_{N-1} \\ V_N \end{bmatrix} \quad \text{or} \quad \mathbf{F}_5 \mathbf{V} = \lambda \mathbf{V}$$

(b) Verify that the eigenvalues of the matrix $\mathbf{F}_5$ are given by

$$\lambda_n = 2 - 2 \cos \frac{(2n - 1)\pi}{2N + 1}, \qquad n = 1, 2, \ldots, N$$

and the corresponding eigenvectors are

$$\mathbf{V}_n = \begin{bmatrix} \cos \dfrac{(2n - 1)\pi x_1}{2} \\ \cos \dfrac{(2n - 1)\,\pi x_2}{2} \\ \vdots \\ \cos \dfrac{(2n - 1)\pi x_N}{2} \end{bmatrix} = \cos \frac{(2n - 1)\pi x}{2}, \qquad n = 1, 2, \ldots, N$$

(c) Without doing any calculation, show $\mathbf{V}_n \cdot \mathbf{V}_m = 0$ for $n \neq m$.

(d) Show that $\mathbf{V}_n \cdot \mathbf{V}_n = \frac{1}{4}(2N + 1)$.

15. For the case $N = 3$ write expressions for
  (a) the eigenvalues λ_n, $n = 1, 2, 3$, of the matrix $\mathbf{F}_5$
  (b) the eigenvectors $\mathbf{V}_n$, $n = 1, 2, 3$, of the matrix $\mathbf{F}_5$

16. Use an eigenvector expansion to solve

$$\mathbf{U}^{j+1} = (\mathbf{I} - r\mathbf{F}_1)\mathbf{U}^j + k\mathbf{S}$$
$$\mathbf{U}^0 = 0$$

8.5 STABILITY OF FINITE-DIFFERENCE METHODS

Continuous Dependence on Data

In Section 8.3 we saw that the forward difference method fails to produce a useful numerical approximation when the condition $r \leq \frac{1}{2}$ is violated. Again, in Section 8.4, we observed that the analytical solution to a forward difference discretization fails to remain bounded when $r > \frac{1}{2}$. In each of these examples, the stability of the finite-difference method determines the nature of the numerical solution.

The stability of a finite-difference method is closely related to the notion of continuous dependence of a solution on the problem data. We have already encountered the idea of continuous dependence of a solution on problem data in Sections 1.5, 6.2, and 6.3.

To make the present discussion specific, consider a linear, constant-coefficient initial-boundary-value problem with homogeneous boundary conditions. For example,

$$L[u] = u_t - a^2 u_{xx} = 0, \qquad 0 < x < 1, 0 < t \le T \tag{8.5.1}$$
$$u(x, 0) = f(x), \qquad 0 < x < 1 \tag{8.5.2}$$
$$u(0, t) = 0 = u(1, t), \qquad 0 < t \le T \tag{8.5.3}$$

The solution of (8.5.1)–(8.5.3) is said to *depend continuously on the data function* $f(x)$ provided there exists a constant C independent of f such that

$$\|u(x, t)\| \le C\|f(x)\|, \qquad 0 \le t \le T \tag{8.5.4}$$

Here $\|\cdot\|$ denotes a general norm that could be, for example, the *maximum norm*

$$\|f(x)\| = \max_{0 \le x \le 1} |f(x)|$$

or the *energy norm*

$$\|f(x)\| = \left\{ \int_0^1 f(x)^2 \, dx \right\}^{1/2}$$

The maximum norm is also called the L^∞ norm, and the energy norm is called the L^2 norm. See also Section 2.2.

To gain a clearer understanding of the notion of continuous dependence on the data, consider the following "perturbation" of the problem (8.5.1)–(8.5.3):

$$u_t - a^2 u_{xx} = 0, \qquad 0 < x < 1, 0 < t \le T$$
$$u(x, 0) = f(x) + e(x), \qquad 0 < x < 1$$
$$u(0, t) = 0 = u(1, t), \qquad 0 < t \le T$$

Let $u(x, t)$ denote the solution of (8.5.1)–(8.5.3) and $\hat{u}(x, t)$ denote the solution of the perturbed problem. Then, if the solution depends continuously on the data, it follows that

$$\|u(x, t) - \hat{u}(x, t)\| \le C\|e(x)\|, \qquad 0 \le t \le T$$

If we think of $e(x)$ as an error in representing the initial condition (8.5.2), then *continuous dependence on the data implies that small errors in the data result in small changes in the solution.*

Stability

To discretize the continuous initial-value problem (8.5.1)–(8.5.3), introduce a uniform x grid on the interval $0 \le x \le 1$,

$$x_n = nh, \qquad n = 1, 2, \ldots, N, h = 1/(N + 1)$$

Also, introduce a uniform time grid

$$t_j = jk, \quad j = 0, 1, 2, \ldots, J, k = T/J$$

Let

$$D[U_n^j] = 0, \qquad n = 1, 2, \ldots, N, j = 1, 2, \ldots, J \qquad (8.5.5)$$
$$U_n^0 = f_n, \qquad n = 1, 2, \ldots, N \qquad (8.5.6)$$
$$U_0^j = 0 = U_{N+1}^j, \qquad j = 1, 2, \ldots, J \qquad (8.5.7)$$

represent a finite-difference approximation to the initial-value problem (8.5.1)–(8.5.3). For example, $D[U_n^j] = 0$ could represent the forward difference, the backward difference, or the Crank–Nicolson discretization. Let $\mathbf{U}^j$ and $\mathbf{f}$ denote column vectors defined by $\mathbf{U}^j = [U_1^j, U_2^j, \ldots, U_N^j]^T$ and $\mathbf{f} = [f_1, f_2, \ldots, f_N]^T$. The finite-difference method (8.5.5)–(8.5.7) is said to be (unconditionally) *stable* if, for any h and k, there exists a constant, C, independent of h, k, and $\mathbf{f}$ such that

$$\|\mathbf{U}^j\| \le C\|\mathbf{f}\|, \qquad 0 \le jk \le T \qquad (8.5.8)$$

If k must be functionally related to h for (8.5.8) to hold, the difference method is called *conditionally stable*.

The norm in (8.5.8) is general. For example,

$$\|\mathbf{f}\| = \max_{1 \le n \le N} |f_n|$$

or

$$\|\mathbf{f}\| = \left\{ \sum_{n=1}^{N} f_n^2 h \right\}^{1/2}$$

Let $\hat{U}_n^j$ denote the solution to a perturbation of the difference system (8.5.5)–(8.5.7) that differs only in the initial condition. Namely,

$$D[U_n^j] = 0, \qquad n = 1, 2, \ldots, N, j = 1, 2, \ldots$$
$$\hat{U}_n^0 = f_n + E_n, \qquad n = 1, 2, \ldots, N$$
$$\hat{U}_0^j = 0 = \hat{U}_{N+1}^j, \qquad j = 1, 2, \ldots, J$$

where E_n could be thought of as an error superimposed on the initial condition f_n. If the difference method (8.5.5)–(8.5.7) is stable, then it follows that

$$\|\hat{\mathbf{U}}^j - \mathbf{U}^j\| \le C\|\mathbf{E}\|, \qquad jk \le T$$

where $\hat{\mathbf{U}}^j = [\hat{U}_1^j, \hat{U}_2^j, \ldots, \hat{U}_N^j]^T$ and $\mathbf{E} = [E_1, E_2, \ldots, E_N]^T$.

From the last inequality, we see that *stability of the finite-difference method implies that small errors in the initial condition result in small errors in the solution to the difference equations*. In most practical applications, errors due to computer rounding or truncation are introduced not only in the initial condition, but also at each step of a finite-difference calculation. More generally, if at each time step errors are introduced into a stable finite-difference calculation, then at the very worst these errors will accumulate in an additive fashion. It is the exponential growth of errors that must be avoided to maintain stability in approximating a bounded solution. We now present two methods for analyzing the stability properties of a given finite-difference method.

Spectral Method

Consider an initial-boundary-value problem of the form

$$u_t - a^2 u_{xx} = 0, \qquad 0 < x < 1, 0 < t < T \qquad (8.5.9a)$$
$$u(x, 0) = f(x), \qquad 0 < x < 1 \qquad (8.5.9b)$$
$$\alpha_0 u(0, t) + \beta_0 u_x(0, t) = 0, \qquad \alpha_1 u(1, t) + \beta_1 u_x(1, t) = 0 \qquad (8.5.10)$$

Using the notation of Section 8.4, formulate a two-level, homogeneous finite-difference system of the form

$$A(\mathbf{F})\mathbf{U}^{j+1} = B(\mathbf{F})\mathbf{U}^j \qquad (8.5.11)$$
$$\mathbf{U}^0 = \mathbf{f} \qquad (8.5.12)$$

where $A(\mathbf{F})$ and $B(\mathbf{F})$ are $N \times N$ matrices that are polynomial functions of the $N \times N$ matrix $\mathbf{F}$. Recall that the matrix $\mathbf{F}$ incorporates the boundary conditions (8.5.10). We assume that $\mathbf{F}$ satisfies the hypotheses of Theorem 8.4.2.

As we saw in Section 8.4, if we look for a solution of the difference system in the form of an eigenvector expansion,

$$\mathbf{U}^j = \sum_{n=0}^{N} c_n^j \mathbf{V}_n \qquad (8.5.13)$$

with the $\mathbf{V}_n$ eigenvectors of the matrix $\mathbf{F}$ and the $\mathbf{W}_n$ eigenvectors of $\mathbf{F}^T$, then it follows that

$$c_n^j = \left[\frac{B(\lambda_n)}{A(\lambda_n)} \right]^j c_n^0 \quad \text{and} \quad c_n^0 = \frac{\mathbf{f} \cdot \mathbf{W}_n}{\mathbf{V}_n \cdot \mathbf{W}_n}$$

where λ_n is an eigenvalue of $\mathbf{F}$. The spectral stability condition is the requirement that no component in the eigenvector expansion grow without bound as $j \to \infty$. From the recursion formula for the coefficients c_n^j, it follows that

$$\frac{c_n^{j+1}}{c_n^j} = \frac{B(\lambda_n)}{A(\lambda_n)}, \qquad n = 1, 2, \ldots, N$$

If, for some value $n = p$, the magnitude of the ratio c_p^{j+1}/c_p^j exceeds 1, then the $\mathbf{V}_p$ component of the solution (8.5.13) will grow without bound as $j \to \infty$. Thus, the spectral stability condition can be stated as follows.

Spectral Stability Condition. The finite-difference system (8.5.11) and (8.5.12) is said to satisfy the spectral stability condition provided

$$\left| \frac{B(\lambda_n)}{A(\lambda_n)} \right| \leq 1, \qquad n = 1, 2, \ldots, N \qquad (8.5.14)$$

holds for arbitrary N.

The spectral stability condition is also known as the *matrix stability condition*. The following theorem relates the spectral stability condition (8.5.14) to the definition of stability (8.5.8).

Theorem 8.5.1. Given the finite-difference method (8.5.11) and (8.5.12) with $\mathbf{F}$ satisfying the hypotheses of Theorem 8.4.2, then:

1. The spectral stability condition is always necessary for the difference method to be stable in the sense of (8.5.8).
2. If $\mathbf{F}$ is symmetric, then the spectral stability condition is both necessary and sufficient for the difference method to be stable in the sense of (8.5.8).

The proof of part 1 of Theorem 8.5.1 follows from the preceding development of the spectral stability condition. Exercise 6 outlines the method of proof of part 2 of the theorem.

For a fixed x grid, the spectral stability condition is both necessary and sufficient to prevent unbounded error growth. When $\mathbf{F}$ is not symmetric, there is a problem in exhibiting a fixed constant C that can be used in (8.5.8) as $N \rightarrow \infty$.

Using the formulas of Section 8.4 for the eigenvalues of the matrices $\mathbf{F}_1$, $\mathbf{F}_2$, $\mathbf{F}_3$, and $\mathbf{F}_4$, we can analyze the spectral stability condition for a number of finite-difference systems.

EXAMPLE 8.5.1 ————————————————————————————

If we discretize the initial-boundary-value problem

$$u_t - u_{xx} = 0, \qquad 0 < x < 1, t > 0$$
$$u(x, 0) = f(x), \qquad 0 < x < 1$$
$$u(0, t) = 0 = u(1, t), \qquad t > 0$$

using the forward-in-time method (8.2.6), then, on the grid of Theorem 8.4.1, the resulting difference system is

$$\mathbf{U}^{j+1} = (\mathbf{I} - r\mathbf{F}_1)\mathbf{U}^j, \qquad r = k/h^2$$
$$\mathbf{U}^0 = \mathbf{f}$$

From Theorem 8.4.1, we know the eigenvalues of $\mathbf{F}_1$ are given by

$$\lambda_n = 2 - 2 \cos \frac{n\pi}{N+1} = 4 \sin^2 \frac{n\pi}{2(N+1)}, \qquad n = 1, 2, \ldots, N$$

Since $A(\mathbf{F}) = \mathbf{I}$ and $B(\mathbf{F}) = \mathbf{I} - r\mathbf{F}$, it follows that, for this example, the spectral stability condition is $|B(\lambda_n)| \leq 1$ or

$$\left| 1 - 4r \sin^2 \frac{n\pi}{2(N+1)} \right| \leq 1, \qquad n = 1, 2, \ldots, N$$

For the preceding inequality to hold for arbitrary N, we must have $r \leq \frac{1}{2}$. Thus, our spectral stability analysis shows that the forward-in-time difference method (8.2.6) is conditionally stable with the stability restriction given by

$$r = k/h^2 \leq \frac{1}{2} \qquad\qquad\qquad ∎∎$$

EXAMPLE 8.5.2 ————————————————————————————————————

Given the initial-boundary-value problem

$$u_t - a^2 u_{xx} = 0, \qquad 0 < x < 1, t > 0$$
$$u(x, 0) = f(x), \qquad 0 < x < 1$$
$$u_x(0, t) = 0 = u_x(1, t), \qquad t > 0$$

if the Crank–Nicolson method (8.2.8) is used to discretize on the grid $x_n = nh$, $h = 1/(N - 1)$, then the resulting system of finite-difference equations can be written in the form

$$\left(I + \frac{r}{2} F_2\right) U^{j+1} = \left(I - \frac{r}{2} F_2\right) U^j, \qquad r = \frac{a^2 k}{h^2}$$

and we see that the $N \times N$ matrices $A(F_2) = I - (r/2)F_2$ and $B(F_2) = I + (r/2)F_2$. The spectral stability condition for this example is

$$\left| \frac{1 - (r/2)\lambda_n}{1 + (r/2)\lambda_n} \right| \le 1, \qquad n = 1, 2, \dots, N$$

where λ_n is an eigenvalue of F_2. Using Theorem 8.4.3 and the identity $1 - \cos \beta = 2 \sin^2(\beta/2)$, we can write the spectral stability condition in the form

$$\left| \frac{1 - 2r \sin^2(\beta_n/2)}{1 + 2r \sin^2(\beta_n/2)} \right| \le 1, \qquad n = 1, 2, \dots, N$$

where $\beta_n = (n - 1)\pi/(N - 1)$. This shows that the spectral stability condition holds unconditionally for this Crank–Nicolson finite-difference system. Notice that for $n = 1$ equality holds in the spectral stability condition. Thus, we cannot expect any damping of errors in the V_1 component. What can you say about the propagation of the V_N component in the case that $r = \frac{1}{2}$? ∎ ∎

As the preceding examples show, when the eigenvalues λ_n of the matrix F in (8.5.11) are known, then the analysis of the spectral stability condition is straightforward. In many finite-difference systems of the form (8.5.11) and (8.5.12) we encounter a matrix F whose eigenvalues cannot be explicitly displayed. In these cases, the following theorem can often be used to determine sufficient conditions for the spectral stability condition to hold.

Theorem 8.5.2 Gerschgorin Circle Theorem. Let $A = [a_{nj}]$ be an $N \times N$ matrix and let C_n denote the circular region in the complex plane with center a_{nn} and radius equal to the sum of the magnitudes of the off-diagonal entries in the ith row of A. That is,

$$C_n = \left\{ z : |z - a_{nn}| \le \sum_{\substack{j=1 \\ j \ne n}}^{N} |a_{ij}| \right\}, \qquad n = 1, 2, \dots, N$$

The eigenvalues of A are contained in the union of the C_n.

EXAMPLE 8.5.3 ――

Given the matrix

$$A = \begin{bmatrix} 1 & 0 & 1 \\ -1 & 2 & 1 \\ -1 & \frac{1}{2} & 6 \end{bmatrix}$$

then $C_1 = \{z : |z - 1| \leq 1\}$, $C_2 = \{z : |z - 2| \leq 2\}$, $C_2 = \{z : |z - 6| \leq \frac{3}{2}\}$. The Gerschgorin theorem implies that the eigenvalues of $\mathbf{A}$ are contained in the union of the circular regions C_1, C_2, and C_3. ■ ■

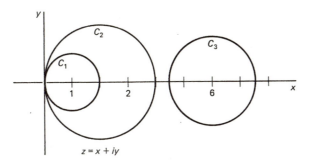

EXAMPLE 8.5.4 ――

Consider the initial-boundary-value problem

$$\begin{aligned} u_t - u_{xx} &= 0, & 0 < x < 1, t > 0 \\ u(x, 0) &= f(x), & 0 < x < 1 \\ 3u(0, t) - u_x(0, t) &= 0, & t > 0 \\ u(1, t) &= 0, & t > 0 \end{aligned}$$

If we apply the forward difference method (8.2.6) on the grid $x_n = (n - 1)h$, $n = 1, 2, \ldots, N$, $h = 1/N$,

$$\begin{array}{ccccccccc} & 0 & & & & & & & 1 \\ \cdots & \cdot & - & \cdot & - & \cdots & \cdot & - & \cdot & - & \cdots & - & \cdot \\ & x_0 & x_1 & x_2 & & & x_{n-1}\, x_n & x_{n+1} & & x_N & x_{N+1} \end{array}$$

the resulting difference system is

$$U_n^{j+1} - U_n^j = r\,\delta_x^2 U_n^j, \qquad n = 1, 2, \ldots, N, r = k/h^2$$

$$U_n^j = f_n, \qquad n = 1, 2, \ldots, N$$

$$3U_1^j - \frac{U_2^j - U_0^j}{2h} = 0$$

$$U_{N+1}^j = 0$$

Using the discretization of the mixed, left boundary condition, we can solve for U_0^j in terms of U_1^j and U_2^j. Incorporating

$$U_0^j = U_2^j - 6hU_1^j$$

in the finite-difference equation leads to the matrix formulation

$$\mathbf{U}^{j+1} = (I - r\mathbf{F})\mathbf{U}^j$$
$$\mathbf{U}^0 = \mathbf{f}$$

where the matrix $\mathbf{F}$ is given by

$$\mathbf{F} = \begin{bmatrix} 2 + 6h & -2 & & & & & \\ -1 & 2 & -1 & & & & \\ & -1 & 2 & -1 & & & \\ & & & \ddots & \ddots & & \\ & & & & -1 & 2 & -1 \\ & & & & & -1 & 2 \end{bmatrix}$$

From the Gershgorin circle theorem, we know that the eigenvalues λ_n of the matrix $\mathbf{F}$ are contained in the union of the three circles:

$$C_1 = \{z : |z - (2 + 6h)| \leq 2\}$$
$$C_2 = \{z : |z - 2| \leq 2\}$$
$$C_3 = \{z : |z - 2| \leq 1\}$$

In the union of C_1, C_2, and C_3, the point that is farthest away from the point $z = 1$ is the point $z = 4 + 6h$. Now, it is not difficult to see that a sufficient condition for the spectral stability condition $|1 - r\lambda_n| \leq 1$ to hold is that r satisfy the inequality

$$r \leq \frac{2}{4 + 6h} \qquad \blacksquare \blacksquare$$

The matrix or spectral stability analysis applies to initial-boundary-value problems given on a bounded spatial region. Thus, the spectral stability analysis cannot be applied to a pure initial-value problem. Also, a spectral stability analysis requires that we formulate the difference system as a matrix equation. A matrix stability analysis does not apply to a difference system in its coordinate formulation in terms of U_n^j.

von Neumann Method

The von Neumann method for studying the stability of a finite-difference scheme is closely related to the spectral method. A discrete Fourier series is used to represent the initial condition for the finite-difference system. This representation of the initial condition by a discrete Fourier series is a direct analogy with the eigenvector expansions used in the spectral method. To apply the von Neumann method, we examine how each Fourier mode is propagated by the finite-difference equation. Put in general terms, the von Neumann stability criterion is that the finite-difference equation does not permit any Fourier mode to increase as the finite-difference solution is advanced from t_j to t_{j+1}.

A von Neumann stability analysis is carried out on the component form of the finite-difference equation rather than the matrix formulation of the difference system. The following example illustrates the use of the von Neumann method of examining stability applied to a pure initial-value problem.

EXAMPLE 8.5.5 _____

Consider the initial-boundary-value problem

$$u_t - a^2 u_{xx} = 0, \qquad -\infty < x < \infty, \, t > 0$$
$$u(x, 0) = f(x), \qquad -\infty < x < \infty$$

and assume that both the initial condition f and the solution $u(x, t)$ are periodic. Since a rescaling of the x axis leaves the form of the problem invariant, we can take f and u to be 2π periodic. Introduce a uniform x grid defined by

$$x_n = -\pi + nh, \qquad n = 0, \pm 1, \pm 2, \ldots, \qquad h = 2\pi/N$$

If we discretize this problem with the forward-in-time difference method (8.2.6), the resulting difference system is

$$U_n^{j+1} - U_n^j = r\delta_x^2 U_n^j, \qquad n = 0, \pm 1, \pm 2, \ldots, \qquad r = ka^2/h^2$$
$$U_n^j = f_n, \qquad n = 0, \pm 1, \pm 2, \ldots$$

where f_n is 2π periodic in the sense that the values of f_n satisfy

$$f_{n+N} = f_n, \qquad n = 0, \pm 1, \pm 2, \ldots$$

and the f_n values at only the grid points

$$x_n = -\pi + nh, \qquad n = 0, 1, 2, \ldots, N - 1, \qquad h = 2\pi/N$$

are sufficient to fix all the values of f_n. If f_n is 2π periodic, then for each j we can show that U_n^j is 2π periodic. Now, using the results of Section 2.4, discrete Fourier series can be used to represent f_n and U_n^j:

$$f_n = \sum_{m=0}^{N-1} \xi_m e^{imx_n}, \qquad \xi_m = \frac{1}{N} \sum_{n=0}^{N-1} f_n e^{-imx_n}$$

$$U_n^j = \sum_{m=0}^{N-1} \xi_m^j e^{imx_n}, \qquad \xi_m^j = \frac{1}{N} \sum_{n=0}^{N-1} U_n^j e^{-imx_n}$$

The initial condition requires $\xi_m^0 = \xi_m$. Putting U_n^j into the difference equation yields

$$\sum_{m=0}^{N-1} \xi_m^{j+1} e^{imx_n} - \sum_{m=0}^{N-1} \xi_m^j e^{imx_n} = r\delta_x^2 \sum_{m=0}^{N-1} \xi_m^j e^{imx_n}$$

or

$$\sum_{m=0}^{N-1} (\xi_m^{j+1} - \xi_m^j) e^{imx_n} = \sum_{m=0}^{N-1} \xi_m^j r\delta_x^2 e^{imx_n}$$

$$= \sum_{m=0}^{N-1} \xi_m^j r(e^{imx_{n-1}} - 2e^{imx_n} + e^{imx_{n+1}})$$

from which we see

$$\sum_{m=0}^{N-1} [\xi_m^{j+1} - \xi_m^j - \xi_m^j r(e^{-imh} - 2 + e^{imh})] e^{imx_n} = 0$$

Since the complex exponentials e^{imx_n}, $n = 0, 1, \ldots, N - 1$, are linearly independent, the bracketed quantity in the last equation must be zero. From

$$\xi_m^{j+1} - \xi_m^j - \xi_m^j r(e^{-imh} - 2 + e^{imh}) = 0$$

we can see that the ratio of the Fourier coefficients at time levels t_{j+1} and t_j is given by

$$\xi_m^{j+1}/\xi_m^j = 1 + r(e^{-imh} - 2 + e^{imh})$$
$$= 1 - 2r(1 - \cos mh)$$
$$= 1 - 4r \sin^2(mh/2)$$

For the calculation to be stable, no Fourier mode can be allowed to grow without bound as $j \to \infty$. Thus, the von Neumann stability condition is that the inequality

$$|\xi_m^{j+1}/\xi_m^j| \le 1, \qquad m = 0, 1, \ldots, N - 1$$

must hold for arbitrary N. Since $h = 2\pi/N$, it follows that as $N \to \infty$, the quantity mh ranges over the interval 2π. For this example, which deals with the the forward-in-time difference method, the von Neumann stability condition is seen to hold if and only if $r \le \frac{1}{2}$.   ■ ■

An examination of this example reveals that the important steps in the von Neumann stability analysis can be summarized as follows.

1. Assume the solution of the finite-difference system is expressed as a superposition of Fourier modes with a typical Fourier mode having the form

$$\xi^j e^{i\beta n}, \qquad 0 \le \beta < 2\pi$$

To reconcile this form of expressing a typical Fourier mode with the form used in the example, note that

$$\xi_m^j e^{imx_n} = \xi_m^j e^{im(-\pi + nh)} = \xi_m^j (-1)^m e^{imhn} \equiv \xi^j e^{i\beta n}$$

where we have made the assignment $\beta = mh$. As $N \to \infty$, mh ranges over the interval $[0, 2\pi)$.

2. Test the behavior of the finite-difference method on a typical Fourier mode by putting $\xi^j e^{i\beta n}$ into the finite-difference equation and solving for the ratio ξ^{j+1}/ξ^j. The magnitude of this ratio is called the *magnification factor* associated with the β Fourier mode. By testing the finite-difference method on a single, arbitrary Fourier mode, we are assuming that there is no interaction between the various Fourier modes. This essentially restricts the von Neumann stability analysis to constant-coefficient linear difference equations.

3. Impose the requirement that the magnification factor satisfy the von Neumann stability condition,

$$|\xi^{j+1}/\xi^j| \le 1 \quad \text{for all } \beta \in [0, 2\pi)$$

So that we can describe the von Neumann stability analysis for more general finite-difference methods, we introduce some notation. Let S_+ and S_- denote *shift operators* defined by

$$S_+ U_n^j = U_{n+1}^j; \qquad S_- U_n^j = U_{n-1}^j$$

Also, let $P(S_+, S_-)$ and $Q(S_+, S_-)$ denote polynomials in S_+ and S_-. Then, any two-level, homogeneous, linear, constant-coefficient initial-value finite-difference problem can be written in the form

$$P(S_+, S_-)U_n^{j+1} = Q(S_+, S_-)U_n^j \qquad (8.5.15)$$

EXAMPLE 8.5.6 ──

Let $P(S_+, S_-) = 1$ and $Q(S_+, S_-) = r(S_- - 2 + S_+)$ be polynomials in S_+ and S_-. Moreover,

$$Q(S_+, S_-)U_n^j = r(U_{n-1}^j - 2U_n^j + U_{n+1}^j) = r\delta_x^2 U_n^j$$

Therefore, with this choice of P and Q, the forward difference method (8.2.6) for $u_t - a^2 u_{xx} = 0$ on the grid $(x_n, t_j) = (nh, jk)$ can be expressed in the form

$$P(S_+, S_-)U_n^{j+1} = Q(S_+, S_-)U_n^j, \qquad r = a^2 k/h^2 \qquad ■■$$

Assume a general harmonic in a Fourier expansion of U_n^j has the form

$$\xi^j e^{i\beta n} \qquad (8.5.16)$$

The von Neumann stability condition is that no harmonic be magnified from time t_j to time t_{j+1}. To examine how the difference equation acts on a harmonic of this form, we consider

$$P(S_+, S_-)\xi^{j+1}e^{i\beta n} = Q(S_+, S_-)\xi^j e^{i\beta n}$$

The linearity of the difference operators $P(S_+, S_-)$ and $Q(S_+, S_-)$ allows us to write the last equation as

$$\xi^{j+1}P(S_+, S_-)e^{i\beta n} = \xi^j Q(S_+, S_-)e^{i\beta n}$$

Since

$$S_+ e^{i\beta n} = e^{i\beta(n+1)} \quad \text{and} \quad S_- e^{i\beta n} = e^{i\beta(n-1)}$$

we have

$$\xi_m^{j+1}P(e^{i\beta}, e^{-i\beta}) = \xi_m^j Q(e^{i\beta}, e^{-i\beta})$$

From the last equation, it follows that from t_j to t_{j+1} the Fourier mode $e^{i\beta n}$ is multiplied by the *magnification factor*

$$\frac{\xi^{j+1}}{\xi^j} = \frac{Q(e^{i\beta}, e^{-i\beta})}{P(e^{i\beta}, e^{-i\beta})}$$

A necessary and sufficient condition for no Fourier mode of U_n^j to grow without bound as $j \to \infty$ is

$$|\xi^{j+1}/\xi^j| \leq 1$$

Therefore, the von Neumann stability condition is stated as follows.

Von Neumann Stability Condition. The finite-difference equation (8.5.15) is said to satisfy the von Neumann stability condition provided

$$\left| \frac{Q(e^{i\beta}, e^{-i\beta})}{P(e^{i\beta}, e^{-i\beta})} \right| \leq 1, \qquad 0 \leq \beta < 2\pi \tag{8.5.17}$$

As an illustration of the von Neumann stability analysis described in the preceding, let us reconsider the forward-in-time difference method.

EXAMPLE 8.5.7 _____

Given $u_t - a^2 u_{xx} = 0$, we have seen that if we use the forward difference method (8.2.6), the resulting difference equation is

$$P(S_+, S_-)U_n^{j+1} = Q(S_+, S_-)U_n^j$$

where

$$P(S_+, S_-) = 1 \quad \text{and} \quad Q(S_+, S_-) = rS_- + (1 - 2r) + rS_-, \qquad r = a^2 k/h^2$$

Therefore, the von Neumann stability condition is

$$\left| re^{-i\beta} + (1 - 2r) + re^{i\beta} \right| \leq 1, \qquad 0 \leq \beta < 2\pi$$

Using the identities

$$re^{i\beta} + re^{-i\beta} = 2r \cos \beta$$

and

$$1 - \cos \beta = 2 \sin^2 \frac{\beta}{2}$$

allows the von Neumann stability condition to be rewritten as

$$\left| 1 - 4r \sin^2 \frac{\beta}{2} \right| \leq 1, \qquad 0 \leq \beta < 2\pi$$

From the last inequality, it is easy to see that the von Neumann stability condition holds if and only if $r \leq \frac{1}{2}$. ∎ ∎

The following theorem relates the von Neumann stability condition (8.5.17) to the definition of stability (8.5.8).

Theorem 8.5.3. Given the finite-difference method (8.5.15):

1. The von Neumann stability condition is always necessary for the difference method to be stable in the sense of (8.5.8).
2. For an initial-value problem with a periodic solution, the von Neumann stability condition is both necessary and sufficient for the difference method to be stable in the sense of (8.5.8).

Exercise 8 outlines the proof of part 2 of Theorem 8.5.3. Conclusion 2 of Theorem

8.5.3 also holds for initial-value problems whose solutions have finite l^2 energy. See Exercise 9 for a definition of l^2 and an outline of the proof.

The von Neumann stability analysis is not restricted to pure initial-value problems. For example, it can be used to analyze the stability of a difference method for an initial-boundary-value problem whose solution has a smooth periodic extension to the entire x axis. The initial-boundary-value problem (8.5.1)–(8.5.3) has this property.

In practice, the von Neumann stability analysis method is applied to general initial-boundary-value problems by disregarding the boundary conditions. Given an initial-boundary-value problem, suppose we apply the spectral stability method and the von Neumann method with the boundary conditions disregarded. Generally these two approaches lead to the same stability restrictions in the limit as the x grid spacing approaches zero. On a fixed grid, the spectral method will sometimes provide information about how the boundary conditions enter the stability analysis. However, it is the difference equation, not the boundary conditions, that dominates the stability analysis.

Lax Equivalence Theorem

Now that we have discussed the notion of a stable finite-difference method in some detail, let us review two other definitions that were given in Section 8.1. Recall that the local truncation error for a difference method is defined as the amount by which the solution of the continuous partial differential equation problem fails to satisfy the finite-difference system. We say that a difference method is *consistent* with a continuous problem provided the local truncation error goes to zero in the limit as the grid spacings approach *zero*. Recall also that the finite-difference method is called *convergent* if the solution U_n^j of the difference system and the solution u_n^j of the continuous problem can be made arbitrarily close by choosing the grid spacings sufficiently small. Our final result of this section provides a connection between the notions of consistency, convergence, and stability.

Theorem 8.5.4 Lax Equivalence Theorem. Given a well-posed linear initial-value problem or initial-boundary-value problem and a finite-difference system consistent with it, stability is both necessary and sufficient for convergence.

Exercises

1. Determine the restriction, if any, on h and k for the spectral stability condition (8.5.15) to hold if the forward difference method of Table 8.2.2 is used for the problem

   (a) $u_t - a^2 u_{xx} = 0, 0 < x < L, t > 0$
   $u(x, 0) = f(x), 0 < x < L$
   $u(0, t) = 0 = u(L, t), t > 0$

   (b) $u_t - a^2 u_{xx} = 0, 0 < x < L, t > 0$
   $u(x, 0) = f(x), 0 < x < L$
   $u_x(0, t) = 0 = u_x(L, t), t > 0$

   (c) $u_t - a^2 u_{xx} = 0, 0 < x < L, t > 0$
   $u(x, 0) = f(x), 0 < x < L$
   $u_x(0, t) = 0 = u(L, t), t > 0$

2. Repeat Exercise 1 using the Backward difference method of Table 8.2.2.

3. Repeat Exercise 1 using the Crank–Nicolson method of Table 8.2.2

4. Given the initial-value problem,

$$u_t - a^2 u_{xx} = 0, \qquad -\infty < x < \infty, \, t > 0$$
$$u(x, 0) = f(x), \qquad -\infty < x < \infty,$$

show that the von Neumann stability condition is satisfied for all h and k if
(a) the backward difference method (8.2.7) is used.
(b) the Crank–Nicolson method (8.2.8) is used.

5. Given the initial-value problem

$$u_t - a^2 u_{xx} = cu, \qquad -\infty < x < \infty, \, t > 0, \, c > 0$$
$$u(x, 0) = f(x), \qquad -\infty < x < \infty$$

(a) Use a Fourier integral transform, or another method if you prefer, to obtain the exact solution.

(b) Show that the exact solution grows exponentially in time. When the PDE has a solution that is unbounded in time, the von Neumann stability condition is relaxed to allow errors to grow with, but not faster than, the solution of the FDE. This relaxed von Neumann stability condition is

$$\left| \frac{Q(e^{i\beta}, e^{-i\beta})}{P(e^{i\beta}, e^{-i\beta})} \right| \leq 1 + O(k), \qquad 0 \leq \beta < 2\pi$$

(c) Use the forward difference method (8.2.6) to difference the initial-value problem and show that the relaxed von Neumann stability condition of part (b) is satisfied provided $r \leq \frac{1}{2}$.

6. Rework Problem 1 using the Gerschgorin circle theorem to estimate the eigenvalues of the difference matrices $\mathbf{F}_1, \mathbf{F}_2, \mathbf{F}_3$.

7. Let λ_n and $\mathbf{V}_n$, $n = 1, 2, \ldots, N$, be the eigenvalues and eigenvectors of the $N \times N$ matrix $\mathbf{F}_1$. Let $f(x)$ be a continuous function on $0 \leq x \leq 1$ and set $f_n = f(nh)$, $n = 1, 2, \ldots, N$ where $h = 1/(N + 1)$.
(a) If $f = \sum_1^N c_n \mathbf{V}_n$, find an expression for c_n.
(b) Show that $\mathbf{f} \cdot \mathbf{f} = [(N + 1)/2] \sum_1^N c_n^2$.
(c) If $\|\mathbf{f}\|$ denotes the energy norm of $\mathbf{f}$ show that

$$\|\mathbf{f}\| = \text{const} \sum_1^N c_n^2$$

and determine the value of the constant.
(d) If $\mathbf{U}^j$ is a solution of the difference system (8.5.12)–(8.5.13) given by $\mathbf{U}^j = \sum_1^N c_n^j \mathbf{V}_n$, show $\|\mathbf{U}^j\| \leq \|\mathbf{f}\|$ provided the spectral stability condition holds.

8. Given the Fourier representation

$$U_n^j = \sum_{m=0}^{N-1} \xi_m^j e^{imx_n}, \qquad \xi_m^j = \frac{1}{N} \sum_{n=0}^{N-1} U_n^j e^{-imx_n}$$

$$x_n = -\pi + nh, \qquad n = 0, 1, 2, \ldots, N - 1, \qquad h = 2\pi/N$$

(a) Establish the discrete Parseval equation

$$\|U^j\| = (2\pi)^{1/2} \left\{ \sum_{n=0}^{N-1} (\xi_n^j)^2 \right\}^{1/2}$$

for the energy norm

$$\|\mathbf{f}\| = \left\{ \sum_{n=0}^{N-1} f_n^2 h \right\}^{1/2}$$

In connection with 8(a), see Theorem 2.5.2.

(b) Assuming that the von Neumann stability condition,

$$\left| \frac{\xi_n^{j+1}}{\xi_n^j} \right| \le 1$$

show that $\|\mathbf{U}^j\| \le \|\mathbf{U}^{j-1}\|$.

(c) Establish (2) of Theorem 8.5.3.

9. A sequence $\{f_n\}$, $n = 0, \pm 1, \pm 2, \ldots$, is said to be in l^2 provided

$$\sum_{n=-\infty}^{\infty} |f_n|^2 < \infty$$

A function $\xi(\beta)$ is said to be in $L^2[a, b]$ provided

$$\int_a^b |\xi(\beta)|^2 \, d\beta < \infty$$

Given that to each sequence $\{f_n\}$ in l^2 there corresponds a function $\xi(\beta)$ in $L^2[-\pi, \pi]$ such that

$$\xi(\beta) = \sum_{n=-\infty}^{\infty} f_n e^{i\beta n}$$

(a) Show

$$f_n = \frac{1}{2\pi} \int_{-\pi}^{\pi} \xi(\beta) e^{-i\beta n} \, d\beta$$

$$\left(\text{Hint: } \int_{-\pi}^{\pi} e^{i\beta n} e^{-i\beta m} \, d\beta = \begin{cases} 0, & n \ne m \\ 2\pi, & n = m \end{cases} \right)$$

(b) Show

$$\sum_{n=-\infty}^{\infty} |f_n|^2 = \frac{1}{2\pi} \int_{-\pi}^{\pi} |\xi(\beta)|^2 \, d\beta$$

10. Let U_n^j represent a solution to the finite difference equation $P(S_+, S_-)U_n^{j+1} = Q(S_+, S_-)U_n^j$. Assume that for each $j = 0, 1, 2, \ldots$, U_n^j is in l^2.

(a) Use the last problem to show

$$U_n^j = \frac{1}{2\pi} \int_{-\pi}^{\pi} \xi^j(\beta) e^{-i\beta n} \, d\beta; \qquad \xi^j(\beta) = \sum_{n=-\infty}^{\infty} U_n^j e^{i\beta n}$$

and

$$\|\mathbf{U}^j\| \equiv \sum_{n=-\infty}^{\infty} |U_n^j|^2 = \frac{1}{2\pi} \int_{-\pi}^{\pi} |\xi^j(\beta)|^2 \, d\beta$$

(b) Using the finite difference equation shows that

$$\frac{1}{2\pi} \int_{-\pi}^{\pi} \xi^{j+1} P(e^{i\beta}, e^{-i\beta}) e^{i\beta n} \, d\beta = \frac{1}{2\pi} \int_{-\pi}^{\pi} \xi^j Q(e^{i\beta}, e^{-i\beta}) e^{i\beta n} \, d\beta$$

(c) Using part (b) show

$$\xi^{j+1} = \frac{Q(e^{i\beta}, e^{-i\beta})}{P(e^{i\beta}, e^{-i\beta})} \xi^j$$

(d) Show that if the von Neumann stability condition holds, then

$$\|\mathbf{U}^{j+1}\| \le \|\mathbf{U}^j\| \le \cdots \le \|\mathbf{U}^1\| \le \|\mathbf{f}\|$$

so the method is stable in the sense of (8.5.3).

11. Let $\mathbf{V}_n$ be an eigenvalue of the $N \times N$ matrix $\mathbf{F}_1$.

(a) Show that if an N vector $\mathbf{U}^j$ has the eigenvector expansion

$$\mathbf{U}^j = \sum_{n=0}^{N} c_n^j \mathbf{V}_n, \qquad c_n^j = \frac{\mathbf{U}^j \cdot \mathbf{V}_n}{\mathbf{V}_n \cdot \mathbf{V}_n}$$

then, in the discrete energy norm, the following discrete Parseval equation holds.

$$\|\mathbf{U}^j\| = (2)^{-1/2} \left\{ \sum_{n=1}^{N} |c_n^j|^2 \right\}^{1/2}$$

(b) Use part (a) to show that if the spectral stability condition (8.5.16) holds and if $\mathbf{U}^j$ satisfies the difference system (8.5.12)–(8.5.13), then $\|\mathbf{U}^j\| \le \|\mathbf{U}^{j-1}\|$.

(c) Show that if $F = F_1$ in (8.5.12) then the spectral stability condition is necessary and sufficient for the difference method to be stable in the sense of (8.5.3)

8.6 DIFFERENCE METHODS IN TWO SPACE DIMENSIONS

In this section we present some finite-difference methods for dealing with parabolic equations in two space dimensions. We begin with the constant-coefficient diffusion equation for an unknown function $u = u(x, y, t)$. Consider the following initial-boundary value problem on a rectangular region Ω (see Figure 8.6.1)

$$u_t - a^2(u_{xx} + u_{yy}) = S(x, y, t), \qquad \begin{array}{l} 0 < x < X, \\ 0 < y < Y, \end{array} t > 0 \qquad (8.6.1)$$

$$u(x, y, 0) = f(x, y), \qquad 0 < x < X, \, 0 < y < Y \qquad (8.6.2)$$

$$\begin{array}{ll} u(0, y, t) = px(y, t), & \\ u(X, y, t) = qx(y, t) & 0 < y < Y, \\ u(x, 0, t) = py(x, t), & t > 0, \\ u(x, Y, t) = qy(x, t) & 0 < x < X \end{array} \qquad (8.6.3)$$

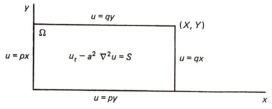

Figure 8.6.1 Boundary conditions for (8.6.1)–(8.6.3).

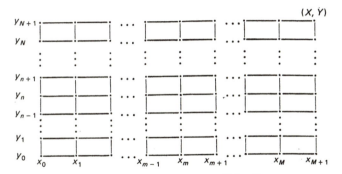

Figure 8.6.2 Rectangular finite-difference grid for Ω.

The exact solution of problem (8.6.1)–(8.6.3) in terms of Fourier series is discussed in Example 3.2.4.

To approximate the solution of (8.6.1)–(8.6.3), let us introduce a rectangular, uniform grid on the region Ω by defining

$$x_m = m \cdot hx, \qquad m = 0, 1, 2, \ldots, M + 1; \qquad hx = X/(M + 1)$$

and

$$y_n = n \cdot hy, \qquad n = 0, 1, 2, \ldots, N + 1; \qquad hy = Y/(N + 1)$$

See Figure 8.6.2. Using the grid of Figure 8.6.2, we extend our subscript–superscript notation by setting

$$u^j_{mn} = u(x_m, y_n, t_j)$$

Similarly, we let U^j_{mn} denote a numerical approximation of u^j_{mn}. For centered second difference we adopt the notation

$$\delta^2_x U^j_{mn} = U^j_{m-1,n} - 2U^j_{mn} + U^j_{m+1,n} \tag{8.6.4}$$

and

$$\delta^2_y U^j_{mn} = U^j_{m,n-1} - 2U^j_{mn} + U^j_{m,n+1} \tag{8.6.5}$$

Explicit Difference Method

Using the above notation, it is easy to construct an explicit, forward-in-time difference equation for (8.6.1). Namely,

$$\frac{U^{j+1}_{mn} - U^j_{mn}}{k} - a^2 \left[\frac{\delta^2_x U^j_{mn}}{hx^2} + \frac{\delta^2_y U^j_{mn}}{hy^2} \right] = S^j_{mn} \tag{8.6.6}$$

where we have used S^j_{mn} to denote $S(x_m, y_n, t_j)$. If we set

$$rx = ka^2/(hx)^2 \quad \text{and} \quad ry = ka^2/(hy)^2$$

then (8.6.6) can be solved for U^j_{mn} to obtain

$$U^{j+1}_{mn} = U^j_{mn} + rx \cdot \delta^2_x U^j_{mn} + ry \cdot \delta^2_y U^j_{mn} + k \cdot S^j_{mn} \tag{8.6.7}$$

Together with the initial condition

$$U^0_{mn} = f_{mn}, \qquad m = 1, 2, \ldots, M, \, n = 1, 2, \ldots, N \qquad (8.6.8)$$

and the boundary conditions

$$U^j_{on} = px^j_n, \qquad U^j_{Mn} = qx^j_n, \qquad \begin{matrix} n = 1, 2, \ldots, N, \\ j = 1, 2, \ldots \end{matrix}$$

$$U^j_{mo} = py^j_m, \qquad U^j_{mN} = qy^j_m, \qquad \begin{matrix} m = 1, 2, \ldots, M, \\ j = 1, 2, \ldots \end{matrix} \qquad (8.6.9)$$

the difference equations (8.6.7) can be used to obtain a numerical approximation to the solution of the initial-boundary-value problem (8.6.1)–(8.6.3). To advance the solution from t_j to t_{j+1} requires we use five values of U^j_{mn}, U^j_{mn}, $U_{m\pm 1,n}$, $U_{m,n\pm 1}$ and the value of S^j_{mn}. See Figure 8.6.3.

The forward-in-time difference method given by (8.6.7)–(8.6.9) is a straightforward method for obtaining an approximation to the solution of the initial-boundary-value problem (8.6.1)–(8.6.3). However, to obtain a stable calculation using this finite-difference method requires that the time step be chosen sufficiently small. In fact, the stability condition for (8.6.7) is more restrictive than the stability condition for its one-space-variable counterpart.

We can perform a von Neumann stability analysis of the finite-difference method (8.6.7) by supposing the solution to be expressed as a discrete double Fourier series. A typical harmonic, in complex notation, has the form

$$\xi^j e^{im \cdot hx + n \cdot hy}, \qquad i = \sqrt{-1}$$

If we put this expression into (8.6.7) with S^j_{mn} set to zero, then we can solve for the magnification factor associated with this harmonic and find (see Exercise 5)

$$\frac{\xi^{j+1}}{\xi^j} = \left(1 - 4rx \sin^2 \frac{m \cdot hx}{2} - 4ry \sin^2 \frac{n \cdot hy}{2}\right)$$

For the von Neumann stability condition

$$|\xi^{j+1}/\xi^j| \le 1$$

to hold for all choices of $m \cdot hx$ and $n \cdot hy$, it follows that we must restrict rx and ry to satisfy $rx + ry \le \frac{1}{2}$. In applications, this stability restriction sometimes makes it impractical to use the explicit difference method (8.6.7). For example, on the region

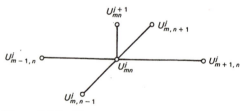

Figure 8.6.3 Five-point computational stencil for forward difference method (8.6.7).

$0 < x < 1, 0 < y < 1$, if we choose $hx = hy = 0.01$, then for stability we must restrict the time step k to satisfy $k \leq 2.5 \times 10^{-5}$. In this case we would require 40,000 time steps to advance the solution from $t = 0$ to $t = 1$. We can summarize our discussion of the explicit forward-in-time method for dealing with (8.6.1)–(8.6.3) as follows.

Theorem 8.6.1 The forward-in-time finite-difference method (8.6.6) has local truncation error $O(k + hx^2 + hy^2)$ and it is stable if and only if $rx + ry \leq \frac{1}{2}$.

Exact Solution by Fourier's Method

Following the approach of Section 8.4, we can obtain the exact solution to the difference equation (8.6.7) in the case that S is zero and the boundary conditions are homogeneous. This exact solution clearly displays the role of the stability condition of Theorem 8.6.1. Also, the exact solution of the difference system can be used to verify computer code written to implement the method (8.6.7)–(8.6.9).

Assume the boundary conditions are homogeneous,

$$px_{mn}^j = qx_{mn}^j = 0; \qquad py_{mn}^j = qy_{mn}^j = 0$$

and $S_{mn}^j = 0$. Following the solution of the one-dimensional difference equation in Section 8.4, we look for solutions to

$$U_{mn}^{j+1} - U_{mn}^j - rx\,\delta_x^2 U_{mn}^j - ry\,\delta_y^2 U_{mn}^j = 0$$

in the form

$$U_{mn}^j = T^j \Phi_m \Psi_n, \qquad \Phi_0 = \Phi_{M+1} = 0, \qquad \Psi_0 = \Psi_{N+1} = 0$$

Putting $T^j \Phi_m \Psi_n$ into the difference equation and separating indices we conclude

$$\frac{T^{j+1} - T^j}{T^j} - \frac{rx\,\delta_x^2 \Phi_m}{\Phi_m} = \frac{ry\,\delta_y^2 \Psi_n}{\Psi_n}$$

Since the left side of the last equation depends only on the indices j and m and the right side depends only on n, it follows that

$$\frac{ry\,\delta_y^2 \Psi_n}{\Psi_n} = \text{const}$$

From the homogeneous boundary conditions, $\Psi_0 = \Psi_{N+1} = 0$, it follows from Theorem 8.4.1 that for Ψ_n to not be identically zero on the grid, it must be that

$$\frac{\text{const}}{ry} = -\lambda_q = -2 + 2\cos\frac{q\pi}{N+1} = -4\sin^2\frac{q\pi}{2(N+1)}, \qquad q = 1, 2, \ldots, N$$

and

$$\Psi_n = \beta_n \sin q\pi y_n, \qquad q = 1, 2, \ldots, N, \qquad \beta_n = \text{const}$$

Similar arguments show the x-eigenvector components and corresponding eigenvalues to be given by

$$\Phi_m = \alpha_m \sin p\pi x_m, \qquad p = 1, 2, \ldots, M, \qquad \alpha_m = \text{const}$$

and

$$-\lambda_p = -2 + 2 \cos \frac{p\pi}{M + 1} = -4 \sin^2 \frac{p\pi}{2(M + 1)}, \qquad p = 1, 2, \ldots, M$$

Also, it is easy to see that the transient part of the solution satisfies the recursion formula

$$T^j = (1 + rx\lambda_p + ry\lambda_q)T^{j-1} = (1 + rx\lambda_p + ry\lambda_q)^j T^0, \qquad j = 1, 2, \ldots$$

We now use superposition to express the solution of the difference system in the form

$$U^j_{mn} = \sum_{p=1}^{M} \sum_{q=1}^{N} \left(1 - 4rx \sin^2 \frac{p\pi}{2(M + 1)} - 4ry \sin^2 \frac{q\pi}{2(N + 1)} \right)^j$$
$$\times T^0 \alpha_m \beta_n \sin p\pi x_m \sin q\pi y_n$$

Finally, we use the initial condition and the orthogonality result from Theorem 8.4.1 to fix the values of $T^0 \alpha_m \beta_n$ to be

$$T^0 \alpha_m \beta_n = \frac{4}{(M + 1)(N + 1)} \sum_{p=1}^{M} \sum_{q=1}^{N} f_{mn} \sin p\pi x_m \sin q\pi y_n$$

This exact solution remains bounded for all j if and only if

$$\left| 1 - 4rx \sin^2 \frac{p\pi}{2(M + 1)} - 4ry \sin^2 \frac{q\pi}{2(N + 1)} \right| \leq 1$$

for $p = 1, 2, \ldots, M$, $q = 1, 2, \ldots, N$, with M and N arbitrary. Thus, for stability $rx + ry \leq \frac{1}{2}$ must hold.

Implicit Difference Methods

A backward-in-time, implicit difference method for dealing with (8.6.1)–(8.6.3) can be constructed by simply evaluating the spatial differences in (8.6.7) at t_{j+1} rather than t_j. Namely,

$$U^{j+1}_{mn} - [rx \, \delta^2_x U^{j+1}_{mn} + ry \, \delta^2_y U^{j+1}_{mn}] = U^j_{mn} + kS^{j+1}_{mn} \qquad (8.6.10)$$
$$U^0_{mn} = f_{mn}, \qquad m = 1, 2, \ldots, M, n = 1, 2, \ldots, N \qquad (8.6.11)$$

$$U^j_{on} = px^j_n, \qquad U^j_{Mn} = qx^j_n, \qquad \begin{array}{l} n = 1, 2, \ldots, N, \\ j = 1, 2, \ldots \end{array}$$

$$U^j_{mo} = py^j_m, \qquad U^j_{mN} = qy^j_m, \qquad \begin{array}{l} m = 1, 2, \ldots, M, \\ j = 1, 2, \ldots \end{array} \qquad (8.6.12)$$

The finite-difference system (8.6.10)–(8.6.12) is the two-dimensional extension of (8.2.7). In a similar fashion, we can extend the Crank–Nicolson method (8.2.8) to two space dimensions as follows:

$$[1 - \tfrac{1}{2}rx \, \delta^2_x - \tfrac{1}{2}ry \, \delta^2_y]U^{j+1}_{mn} = [1 + \tfrac{1}{2}rx \, \delta^2_x + \tfrac{1}{2}ry \, \delta^2_y] U^j_{mn}$$

Theorem 8.6.2. The backward-in-time method (8.6.10) has local truncation error $O(k + hx^2 + hy^2)$ and it is unconditionally stable.

Theorem 8.6.3.   The Crank–Nicolson method has local truncation error $O(k^2 + hx^2 + hy^2)$ and it is unconditionally stable.

A von Neumann stability analysis shows that the difference method (8.6.10)–(8.6.12) is unconditionally stable. However, to obtain this stability, we have sacrificed computational simplicity. From Figure 8.6.4 we see that, in general, each of the linear equations defined by (8.6.10) involves five unknown U_{mn} values at time t_{j+1}. If we arrange the unknowns that appear in (8.6.10) in the order

$$U_{11}^{j+1}, U_{21}^{j+1}, \ldots, U_{M1}^{j+1}, U_{12}^{j+1}, U_{22}^{j+1}, \ldots, U_{M2}^{j+1}, \ldots, U_{1N}^{j+1}, U_{2N}^{j+1}, \ldots, U_{MN}^{j+1}$$

then it follows that the linear equations defined by (8.6.10) have the form

$$\mathbf{A}\mathbf{U}^{j+1} = \mathbf{c}$$

or

$$
\begin{bmatrix}
\mathbf{B} & -ry\mathbf{I} & & & & & & & \\
-ry\mathbf{I} & \mathbf{B} & -ry\mathbf{I} & & & & & & \\
& \ddots & \ddots & \ddots & & & & & \\
& & -ry\mathbf{I} & \mathbf{B} & -ry\mathbf{I} & & & & \\
& & & -ry\mathbf{I} & \mathbf{B} & -ry\mathbf{I} & & & \\
& & & & -ry\mathbf{I} & \mathbf{B} & -ry\mathbf{I} & & \\
& & & & & \ddots & \ddots & \ddots & \\
& & & & & & -ry\mathbf{I} & \mathbf{B} & -ry\mathbf{I} \\
& & & & & & & -ry\mathbf{I} & \mathbf{B}
\end{bmatrix}
\begin{bmatrix}
\{\vec{\mathbf{U}}_1^{j+1}\} \\
\{\vec{\mathbf{U}}_2^{j+1}\} \\
\vdots \\
\{\vec{\mathbf{U}}_{n-1}^{j+1}\} \\
\{\vec{\mathbf{U}}_n^{j+1}\} \\
\{\vec{\mathbf{U}}_{n+1}^{j+1}\} \\
\vdots \\
\{\vec{\mathbf{U}}_{N-1}^{j+1}\} \\
\{\vec{\mathbf{U}}_N^{j+1}\}
\end{bmatrix}
=
\begin{bmatrix}
\{\vec{\mathbf{C}}_1^j\} \\
\{\vec{\mathbf{C}}_2^j\} \\
\vdots \\
\{\vec{\mathbf{C}}_{n-1}^j\} \\
\{\vec{\mathbf{C}}_n^j\} \\
\{\vec{\mathbf{C}}_{n+1}^j\} \\
\vdots \\
\{\vec{\mathbf{C}}_{N+1}^j\} \\
\{\vec{\mathbf{C}}_N^j\}
\end{bmatrix}
\quad (8.6.13)
$$

where we have used the notation

$$
\{\vec{\mathbf{U}}_n^{j+1}\} =
\begin{bmatrix}
U_{1,n}^{j+1} \\
\vdots \\
U_{m-1,n}^{j+1} \\
U_{m,n}^{j+1} \\
U_{m+1,n}^{j+1} \\
\vdots \\
U_{M,n}^{j+1}
\end{bmatrix}
; \quad
\mathbf{B} =
\begin{bmatrix}
b & -rx & & & & & \\
& \ddots & \ddots & & & & \\
-rx & b & -rx & & & \\
& -rx & b & -rx & & \\
& & -rx & b & -rx & \\
& & & \ddots & \ddots & \ddots \\
& & & & -rx & b
\end{bmatrix}
$$

$$b = (1 + 2rx + 2ry) \tag{8.6.14}$$

The determination of the components of the right side of (8.6.13) is left to Exercise 14.

The coefficient matrix $\mathbf{A}$ in (8.6.13) is seen to be an $N \times N$ *block tridiagonal matrix* with $M \times M$ blocks defined by $\mathbf{B}$ and $\mathbf{I}$. This coefficient matrix can also be described as *five banded* since all of the nonzero entries of $\mathbf{A} = [a_{ij}]$ are confined to just the five bands defined by a_{pp}, $a_{p,p\pm1}$, and $a_{p,p\pm M}$.

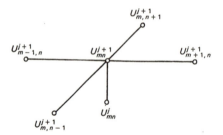

Figure 8.6.4 Computational stencil for backward-in-time method (8.6.10).

Linear systems of the form (8.6.13) occur frequently in the numerical solution of partial differential equations in two space variables. A number of methods for dealing with systems of this form have been developed. One direct method for solving (8.6.13) is a version of Gaussian elimination that exploits the five bandedness of the coefficient matrix. The logic of a *five-banded linear solver* is very similar to that used in our tridiagonal solver. However, in the forward sweep that eliminates the subdiagonal entries, the bands between $a_{p,p+1}$ and $a_{p,p+M}$ are generally filled with nonzero entries. If one is willing to store **A**, including all its nonzero entries, in an $MN \times MN$ array, it is not difficult to construct a five-banded linear solver. However, the storage requirements in this approach are often unacceptable. As an alternative to developing code to solve (8.6.13), a number of computer library routines are available. For example, IMSL and LINPAK have very efficient and well-tested routines for solving (8.6.13).

In the case that the initial-boundary-value problem (8.6.1)–(8.6.3) has $S = 0$ and homogeneous boundary conditions, it is possible to construct discrete double Fourier series to provide the exact solution of the difference equations (8.6.10)–(8.6.12). This approach closely follows the development given above and in Section 8.4. See Exercise 17.

Another approach to solving (8.6.13) is to construct a sequence of approximations to its exact solution. This approach is known as an *indirect method* or an *iterative method* for solving (8.6.13). One popular iterative method is successive overrelaxation (SOR). Iterative methods are discussed in Section 10.4. It is possible to proceed from this point to that section with no loss of continuity. In general, the computations required to implement an iterative method are quite easy to code.

Alternating-Direction Implicit Methods

Alternating-direction implicit (ADI) finite-difference methods for (8.6.1)–(8.6.3) have been developed to overcome the stability restriction encountered in using explicit methods and to avoid the five-banded matrices that arise in backward-in-time or a Crank–Nicolson difference system. Let us begin with the Crank–Nicolson method for (8.6.1),

$$[1 - \tfrac{1}{2}r\,\delta_x^2 - \tfrac{1}{2}r\,\delta_y^2]U_{mn}^{j+1} = [1 + \tfrac{1}{2}r\,\delta_x^2 + \tfrac{1}{2}r\,\delta_y^2]U_{mn}^j \qquad (8.6.15)$$

in which we have, for simplicity, set $S(x, y, t) = 0$ and selected a square grid with $hx = hy = h$ and

$$rx = ry = r = a^2k/h^2$$

Recall that the Crank–Nicolson method has local truncation error $O(k^2 + h^2)$. Since

$$r^2 \, \delta_x^2 \, \delta_y^2 (u_{mn}^{j+1} - u_{mn}^j) = k^3 u_{xxyyt}(x_m, y_n, t_{j+1/2}) + O(k^4 + k^3 h^2)$$

it follows that the difference method

$$[1 - \tfrac{1}{2}r \, \delta_x^2][1 - \tfrac{1}{2}r \, \delta_y^2] U_{mn}^{j+1} = [1 + \tfrac{1}{2}r \, \delta_x^2][1 + \tfrac{1}{2}r \, \delta_y^2] U_{mn}^j \qquad (8.6.16)$$

is equivalent to the Crank–Nicolson method up to the order of the truncation error of the Crank–Nicolson method. Now Equation (8.6.16) suggests a difference method of the form

$$[1 - \tfrac{1}{2}r \, \delta_x^2] U_{mn}^* = \mathcal{A} U_{mn}^j \qquad (8.6.17a)$$

$$[1 - \tfrac{1}{2}r \, \delta_y^2] U_{mn}^{j+1} = \mathcal{B} U_{mn}^* \qquad (8.6.17b)$$

where $\mathcal{A}$ and $\mathcal{B}$ are difference operators to be chosen so that, when U_{mn}^* is eliminated from (8.6.17a)–(8.6.17b), the resulting difference equation is equivalent to (8.6.16) up to the order of the truncation error of (8.6.16). Several choices of $\mathcal{A}$ and $\mathcal{B}$ are possible. Selecting

$$\mathcal{A} = [1 + \tfrac{1}{2}r \, \delta_y^2] \quad \text{and} \quad \mathcal{B} = [1 + \tfrac{1}{2}r \, \delta_x^2]$$

gives the Peaceman–Rachford ADI method,

$$[1 - \tfrac{1}{2}r \, \delta_x^2] U_{mn}^* = [1 + \tfrac{1}{2}r \, \delta_y^2] U_{mn}^j \qquad (8.6.18a)$$

$$[1 - \tfrac{1}{2}r \, \delta_y^2] U_{mn}^{j+1} = [1 + \tfrac{1}{2}r \, \delta_x^2] U_{mn}^* \qquad (8.6.18b)$$

As Algorithm 8.7 shows, the implementation of the Peaceman–Rachford method requires only that we solve tridiagonal systems of linear equations.

In setting up the tridiagonal system defined by (8.6.18a), boundary conditions are required for $U_{o,n}^*$ and $U_{M,n}^*$. By adding the right side of (8.6.18b) to the left side of (8.6.18a), we can see that U_{mn}^* satisfies

$$2U_{mn}^* = [1 - \tfrac{1}{2}r \, \delta_y^2] U_{mn}^{j+1} + [1 + \tfrac{1}{2}r \, \delta_y^2] U_{mn}^j$$

Thus on the boundary we should assign U_{mn}^* the following values

$$U_{on}^* = \tfrac{1}{2}[1 - \tfrac{1}{2}r \, \delta_y^2] px(y_n, t_{j+1}) + \tfrac{1}{2}[1 + \tfrac{1}{2}r \, \delta_y^2] px(y_n, t_j)$$

and

$$U_{mmax,n}^* = \tfrac{1}{2}[1 - \tfrac{1}{2}r \, \delta_y^2] qx(y_n, t_{j+1}) + \tfrac{1}{2}[1 + \tfrac{1}{2}r \, \delta_y^2] qx(y_n, t_j)$$

For the case of time independent boundary conditions the above equations for the boundary values of U^* reduce to

$$U_{on}^* = px(y_n) \quad \text{and} \quad U_{m,max,n}^* = qx(y_n)$$

Theorem 8.6.4. The Peaceman–Rachford method (8.6.18a) and (8.6.18b) has local truncation error $O(k^2 + hx^2 + hy^2)$ and it is unconditionally stable.

ALGORITHM 8.7   Peaceman–Rachford Method: Dirichlet Initial-Boundary-Value Problem

Step 1. Document

This algorithm uses the Peaceman–Rachford method (8.6.18) to approximate the solution to the initial-boundary value problem

$$u_t - a^2(u_{xx} + u_{yy}) = 0, \qquad \begin{matrix} 0 < x < X \\ 0 < y < Y \end{matrix} \, t > 0$$

$$u(x, y, 0) = f(x, y), \qquad \begin{matrix} 0 < x < X \\ 0 < y < Y \end{matrix}$$

$$u(0, y, t) = px(y); \; u(X, y, t) = qx(y), \qquad 0 < y < Y, t > 0$$

$$u(x, 0, t) = py(x); \; u(x, Y, t) = qy(x), \qquad 0 < x < X, t > 0$$

INPUT

Real a^2, diffusivity
Reals X, Y, x and y dimensions
Real k, time step
Integer $jmax$, number of time steps
Integers $mmax$, $nmax$, number of nodes in x, y directions
Function $f(x, y)$, initial condition
Functions $px(y)$, $qx(y)$, $py(x)$, $qy(x)$, boundary conditions

OUTPUT

time t and approximation U_{mn}^j to $u(x_m, y_n, t_j)$ for
$m = 0, 1, \ldots, nmax + 1; n = 0, 1, \ldots, nmax + 1$ and
$j = 0, 1, \ldots, jmax$

Step 2. Define a grid

Set $hx = X/(mmax + 1)$
  $hy = Y/(nmax + 1)$
  $rx = a^2k/(hx)^2$
  $ry = a^2k/(hy)^2$

Step 3. Initialize numerical solution

Set $t = 0$
  $x_0 = 0$
  $y_0 = 0$
For $m = 1, 2, \ldots, mmax$
  For $n = 1, 2, \ldots, nmax$ set

$$x_m = x_{m-1} + hx$$
$$y_n = y_{n-1} + hy$$
$$U_{mn} = f(x_m, y_n)$$
$$x_{mmax+1} = X$$
$$y_{nmax+1} = Y$$

For $m = 1, 2, \ldots, mmax$ set
$$U_{m,0} = py(x_m)$$
$$U_{m,nmax+1} = qy(x_m)$$

For $n = 1, 2, \ldots, nmax$ set
$$U_{0,n} = px(y_n)$$
$$U_{mmax+1,n} = qx(y_n)$$

Output t

For $n = nmax + 1, nmax, \ldots, 2, 1, 0$

    For $m = 0, 1, \ldots, mmax + 1$

        Output x_m, y_n, U_{mn}

Step 4. Begin time stepping

For $j = 1, 2, \ldots, jmax$
    Do steps 5–11

Step 5 Begin y sweep

For $n = 1, 2, \ldots, nmax$
    Do steps 6–7

Step 6. Define tridiagonal system for (8.6.18a)

For $m = 1, 2, \ldots, mmax$ set
$$a_m = -rx/2$$
$$b_m = (1 + rx)$$
$$c_m = -rx/2$$
$$d_m = \left(\frac{ry}{2}\right) U_{m,n-1} + (1 - ry)U_{mn} + \left(\frac{ry}{2}\right)U_{m,n+1}$$

Set $d_1 = d_1 + \left(\frac{rx}{2}\right) \cdot U_{0,n}$

$$d_{mmax} = d_{mmax} + \left(\frac{rx}{2}\right) \cdot U_{mmax+1,n}$$

Step 7. Solve for $V_{mn} = U_{mn}^*$

CALL TRIDI($mmax, a, b, c, d$)
For $m = 1, 2, \ldots, mmax$ set
$$V_{mn} = d_m$$

Step 8. Begin x sweep

For $m = 1, 2, \ldots, mmax$
    Do steps 9–10

Step 9. Define tridiagonal system for (8.6.18b)

For $n = 1, 2, \ldots, nmax$ set
$$a_n = -ry/2$$
$$b_n = (1 + ry)$$
$$c_n = -ry/2$$
$$d_n = \left(\frac{rx}{2}\right) V_{m-1,n} + (1 - rx)V_{mn} + \left(\frac{rx}{2}\right) V_{m+1,n}$$

Set $d_1 = d_1 + \left(\frac{ry}{2}\right) \cdot U_{m,0}$

$$d_{nmax} = d_{nmax} + \left(\frac{ry}{2}\right) \cdot U_{m,nmax+1}$$

Step 10. Solve for U_{mn}^{j+1}

CALL TRIDI($nmax, a, b, c, d$)
For $n = 1, 2, \ldots, nmax$ set
$$U_{mn} = d_n$$

Step 11. Output numerical solution

Set $t = t + k$
Output t
For $n = nmax + 1, nmax, \ldots, 2, 1$
    For $m = 0, 1, \ldots, mmax + 1$
        Output x_{mn}, U_{mn}

EXAMPLE 8.6.1 _____

Consider the initial-boundary-value problem

$$u_t - (u_{xx} + u_{yy}) = 0, \quad 0 < x < 1, 0 < y < 1, t > 0$$
$$u(x, y, 0) = 100 \sin \pi x \sin \pi y, \quad 0 < x < 1, 0 < y < 1$$
$$u(0, y\ t) = 0 = u(1, y, t), \quad 0 < y < 1, t > 0$$
$$u(x, 0, t) = 0 = u(x, 1, t), \quad 0 < x < 1, t > 0$$

whose exact solution is given by

$$u(x, y, t) = 100e^{-2\pi^2 t} \sin \pi x \sin \pi y$$

TABLE 8.6.1 COMPARISON OF NUMERICAL AND EXACT SOLUTIONS FOR EXAMPLE 8.6.1
(T = 0.10)

Numerical Solution

					m				
n	1	2	3	4	5	6	7	8	9
9	1.35	2.56	3.52	4.14	4.36	4.14	3.52	2.56	1.35
8	2.56	4.87	6.70	7.88	8.29	7.88	6.70	4.87	2.56
7	3.52	6.70	9.23	10.85	11.40	10.85	9.23	6.70	3.52
6	4.14	7.88	10.85	12.75	13.41	12.75	10.85	7.88	4.14
5	4.36	8.29	11.40	13.41	14.10	13.41	11.40	8.29	4.36
4	4.14	7.88	10.85	12.75	13.41	12.75	10.85	7.88	4.14
3	3.52	6.70	9.23	10.85	11.40	10.85	9.23	6.70	3.52
2	2.56	4.87	6.70	7.88	8.29	7.88	6.70	4.87	2.56
1	1.35	2.56	3.52	4.14	4.36	4.14	3.52	2.56	1.35

Exact Solution

					m				
n	1	2	3	4	5	6	7	8	9
9	1.33	2.52	3.47	4.08	4.29	4.08	3.47	2.52	1.33
8	2.52	4.80	6.61	7.77	8.16	7.77	6.61	4.80	2.52
7	3.47	6.61	9.09	10.69	11.24	10.69	9.09	6.61	3.47
6	4.08	7.77	10.69	12.56	13.21	12.56	10.69	7.77	4.08
5	4.29	8.16	11.24	13.21	13.89	13.21	11.24	8.16	4.29
4	4.08	7.77	10.69	12.56	13.21	12.56	10.69	7.77	4.08
3	3.47	6.61	9.09	10.69	11.24	10.69	9.09	6.61	3.47
2	2.52	4.80	6.61	7.77	8.16	7.77	6.61	4.80	2.52
1	1.33	2.52	3.47	4.08	4.29	4.08	3.47	2.52	1.33

Choosing X and $Y = 1$, *mmax* and *nmax* $= 9$ (so that $hx = hy = 0.1$), $k = 0.01$, and *jmax* $= 10$ in Algorithm 8.7 produces the numerical solution displayed in Table 8.6.1.   ■ ■

Another popular ADI method can be developed from the backward-in-time difference method

$$[1 - r\,\delta_x^2 - r\,\delta_y^2]U_{mn}^{j+1} = U_{mn}^j \qquad (8.6.19)$$

The local truncation error for this method is $O(k + h^2)$. It can be shown that

$$[1 - r\,\delta_x^2][1 - r\,\delta_y^2]U_{mn}^{j+1} = U_{mn}^j \qquad (8.6.20)$$

is equivalent to (8.6.19) up to the order of the local truncation error. The following ADI method is due to Douglas and Rachford:

$$[1 - r\,\delta_x^2]U_{mn}^* = [1 + r\,\delta_y^2]U_{mn}^j \tag{8.6.21a}$$

$$[1 - r\,\delta_y^2]U_{mn}^{j+1} = U_{mn}^* - r\,\delta_y^2\,U_{mn}^j \tag{8.6.21b}$$

By eliminating U_{mn}^* from (8.6.21a) and (8.6.21b), we see that

$$[1 - r\,\delta_x^2][1 - r\,\delta_y^2]U_{mn}^{j+1} = [1 + r^2\,\delta_x^2\,\delta_y^2]U_{mn}^j$$

which can be shown to be equivalent to (8.6.20) up to $O(k + h^2)$. Thus we conclude that, up to order $k + h^2$, the Douglas–Rachford method is equivalent to the backward-in-time method (8.6.19).

From (8.6.21b), the boundary conditions for U^* are seen to be

$$U_{0,n}^* = [1 - r\,\delta_y^2]px(y_n, t_{j+1}) + r\,\delta_y^2 px(y_n, t_j)$$

and

$$U_{mmax,n}^* = [1 - r\,\delta_y^2]qx(y_n, t_{j+1}) + r\,\delta_y^2 qx(y_n, t_j)$$

In the time-independent case these boundary conditions, for U^* reduce to

$$U_{0,n}^* = px(y_n) \quad \text{and} \quad U_{mmax,n}^* = qx(y_n)$$

Theorem 8.6.5. The Douglas–Rachford method (8.6.21a) and (8.6.21b) has local truncation error $O(k + hx^2 + hy^2)$ and it is unconditionally stable.

The Peaceman–Rachford method is restricted to partial differential equations in *two* space variables. The Douglas–Rachford method can be generalized to three space variables. Given the partial differential equation

$$u_t = a^2(u_{xx} + u_{yy} + u_{zz}) = 0, \qquad 0 < x < X, 0 < y < Y, 0 < z < Z, t > 0$$

let $(x_m, y_n, z_p) = (m \cdot h, n \cdot h, p \cdot h)$ define a uniform grid and let U_{mnp}^j denote an approximation to $u(x_m, y_n, z_p, t_j)$. Then the Douglas–Rachford method takes the form

$$[1 - r\,\delta_x^2]U_{mnp}^* = [1 + r(\delta_y^2 + \delta_z^2)]U_{mnp}^j \tag{8.6.22a}$$

$$[1 - r\,\delta_y^2]U_{mnp}^{**} = U_{mnp}^* - r\,\delta_y^2 U_{mnp}^j \tag{8.6.22b}$$

$$[1 - r\,\delta_z^2]U_{mnp}^{j+1} = U_{mnp}^{**} - r\,\delta_z^2 U_{mnp}^j \tag{8.6.22c}$$

Exercises

1. Given the initial-boundary-value problem

$$u_t - (u_{xx} + u_{yy}) = -e^{-t}xy, \qquad 0 < x < 1, 0 < y < 1, t > 0$$
$$u(x, y, 0) = xy$$
$$u(0, y, t) = 0, \qquad u(1, y, t) = ye^{-t}$$
$$u(x, 0, t) = 0, \qquad u(x, 1, t) = xe^{-t}$$

Write the difference equations for
(a) explicit, forward-in-time difference method (8.6.7)

(b) implicit, backward-in-time difference method (8.6.10)
(c) Crank–Nicolson method

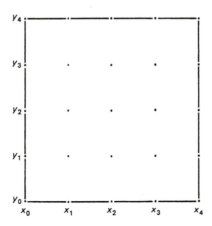

2. In Exercise 1(a) what restriction must k satisfy for the difference method to be stable?

3. In Exercise 1(a) choose $k = 0.01$ and determine the numerical solution at
   (a) $t = 0.01$
   (b) $t = 0.05$
   (c) $t = 0.5$
   and compare the numerical solution with the exact solution

   $$u(x, y, t) = xye^{-t}$$

4. Write the Crank–Nicolson method for the initial-boundary-value problem (8.6.1)–(8.6.3) as a system of linear algebraic equations using matrix notation.

5. In the von Neumann stability analysis of the explicit, forward-in-time difference method (8.6.7) verify that the magnification factor is given by

   $$\frac{\xi^{j+1}}{\xi^j} = \left(1 - 4rx \sin^2 \frac{mhx}{2} - 4ry \sin^2 \frac{nhy}{2}\right)$$

6. Use von Neumann's method to establish the stability result stated in
   (a) Theorem 8.6.2.
   (b) Theorem 8.6.3.

7. Using Equations (8.6.17a)–(8.6.17b) derive the D'Yakonov ADI method,

   $$[1 - \tfrac{1}{2}r \, \delta_x^2]U_{mn}^* = [1 + \tfrac{1}{2}r \, \delta_x^2][1 + \tfrac{1}{2}r \, \delta_y^2]U_{mn}^j$$
   $$[1 - \tfrac{1}{2}r \, \delta_y^2]U_{mn}^{j+1} = U_{mn}^*$$

8. Show that for the nonhomogeneous equation

   $$u_t - a^2(u_{xx} + u_{yy}) = S(x, y, t)$$

   the D'Yakonov method of Exercise 7 takes the form

   $$[1 - \tfrac{1}{2}r \, \delta_x^2]U_{mn}^* = [1 + \tfrac{1}{2}r \, \delta_x^2][1 + \tfrac{1}{2}r \, \delta_y^2]U_{mn}^j + \tfrac{1}{2}k(S_{mn}^j + S_{mn}^{j+1})$$
   $$[1 - \tfrac{1}{2}r \, \delta_y^2]U_{mn}^{j+1} = U_{mn}^*$$

9. Show that if in (8.6.17a) and (8.6.17b) we choose

$$\mathcal{A} = [1 + \tfrac{1}{2}r\,\delta_x^2] \quad \text{and} \quad \mathcal{B} = [1 + \tfrac{1}{2}r\,\delta_y^2]$$

then the result is a *locally one-dimensional* (LOD) ADI method in which the first step involves only the x variable and the second step involves only the y variable. Then show that this LOD method satisfies the von Neumann stability condition for all $r > 0$.

10. Investigate the stability of the ADI method that results from choosing

$$\mathcal{A} = 1 \quad \text{and} \quad \mathcal{B} = [1 + \tfrac{1}{2}r\,\delta_x^2][1 + \tfrac{1}{2}r\,\delta_y^2]$$

in (8.6.17a) and (18.6.17b).

11. Given the initial-boundary-value problem

$$\begin{aligned}
u_t - a^2(u_{xx} + u_{yy}) &= 0, & 0 &< x < X, 0 < y < Y, t > 0 \\
(u(x, y, 0) &= f(x, y), & 0 &< x < X, 0 < y < Y \\
u_x(0, y, t) &= 0 = u_x(X, y, t), & 0 &< y < Y, t > 0 \\
u_y(x, 0, t) &= 0 = u_y(x, Y, t), & 0 &< x < X, t > 0
\end{aligned}$$

   (a) Formulate an explicit system of difference equations using centered differences to approximate all spatial derivatives.
   (b) Use Fourier's method to obtain the exact solution to the finite-difference system in (a).

12. Repeat Exercise 11 for the boundary conditions

$$\begin{aligned}
u_x(0, y, t) &= 0 = u_x(X, y, t), & 0 &< y < Y, t > 0 \\
u(x, 0, t) &= 0 = u(x, Y, t), & 0 &< x < X, t > 0
\end{aligned}$$

13. Repeat Exercise 11 for the boundary conditions

$$\begin{aligned}
u_x(0, y, t) &= 0 = u(X, y, t), & 0 &< y < Y, t > 0 \\
u_y(x, 0, t) &= 0 = u(x, Y, t), & 0 &< x < X, t > 0
\end{aligned}$$

14. Obtain the right side of the matrix difference equation (8.6.13).

15. For the case $\mathbf{c} = \mathbf{0}$, use Fourier's method to solve (8.6.13) subject to the initial condition $\mathbf{U} = \mathbf{f}$.

16. List four methods for solving the backward in time-difference equation (8.6.13) and list and advantage and a disadvantage of each method.

17. Use a von Neumann stability analysis to show that the Peaceman–Rachford method (8.6.18a) and (8.6.18b) is unconditionally stable. Show also that neither equation (8.6.18a) or (8.6.18b) alone provides an unconditionally stable difference method.

18. (a) Use Algorithm 8.7 to approximate the solution of the problem

$$\begin{aligned}
u_t - (u_{xx} + u_{yy}) &= 0, & 0 &< x < 1, 0 < y < 1, t > 0 \\
u(x, y, 0) &= 100 \sin 2\pi x \sin \pi y \\
u(0, y, t) &= 0 = u(1, y, t) \\
u(x, 0, t) &= 0 = u(x, 1, t)
\end{aligned}$$

and compare the numerical solution with the exact solution

$$u(x, y, 0) = 100e^{-5\pi^2 t} \sin 2\pi x \sin \pi y$$

   (b) Use the Douglas–Rachford method to approximate the solution of 18(a).

19. Use Algorithm 8.7 to approximate the solution of the problem

$$u_t - (u_{xx} + u_{yy}) = 0, \qquad 0 < x < 1, 0 < y < 1, t > 0$$
$$u(x, y, 0) = 0$$
$$u(0, y, t) = y^2 \qquad u(1, y, t) = 1 - y^2$$
$$u(x, 0, t) = x^2 \qquad u(x, 1, t) = 1 - x^2$$

and, for large time, compare the numerical solution with the steady-state solution $u(x, y) = x^2 - y^2$.

20. (a) Modify Algorithm 8.7 to handle the time-dependent boundary conditions

$$u(0, y, t) = px(y, t), \qquad u(1, y, t) = qx(y, t)$$
$$u(x, 0, t) = py(x, t), \qquad u(x, 1, t) = qy(y, t)$$

and a source/sink term $S(x, y, t)$ in $u_t - (u_{xx} + u_{yy}) = S(x, y, t)$.
[*Hint*: Put half the effect of $S(x, y, t)$ into each of (8.6.18a) and (8.6.18b).]

(b) Test your modification of Algorithm 8.7 on the initial-boundary-value problem of Exercise 1.

8.7 CONSERVATION LAW DIFFERENCE EQUATIONS

Parabolic Conservation Laws in One Space Dimension

In Sections 7.3 and 7.5, we studied one-dimensional conservation laws of the form

$$u_t + F_x = 0 \tag{8.7.1}$$

where $u = u(x, t)$ denotes the linear density of some quantity and F the flux in the x direction of that quantity. When we assume that the flux is a differentiable function of u, $F = F(u)$, then Equation (8.7.1) can be written as a first-order hyperbolic equation in u. Namely,

$$u_t + a(u)u_x = 0, \qquad a(u) = F'(u)$$

A large class of physical problems can be described in terms of a slightly more general conservation law. Suppose that $\rho = \rho(x, t)$ is the linear density of some material quantity M and that q is the flux in the x direction of the material M. Then the one-dimensional conservation law (8.7.1) takes the form

$$\rho_t + \dot{q}_x = 0 \tag{8.7.2}$$

Further, suppose that $u = u(x, t)$ is some observable function that determines the density ρ and the flux q through relationships of the form

$$\rho(x, t) = R(x, u(x, t)) \tag{8.7.3}$$

and

$$q(x, t) = -\sigma u_x(x, t), \qquad \sigma = \sigma(x, u(x, t)) > 0 \tag{8.7.4}$$

If we put the expressions (8.7.3) and (8.7.4) into the conservation law (8.7.2), the result is the *parabolic conservation law*

$$cu_t - (\sigma u_x)_x = 0 \tag{8.7.5}$$

where c denotes $R_u(x, u)$. If ρ is a linear function of u,

$$\rho(x, t) = cu(x, t) + b, \quad c, b = \text{const}, c > 0$$

and if σ is a constant, then (8.7.5) reduces to the constant-coefficient diffusion equation

$$u_t - a^2 u_{xx} = 0, \quad a^2 = \sigma/c$$

EXAMPLE 8.7.1 _____

Suppose that $u = u(x, t)$ represents a one-dimensional population density and assume that the individuals in this population move in the x direction so as to avoid crowding. That is, the flux of population is proportional to the negative of the gradient of population density, $q = -\sigma u_x$, $\sigma > 0$. In the flux expression, the proportionality factor σ determines the rate at which population will relocate in response to a unit gradient in population density. For this example, we have $\rho(x, t) = u(x, t)$ so the conservation law (8.7.5) takes the form

$$u_t - (\sigma u_x)_x = 0$$

For different physical settings, a variety of different expressions for σ might be assumed. For example, if the rate at which individuals respond to crowding is independent of position, time, and population density, then we could assume that σ is a constant. If individuals in the population move at different, but known, rates at some positions x or at some times t, then we could take $\sigma = \sigma(x, t)$. If individuals respond differently to crowding at low population densities than at high population densities, then it would be appropriate to make σ a function of density, $\sigma = \sigma(u)$. ■■

EXAMPLE 8.7.2 _____

Consider a one-dimensional heat-conducting material with density d and specific heat s. See Section 1.2. The specific heat represents the number of calories required to raise a unit mass of the material a unit degree of temperature. The units of d are mass per length and the units for s are calories per degree per mass. For this example we assume that s and d are functions of position: $s = s(x)$ and $d = d(x)$.

Let ρ denote the heat density measured in calories per length. Finally, let $q = -\sigma u_x$ be the heat flux vector. The units of q are calories per time. Here, σ is called the thermal conductivity, and it represents the rate at which heat moves from a hot region to a cold region in response to a unit gradient of temperature. If we assume that σ is a function of position and temperature, $\sigma = \sigma(x, u)$, then the conservation law $\rho_t + q_x = 0$ with

$$\rho = R(x, u) = d(x)s(x)u(x, t)$$

and

$$q = -\sigma(x, u)u_x$$

leads to the parabolic conservation law

$$c(x)u_t - (\sigma(x, u)u_x)_x = 0, \qquad c(x) = d(x)s(x) \qquad \blacksquare\ \blacksquare$$

Material Balance Difference Equations in One Dimension

It is possible to discretize Equation (8.7.5) using the finite-difference expressions that were developed in Section 8.1 from Taylor series. However, a more intuitive and physically based system of finite-difference equations for (8.7.5) can be obtained from a material balance argument.

Consider the following conservation law initial-boundary-value problem

$$cu_t - (\sigma u_x)_x = 0, \qquad 0 < x < 1, t > 0 \qquad (8.7.6)$$

$$u(x, 0) = f(x), \qquad 0 < x < 1 \qquad (8.7.7)$$

$$u(0, t) = \alpha(t), \qquad u(1, t) = \beta(t) \qquad (8.7.8)$$

We assume that (8.7.3) and (8.7.4) hold so that

$$cu_t = \rho_t \quad \text{and} \quad q = -\sigma u_x$$

allows Equation (8.7.6) to be written in the equivalent form

$$\rho_t + q_x = 0 \qquad (8.7.6a)$$

Define a uniform x-grid,

$$x_n = nh, \qquad n = 0, 1, 2, \ldots, N + 1$$

with $x_0 = 0$ and $x_{N+1} = 1$. Also define a set points on the x axis by

$$\xi_n = -\tfrac{1}{2}h + nh, \qquad n = 0, 1, 2, \ldots, N + 2$$

so that x_n is the center of the *finite-difference block* (ξ_n, ξ_{n+1}). See Figure 8.7.1. Finally, define a uniform time grid by

$$t_j = jk, \qquad j = 0, 1, \ldots$$

Now, consider the region in the xt plane defined by

$$\xi_n < x < \xi_{n+1}, \qquad t_j < t < t_{j+1}$$

If we integrate the differential equation conservation law (8.7.6) over this region, we can recover the integral equation conservation law.

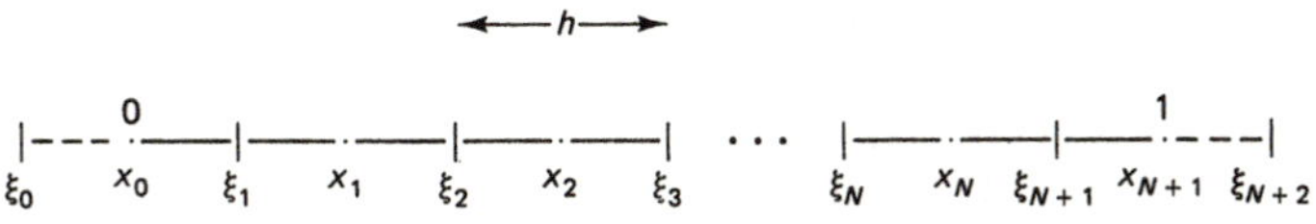

Figure 8.7.1 Block-centered finite-difference grid on $0 < x < 1$ with uniform grid spacing h.

$$\int_{t_j}^{t_{j+1}} \int_{\xi_n}^{\xi_{n+1}} (\rho_t + q_x)\, dx\, dt = \int_{\xi_n}^{\xi_{n+1}} \int_{t_j}^{t_{j+1}} \rho_t\, dt\, dx + \int_{t_j}^{t_{j+1}} \int_{\xi_n}^{\xi_{n+1}} q_x\, dx\, dt$$

$$= \int_{\xi_n}^{\xi_{n+1}} [\rho(x, t_{j+1}) - \rho(x, t_j)]\, dx$$

$$+ \int_{t_j}^{t_{j+1}} [q(\xi_{n+1}, t) - q(\xi_n, t)]\, dt = 0$$

or

$$\int_{\xi_n}^{\xi_{n+1}} [\rho(x, t_{j+1}) - \rho(x, t_j)]\, dx = \int_{t_j}^{t_{j+1}} [q(\xi_n, t) - q(\xi_{n+1}, t)]\, dt \qquad (8.7.9)$$

Equation (8.7.9) is an integral form of the conservation law (8.7.6a) or (8.7.6). In physical terms, Equation (8.7.9) states that the difference in the amount of material M in the block (ξ_n, ξ_{n+1}) from time t_j to time t_{j+1} is equal the flow of M into the block across $x = \xi_n$ minus the flow of M out of the block across $x = \xi_{n+1}$ during the time interval $t_j < t < t_{j+1}$. See Figure 8.7.2.

Using the integral equation conservation law (8.7.9), we can derive a number of finite-difference methods for the initial-boundary-value problem (8.7.6)–(8.7.8). If we use the midpoint quadrature rule to approximate the integral of the densities in (8.7.6), then

$$\int_{\xi_n}^{\xi_{n+1}} [\rho(x, t_{j+1}) - \rho(x, t_j)]\, dx \simeq [\rho(x_n, t_{j+1}) - \rho(x_n, t_j)]h$$

The choice of the midpoint rule for the last integral relates the densities to the x grid points.

To approximate the integral of the fluxes in (8.7.9), we first choose the left endpoint quadrature rule to obtain

$$\int_{t_j}^{t_{j+1}} [q(\xi_n, t) - q(\xi_{n+1}, t)]\, dt \simeq [q(\xi_n, t_j) - q(\xi_{n+1}, t_j)]k \qquad (8.7.10)$$

The above approximation of the fluxes at time t_j will lead to an explicit difference method that is an extension of (8.2.6). The use of the right-end rule for the the integral (8.7.10)

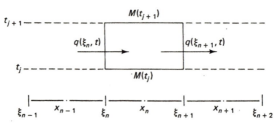

Figure 8.7.2 Material balance diagram showing that change in $M(t) = \displaystyle\int_{\xi_n}^{\xi_{n+1}} \rho(x, t)\, dx$ equals flow in minus flow out.

and the resulting implicit difference equation (8.7.10) is discussed below. The trapezoidal rule approximation of (8.7.10) is considered in Exercise 7.

We now have a difference equation that relates the density function ρ to the flux q. Namely,

$$[\rho(x_n, t_{j+1}) - \rho(x_n, t_j)]h = [q(\xi_n, t_j) - q(\xi_{n+1}, t_j)]k \qquad (8.7.11)$$

Equation (8.7.11) represents a single difference equation for the two quantities ρ and q. To derive a useful difference equation for dealing with the problem (8.7.6)–(8.7.8), we need to relate ρ and q to u. From (8.7.3) we have

$$\rho_n^{j+1} - \rho_n^j = R(x_n, U_n^{j+1}) - R(x_n, U_n^j)$$

From the mean-value theorem it follows that

$$R(x_n, u_n^{j+1}) - R(x_n, u_n^j) = R_u(x_n, V_n^j)[U_n^{j+1} - U_n^j]$$
$$= c(x_n, V_n^j)[U_n^{j+1} - U_n^j]$$

where V_n^j is between U_n^j and U^{j+1}. If we make the approximation $V_n^j = U_n^j$, then the left side of the conservation law difference equation (8.7.11) can be replaced as follows:

$$[\rho_n^{j+1} - \rho_n^j]h \simeq c(x_n, U_n^j)[U_n^{j+1} - U_n^j]h \qquad (8.7.12)$$

In some circumstances it may be desirable to estimate V_n^j by U_n^{j+1} or by an average of U_n^j and U_n^{j+1}.

It remains to find a replacement for $q(\xi_n, t_j)$ and $q(\xi_{n+1}, t_j)$ in terms of U_n^j. When σ is independent of u, it follows from (8.7.4) that $q(\xi_n, t_j)$ can be approximated by

$$q(\xi_n, t_j) = -\sigma(\xi_n) \frac{U_n^j - U_{n-1}^j}{h}$$

and $q(\xi_{n+1}, t_j)$ can be approximated by

$$q(\xi_{n+1}, t_j) = -\sigma(\xi_{n+1}) \frac{U_{n+1}^j - U_n^j}{h}$$

Combining these *numerical fluxes* into Equations (8.7.11) and (8.7.12) yields the difference equation

$$c(x_n, U_n^j)[U_n^{j+1} - U_n^j] = r\sigma(\xi_n)U_{n-1}^j - r(\sigma(\xi_n) + \sigma(\xi_{n+1}))U_n^j + r\sigma(\xi_{n+1})U_{n+1}^j$$

or

$$c_n^j[U_n^{j+1} - U_n^j] = r\sigma_{n-1/2}U_{n-1}^j - r(\sigma_{n-1/2} + \sigma_{n+1/2})U_n^j \qquad (8.7.13)$$
$$+ r\sigma_{n+1/2}U_{n+1}^j, \qquad r = k/h^2$$

EXAMPLE 8.7.3 ———————————————————————————————————

Consider the initial-value problem

$$u_t - (xu_x)_x, \qquad 0 < x < 1, t > 0$$
$$u(x, 0) = f(x), \qquad 0 < x < 1$$
$$u(0, t) = 0 = u(1, t), \qquad t > 0$$

on the grid

$$
\begin{array}{c}
0 \qquad \frac{1}{4} \qquad \frac{1}{2} \qquad \frac{3}{4} \qquad 1 \\
|\cdots\cdot-|-\cdot-|-\cdot-|-\cdot-|-\cdot\cdots| \\
\frac{1}{8} \qquad \frac{3}{8} \qquad \frac{5}{8} \qquad \frac{7}{8}
\end{array}
$$

According to (8.7.13), the difference system for this problem is

$$
U_1^{j+1} - U_1^j = r \cdot \tfrac{1}{8} U_0^j - r(\tfrac{1}{8} + \tfrac{3}{8}) U_1^j + r \cdot \tfrac{3}{8} U_2^j
$$
$$
U_2^{j+1} - U_2^j = r \cdot \tfrac{3}{8} U_1^j - r(\tfrac{3}{8} + \tfrac{5}{8}) U_2^j + r \cdot \tfrac{5}{8} U_3^j
$$
$$
U_3^{j+1} - U_3^j = r \cdot \tfrac{5}{8} U_2^j - r(\tfrac{5}{8} + \tfrac{7}{8}) U_3^j + r \cdot \tfrac{7}{8} U_4^j
$$
$$
U_n^0 = f(x_n), \qquad n = 1, 2, 3, \qquad U_0^0 = f(0)/2, \qquad U_4^0 = f(1)/2,
$$
$$
U_0^j = 0 = U_4^j, \qquad j = 1, 2, \ldots
$$

where $r = k/h^2$ and $h = \tfrac{1}{4}$. ■ ■

Let us now consider the flux $q(\xi_n, t_j)$ in the case that σ is either u dependent or only known at the points x_n. From Equation (8.7.4), we have

$$
q(x, t_j) = -\sigma(x, u(x, t_j)) u_x(x, t_j)
$$

which implies

$$
u_x(x, t_j) = -\frac{q(x, t_j)}{\sigma(x, u(x, t_j))}
$$

Integrating this last equation from x_{n-1} to x_n yields

$$
u_n^j - u_{n-1}^j = \int_{x_{n-1}}^{x_n} u_x(x, t_j)\, dx = -\int_{x_{n-1}}^{x_n} \frac{q(x, t_j)}{\sigma(x, u(x, t_j))}\, dx
$$
$$
\simeq -q(\xi_n, t_j) \int_{x_{n-1}}^{x_n} \frac{1}{\sigma(x, u(x, t_j))}\, dx
$$
$$
= -q(\xi_n, t_j) \left\{ \int_{x_{n-1}}^{\xi_n} \frac{1}{\sigma(x, u(x, t_j))}\, dx + \int_{\xi_n}^{x_n} \frac{1}{\sigma(x, u(x, t_j))}\, dx \right\}
$$
$$
\simeq -q(\xi_n, t_j) \left\{ \frac{1}{\sigma(x_{n-1}, u(x_{n-1}, t_j))} \frac{h}{2} + \frac{1}{\sigma(x_n, u(x_n, t_j))} \frac{h}{2} \right\}
$$
$$
= -q(\xi_n, t_j) \left\{ \frac{1}{\sigma_{n-1}^j} \frac{h}{2} + \frac{1}{\sigma_n^j} \frac{h}{2} \right\} = -q(\xi_n, t_j) \left\{ \frac{h\sigma_n^j + h\sigma_{n-1}^j}{2\sigma_n^j \sigma_{n-1}^j} \right\}
$$

Now we can write

$$
q(\xi_n, t_j) \simeq \frac{-2\sigma_n^j \sigma_{n-1}^j}{(\sigma_n^j + \sigma_{n-1}^j)} \frac{U_n^j - U_{n-1}^j}{h} \tag{8.7.14}
$$

Similarly, we have

$$
q(\xi_{n+1}, t_j) \simeq \frac{-2\sigma_n^j \sigma_{n+1}^j}{(\sigma_n^j + \sigma_{n+1}^j)} \frac{U_{n+1}^j - U_n^j}{h} \tag{8.7.15}
$$

In Equations (8.7.14) and (8.7.15), the coefficient involving σ is called a *harmonic average* or *harmonic mean*. For a more compact notation, we use

$$\mu(\sigma_m, \sigma_n) = \frac{2\sigma_m\sigma_n}{\sigma_m + \sigma_n}$$

to denote the harmonic mean of σ_m and σ_n. Also, to avoid a division by zero, we define $\mu(0, 0) = 0$. For example, in terms of this harmonic mean notation, Equations (8.7.14) and (8.7.15) take the form

$$q(\xi_n, t_j) = \mu(\sigma_n^j, \sigma_{n-1}^j)[U_n^j - U_{n-1}^j]/h$$

and

$$q(\xi_{n+1}, t_j) = \mu(\sigma_n^j, \sigma_{n+1}^j)[U_{n+1}^j - U_n^j]/h$$

Now, if we put Equations (8.7.14) and (8.7.15) into (8.7.11), then the difference equation for $cu_t - (\sigma u_x)_x = 0$ that results is

$$c(x_n, U_n^j)[U_n^{j+1} - U_n^j]h = \{\mu(\sigma_n^j, \sigma_{n-1}^j)[U_n^j - U_{n-1}^j]/h - \mu(\sigma_n^j, \sigma_{n+1}^j)[U_{n+1}^j - U_n^j]/h\}k$$

Introducing the notation

$$c_n^j = c(x_n, U_n^j), \qquad \mu_{n-1}^j = \mu(\sigma_n^j, \sigma_{n-1}^j), \qquad \mu_{n+1}^j = \mu(\sigma_n^j, \sigma_{n+1}^j)$$

allows the finite-difference system for (8.7.6)–(8.7.8) to be expressed in the form

$$c_n^j[U_n^{j+1} - U_n^j] = r[\mu_{n-1}^j U_{n-1}^j - (\mu_{n-1}^j + \mu_{n+1}^j)U_n^j + \mu_{n+1}^j U_{n+1}^j]$$
$$r = k/h^2 \tag{8.7.16}$$
$$U_n^0 = f(x_n), \qquad n = 1, 2, \ldots, N \tag{8.7.17}$$
$$U_0^j = \alpha(t_j), \qquad U_{N+1}^j = \beta(t_j), \qquad j = 1, 2, \ldots \tag{8.7.18}$$

If all the coefficients $c_n^j = $ const and $\mu_n^j = $ const in (8.7.16), then (8.7.16) is the forward-in-time difference method.

EXAMPLE 8.7.4 _____

Consider the initial-boundary-value problem

$$u_t - (2u^2 u_x)_x, \qquad 0 < x < 1, t > 0$$
$$u(x, 0) = (1 - x)^{1/2}, \qquad 0 < x < 1$$
$$u(0, t) = (1 + t)^{1/2}; \qquad u(1, t) = t^{1/2}, \qquad t > 0.$$

Then (8.7.16)–(8.7.18) is a finite-difference discretization of this problem provided we set

$$c_n^j = 1, \qquad U_n^0 = (1 - x_n)^{1/2}$$
$$U_0^j = (1 + t_j)^{1/2}, \qquad U_{N+1}^j = t^{1/2}$$
$$\mu_{n-1} = \frac{2 \cdot 2(U_n^j)^2 \cdot 2(U_{n-1}^j)^2}{2(U_n^j)^2 + 2(U_{n-1}^j)^2}, \qquad \mu_{n+1} = \frac{2 \cdot 2(U_n^j)^2 \cdot 2(U_{n+1}^j)^2}{2(U_n^j)^2 + 2(U_{n+1}^j)^2} \qquad ■ ■$$

The finite-difference equation (8.7.14) is explicit in the sense that it involves only one value of the numerical solution at the level t_{j+1}. If we had used a right-hand rectangular quadrature rule in approximating the integral of the fluxes in (8.7.9), then we would have obtained the following backward-in-time difference equation for (8.7.6):

$$c_n^j[U_n^{j+1} - U_n^j] = r[\mu_{n-1}^{j+1}U_{n-1}^{j+1} - (\mu_{n-1}^{j+1} + \mu_{n+1}^{j+1})U_n^{j+1} \quad (8.7.19a)$$
$$+ \mu_{n+1}^{j+1}U_{n+1}^{j+1}], \quad r = k/h^2$$

Since (8.7.19a) involves U_{n-1}^{j+1}, U_n^{j+1}, and U_{n+1}^{j+1}, we classify it as an implicit difference equation. Moreover, when σ depends on u, then (8.7.19a) is nonlinear in the unknowns U_{n-1}^{j+1}, U_n^{j+1}, and U_{n+1}^{j+1}. One common technique for avoiding having to solve a nonlinear system of difference equations is to *lag the nonlinearities* by evaluating the coefficients at t_j rather than t_{j+1}. If we lag the nonlinearities in (8.7.19a) then the resulting linearized backward difference equation is

$$c_n^j[U_n^{j+1} - U_n^j] = r[\mu_{n-1}^j U_{n-1}^{j+1} - (\mu_{n-1}^j + \mu_{n+1}^j)U_n^{j+1} + \mu_{n+1}^j U_{n+1}^{j+1}], \quad (8.7.19b)$$
$$r = k/h^2$$

None of our techniques for analyzing the stability of a finite-difference method applies to difference equations that arise from partial differential equations with variable coefficients. However, there is computational evidence that the forward difference method (8.7.16) becomes unstable unless

$$r(\mu_{n-1}^j + \mu_{n+1}^j) \leq c_n^j, \quad n = 1, 2, \ldots, N, j = 1, 2, \ldots$$

Numerical experiments also indicate that the linearized, backward difference, and Crank–Nicolson methods are unconditionally stable.

Exercises

1. Given the initial-value problem

$$cu_t - (e^x u_x)_x = 0, \quad 0 < x < 1, t > 0, c = \text{const}, c > 0$$
$$u(x, 0) = x(x - 1), \quad 0 < x < 1$$
$$u(0, t) = 0 = u(1, t)$$

and the block-centered grid of Example 8.7.3, use (8.7.13) to develop a system of finite-difference equations.

2. Given the initial-value problem

$$cu_t - (e^u u_x)_x = 0, \quad 0 < x < 1, t > 0$$
$$u(x, 0) = \sin \pi x, \quad 0 < x < 1$$
$$u(0, t) = 0 = u(1, t)$$

and the block-centered grid of Example 8.7.3, use (8.7.16) to develop a system of finite-difference equations.

3. Given the initial-boundary-value problem

$$cu_t - (\sigma u_x)_x = 0, \quad 0 < x < 1, t > 0$$
$$u(x, 0) = f(x), \quad 0 < x < 1$$
$$u_x(0, t) = 0 = u_x(1, t)$$

Show that the homogeneous Neumann boundary conditions can be incorporated into a difference scheme by ensuring that

$$q(\xi_1, t) = 0 = q(x_{N+1}, t)$$

4. Repeat Exercise 1 for the Neumann boundary conditions

$$u_x(0, t) = 0 = u_x(1, t)$$

5. Repeat Exercise 2 for the Neumann boundary conditions

$$u_x(0, t) = 0 = u_x(1, t)$$

6. (a) Show that a material balance equation that accounts for a source–sink term can be written in the form

$$cu_t - (\sigma u_x)_x = S(x, t)$$

and specify the units associated with c, σ, and the source–sink term S.

  (b) Modify the difference equation so that it applies to the differential equation of (a).

7. (a) Using the trapezoidal rule to evaluate the integral of the fluxes in Figure 8.7.2, derive a Crank–Nicolson method for the conservation law $c(x, t)u_t - (\sigma(x, t)u_x)_x = 0$.

  (b) In the case that $c = c(u)$ and $\sigma = \sigma(u)$ are u dependent, describe how to lag the nonlinearities in the difference equation.

8. Use the weighted mean-value theorem from calculus to support the estimate

$$\int_{x_{n-1}}^{x_n} \frac{q(x, t_j)}{\sigma(x, u(x, t_j))} \, dx \simeq q(\xi_n, t_j) \int_{x_{n-1}}^{x_n} \frac{1}{\sigma(x, u(x, t_j))} \, dx$$

9. Given a nonuniform block centered grid,

$$\leftarrow h_0 \rightarrow \leftarrow h_1 \rightarrow \leftarrow h_2 \rightarrow \qquad \leftarrow h_N \rightarrow \leftarrow h_{N+1} \rightarrow$$

$$
\begin{array}{cccccccc}
 & 0 & & & & & 1 & \\
| \cdots & \dfrac{\cdot}{x_0} & \dfrac{\cdot}{x_1} & \dfrac{\cdot}{x_2} & \cdots & \dfrac{\cdot}{x_N} & \dfrac{\cdot}{x_{N+1}} & \cdots | \\
\xi_0 & \xi_1 & \xi_2 & \xi_3 \ \ \xi_N & & \xi_{N+1} & & \xi_{N+2}
\end{array}
$$

write a formula for x_n in terms of the block sizes $h_0, h_1, \ldots, h_n$.

10. With reference to the nonuniform grid of Exercise 9 use the equation

$$u_n^j - u_{n-1}^j = \int_{x_{n-1}}^{x_n} u_x(x, t_j) \, dx = -\int_{x_{n-1}}^{x_n} \frac{q(x, t_j)}{\sigma(x, u(x, t_j))} \, dx$$

to develop the following formula for the numerical flux:

$$q(\xi_n, t_j) = \frac{-2\sigma_n^j \sigma_{n-1}^j}{\sigma_n^j h_{n-1} + \sigma_{n-1}^j h_n} [U_n^j - U_{n-1}^j]$$

11. Using the grid of Exercise 9, develop a finite-difference equation for (8.7.5).

12. The flux q given by (8.7.4) is often called a *diffusive flux*. When the flux is partly diffusive and partly convective, then $q = -\sigma u_x + vu$, where v is the convective velocity. Derive the convective–diffusive analogue of the conservation law (8.7.5),

$$cu_t = (\sigma u_x)_x - (vu)_x$$

13. Using centered difference formulas for all spatial derivatives, derive a difference formula for the linear convective–diffusive conservation law $cu_t = (\sigma u_x)_x - (vu)_x$ using

  (a) forward-in-time difference method

  (b) backward-in-time difference method

  (c) Crank–Nicolson method

14. Given the difference equation

$$c_n[U_n^{j+1} - U_n^j] = r[U_{n-1} - 2U_n^j + U_{n+1}^j], \qquad r = k/h^2, \ n = 1, 2, 3, 4, 5$$

where c_n is the capacity coefficient of the difference block centered at x_n. The capacity coefficient fixes the time scale on which the M content of a block changes in response to a given flux. To see the effect of the capacity coefficient, apply (8.7.13) for one time step to the following initial-value problem:

$$cu_t - u_{xx} = 0, \quad 0 < x < 1, t > 0$$
$$u(x, 0) = x(1 - x), \quad 0 < x < 1$$
$$u(0, t) = 0 = u(1, t), \quad t > 0$$

Use $r = 0.25$ and choose $c = 0.01$, $c = 1$, and $c = 100$.

8.8 MATERIAL BALANCE DIFFERENCE EQUATIONS IN TWO SPACE VARIABLES

Two-Dimensional Conservation Laws

Suppose that $\rho = \rho(x, y, t)$ represents the areal density of some material quantity M and that $\mathbf{q} = (qx, qy)$ is the flux vector of the material M. Then in a region ω that is free of sources and sinks, the time rate of increase of material M in ω must be balanced by the net flow of material M inward across s, the boundary of ω. That is,

$$\frac{d}{dt} \int_\omega \rho(x, y, t) \, dx \, dy = -\int_s \mathbf{q} \cdot \mathbf{n} \, ds \tag{8.8.1}$$

Equation (8.8.1) is an integral equation conservation law for the material M. The appearance of the minus sign on the right side of (8.8.1) is due to the outward orientation of $\mathbf{n}$, the normal to s. An inward flux of M across s is given by $-\mathbf{q} \cdot \mathbf{n}$. See Figure 8.8.1. Using the divergence theorem to express the integral over s as a surface integral, we find

$$\int_s \mathbf{q} \cdot \mathbf{n} \, ds = \int_\omega \nabla \cdot \mathbf{q} \, dx \, dy$$

Thus, we conclude that if (8.8.1) holds for an arbitrary subregion ω of some larger region Ω and if ρ, qx, and qy are differentiable, then the following two-dimensional conservation law must hold throughout Ω:

$$\rho_t + \nabla \cdot \mathbf{q} = 0 \quad \text{or} \quad \rho_t + qx_x + qy_y = 0 \tag{8.8.2}$$

Suppose that $u = u(x, y, t)$ is some observable function that determines values for the density ρ and the flux components qx and qy through relationships of the form

$$\rho(x, y, t) = R(x, y, u(x, y, t)) \tag{8.8.3}$$

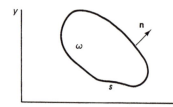

Figure 8.8.1 Control region ω with boundary s and outward normal $\mathbf{n}$.

$$qx(x, y, t) = -\alpha u_x(x, y, t), \qquad \alpha = \alpha(x, y, u(x, y, t)) > 0 \qquad (8.8.4)$$

and

$$qy(x, y, t) = -\beta u_y(x, y, t), \qquad \beta = \beta(x, y, u(x, y, t)) > 0 \qquad (8.8.5)$$

If we put Equations (8.8.3)–(8.8.5) into (8.8.2), the result is the two-dimensional parabolic partial differential equation

$$cu_t - (\alpha u_x)_x - (\beta u_y)_y = 0 \qquad (8.8.6)$$

where c denotes $R_u(x, y, u)$. If ρ is a linear function of u,

$$\rho(x, y, t) = cu(x, y, t) + b, \qquad c, b\text{-const}, c > 0$$

and if α and β have a common, constant value, say σ, then (8.8.6) reduces to the constant coefficient two-dimensional diffusion equation

$$u_t - a^2(u_{xx} + u_{yy}) = 0, \qquad a^2 = \sigma/c$$

EXAMPLE 8.8.1 ——

An aquifer is a porous subsurface geological formation that contains water in the pore spaces between the solid grains that make up the formation. The water in an aquifer is referred to as groundwater. In many areas, groundwater provides the major source of water for both agricultural and municipal use. Conservation law partial differential equations of the form (8.8.6) are used as mathematical models of groundwater flows.

Consider an xy coordinate system on a region and suppose that below this region there is an aquifer with uniform thickness that is confined between two impermeable layers (Figure 8.8.2).

The pressure or head in an aquifer at a position (x, y) at time t can be visualized as follows. Suppose that at time t at position (x, y) a well is drilled from the ground surface, through the upper confining layer, into the aquifer. Then in response to the pressure field in the aquifer, water would rise a certain distance $u(x, y, t)$, measured from, say, the lower confining layer up the well bore. This function $u(x, y, t)$ is known as the pressure head, or simply the head distribution in the aquifer.

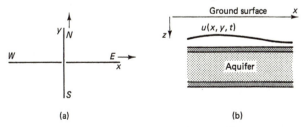

(a)   (b)

Figure 8.8.2 (a) Plan (map) view of xy coordinate system on ground surface. (b) Vertical section showing aquifer between confining layers.

Let $\rho = \rho(x, y, t)$ denote the areal density of the mass of water in the aquifer. For many groundwater flows, it can be argued that ρ is related to u linearly,

$$\rho = c(x, y)u(x, y, t) + b(x, y), \qquad c \geq 0$$

where c is the water capacity function. Note that if c is small, then large changes in head u can correspond to small changes in water content ρ. Conversely if c is large, then water content in the aquifer can vary greatly and the head can remain nearly constant. The xy variation in b and c are due to inhomogeneities in aquifer properties such as porosity or compressibility.

Darcy's law is based on the observation that the groundwater will move from a region of high head (high pressure) to a region of lower head. For the case of horizontal flow, one version of Darcy's law states that the flux of groundwater is given by

$$\mathbf{q} = (qx, qy) = (-\alpha u_x, -\beta u_y)$$

where $\alpha = \alpha(x, y)$ and $\beta = \beta(x, y)$ are nonnegative functions. The functions α and β are called the hydraulic conductivities of the aquifer in the x and y directions, respectively. The hydraulic conductivity in a given direction represents the flux of groundwater in that direction in response to a unit head gradient. For certain types of geologically stratified aquifers, it is easier for water to flow in one direction than in an orthogonal direction. Thus, it is possible that the hydraulic conductivities in the x and y directions be different.

Let Ω be a region on the ground surface and let ω be any subregion of Ω. Let s denote the boundary of ω and consider the right cylinder formed by extending a vertical line through each point of s downward through the aquifer. Assume that no sources or sinks of groundwater exist within this cylinder. The rate of increase of groundwater content in the portion of the aquifer below ω is equal to the net flow across the cylindrical boundary s. Thus, we conclude that the integral conservation law (8.8.1) holds for ω and the differential equation conservation law (8.8.6) holds in Ω.   ■ ■

Material Balance Difference Equations in Two Dimensions

Consider the initial-boundary-value problem

$$cu_t - (\alpha u_x)_x - (\beta u_y)_y = 0, \qquad 0 < x < X, 0 < y < Y, t > 0 \qquad (8.8.7)$$
$$u(x, y, 0) = f(x, y), \qquad 0 < x < X, 0 < y < Y \qquad (8.8.8)$$
$$u(0, y, t) = gx(y, t), \qquad u(X, y, t) = Gx(y, t)$$
$$u(x, 0, t) = gy(y, t), \qquad u(x, Y, t) = Gy(y, t) \qquad (8.8.9)$$

Let us introduce a two-dimensional, block-centered finite-difference grid as indicated in Figure 8.8.3.

In the integral form of the conservation law, (8.8.1), suppose that the region ω is the finite-difference block defined by $\xi_m < x < \xi_{m+1}$, $\eta_n < y < \eta_{n+1}$. If we use the midpoint rule to approximate all integrals in (8.8.1) and evaluate the fluxes at time t_j, then we obtain

$$\frac{d}{dt} \rho(x_m, y_n, t)(hx)(hy) = [qx(\xi_m, y_n, t) - qx(\xi_{m+1}, y_n, t)]hy \qquad (8.8.10)$$
$$+ [qy(x_m, \eta_n, t) - qy(x_m, \eta_{n+1}, t)]hx$$

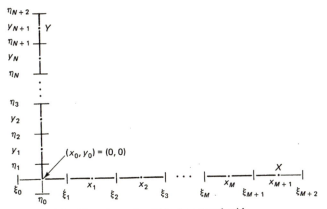

Figure 8.8.3 Two dimensional block centered grid.

The last equation states that the time rate of change of material M in the difference block centered at (x_m, y_n) is accounted for by the net flux of M into the block across the boundary of the block. See Figure 8.8.4. To obtain a difference equation for ρ, qx, and qy, we integrate Equation (8.8.10) from t_j to t_{j+1}. The time integral of the fluxes can be approximated in several ways. As we shall see, if we use a left endpoint rule and evaluate qx and qy at t_j an explicit, forward-in-time difference method will result. Evaluating qx and qy at t_{j+1} leads to a backward-in-time, implicit method. The use of the trapezoid rule in integrating the fluxes from t_j to t_{j+1} gives a Crank–Nicolson method. For now let us select the left endpoint rule. Then, after integrating (8.8.10) from t_{j+1} to t_j, we find

$$[\rho_{mn}^{j+1} - \rho_{mn}^{j}](hx)(hy) = \{[qx(\xi_m, y_n, t) - qx(\xi_{m+1}, y_n, t_j)]hy \qquad (8.8.11)$$
$$+ [qy(x_m, \eta_n, t) - qy(x_m, \eta_{n+1}, t_j)]hx\}k$$

where $k = t_{j+1} - t_j$.

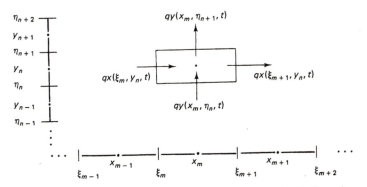

Figure 8.8.4 Net flow of M into block centered at (x_m, y_n) with dimension hx by hy.

Paralleling the development used in the case of one space variable, we conclude

$$[\rho_{mn}^{j+1} - \rho_{mN}^{j}] \simeq c(x_m, y_n, U_{mn}^{j})[U_{mn}^{j+1} - U_{mn}^{j}]$$

$$qx(\xi_m, y_n, t_j) \simeq \frac{-2\alpha_{mn}^{j}\alpha_{m-1,n}^{j}}{(\alpha_{mn}^{j} + \alpha_{m-1,n}^{j})} \frac{[U_{mn}^{j} - U_{m-1,n}^{j}]}{hx}$$

$$qx(\xi_{m+1}, y_n, t_j) \simeq \frac{-2\alpha_{mn}^{j}\alpha_{m+1,n}^{j}}{(\alpha_{mn}^{j} + \alpha_{m+1,n}^{j})} \frac{[U_{m+1,n}^{j} - U_{mn}^{j}]}{hx}$$

$$qy(x_m, \eta_n, t_j) \simeq \frac{-2\beta_{mn}^{j}\beta_{m,n-1}^{j}}{(\beta_{mn}^{j} + \beta_{m,n-1}^{j})} \frac{[U_{mn}^{j} - U_{m,n-1}^{j}]}{hy}$$

$$qy(x_m, \eta_{n+1}, t_j) \simeq \frac{-2\beta_{mn}^{j}\beta_{m,n+1}^{j}}{(\beta_{mn}^{j} + \beta_{m,n+1}^{j})} \frac{[U_{m,n+1}^{j} - U_{mn}^{j}]}{hy}$$

To avoid having to write out the expressions for the harmonic means that appear in the flux expressions, we adopt the notation

$$\mu_{m-1,n}^{j} = \frac{2\alpha_{mn}^{j}\alpha_{m-1,n}^{j}}{(\alpha_{mn}^{j} + \alpha_{m-1,n}^{j})}, \qquad \mu_{m+1,n}^{j} = \frac{2\alpha_{mn}^{j}\alpha_{m+1,n}^{j}}{(\alpha_{mn}^{j} + \alpha_{m+1,n}^{j})}$$

and

$$\mu_{m,n-1}^{j} = \frac{2\beta_{mn}^{j}\beta_{m,n-1}^{j}}{(\beta_{mn}^{j} + \beta_{m,n-1}^{j})}, \qquad \mu_{m,n+1}^{j} = \frac{2\beta_{mn}^{j}\beta_{m,n+1}^{j}}{(\beta_{mn}^{j} + \beta_{m,n+1}^{j})}$$

Then an explicit, forward-in-time material balance difference equation for (8.8.7) is

$$c_{mn}^{j}[U_{mn}^{j+1} - U_{mn}^{j}] = rx[\mu_{m-1,n}^{j}U_{m-1,n}^{j} - (\mu_{m-1,n}^{j} + \mu_{m+1,n}^{j})U_{mn}^{j} + \mu_{m+1,n}^{j}U_{m+1,n}^{j}]$$
$$+ ry[\mu_{m,n-1}^{j}U_{m,n-1}^{j} - (\mu_{m,n-1}^{j} + \mu_{m,n+1}^{j})U_{mn}^{j} + \mu_{m,n+1}^{j}U_{m,n+1}^{j}]$$
$$\tag{8.8.12}$$

where $rx = k/(hx)^2$ and $ry = k/(hy)^2$. Backward-in-time and Crank–Nicolson method similar to (8.8.11) can also be derived.

For simplicity, we have presented the development of material balance difference equations for the case that no sources or sinks of material M are present. It is easy to modify the above discussion so that it applies to a conservation law of the form

$$cu_t - (\alpha u_x)_x - (\beta u_y)_y = S(x, y, t)$$

where $S(x, y, t)$ represents the time rate of production or removal of material M per unit area. In this case the corresponding integral equation conservation law is

$$\frac{d}{dt} \int_\omega \rho(x, y, t) \, dx \, dy = -\int_s \mathbf{q} \cdot \mathbf{n} \, ds + \int_\omega S(x, y, t) \, dx \, dy$$

and the counterpart to (8.8.12) is

$$c_{mn}^{j}[U_{mn}^{j+1} - U_{mn}^{j}] = rx[\mu_{m-1,n}^{j}U_{m-1,n}^{j} - (\mu_{m-1,n}^{j} + \mu_{m+1,n}^{j})U_{mn}^{j} + \mu_{m+1,n}^{j} U_{m+1,n}^{j}]$$
$$+ ry[\mu_{m,n-1}^{j}U_{m,n-1}^{j} - (\mu_{m,n-1}^{j} + \mu_{m,n+1}^{j})U_{mn}^{j} + \mu_{m,n+1}^{j}U_{m,n+1}^{j}]$$
$$+ kS_{mn}^{j}$$

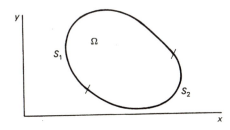

Figure 8.8.5 Nonrectangular region for initial-boundary-value problem (8.7)–(8.8.13-15).

Irregular Regions

We have developed finite-difference methods for approximating the solution to a parabolic initial-boundary-value problem in two space variables on a rectangular region. We now show how these methods can be modified to apply to problems on irregularly shaped regions. Consider an initial-value problem in the conservation law form

$$cu_t - (\alpha u_x)_x - (\beta u_y)_y = 0, \qquad (x, y) \in \Omega, \, t > 0 \qquad (8.8.13)$$
$$u(x, y, 0) = f(x, y), \qquad (x, y) \in \Omega \qquad (8.8.14)$$
$$u(x, y, t) = g(x, y), \qquad (x, y) \in S_1$$
$$\frac{\partial u}{\partial n}(x, y, t) = 0, \qquad (x, y) \in S_2 \qquad (8.8.15)$$

where S_1 and S_2 are complementary subsets of S, the boundary of Ω. See Figure 8.8.5.

Let us now introduce a finite-difference grid that contains the region Ω as a proper subset. See Figure 8.8.6. Next, we attach to each finite-difference block one of the integer labels 0, 1, or 2. To each block whose center lies in Ω, we assign the value 1. To each block that is not in Ω, but is adjacent to an interior block through which S_1 passes, we assign the value 2. All other blocks are given the value 0.

Blasting is a convenient device for holding the u values in certain finite-difference blocks constant during the numerical simulation. Suppose that all the data, c, α, β, f, and g, in problem (8.8.13)–(8.8.15) is of order 1. Define a *blasting constant* $\mathcal{B}$ to be a

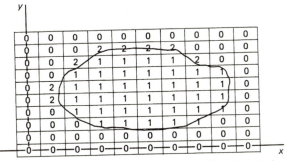

Figure 8.8.6 Block-centered grid on region: 0, no flow block; 1, interior block; 2, Dirichlet boundary condition block.

fixed positive constant that is several orders of magnitude larger than the problem data. For example, if c, α, β, f, and g have the form $z \times 10^0$ with $|z| < 10$, we could choose $\mathscr{B} = 10^{10}$. Then any calculation involving a division of $\mathscr{B}$ into one of the data will produce a value that is essentially zero. In particular, if we assign c_{pq}^j the value $\mathscr{B}$, then the difference method (8.6.30) will produce the results

$$U_{pq}^{j+1} \simeq U_{pq}^{j} \simeq U_{pq}^{j-1} \simeq \cdots \simeq U_{pq}^{1} \simeq U_{pq}^{0}$$

where the approximate equalities are accurate to about 10 decimal places.

In setting up our material balance difference equations, we have used the harmonic mean to calculate the fluxes across the faces of the finite-difference blocks. It follows from this use of harmonic averages that the flow across any vertical face of a finite-difference block will be zero if α_{mn} is zero in either of the blocks that meet on this face. Similarly, the flow across any horizontal face common to two finite-difference blocks is zero if β_{mn} is zero in either of the adjoining blocks. See Figure 8.8.7.

Now, to embed the finite-difference problem for (8.8.13)–(8.8.15) in the rectangular finite difference grid of Figure 8.8.6, we proceed as follows.

In the difference cells with label 1 assign

$$\alpha_{mn} = \alpha(x_m, y_n)$$
$$\beta_{mn} = \beta(x_m, y_n)$$
$$c_{mn}^j = c(x_m, y_n, U_{mn}^j)$$
$$U_{mn}^0 = f(x_m, y_n)$$

In the difference cells with label 0 assign

$\alpha_{mn} = 0$

$\beta_{mn} = 0$

$c_{mn}^j = \mathscr{B}$ ($\mathscr{B}$ a blasting constant)

$U_{mn}^0 = U_{\text{flag}}$ (U_{flag} a no-flow cell identifier value, e.g., $U_{\text{flag}} = -99$)

In the difference cells with label 2 assign

$$\alpha_{mn} = \alpha(x_m, y_n)$$
$$\beta_{mn} = \beta(x_m, y_n)$$
$$c_{mn}^j = \mathscr{B} \quad (\mathscr{B} \text{ a blasting constant})$$
$$U_{mn}^0 = g(x_m, y_n)$$

Algorithm 8.8 incorporates the above method for dealing with irregular boundaries into the Peaceman–Rachford method of Algorithm 8.7. Other methods, such as the Douglas–Rachford method, for solving the difference equations can easily be substituted into Algorithm 8.8.

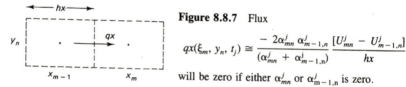

Figure 8.8.7 Flux

$$qx(\xi_m, y_n, t_j) \cong \frac{-2\alpha_{mn}^j \, \alpha_{m-1,n}^j}{(\alpha_{mn}^j + \alpha_{m-1,n}^j)} \frac{[U_{mn}^j - U_{m-1,n}^j]}{hx}$$

will be zero if either α_{mn}^j or $\alpha_{m-1,n}^j$ is zero.

ALGORITHM 8.8   Peaceman–Rachford Method: Irregular Region

Step 1. Document

The following algorithm uses the Peaceman–Rachford method (8.6.18) and the material balance difference equation (8.8.13) to approximate the solution to a Dirichlet intitial-boundary-value problem

$$\gamma u_t - (\alpha u_x)_x - (\beta u_y)_y = 0 \text{ in } \Omega,\ t > 0$$
$$u(x, y, 0) = f(x, y) \text{ in } \Omega$$
$$u(x, y, t) = g(x, y) \text{ on } S$$

INPUT

Array $l(m, n)$, label of 0, 1, or 2 for each block
Reals X, Y, x, y dimensions of rectangular region
Integers $mmax, nmax$, number of nodes in x, y directions
Real k, time step
Real blast, blasting constant
Integer $jmax$, number of time steps
Functions $f(x, y)$, $g(x, y)$ initial, boundary condition
Functions $\gamma(x, y)$, $\alpha(x, y)$, $\beta(x, y)$, coefficients in PDE

OUTPUT

Time t and approximation U_{mn}^j to $u(x_m, y_n, t_j)$ for
$m = 0, 1, \ldots, nmax + 1$; $n = 0, 1, \ldots, nmax + 1$ and
$j = 0, 1, \ldots, jmax$

Step 2. Define a grid

Set $hx = X/(mmax + 1)$
$\quad hy = Y/(nmax + 1)$
$\quad rx = k/(hx)^2$
$\quad ry = k/(hy)^2$

Step 3. Initialize numerical solution

Set $t = 0$
$\quad x_0 = 0$
$\quad y_0 = 0$
For $n = 1, 2, \ldots, nmax$ set
$\quad y_n = y_{n-1} + hy$

For $m = 1, 2, \ldots mmax$ set
$$x_m = x_{m-1} + hx$$
Set $x_{mmax+1} = X$
$\quad y_{nmax+1} = Y$
For $n = 0, 1, 2, \ldots, nmax + 1$
For $m = 0, 1, 2, \ldots, mmax + 1$
If $l(m, n) = 0$ then set
$\quad al_{mn} = 0$
$\quad bt_{mn} = 0$
$\quad cp_{mn} = $ blast
$\quad u_{mn} = $ uflag
$\quad v_{mn} = $ uflag
else if $l(m, n) = 1$ then set
$\quad al_{mn} = \alpha(x_m, y_n)$
$\quad bt_{mn} = \beta(x_m, y_n)$
$\quad cp_{mn} = \gamma(x_m, y_n)$
$\quad u_{mn} = f(x_m, y_n)$
$\quad v_{mn} = f(x_m, y_n)$
else if $l(m, n) = 2$ then set
$\quad al_{mn} = \alpha(x_m, y_n)$
$\quad bt_{mn} = \beta(x_m, y_n)$
$\quad cp_{mn} = $ blast
$\quad u_{mn} = g(x_m, y_n)$
$\quad v_{mn} = g(x_m, y_n)$
Output t
For $n = nmax + 1, nmax, \ldots, 2, 1$
$\quad$ For $m = 0, 1, \ldots, mmax + 1$
$\quad\quad$ Output U_{mn}

Step 4. Define harmonic mean on each face of interior blocks

For $n = 1, 2, \ldots, nmax$
$\quad$ For $m = 1, 2, \ldots, mmax$
$\quad denw = al_{mn} + al_{m-1,n}$
$\quad$ If $denw = 0$ then set
$\quad hw_{mn} = 0$
$\quad$ else set
$\quad hw_{mn} = 2 \cdot al_{mn} \cdot al_{m-1,n} / denw$
$\quad dene = al_{mn} + al_{m+1,n}$
$\quad$ If $dene = 0$ then set
$\quad he_{mn} = 0$

else set
$$he_{mn} = 2 \cdot al_{mn} \cdot al_{m+1,n}/dene$$
$$denn = bt_{mn} + bt_{m,n+1}$$
If $denn = 0$ then set
$$hn_{mn} = 0$$
else set
$$hn_{mn} = 2 \cdot bt_{mn} \cdot bt_{m,n+1}/denn$$
$$dens = bt_{mn} + bt_{m,n-1}$$
If $dens = 0$ then set
$$hs_{mn} = 0$$
else set
$$hs_{mn} = 2 \cdot bt_{mn} \cdot bt_{m,n-1}/dens$$

Step 5. Begin time stepping

For $j = 1, 2, \ldots, jmax$
    Do steps 6–12

Step 6. Begin y sweep

For $n = 1, 2, \ldots, nmax$
    Do steps 7–8

Step 7. Define tridiagonal system

For $m = 1, 2, \ldots, mmax$ set
$$a_m = -0.5rx \cdot hw_{mn}$$
$$c_m = -0.5rx \cdot he_{mn}$$
$$b_m = (cp_{mn} - a_m - c_m)$$
$$zs = 0.5 \cdot ry \cdot hs_{mn}$$
$$zn = 0.5 \cdot ry \cdot hn_{mn}$$
$$d_m = zs \cdot U_{m,n-1} + (cp_{mn} - zs - zn)U_{mn} + zn \cdot U_{m,n+1}$$

Step 8. Solve for $V_{mn} = U^*_{mn}$

CALL TRIDI($mmax, a, b, c, d$)
For $m = 1, 2, \ldots, mmax$ set
    $V_{mn} = d_m$

Step 9. Begin x sweep

For $m = 1, 2, \ldots, mmax$
    Do steps 10–11

Step 10. Define tridiagonal system

For $n = 1, 2, \ldots, nmax$ set

$$a_n = -0.5 ry \cdot hs_{mn}$$
$$c_n = -0.5 ry \cdot hn_{mn}$$
$$b_n = (cp_{mn} - a_n - c_n)$$
$$zw = 0.5 \cdot rx \cdot hw_{mn}$$
$$ze = 0.5 \cdot rx \cdot he_{mn}$$
$$d_n = zw \cdot V_{m-1,n} + (cp_{mn} - zw - ze)V_{mn} + ze \cdot V_{m+1,n}$$

Step 11. Solve for U_{mn}^{j+1}

CALL TRIDI($nmax, a, b, c, d$)
For $n = 1, 2, \ldots, nmax$ set
$$U_{mn} = d_n$$

Step 12. Output numerical solution

Set $t = t + k$
Output t
For $n = nmax, nmax - 1, \ldots, 2, 1$
    For $m = 1, 2, \ldots, nmax$
        Output u_{mn}

EXAMPLE 8.8.2

Consider the initial-boundary-value problem

$$u_t - (u_{xx} + u_{yy}) = 0, \qquad (x, y) \in \Omega, t > 0$$
$$u(x, y, 0) = f(x, y), \qquad (x, y) \in \Omega$$
$$u(x, y, t) = 0, \qquad (x, y) \in S, t > 0$$

where Ω is the quarter disk that is given in polar coordinates by

$$\Omega = \{(r, \theta) : 0 < r < 1, 0 < \theta < \pi/2\}$$

and S denotes the boundary of this disk.

If we choose the initial condition to be

$$f(x, y) = F(r, \theta) = 100 \sin 2\theta J_2(\mu r)$$

where μ (≈ 5.13562) is the first positive zero of the Bessel function J_2, then the exact solution of the example problem is known to be

$$u(x, y) = U(r, \theta) = 100 \exp(-\mu^2 t)\sin 2\theta J_2(\mu r)$$

We use this known, exact solution to test Algorithm 8.8. See Figure 8.8.8.

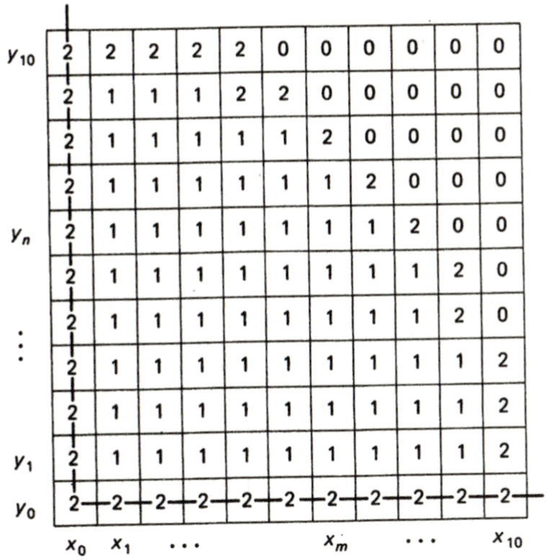

y_{10}

2	2	2	2	2	0	0	0	0	0	0
2	1	1	1	2	2	0	0	0	0	0
2	1	1	1	1	1	2	0	0	0	0
2	1	1	1	1	1	1	2	0	0	0
2	1	1	1	1	1	1	1	2	0	0
2	1	1	1	1	1	1	1	1	2	0
2	1	1	1	1	1	1	1	1	2	0
2	1	1	1	1	1	1	1	1	1	2
2	1	1	1	1	1	1	1	1	1	2
2	1	1	1	1	1	1	1	1	1	2
2	2	2	2	2	2	2	2	2	2	2

y_n (row 5), $\vdots$, y_1 (row 10), y_0 (bottom row)

x_0 x_1 $\cdots$ x_m $\cdots$ x_{10}

Figure 8.8.8 Block-centered finite difference for Example 8.8.2.

Note that the differential equation can be written in conservation law form as

$$cu_t - (\alpha u_x)_x - (\beta u_y)_y = 0$$

with the understanding that, at each point of Ω,

$$c = 1, \qquad \alpha = \beta = 1$$

For this example problem, in the difference cells with label 1 assign

$$\alpha_{mn} = 1$$
$$\beta_{mn} = 1$$
$$c^j_{mn} = 1$$
$$U^0_{mn} = f(x_m, y_n)$$

In the diffference cells with label 0 assign

$$\alpha_{mn} = 0$$
$$\beta_{mn} = 0$$
$$c^j_{mn} = \mathscr{B} \qquad (\mathscr{B} \text{ a blasting constant})$$
$$U^0_{mn} = U_{\text{flag}} \qquad (U_{\text{flag}} \text{ a no-flow cell identifier value, } U_{\text{flag}} = -99999)$$

In the difference cells with label 2 assign

$$\alpha_{mn} = 1$$
$$\beta_{mn} = 1$$
$$c^j_{mn} = \mathscr{B} \qquad (\mathscr{B} \text{ a blasting constant, } \mathscr{B} = 10^{10})$$
$$U^0_{mn} = 0$$

TABLE 8.8.1 COMPARISON OF NUMERICAL AND EXACT SOLUTIONS
FOR EXAMPLE 2 ON THE GRID OF FIGURE 8.8.8[a]

Numerical solution
```
0.00   0.00   0.00   0.00
0.00   0.30   0.49   0.46   0.00   0.00
0.00   0.61   1.08   1.26   1.10   0.74   0.00
0.00   0.92   1.66   2.08   2.09   1.72   0.97   0.00
0.00   1.17   2.14   2.76   2.92   2.63   1.93   0.98   0.00
0.00   1.31   2.43   3.17   3.46   3.25   2.63   1.73   0.77   0.00
0.00   1.32   2.45   3.23   3.58   3.46   2.93   2.10   1.10   0.00
0.00   1.17   2.17   2.89   3.23   3.18   2.76   2.08   1.27   0.48   0.00
0.00   0.87   1.63   2.17   2.45   2.43   2.15   1.67   1.08   0.50   0.00
0.00   0.47   0.87   1.17   1.32   1.32   1.17   0.92   0.62   0.30   0.00
0.00   0.00   0.00   0.00   0.00   0.00   0.00   0.00   0.00   0.00   0.00
```
Numerical solution minus exact solution
```
0.00   0.00   0.00   0.00
0.00   0.03   0.07   0.07   0.00   0.00
0.00   0.04   0.06   0.06   0.03   0.10   0.00
0.00   0.04   0.07   0.08   0.08   0.07   0.01   0.00
0.00   0.04   0.08   0.10   0.10   0.08   0.04   0.00   0.00
0.00   0.04   0.08   0.11   0.11   0.11   0.09   0.08   0.12   0.00
0.00   0.04   0.08   0.11   0.12   0.12   0.10   0.08   0.04   0.00
0.00   0.04   0.07   0.09   0.11   0.11   0.10   0.09   0.07   0.09   0.00
0.00   0.03   0.05   0.07   0.08   0.08   0.08   0.08   0.07   0.08   0.00
0.00   0.01   0.03   0.04   0.04   0.05   0.05   0.04   0.04   0.04   0.00
0.00   0.00   0.00   0.00   0.00   0.00   0.00   0.00   0.00   0.00   0.00
```

[a]$T = 0.10$; time step $k = 0.01$; space steps $hx = hy = 0.1$

Table 8.8.1 displays the results obtained when Algorithm 8.8 is applied to this example problem. ■ ■

The procedure for incorporating irregular boundaries in a finite-difference method illustrated in Example 8.8.2 is only one of several possible approaches. For example, the difference equations for Example 8.8.2 could have been formulated in polar coordinates following an approach illustrated elsewhere (P. DuChateau and D. Zachmann, *Schaum's Outline of Partial Differential Equations*, McGraw-Hill, New York, 1986, pp. 176–177). Alternatively, at grid points adjacent to the boundary of Ω, the computational stencil can be modified. (For more on this approach, see Exercise 6 and P. DuChateau and D. Zachmann, *Schaum's Outline of Partial Differential Equations*, McGraw-Hill, New York, 1986, p. 175, Mitchell and Griffiths, *The Finite Difference Method in Partial Differential Equations*, Wiley, New York, pp. 56 and 111–114, and Sod, *Numerical Methods in Fluid Dynmics*, Cambridge Univ. Press Cambridge, 1985, pp. 130–134.)

Finally, a popular and very effective numerical method for dealing with irregular geometries is the finite-element method. See Mitchell and Wait, *The Finite Element Method in Partial Differential Equations*, Wiley, New York, 1977.

Exercises

1. Given the block-centered difference grid indicated below and the initial-boundary-value problem

$$cu_t - (\alpha(x, y)u_x)_x - (\beta(x, y)u_y)_y = 0, \qquad 0 < x < 1, 0 < y < 1, t > 0$$
$$u(x, y, 0) = 4, \qquad 0 < x < 1, 0 < y < 1$$
$$u(0, y, t) = 0 = u(1, y, t), \qquad u(x, 0, t) = 0 = u(x, 1, t)$$

   (a) Write the difference equations (8.8.13) assuming $c_{mn} = 1$, $j = 0, 1, 2, \ldots;$ $\alpha_{mn} = 1 = \beta_{mn}$, for all (m, n) except $\alpha_{22} = 0 = \beta_{22}$.

   (b) What are the numerical fluxes across each face of the block with center at (x_2, y_2)?

   (c) Does the value of U_{22}^j affect the other U_{mn} values?

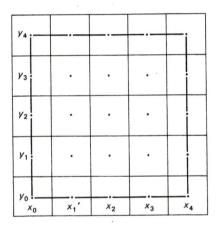

2. Given the boundary-value problem and grid of Exercise 1,

   (a) Write the difference equations (8.8.12) assuming $c_{mn}^j = 1$, $j = 0, 1, 2, \ldots$, for all (m, n) except $c_{22} = 10^6$ and $\alpha_{mn} = 1 = \beta_{mn}$.

   (b) What is the value of U_{22}^j for $j = 0, 1, 2, \ldots$?

   (c) Does the value of U_{22}^j affect the other U_{mn} values?

3. Given the initial-boundary-value problem

$$u_t - (u_{xx} + u_{yy}) = 0, \qquad (x, y) \in \Omega, t > 0$$
$$u(x, y, 0) = f(x, y), \qquad (x, y) \in \Omega$$
$$u(x, y, t) = 0, \qquad (x, y) \in S, t > 0$$

   where Ω is the quarter disk that is given in polar coordinates by

$$\Omega = \{(r, \theta) : 0 < r < 1, 0 < \theta < \pi/2\}$$

   and S denotes the boundary of this disk, let

$$f(x, y) = F(r, \theta) = 100 \sin 4\theta J_4(\mu r)$$

where μ is the first positive zero of the Bessel function J_4. Use the method of Example 8.8.2 to approximate the solution and compare the numerical and exact solution.

4. Given the block-centered difference grid indicated in the accompanying figure and the initial-boundary value problem

$$u_t - (e^x u_x)_x - (e^{-y} u_y)_y = 0, \qquad 0 < x < 1, 0 < y < 1$$
$$u(x, y, 0) = \sin \pi x \sin \pi y, \qquad 0 < x < 1, 0 < y < 1$$
$$u(0, y, t) = u(1, y, t) = 0, \qquad 0 < y < 1$$
$$u(x, 0, t) = u(x, 1, t) = 0, \qquad 0 < x < 1$$

formulate a finite-difference system of equations in which e^x and e^{-y} are evaluated on block faces.

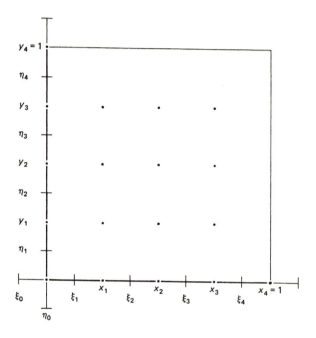

5. Using the grid and the initial-boundary-value conditions of Exercise 4, apply (8.8.13) to obtain a difference system for

$$u_t - (e^u u_x)_x - (e^{-u} u_y)_y = 0, \qquad 0 < x < 1, 0 < y < 1$$

subject to the same initial and boundary conditions.

6. Suppose $u(x, y)$ satisfies Laplace's equation. Use Taylor's series expansions for $u(x - h, y)$, $u(x, y - h)$, $u(x + he, y)$, and $u(x, y + hn)$ and the accompanying figure to obtain a finite-difference stencil relating U_{mn}, $U_{m-1,n}$, $U_{m,n-1}$, and the boundary values

$$ge = g(x + he, y) \quad \text{and} \quad gn = g(x, y + hn)$$

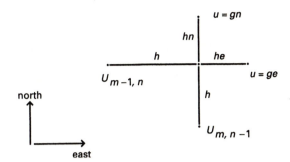

Chapter 9

Numerical Solutions of Hyperbolic Equations

In Chapter 8 we saw how finite-difference methods can be used to approximate the solution to parabolic partial differential equations. In this chapter we continue our discussion of finite-difference methods with applications to hyperbolic differential equations.

Parabolic equations generally model diffusive physical processes while hyperbolic equations usually arise in connection with a purely convective phenomenon. Partly because of the different physical origins of parabolic and hyperbolic equations, different types of finite-difference methods are required for the two classes of equations.

In Chapter 7 we saw that the method of characteristics is a powerful method for obtaining the exact solution to a hyperbolic partial differential equation. However, in many problems the characteristic equations cannot be integrated in closed form. In those cases a numerical method is called for. In this chapter we also introduce a numerical approach for applying the method of characteristics.

9.1 DIFFERENCE METHODS FOR A SCALAR INITIAL-VALUE PROBLEM

Introduction

In Chapter 7 we saw that a number of interesting physical problems can be modeled mathematically with a single hyperbolic partial differential equation of the form

$$a(x, t, u)u_x + b(x, t, u)u_t = c(x, t, u), \qquad -\infty < x < \infty, t > 0 \qquad (9.1.1)$$

subject to an initial condition

$$u(x, 0) = f(x), \qquad -\infty < x < \infty \qquad (9.1.2)$$

For example, in Section 7.5 equations of the form (9.1.1) were used to describe the flow of traffic and the transport of fluid in a porous medium.

In Section 7.2 we discussed the method of characteristics for solving the initial-value problem (9.1.1) and (9.1.2). Recall that the basis of the method of characteristics is the observation that along characteristics the first-order partial differential equation (9.1.1) can be formulated as an equivalent system of first-order ordinary differential equations. These ordinary differential equations, called the characteristic equations, are

$$\frac{dx}{dr} = a, \qquad \frac{dt}{dr} = b, \qquad \frac{du}{dr} = c \qquad (9.1.3)$$

where r is a parameter that determines the position along a characteristic. To solve (9.1.1) and (9.1.2) by the method of characteristics of Section 7.2, we must obtain an exact solution of the characteristic equations. For this reason, we refer to the method presented in Section 7.2 as the *exact method of characteristics*.

In many cases the form of the functions a, b, and c in (9.1.3) does not permit an exact solution of the characteristic equations. Then some numerical method for dealing with the problem (9.1.1) and (9.1.2) is required. In this chapter, two numerical methods for approximating the solution of (9.1.1) and (9.1.2) will be described.

In this section we show how *finite differences* can be applied to the problem (9.1.1) and (9.1.2). Then in Section 9.7 we consider a *numerical method of characteristics* that is based on obtaining an approximate solution to the characteristic equations (9.1.3).

Advantages of the finite-difference method are that (1) it is relatively easy to implement and (2) it produces the numerical approximation to the solution of (9.1.1) and (9.1.2) on a rectangular grid in the xt plane. A disadvantage of the finite-difference method is that it generally does not directly incorporate the fact that information about the solution of (9.1.1) is propagated along characteristics. As a consequence, finite-difference methods often produce poor results when the solution of (9.1.1) and (9.1.2) has large gradients.

The major strength of the numerical method of characteristics is that it tracks information about the solution of (9.1.1) along approximations to the characteristics. This makes the numerical method of characteristics generally better at dealing with a solution that has sharp fronts. Disadvantages of the numerical method of characteristics are that (1) it produces an approximation to the solution on an irregularly spaced set of points in the xt plane and (2) it is somewhat more complicated to implement than a finite-difference method.

Explicit Finite-Difference Methods

To begin our discussion of finite-difference methods for first-order differential equations of the form (9.1.1), we consider the simple equation

$$au_x + u_t = 0, \qquad -\infty < x < \infty \qquad (9.1.4)$$

and assume that a is a nonzero constant. Consider a grid on the upper half of the xt plane defined by

$$x_n = nh, \qquad n = 0, \pm 1, \pm 2, \ldots; \qquad t_j = jk, \qquad j = 0, 1, 2, \ldots$$

TABLE 9.1.1   EXPLICIT FINITE-DIFFERENCE METHODS AND LOCAL TRUNCATION
ERRORS (LTEs) FOR $au_x + u_t = 0$

Forward in time–forward in space (FTFS), LTE $= O(k + h)$:

$$a\frac{U_{n+1}^j - U_n^j}{h} + \frac{U_n^{j+1} - U_n^j}{k} = 0$$

or

$$U_n^{j+1} = (1 + sa)U_n^j - saU_{n+1}^j, \qquad s = k/h \qquad\qquad (9.1.5)$$

Forward in time–backward in space (FTBS), LTE $= O(k + h)$:

$$a\frac{U_n^j - U_{n-1}^j}{h} + \frac{U_n^{j+1} - U_n^j}{k} = 0$$

or

$$U_n^{j+1} = (1 - sa)U_n^j + saU_{n-1}^j, \qquad s = k/h \qquad\qquad (9.1.6)$$

Forward in time–centered in space (FTCS), LTE $= O(k + h^2)$:

$$a\frac{U_{n+1}^j - U_{n-1}^j}{2h} + \frac{U_n^{j+1} - U_n^j}{k} = 0$$

or

$$U_n^{j+1} = U_n^j - \frac{sa}{2}(U_{n+1}^j - U_{n-1}^j), \qquad s = k/h \qquad\qquad (9.1.7)$$

Lax–Friedrichs method, LTE $= O(k + h)$:

$$a\frac{U_{n+1}^j - U_{n-1}^j}{2h} + \frac{U_n^{j+1} - \frac{1}{2}(U_{n-1}^j + U_{n+1}^j)}{k} = 0$$

or

$$U_n^{j+1} = \tfrac{1}{2}(U_{n+1}^j + U_{n-1}^j) - \tfrac{1}{2}sa(U_{n+1}^j - U_{n-1}^j), \qquad s = k/h \qquad\qquad (9.1.8)$$

Centered in time–centered in space (CTCS or leapfrog), LTE $= O(k^2 + h^2)$:

$$a\frac{U_{n+1}^j - U_{n-1}^j}{2h} + \frac{U_n^{j+1} - U_n^{j-1}}{2k} = 0$$

or

$$U_n^{j+1} = U_n^{j-1} - sa(U_{n+1}^j - U_{n-1}^j), \qquad s = k/h \qquad\qquad (9.1.9)$$

Lax–Wendroff method, LTE $= O(k^2 + h^2)$:

$$U_n^{j+1} = U_n^j - \tfrac{1}{2}sa(U_{n+1}^j - U_{n-1}^j) + \tfrac{1}{2}s^2a^2(U_{n-1}^j - 2U_n^j + U_{n+1}^j), \qquad s = k/h \qquad\qquad (9.1.10)$$

As before, let $u_n^j = u(x_n, t_j)$ and let U_n^j denote an approximation to u_n^j. Using the difference formulas of Table 8.1.1, it is easy to write down a number of finite-difference approximations to (9.1.4). See Table 9.1.1.

With the exception of (9.1.8) and (9.1.10), the difference formulas of Table 9.1.1 follow directly from the difference quotient expressions of Table 8.1.1. The order of LTE for each method is easy to obtain from Taylor's Theorem 8.1.1.

The Lax–Friedrichs method (9.1.8) can be gotten from the FTCS method by replacing the U_n^j that appears in the time derivative by the average of its left and right neighbors, U_{n-1}^j and U_{n+1}^j.

One way to obtain the Lax–Wendroff method is to use the Taylor series expansion

$$u(x, t + k) = u(x, t) + u_t(x, t)k + u_{tt}(x, t)\frac{k^2}{2} + O(k^3) \qquad (9.1.11)$$

From the differential equation $au_x + u_t = 0$, we have

$$u_t = -au_x \quad \text{and} \quad u_{tt} = (-au_x)_t = -a(u_t)_x = -a(-au_x)_x = a^2u_{xx}$$

Putting $u_t = -au_x$ and $u_{tt} = a^2u_{xx}$ into the Taylor series (9.1.11) yields

$$u(x, t + k) = u(x, t) - au_x(x, t)k + \tfrac{1}{2}a^2k^2u_{xx}(x, t) + O(k^3)$$

Now, if we use centered differences to approximate $u_x(x, t)$ and $u_{xx}(x, t)$ at $(x, t) = (x_n, t_j)$,

$$u_x(x_n, t_j) = \frac{U_{n+1}^j - U_{n-1}^j}{2h} + O(h^2)$$

$$u_{xx}(x_n, t_j) = \frac{U_{n-1}^j - 2U_n^j + U_{n+1}^j}{h^2} + O(h^2)$$

then the Lax–Wendroff method (9.1.10) results. In Section 9.4 we describe another method for obtaining the difference formulas (9.1.8) and (9.1.10).

It is not difficult to extend the difference methods of Table 9.1.1 to the nonhomogeneous equation

$$au_x + u_t = c(x, t)$$

For the FTFS, FTBS, and Lax–Friedrichs methods, we need only add the term kc_n^j to the right side of the difference equation. For the leapfrog method we add $2kc_n^j$. The easiest way to modify the Lax–Wendroff method and maintain a LTE $= O(k^2 + h^2)$ is to use both the function $c(x, t)$ and its partial derivatives $c_t(x, t)$ and $c_x(x, t)$. Then the Lax–Wendroff method for the nonhomogeneous equation $au_x + u_t = c(x, t)$ is

$$U_n^{j+1} = U_n^j - \tfrac{1}{2}sa(U_{n+1}^j - U_{n-1}^j) + \tfrac{1}{2}s^2a^2(U_{n-1}^j - 2U_n^j + U_{n+1}^j)$$
$$+ kc_n^j + \tfrac{1}{2}k^2(c_t - ac_x)_n^j$$

Stability

Because of stability considerations, not all of the methods given in Table 9.1.1 can be used to appoximate the solution of the initial-value problem

$$au_x + u_t = 0, \qquad -\infty < x < \infty \qquad (9.1.12)$$
$$u(x, 0) = f(x), \qquad -\infty < x < \infty \qquad (9.1.13)$$

The FTCS method (9.1.7) is perhaps the most "natural" of the finite-difference methods of Table 9.1.1. However, as we show in Example 9.1.2, it is unconditionally unstable. This illustrates that care must be exercised in choosing a finite-difference approximation for even an equation as simple as (9.1.12).

Using the notation of Section 8.5, we see that, with the exception of the leapfrog method, each of the difference equations of Table 9.1.1 can be expressed in the form

$$U_n^{j+1} = Q(S_+, S_-)U_n^j \qquad (9.1.14)$$

where Q denotes a polynomial in the shift operators S_+ and S_-. The von Neumann stability condition for a difference equation of the form (9.1.14) is

$$|Q(e^{i\beta}, e^{-i\beta})| \leq 1, \qquad 0 \leq \beta < 2\pi$$

EXAMPLE 9.1.1 _____

To put the FTFS method (9.1.5) in the form (9.1.14), we define Q by

$$Q(S_+, S_-) = (1 + sa) - saS_+$$

Thus, the von Neumann stability condition is that

$$|1 + sa - sae^{i\beta}| \leq 1, \qquad 0 \leq \beta < 2\pi$$

or equivalently

$$|1 + sa - sae^{i\beta}|^2 \leq 1, \qquad 0 \leq \beta < 2\pi$$

The square of the magnitude of the complex number

$$1 + sa - sae^{i\beta} = 1 + sa(1 - \cos \beta) - isa \sin \beta$$

is the sum of the squares of its real and imaginary parts. Thus,

$$\begin{aligned}
|1 + sa - sae^{i\beta}|^2 &= [1 + sa(1 - \cos \beta)]^2 + [sa \sin \beta]^2 \\
&= 1 + 2sa(1 + sa)(1 - \cos \beta) \\
&= 1 + 4sa(1 + sa)\sin^2(\beta/2)
\end{aligned}$$

From the last expression, we see that the difference method (9.1.8) for the equation $au_x + u_t = 0$ satisfies the von Neumann stability condition if and only if a is negative and the grid spacings are chosen so that $|sa| \leq 1$, that is, $-1 \leq ak/h \leq 0$. ∎ ∎

EXAMPLE 9.1.2 _____

For the FTCS method (9.1.7) we have

$$Q(S_+, S_-) = 1 - \tfrac{1}{2}sa(S_+ - S_-)$$

so the von Neumann stability condition is that

$$|1 - \tfrac{1}{2}sa(e^{i\beta} - e^{-i\beta})| \leq 1, \qquad 0 \leq \beta < 2\pi$$

or

$$|1 - (sa \sin \beta)i| \leq 1, \qquad 0 \leq \beta < 2\pi$$

or

$$1 + (sa)^2\sin^2\beta \leq 1, \qquad 0 \leq \beta < 2\pi$$

Therefore, the only way the von Neumann stability condition can hold is if $sa = 0$. This shows that the FTCS method is unconditionally unstable. ∎ ∎

Each of the two-level difference methods of Table 9.1.1 can be tested for stability using the von Neumann method illustrated in the preceding examples. A slightly more general von Neumann stability analysis applies to the leapfrog method. These stability results are summarized in the following theorem.

Theorem 9.1.1. Given the initial-boundary-value problem

$$au_x + u_t = 0, \qquad -\infty < x < \infty, t > 0, a = \text{const}$$
$$u(x, 0) = f(x), \qquad -\infty < x << \infty$$

then:

1. The FTFS method (9.1.5) is stable if and only if $a < 0$ and $|sa| \leq 1$. It has LTE $= O(k + h)$ provided $u(x, t)$ is C^2 in both x and t.
2. The FTBS method (9.1.6) is stable if and only if $a > 0$ and $|sa| \leq 1$. It has LTE $= O(k + h)$ provided $u(x, t)$ is C^2 in both x and t.
3. The FTCS method (9.1.7) is unstable.
4. The Lax–Friedrichs method (9.1.8) is stable if and only if $|sa| \leq 1$. It has LTE $= O(k + h)$ provided $u(x, t)$ is C^4 in x and t.
5. The leapfrog method (9.1.9) is stable if and only if $|sa| \leq 1$. It has LTE $= O(k^2 + h^2)$ provided $u(x, t)$ is C^4 in both x and t.
6. The Lax–Wendroff method (9.1.10) is stable if and only if $|sa| \leq 1$. It has LTE $= O(k^2 + h^2)$ provided $u(x, t)$ is C^4 in both x and t.

The Courant–Friedrichs–Lewy (CFL) condition provides a necessary condition for stability of an explicit finite-difference method for a hyperbolic partial differential equation. The CFL condition imposes a restriction on the relationship of the characteristics of the partial differential equation to the grid spacings in the difference equation. For a two-level difference method, the CFL condition can be stated as follows.

CFL Stability Condition: A necessary condition for the stability of an explicit finite-difference method is that the characteristics through the point (x_n, t_{j+1}) must intersect the line $t = t_j$ within the span of the grid points used to define U_n^{j+1}.

The CFL condition is often stated by saying that the *numerical domain of dependence must contain the analytical domain of dependence*. See Figure 9.1.1. In Figure 9.1.1a, b, the CFL condition holds for the unconditionally unstable FTCS method. This demonstrates that the CFL condition is only a necessary condition, not a sufficient condition, for stability.

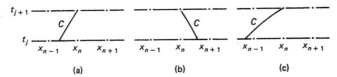

Figure 9.1.1 The C is characteristic of PDE $au_x + u_t = 0, a > 0$.
(a) CFL condition holds for all methods of Table 9.1.1 except FTFS.
(b) CFL condition holds for all methods of Table 9.1.1 except BTBS.
(c) CFL condition does not hold for any method of Table 9.1.1.

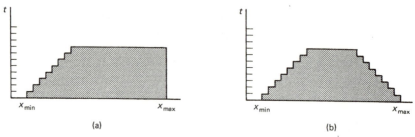

Figure 9.1.2 Computational region for (a) FTBS method and (b) Lax–Wendroff method.

Computations

The implementation of the difference methods of Table 9.1.1 is straightforward. For the two-level methods, only two arrays are required, one to store the solution at time t_j and one to store the solution at time t_{j+1}. Starting values at time t_0 are provided by the initial condition $u(x, 0) = f(x)$.

The leapfrog method can also be implemented with two arrays, but it is easier to use one array for each of the time levels t_{j+1}, t_j, and t_{j-1}. The leapfrog method requires starting values for U_n^0 and U_n^1. The U_n^0 values are obtained from the initial condition. Some two-level method with LTE $= O(k^2 + h^2)$, the same LTE as the leapfrog method, is used to set values of U_n^1. For example, the Lax–Wendroff method can be used to start the leapfrog method.

When using any of the methods of Table 9.1.1 to advance the numerical solution of an initial-value problem, one or two of the grid points must be dropped from the computation. For example, if we use the FTBS method to advance a solution initialized on the interval $x_{\min} < x < x_{\max}$, then the computation proceeds on the region indicated in Figure 9.1.2a. If the Lax–Wendroff method is used, the region covered by the numerical solution is indicated in Figure 9.1.2b. To construct a numerical solution U_n^j on the grid points between x_p and x_q at time t_j by the Lax–Wendroff method, an initial condition must be given on the grid points between $x_{\min}$ and $x_{\max}$ where min $= p - J$ and max $= q + J$.

We now illustrate how to apply the Lax–Wendroff method. It is easy to modify Algorithm 9.1 to fit one of the other methods of Table 9.1.1.

ALGORITHM 9.1   Lax–Wendroff Method: Initial-Value Problem

Step 1. Document

This algorithm uses the Lax–Wendroff method
(9.1.10) to approximate the solution of the initial value problem

$$au_x + u_t = c(x, t), \qquad -\infty < x < \infty, t > 0$$
$$u(x, 0) = f(x), \qquad -\infty < x < \infty$$

for $t_j \leq t_{jmax}$ and $x_p \leq x_n \leq x_q$.

INPUT

Integer p, lower index for solution at time t_{jmax}
Integer q, upper index for solution at time t_{jmax}
Integer $jmax$, number of time steps
Real k, time step $(k \cdot jmax = t_{jmax})$
Real h, space step
Real a, coefficient in equation
Function $f(x)$, initial condition
Function $c(x, t)$, right side of equation
Functions $c_x(x, t)$; $c_t(x, t)$, x and t derivatives of $c(x, t)$

OUTPUT

Time t and approximation U_n^j to $u(x_n, t_j)$ for
$n = p, p + 1, \ldots, q$ and $j = 0, 1, \ldots, jmax$

Step 2. Define grid ratio

Set $s = k/h$
If $|sa| > 1$ output message that computation is unstable.

Step 3. Initialize numerical solution

Set $t = 0$
   $nmin = p - jmax$
   $nmax = q + jmax$
For $n = nmin, nmin + 1, \ldots, nmax$ set
   $x_n = n \cdot h$
   $V_n = f(x_n)$
Output t
For $n = p, p + 1, \ldots, q$
   Output x_n, V_n

Step 4. Begin time stepping

For $j = 1, 2, \ldots, jmax$
   Do steps 5–7

Step 5. Advance solution one time step

Set $nmin = nmin + 1$
   $nmax = nmax - 1$

For $n = nmin, nmin + 1, \ldots, nmax$ set

$$U_n = V_n - \frac{sa}{2}(V_{n+1} - V_{n-1}) + \frac{s^2 a^2}{2}(V_{n-1} - 2V_n + V_{n+1})$$
$$+ kc(x_n, t) + \frac{k^2}{2}(c_t(x_n, t) - ac_x(x_n, t))$$

Step 6. Prepare for next time step

Set $t = t + k$
For $n = nmin, nmin + 1, \ldots, nmax$ set
   $V_n = U_n$

Step 7. Output numerical solution

Output t
For $n = p, p + 1, \ldots, q$
   Output x_n, U_n

EXAMPLE 9.1.3 ───

Given the initial-value problem

$$u_x + u_t = -x^2 t, \qquad -\infty < x < \infty, t > 0$$
$$u(x, 0) = 2 + \sin \pi x, \qquad -\infty < x < \infty$$

the method of characteristics of Section 7.2 can be used to develop the exact solution

$$u(x, t) = 2 + \sin \pi(x - t) - \tfrac{1}{3}x^3 t + \tfrac{1}{12}x^4 - \tfrac{1}{12}(x - t)^4$$

TABLE 9.1.2 COMPARISON OF NUMERICAL
AND EXACT SOLUTIONS
FOR EXAMPLE 9.1.3

$T = 1.00$	Numerical	Exact
$X = 0.0$	1.879359	1.916667
$X = 0.1$	1.602068	1.635983
$X = 0.2$	1.348416	1.375548
$X = 0.3$	1.145018	1.162650
$X = 0.4$	1.012593	1.018943
$X = 0.5$	0.963931	0.958333
$X = 0.6$	1.002646	0.985610
$X = 0.7$	1.122819	1.095983
$X = 0.8$	1.309579	1.275548
$X = 0.9$	1.540557	1.502650
$X = 1.0$	1.788079	1.750000

Let us choose an x grid spacing of $h = 0.1$ and a time grid spacing of $k = 0.05$ and use the Lax–Wendroff method to approximate this solution on the interval $0 \le x \le 1$ at time $t = 1$. In terms of Algorithm 9.1, we have $p = 0$ and $q = 10$. Since 20 time steps of $k = 0.05$ are required to advance the numerical solution to time $t = 1.0$, we must set $nmin = -20$ and $nmax = 30$. Table 9.1.2 shows a comparison of the numerical and exact solutions.

Exercises

1. Modify Algorithm 9.1 to implement
   (a) FTBS method
   (b) FTFS method
   (c) Lax–Friedrichs method
   (d) leapfrog method

2. Approximate the solution of the initial-value problem of Example 9.1.3 on the interval $0 \le x \le 1$ for $0 \le t_j \le 1.5$ with $h = 0.1$ and $k = 0.075$ using
   (a) FTBS method
   (b) Lax–Friedrichs method
   (c) leapfrog method

3. Repeat Exercise 2 with $h = 0.1$ and $k = 0.1$.

4. Given the initial-value problem

$$2u_x + u_t = 3\cos(x + t), \qquad -\infty < x < \infty, t > 0$$
$$u(x, 0) = 4 + \sin x, \qquad -\infty < x < \infty$$

   obtain a numerical approximation to the solution on the region $0 \le x \le 1, 0 \le t \le 1$ with $h = 0.1$ using
   (a) FTBS method
   (b) Lax–Wendroff method
   (c) Lax–Friedrichs method
   (d) leapfrog method
   Compare the numerical results with the exact solution,

$$u(x, t) = 4 + \sin(x + t)$$

5. Consider the initial-value problem

$$u_x + u_t = 2xt + x^2, \qquad -\infty < x < \infty, t > 0$$
$$u(x, 0) = 0, \qquad -\infty < x < \infty$$

   whose exact solution is $u(x, t) = x^2 t$. Find the local truncation error for this problem when using
   (a) FTBS method
   (b) Lax–Wendroff method

6. Write each of the following methods in the form

$$U_n^{j+1} = Q(S_+, S_-)U_n^j$$

   where Q is a polynomial in the shift operators S_+ and S_-:
   (a) FTBS method
   (b) Lax–Wendroff method
   (c) Lax–Friedrichs method

7. Using the results of Exercise 6 show that the von Neumann stability condition is satisfied for
   (a) FTBS method if and only if $a > 0$ and $0 \le sa \le 1$
   (b) Lax–Wendroff method if and only if $-1 \le sa \le 1$
   (c) Lax–Friedrichs method if and only if $-1 \le sa \le 1$

8. Consider the initial-value problem

$$u_x + u_t = 0, \qquad -\infty < x < \infty, t > 0$$

$$u(x, 0) = \begin{cases} 1, & |x| < 0.5, \\ 0, & |x| \ge 0.5, \end{cases} \quad -\infty < x < \infty$$

which has the (generalized) solution

$$u(x, t) = \begin{cases} 1, & |x - t| < 0.5 \\ 0, & |x - t| \ge 0.5 \end{cases} \text{ where } -\infty < x < \infty, t > 0.$$

Approximate the solution on $-2 < x < 2, 0 < t < 1$, with $h = 0.1$ using
   (a) FTBS method
   (b) Lax–Wendroff method
   (c) Lax–Friedrichs method
   (d) Leapfrog method
   Experiment with the time step and graph some of the numerical results superimposed on the exact solution.

9. Approximate the solution to the initial-boundary-value problem of Example 9.1.3 using
   (a) FTFS method with $h = 0.1$ and $k = 0.05$
   (b) Lax–Wendroff method with $h = 0.05$ and $k = 0.1$
   and in each case explain the poor performance of the method.

9.2 DIFFERENCE METHODS FOR A SCALAR INITIAL-BOUNDARY-VALUE PROBLEM

If a finite-difference equation for advancing a numerical solution from t_j to t_{j+1} involves only one term evaluated at the time level t_{j+1}, then the difference equation is called *explicit in time*. All of the methods of Table 9.1.1 are explicit in time. An *implicit difference method* is one that involves more than one term at the most advanced time level.

Explicit methods are the only kind of difference methods that apply directly to pure initial-value problems. If we try to apply an implicit method to a pure initial-value problem, we are led to a system of equations with an infinite number of unknowns. When dealing with an initial-boundary-value problem, both explicit and implicit finite-difference methods are available. Generally, implicit difference methods are unconditionally stable.

Consider the initial-boundary-value problem

$$au_x + u_t = 0, \qquad 0 < x < \infty, t > 0 \tag{9.2.1}$$

$$u(x, 0) = f(x), \qquad 0 < x < \infty \tag{9.2.2}$$

$$u(0, t) = g(t), \qquad t > 0 \tag{9.2.3}$$

in which we assume a to be a positive constant. Using the results of Section 7.1, this problem can be solved exactly by the method of characteristics to give

$$u(x, t) = \begin{cases} f(x - at), & x > at \\ g\left(\dfrac{at - x}{a}\right), & 0 < x < at \end{cases} \qquad (9.2.4)$$

The initial-boundary-value problem (9.2.1) and (9.2.3) with its known solution (9.2.4) will be used as a model problem in developing finite-difference methods for more general problems of this type.

Explicit Methods

With only slight modifications to account for the boundary condition $u(x, 0) = g(t)$, the explicit FTBS, Lax–Friedrichs, leapfrog, and Lax–Wendroff methods of Table 9.1.1 apply to the initial-boundary-value problem. The explicit FTFS method does not apply since we have assumed a to be positive.

EXAMPLE 9.2.1 ──

Consider the initial-boundary-value problem

$$\begin{aligned} 2u_x + u_t &= 0, & 0 < x < \infty, \, t > 0 \\ u(x, 0) &= x^2, & 0 < x < \infty \\ u(0, t) &= \sin t, & t > 0 \end{aligned}$$

whose exact solution is

$$u(x, t) = \begin{cases} (x - 2t)^2, & x > 2t \\ \sin\left(\dfrac{2t - x}{2}\right), & 0 < x < 2t \end{cases}$$

To illustrate an application of an explicit difference method to this problem, let us use the FTBS method to approximate the solution on $0 < x \le 1, \, 0 < t \le 0.5$, with

$$x_n = nh, \qquad n = 0, 1, \ldots, 10, \, h = 0.1$$

Since $a = 2$ and $h = 0.1$, the stability restriction $0 < 2k/h \le 1$ implies that the largest time step that can be used is $k = 0.05$. If we choose $k = 0.05$, the FTBS method will reduce to the method of characteristics and produce the exact solution. If a were not a constant, then we would not be able to make the difference method behave as the method of characteristics. To demonstrate how the difference method behaves when it does not coincide with the method of characteristics, let us choose $k = 0.025$.

The FTBS can be easily implemented on this example problem by using

$$\begin{aligned} U_n^0 &= (n \cdot h)^2, & n &= 0, 1, \ldots, 10 \\ U_0^j &= \sin(j \cdot k), & j &= 1, 2, \ldots, 20 \\ U_n^{j+1} &= (1 - sa)U_n^j + saU_{n-1}^j, & n &= 1, 2, \ldots, 10, \, s = 0.025/.1, \\ & & j &= 1, 2, \ldots, 20, \, a = 2 \end{aligned}$$

TABLE 9.2.1 NUMERICAL AND EXACT SOLUTIONS FOR EXAMPLE 1 USING FTBS METHOD

$T = 0.30$	Numerical ($k = 0.025$)	Numerical ($k = 0.1$)	Exact
$X = 0.0$	0.295520	0.295520	0.295520
$X = 0.1$	0.247279	0.187672	0.247404
$X = 0.2$	0.198662	0.419334	0.198669
$X = 0.3$	0.150638	0.030000	0.149438
$X = 0.4$	0.105904	-0.020000	0.099833
$X = 0.5$	0.069669	-0.050000	0.049979
$X = 0.6$	0.048790	-0.060000	0
$X = 0.7$	0.049128	-0.050000	0.010000
$X = 0.8$	0.073298	-0.020000	0.040000
$X = 0.9$	0.120835	0.030000	0.090000
$X = 1.0$	0.190132	0.100000	0.160000

TABLE 9.2.2 IMPLICIT FINITE-DIFFERENCE METHODS AND LTE FOR $au_x + u_t = S(x, t)$, $a > 0$

Backward in time–backward in space (BTBS), LTE $= O(k + h)$:

$$a \frac{U_{n+1}^{j+1} - U_n^{j+1}}{h} + \frac{U_{n+1}^{j+1} - U_{n+1}^{j}}{k} = S_{n+1}^{j+1}$$

or

$$(1 + sa)U_{n+1}^{j+1} - saU_n^{j+1} = U_{n+1}^{j} + kS_{n+1}^{j+1}, \qquad s = k/h \qquad (9.2.5)$$

Backward in time–centered in space (BTCS), LTE $= O(k + h^2)$:

$$a \frac{U_{n+1}^{j+1} - U_{n-1}^{j+1}}{2h} + \frac{U_n^{j+1} - U_n^{j}}{k} = S_n^{j+1}$$

or $(9.2.6)$

$$U_n^{j+1} + \tfrac{1}{2}sa\{U_{n+1}^{j+1} - U_{n-1}^{j+1}\} = U_n^{j} + kS_n^{j+1}, \qquad s = k/h$$

Crank–Nicolson, LTE $= O(k^2 + h^2)$:

$$\frac{1}{2}\left\{ a \frac{U_{n+1}^{j+1} - U_{n-1}^{j+1}}{2h} + a \frac{U_{n+1}^{j} - U_{n-1}^{j}}{2h} \right\} + \frac{U_n^{j+1} - U_n^{j}}{k} = S_n^{j+1/2}$$

or

$$U_n^{j+1} + \tfrac{1}{4}sa\{U_{n+1}^{j+1} - U_{n-1}^{j+1}\} = U_n^{j} - \tfrac{1}{4}sa\{U_{n+1}^{j} - U_{n-1}^{j}\} + kS_n^{j+1/2} \qquad (9.2.7)$$

Wendroff method, LTE $= O(k^2 + h^2)$:

$$\frac{a}{2}\left\{ \frac{U_{n+1}^{j+1} - U_n^{j+1}}{h} + \frac{U_{n+1}^{j} - U_n^{j}}{h} \right\} + \frac{1}{2}\left\{ \frac{U_{n+1}^{j+1} - U_{n+1}^{j}}{k} + \frac{U_n^{j+1} - U_n^{j}}{k} \right\} = S_{n+1/2}^{j+1/2}$$

or

$$(1 + sa)U_{n+1}^{j+1} + (1 - sa)U_n^{j+1} = (1 - sa)U_{n+1}^{j} + (1 + sa)U_n^{j} + 2kS_{n+1/2}^{j+1/2}, \qquad s = k/h \qquad (9.2.8)$$

Comparisons of the exact and numerical solution are shown in Table 9.2.1. Observe that on this fairly coarse grid the FTBS method does not do a very good job of approximating the exact solution. For the choice $k = 0.1$, which makes the FTBS method unstable, the numerical solution does not even resemble the exact solution. ■ ■

Implicit Methods

We next show how implicit difference methods can be applied to the model initial-boundary-value problem (9.2.1)–(9.2.3). Examples of implicit difference methods can be found in Table 9.2.2.

A von Neumann stability analysis can be used to show that all of the implicit finite-difference methods of Table 9.2.1 are unconditionally stable. The BTBS method (9.2.5) and the Wendroff method each involve two adjacent grid points at the time level t_{j+1}. This computational stencil permits these methods to be implemented without having to solve a set of simultaneous equations. For example, to calculate U_{n+1}^{j+1} using the BTBS method, we need only reach backward in time for U_{n+1}^{j} and backward in space for U_{n}^{j+1}. The value of U_{n+1}^{j} is available from a previous calculation or from the initial condition. Similarly, the value of U_{n}^{j+1} is available from a previous calculation or from the boundary condition. Algorithm 9.2 shows how to implement the Wendroff implicit method.

ALGORITHM 9.2   Wendroff Implicit Method
Initial-Boundary-Value Problem

Step 1. Document

This algorithm uses the Wendroff method (9.2.8) to approximate the solution of the initial-boundary-value problem

$$au_x + u_t = c(x, t), \qquad x > 0, 0 < t < t_{jmax}$$
$$u(x, 0) = f(x), \qquad x > 0$$
$$u(0, t) = g(t), \qquad 0 < t < t_{jmax}$$

INPUT

Integer n max, maximum x subscript for solution at time t_{jmax}
Integer j max, number of time steps
Reak k, time step ($k \cdot jmax = t_{jmax}$)
Real h, space step
Real a, coefficient in equation
Function $f(x)$, initial condition
Function $g(t)$, boundary condition
Function $c(x, t)$, G right side of PDE

OUTPUT

Time t_j and approximation U_n^j to $u(x_n, t_j)$ for $n = 0, 1, \ldots, nmax$
and $j = 0, 1, \ldots, jmax$

Step 2. Define grid ratio

Set $s = k/h$

Step 3. Initialize numerical solution

Set $t = 0$
For $n = 0, 1, \ldots, nmax$ set
  $x_n = n \cdot h$
  $V_n = f(x_n)$
Output t
For $n = 0, 1, 2, \ldots, nmax$
  Output x_n, V_n

Step 4. Begin time stepping

For $j = 1, 2, \ldots, jmax$
  Do steps 5–7

Step 5. Advance solution one time step

Set $t = t + k$
  $U_0 = g(t)$
  $q = (1 - sa)/(1 + sa)$
For $n = 0, 1, \ldots, nmax - 1$ set

$$U_{n+1} = V_n + qV_{n+1} - qU_n + 2kc\left(x_n + \frac{h}{2}, t - \frac{k}{2}\right)/(1 + sa)$$

Step 6. Prepare for next time step

For $n = 0, 1, \ldots, nmax$ set
  $V_n = U_n$

Step 7. Output numerical solution

Output t
For $n = 0, 1, \ldots, nmax$
  Output x_n, U_n

EXAMPLE 9.2.2 ————————————————————————————————

Consider, once again, the initial-boundary-value problem

$$2u_x + u_t = 0, \qquad 0 < x < \infty, \, t > 0$$
$$u(x, 0) = x^2, \qquad 0 < x < \infty$$
$$u(0, t) = \sin t, \qquad t > 0$$

whose exact solution is

$$u(x, t) = \begin{cases} (x - 2t)^2, & x > 2t \\ \sin\left(\dfrac{2t - x}{2}\right), & 0 < x < 2t \end{cases}$$

To illustrate an application of an implicit difference method to this problem, use Wendroff's method to approximate the solution on $0 < x \le 1$, $0 < t \le 1$, with

$$x_n = nh, \qquad n = 0, 1, \ldots, 10, \, h = 0.1$$

First, let us choose $k = 0.025$ so that the performance of Wendroff's method can be compared with that of the FTBS method used in Example 9.2.1. Using Algorithm 9.2 and $k = 0.025$ to implement Wendroff's method, we obtain the results shown in Table 9.2.2.

Recall that when the explicit FTBS method was applied to this example problem, the stability restriction was that the time step satisfy $k \le 0.05$ for $h = 0.1$. To illustrate that the implicit Wendroff method is not subject to this stability restriction, we also compute the numerical solution with $k = 0.1$ and $h = 0.1$. Comparisons of the exact and numerical solutions with the choice of $k = 0.1$ are also displayed in Table 9.2.3. Compare Tables 9.2.1 and 9.2.2. ■ ■

TABLE 9.2.3 NUMERICAL AND EXACT SOLUTIONS FOR EXAMPLE 9.2.2 USING ALGORITHM 9.2

$T = 0.30$	Numerical ($k = 0.025$)	Numerical ($k = 0.1$)	Exact
$X = 0.0$	0.2295520	0.295520	0.295520
$X = 0.1$	0.247412	0.245151	0.247404
$X = 0.2$	0.198689	0.211208	0.198669
$X = 0.3$	0.149522	0.140436	0.149438
$X = 0.4$	0.100436	0.073187	0.099833
$X = 0.5$	0.053635	0.028272	0.049979
$X = 0.6$	0.016248	0.009077	0
$X = 0.7$	0.003026	0.014216	0.010000
$X = 0.8$	0.029240	0.041867	0.040000
$X = 0.9$	0.091608	0.090798	0.090000
$X = 1.0$	0.166525	0.160332	0.160000

Numerical Boundary Treatments

For ease of implementation, the BTBS and Wendroff methods are usually preferred over the BTCS and Crank–Nicolson methods of Table 9.2.1. To apply either the BTCS or the Crank–Nicolson method to the problem of Example 9.2.2 requires that we manufacture some sort of condition at the last downstream point of interest. This last downstream point is $x = 1$ in Example 9.2.2. Otherwise the tridiagonal systems of equations defined by (9.2.6) and (9.2.7) will contain an infinite number of unknowns.

Methods for setting a value of U_n^{j+1} at the downstream boundary include using (i) the numerical method of characteristics to look backward along the characteristic through the downstream boundary point, (ii) an explicit method to fix the value, and (iii) physical reasoning to obtain a problem-specific downstream boundary condition. If such a numerical treatment of the downstream boundary introduces an error, the backward sweep of the tridiagonal solver used in BTCS and Crank–Nicolson methods will carry this error back upstream, against the flow of the characteristics, and make that error felt at all x nodes.

Although it is usually best to avoid having to provide a special method for dealing with a downstream node, the following example illustrates one situation in which a numerical treatment of a downstream boundary may be desirable.

EXAMPLE 9.2.3 ───

Consider using the Lax–Wendroff method to estimate the solution of the problem

$$\begin{aligned}
10u_x + u_t &= 0, & 0 &< x < \infty, t > 0 \\
u(x, 0) &= e^{-x^2}, & 0 &< x < \infty \\
u(0, t) &= e^{-t}, & t &> 0
\end{aligned} \qquad (9.2.9)$$

Suppose that we are only interested in the solution on the interval $0 \le x \le 1$ for $0 \le t \le 5$, and that we would like to use an x grid spacing of $h = 0.01$. The stability restriction for the Lax–Wendroff method, $|ka/h| \le 1$, requires $k \le 10^{-3}$, since $a = 10$ and $h = 0.01$. If we use the Lax–Wendroff method as it was given in Section 9.1, then we find we must initialize the numerical solution U_n^0 for $n = 0, 1, \ldots, 5100$. Recall that we must drop the rightmost grid point at each time step. Now a direct application of Lax–Wendroff requires two arrays of dimension over 5000 and more than 5000 grid updates for the first several time steps. ■ ■

To illustrate a numerical boundary treatment, consider the problem

$$\begin{aligned}
au_x + u_t &= 0, & 0 &< x < 1, t > 0, a > 0 & (9.2.10) \\
u(x, 0) &= f(x), & 0 &< x < 1 & (9.2.11) \\
u(0, t) &= g(t), & t &> 0 & (9.2.12)
\end{aligned}$$

Let $x_n = nh$, $n = 0, 1, \ldots, N$, $h = 1/N$, denote a finite-difference grid for the problem (9.2.10)–(9.2.12). By Taylor's theorem,

$$u(x_N, t + k) = u(x_N, t) + u_t(x_N, t)k + u_{tt}(x_N, t)\frac{k^2}{2} + O(k^3) \qquad (9.2.13)$$

The differential equation $au_x + u_t = 0$ implies

$$u_t = -au_x \quad \text{and} \quad u_{tt} = (-au_x)_t = -a(u_t)_x = -a(-au_x)_x = a^2 u_{xx}$$

Putting $u_t = -au_x$ and $u_{tt} = a^2 u_{xx}$ into (9.2.13) yields

$$u(x_N, t + k) = u(x_N, t) - au_x(x_N, t)k + \tfrac{1}{2}a^2k^2u_{xx}(x_N, t) + O(k^3)$$

Now, if we use one-sided differences to approximate $u_x(x_N, t)$ and $u_{xx}(x_N, t)$ at $t = t_j$,

$$u_x(x_N, t_j) = \frac{3U_N^j - 4U_{N-1}^j + U_{N-2}^j}{2h} + O(h^2)$$

$$u_{xx}(x_N, t_j) = \frac{U_N^j - 2U_{N-1}^j + U_{N-2}^j}{h^2} + O(h^2)$$

then, instead of the centered Lax–Wendroff method, we can compute at the downstream boundary with

$$\begin{aligned}U_N^{j+1} = U_N^j &- \tfrac{1}{2}sa(3U_N^j - 4U_{N-1}^j + U_{N-2}^j) \\ &+ \tfrac{1}{2}s^2a^2(U_N^j - 2U_{N-1}^j + U_{N-2}^j)\end{aligned} \qquad (9.2.14)$$

Now the Lax–Wendroff method (9.1.10) applies at the grid points x_n, $n = 1, 2, \ldots,$ $N - 1$, and (9.2.14) can be used at the downstream boundary node x_N. It can be shown that this method of using (9.1.10) and (9.2.14) to approximate the solution of the initial-boundary-value problem (9.2.10)–(9.2.12) is stable provided $|sa| \le 1$.

Considerable care must be used in choosing methods to "close" a system of difference equations by introducing into a finite-difference calculation a boundary condition that is not present in the associated partial differential equation problem. With such numerical boundary treatments, there is a risk of introducing an instability in the difference calculation and/or violating the physics that the partial differential equation models.

Exercises

1. Modify Algorithm 9.1 so that it applies to the initial-boundary-value problem (9.2.1)–(9.2.3).
2. Write an algorithm that uses
   (a) FTBS method
   (b) Lax–Friedrichs method
   to approximate the solution of (9.2.1)–(9.2.3).
3. For $0 < x < 1$, $0 < t < 1$, approximate the solution to the initial-boundary-value problem of Example 9.2.1 using $h = 0.1$ and $k = 0.04$ and
   (a) Lax–Wendroff method
   (b) Lax–Friedrichs method
4. Repeat Exercise 3 with $h = 0.1$ and $k = 0.1$. Explain your results.
5. Show that the von Neumann stability condition is satisfied for any choice of $s = k/h$ for
   (a) BTBS method (9.2.5)
   (b) BTCS method (9.2.6)
   (c) Crank–Nicolson method (9.2.7)
   (d) Wendroff method (9.2.8)

6. Modify Algorithm 9.2 to implement
   (a) BTBS method (9.2.5)
   (b) BTCS method (9.2.6)
   (c) Crank–Nicolson method (9.2.7)

7. Using $h = 0.1$ and $k = 0.04$ for $0 \le x \le 1$, $0 \le t \le 1$, approximate the solution to the initial-boundary-value problem of Example 9.2.2 using
   (a) BTBS method (9.2.5)
   (b) BTCS method (9.2.6)
   (c) Crank–Nicolson method (9.2.7)

8. For $0 \le x \le 1$, $0 \le t \le 1$, using $h = 0.1$, use Wendroff's method to approximate the solution to each of the following initial-boundary-value problems and compare the numerical solution with the exact solution
   (a) $u_t + u_x = -x^2 t$, $x > 0$, $t > 0$
       $u(x, 0) = 2 + \sin x$, $x > 0$
       $u(0, t) = 2 - \sin t - \frac{1}{12}t^4$, $t > 0$
       $u(x, t) = 2 + \sin(x - t) + \frac{1}{12}x^4 - \frac{1}{3}x^3 t - \frac{1}{12}(x - t)^4$
   (b) $u_x + u_t = (2x - x^2)e^{-t}$, $x > 0$, $t > 0$
       $u(x, 0) = x^2$, $x > 0$
       $u(0, t) = 0$, $t > 0$
       $u(x, t) = x^2 e^{-t}$
   (c) $u_x + u_t = \sin t + x \cdot \cos t$, $x > 0$, $t > 0$
       $u(x, 0) = 0$, $x > 0$
       $u(0, t) = 0$, $t > 0$
       $u(x, t) = x \sin t$
   (d) $u_x + u_t = 0$, $x > 0$, $t > 0$
       $u(x, 0) = 2 + \sin x$, $x > 0$
       $u(0, t) = 2 - \sin t$, $t > 0$
       $u(x, t) = 2 + \sin(x - t)$

9. Repeat Exercise 8 using
   (a) BTBS method
   (b) Lax–Wendroff method
   (c) FTBS method

10. Show that the exact solution of Examples 9.2.1 and 9.2.2 is continuous but not continuously differentiable across the line $x = 2t$. Explain the poor numerical approximation in Table 9.2.2 near the grid point $x = 0.6$.

11. For each of the methods listed do the following:
    (a) Sketch the computational stencil.
    (b) Develop the method by replacing u_x and u_t by appropriate difference quotient approximations.
    (c) Expand $u(x, t)$ in Taylor series about the indicated point $(\bar{x}, \bar{t})$ and find the LTE. The methods to use are
        (i)   BTBS, $(\bar{x}, \bar{t}) = (x_{n+1}, t_{j+1})$
        (ii) BTCS, $(\bar{x}, \bar{t}) = (x_n, t_{j+1})$
        (iii) Crank–Nicolson, $(\bar{x}, \bar{t}) = (x_n, t_{j+1/2})$
        (iv) Wendroff, $(\bar{x}, \bar{t}) = (x_{n+1/2}, t_{j+1/2})$

9.3 SCALAR CONSERVATION LAWS

Difference Equations in Conservation Law Form

The first-order conservation law

$$F(u)_x + u_t = 0 \tag{9.3.1}$$

or its equivalent

$$a(u)u_x + u_t = 0, \qquad a(u) = F'(u) \tag{9.3.2}$$

occurs frequently in applications. See Sections 7.3–7.5. Our goal in this section is to develop some finite-difference methods that are specifically tailored to conservation law partial differential equations.

As before, let h denote the x grid spacing and k the t grid spacing. In direct analogy with (9.3.1), we say that an explicit finite-difference method is in *conservation law form* if it can be expressed in the form

$$\frac{Q_{n+1/2}^j - Q_{n-1/2}^j}{h} + \frac{U_n^{j+1} - U_n^j}{k} = 0 \tag{9.3.3}$$

where $Q_{n-1/2}^j$ and $Q_{n+1/2}^j$ represent numerical approximations to the fluxes $F(u_{n-1/2}^j)$ and $F(u_{n+1/2}^j)$, respectively, $Q_{n-1/2}^j$ and $Q_{n+1/2}^j$ the *numerical fluxes*. A number of conservation law finite-difference methods will be obtained by expressing $Q_{n-1/2}^j$ and $Q_{n+1/2}^j$ in the form

$$Q_{n-1/2}^j = Q(U_{n-1}^j, U_n^j) \quad \text{and} \quad Q_{n+1/2}^j = Q(U_n^j, U_{n+1}^j)$$

where the function Q that defines the numerical fluxes varies from method to method. We require that the numerical flux Q be *consistent* with the continuous flux F in the sense that

$$Q(U, U) = F(U)$$

Observe that (9.3.3) can be rearranged to read

$$U_n^{j+1} \cdot h = U_n^j \cdot h + Q_{n-1/2}^j \cdot k - Q_{n+1/2}^j \cdot k \tag{9.3.4}$$

In a conservation law of the form (9.3.1), u represents the linear density of some material M and F the flux of that material. Now, with the help of Figure 9.3.1, we see that the conservation law difference equation (9.3.4) states that the M content of the interval $(\xi_n, \xi_{n+1}) = (x_n - h/2, x_n + h/2)$ at time t_{j+1} is equal to the M content of that interval at time t_j adjusted for the net flow of material M across the boundaries of the interval, $x = \xi_n$ and $x = \xi_{n+1}$, from time t_j to time t_{j+1}.

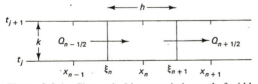

Figure 9.3.1 Change in M content in interval of width h centered at x_n equals flow in minus flow out.

Our reason for formulating difference methods for (9.1.1) in conservation law form is given in the following result.

Theorem 9.3.1. If the solution U_n^j of the conservation law difference equation (9.3.4) converges to a bounded function $u(x, t)$, then $u(x, t)$ is a generalized solution of the conservation law (9.1.1).

Theorem 9.3.1 is due to P.D. Lax and B. Wendroff, "Systems of Conservation Laws," *Commun. Pure Appl. Math.*, **13**, 217 (1960). A proof can also be found in G. A. Sod, *Numerical Methods in Fluid Dynamics*, Cambridge University Press, Cambridge, 1985.

Conservation Law Difference Equations

Probably, the first conservation law difference equation that comes to mind is the one in which the numerical fluxes are defined so that the flux across the left (right) boundary of the interval $(x_n - h, x_n + h)$ is the flux value at the node to the left (right). That is, set

$$Q_{n-1}^j = F(U_{n-1}^j) = F_{n-1}^j \quad \text{and} \quad Q_{n+1}^j = F(U_{n+1}^j) = F_{n+1}^j$$

to obtain

$$U_n^{j+1} = U_n^j - \tfrac{1}{2}s(F_{n+1}^j - F_{n-1}^j), \qquad s = k/h \tag{9.3.5}$$

However, we must reject the difference method (9.3.5) because it is unconditionally unstable for even the simple case in which the flux is linear with $F(u) = au$, $a = $ const. In this case the method (9.3.5) reduces to the FTCS method of Section 9.1.

In Section 9.1, we saw that the FTCS method can be modified to produce a stable method called the Lax–Friedrichs method. This modification consists of replacing the term U_n^j in the FTCS method by the arithmetic mean of its left and right neighbor values. Using the same modification in (9.3.5), we obtain the Lax–Friedrichs method for the conservation law (9.3.1).

Lax–Friedrichs Method

$$U_n^{j+1} = \tfrac{1}{2}(U_{n+1}^j + U_{n-1}^j) - \tfrac{1}{2}s(F_{n+1}^j - F_{n-1}^j), \qquad s = k/h \tag{9.3.6}$$

With numerical fluxes at $x_n + h$ and $x_n - h$ defined by

$$Q_{n+1}^j = F_{n+1}^j + \frac{1}{s}(U_n^j - U_{n+1}^j) \equiv Q(U_n^j, U_{n+1}^j) \tag{9.3.7a}$$

$$Q_{n-1}^j = F_{n-1}^j + \frac{1}{s}(U_{n-1}^j - U_n^j) \equiv Q(U_{n-1}^j, U_n^j) \tag{9.3.7b}$$

the Lax–Friedrichs method is seen to be a conservation law difference equation. Moreover, the function Q in (9.3.7a) and (9.3.7b) is consistent with F since $Q(U_n^j, U_n^j) = F(U_n^j)$.

When $F(u)$ is linear with $F(u) = au$, $a = $ const, then the conservation law Lax–Friedrichs method (9.3.6) reduces to the Lax–Friedrichs method of Section 9.1. In this linear case the stability restriction is $|sa| \leq 1$. When $F(u)$ is nonlinear, the von Neumann stability analysis does not apply because the superposition principle is no longer valid. However, by considering a linearization of (9.3.6) about the numerical solution, a linearized stability analysis can be performed. This linearized stability analysis predicts that the method (9.3.6) is stable provided

$$|s \cdot a(U_n^j)| \leq 1 \quad \text{for all } n, j \tag{9.3.8}$$

Lax–Wendroff Method

The Lax–Wendroff method for the conservation law can be obtained from an argument similar to that used in Section 9.1. By Taylor's Theorem 8.1.1,

$$u(x_n, t_{j+1}) = u(x_n, t_j) + u_t(x_n, t_j)k + u_{tt}(x_n, t_j) \frac{k^2}{2} + O(k^3) \tag{9.3.9}$$

From the differential equation $F(u)_x + u_t = 0$, we have

$$u_t = -F(u)_x$$

and

$$u_{tt} = (-F(u)_x)_t = -(F(u)_t)_x = -(F'(u)u_t)_x$$
$$= -(a(u)u_t)_x = (a(u)F(u)_x)_x$$

Now we approximate $u_t(x_n, t_j)$ by

$$u_t(x_n, t_j) = -\frac{F_{n+1}^j - F_{n-1}^j}{2h} + O(h^2) \tag{9.3.10}$$

To obtain a finite-difference replacement for $u_{tt}(x_n, t_j)$, recall that the difference operator δ_x is defined by

$$\delta_x U_n^j = U_{n+1/2}^j - U_{n-1/2}^j$$

Therefore, a centered difference formula for $F(u)_x$ using a step of size $h/2$ is given by

$$F(u)_x = \frac{1}{h} \delta_x F_n^j + O(h^2)$$

Now we have,

$$u_{tt}(x_n, t_j) = \frac{1}{h^2} \delta_x(a(U_n^j)\delta_x F(U_n^j)) + O(h^2)$$

$$= \frac{1}{h^2} \delta_x(a_n^j(F_{n+1/2}^j - F_{n-1/2}^j)) + O(h^2)$$

$$= \frac{1}{h^2} [a_{n+1/2}^j(F_{n+1}^j - F_n^j) - a_{n-1/2}^j(F_n^j - F_{n-1}^j)] + O(h^2) \tag{9.3.11}$$

The coefficients $a_{n\pm1/2}^j$ in (9.3.6) can be evaluated using either

$$a_{n+1/2}^j = a\left(\frac{U_n^j + U_{n+1}^j}{2}\right) \quad \text{and} \quad a_{n-1/2}^j = a\left(\frac{U_n^j + U_{n-1}^j}{2}\right)$$

or

$$a^j_{n+1/2} = \frac{a(U^j_n) + a(U^j_{n+1})}{2} \quad \text{and} \quad a^j_{n-1/2} = \frac{a(U^j_n) + a(U^j_{n-1})}{2}$$

Using the difference formulas (9.3.10) and (9.3.11) to replace u_t and u_{tt} in (9.3.9), we obtain the *conservation law form of the Lax–Wendroff method.*

$$U^{j+1}_n = U^j_n - \tfrac{1}{2}s(F^j_{n+1} - F^j_{n-1}) + \tfrac{1}{2}s^2[(a^j_{n+1/2})(F^j_{n+1} - F^j_n) \quad (9.3.12)$$
$$- (a^j_{n-1/2})(F^j_n - F^j_{n-1})], \quad s = k/h$$

Notice that in the case of a linear conservation law, $F(u) = au$, $a = \text{const}$, the conservation law Lax–Wendroff method reduces to the Lax–Wendroff method of Section 9.1. With numerical fluxes defined by

$$Q^j_{n+1} = F^j_{n+1} + sa_{n+1/2}(F^j_n - F^j_{n+1}) \equiv Q(U^j_n, U^j_{n+1}) \quad (9.3.13a)$$
$$Q^j_{n-1} = F^j_{n-1} + sa_{n-1/2}(F^j_n - F^j_{n-1}) \equiv Q(U^j_n, U^j_{n-1}) \quad (9.3.13b)$$

the Lax–Wendroff method is seen to be a conservation law difference equation with consistent numerical fluxes. A linearized stability analysis indicates that the conservation law Lax–Wendroff method is subject to the stability condition (9.3.8).

To avoid the calculation of the quantities $a_{n\pm 1/2}$ in the Lax–Wendroff method, a number of two-step modifications have been developed. Two examples of two-step modifications of the Lax–Wendroff method are the Richtmeyer method and the MacCormack method in Table 9.3.1. Each of these two-step replacements have the property that they reduce to the Lax–Wendroff method in the linear case $F(u) = au$, $a = \text{const}$.

Monotone Methods

Since the Lax–Wendroff method has local truncation error $O(k^2 + h^2)$, it is said to be second order in both time and space, or simply a second-order method. Second-order finite-difference methods generally do a good job of approximating the solution of the conservation law (9.3.1) when that solution is smooth. However, when the solution of (9.3.1) has a discontinuity or a very sharp front, then a numerical solution obtained by a second-order method will often exhibit oscillations even if the exact solution is monotone. In many cases these oscillations in the numerical solution are merely an annoyance, but in some instances the oscillations can produce an instability in the calculation.

Consider a finite-difference equation of the form

$$U^{j+1}_n = R(U^j_{n-1}, U^j_n, U^j_{n+1}) \quad (9.3.14)$$

We say that the finite-difference method (9.3.14) is *monotone* provided

$$\frac{\partial R}{\partial U^j_i} \geq 0 \quad \text{for } i = n - 1, n, n + 1$$

EXAMPLE 9.3.1 _____

Since the Lax–Friedrichs method has the form

$$U^{j+1}_n = \tfrac{1}{2}(U^j_{n+1} + U^j_{n-1}) - \tfrac{1}{2}s(F^j_{n+1} - F^j_{n-1}) \equiv R(U^j_{n-1}, U^j_n, U^j_{n+1})$$

it follows that

$$\frac{\partial R}{\partial U_{n-1}^j} = \tfrac{1}{2} + \tfrac{1}{2}sa(U_{n-1}^j)$$

$$\frac{\partial R}{\partial U_n^j} = 0$$

$$\frac{\partial R}{\partial U_{n+1}^j} = \tfrac{1}{2} - \tfrac{1}{2}sa(U_{n+1}^j)$$

Therefore, it follows that the Lax–Friedrichs method is monotone provided (9.3.8) holds. ∎ ∎

We say that the numerical solution U_n^j is *monotone increasing* if

$$m < n \quad \text{implies} \quad U_m^j \le U_n^j$$

Similarly the numerical solution U_n^j is *monotone decreasing* if

$$m < n \quad \text{implies} \quad U_m^j \ge U_n^j$$

We say U_n^j is monotone if it is either monotone increasing or monotone decreasing. The finite-difference operator R of (9.3.14) is said to be *monotonicity preserving* provided all solutions of (9.3.14) have the property that if U_n^j is monotone, then U_n^{j+1} is monotone of the same type. The next theorem shows that a monotone finite-difference method applied to a monotone initial condition produces a numerical solution that is monotone of the same type.

Theorem 9.3.2. Every monotone finite-difference method of the form (9.3.14) is monotonicity preserving.

Proof. From (9.3.14) we have

$$U_n^{j+1} = R(U_{n-1}^j, U_n^j, U_{n+1}^j)$$

and

$$U_{n+1}^{j+1} = R(U_n^j, U_{n+1}^j, U_{n+2}^j)$$

From the mean-value theorem, there exists a constant γ, $0 < \gamma < 1$, such that

$$
\begin{aligned}
U_{n+1}^{j+1} - U_n^{j+1} &= R(U_n^j, U_{n+1}^j, U_{n+2}^j) - R(U_{n-1}^j, U_n^j, U_{n+1}^j)\\[4pt]
&= \frac{\partial R}{\partial U_{n-1}^j} (\overline{U}_{n-1}^j, \overline{U}_n^j, \overline{U}_{n+1}^j)[U_n^j - U_{n-1}^j]\\[4pt]
&\quad + \frac{\partial R}{\partial U_n^j} (\overline{U}_{n-1}^j, \overline{U}_n^j, \overline{U}_{n+1}^j)[U_{n+1}^j - U_n^j]\\[4pt]
&\quad + \frac{\partial R}{\partial U_{n+1}^j} (\overline{U}_{n-1}^j, \overline{U}_n^j, \overline{U}_{n+1}^j)[U_{n+2}^j - U_{n+1}^j]
\end{aligned}
$$

where

$$
\begin{aligned}
(\overline{U}_{n-1}^j, \overline{U}_n^j, \overline{U}_{n+1}^j) = (\gamma U_n^j + (1-\gamma)U_{n-1}^j,\\
\gamma U_{n+1}^j + (1-\gamma)U_n^j, \gamma U_{n+2}^j - (1-\lambda)U_{n+1}^j)
\end{aligned}
$$

Since R is monotone, all of its partial derivatives are nonnegative. Since U_n^j is monotone, $[U_n^j - U_{n-1}^j]$, $[U_{n+1}^j - U_n^j]$, and $[U_{n+2}^j - U_{n+1}^j]$ are all either nonpositive or nonnegative. Therefore, $[U_{n+1}^{j+1} - U_n^{j+1}]$ has the same sign as $[U_{n+1}^j - U_n^j]$, and the proof is complete.

A. Harten, J. M. Hyman, and P. D. Lax ["On Finite Difference Approximations and Entropy Conditions for Shocks," *Commun. Pure Appl. Math.*, **29**, 297, (1976)] have constructed an example conservation law whose solution contains three shocks. Using the Lax–Wendroff method (9.3.12) to approximate the solution of this problem, they obtain a numerical solution that satisfies the entropy condition across only two of the three discontinuities of the numerical solution. This example illustrates that it is possible for a finite-difference method to produce nonphysical numerical solutions. Recall that by requiring the entropy condition, we rule out all discontinuities in the solution of (9.3.1) except those that are essential for conservation of material M. The failure of the Lax–Wendroff numerical solution to satisfy the entropy condition can be attributed to the fact that the Lax–Wendroff method is not monotone.

From Theorem (9.3.2), we already know that montone finite-difference methods will not introduce oscillations into the approximation of a montone solution of (9.3.1). The next theorem shows that monotone difference methods in conservation form will produce numerical solutions that satisfy the entropy condition.

Theorem 9.3.3. Consider a monotone finite-difference method in conservation form for the conservation law (9.3.1). If the solution of this difference method converges to some function $u(x, t)$ as k and h approach zero with $s = k/h$ fixed, then $u(x, t)$ is a generalized solution of (9.3.1) and the entropy condition (7.4.7) is satisfied across all discontinuities of u.

Theorems 9.3.2 and 9.3.3 give two advantages of monotone difference methods in conservation form. The next result shows that to develop a method that behaves in the favorable manner described in Theorems 9.3.2 and 9.3.3, we must sacrifice something with regard to local truncation error.

Theorem 9.3.4. A monotone finite-difference method in conservation form has local truncation error $O(k + h)$.

The proof of Theorem 9.3.3 can be found in G. A. Sod, *Numerical Methods in Fluid Dynamics*, Cambridge University Press, Cambridge, 1985, pp. 298, 299.

Theorems 9.3.2–9.3.4 can be used as a guide in selecting a finite-difference method to apply to (9.3.1). If we expect the solution to contain one or more shocks, we should consider using a monotone difference method in conservation form. If we believe the solution of (9.3.1) is smooth, then a higher order scheme such as the Lax–Wendroff method would probably be appropriate.

Upwind Methods

We have already seen that the Lax–Friedrichs method is a monotone method in conservation form. Other monotone difference methods in conservation from can be obtained by using one-sided differences to approximate the term $F(u)_x$ in (9.3.1).

A physical interpretation of this one-sided differencing is the following. We know that the sign of $a(u)$ determines the direction of the characteristic. If $a > 0$ along a characteristics, then the characteristic equation $dx/dt = a$ shows that the characteristic points forward in x with increasing time. Similarly, if $a < 0$, a characteristic points backward in x with increasing time. Thus, the characteristics of (9.3.1) define a flow field in the xt plane. The close connection of conservation laws and fluid mechanics has led to the terms *stream* and *wind* to describe this flow field. The direction one must go to remain on a characteristic with increasing time is called *downstream* or *downwind*. Of course, the opposite directions are called *upstream* or *upwind*. If $a > 0$, then during the time interval t_j to t_{j+1}, we can argue that the downstream flux at position x_{n+1} will not have a direct effect on the M content of the interval $(x_n - h/2, x_n + h/2)$. Since u is constant along characteristics, the flux $F(u)$ is also constant along characteristics. Thus, from Figure 9.3.2, we see that the upstream flux F_{n-1}^j approximates the flow into the interval, F_n^j approximates the flow out of the region, while F_{n+1}^j has no impact on the M balance for the interval. Reasoning that we should look upstream to see what flux will be affecting the interval during this time period, we are led to develop two *upwind* or *upstream difference methods*:

FTBS upwind method:

$$U_n^{j+1} = U_n^j + s(F_{n-1}^j - F_n^j), \qquad a(u) > 0 \qquad (9.3.15a)$$

FTFS upwind method:

$$U_n^{j+1} = U_n^j - s(F_{n+1}^j - F_n^j), \qquad a(u) < 0 \qquad (9.3.15b)$$

It is not difficult to show that the difference methods (9.3.15a) and (9.3.15b) are in conservation law form. In (9.3.15a) the numerical fluxes are given by

$$Q_{n-1/2}^j = F_{n-1}^j \quad \text{and} \quad Q_{n+1/2}^j = F_n^j$$

For (9.3.15b) they are

$$Q_{n-1/2}^j = F_n^j \quad \text{and} \quad Q_{n+1/2}^j = F_{n+1}^j$$

Also, if

$$0 \le sa(U_n^j) \le 1 \quad \text{for all } n, j$$

then the upwind method (9.3.15a) is monotone. Similarly, if

$$-1 \le sa(U_n^j) \le 0 \quad \text{for all } n, j$$

then the method (9.3.15b) is monotone.

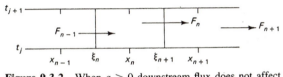

Figure 9.3.2 When $a > 0$ downstream flux does not affect M balance of interval (ξ_{n-1}, ξ_n). Characteristics are curves (———).

When $a(u)$ is not of one sign, we need to combine methods (9.3.15a) and (9.3.15b) so that the flux across a boundary, ξ_n or ξ_{n+1}, is automatically computed from the upwind direction. Also, it would be desirable to have a difference method that does not require the function $a(u)$.

Since $a(u) = F'(u)$, from the mean-value theorem it follows that

$$\frac{F(U_{n-1}^j) - F(U_n^j)}{U_{n-1}^j - U_n^j} = a(\overline{U}_n^j), \qquad U_{n-1}^j < \overline{U}_n^j < U_n^j \tag{9.3.16}$$

Equation (9.3.16) supports using the difference quotient on the left of (9.3.16) to fix the sign of $a(U_{n-1/2}^j)$. The sign of $a(U_{n-1/2}^j)$ determines whether the characteristic passes through $x = \xi_n$ from left to right or from right to left. Let us define an indicator function by

$$\delta(V, W) = \begin{cases} 1, & \dfrac{F(V) - F(W)}{V - W} \geq 0 \\[4mm] 0, & \dfrac{F(V) - F(W)}{V - W} < 0 \end{cases}$$

Also define

$$\delta_{n-1/2}^j = \delta(U_{n-1}^j, U_n^j) \quad \text{and} \quad \delta_{n+1/2}^j = \delta(U_n^j, U_{n+1}^j)$$

Now, numerical fluxes satisfying

$$Q_{n-1/2}^j = Q(U_{n-1}^j, U_n^j) = \begin{cases} F_{n-1}^j & \text{if } \delta_{n-1/2}^j \geq 0 \\ F_n^j & \text{if } \delta_{n-1/2}^j < 0 \end{cases}$$

and

$$Q_{n+1/2}^j = Q(U_n^j, U_{n+1}^j) = \begin{cases} F_n^j & \text{if } \delta_{n+1/2}^j \geq 0 \\ F_{n+1}^j & \text{if } \delta_{n+1/2}^j < 0 \end{cases}$$

can be defined by

$$Q_{n-1/2}^j = F_{n-1}^j + [1 - \delta_{n-1/2}^j](F_n^j - F_{n-1}^j)$$

and

$$Q_{n+1/2}^j = F_n^j + [1 - \delta_{n+1/2}^j](F_{n+1}^j - F_n^j)$$

Putting these numerical fluxes into (9.3.4), we obtain a monotone difference method in conservation law form due to Harten:

Harten upwind method:

$$U_n^{j+1} = U_n^j + s\{F_{n-1}^j + [1 - \delta_{n-1/2}^j](F_n^j - F_{n-1}^j) \tag{9.3.17}$$
$$- F_n^j + [1 - \delta_{n+1/2}^j](F_{n+1}^j - F_n^j)\}$$

TABLE 9.3.1 CONSERVATION LAW DIFFERENCE METHODS FOR $F(u)_x + u_t = 0$

Centered in time–centered in space (unstable):

$$U_n^{j+1} = U_n^j - \tfrac{1}{2}s(F_{n+1}^j - F_{n-1}^j), \qquad s = k/h \tag{9.3.18}$$

Lax–Friedrichs:

$$U_n^{j+1} = \tfrac{1}{2}(U_{n+1}^j + U_{n-1}^j) - \tfrac{1}{2}s(F_{n+1}^j - F_{n-1}^j), \qquad s = k/h \tag{9.3.19}$$

Lax–Wendroff:

$$U_n^{j+1} = U_n^j - \tfrac{1}{2}s(F_{n+1}^j - F_{n-1}^j) + \tfrac{1}{2}s^2[(a_{n+1/2}^j)(F_{n+1}^j - F_n^j) \tag{9.3.20}$$
$$- (a_{n-1/2}^j)(F_n^j - F_{n-1}^j)], \; s = k/h$$

Richtmeyer method:

$$U_{n+1/2}^* = \tfrac{1}{2}(U_{n+1}^j + U_n^j) - \tfrac{1}{2}s(F_{n+1}^j - F_n^j) \tag{9.3.21a}$$
$$U_n^{j+1} = U_n^j - s(F_{n+1/2}^* - F_{n-1/2}^*) \tag{9.3.21b}$$

MacCormack method:

$$U_n^* = U_n^j - s(F_{n+1}^j - F_n^j) \tag{9.3.22a}$$
$$U_n^{j+1} = \tfrac{1}{2}[U_n^j + U_n^* - s(F_n^* - F_{n-1}^*)] \tag{9.3.22b}$$

FTBS upwind method:

$$U_n^{j+1} = U_n^j + s(F_{n-1}^j - F_n^j), \qquad a(u) > 0 \tag{9.3.23}$$

FTFS upwind method:

$$U_n^{j+1} = U_n^j - s(F_{n+1}^j - F_n^j), \qquad a(u) < 0 \tag{9.3.24}$$

Harten upwind method:

$$U_n^{j+1} = U_n^j + s\{F_{n-1}^j + [1 - \delta_{n-1/2}^j](F_n^j - F_{n-1}^j) \tag{9.3.25}$$
$$- F_n^j + [1 - \delta_{n+1/2}^j](F_{n+1}^j - F_n^j)\}$$

For convenient reference, the conservation law difference methods of this section are listed in Table 9.3.1.

EXAMPLE 9.3.2 ───

Consider following initial-value problem

$$(u^2/2)_x + u_t = 0, \qquad -\infty < x < \infty, t > 0$$
$$u(x, 0) = x, \qquad -\infty < x < \infty$$

Since $a(u) = u$ and $f(x) = x$ produce $a'(u) = 1$ and $f'(x) = 1$, the product $f'a'$ is always positive. Thus we know from Theorem 7.3.1 that this problem has a smooth solution. In fact, Theorem 7.3.1 implies the exact solution is

$$u(x, t) = x/(1 + t), \qquad t > 0$$

Since we anticipate a smooth solution, let us approximate the solution by the second-order Lax–Wendroff method. A slight modification of Algorithm 9.1 produces the results shown in Table 9.3.2. ■ ■

TABLE 9.3.2 NUMERICAL AND EXACT SOLUTION
FOR EXAMPLE 9.3.2 USING
LAX–WENDROFF METHOD (9.3.20)

$T = 0.50$	Numerical ($k = 0.05$)	Exact
$X = 0.0$	0	0
$X = 0.1$	0.066707	0.066667
$X = 0.2$	0.133414	0.133333
$X = 0.3$	0.200121	0.200000
$X = 0.4$	0.266828	0.266667
$X = 0.5$	0.333535	0.333333
$X = 0.6$	0.400242	0.400000
$X = 0.7$	0.466949	0.466667
$X = 0.8$	0.533656	0.533333
$X = 0.9$	0.600363	0.600000
$X = 1.0$	0.667070	0.666667

EXAMPLE 9.3.3 _____

To illustrate a difference method on a problem that has a shock in its solution, consider
the initial-value problem

$$(u^2/2)_x + u_t = 0, \qquad -\infty < x < \infty, \, t > 0$$
$$u(x, 0) = \begin{cases} 2, & x < 0 \\ 0, & x > 0 \end{cases}$$

From Section 7.4, we know the (generalized) solution to this problem to be

$$u(x, t) = \begin{cases} 2, & x < t, \, t > 0 \\ 0, & x > t, \, t > 0 \end{cases}$$

Using the FTBS upwind method (9.3.23) with $h = 0.02$ and $k = 0.005$, we find the
numerical solution (○) and the exact solution (———) at time $t = 0.5$ to be as shown in
Figure 9.3.3. Note that the shock in the numerical solution is rounded off and lags the
shock in the exact solution.   ■ ■

Exercises

1. Use (9.3.7a) and (9.3.7b) to verify that the Lax–Friedrichs method (9.3.6) can be written in
the conservation law form (9.3.3).

2. Verify that the Lax–Wendroff method (9.3.12) can be written in the conservation law form
(9.3.3).

3. Show that the Lax–Wendroff method is not monotone. [*Hint*: Consider (9.1.10).]

4. Show that the upwind method (9.3.15a) can be written in conservation law form and that it
is monotone provided a "stability condition" of the form $0 \leq sa(U_n^j) \leq 1$ is satisfied.

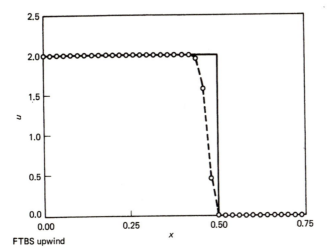

FTBS upwind

Figure 9.3.3

5. In applying the Lax–Wendroff method to the problem of Example 9.1.1, what is the largest time step you would use? Justify your answer.

6. Rework Example 9.1.1 using
   (a) Richtemeyer method (9.3.21)
   (b) MacCormack method (9.3.22)
   (c) Lax–Friedrichs method (9.3.19)

7. Show that when $F(u) = au$ with $a = $ const, the following reduces to the Lax–Wendroff method for $au_x + u_t = 0$:
   (a) MacCormack method (9.3.22)
   (b) Richtmeyer method (9.3.21)

8. For $0 \leq x \leq 1, 0 < t \leq 1$, use the FTBS upwind method (9.3.23) to approximate the solution to the initial value problem

$$(e^u)_x + u_t = 0, \qquad -\infty < x < \infty, t > 0$$

$$u(x, 0) = \begin{cases} 1, & x < 0, \ -\infty < x < \infty \\ 0, & x > 0, \ -\infty < x < \infty \end{cases}$$

and compare the numerical solution with the exact solution $(t > 0)$

$$u(x, t) = \begin{cases} 1, & x < (e - 1)t \\ 0, & x > (e - 1)t \end{cases}$$

9. Repeat Exercise 7 using
   (a) Lax–Friedrichs method
   (b) Lax–Wendroff method
   (c) Richtmeyer method
   (d) MacCormack method

10. For $0 \le x \le 1$, $0 < t \le 1$, use the FTBS upwind method (9.3.23) to approximate the solution to the initial-value problem

$$2uu_x + u_t = 0$$

$$u(x, 0) = \begin{cases} \sqrt{|x|}, & x < 0 \\ 0, & x > 0 \end{cases}$$

and compare the numerical solution with the exact solution

$$u(x, t) = \begin{cases} t + \sqrt{t^2 - x}, & x < 3t^2/4 \\ 0, & x > 3t^2/4 \end{cases}$$

11. For $0 \le x \le 1$, $0 < t \le 1$, approximate the solution to each of the following initial-value problems:

(a) $(u^2)_x + u_t = 0$, $u(x, 0) = \begin{cases} 1, x < 0 \\ 3, x > 0 \end{cases}$

(b) $(e^u)_x + u_t = 0$, $u(x, 0) = \begin{cases} 0, x < 0 \\ 1, x > 0 \end{cases}$

(c) $(u^4)_x + u_t = 0$, $u(x, 0) = \begin{cases} 0, x < 0 \\ 1, x > 0 \end{cases}$

using the Lax–Wendroff method. Compare the numerical and exact solutions.

12. Repeat Exercise 10 using
   (a) Lax–Friedrichs method
   (b) FTBS upwind method
   (c) Richtmeyer method
   (d) MacCormack method

9.4 DISPERSION AND DISSIPATION

Dispersion and Dissipation for Partial Differential Equations

Using Fourier series or the Fourier integral theorem, the solution to a homogeneous linear partial differential equation in x and t,

$$L[u] = 0 \tag{9.4.1}$$

can often be expressed as a superposition of Fourier modes of the form

$$\exp[i(\beta x - \omega t)] \tag{9.4.2}$$

To see examples of solutions expressed in this form, see Section 5.1. In (9.4.2), ω is the *frequency* of the Fourier mode, β is the *wave number*, and $\Lambda = 2\pi/\beta$ is the *wavelength*.

If we substitute $\exp[i(\beta x - \omega t)]$ into the partial differential equation (9.4.1) and divide out the common exponential factor, then a relationship between the frequency and the wave number results. This relationship, which can be written in the form $\omega = \omega(\beta)$, is the *dispersion relation* for the partial differential equation (9.4.1). If ω is real for all real values of β and if $\omega(\beta)$ is linear in β, then the partial differential equation is said to be *nondispersive*. If $\omega'' \neq 0$, then the partial differential equation is *dispersive*. In Example 5.1.3 we have encountered the phenomenon of dispersion in connection with the second-order wave equation.

If $\omega(\beta)$ is real for all real β, then the solutions of $L[u] = 0$ are undamped waves and the differential equation (9.4.1) is *nondissipative* or of *conservative type*. If $\omega(\beta)$ is complex and the imaginary part of ω is negative, then the partial differential equation is *dissipative*.

EXAMPLE 9.4.1 —————————————————————————————

Consider the diffusion equation

$$u_t - a^2 u_{xx} = 0$$

If we put the expression $\exp[i(\beta x - \omega t)]$ into this differential equation, then we find

$$-i\omega \exp[i(\beta x - \omega t)] - a^2(i\beta)^2 \exp[i(\beta x - \omega t)] = 0$$

Dividing out the common exponential factor and using the facts that $i^2 = -1$ and $1/i = -i$, we obtain the dispersion relation

$$\omega = -ia^2\beta^2$$

Since ω is complex with negative imaginary part, the diffusion equation is seen to be dissipative. Since $\omega''(\beta) \neq 0$ the diffusion equation is also dispersive. ■■

EXAMPLE 9.4.2 —————————————————————————————

Given the advection equation $au_x + u_t = 0$, $a = \text{const}$, if we put $u = \exp[i(\beta x - \omega t)]$, then it follows that

$$ai\beta \exp[i(\beta x - \omega t)] - i\omega \exp[i(\beta x - \omega t)] = 0$$

This shows the dispersion relation for this advection equation to be

$$\omega = a\beta$$

Since ω is real for all real β, the partial differential equation is nondissipative or conservative. Also, since $\omega''(\beta) = 0$, the advection equation is nondispersive. Note that the signal speed, or celerity, $dx/dt = a$ is related to the wave number and frequency by $a = \omega/\beta$. ■■

EXAMPLE 9.4.3 —————————————————————————————

For the second-order wave equation

$$u_{tt} - c^2 u_{xx} = 0$$

the dispersion relation is $\omega^2 = c^2\beta^2$, which has the two linear solutions $\omega = \omega_1(\beta) = c\beta$ and $\omega = \omega_2(\beta) = -c\beta$. Since ω is real, the partial differential equation is nondissipative. Since ω_1 and ω_2 are linear functions of β, the partial differential equation is also nondispersive. ■■

So that we may better understand the notions of dispersion and dissipation, consider a Fourier analysis of the simple advection equation initial-value problem

$$au_x + u_t = 0, \qquad a = \text{const}, \qquad -\infty < x < \infty \qquad (9.4.3)$$
$$u(x, 0) = f(x), \qquad\qquad\qquad -\infty < x < \infty \qquad (9.4.4)$$

Suppose that the initial condition $f(x)$ can be expressed as a superposition of Fourier modes using either a Fourier series or a Fourier integral representation. For purposes of discussion, suppose that $f(x)$ is 2π periodic and that it has a Fourier series representation

$$f(x) = \frac{a_0}{2} + \sum_{m=1}^{\infty} (a_m \cos mx + b_m \sin mx) \qquad (9.4.5)$$

where

$$a_m = \frac{1}{\pi} \int_{-\pi}^{\pi} f(x)\cos mx \; dx \quad \text{and} \quad b_m = \frac{1}{\pi} \int_{-\pi}^{\pi} f(x)\sin mx \; dx$$

An alternative way of writing the Fourier series of f is the complex form

$$f(x) = \sum_{m=-\infty}^{\infty} c_m e^{imx} \qquad (9.4.6)$$

where

$$c_m = \frac{1}{2\pi} \int_{-\pi}^{\pi} f(x)e^{-imx} \; dx$$

Recall that Euler's formulas,

$$\cos \theta = \frac{e^{i\theta} + e^{-i\theta}}{2} \quad \text{and} \quad \sin \theta = \frac{e^{i\theta} - e^{-i\theta}}{2i}$$

can be used to establish the equivalence of the representation (9.4.5) and (9.4.6). We refer to a typical term in (9.4.6) as a *Fourier mode* or a *harmonic* of the function $f(x)$. The low harmonics are those that have a low wave number, such as $m = 0$ and $m = 1$. Large values of m are associated with high harmonics in the representation of f.

From Section 7.1, we know that the solution to the initial-value problem (9.4.3) and (9.4.4) is given by $u(x, t) = f(x - at)$. Thus, $u(x, t)$ has the Fourier representation

$$u(x, t) = \sum_{m=-\infty}^{\infty} c_m e^{im(x - at)} \qquad (9.4.7)$$

Calculating the magnitude of the mth harmonic of the solution, we find

$$\left| c_m e^{im(x - at)} \right| = \left| c_m \right| \left| e^{im(x - at)} \right| = \left| c_m \right|$$

This shows that the magnitude of the mth harmonic of the solution is the same as the magnitude of the mth harmonic of the initial condition. This exemplifies the nondissipative character of the differential equation (9.4.3). Also observe that each harmonic in the representation (9.4.7) is advected, or translated, by the constant amount a for each unit increase in time. The fact that all harmonics propagate at the same speed illustrates the nondispersive nature of the differential equation (9.4.3).

Dispersion and Dissipation for Finite-Difference Equations

When choosing a finite-difference method to approximate the solution of a partial differential equation, we would like the difference method to have the same dissipative and dispersive nature as the partial differential equation. Unfortunately, some compromises must generally be made. For example, to obtain a stable difference method, it is often necessary for the difference method to contain a measure of dissipation even if none is present in the partial differential equation. Similarly, we shall see in what follows that some degree of numerical dispersion is usually present in a difference method even if the partial differential equation is nondispersive. When interpreting the results of a finite-difference method, it is important to understand the mechanisms of numerical dissipation and dispersion.

We next analyze the dispersive and dissipative nature of finite-difference methods for the advection equation initial-value problem (9.4.3)–(9.4.5). A dispersion–dissipation analysis of a finite-difference equation follows the same lines as the von Neumann stability analysis.

Following the preceding discussion of dispersion and dissipation for the initial-boundary-value problem (9.4.3) and (9.4.4), let us continue to assume that the initial condition is 2π periodic. A discrete Fourier representation of the initial condition is most convenient when an even number of grid points is used. Therefore, define a finite-difference grid on the interval $[-\pi, \pi)$ by

$$x_n = -\pi + nh, \qquad n = 0, 1, 2, \ldots, 2p - 1, \qquad h = 2\pi/2p = \pi/p$$

Now, using the results of Section 2.1, we can represent the initial condition by

$$f_n = \frac{A_0 + A_p \cos px_n}{2} + \sum_{m=1}^{p} (A_m \cos mx_n + B_m \sin mx_n) \tag{9.4.8}$$

where

$$A_m = \frac{1}{p} \sum_{n=0}^{2p-1} f_n \cos mx_n \quad \text{and} \quad B_m = \frac{1}{p} \sum_{n=0}^{2p-1} f_n \sin mx_n$$

The real Fourier representation of the initial condition (9.4.8) proves to be inconvenient for a dispersion–dissipation analysis. Therefore, we consider the complex equivalent of (9.4.8),

$$f_n = \sum_{m=-p}^{p} C_m e^{imx_n} \tag{9.4.9}$$

The complex Fourier coefficients C_m are related to the A_m and B_m by

$$C_p = C_{-p} = \tfrac{1}{4}A_p; \qquad C_0 = \tfrac{1}{2}A_0$$
$$C_m = \tfrac{1}{2}(A_m - iB_m); \qquad C_{-m} = \tfrac{1}{2}(A_m + iB_m), \qquad m = 1, 2, \ldots, p - 1$$

A slightly different Fourier representation of f_n was used in the von Neumann stability analysis of Section 8.5. We have chosen the complex representation (9.4.9) for f_n because it clearly shows that the low harmonics correspond to $m = 0, 1, \ldots$ and the high harmonics are obtained at $|m| = p, p - 1, \ldots$.

Given a finite-difference equation for $au_x + u_t = 0$, we proceed with a dispersion and dissipation analysis as follows.

1. Assume the difference equation has a solution expressible in the form

$$U_n^j = \sum_{m=-p}^{p} C_m e^{im(x_n - \alpha t_j)}$$

2. Put a general harmonic $e^{im(x_n - \alpha t_j)}$ into the finite-difference equation and divide out a common exponential factor to obtain a *discrete dispersion relation* of the form

$$\alpha = \alpha(m, h, k) \tag{9.4.10}$$

where k and h represent the time and space grid spacings. In general, α is complex with

$$\alpha = \mathcal{R}(\alpha) + i\mathcal{I}(\alpha)$$

The real part of α determines the dispersive nature of the difference method and the imaginary part of α dictates the dissipative character.

3. With α given by (9.4.10), a dissipation analysis can be performed by inspecting the magnification factor

$$|\xi| = \left| \frac{e^{im(x_n - \alpha t_{j+1})}}{e^{im(x_n - \alpha t_j)}} \right| = |e^{-im\alpha k}| = e^{m\mathcal{I}(\alpha)k} \tag{9.4.11}$$

If $|\xi| > 1$ $[\mathcal{I}(\alpha) > 0]$ for some m, the difference method is *unstable*. If $|\xi| = 1$ $[\mathcal{I}(\alpha) = 0]$ for all m, the difference method is *nondissipative*. If $|\xi| \leq 1$ for all m and $|\xi| < 1$ for at least one m, the difference method is *dissipative*. If there exists a positive constant c and a positive integer s such that

$$|\xi| \leq 1 - c(mh)^{2s}, \qquad 0 \leq mh \leq \pi$$

we say the difference method is *dissipative of order 2s*. When the method is dissipative, we often display the amount of dissipation by plotting $|\xi|$ as a function of m for various choices of h and k.

4. If the $\mathcal{R}(\alpha)$ given by (9.4.10) is not equal to the constant a, then the difference method is *dispersive*. There are many ways to display the amount of dispersion exhibited by a difference scheme. One method is to plot the ratio of $\mathcal{R}(\alpha)/a$ as a function of the wave number m for various choices of h and k.

EXAMPLE 9.4.4 _____

To illustrate a dispersion–dissipation analysis of a finite-difference method, consider the FTBS method for (9.4.3) and (9.4.4). Namely,

$$a \frac{U_n^j - U_{n-1}^j}{h} + \frac{U_n^{j+1} - U_n^j}{k} = 0, \qquad a > 0 \tag{9.4.12}$$

Putting a general harmonic of the form $e^{im(x_n - \alpha t_j)}$ into the difference equation (9.4.12) and dividing by a common exponential factor gives

$$a \frac{1 - e^{-imh}}{h} + \frac{e^{-im\alpha k} - 1}{k} = 0 \tag{9.4.13}$$

which implicitly defines the discrete dispersion relation (9.4.10). We do not need to express α explicitly as a function of m, h, and k to perform a dissipation analysis. From (9.4.11) and (9.4.13), we have

$$\xi = e^{-im\alpha k} = 1 - sa(1 - e^{-imh}), \qquad s = k/h$$

Euler's formula for the sine allows us to write

$$\xi = 1 - 2sa i e^{-imh/2} \left(\sin \frac{mh}{2} \right)$$

$$= 1 - 2sa \sin^2 \frac{mh}{2} - i \left(2sa \cos \frac{mh}{2} \sin \frac{mh}{2} \right)$$

so the square of the magnification factor is given by

$$|\xi|^2 = 1 - 4sa(1 - sa)\sin^2 \frac{mh}{2} \tag{9.4.14}$$

We can gain a good deal of information about the FTBS method by studying equation (9.4.14). First, if sa does not satisfy $0 \le sa \le 1$, then $|\xi| > 1$ and the difference method is unstable. Second, if $sa = 1$, in which case the characteristics of (9.4.3) pass through the grid points, then $|\xi| = 1$ and the method is nondissipative. Third, for any fixed choice of mh, maximum dissipation occurs when $sa = \frac{1}{2}$.

Recall that the x grid spacing is given by $h = \pi/p$ and that as m varies from 0 to p, all harmonics in the discrete solution are represented. The highest harmonics correspond to $m = p$. These high harmonics represent the high-wave-number, most oscillatory, components of the difference solution. Putting $h = \pi/p$ in (9.4.14) and allowing m to vary from 0 to p, we see that for any fixed choice of sa, $0 < sa < 1$, the highest harmonics are the ones that are most rapidly dissipated or damped.

In Figure 9.4.1, the magnification factor $|\xi|$ is plotted against m/p. The choice $sa = 0.4$ ($\triangle$) is seen to be more dissipative than $sa = 0.2$ ($\bigcirc$). The plot of the magnification factor for $sa = 0.6$ coincides with that for $sa = 0.4$ and the one for $sa = 0.8$ with $sa = 0.2$.

To carry out a dispersion analysis, we must use (9.4.13) to express $\Re(\alpha)$ as a function of m, h, and k. From

$$e^{-i\alpha mk} = \xi$$

it follows that

$$\Re(\alpha) = \frac{-1}{mk} \tan^{-1} \left[\frac{\Im(\xi)}{\Re(\xi)} \right]$$

From

$$\xi = 1 - 2sa \sin^2 \frac{mh}{2} - i \left(2sa \cos \frac{mh}{2} \sin \frac{mh}{2} \right)$$

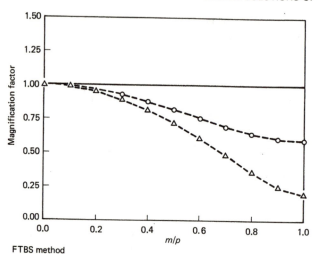

FTBS method

Figure 9.4.1

we have

$$\mathcal{R}(\xi) = 1 - 2sa \sin^2 \frac{mh}{2} \quad \text{and} \quad \mathcal{I}(\xi) = -2sa \cos \frac{mh}{2} \sin \frac{mh}{2}$$

Therefore, the discrete dispersion relation is

$$\mathcal{R}(\alpha) = \frac{1}{mk} \tan^{-1} \left[\frac{2sa \cos(mh/2)\sin(mh/2)}{1 - 2sa \sin^2(mh/2)} \right]$$

Using the trigonometric identities

$$2 \cos \frac{mh}{2} \sin \frac{mh}{2} = \sin mh$$

and

$$2 \sin^2 \frac{mh}{2} = (1 - \cos mh)$$

the discrete dispersion relation can be written in the form

$$\mathcal{R}(\alpha) = \frac{1}{mk} \tan^{-1} \left[\frac{sa \sin mh}{1 - sa(1 - \cos mh)} \right] \tag{9.4.15}$$

From the dispersion relation (9.4.15), we see that if $sa = 1$, in which case the discrete signal speed h/k equals the signal speed a of the differential equation, then

$$\mathcal{R}(\alpha) = \frac{1}{mk} \tan^{-1}[\tan mh] = \frac{h}{k} = \frac{1}{s} = a$$

and in the case $sa = 1$ the method is nondispersive.

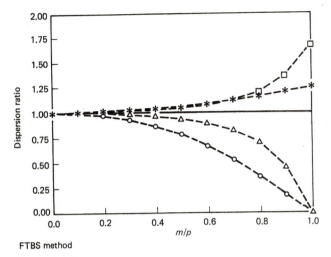

FTBS method

Figure 9.4.2

For values of sa other that $sa = 1$, it is interesting to plot the ratio $\mathcal{R}(\alpha)/a$ as a function of the wave number m. If $\mathcal{R}(\alpha)/a > 1$, then the numerical solution is propagating faster than the exact solution, while a value of $\mathcal{R}(\alpha)/a < 1$ indicates that the numerical solution is lagging the exact solution. If we divide Equation (9.4.15) by a and put $h = \pi/p$, then we obtain

$$\frac{\mathcal{R}(\alpha)}{a} = \frac{1}{sa \cdot m\pi/p} \tan^{-1}\left[\frac{sa\,\sin(m\pi/p)}{1 - sa[1 - \cos(m\pi/p)]}\right] \qquad (9.4.16)$$

Expressing the discrete dispersion relation in the form (9.4.16) displays the amount of dispersion as a function of the wave number m, $n = 0, 1, \ldots, p$, for any fixed value of sa. Figure 9.4.2 shows the dispersion ratio (9.4.16) for the cases

$$sa = 0.2 \quad (\bigcirc), \qquad sa = 0.4 \quad (\triangle)$$
$$sa = 0.6 \quad (\square), \text{ and} \qquad sa = 0.8 \quad (*)$$

The FTBS method is nondispersive for the choices $sa = 0.5$ and $sa = 1$. ∎ ∎

Artificial Dissipation

The dissipative and dispersive nature of a finite-difference method for (9.4.3) is linked to the form of the local truncation error for that method. If the truncation error has either of the forms

$$\nu u_{xx} + \gamma u_{xxx} + \text{ higher order terms}$$

or

$$\gamma u_{xxx} + \nu u_{xxxx} + \text{ higher order terms}$$

then ν is the *coefficient of numerical dissipation* and γ the *coefficient of numerical dispersion.* That is, even-order derivatives in the truncation error account for dissipation and odd-order derivatives account for dispersion.

This observation has been used in the construction of a number of finite-difference methods. Given an unstable or marginally stable finite-difference method, one can often stabilize the method by introducing a small amount of dissipation into the method. In computational fluid dynamics such a dissipative term is called an *artificial viscosity term*.

To illustrate, consider an advection equation of the form (9.4.3) to which we add an artificial dissipation term vu_{xx}. Namely,

$$au_x + u_t = vu_{xx} \tag{9.4.17}$$

If we use centered differences to approximate both x derivatives to second order in h and a forward difference for the t derivative, the result is

$$a\frac{U_{n+1}^j - U_{n-1}^j}{2h} + \frac{U_n^{j+1} - U_n^j}{k} = v\frac{\delta_x^2 U_n^j}{h^2}$$

or

$$U_n^{j+1} = U_n^j - \tfrac{1}{2}sa(U_{n+1}^j - U_{n-1}^j) + \frac{kv}{h^2}\delta_x^2 U_n^j \tag{9.4.18}$$

In the case $v = 0$, we recognize (9.4.18) as the FTCS method of Section 9.1. The FTCS method was shown to be unconditionally unstable. A von Neumann stability analysis shows that for (9.4.18) to be stable, then

$$(sa)^2/2 \le kv/h^2 \le 1/2$$

must hold. This imposes upper and lower bounds on the amount of numerical dissipation we are free to introduce in our effort to stabilize the FTCS method. It is not difficult to verify that (9.4.18) with

    (i) $kv/h^2 = 1/2$ yields the Lax–Friedrichs method;
    (ii) $kv/h^2 = (sa)^2/2$ yields the Lax–Wendroff method;
    (iii) $kv/h^2 = sa/2$ yields the FTBS method.

Exercises

1. Show that the partial differential equation $u_x + u_t + cu = 0$, $c = $ const, is
   (a) nondispersive
   (b) dissipative if $c > 0$

2. Following Example 9.4.4, carry out a dissipation analysis of
   (a) Lax–Wendroff method (9.1.10)
   (b) Lax–Friedrichs method (9.1.8)
   (c) BTBS method (9.2.5)
   (d) Wendroff implicit method (9.2.8)

3. Repeat Exercise 2 with a dispersion analysis.

4. Show that the Wendroff implicit method (9.2.8) is nondissipative.

5. Using (9.4.14), show that the FTBS method is most dissipative when $sa = \tfrac{1}{2}$.

6. Use the discrete dispersion relation (9.4.15) to show that the FTBS method is nondispersive for $sa = \tfrac{1}{2}$.

7. Show that, in (9.4.18), the choice
   (a) $kv/h^2 = 1/2$ yields the Lax–Wendroff method;
   (b) $kv/h^2 = (sa)^2/2$ yields the Lax–Friedrichs method;
   (c) $kv/h^2 = sa/2$ yields the FTBS method.

8. Find the coefficient of numerical dissipation for
   (a) Lax–Wendroff method (9.1.10)
   (b) Lax–Friedrichs method (9.1.8)
   (c) BTBS method (9.2.5)
   (d) Wendroff implicit method (9.2.8)
   (e) FTBS method (9.1.6)

9. Repeat Exercise 8 for the coefficient of numerical dispersion.

9.5 SYSTEMS OF EQUATIONS

Linear Systems

Consider the hyperbolic system of M linear first-order hyperbolic equations

$$\mathbf{Au}_x + \mathbf{u}_t = 0 \tag{9.5.1}$$

where $\mathbf{A}$ is a constant $M \times M$ matrix and $\mathbf{u}$ is an M vector,

$$\mathbf{u} = [u_1(x, t), u_2(x, t), \ldots, u_M(x, t)]^T$$

Recall that for (9.5.1) to be hyperbolic, the matrix $\mathbf{A}$ must have M real eigenvalues, $\lambda_1, \lambda_2, \ldots, \lambda_M$. The explicit finite-difference methods (9.1.5)–(9.1.10) and the implicit methods (9.2.5)–(9.2.8) for the scalar, hyperbolic equation $au_x + u_t = 0$ extend to finite-difference methods for (9.5.1) when a is replaced by $\mathbf{A}$, $1 \pm sa$ by $\mathbf{I} \pm s\mathbf{A}$, and U_n^j by $\mathbf{U}_n^j$. In this context, $\mathbf{U}_n^j$ denotes an approximation to $\mathbf{u}(x_n, t_j)$. Table 9.5.1 displays some difference method for (9.5.1)

TABLE 9.5.1[a] FINITE-DIFFERENCE METHODS AND LTEs FOR $\mathbf{Au}_x + \mathbf{u}_t = 0$

Forward in time–forward in space (FTFS), LTE $= O(k + h)$; stability restriction, $-1 \le s\lambda_p \le 0$, $p = 1, 2, \ldots, M$:

$$\mathbf{U}_n^{j+1} = (\mathbf{I} + s\mathbf{A})\mathbf{U}_n^j - s\mathbf{A}\mathbf{U}_{n+1}^j, \qquad s = k/h \tag{9.5.2}$$

Forward in time–backward in space (FTBS), LTE $= O(k + h)$; stability restriction, $0 \le s\lambda_p \le 1$, $p = 1, 2, \ldots, M$:

$$\mathbf{U}_n^{j+1} = (\mathbf{I} - s\mathbf{A})\mathbf{U}_n^j + s\mathbf{A}\mathbf{U}_{n-1}^j, \qquad s = k/h \tag{9.5.3}$$

Forward in time–centered in space (FTCS), LTE $= O(k + h^2)$; unstable:

$$\mathbf{U}_n^{j+1} = \mathbf{U}_n^j - \tfrac{1}{2}s\mathbf{A}(\mathbf{U}_{n+1}^j - \mathbf{U}_{n-1}^j), \qquad s = k/h \tag{9.5.4}$$

Lax–Friedrichs method, LTE $= O(k + h)$; stability restriction, $|s\lambda_p| \le 1$, $p = 1, 2, \ldots, M$:

$$\mathbf{U}_n^{j+1} = \tfrac{1}{2}(\mathbf{U}_{n+1}^j + \mathbf{U}_{n-1}^j) - \tfrac{1}{2}s\mathbf{A}(\mathbf{U}_{n+1}^j - \mathbf{U}_{n-1}^j) \tag{9.5.5}$$

Centered in time–centered in space (CTCS or leapfrog), LTE $= O(k^2 + h^2)$; stability restriction, $|s\lambda_p| \le 1$, $p = 1, 2, \ldots, M$:

$$\mathbf{U}_n^{j+1} = \mathbf{U}_n^{j-1} - s\mathbf{A}(\mathbf{U}_{n+1}^j - \mathbf{U}_{n-1}^j), \qquad s = k/h \tag{9.5.6}$$

TABLE 9.5.1　*(cont.)*

Lax–Wendroff method, LTE $= O(k^2 + h^2)$; stability restriction, $|s\lambda_p| \le 1$, $p = 1, 2, \ldots, M$:

$$\mathbf{U}_n^{j+1} = \mathbf{U}_n^j - \tfrac{1}{2}s\mathbf{A}(\mathbf{U}_{n+1}^j - \mathbf{U}_{n-1}^j) + \tfrac{1}{2}s^2\mathbf{A}^2(\mathbf{U}_{n-1}^j - 2\mathbf{U}_n^j + \mathbf{U}_{n+1}^j), \qquad s = k/h \qquad (9.5.7)$$

Backward in time–backward in space (BTBS), LTE $= O(k + h)$; stability restriction, none, unconditionally stable:

$$(\mathbf{I} + s\mathbf{A})\mathbf{U}_n^{j+1} - s\mathbf{A}\mathbf{U}_{n-1}^{j+1} = \mathbf{U}_n^j, \qquad s = k/h \qquad (9.5.8)$$

Backward in time–centered in space (BTCS), LTE $= O(k + h^2)$; stability restriction, none, unconditionally stable:

$$\mathbf{U}_n^{j+1} + \tfrac{1}{2}s\mathbf{A}\{\mathbf{U}_{n+1}^{j+1} - \mathbf{U}_{n-1}^{j+1}\} = \mathbf{U}_n^j, \qquad s = k/h \qquad (9.5.9)$$

Crank–Nicolson method, LTE $= O(k^2 + h^2)$; stability restriction, none, unconditionally stable:

$$\mathbf{U}_n^{j+1} + \tfrac{1}{4}s\mathbf{A}\{\mathbf{U}_{n+1}^{j+1} - \mathbf{U}_{n-1}^{j+1}\} = \mathbf{U}_n^j - \tfrac{1}{4}s\mathbf{A}\{\mathbf{U}_{n+1}^j - \mathbf{U}_{n-1}^j\} \qquad (9.5.10)$$

Wendroff method, LTE $= O(k^2 + h^2)$; stability restriction, none, unconditionally stable:

$$(\mathbf{I} + s\mathbf{A})\mathbf{U}_n^{j+1} + (\mathbf{I} - s\mathbf{A})\mathbf{U}_{n-1}^{j+1} = (\mathbf{I} - s\mathbf{A})\mathbf{U}_n^j + (\mathbf{I} + s\mathbf{A})\mathbf{U}_{n-1}^j, \qquad s = k/h \qquad (9.5.11)$$

---

[a]Eigenvalue λ_p represents a general eigenvalue of $\mathbf{A}$.

　　The difference methods of Table 9.5.1 apply to the hyperbolic system (9.5.1) in the case that $\mathbf{A}$ is a constant matrix. If $\mathbf{A}$ is a function of only x, then the methods (9.5.3)–(9.5.10) apply with $\mathbf{A}$ replaced by $\mathbf{A}_n = \mathbf{A}(x_n)$.

　　If $\mathbf{A}$ is a function of both x and t, then the methods with LTE $= O(k^2 + h^2)$ require modification if they are to retain their second-order accuracy. For example, when $\mathbf{A}$ is x and t dependent, then the Lax–Wendroff method takes the form

$$\begin{aligned}
\mathbf{U}_n^{j+1} = \ & \mathbf{U}_n^j - \tfrac{1}{2}s\mathbf{A}_n^{j+1/2}(\mathbf{U}_{n+1}^j - \mathbf{U}_{n-1}^j) \\
& + \tfrac{1}{4}s^2[\mathbf{A}_n^{j+1/2}\triangle_x\mathbf{A}_n^{j+1/2}\nabla_x + \mathbf{A}_n^{j+1/2}\triangle_x\mathbf{A}_n^{j+1/2}\nabla_x]\mathbf{U}_n^j \qquad (9.5.12)
\end{aligned}$$

where $\triangle_x$ and ∇_x denote difference operators defined by

$$\triangle_x\mathbf{U}_n^j = \mathbf{U}_{n+1}^j - \mathbf{U}_n^j \quad \text{and} \quad \nabla_x\mathbf{U}_n^j = \mathbf{U}_n^j - \mathbf{U}_{n-1}^j$$

Similarly, we can show that the Wendroff implicit method takes the form

$$\begin{aligned}
(1 + s\mathbf{A}_{n+1/2}^{j+1/2})\mathbf{U}_n^{j+1} + (1 - s\mathbf{A}_{n+1/2}^{j+1/2})\mathbf{U}_{n-1}^{j+1} = \ & (1 - s\mathbf{A}_{n+1/2}^{j+1/2})\mathbf{U}_n^j \\
& + (1 + s\mathbf{A}_{n+1/2}^{j+1/2})\mathbf{U}_{n-1}^j \qquad (9.5.13)
\end{aligned}$$

Stability

　　The von Neumann method can be used to examine the stability of finite-difference methods for (9.5.1) when the matrix $\mathbf{A}$ is constant. To perform a von Neumann stability analysis, we assume that each component of the solution, $\mathbf{U}_n^j$, of the difference equation can be expanded in a Fourier series. A typical harmonic in this expansion has the form

$$\mathbf{U}_n^j = \boldsymbol{\xi}^j e^{i\beta x_n}, \qquad 0 < \beta \le 2\pi \qquad (9.5.14)$$

where $\boldsymbol{\xi}$ depends on β.

　　If we substitute (9.5.14) into one of the difference methods of Table 9.5.1 and divide out a common factor of $e^{i\beta x_n}$, the result is an equation of the form

$$\boldsymbol{\xi}^{j+1} = \mathbf{G}\boldsymbol{\xi}^j \qquad (9.5.15)$$

where $\mathbf{G}$ is a constant $M \times M$ matrix called the *amplification matrix* for the difference method. If $\mu_1, \mu_2, \ldots, \mu_M$ denote the M eigenvalues of $\mathbf{G}$, then the von Neumann necessary condition for stability of the difference method is

$$|\mu_p| \leq 1, \qquad p = 1, 2, \ldots, M \tag{9.5.16}$$

EXAMPLE 9.5.1 ───

Consider the Lax–Friedrichs method

$$\mathbf{U}_n^{j+1} = \tfrac{1}{2}(\mathbf{U}_{n+1}^j + \mathbf{U}_{n-1}^j) - \tfrac{1}{2}s\mathbf{A}(\mathbf{U}_{n+1}^j - \mathbf{U}_{n-1}^j)$$

If we put

$$\mathbf{U}_n^j = \boldsymbol{\xi}^j e^{i\beta x_n}$$

and use

$$e^{i\beta x_{n+1}} = e^{i\beta x_n}e^{i\beta h} \quad \text{and} \quad e^{i\beta x_{n-1}} = e^{i\beta x_n}e^{-i\beta h}$$

then, after division by $e^{i\beta x_n}$, we find

$$\boldsymbol{\xi}^{j+1} = (\cos \beta h\mathbf{I} - is \sin \beta h\mathbf{A})\boldsymbol{\xi}^j$$

Thus, we see that the amplification matrix for the Lax–Friedrichs method is

$$\mathbf{G} = \cos \beta h\mathbf{I} - is \sin \beta h\mathbf{A}$$

We can relate the eigenvalues of $\mathbf{G}$ to those of $\mathbf{A}$ as follows. The matrix $\mathbf{G}$ has eigenvalue μ if and only if

$$\det[\mathbf{G} - \mu\mathbf{I}] = 0$$

or

$$\det[(\cos \beta h - u\mathbf{I}) - is\mathbf{A} \sin \beta h] = 0$$

or

$$\det\left[\mathbf{A} - \frac{\cos \beta h - \mu}{is \sin \beta h}\mathbf{I}\right] = 0$$

The last equation shows that

$$\frac{\cos \beta h - \mu}{is \sin \beta h} = \lambda$$

where λ is an eigenvalue of $\mathbf{A}$. Solving for μ, we find

$$\mu = \cos \beta h - is\lambda \sin \beta h$$

from which it follows that

$$|\mu|^2 = \cos^2\beta h + s^2\lambda^2\sin^2\beta h$$
$$= 1 + (s^2\lambda^2 - 1)\sin^2\beta h$$

From the preceding expression for $|\mu|^2$, we see that all eigenvalues of the amplification matrix $\mathbf{G}$ satisfy $|\mu| \leq 1$ if and only if $s^2\lambda^2 \leq 1$. Thus, the von Neumann stability condition for the Lax–Friedrichs method is

$$|s\lambda| \leq 1 \quad \text{for all eigenvalues } \lambda \text{ of } \mathbf{A} \qquad \blacksquare\,\blacksquare$$

Conservation Law Systems

A conservation law in M dependent variables,

$$\mathbf{u}(x, t) = [u_1(x, t), u_2(x, t), \ldots, u_M(x, t)]^T$$

can be written in the form

$$[F(\mathbf{u})]_x + \mathbf{u}_t = 0 \tag{9.5.17}$$

where

$$F(\mathbf{u}) = [F_1(\mathbf{u}), F_2(\mathbf{u}), \ldots, F_M(\mathbf{u})]^T$$

If we use the chain rule to expand the x derivative in (9.5.17), then (9.5.13) takes the form

$$\mathbf{A}(\mathbf{u})\mathbf{u}_x + \mathbf{u}_t = 0 \tag{9.5.18}$$

TABLE 9.5.2 CONSERVATION LAW DIFFERENCE METHODS FOR $\mathbf{F(u)}_x + \mathbf{u}_t = 0$

Centered in time–centered in space (unstable):

$$\mathbf{U}_n^{j+1} = \mathbf{U}_n^j - \tfrac{1}{2}s(\mathbf{F}_{n+1}^j - \mathbf{F}_{n-1}^j), \qquad s = k/h \tag{9.5.19}$$

Lax–Friedrichs method:

$$\mathbf{U}_n^{j+1} = \tfrac{1}{2}(\mathbf{U}_{n+1}^j + \mathbf{U}_{n-1}^j) - \tfrac{1}{2}s(\mathbf{F}_{n+1}^j - \mathbf{F}_{n-1}^j), \qquad s = k/h \tag{9.5.20}$$

Lax–Wendroff method:

$$\begin{aligned}
\mathbf{U}_n^{j+1} = \mathbf{U}_n^j &- \tfrac{1}{2}s(\mathbf{F}_{n+1}^j - \mathbf{F}_{n-1}^j) \\
&+ \tfrac{1}{2}s^2[(\mathbf{A}_{n+1/2}^j)(\mathbf{F}_{n+1}^j - \mathbf{F}_n^j) - (\mathbf{A}_{n-1/2}^j)(\mathbf{F}_n^j - \mathbf{F}_{n-1}^j)], \qquad s = k/h
\end{aligned} \tag{9.5.21}$$

Richtmeyer method:

$$\mathbf{U}_{n+1/2}^* = \tfrac{1}{2}(\mathbf{U}_{n+1}^j + \mathbf{U}_n^j) - \tfrac{1}{2}s(\mathbf{F}_{n+1}^j - \mathbf{F}_n^j) \tag{9.5.22a}$$

$$\mathbf{U}_n^{j+1} = \mathbf{U}_n^j - s(\mathbf{F}_{n+1/2}^* - \mathbf{F}_{n-1/2}^*) \tag{9.5.22b}$$

MacCormack method:

$$\mathbf{U}_n^* = \mathbf{U}_n^j - s(\mathbf{F}_{n+1}^j - \mathbf{F}_n^j) \tag{9.5.23a}$$

$$\mathbf{U}_n^{j+1} = \tfrac{1}{2}[\mathbf{U}_n^j + \mathbf{U}_n^* - s(\mathbf{F}_n^* - \mathbf{F}_{n-1}^*)] \tag{9.5.23b}$$

FTBS upwind method:

$$\mathbf{U}_n^{j+1} = \mathbf{U}_n^j + s(\mathbf{F}_{n-1}^j - \mathbf{F}_n^j), \qquad \lambda_p \geq 0, p = 1, 2, \ldots, M \tag{9.5.24}$$

FTFS upwind method:

$$\mathbf{U}_n^{j+1} = \mathbf{U}_n^j - s(\mathbf{F}_{n+1}^j - \mathbf{F}_n^j), \qquad \lambda_p \leq 0, p = 1, 2, \ldots, M \tag{9.5.25}$$

where the $M \times M$ matrix $\mathbf{A}(u)$ is the Jacobian matrix of the derivatives of $\mathbf{f}$ with respect to the components of $\mathbf{u}$. That is,

$$\mathbf{A(u)} = \begin{bmatrix} a_{11} & a_{12} & \cdots & a_{1M} \\ a_{21} & a_{22} & \cdots & a_{2M} \\ \vdots & \vdots & & \vdots \\ a_{M1} & a_{M2} & \cdots & a_{MM} \end{bmatrix}$$

with $a_{pq} = a_{pq}(\mathbf{u}) = \partial F_p / \partial u_q$, $p, q = 1, 2, \ldots, M$.

In Section 9.3 several difference methods were developed for a scalar conservation law of the form $F(u)_x + u_t = 0$. With the exception of (9.3.25), each of the methods in Table 9.3.1 generalizes to (9.5.17) and (9.5.18). See Table 9.5.2.

To apply one of the upwind methods of Table 9.5.2, all of the eigenvalues of the matrix $\mathbf{A}$, $\lambda_1, \lambda_2, \ldots, \lambda_M$, must be of one sign. Recall that the characteristics of (9.5.14) are the curves defined by

$$\frac{dx}{dt} = \lambda_p, \qquad p = 1, 2, \ldots, M$$

Information about the solution of (9.5.17) or (9.5.18) is propagated "downwind" along these characteristics.

The two-step methods (9.5.22) and (9.5.23) reduce to the Lax–Wendroff method in the linear case in which $F(\mathbf{u}) = \mathbf{Au}$, with $\mathbf{A}$ a constant matrix. Such two-step variations of the Lax–Wendroff method are popular since they do not require the calculation of the Jacobian matrix of $\mathbf{F}$.

Computations

If a particular method has been coded for the scalar equation $au_x + u_t = 0$, then it is usually an easy matter to adapt that code to apply to the system (9.5.1). In the case of an initial-boundary-value problem, care must be taken to ensure that on each boundary information is specified only on characteristics that point away from that boundary.

Algorithm 9.3 illustrates the use of the Lax–Wendroff to solve an initial-value problem for a constant-coefficient, linear pair of first-order equations. Following the outline of Algorithm 9.3, it is not hard to incorporate other methods from Tables 9.5.1 and 9.5.2.

EXAMPLE 9.5.2 _____

To illustrate Algorithm 9.3, consider the initial-value problem

$$4 \frac{\partial u_1}{\partial x} - 6 \frac{\partial u_2}{\partial x} + \frac{\partial u_1}{\partial t} = 0$$

$$\frac{\partial u_1}{\partial x} - 3 \frac{\partial u_2}{\partial x} + \frac{\partial u_2}{\partial t} = 0, \qquad -\infty < x < \infty, \, t > 0$$

$$u(x, 0) = \sin x, \qquad v(x, 0) = \cos x, \qquad -\infty < x < \infty$$

Written in the form $\mathbf{Au}_x + \mathbf{u}_t = 0$ with $\mathbf{u} = [u, v]^T$, we see that

$$\mathbf{A} = \begin{bmatrix} 4 & -6 \\ 1 & -3 \end{bmatrix}$$

ALGORITHM 9.3   Lax–Wendroff Method for a Pair of Equations

Step 1. Document

This algorithm uses the Lax–Wendroff method (9.5.7)
to approximate the solution of the initial-value problem

$$a_{11}u_x + a_{12}v_x + u_t = 0,$$
$$a_{21}u_x + a_{22}v_x + v_t = 0, \qquad -\infty < x < \infty, \; t > 0$$
$$u(x, 0) = f(x), \qquad v(x, 0) = g(x)$$

for $x_p \le x \le x_q$ and $0 < t \le t_{max}$

INPUT

Integer p, lower index for solution at time t_{jmax}
Integer q, upper index for solution at time t_{jmax}
Integer $jmax$, number of time steps
Real k, time step ($k \cdot jmax = t_{jmax}$)
Real h, space step
Array a_{ij}, $i, j = 1, 2$, coefficient matrix
Array b_{ij}, $i, j = 1, 2$, coefficient matrix squared
Function $f(x)$, initial condition for u
Function $g(x)$, initial condition for v

OUTPUT

Approximation (U_n^j, V_n^j) to $(u(x_n, t_j), v(x_n, t_j))$
for $n = p, p + 1, \ldots, q$ and $j = 0, 1, \ldots, jmax$

Step 2. Define grid ratio

Set $s = k/h$

Step 3. Initialize numerical solution

Set $t = 0$
    $nmin = p - jmax$
    $nmax = q + jmax$
For $n = nmin, nmin + 1, \ldots, nmax$ set
    $x_n = n \cdot h$
    $U_n = f(x_n)$
    $V_n = g(x_n)$
    Output x_n, w_n

Output t
For $n = p, p + 1, \ldots, q$
    Output x_n, U_n, V_n

Step 4. Begin time stepping

For $j = 1, 2, \ldots, jmax$
    Do steps 5–7

Step 5. Advance solution one time step

Set $nmin = nmin + 1$
    $nmax = nmax - 1$
For $n = nmin, nmin + 1, \ldots, nmax$ set
$$\begin{aligned}
UNEW_n &= U_n - 0.5 \cdot s \cdot a_{11} \cdot (U_{n+1} - U_{n-1}) \\
&\quad - 0.5 \cdot s \cdot a_{12} \cdot (V_{n+1} - V_{n-1}) \\
&\quad + 0.5 \cdot s^2 \cdot b_{11} \cdot (U_{n-1} - 2U_n + U_{n+1}) \\
&\quad + 0.5 \cdot s^2 \cdot b_{12} \cdot (V_{n-1} - 2V_n + V_{n+1}) \\
VNEW_n &= V_n - 0.5 \cdot s \cdot a_{21} \cdot (U_{n+1} - U_{n-1}) \\
&\quad - 0.5 \cdot s \cdot a_{22} \cdot (V_{n+1} + V_{n-1}) \\
&\quad + 0.5 \cdot s^2 \cdot b_{21} \cdot (U_{n-4} - 2U_n + U_{n+1}) \\
&\quad + 0.5 \cdot s^2 \cdot b_{22} \cdot (V_{n-1} - 2V_n + V_{n+1})
\end{aligned}$$

Step 6. Prepare for next time step

Set $t = t + k$
For $n = nmin, nmin + 1, \ldots, nmax$ set
$U_n = UNEW_n$
$V_n = VNEW_n$

Step 7. Output numerical solution

Output t
For $n = p, p + 1, \ldots, q$
    Output x_n, U_n, V_n

The eigenvalues of A are $\lambda = -2$ and $\lambda = 3$. Using the results of Section 7.6, we find the exact solution to be

$$u_1(x, t) = \tfrac{1}{5}[6 \sin(x - 3t) - 6 \cos(x - 3t) - \sin(x + 2t) - 6 \cos(x + 2t)]$$
$$u_2(x, t) = \tfrac{1}{5}[\sin(x - 3t) - \cos(x - 3t) - \sin(x + 2t) + 6 \cos(x + 2t)]$$

Since the eigenvalues of A are $\lambda = -2$ and $\lambda = 3$, the stability restriction for the Lax–Wendroff method requires $3k/h \leq 1$. Choosing $h = 0.05$ and $k = 0.0125$ to approximate the solution for $0 \leq x \leq 1$, $0 < t \leq 0.5$, we obtain the results shown in Table 9.5.3. ■ ■

TABLE 9.5.3 COMPARISON OF NUMERICAL AND EXACT SOLUTIONS
FOR EXAMPLE 9.5.2

$t = 0.5$ x	Numerical u	Exact u	Numerical v	Exact v
0.0	-0.801758	-0.801810	0.266720	0.266422
0.1	-1.020323	-1.020427	0.135308	0.134991
0.2	-1.228694	-1.228847	0.002545	0.002211
0.3	-1.424788	-1.424989	-0.130245	-0.130592
0.4	-1.606646	-1.606893	-0.261732	-0.262090
0.5	-1.772451	-1.772742	-0.390605	-0.390969
0.6	-1.920546	-1.920878	-0.515574	-0.515941
0.7	-2.049451	-2.049821	-0.635392	-0.635758
0.8	-2.157879	-2.158283	-0.748862	-0.749224
0.9	-2.244747	-2.245180	-0.854849	-0.855203
1.0	-2.309186	-2.309645	-0.952295	-0.952637

Exercises

1. Consider the initial-value problem

$$2\frac{\partial u_1}{\partial x} + \frac{\partial u_2}{\partial x} + \frac{\partial u_1}{\partial t} = 0$$

$$-\infty < x < \infty, t > 0$$

$$2\frac{\partial u_1}{\partial x} + 3\frac{\partial u_2}{\partial x} + \frac{\partial u_2}{\partial t} = 0,$$

$$u_1(x, 0) = \sin x, \qquad u_2(x, 0) = x^2, \qquad -\infty < x < \infty$$

(a) Develop the exact solution

$$u_1(x, t) = \tfrac{1}{3}[2 \sin(x - t) - (x - t)^2 + \sin(x - 4t) + (x - 4t)^2]$$
$$u_2(x, t) = \tfrac{1}{3}[-2 \sin(x - t) + (x - t)^2 + 2 \sin(x - 4t) + 2(x - 4t)^2]$$

(b) For $0 \le x \le 1$, $0 < t \le 1$, approximate the solution using the Lax–Wendroff method. Compare the numerical solution with the exact solution.

2. Repeat Exercise 1(b) using the
   (a) FTBS method (9.5.3)
   (b) Lax–Friedrichs method (9.5.5)
   (c) CTCS leapfrog method (9.5.6)

3. Consider the initial-boundary-value problem

$$2\frac{\partial u_1}{\partial x} + \frac{\partial u_2}{\partial x} + \frac{\partial u_1}{\partial t} = 0$$

$$x > 0, t > 0$$

$$2\frac{\partial u_1}{\partial x} + 3\frac{\partial u_2}{\partial x} + \frac{\partial u_2}{\partial t} = 0,$$

$$u_1(x, 0) = \sin x, \qquad u_2(x, 0) = x^2, \qquad x > 0$$
$$u_1(0, t) = \tfrac{1}{3}[-2 \sin t - \sin(4t) + 15t^2], \qquad t > 0$$
$$u_2(0, t) = \tfrac{1}{3}[2 \sin t - 2 \sin(4t) + 33t^2]$$

whose exact solution is given in Exercise 1(a). For $0 \le x \le 1$, $0 < t \le 1$, approximate the solution using the Wendroff method (9.5.11). Compare the numerical solution with the exact solution.

4. Repeat Exercise 3 using the BTBS method (9.5.8).

5. Consider the initial-boundary-value problem

$$\frac{\partial u_1}{\partial x} - 0\frac{\partial u_2}{\partial x} + \frac{\partial u_1}{\partial t} = 0$$

$$0 < x, t > 0$$

$$\frac{\partial u_1}{\partial x} + 2\frac{\partial u_2}{\partial x} + \frac{\partial u_2}{\partial t} = 0,$$

$$u_1(x, 0) = x + 1, \qquad u_2(x, 0) = x + 2, \quad 0 < x < \infty$$

$$u_1(0, t) = 1 - t, \qquad u_2(0, t) = 2 - 3t, \quad t > 0$$

(a) Show that the eigenvalues for the matrix **A** of this system are $\lambda = 1$ and $\lambda = 2$. Sketch a few characteristics in the region $x > 0$, $t > 0$.

(b) Obtain the exact solution

$$u_1(x, t) = 1 + x - t$$
$$u_2(x, t) = 2 + x - 3t$$

(c) Use the Wendroff implicit method to approximate the solution on the region $0 < x < 1$, $0 < t < 1$.

6. Repeat Exercise 5(c) using the BTBS method.

7. Consider the initial-value problem

$$4\frac{\partial u_1}{\partial x} + \frac{\partial u_2}{\partial x} + \frac{\partial u_1}{\partial t} = 0$$

$$-\infty < x < \infty, t > 0$$

$$\frac{\partial u_1}{\partial x} + 4\frac{\partial u_2}{\partial x} + \frac{\partial u_2}{\partial t} = 0,$$

$$u_1(x, 0) = \sin x, \qquad u_2(x, 0) = \cos x, \qquad -\infty < x < \infty$$

(a) Find the exact solution.

(b) For $0 \le x \le 1$, $0 < t \le 1$, approximate the solution using the Lax–Wendroff method. Compare the numerical solution with the exact solution.

8. Repeat Exercise 7(b) using the
(a) FTBS method (9.5.3)
(b) Lax–Friedrichs method (9.5.5)
(c) CTCS leapfrog method (9.5.6)

9. The flow of water in an open channel of rectangular cross section can be described by the pair of conservation laws

$$(uv)_x + u_t = 0$$
$$(gu + v^2/2)_x + v_t = 0$$

where $u = u(x, t)$ denotes the flow depth and $v = v(x, t)$ the flow velocity. x measures distance along the channel, t represents time, and g is the gravitational constant; $g = 9.8$ m/s^2 in the mks system. Suppose that, for $t < 0$, the water in the channel is not moving and that a dam at $x = 0$ maintains a water of depth 2 m in the region $x < 0$ while for $x > 0$ the water depth is 1 m. At time $t = 0$ the dam is suddenly removed producing the initial-value problem

$$(uv)_x + u_t = 0 \qquad -\infty < x < \infty, t > 0$$
$$(gu + v^2/2)_x + v_t = 0,$$

$$u(x, 0) = \begin{cases} 2, & x < 0 \\ 1, & x > 0 \end{cases} \qquad v(x, 0) = 0$$

Approximate the solution to this problem using

(a) Richtmeyer method
(b) Lax–Friedrichs method
(c) FTBS method
(d) MacCormack method

10. Using the fact that the disturbance caused by the removal of the dam travels with finite speed to impose boundary conditions

$$
\begin{aligned}
u(x, 0) &= 2, & x &\ll 0 \\
u(x, 0) &= 1, & x &\gg 0 \\
v(x, 0) &= 0, & |x| &\gg 0
\end{aligned}
$$

approximate the solution to the initial-value problem of Exercise 9 using a BTBS upwind method.

11. For each of the following methods (i) obtain the amplification matrix $\mathbf{G}$; (ii) relate the eigenvalues of $\mathbf{G}$ to those of $\mathbf{A}$; (iii) obtain the von Neumann stability restriction:

(a) (9.5.2)   (f) (9.5.8)
(b) (9.5.3)   (g) (9.5.9)
(c) (9.5.4)   (h) (9.5.10)
(d) (9.5.6)   (i) (9.5.11)
(e) (9.5.7)

12. Show that to apply the Wendroff implicit method to an initial-boundary-value problem with no numerical boundary conditions, the matrix $\mathbf{A}$ must be definite, that is, all the eigenvalues of $\mathbf{A}$ must be of one sign.

9.6 SECOND-ORDER EQUATIONS

The simplest second-order hyperbolic partial differential equation is the wave equation

$$
u_{tt} - c^2 u_{xx} = 0, \qquad c = \text{const} \tag{9.6.1}
$$

In the same way that we used the heat equation as a prototype in our discussion of parabolic equations, Equation (9.6.1) will serve as a model equation for discussing finite-difference methods for hyperbolic equations of second order in x and t.

The characteristics for (9.6.1) are the straight lines in the xt plane defined by

$$
\frac{dx}{dt} = \pm c
$$

From Section 5.1, the initial-value problem consisting of (9.6.1) subject to

$$
\begin{aligned}
u(x, 0) &= f(x) \tag{9.6.2} \\
u_t(x, 0) &= g(x) \tag{9.6.3}
\end{aligned}
$$

has the D'Alembert solution

$$
u(x, t) = \tfrac{1}{2}[f(x + ct) + f(x - ct)] + \frac{1}{2c} \int_{x-ct}^{x+ct} g(\xi)\, d\xi \tag{9.6.4}
$$

For the initial-value problem (9.6.1)–(9.6.3), the D'Alembert formula provides an exact solution for comparison with the various numerical approximations discussed in what

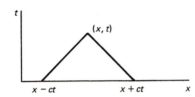

Figure 9.6.1 Characteristic diagram showing interval of dependence, $[x - ct, x + ct]$, for solution of (9.6.1)–(9.6.3) at position x and time t.

follows. Also, the D'Alembert solution displays the role of the characteristics in relating the initial data function f and g to the solution $u(x, t)$. See Figure 9.6.1.

Given the elegant, D'Alembert exact solution to the initial-value problem (9.6.1)–(9.6.3), in the case that c is a constant, there is no need for numerical methods for approximating a solution such as (9.6.4). All that might be required in using (9.6.4) to provide $u(x, t)$ is a numerical method for evaluation of the integral in (9.6.4). However, when c is allowed to depend on x, on x and t, or even on u, a need for numerical methods arises. Also, other second-order hyperbolic partial differential equations may not have an easily expressible exact solution.

Explicit Method

Let $x_n = nh$, $n = 0, \pm 1, \pm 2, \ldots$, $t_j = jk$, $j = 0, 1, 2, \ldots$, denote a uniform difference grid on the region $-\infty < x < \infty$, $t \geq 0$. A finite-difference method for (9.6.1) can be obtained by simply replacing the second derivatives u_{tt} and u_{xx} by centered difference approximations to obtain

$$\frac{\delta_t^2 U_n^j}{k^2} - c^2 \frac{\delta_x^2 U_n^j}{h^2} = 0 \qquad (9.6.5)$$

or

$$U_n^{j-1} - 2U_n^j + U_n^{j+1} - s^2 c^2 \{U_{n-1}^j - 2U_n^j + U_{n+1}^j\} = 0 \qquad s = k/h$$

The difference method (9.6.5) has local truncation error $O(k^2 + h^2)$ provided u is C^4 in x and t. Equation (9.6.5) is an explicit method since it involves only one value of the numerical solution at the most advanced time level, $t = t_{j+1}$. The computational stencil for (9.6.5) is shown in Figure 9.6.2.

Since the difference method (9.6.5) involves three time levels, the values of U_n^0 and U_n^1 are required to begin stepping the numerical solution forward in time. Clearly, we should initialize U_n^0 in accordance with (9.6.2) to obtain

$$U_n^0 = f_n = f(x_n) \qquad (9.6.6)$$

The starting values for U_n^0 and U_n^1 should represent the initial data with an error no worse than the local truncation error for method (9.6.5). There is no error associated with the

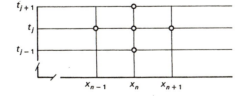

Figure 9.6.2 Computational stencil showing grid points (○) involved in method (9.6.5).

assignment (9.6.6). However, if we were to use a convenient, one-sided difference formula to approximate the time derivative in (9.6.3), then a truncation error $O(k)$ would be introduced into the computation.

To avoid "contaminating" the numerical solution with a poor numerical initial condition for U_n^1, we proceed as follows. Assume that the differential equation (9.6.1) holds at time $t = 0$ and that the function $f(x)$ is C^2. Then, Taylor's theorem gives

$$u(x_n, t_1) = u(x_n, 0) + ku_t(x_n, 0) + \tfrac{1}{2}k^2 u_{tt}(x_n, 0) + O(k^3)$$

$$= u(x_n, 0) + kg(x_n) + \tfrac{1}{2}k^2 c^2 f''(x_n) + O(k^3)$$

$$= u(x_n, 0) + kg(x_n) + \tfrac{1}{2}k^2 c^2 \frac{\delta_x^2 f_n}{h^2} + O(k^3 + k^2 h^2) \qquad (9.6.7)$$

where in the last equation, $f''(x_n)$ has been approximated to $O(h^2)$ with a centered difference formula. From (9.6.7), we see that the initial condition (9.6.3) is approximated to $O(kh^2 + k^2)$ by

$$\frac{U_n^1 - U_n^0}{k} = g(x_n) + \tfrac{1}{2}c^2 s^2 [f(x_{n-1}) - 2f(x_n) + f(x_{n+1})]$$

Thus, it follows that the starting values for U_n^1 should be defined by

$$U_n^1 = U_n^0 + kg_n + \tfrac{1}{2}kc^2 s^2 [f_{n-1} - 2f_n + f_{n+1}] \qquad (9.6.8)$$

The following algorithm implements the explicit, forward-in-time difference method with the starting values given by (9.6.8). This algorithm applies to an initial-boundary-value problem with the values of the solution specified at $x = 0$ and $x = x$ max. A similar algorithm can be applied to a pure initial-value problem or an initial-boundary-value problem using the approach illustrated in Algorithm 9.1. See also Figure 9.1.2.

ALGORITHM 9.4   Explicit Difference Method: Dirichlet Initial-Boundary-Value Problem

Step 1. Document

This algorithm uses the forward difference method (9.6.5) to approximate the solution to the initial-boundary-value problem (9.6.1)–(9.6.3).

INPUT

Real c, celerity
Real $xmax$, length of x interval
Function $f(x)$, initial condition for u
Function $g(x)$, initial condition for u_t
Function $p(t)$, left boundary condition
Function $q(t)$, right boundary condition

Real k, time step
Integer $jmax$, number of time steps
Integer $nmax$, number of x nodes

OUTPUT

Approximation U_n^j to $u(x_n, t_j)$ for $n = 0, 1, \ldots, nmax + 1$
and $j = 0, 1, \ldots, jmax$

Step 2. Define a grid

Set $h = xmax/(nmax + 1)$
  $s = k/h$
  $r = (c \cdot s)^2$
If $r > 1$, output message that method is unstable

Step 3. Initialize numerical solution at $t = 0$

Set $t = 0$
  $x_0 = 0$
  $w_0 = (p(0) + f(0))/2$
For $n = 1, 2, \ldots, nmax$ set
  $x_n = x_{n-1} + h$
  $w_n = f(x_n)$
Set $w_{nmax+1} = (q(0) + f(xmax))/2$
  $x_{nmax+1} = xmax$
Output t
For $n = 0, 1, \ldots, nmax + 1$
  Output x_n, w_n

Step 4. Initialize numerical solution at $t = k$

Set $t = k$
  $v_0 = p(k)$
For $n = 1, 2, \ldots, nmax$ set
  $v_n = w_n + k \cdot g(x_n) + r \cdot [w_{n-1} - 2w_n + w_{n+1}]/2$
Set $v_{nmax+1} = q(k)$
Output t
For $n = 0, 1, \ldots, nmax + 1$
  Output x_n, v_n

Step 5. Begin time stepping

For $j = 1, 2, \ldots, jmax$
  Do steps 6–8

Step 6. Advance solution one time step

For $n = 1, 2, \ldots, nmax$ set
$$u_n = 2 \cdot v_n - w_n + r[v_{n-1} - 2v_n + v_{n+1}]$$
Set $t = t + k$
$$u_0 = p(t)$$
$$u_{nmax+1} = q(t)$$

Step 7. Output numerical solution

Output t
For $n = 0, 1, \ldots, nmax + 1$
    Output x_n, u_n

Step 8. Prepare for next time step

For $n = 0, 1, \ldots, nmax + 1$ set
$$w_n = v_n$$
$$v_n = u_n$$

EXAMPLE 9.6.1 _____

Consider the initial-value problem

$$u_{tt} - u_{xx} = 0, \qquad -\infty < x < \infty, \, t > 0$$
$$u(x, 0) = \sin x, \qquad -\infty < x < \infty$$
$$u_t(x, 0) = e^x, \qquad -\infty < x < \infty$$

By (9.6.4), the exact solution is

$$u(x, t) = \sin x \cos t + \tfrac{1}{2}[e^{x+t} + e^{x-t}]$$

Since we know the exact solution, we can set

$$p(t) = u(0, t) \quad \text{and} \quad q(t) = u(1, t)$$

and use Algorithm 9.4 to estimate the solution. With $h = 0.1$ and $k = 0.075$, we obtain the numerical solution first using (9.6.8) for $O(k^2)$ accurate starting values for U_n^1. Then for comparison we use the $O(k)$ approximation $U_n^1 = U_n^0 + kg_n$ to start the calculation. After 10 time steps the numerical results with the $O(k^2)$ starting values, the numerical results with the $O(k)$ starting values, and the exact solution are compared. See Table 9.6.1. ■ ■

Stability

As with any explicit difference method, we must be concerned with the stability of (9.6.5). To perform a von Neumann stability analysis, we set

$$U_n^j = \xi^j e^{i\beta x_n}$$

TABLE 9.6.1 COMPARISON OF NUMERICAL AND EXACT
SOLUTION FOR EXAMPLE 9.6.1

$T = 1.50$	Numerical $O(k^2)$	Exact	Numerical $O(k)$
$X = 0.0$	2.129279	2.129279	2.129279
$X = 0.1$	2.360488	2.360280	2.358690
$X = 0.2$	2.615148	2.614761	2.611845
$X = 0.3$	2.895760	2.895131	2.890115
$X = 0.4$	3.204905	3.204058	3.197198
$X = 0.5$	3.545390	3.544502	3.537585
$X = 0.6$	3.920561	3.919741	3.913671
$X = 0.7$	4.334046	4.333412	4.328770
$X = 0.8$	4.789969	4.789542	4.786393
$X = 0.9$	5.292819	5.292593	5.291026
$X = 1.0$	5.847505	5.847505	5.847505

in (9.6.5) to find that the magnification factor $\xi = \xi^{j+1}/\xi^j$ satisfies

$$\xi - 2 + \xi^{-1} = c^2 s^2 (e^{i\beta h} - 2 + e^{-i\beta h})$$

which is equivalent to

$$\xi^2 + 2 \left(2c^2 s^2 \sin^2 \frac{\beta h}{2} - 1 \right) \xi + 1 = 0 \qquad (9.6.9)$$

Because the product of the roots of (9.6.9) is 1, for stability, both roots must satisfy $|\xi| = 1$. Otherwise, the magnitude of one of the roots of (9.6.9) would exceed 1. From the quadratic formula, the condition that the roots of (9.6.9) have magnitude 1 is

$$\left| 2c^2 s^2 \sin^2 \frac{\beta h}{2} - 1 \right| \leq 1$$

which is satisfied for arbitrary βh if and only if the von Neumann stability condition

$$c^2 s^2 \leq 1 \qquad (9.6.10)$$

holds. If (9.6.10) holds, then $|\xi| = 1$ and the difference method (9.6.5) is nondissipative.

Associated with the numerical method (9.6.5) is a discrete signal speed defined by h/k. An explicit method cannot propagate information more than one space grid point per time step. Recall that for the partial differential equation the signal speeds are given by the characteristic equation $dx/dt = \pm c$. The stability requirement (9.6.10) states that the finite-difference method must be capable of propagating information at least as fast as information is propagated by the partial differential equation. This is also the content of the Courant–Friedrichs–Lewy (CFL) condition.

Implicit Methods

A number of implicit finite-difference methods for (9.6.1) can be obtained by approximating the u_{tt} term by a centered difference and using a weighted average of centered x

differences at times t_{j-1}, t_j, and t_{j+1} to approximate the u_{xx} term in (9.6.1). Namely,

$$\frac{\delta_t^2 U_n^j}{k^2} - c^2 \left\{ \omega \frac{\delta_x^2 U_n^{j-1}}{h^2} + (1 - 2\omega) \frac{\delta_x^2 U_n^j}{h^2} + \omega \frac{\delta_x^2 U_n^{j+1}}{h^2} \right\} = 0 \qquad (9.6.11)$$

where ω is a weight to be assigned a nonnegative value. If ω is zero, (9.6.11) reduces to the explicit method (9.6.5). A von Neumann stability analysis shows that (9.6.11) is unconditionally stable provided $\omega \geq \frac{1}{4}$. Two common choices of ω that lead to symmetric difference stencils are $\omega = \frac{1}{4}$ and $\omega = \frac{1}{2}$.

The implicit method (9.6.11) with $\omega > 0$ involves three unknowns at the most advanced time level, $t = t_{j+1}$. Therefore, the difference method (9.6.11) is directly applicable only to initial-boundary-value problems on bounded intervals. The tridiagonal system of linear equations defined by (9.6.11) can be solved using Algorithm 8.2.

The implicit method (9.6.11) is unconditionally stable for $\omega \geq \frac{1}{4}$. However, if a large value of ck/h is chosen, the numerical solution will often be very poor. A poor performance of the implicit method can often be traced to a value of ck/h, which introduces a large degree of numerical dissipation or numerical dispersion. Thus, although we are free to violate the condition $|ck/h| \leq 1$ without introducing computational instabilities, accuracy considerations generally dictate that we not violate this condition by too much.

ALGORITHM 9.5   Implicit Difference Method: Dirichlet Initial-Boundary-Value Problem

Step 1. Document

The following algorithm uses the implicit difference method (9.6.11) with $\omega = \frac{1}{2}$ to approximate the solution to the initial-boundary-value problem (9.6.1)–(9.6.3) subject to $u(0, t) = p(t)$, $u(xmax, t) = q(t)$.

INPUT

Real cc, celerity ($cc = c$)
Real $xmax$, length of x interval
Function $f(x)$, initial condition for u
Function $g(x)$, initial condition for u_t
Function $p(t)$, left boundary condition
Function $q(t)$, right boundary condition
Real k, time step
Integer $jmax$, number of time steps
Integer $nmax$, number of x nodes

OUTPUT

Approximation U_n^j to $u(x_n, t_j)$ for $n = 0, 1, \ldots, nmax + 1$ and $j = 0, 1, \ldots, jmax$

Step 2. Define a grid

Set $h = xmax/(nmax + 1)$
$\quad s = k/h$
$\quad r = (cc{\cdot}s)^2$

Step 3. Initialize numerical solution at *t* = 0

Set $t = 0$
$\quad x_0 = 0$
$\quad w_0 = (p(0) + f(0))/2$
For $n = 1, 2, \ldots, nmax$ set
$\quad x_n = x_{n-1} + h$
$\quad w_n = f(x)$
Set $w_{nmax+1} = (q(0) + f(xmax))/2$
$\quad x_{nmax+1} = xmax$
Output t
For $n = 0, 1, \ldots, nmax + 1$
$\quad$ Output x_n, w_n

Step 4. Initialize numerical solution at *t* = *k*

Set $t = k$
$\quad v_0 = p(k)$
For $n = 1, 2, \ldots, nmax$ set
$\quad v_n = w_n + k \cdot g(x_n) + r \cdot \tfrac{1}{2}[w_{n-1} - 2w_n + w_{n+1}]$
Set $v_{nmax+1} = q(k)$
Output t
For $n = 0, 1, \ldots, nmax + 1$
$\quad$ Output x_n, v_n

Step 5. Begin time stepping

For $j = 1, 2, \ldots, jmax$
$\quad$ Do steps 6–9

Step 6. Define tridiagonal system

Set $t = t + k$
For $n = 1, 2, \ldots, nmax$ set
$\quad a_n = -r$
$\quad b_n = 2 + 2r$
$\quad c_n = -r$
$\quad d_n = 4v_n - 2w_n + r[w_{n-1} - 2w_n + w_{n+1}]$
Set $d_1 = d_1 + r \cdot p(t)$
$\quad d_{nmax} = d_{nmax} + r \cdot q(t)$

Step 7. Advance solution one time step

CALL TRIDI($nmax, a, b, c, d$)
For $n = 1, 2, \ldots, nmax$ set
$\quad U_n = d_n$
Set $U_0 = p(t)$
$\quad U_{nmax+1} = q(t)$

Step 8. Prepare for next time step

For $n = 0, 1, \ldots, nmax + 1$ set
$\quad w_n = v_n$
$\quad v_n = u_n$

Step 9. Output numerical solution

Output t
For $n = 0, 1, \ldots, nmax + 1$
$\quad$ Output x_n, u_n

EXAMPLE 9.6.2

To illustrate an implicit difference method, reconsider the initial-boundary-value problem of Example 9.6.3. So that we can compare the implicit and explicit numerical results, choose the h and k values to be those used in Example 9.6.3. Table 9.6.2 displays a comparison of the exact solution and the numerical solution obtained by (9.6.11) with $\omega = \frac{1}{2}$. The choice of $k = 0.15$ for the implicit method yields a good approximation to the exact solution. This choice of k in the explicit method would result in an unstable computation. ∎ ∎

TABLE 9.6.2 COMPARISON OF NUMERICAL AND EXACT
SOLUTION FOR EXAMPLE 9.6.2

$T = 1.50$	Numerical ($k = 0.075$)	Exact	Numerical ($k = 0.150$)
$X = 0.0$	2.129279	2.129279	2.129279
$X = 0.1$	2.361019	2.360280	2.363176
$X = 0.2$	2.616083	2.614761	2.620119
$X = 0.3$	2.896857	2.895131	2.901943
$X = 0.4$	3.206113	3.204058	3.211555
$X = 0.5$	3.546735	3.544502	3.551993
$X = 0.6$	3.921860	3.919741	3.926723
$X = 0.7$	4.335243	4.333412	4.339714
$X = 0.8$	4.790991	4.789542	4.794846
$X = 0.9$	5.293408	5.292593	5.295480
$X = 1.0$	5.847505	5.847505	5.847505

Reduction to a First-Order System

Another way to apply finite-difference methods to approximate the solution of (9.6.1)–(9.6.3) is to first write (9.6.1) as an equivalent system of first-order equations.

EXAMPLE 9.6.3 ───

Given $w_{tt} - c^2 w_{xx} = 0$, if we set $u = w_x$ and $v = w_t$ and assume that $w_{xt} = w_{tx}$, then it follows that

$$u_t = v_x$$

From $w_{tt} = -c^2 w_{xx} = 0$, it follows that

$$c^2 u_x = v_t$$

The last two first-order equations for u and v can be written as the first-order system

$$\begin{bmatrix} 0 & -1 \\ -c^2 & 0 \end{bmatrix} \begin{bmatrix} u_x \\ v_x \end{bmatrix} + \begin{bmatrix} u_t \\ v_t \end{bmatrix} = \begin{bmatrix} 0 \\ 0 \end{bmatrix} \qquad (9.6.12)$$

With

$$\mathbf{A} = \begin{bmatrix} 0 & -1 \\ -c^2 & 0 \end{bmatrix} \quad \text{and} \quad \mathbf{u} = \begin{bmatrix} u \\ v \end{bmatrix}$$

Equation (9.6.12) is seen to have the form $\mathbf{A}\mathbf{u}_x + \mathbf{u}_t = \mathbf{O}$, which is Equation (9.5.1). Initial conditions for u and v are provided by $u(x, 0) = f'(x)$ and $v(x, 0) = g(x)$. Thus, in vector notation, the initial condition is

$$\mathbf{u}(x, 0) = \mathbf{f}(x) = \begin{bmatrix} f'(x) \\ g(x) \end{bmatrix}$$

If (9.6.12) is solved for u and v, then we have numerical approximations to w_x and w_t at each grid point. We must still find a numerical approximation to the solution w at the grid points. To this end, consider the Taylor expansion

$$w(x_n, t_{j+1}) = w(x_n, t_j) + k w_t(x_n, t_j) + \tfrac{1}{2} k^2 w_{tt}(x_n, t_j) + O(k^3)$$

Assume that an estimate W_n^j is available. For $j = 0$, we have $W_n^0 = f_n$. Since $v = w_t$, we can use a numerical solution for v to replace $w_t(x_n, t_j)$,

$$w_t(x_n, t_j) = V_n^j$$

to the order of the difference method used to obtain V_n^j. Also, since $w_{tt} = c^2 w_{xx}$ and $u = v_x$, we can approximate $w_{tt}(x_n, t_j)$ as follows:

$$w_{tt}(x_n, t_j) = c^2 w_{xx}(x_n, t_j) = c^2 u_x(x_n, t_j)$$

$$\simeq c^2 \frac{U_{n+1}^j - U_{n-1}^j}{2h}$$

Now, it follows that the solution w can be estimated at the grid points by

$$W_n^{j+1} = W_n^j + k V_n^j + \frac{c^2 k^2}{4h} (U_{n+1}^j - U_{n-1}^j) \qquad (9.6.13)$$

After reducing the scalar, second-order initial-value problem (9.6.1)–(9.6.3) to a system of first-order equations as in Example 9.6.3, most of the difference methods of Section 9.5 are available. However, the one-sided difference methods (9.5.4) and (9.5.8) of Table 9.5.1 cannot be used to approximate the solution of (9.6.12). Each of those methods requires that the eigenvalues of A be of one sign. An easy calculation shows the eigenvalues of the matrix A of (9.6.12) to be $\lambda = \pm c$.   ■ ■

EXAMPLE 9.6.4 ————————————————————————————————————

Given the initial-value problem

$$w_{tt} - 4w_{xx} = 0, \qquad -\infty < x < \infty, \, t > 0$$
$$w(x, 0) = \sin x, \qquad w_t(x, 0) = 0, \qquad -\infty < x < \infty$$

whose exact solution is

$$w(x, t) = \tfrac{1}{2}[\sin(x + 2t) + \sin(x - 2t)] = \sin x \cos 2t$$

let us use the Lax–Wendroff method (9.5.7) as given in Algorithm 9.3 to approximate the solution of the first-order system equivalent of this problem. Namely,

$$\begin{bmatrix} 0 & -1 \\ -2^2 & 0 \end{bmatrix} \begin{bmatrix} u_x \\ v_x \end{bmatrix} + \begin{bmatrix} u_t \\ v_t \end{bmatrix} = \begin{bmatrix} 0 \\ 0 \end{bmatrix} \tag{9.6.14}$$

$$\begin{bmatrix} u(x, 0) \\ v(x, 0) \end{bmatrix} = \begin{bmatrix} \cos x \\ 0 \end{bmatrix} \tag{9.6.15}$$

Table 9.6.3 displays the numerical solution obtained by applying the Lax–Wendroff method to (9.6.14) and (9.6.15) with $h = 0.05$ and $k = 0.02$. In Table 9.6.4, we show the comparison for the numerical and exact solutions to the second-order equations having used (9.6.13) and the results of Table 9.6.3 to recover W_n^j.   ■ ■

TABLE 9.6.3 COMPARISON OF NUMERICAL AND EXACT SOLUTIONS FOR SYSTEM OF EXAMPLE 9.6.4

$T = 0.50$	Numerical u value	Exact u value	Numerical v value	Exact v value
$X = 0.0$	0.540426	0.540302	0	0
$X = 0.1$	0.537726	0.537603	-0.167997	-0.168014
$X = 0.2$	0.529653	0.529532	-0.334315	-0.334349
$X = 0.3$	0.516289	0.516171	-0.497293	-0.497343
$X = 0.4$	0.497765	0.497651	-0.655302	-0.655368
$X = 0.5$	0.474268	0.474160	-0.806764	-0.806845
$X = 0.6$	0.446033	0.445931	-0.950165	-0.950261
$X = 0.7$	0.413341	0.413246	-1.084072	-1.084181
$X = 0.8$	0.376518	0.376432	-1.207147	-1.207269
$X = 0.9$	0.335934	0.335857	-1.318161	-1.318294
$X = 1.0$	0.291993	0.291927	-1.416004	-1.416147

TABLE 9.6.4 COMPARISON OF NUMERICAL AND EXACT SOLUTION FOR $w_{tt} - 4w_{xx} = 0$ OF EXAMPLE 9.6.4

$T = 0.50$	Numerical w value	Exact w value
$X = 0.0$	0	0
$X = 0.1$	0.053935	0.053940
$X = 0.2$	0.107330	0.107341
$X = 0.3$	0.159653	0.159670
$X = 0.4$	0.210381	0.210404
$X = 0.5$	0.259007	0.259035
$X = 0.6$	0.305045	0.305078
$X = 0.7$	0.348035	0.348072
$X = 0.8$	0.387548	0.387589
$X = 0.9$	0.423189	0.423233
$X = 1.0$	0.454601	0.454649

Exercises

1. Use the forward-in-time explicit method (9.6.5) with $h = 0.1$ and $k = 0.113$ to estimate the solution for $0 \le x \le 1$ to the initial-value problem of Example 9.6.1 at time $t = 0.1$. Do not use any boundary conditions.

2. Given the initial-boundary-value problem

$$u_{tt} - 4u_{xx} = 0, \qquad\qquad 0 < x < 1, t > 0$$
$$u(x, 0) = \sin 2\pi x, \qquad 0 < x < 1$$
$$u_t(x, 0) = 0, \qquad\qquad 0 < x < 1$$
$$u(0, t) = 0 = u(1, t), \qquad t > 0$$

(a) Use separation of variables to obtain the exact solution.
(b) Use the explicit method (9.6.5) to estimate the solution for $0 < t < 2$. Take $h = 0.1$.
(c) Use the implicit method (9.6.11) with $\omega = \frac{1}{2}$ and $h = 0.1$ to estimate the solution for $0 < t < 2$.
(d) Rewrite the second-order equation as a pair of first-order equations and apply the Lax–Wendroff method to estimate the solution for $0 < t < 2$. In each case experiment with k.

3. Consider the implicit method (9.6.11) with $\omega = \frac{1}{2}$.
(a) Put $U_n^j = \xi^j e^{i\beta x_n}$ into the difference equation and obtain a quadratic equation for the magnification factor ξ in the form

$$\xi^2 + 2b\xi + 1 = 0$$

(b) Show that the two roots of the quadratic equation

$$\xi^2 + 2b\xi + 1 = 0$$

both satisfy $|\xi| = 1$ if and only if $|b| \le 1$.
(c) Show that (9.6.11) with $\omega = \frac{1}{2}$ is unconditionally stable.
(d) Show that (9.6.11) with $\omega = \frac{1}{2}$ is nondissipative.

4. Consider the initial-value problem

$$u_{tt} - u_{xx} = 0, \qquad -\infty < x < \infty, \, t > 0$$

$$u(x, 0) = \begin{cases} 1 - |x|, & \text{for } |x| < 1 \\ 0, & \text{for } |x| > 1 \end{cases}$$

$$u_t(x, 0) = 0, \qquad 0 < x < 1$$

   (a) Use D'Alembert's formula to obtain the exact solution.
   (b) Use the explicit method to approximate the solution for $|x| < 2$, for $0 < t < 1$.

5. (a) Modify the implicit method, Algorithm 9.5, to accommodate the Neumann boundary conditions $u_x(0, t) = 0$, $u_x(L, t) = 0$.
   (b) Modify the method to accommodate a mixed (elastic) boundary condition,

$$\alpha u(0, t) + \beta u_x(0, t) = 0, \qquad \alpha\beta \neq 0$$

   at $x = 0$ and a Dirichlet boundary condition $u(L, t) = 0$.

6. The air pressure $p(x, t)$ in an organ pipe can be modeled by the wave equation

$$p_{tt} - c^2 p_{xx} = 0, \qquad 0 < x < L, \, t > 0$$

   where L is the length of the pipe and c is a physical constant. Assume $L = 1$ and $c = 2$ and approximate the pressure distribution in the pipe for $0 < t < 1$ if

$$\begin{array}{ll} p(x, 0) = \sin 2\pi x, & 0 < x < L \\ p_t(x, 0) = 2\pi \sin 2\pi x, & 0 < x < L \\ p(0, t) = p(L, t) = 0, & t > 0 \end{array}$$

   (a) Use the explicit method and $h = 0.1$, $h = 0.05$.
   (b) Use the implicit method and $h = 0.05$.
   (c) Reduce the problem to a pair of first-order equations and use the method of Example 9.6.4 with $h = 0.05$. In each case compare with the exact solution

$$u(x, t) = \sin 2\pi x[\cos 2\pi t + \sin 2\pi t]$$

7. Show that the maximum time step that could be used in the solution of Example 9.6.4 with $h = 0.05$ is $k = 0.025$.

9.7 METHOD OF CHARACTERISTICS

Introduction

The numerical method of characteristics is an effective method for approximating the solution to (i) a single first-order equation; (ii) a pair of first-order equations; or (iii) a single second-order equation of hyperbolic type. To apply the method of characteristics, the partial differential equation problem is reduced to an ordinary differential equation problem along characteristic curves. This reduction can be carried out for a single equation in any number of independent variables. However, for a system of equations or for a second-order equation, the characteristics are curves (rather than surfaces) only in the case that just two independent variables are involved.

In Section 7.2 we used the method of characteristics to solve a number of first-order quasi-linear equations. There we found we could construct an exact solution provided we could obtain an exact antiderivative of the characteristic equations. The numerical method of characteristics of this section does not require any exact antiderivatives.

An exact method of characteristics for a constant-coefficient, homogeneous system of first-order equations was given in Section 7.6. The numerical method of characteristics extends those results to the case of a general quasi-linear pair of equations in x and t.

The D'Alembert solution of Section 4.1 for the wave equation $u_{tt} - c^2 u_{xx} = 0$ is an example of an exact method of characteristics solution to a second-order equation. In this section we consider the numerical method of characteristics solution of a general quasi-linear equation of the form $au_{xx} + 2bu_{xy} + cu_{yy} = S$.

A Single First-Order Equation

Consider the quasi-linear first-order equation

$$a(x, t, u)u_x + u_t = c(x, t, u), \qquad -\infty < x < \infty, t > 0 \qquad (9.7.1)$$

subject to the initial condition

$$u(x, 0) = f(x), \qquad -\infty < x < \infty \qquad (9.7.2)$$

From Section 7.2, the characteristic equations of (9.7.1) are known to be

$$\frac{dx}{dr} = a, \qquad \frac{dt}{dr} = 1, \qquad \frac{du}{dr} = c \qquad (9.7.3)$$

where r is a parameter that varies along a characteristic of (9.7.1). If we let $t = r$, then the second of the characteristic equations (9.7.3) is identically satisfied and the remaining two equations take the form

$$\frac{dx}{dt} = a(x, t, u), \qquad \frac{du}{dt} = c(x, t, u) \qquad (9.7.4)$$

The solution $x = X_n(t)$, $u = U_n(t)$, of the system (9.7.4) along the characteristic through the point $(x, t) = (x_n, 0)$ satisfies the initial condition

$$X_n(0) = x_n, \qquad U_n(0) = f(x_n) = f_n \qquad (9.7.5)$$

The solution to an initial-value problem of the form (9.7.4) can be approximated using standard numerical methods for dealing with an ordinary differential equation initial-value problem. For example, the approximate solution of (9.7.4) and (9.7.5) at time $t = k$ by Euler's method is

$$X_n(k) = X_n(0) + a(X_n(0), 0, U_n(0))k$$
$$U_n(k) = U_n(0) + c(X_n(0), 0, U_n(0))k$$

or

$$X_n(k) = x_n + a(x_n, 0, f_n)k$$
$$\dot{U}_n(k) = f_n + c(x_n, 0, f_n)k$$

The Euler approximation to the solution of (9.7.4) and (9.7.5) can be marched forward on a time grid $t_j = jk, j = 0, 1, \ldots$, by the calculations

$$X_n^{j+1} = X_n^j + a(X_n^j, t_j, U_n^j)k, \qquad X_n^0 = x_n \qquad (9.7.6)$$
$$U_n^{j+1} = U_n^j + c(X_n^j, t_j, U_n^j)k, \qquad U_n^0 = f_n \qquad (9.7.7)$$

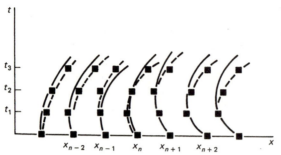

Figure 9.7.1 Characteristics (——) and numerical approximations to characteristics (----). (■) Numerical characteristic grid point.

By allowing n to vary over an x grid as in Figure 9.7.1, Equations (9.7.6) and (9.7.7) can be used to approximate the position of the characteristics of (9.7.1) and the solution to the initial-value problem (9.7.1) and (9.7.2). The numerical solution obtained from (9.7.6) and (9.7.7) generally provides an approximation to the exact solution on an irregularly spaced *characteristic grid* for $t_j > 0$. The point (X_n^j, t_j) is an approximation to the (x, t_j) coordinate of the point on the characteristic through (x_n, t_j). Here, U_n^j is an approximation to the $u(x, t_j)$ on that characteristic.

In the preceding discussion, Euler's method has been used to approximate the solution to the characteristic equations (9.7.4) and (9.7.5). This approach produces an approximation to the exact solution that is accurate to $O(k)$. With little additional computational effort, much better time accuracy can be obtained. As an example of an alternative to Euler's method, we present the following fourth-order, $O(k^4)$, Runge–Kutta method for approximating the solution of (9.7.4) and (9.7.5). A discussion of the Runge–Kutta method can be found in most numerical analysis books. See, for example, R. L. Burden and J. D. Faires, *Numerical Analysis*, 3rd ed., Prindle, Weber & Schmidt, Boston, 1985, pp. 220, 229.

ALGORITHM 9.6   Runge-Kutta Method for a Pair of Ordinary Differential Equations

Step 1. Document

This algorithm uses the Runge-Kutta method of order 4 to approximate the solution to the initial-value problem for the characteristic equations (9.7.4)–(9.7.5),

$$\frac{dx}{dt} = a(x, t, u), \qquad \frac{du}{dt} = c(x, t, u)$$
$$x(0) = x_n, \qquad u(0) = f_n$$

INPUT

Functions $a(x, t, u)$, $c(x, t, u)$, coefficients from PDE

Real x_n, x coordinate of characteristic at $t = 0$
Real f_n, u value at $(x, t) = (x_n, 0)$
Real k, time step
Integer $jmax$, number of time steps

OUTPUT

Time t_j and approximations X_n^j and U_n^j to $x_n(t_j)$ and $u_n(t_j)$ for $j = 0, 1, \ldots, jmax$.

Step 2. Initialize numerical solution

Set $t = 0$
$\quad X = x_n$
$\quad U = f_n$

Step 3. Begin time stepping

For $j = 1, 2, \ldots, jmax$
$\quad$ Do steps 4–6

Step 4. Define Runge-Kutta coefficients

Set $r_1 = ka(X, t, U)$
$\quad s_1 = kc(X, t, U)$

$$r_2 = ka\left(X + \frac{r_1}{2}, t + \frac{k}{2}, U + \frac{s_1}{2}\right)$$

$$s_2 = kc\left(X + \frac{r_1}{2}, t + \frac{k}{2}, U + \frac{s_1}{2}\right)$$

$$r_3 = ka\left(X + \frac{r_2}{2}, t + \frac{k}{2}, U + \frac{s_2}{2}\right)$$

$$s_3 = kc\left(X + \frac{r_2}{2}, t + \frac{k}{2}, U + \frac{s_2}{2}\right)$$

$$r_4 = ka(X + r_3, t + k, U + s_3)$$
$$s_4 = kc(X + r_3, t + k, U + s_3)$$

Step 5. Update numerical solution

Set $t = t + k$
$\quad X = X + \frac{1}{6}(r_1 + 2r_2 + 2r_3 + r_4)$
$\quad U = U + \frac{1}{6}(s_1 + 2s_2 + 2s_3 + s_4)$

Step 6. Output numerical solution

Output t, X, U

By simply embedding Algorithm 9.5 in an n loop that cycles through an x grid, the numerical method of characteristics can be used to generate an approximation to the solution of (9.7.1) and (9.7.2). Initially the x grid satisfies $x_{n+1} - x_n = h$. At each time level, the numerical solution must be tested to determine if $X_{n+1}^j - X_n^j$ is positive for all values of n. If $X_{n+1}^j - X_n^j$ is negative for some n, that indicates that two characteristics of (9.7.1) have intersected and that, possibly, a shock has developed in the solution. With the development of a shock, the basic method of characteristics no longer applies. The shock condition and the entropy condition must now be incorporated into the numerical solution process.

EXAMPLE 9.7.1 ————————————————————————————————————

Consider the quasi-linear initial-value problem

$$uu_x + u_t = -2u^3, \qquad -\infty < x < \infty, t > 0$$
$$u(x, 0) = x, \qquad -\infty < x < \infty$$

Along the characteristic through $x = 1$, we can use the methods of Section 7.2 to show that

$$x = X(t) = \tfrac{1}{2}[(1 + 4t)^{1/2} + 1]$$

and

$$u = U(t) = (1 + 4t)^{-1/2}$$

In Table 9.7.1 we display the numerical approximations to the solution along this characteristic. For comparison, both Euler's method and the Runge–Kutta (RK) method for solving (9.7.4) and (9.7.5) are displayed along with the exact solution. A time step of $k = 0.1$ is used.

TABLE 9.7.1 NUMERICAL AND EXACT SOLUTIONS ALONG CHARACTERISTIC THROUGH $(x, t) = (1, 0)$ in EXAMPLE 9.7.1

Time	RK-X	Exact X	Euler X	RK U	Exact U	Euler U
0.00	1.0000	1.0000	1.0000	1.0000	1.0000	1.0000
0.10	1.0916	1.0916	1.1000	0.8451	0.8452	0.8000
0.20	1.1708	1.1708	1.1800	0.7453	0.7454	0.6976
0.30	1.2416	1.2416	1.2498	0.6742	0.6742	0.6297
0.40	1.3062	1.3062	1.3127	0.6202	0.6202	0.5798
0.50	1.3660	1.3660	1.3707	0.5773	0.5774	0.5408
0.60	1.4219	1.4220	1.4248	0.5423	0.5423	0.5092
0.70	1.4746	1.4747	1.4757	0.5130	0.5130	0.4828
0.80	1.5247	1.5247	1.5240	0.4879	0.4880	0.4603
0.90	1.5723	1.5724	1.5700	0.4662	0.4663	0.4408
1.00	1.6180	1.6180	1.6141	0.4472	0.4472	0.4236

As the results of Table 9.7.1 show, it clearly pays to use the Runge–Kutta method, rather than the Euler method, to integrate along the characteristic. ∎ ∎

A Pair of First-Order Equations

Consider a quasi-linear pair of first-order hyperbolic equations of the form

$$a_{11}u_x + a_{12}v_x + u_t = c_1 \tag{9.7.8a}$$
$$a_{21}u_x + a_{22}v_x + v_t = c_2 \tag{9.7.8b}$$

Using the matrix $\mathbf{A} = [a_{ij}]$ and the vectors

$$\mathbf{u} = \begin{bmatrix} u \\ v \end{bmatrix} \quad \text{and} \quad \mathbf{c} = \begin{bmatrix} c_1 \\ c_2 \end{bmatrix}$$

(9.7.8) can be written in the form

$$\mathbf{A}\mathbf{u}_x + \mathbf{u}_t = \mathbf{c} \tag{9.7.9}$$

Recall from Section 7.2 that (9.7.9) has two families of characteristics defined by

$$\frac{dx}{dt} = \lambda_1 \quad \text{and} \quad \frac{dx}{dt} = \lambda_2$$

where λ_1 and λ_2 are the zeros of $\det(\mathbf{A} - \lambda\mathbf{I}) = 0$. It will be convenient to introduce a pair of parameters α and β that determine the position on the λ_1 and λ_2 characteristics, respectively. Let α be a parameter that varies only along a λ_1 characteristic. Similarly, β is a parameter that varies only along a λ_2 characteristic. See Figure 9.7.2.

Using the characteristic coordinates, we can specify any point (x, t) by giving values of α and β. This correspondence allows us to view x and t as functions of α and β,

$$x = x(\alpha, \beta), \qquad t = t(\alpha, \beta)$$

Along an α characteristic,

$$\frac{dx}{dt} = \frac{x_\alpha(\alpha, \beta)}{t_\alpha(\alpha, \beta)} = \lambda_1 \quad \text{or} \quad x_\alpha = \lambda_1 t_\alpha \tag{9.7.10}$$

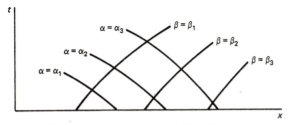

Figure 9.7.2 The α and β characteristics. Along curve $\alpha = \alpha_p = $ const only β varies. Along curve $\beta = \beta_q = $ const only α varies.

Similarly, along a β characteristic

$$\frac{dx}{dt} = \frac{x_\beta(\alpha, \beta)}{t_\beta(\alpha, \beta)} = \lambda_2 \quad \text{or} \quad x_\beta = \lambda_2 t_\beta \tag{9.7.11}$$

If λ_1 and λ_2 are independent of u and v, then (9.7.10)–(9.7.12) can be integrated to determine a characteristic grid. However, for a general quasi-linear equation, λ_1 and λ_2 depend on u and v. Then some knowledge of the solution is required to fix the characteristics. We now proceed to find some conditions on u and v along the characteristics.

If we use the chain rule to differentiate $u = u(x, t)$ along an α characteristic, then we find

$$\frac{\partial u}{\partial \alpha} = \frac{\partial u}{\partial x}\frac{\partial x}{\partial \alpha} + \frac{\partial u}{\partial t}\frac{\partial t}{\partial \alpha} \quad \text{or} \quad u_\alpha = u_x u_\alpha + u_t t_\alpha$$

From the last equation and (9.7.10) it follows that

$$\lambda_1 u_x + u_t = u_\alpha / t_\alpha \tag{9.7.12}$$

Similarly, along an α characteristic, v satisfies

$$\lambda_1 v_x + v_t = v_\alpha / t_\alpha \tag{9.7.13}$$

Now, Equations (9.7.8a), (9.78b), (9.7.12), and (9.7.13) comprise four linear equations in the four unknowns u_x, v_x, u_t, and v_t. In matrix form these four equations are

$$\begin{bmatrix} a_{11} & a_{12} & 1 & 0 \\ a_{21} & a_{22} & 0 & 1 \\ \lambda_1 & 0 & 1 & 0 \\ 0 & \lambda_1 & 0 & 1 \end{bmatrix} \begin{bmatrix} u_x \\ v_x \\ u_t \\ v_t \end{bmatrix} = \begin{bmatrix} c_1 \\ c_2 \\ u_\alpha / t_\alpha \\ v_\alpha / t_\alpha \end{bmatrix} \tag{9.7.13}$$

Since $\det(\mathbf{A} - \lambda_1 \mathbf{I}) = 0$, it follows that the determinant of the coefficient matrix in (9.7.13) is zero. Now, from Cramer's rule, we know that (9.7.13) has a solution if and only if

$$\det \begin{bmatrix} a_{11} & a_{12} & 1 & c_1 \\ a_{21} & a_{22} & 0 & c_2 \\ \lambda_1 & 0 & 1 & u_\alpha / t_\alpha \\ 0 & \lambda_1 & 0 & v_\alpha / t_\alpha \end{bmatrix} = 0 \tag{9.7.14}$$

If we expand the determinant (9.7.14) by minors using the last column, the result is

$$-c_1 \begin{vmatrix} a_{21} & a_{22} & 0 \\ \lambda_1 & 0 & 1 \\ 0 & \lambda_1 & 0 \end{vmatrix} + c_2 \begin{vmatrix} a_{11} & a_{12} & 1 \\ \lambda_1 & 0 & 1 \\ 0 & \lambda_1 & 0 \end{vmatrix} - \frac{u_\alpha}{t_\alpha} \begin{vmatrix} a_{11} & a_{12} & 1 \\ a_{21} & a_{22} & 0 \\ 0 & \lambda_1 & 0 \end{vmatrix} + \frac{v_\alpha}{t_\alpha} \begin{vmatrix} a_{11} & a_{12} & 1 \\ a_{21} & a_{22} & 0 \\ \lambda_1 & 0 & 1 \end{vmatrix} = 0$$

which can be rearranged to read

$$a_{11}^* u_\alpha + a_{12}^* v_\alpha = c_1^* t_\alpha \tag{9.7.15}$$

where

$$a_{11}^* = - \begin{vmatrix} a_{11} & a_{12} & 1 \\ a_{21} & a_{22} & 0 \\ 0 & \lambda_1 & 0 \end{vmatrix}, \qquad a_{12}^* = \begin{vmatrix} a_{11} & a_{12} & 1 \\ a_{21} & a_{22} & 0 \\ \lambda_1 & 0 & 1 \end{vmatrix}$$

and

$$c_1^* = c_1 \begin{vmatrix} a_{21} & a_{22} & 0 \\ \lambda_1 & 0 & 1 \\ 0 & \lambda_1 & 0 \end{vmatrix} - c_2 \begin{vmatrix} a_{11} & a_{12} & 1 \\ \lambda_1 & 0 & 1 \\ 0 & \lambda_1 & 0 \end{vmatrix}$$

A completely parallel argument along the β characteristics yields

$$a_{21}^* u_\beta + a_{22}^* v_\beta = c_2^* t_\beta \qquad (9.7.16)$$

where

$$a_{21}^* = \begin{vmatrix} a_{11} & a_{12} & 1 \\ a_{21} & a_{22} & 0 \\ 0 & \lambda_2 & 0 \end{vmatrix}, \qquad a_{22}^* = \begin{vmatrix} a_{11} & a_{12} & 1 \\ a_{21} & a_{22} & 0 \\ \lambda_2 & 0 & 1 \end{vmatrix}$$

and

$$c_2^* = c_1 \begin{vmatrix} a_{21} & a_{22} & 0 \\ \lambda_2 & 0 & 1 \\ 0 & \lambda_2 & 0 \end{vmatrix} - c_2 \begin{vmatrix} a_{11} & a_{12} & 1 \\ \lambda_2 & 0 & 1 \\ 0 & \lambda_2 & 0 \end{vmatrix}$$

Equations (9.7.10), (9.7.11), (9.7.15), and (9.7.16) are the basis for the numerical method of characteristics. The development leading to Equations (9.7.15) and (9.7.16) offers some new insight into the nature of the characteristics. From the fact that the coefficient matrix in (9.7.13) has zero determinant, we conclude that the *characteristics of the system (9.7.9) are those curves along which the differential equations for u and v and a knowledge of the u and v values are insufficient to uniquely fix the values of the x and t derivatives of u and v.*

With reference to Figure 9.7.3, suppose that the u and v values are known at points Q and R and that we want to estimate the location of the point P and the u and v values at P. To implement the numerical method of characteristics, we simply approximate all α and β derivatives with difference quotients. For example, we approximate x_α by $[x(P - x(Q)]/[\alpha(P) - \alpha(Q)]$, t_α by $[t(P) - t(P)]/[\alpha(P) - \alpha(Q)]$, and so on. Thus, the characteristic equations to be solved numerically are

$$x(P) - x(Q) = \lambda_1[t(P) - t(Q)] \qquad (9.7.17)$$
$$x(P) - x(R) = \lambda_2[t(P) - t(R)] \qquad (9.7.18)$$
$$a_{11}^*[u(P) - u(Q)] + a_{12}^*[v(P) - v(Q)] = c_1^*[t(P) - t(Q)] \qquad (9.7.19)$$
$$a_{21}^*[u(P) - u(R)] + a_{22}^*[v(P) - v(R)] = c_2^*[t(P) - t(R)] \qquad (9.7.20)$$

If we evaluate λ_1, a_{11}^*, a_{12}^*, and c_1^* at Q and evaluate λ_2, a_{21}^*, a_{22}^*, and c_2^* at R, then (9.7.17)–(9.7.20) provide four linear equations for the determination of the four unknowns $x(P)$, $t(P)$, $u(P)$, and $v(P)$.

EXAMPLE 9.7.2 _____

The initial-value problem

$$uu_x + u_t = 2xe^t$$
$$u_x - xv_x + v_t = e^t - 2x^2 - 2t$$
$$u(x, 0) = x, \qquad v(x, 0) = x^2$$

has exact solution $u = xe^t$ and $v = x^2 - t^2$. To illustrate the numerical method of characteristics, we estimate u and v at the first characteristic grid point formed by the intersection of the characteristics through $Q = (1, 0)$ and $R = (1.5, 0)$.

Setting the characteristic polynomial to zero,

$$\begin{vmatrix} u - \lambda & 0 \\ 1 & -x - \lambda \end{vmatrix} = 0$$

we find $\lambda_1 = u$ and $\lambda_2 = -x$. We know to label the roots in this way because λ_1 is positive at both Q and R and λ_2 is negative at both Q and R. See Figure 9.7.3. We evaluate λ_1 at Q and λ_2 at R to obtain $\lambda_1(Q) = 1$ and $\lambda_2(R) = -1.5$. Next, we find

$$a^*_{11}(Q) = - \begin{vmatrix} 1 & 0 & 1 \\ 1 & -1 & 0 \\ 0 & 1 & 0 \end{vmatrix} = -1, \qquad a^*_{12}(Q) = \begin{vmatrix} 1 & 0 & 1 \\ 1 & -1 & 0 \\ 1 & 0 & 1 \end{vmatrix} = 0$$

and

$$c^*_1(Q) = c_1(Q) \begin{vmatrix} 1 & 0 & 0 \\ 1 & 0 & 1 \\ 0 & 1 & 0 \end{vmatrix} - c_2(Q) \begin{vmatrix} 1 & 0 & 1 \\ 1 & 0 & 1 \\ 0 & 1 & 0 \end{vmatrix} = -2$$

Similarly,

$$a^*_{21}(R) = - \begin{vmatrix} 1.5 & 0 & 1 \\ 1 & -1.5 & 0 \\ 0 & -1.5 & 0 \end{vmatrix} = 1.5, \qquad a^*_{22}(R) = \begin{vmatrix} 1.5 & 0 & 1 \\ 1 & -1.5 & 0 \\ -1.5 & 0 & 1 \end{vmatrix} = -4.5$$

and

$$c^*_2(R) = c_1(R) \begin{vmatrix} 1 & -1.5 & 0 \\ -1.5 & 0 & 1 \\ 0 & -1.5 & 0 \end{vmatrix} - c_2(R) \begin{vmatrix} 1.5 & 0 & 1 \\ -1.5 & 0 & 1 \\ 0 & -1.5 & 0 \end{vmatrix} = 4.5^2$$

Putting these coefficients into the numerical characteristic equations (9.7.17)–(9.7.20), we find

$$x(P) - 1 = 1[t(P) - 0]$$
$$x(P) - 1.5 = -1.5[t(P) - 0]$$
$$-1[u(P) - 1] + 0[v(P) - 1^2] = -2[t(P) - 0]$$
$$1.5[u(P) - 1.5] - 4.5[v(P) - 2.25] = 4.5^2[t(P) - 0]$$

Solving these four linear equations, we find the approximate solution by the numerical method of characteristics to be

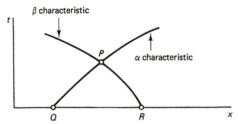

Figure 9.7.3 An α characteristic through Q intersects β characteristic through R at point P.

$$t(P) = 0.2, \quad x(P) = 1.2$$
$$u(P) = 1.4, \quad v(P) = 1.383$$

Now that we have estimated the solution of Example 9.7.2 using the numerical method of characteristics, let us use the exact method of characteristics for comparison. Since our example problem has a known solution, we can find the forward-facing characteristic through Q to be

$$x = \exp(e^t - 1)$$

and the backward-facing characteristic through R to be

$$x = 1.5e^{-t}$$

These characteristics intersect when

$$e^t - 1 + t - \ln 1.5 = 0$$

Using Newton's method, we approximate the t coordinate of the point P to be $t(P) = 0.1928$. Putting this value of t into the equation for either the characteristic through Q or the one through R, we find $x(P) = 1.237$. Evaluating $u(x, t)$ and $v(x, t)$ at $(x, t) = (1.237, 0.1928)$ yields $u(P) = 1.500$ and $v(P) = 1.493$. Comparing the values obtained by the numerical method of characteristics with those obtained by the exact method of characteristics, we find:

	$x(P)$	$t(P)$	$u(P)$	$v(P)$
Numerical	1.2	0.2	1.4	1.383
Exact	1.237	0.1928	1.500	1.493

In Example 9.7.2 Euler's method was used to approximate the solution to the characteristic equations. More accurate estimates to the coefficients in (9.7.17)–(9.7.20) usually yield a method more accurate than Euler's method. For example, we could use a trapezoidal approximation in which we estimate λ_1 by $\frac{1}{2}[\lambda_1(P) + \lambda_1(Q)]$ with similar expressions for the other coefficients. These higher order methods generally result in a nonlinear system of characteristic equations whose solution must be approximated by an iterative procedure.

Second-Order Equations

Let $u = u(x, y)$ satisfy a second-order hyperbolic equation of the form

$$au_{xx} + 2bu_{xy} + cu_{yy} = f \tag{9.7.21}$$

subject to the initial conditions

$$u(x, 0) = g(x) \tag{9.7.22}$$
$$u_y(x, 0) = h(x) \tag{9.7.23}$$

The method of characteristic applies to (9.7.21) only when it is of hyperbolic type. Recall that (9.7.21) is classified as hyperbolic if the discriminant $b^2 - ac$ is positive. The numerical method of characteristics applies equally well to the linear case

$$a = a(x, y), \qquad b = b(x, y), \qquad c = c(x, y), f = d(x, y)u + e(x, y)$$

or to the quasi-linear case with a, b, c, and f allowed to depend on x, y, u, u_x, u_y but not on u_{xx}, u_{xy}, u_{yy}.

The characteristics of (9.7.21) are those curves along which a knowledge of u, u_x, and u_y together with (9.7.21) are insufficient to uniquely fix values for u_{xx}, u_{xy}, and u_{yy}. It is not difficult to show that the characteristics are defined by

$$\frac{dy}{dx} = \frac{b + \sqrt{b^2 - ac}}{a} \equiv \lambda_+ \tag{9.7.24}$$

and

$$\frac{dy}{dx} = \frac{b - \sqrt{b^2 - ac}}{a} \equiv \lambda_- \tag{9.7.25}$$

If (9.7.24) holds along a level curve of a function $F(x, y)$,

$$F(x, y) = \beta, \qquad \beta = \text{const} \tag{9.7.26}$$

we say that the curve (9.7.26) is an α characteristic of (9.7.21). Similarly, if (9.7.25) holds along a level curve of a function $G(x, y)$,

$$G(x, y) = \alpha, \qquad \alpha = \text{const} \tag{9.7.27}$$

the curves (9.7.27) are called β characteristics. We assume that to every pair of Cartesian coordinates (x, y) there corresponds a unique pair of characteristic coordinates (α, β). That is, we view x and y as functions of α and β:

$$x = x(\alpha, \beta), \qquad y = y(\alpha, \beta)$$

Using α as a parameter that varies along a curve defined by (9.7.24) or (9.7.26) to write

$$\frac{dy}{dx} = \frac{y_\alpha}{x_\alpha}$$

we can express (9.7.24) in the form

$$y_\alpha = \lambda_+ x_\alpha$$

Similarly, using β as a parameter that varies along a curve defined by (9.7.25) or (9.7.27), we can express (9.7.25) in the form

$$y_\beta = \lambda_- x_\beta$$

The chain rule can be used to show

$$u_\alpha = u_x x_\alpha + u_y y_\alpha, \qquad\qquad u_\beta = u_x x_\beta + u_y y_\beta$$
$$\lambda_+ a(u_x)_\alpha + c(u_y)_\alpha = f y_\alpha, \qquad \lambda_- a(u_x)_\beta + c(u_y)_\beta = f y_\beta$$

Collecting the preceding results, we have five essentially ordinary differential equations for dealing with the initial-value problem (9.7.21)–(9.7.23) in the $\alpha\beta$ coordinate system. This characteristic system is

$$y_\alpha = \lambda_+ x_\alpha \qquad\qquad (9.7.28)$$
$$y_\beta = \lambda_- x_\beta \qquad\qquad (9.7.29)$$
$$\lambda_+ a(u_x)_\alpha + c(u_y)_\alpha = f y_\alpha \qquad\qquad (9.7.30)$$
$$\lambda_- a(u_x)_\beta + c(u_y)_\beta = f y_\beta \qquad\qquad (9.7.31)$$
$$u_\alpha = u_x x_\alpha + u_y y_\alpha \quad \text{or} \quad u_\beta = u_x x_\beta + u_y y_\beta \qquad (9.7.32)$$

The numerical method of characteristics now proceeds by replacing all α and β derivatives in (9.7.28)–(9.7.32) by difference quotients. For example, with reference to Figure 9.7.3, we have

$$x_\alpha \simeq \frac{x(P) - x(Q)}{\Delta\alpha}, \qquad x_\beta \simeq \frac{x(P) - X(R)}{\Delta\beta}, \qquad (u_x)_\alpha \simeq \frac{u_x(P) - u_x(Q)}{\Delta\alpha}$$

and so on. Then Equations (9.7.28)–(9.2.31) and one of the equations from (9.7.32) produce five algebraic equations in the five unknowns $x(P)$, $y(P)$, $u_x(P)$, $u_y(P)$, $u(P)$.

Exercises

1. Use the exact method of characteristics to find the solution to each of the following initial-value problems along the characteristic that passes through $(x, t) = (1, 0)$. Then use the numerical method of characteristics to estimate the position of that characteristic and the solution on that characteristic at time $t = 0.1, 0.2, \ldots, 1.0$. Use Euler's method to integrate the characteristic equations.

(a) $u_x + u_t = u$, $u(x, 0) = \cos x$

(b) $u_x + xu_t = u^2$, $u(x, 0) = \begin{cases} 1, & 1 < x < 2 \\ 0, & \text{otherwise} \end{cases}$, $x > 0$

(c) $2xtu_t + u_t - u = 0$, $u(x, 0) = x$

(d) $tu_x + u_t = x$, $u(x, 0) = x^2$

(e) $u_x + xu_t = u^2$, $u(x, 0) = 1$

(f) $u_x + u_t = -u$, $u(x, 0) = 1 + \cos x$

2. (a) Repeat Problem 1 using the Runge–Kutta method to integrate the characteristic equations.

(b) Repeat Exercise 1 using a finite-difference method.

3. Consider the initial-value problem

$$2\frac{\partial u_1}{\partial x} + \frac{\partial u_2}{\partial x} + \frac{\partial u_1}{\partial t} = 0$$

$$-\infty < x < \infty, t > 0$$

$$2\frac{\partial u_1}{\partial x} + 3\frac{\partial u_2}{\partial x} + \frac{\partial u_2}{\partial t} = 0,$$

$$u_1(x, 0) = x, \qquad u_2(x, 0) = x^2, \qquad -\infty < x < \infty$$

(a) Find the exact solution

$$u_1(x, t) = \tfrac{1}{3}[2(x - t) - 2(x - t)^2 + (x - 4t) + 2(x - 4t)^2]$$
$$u_2(x, t) = \tfrac{1}{3}[(x - 4t) + 2(x - 4t)^2 - (x - t) + (x - t)^2]$$

(b) Sketch the characteristics through $Q = (1, 0)$ and $R = (2, 0)$.
(c) Use the mumerical method of characteristics to estimate the solution at the intersection of the characteristic of (b).

4. Consider the initial-value problem

$$4\frac{\partial u_1}{\partial x} + \frac{\partial u_2}{\partial x} + \frac{\partial u_1}{\partial t} = 0$$

$$x > 0, t > 0$$

$$\frac{\partial u_1}{\partial x} + 4\frac{\partial u_2}{\partial x} + \frac{\partial u_2}{\partial t} = 0,$$

$$u_1(x, 0) = \sin x, \qquad u_2(x, 0) = x^2, \qquad x > 0$$

(a) Find the exact solution.
(b) Sketch the characteristics through $Q = (1, 0)$ and $R = (2, 0)$.
(c) Use the numerical method of characteristics to estimate the solution at the intersection of the characteristic of (b).

5. Embed the Runge–Kutta Algorithm 9.6 in a DO LOOP that covers an x grid and obtain an algorithm that develops the solution to a quasi-linear first-order equation on a characteristic grid. Then test the algorithm on a grid, $x_n = nh$, $n = 1$, $10h = 0.1$, by reworking Example 9.7.1.

6. Consider the initial-boundary-value problem

$$x^2 u_{xx} - y^2 u_{yy} = 1, \qquad x > 1, y > 1$$
$$u(x, 1) = \log x, \qquad u_y(x, 1) = 2, \qquad x > 1$$
$$u(1, y) = 2 \log y, \qquad y > 1$$

(a) Find the exact solution. [*Hint:* Set $u = X(x) + Y(y)$.]
(b) Sketch the characteristics through $Q = (2, 1)$ and $R = (2.1, 1)$.
(c) Use the numerical method of characteristics to estimate the solution at the intersection of the characteristic of (b).

7. (a) Verify that $u = xy$ solves

$$u_{xx} - u^2 u_{yy} = 0, \qquad x > 0, y > 2$$
$$u(x, 2) = 2x, \qquad x > 0$$
$$u_y(x, 2) = x, \qquad x > 0$$

(b) Determine the characteristics and the first characteristic grid point P between $Q = (1, 2)$ and $R = (2, 2)$.
(c) Use the numerical method of characteristics to obtain the initial approximation to the solution at point P of (b).

Chapter 10

Finite-Difference Methods for Elliptic Equations

In this chapter we discuss the finite-difference method for elliptic partial differential equations. An elliptic equation generally arises in connection with a physical problem involving a system in equilibrium. Thus, the time variable is not present, and the stability of a difference method will not play the central role that it did in Chapters 8 and 9.

When a linear, elliptic partial differential equation is discretized, the immediate result is a system of linear, algebraic equations. In this chapter, our main goal is to provide efficient numerical methods for dealing with systems of finite-difference equations for elliptic partial differential equations.

10.1 DIFFERENCE EQUATIONS FOR ELLIPTIC EQUATIONS

Laplace and Poisson Equations

Laplace's equation, $\nabla^2 u = 0$, is one of the most frequently encountered partial differential equations. In Chapter 1, we saw that Laplace's equation can be used to describe the equilibrium temperature distribution in a heat-conducting medium. Laplace's equation is also encounterd in such diverse areas as ideal-fluid flow, electrostatic and magnetic potential fields, stress and strain analysis of elastic solids, and groundwater transport.

The nonhomogeneous counterpart of Laplace's equation, $\nabla^2 u = f$, is known as *Poisson's equation*. A nonzero right side in Poisson's equation often corresponds to a distributed source or sink in the underlying physical problem.

EXAMPLE 10.1.1 ───────────────────────────────────

Let $u = u(x, y)$ represent the equilibrium temperature distribution in a two-dimensional heat-conducting medium Ω. Fourier's law states that the heat flux vector is given by

$$\mathbf{q} = (qx, qy) = -K \nabla u = (-Ku_x, -Ku_y)$$

where K is known as the thermal conductivity.

Let ω be an arbitrary subregion of Ω and let σ denote the boundary of ω. If there are no heat sources or sinks interior to Ω, then the net flux of heat through σ must be zero. That is,

$$\int_\sigma \mathbf{q} \cdot \mathbf{n} \, d\sigma = 0$$

where $\mathbf{n}$ denotes a unit outward normal to σ. From the divergence theorem,

$$\int_\sigma \mathbf{q} \cdot \mathbf{n} \, d\sigma = \int_\omega \nabla \cdot \mathbf{q} \, d\omega$$

Since ω is arbitrary, we conclude that if $\nabla \cdot \mathbf{q}$ is continuous, then it must be zero throughout Ω. Thus,

$$\nabla \cdot \mathbf{q} = (-Ku_x)_x + (-Ku_y)_y = 0 \quad \text{in } \Omega$$

If K is assumed to be a constant, then it follows that u satisfies Laplace's equation

$$\nabla^2 u = u_{xx} + u_{yy} = 0 \quad \text{in } \Omega$$

If a distributed heat source term $S(x, y)$ adds heat to Ω in such a way as to maintain thermal equilibrium, then the flow of heat out of ω through σ must balance the heat produced by $S(x, y)$ inside ω. That is,

$$\int_\sigma \mathbf{q} \cdot \mathbf{n} \, d\sigma = \int_\omega S(x, y) \, d\omega$$

A negative value of S corresponds to a heat sink instead of a source. Using the divergence theorem to transform the σ integral in the last equation to an integral over ω yields

$$\int_\omega \nabla \cdot \mathbf{q} \, d\omega = \int_\omega S(x, y) \, d\omega \quad \text{or} \quad \int_\omega [\nabla \cdot q - S(x, y)] \, d\omega = 0$$

If $[\nabla \cdot \mathbf{q} - S(x, y)]$ is continuous, we conclude it must be zero throughout Ω. This argument gives

$$(-Ku_x)_x + (-Ku_y) = S(x, y) \quad \text{in } \Omega$$

If we assume K to be a constant, then u is seen to satisfy Poisson's equation,

$$u_{xx} + u_{yy} = f(x, y) \equiv -S(x, y)/K \quad \text{in } \Omega \qquad \blacksquare\blacksquare$$

To begin our discussion of finite-difference methods for Laplace's or Poisson's equation, define a square grid on the xy plane

$$(x_m, y_n) = (mh, nh), \qquad m = 0, 1, \ldots, M + 1, n = 0, 1, \ldots, N + 1$$

Let $X = (M + 1)h$ and $Y = (N + 1)h$ and let Ω be the rectangular region defined by

$$\Omega = \{(x, y) : 0 < x < X, 0 < y < Y\}$$

Consider the Dirichlet boundary value problem

$$u_{xx} + u_{yy} = f(x, y) \quad \text{in } \Omega \tag{10.1.1}$$
$$u = g(x, y) \quad \text{on } S \tag{10.1.2}$$

where S denotes the boundary of Ω. See Figure 10.1.1.

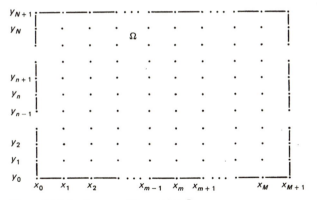

Figure 10.1.1 Square grid on region Ω.

To obtain a system of finite-difference equations for (10.1.1) and (10.1.2) on the grid of Figure 10.1.1, we approximate the derivatives u_{xx} and u_{yy} by centered difference formulas of the form (8.1.11) to obtain

$$\frac{U_{m-1,n} - 2U_{mn} + U_{m+1,n}}{h^2} + \frac{U_{m,n-1} - 2U_{mn} + U_{m,n+1}}{h^2} = f(x_m, y_n) \quad (10.1.3)$$

Using the difference operators δ_x and δ_y, (10.1.3) can be written in the more compact form

$$\delta_x^2 U_{mn} + \delta_y^2 U_{mn} = h^2 f_{mn}, \quad m = 1, 2, \ldots, M, n = 1, 2, \ldots, N \quad (10.1.4)$$

Equation (10.1.4) displays the *computational stencil* for the discrete Laplace difference operator $(\delta_x^2 + \delta_y^2)$. See Figure 10.1.2.

Applying the computational stencil of Figure 10.1.2 to each interior grid point of Figure 10.1.1 and using the boundary condition $u = g$ on S where the stencil touches the boundary yields a set of MN linear equations for the unknown U values,

$$U_{mn}, \quad m = 1, 2, \ldots, M, n = 1, 2, \ldots, N$$

By introducing a single subscript labeling of the unknown U values, we can write this linear system in the form

$$\mathbf{AU} = \mathbf{b} \quad (10.1.5)$$

where $\mathbf{A}$ is a square matrix of order $P = MN$ and $\mathbf{U}$ and $\mathbf{b}$ are column vectors of length P. The U_{mn} can be ordered according to a variety of labeling schemes. For example, we

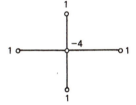

Figure 10.1.2 Five-point computational stencil for discrete Laplace operator.

can label from left to right, bottom to top; from top to bottom, right to left; and so on. In Section 10.5 we discuss a *red–black*, or *checkerboard*, labeling. For the present discussion, let us order the U_{mn} according to the FORTRAN convention that the leftmost subscript varies most rapidly. This provides the singly subscripted vector $\mathbf{U} = [U_1, U_2, \ldots, U_M, U_{M+1}, \ldots, U_{MN}]^T = [U_{11}, U_{21}, \ldots, U_{M1}, U_{12}, \ldots, U_{MN}]^T$, which corresponds to labeling left to right across the row $n = 1$, followed by left to right across row $n = 2$, and so on, left to right across row $n = N$.

EXAMPLE 10.1.2 —————————————————————————————————————

In Figure 10.1.1, suppose that $M = N = 3$. Then there are $P = 9$ interior grid points that can be ordered to form the vector

$$\mathbf{U} = [U_{11}, U_{21}, U_{31}, U_{12}, U_{22}, U_{32}, U_{13}, U_{23}, U_{33}]^T$$
$$= [U_1, U_2, U_3, U_4, U_5, U_6, U_7, U_8, U_9]^T$$

Using the five-point stencil of Figure 10.1.2 to form the linear system (10.1.5), we find

$$
\begin{bmatrix}
-4 & 1 & 0 & 1 & 0 & 0 & 0 & 0 & 0 \\
1 & -4 & 1 & 0 & 1 & 0 & 0 & 0 & 0 \\
0 & 1 & -4 & 0 & 0 & 1 & 0 & 0 & 0 \\
1 & 0 & 0 & -4 & 1 & 0 & 1 & 0 & 0 \\
0 & 1 & 0 & 1 & -4 & 1 & 0 & 1 & 0 \\
0 & 0 & 1 & 0 & 1 & -4 & 0 & 0 & 1 \\
0 & 0 & 0 & 1 & 0 & 0 & -4 & 1 & 0 \\
0 & 0 & 0 & 0 & 1 & 0 & 1 & -4 & 1 \\
0 & 0 & 0 & 0 & 0 & 1 & 0 & 1 & -4
\end{bmatrix}
\begin{bmatrix}
U_1 \\ U_2 \\ U_3 \\ U_4 \\ U_5 \\ U_6 \\ U_7 \\ U_8 \\ U_9
\end{bmatrix}
=
\begin{bmatrix}
b_1 \\ b_2 \\ b_3 \\ b_4 \\ b_5 \\ b_6 \\ b_7 \\ b_8 \\ b_9
\end{bmatrix}
$$

where

$$
\begin{bmatrix}
b_1 \\ b_2 \\ b_3 \\ b_4 \\ b_5 \\ b_6 \\ b_7 \\ b_8 \\ b_9
\end{bmatrix}
=
\begin{bmatrix}
h^2 f_{11} - g_{01} - g_{10} \\
h^2 f_{21} - g_{20} \\
h^2 f_{31} - g_{41} - g_{30} \\
h^2 f_{12} - g_{02} \\
h^2 f_{22} \\
h^2 f_{32} - g_{42} \\
h^2 f_{13} - g_{03} - g_{14} \\
h^2 f_{23} - g_{24} \\
h^2 f_{33} - g_{43} - g_{34}
\end{bmatrix}
$$

For the case $M = N = 3$, the preceding linear system of equations represents the finite-difference system for the Poisson equation (10.1.1) subject to the Dirichlet boundary condition (10.1.2).

The coefficient matrix in the preceding difference system is *block tridiagonal* with $\mathbf{A}$ of the form

$$
\mathbf{A} = [\mathbf{I}; \mathbf{B}; \mathbf{I}] =
\begin{bmatrix}
\mathbf{B} & \mathbf{I} & \mathbf{0} \\
\mathbf{I} & \mathbf{B} & \mathbf{I} \\
\mathbf{0} & \mathbf{I} & \mathbf{B}
\end{bmatrix}
$$

where $\mathbf{I}$ denotes the 3×3 identity matrix and

$$\mathbf{B} = \begin{bmatrix} -4 & 1 & 0 \\ 1 & -4 & 1 \\ 0 & 1 & -4 \end{bmatrix}$$

■ ■

We can make a number of observations regarding the form of the finite-difference system (10.1.5) for the Dirichlet boundary value problem (10.1.1) and (10.1.2).

1. The dimension of $\mathbf{U}$ and $\mathbf{b}$ is equal to the number of finite-difference nodes interior to Ω.
2. The vector $\mathbf{b}$ is determined by the values of the boundary data and the source/sink term in (10.1.1).
3. The matrix $\mathbf{A}$ is square and contains at most five nonzero entries per row.
4. The matrix $\mathbf{A} = [a_{ij}]$ is banded in the sense that there is an integer q such that all of the nonzero entries of $\mathbf{A}$ are confined to the entries a_{ij} for which $|i - j| \leq q$.
5. For a fixed region Ω, the order of the matrix $\mathbf{A}$ and the length of $\mathbf{U}$ and $\mathbf{b}$ increase as the mesh spacing is decreased.
6. The matrix $\mathbf{A}$ is block tridiagonal with the number of diagonal blocks equal to the number of interior nodes in the y direction and the order of each block equal to the number on interior nodes in the x direction.

As Example 10.1.1 illustrates, the finite-difference system for the Dirichlet boundary-value problem (10.1.1) and (10.1.2) consists of a system of linear algebraic equations of the form (10.1.3). This system of difference equations can be uniquely solved provided the coefficient matrix $\mathbf{A}$ is nonsingular.

Theorem 10.1.1. The linear system of finite-difference equations $\mathbf{AU} = \mathbf{b}$ for the Dirichlet boundary-value problem (10.1.1) and (10.1.2) is uniquely solvable.

Proof. From linear algebra, we know that $\mathbf{A}$ is singular if and only if $\lambda = 0$ is an eigenvalue of $\mathbf{A}$. In Section 10.3, we show that all the eigenvalues of the matrix $\mathbf{A}$ are negative.

Derivative Boundary Conditions

Consider the Neumann boundary-value problem

$$u_{xx} + u_{yy} = f(x, y) \quad \text{on } \Omega \qquad (10.1.6a)$$

$$\frac{\partial u}{\partial n} = g(x, y) \quad \text{on } S \qquad (10.1.6b)$$

where Ω is the region shown in Figure 10.1.3 and S is the boundary of Ω. A difference

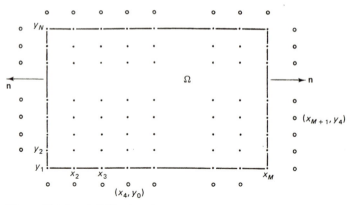

Figure 10.1.3   (•) Grid point where U_{mn} must be found; (○) ghost point.

equation for (10.1.5) is given by

$$\delta_x^2 U_{mn} + \delta_y^2 U_{mn} = h^2 f_{mn}, \qquad m = 1, 2, \ldots, M, \; n = 1, 2, \ldots, N \quad (10.1.7)$$

The truncation error associated with (10.1.7) is of order h^2. If central differences are used for the computation of the boundary derivatives in (10.1.6b), then the overall truncation error for the system of difference equations will be $O(h^2)$. To approximate $\partial u/\partial n$ by central differences requires the ghost points indicated in Figure 10.1.3. The U values at the ghost points can be eliminated using the following discretizations of (10.1.6b):

$$U_{M+1,n} - U_{M-1,n} = 2hg_{Mn}, \qquad n = 0, 1, \ldots, N \qquad (10.1.8)$$
$$U_{m,N+1} - U_{m,N-1} = 2hg_{mN}, \qquad m = 0, 1, \ldots, M \qquad (10.1.9)$$
$$U_{0,n} \qquad - U_{2,n} = 2hg_{1n}, \qquad n = 0, 1, \ldots, N \qquad (10.1.10)$$
$$U_{m,0} \qquad - U_{m,2} = 2hg_{m1}, \qquad m = 0, 1, \ldots, M \qquad (10.1.11)$$

Example 10.1.3 ───

Consider the difference system (10.1.7)–(10.1.11) for the case $M = N = 3$. A difference grid for this problem is shown in Figure 10.1.4. Using the single indexing of Figure 10.1.4, we find the difference system to be

$$\begin{bmatrix} -4 & 2 & 0 & 2 & 0 & 0 & 0 & 0 & 0 \\ 1 & -4 & 1 & 0 & 2 & 0 & 0 & 0 & 0 \\ 0 & 2 & -4 & 0 & 0 & 2 & 0 & 0 & 0 \\ 1 & 0 & 0 & -4 & 2 & 0 & 1 & 0 & 0 \\ 0 & 1 & 0 & 1 & -4 & 1 & 0 & 1 & 0 \\ 0 & 0 & 1 & 0 & 2 & -4 & 0 & 0 & 1 \\ 0 & 0 & 0 & 2 & 0 & 0 & -4 & 2 & 0 \\ 0 & 0 & 0 & 0 & 2 & 0 & 1 & -4 & 1 \\ 0 & 0 & 0 & 0 & 0 & 2 & 0 & 2 & -4 \end{bmatrix} \begin{bmatrix} U_1 \\ U_2 \\ U_3 \\ U_4 \\ U_5 \\ U_6 \\ U_7 \\ U_8 \\ U_9 \end{bmatrix} = \begin{bmatrix} b_1 \\ b_2 \\ b_3 \\ b_4 \\ b_5 \\ b_6 \\ b_7 \\ b_8 \\ b_9 \end{bmatrix}$$

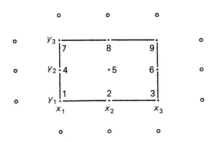

Figure 10.1.4 Difference grid for Example 10.1.3.

where

$$
\begin{bmatrix} b_1 \\ b_2 \\ b_3 \\ b_4 \\ b_5 \\ b_6 \\ b_7 \\ b_8 \\ b_9 \end{bmatrix} = \begin{bmatrix} h^2 f_{11} - 4hg_{11} \\ h^2 f_{21} - 2hg_{21} \\ h^2 f_{31} - 4hg_{31} \\ h^2 f_{12} - 2hg_{12} \\ h^2 f_{22} \\ h^2 f_{32} - 2hg_{32} \\ h^2 f_{13} - 4hg_{13} \\ h^2 f_{23} - 2hg_{23} \\ h^2 f_{33} - 4hg_{33} \end{bmatrix}
$$

The coefficient matrix in this difference system is *block tridiagonal* with A of the form

$$
A = \begin{bmatrix} B & 2I & 0 \\ I & B & I \\ 0 & 2I & B \end{bmatrix}
$$

where I denotes the 3×3 identity matrix and

$$
B = \begin{bmatrix} -4 & 2 & 0 \\ 1 & -4 & 1 \\ 0 & 2 & -4 \end{bmatrix}
$$

■ ■

It is not difficult to show that for the Neumann boundary-value problem, the coefficient matrix in the difference system is singular. One way to see this is by observing that the sum of the entries in each row of A is zero. Therefore, the vector

$$
1 = [1, 1, \ldots, 1]^T
$$

provides a nonzero solution to the homogeneous system $AU = 0$.

We can show that the difference system of Example 10.1.3 has a solution if and only if the *compatibility condition*

$$
\tfrac{1}{2}b_1 + b_2 + \tfrac{1}{2}b_3 + b_4 + 2b_5 + b_6 + \tfrac{1}{2}b_7 + b_8 + \tfrac{1}{2}b_9 = 0
$$

holds. This compatibility condition is the discrete version of the condition

$$
\int_\Omega f(x, y) \, d\Omega = \int_S g(x, y) \, dS
$$

that is required for the solvability of (10.1.6a) and (10.1.6b).

If u is a solution to the Neumann boundary-value problem (10.1.6a) and (10.1.6b), then $u + c$, c = const, is also a solution. Similarly, if U is a solution to the difference system for (10.1.6a) and (10.1.6b), then $U + c\mathbf{1}$, c = const, is also a solution.

Having seen how to approximate a boundary derivative to $O(h^2)$, it is not difficult to incorporate a mixed boundary condition of the form

$$\alpha u + \beta \frac{\partial u}{\partial n} = g \tag{10.1.12}$$

into a difference system for Laplace's or Poisson's equation. A sufficient condition for the unique solvability of (10.1.6a) subject to the mixed condition (10.1.12) is $\alpha\beta > 0$. (See *Schaum's Outline of Partial Differential Equations*, p. 30.)

Material Balance Difference Equations

In many applications involving the time-independent flow of some material, we encounter elliptic boundary-value problems of the form

$$(au_x)_x + (bu_y)_y + cu = f \quad \text{in } \Omega, \qquad ab > 0 \tag{10.1.13a}$$

$$u = g \quad \text{on } S, \tag{10.1.13b}$$

Often the Dirichlet boundary condition (10.1.13b) is replaced by a Neumann or a mixed condition on all or part of S.

EXAMPLE 10.1.4 _____

Consider two-dimensional, horizontal movement of groundwater in an aquifer and assume that the volumetric water flux vector is given by

$$\mathbf{q} = (-a(x, y)u_x, \; -b(x, y)u_y)$$

where a and b represent the hydraulic conductivities in the x and y directions and u is the pressure head in the aquifer. In this connection see Example 8.6.5. If $f(x, y)$ represents the volume of water added or subtracted per unit of plan view area per unit of time, then a mass balance argument leads immediately to (10.1.13a) with $c = 0$.

To illustrate boundary conditions that might accompany such a problem, suppose that the head is observed on three sides of a rectangular flow region Ω while the fourth side is known to be a no-flow boundary. Then the boundary-value problem for the pressure head would be as in Figure (10.1.5). ■ ■

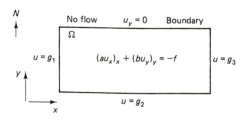

Figure 10.1.5 Schematic of groundwater flow model.

In Section 8.7 we derived material balance difference equations for time-dependent conservation laws involving two space variables. That same development applies to the time-independent conservation law (10.1.13a).

Suppose that the region Ω is covered by a rectangular, cell-centered finite-difference grid

$$(x_m, y_n) = (m \cdot hx, n \cdot hy), \qquad m = 1, 2, \ldots, M, \; n = 1, 2, \ldots, N$$

See Figure 8.6.4. If the values of a and b are available for all values of x and y, then the difference equations for (10.1.13a) are

$$
\begin{aligned}
&[a_{m-1/2,n}U_{m-1,n} - (a_{m-1/2,n} + a_{m+1/2,n})U_{mn} + a_{m+1/2,n}U_{m+1,n}]/hx^2 \\
&+ [b_{m,n-1/2}U_{m,n-1} - (b_{m,n-1/2} + b_{m,n+1/2})U_{mn} \\
&+ b_{m,n+1/2}U_{m,n+1}]/hy^2 + c_{mn}U_{mn} = f_{mn}
\end{aligned}
\tag{10.1.14}
$$

where

$$a_{m\pm1/2,n} = a(x_m \pm hx/2, y_n) \quad \text{and} \quad b_{m,n\pm1/2} = b(x_m, y_n \pm hy/2)$$

If the coefficients a and b are only known at the grid points, then harmonic averages can be used to assign a value of a to each vertical face and a value of b to each horizontal face of each difference cell. Following the approach of Section 8.6, this leads to the difference system

$$
\begin{aligned}
&[\mu_{m-1,n}U_{m-1,n} - (\mu_{m-1,n} + \mu_{m+1,n})U_{mn} + \mu_{m+1,n}U_{m+1,n}]/hx^2 \\
&+ [\mu_{m,n-1}U_{m,n-1} - (\mu_{m,n-1} + \mu_{m,n+1})U_{mn} + \mu_{m,n+1}U_{m,n+1}]/hy^2 + c_{mn}U_{mn} = f_{mn}
\end{aligned}
$$

where

$$\mu_{m-1,n} = \frac{2a_{mn}a_{m-1,n}}{a_{mn} + a_{m-1,n}}, \qquad \mu_{m+1,n} = \frac{2a_{mn}a_{m+1,n}}{a_{mn} + a_{m+1,n}}$$

and

$$\mu_{m,n-1} = \frac{2b_{mn}b_{m,n-1}}{b_{mn} + b_{m,n-1}}, \qquad \mu_{m,n+1} = \frac{2b_{mn}b_{m,n+1}}{b_{mn} + b_{m,n+1}}$$

If $a > 0$, $b > 0$, and $c \leq 0$, then it can be shown that the preceding difference equations for (10.1.13a) supplemented by the boundary condition (10.1.13b) are uniquely solvable provided hx and hy are chosen sufficiently small. (See *Schaum's Outline of Partial Differential Equations*, p. 177.)

Exercises

1. Let Ω be the unit square $0 < x < 1$, $y < 0 < 1$. For each of the following boundary-value problems, obtain a system of nine finite-difference equations. Introduce "ghost points" as necessary and provide a sketch of the grid on which your difference equations hold. A different finite-difference system will result depending on whether a vertex-centered or a cell-centered grid is chosen.

  (a) $\nabla^2 u = 0$ in Ω
    $u(0, y) = 0$, $u(1, y) = 1$, $0 < y < 1$
    $u_y(x, 0) = 0$, $u_y(x, 1) = 5$, $0 < x < 1$
  (b) $\nabla^2 u = f$ in Ω
    $u_x(0, y) = 0$, $u_x(1, y) = 1$, $0 < y < 1$
    $u(x, 0) = 0$, $u(x, 1) = 5$, $0 < x < 1$
  (c) $\nabla^2 u = f$ in Ω
    $u_x(0, y) = 0$, $u(1, y) = 0$, $0 < y < 1$
    $u_y(x, 0) = 0$, $u(x, 1) = 0$, $0 < x < 1$
  (d) $\nabla^2 u = f$ in Ω
    $u_x(0, y) = 0$, $u(1, y) = 1$, $0 < y < 1$
    $u_y(x, 0) = 0$, $u_y(x, 1) = 0$, $0 < x < 1$

2. Using a rectangular grid $(x_m, y_n) = (mhx, nhy)$, write a finite-difference equation for
  (a) Laplace's equation
  (b) Poisson's equation
  (c) $\nabla^2 u + cu = f$

3. Consider the equation $\nabla \cdot [\mathbf{T} \nabla u] = 0$ where $\mathbf{T}$ denotes a constant-coefficient, positive definite 2×2 tensor

$$\mathbf{T} = \begin{bmatrix} t_{11} & t_{12} \\ t_{12} & t_{22} \end{bmatrix}$$

  (a) Show that $u = u(x, y)$ satisfies

$$t_{11} u_{xx} + 2t_{12} u_{xy} + t_{22} u_{yy} = 0$$

  (b) Obtain a finite-difference equation for (a) using a square grid.
  (c) Obtain a finite-difference equation for (a) using a rectangular grid.
  (d) Use the fact that $\mathbf{T}$ is positive definite (both eigenvalues of $\mathbf{T}$ are positive) to show that the partial differential equation of part (a) is elliptic.

4. Write the difference system of Example 10.1.2 in which the unknown U_p are ordered
  (a) by columns with $\mathbf{U} = [U_1, U_4, U_7, U_2, U_5, U_8, U_3, U_6, U_9]^T$
  (b) in a checkerboard, or red–black, fashion with

$$\mathbf{U} = [U_1, U_3, U_5, U_7, U_9, U_2, U_4, U_6, U_8]^T$$

5. Write the coefficient matrix of Example 10.1.2 in block tridiagonal form using only the identity matrix and the matrix $\mathbf{F}_1$ of Section 8.4.

6. Write the coefficient matrix of Example 10.1.3 in block tridiagonal form using only the identity matrix and the matrix $\mathbf{F}_2$ of Section 8.4.

7. Write a system of nine difference equations for the boundary-value problem

$$\begin{aligned}
(e^x u_x)_x + (e^{xy} u_y)_y &= f, & 0 < x < 1, 0 < y < 1 \\
u(0, y) &= 0, & u(1, y) = 0, & \quad 0 < y < 1 \\
u(x, 0) &= 0, & u(x, 1) = 0, & \quad 0 < x < 1
\end{aligned}$$

10.2 DIRECT SOLUTION OF LINEAR EQUATIONS

In Section 10.1, we saw that the finite-difference equations for an elliptic boundary-value problem takes the form of a system of linear algebraic equations,

$$
\begin{aligned}
a_{11}U_1 + a_{12}U_2 + \cdots a_{1P}U_P &= b_1 \\
a_{21}U_1 + a_{22}U_2 + \cdots a_{2P}U_P &= b_2 \\
\vdots \qquad \vdots \qquad \quad \vdots \qquad \quad \vdots \\
a_{P1}U_1 + a_{P2}U_2 + \cdots a_{PP}U_P &= b_P
\end{aligned}
\qquad (10.2.1)
$$

which can be written in terms of the $P \times P$ matrix $\mathbf{A} = [a_{ij}]$ and the column vectors $\mathbf{U}$ and $\mathbf{b}$ as

$$
\mathbf{AU} = \mathbf{b} \qquad (10.2.2)
$$

Thus, the problem of solving the finite-difference equations for an elliptic boundary-value problem reduces immediately to the linear algebra problem of solving a system of linear equations.

Roughly speaking, there are two types of methods for determining the solution of (10.2.1) or (10.2.2). A method for solving (10.2.2) is called a *direct method* if it produces the exact solution of (10.2.2) (up to rounding errors) in a finite number of algebraic operations. The basic direct method is *Gaussian elimination*. Most other direct methods are variations of Gaussian elimination. Examples include Gaussian elimination with pivoting and various $\mathcal{LU}$ factorization approaches that rely on writing $\mathbf{A}$ as a product of a lower triangular matrix $\mathcal{L}$ and an upper triangular matrix $\mathcal{U}$. In this section we discuss some direct methods that are particularly well suited for solving finite-difference systems of the form (10.2.1).

An *iterative method* (or *indirect method*) for (10.2.1) generates a sequence of approximations to the solution of the algebraic equations. In contrast to direct methods, iterative methods generally do not produce the exact solution to the linear system in a finite number of algebraic operations. In the next section, we discuss a number of iterative methods that are widely applied to systems of finite-difference equations.

Direct methods are usually reserved for relatively small systems of linear equations because their computer storage requirements usually exceed those of an iterative method. Also, iterative methods are generally easier to code than direct methods. However, when the exact solution to the difference system is desired, the method of choice is usually a direct method.

Gaussian Elimination

We have already developed one algorithm based on Gaussian elimination, namely, Algorithm 8.2, which solves a system of linear equations in the case that the coefficient matrix is tridiagonal.

Recall that to use Gaussian elimination to solve (10.2.2), we first form the augmented matrix

$$[A; b] = \begin{bmatrix} a_{11} & a_{12} & \cdots & a_{1P} & b_1 \\ a_{21} & a_{22} & \cdots & a_{2P} & b_2 \\ \vdots & \vdots & & \vdots & \vdots \\ a_{P1} & a_{P2} & \cdots & a_{PP} & b_P \end{bmatrix}$$

The Gaussian elimination method consists of two parts. In the first part, multiples of the first row of the augmented matrix are added to the rows 2 through P to eliminate all entries in column 1 below the diagonal of A. Then multiples of the present second row are added to rows 3 through P to eliminate all entries in column 2 below the diagonal. This is continued until all subdiagonal entries of A have been brought to zero. Then, with the coefficient matrix in upper triangular form, it is an easy matter to back substitute and determine the components of the solution vector in the order U_P, $U_{P-1}, \ldots, U_2, U_1$.

The only way the basic Gaussian elimination procedure outlined in the preceding can fail is if, at some stage of the forward substitution, a zero is produced on a diagonal entry of A. If this situation arises, then either the matrix A is singular, in which case (10.2.1) fails to have a unique solution, or the equations (10.2.1) can be reordered to produce a nonzero diagonal entry so that the elimination procedure can proceed. This reordering of the equations by interchanging two rows of the matrix A is called *pivoting*. When no pivoting is required, the implementation of Gaussian elimination is a very straightforward matter. As the next result shows, most of the linear systems that we encounter in applications of finite differences can be solved by Gaussian elimination with no pivoting.

To state the next theorem, we require the notion of a *reducible matrix*. We say the $P \times P$ matrix A of (10.2.2) is *reducible* if some m, $m < P$, components of b uniquely determine m components of the solution U. Thus, we can describe a matrix A as *irreducible* if a change in any component of b affects every component of U. In the context of difference equations for elliptic boundary-value problems, a finite-difference matrix is reducible if some of the U values on the interior nodes are independent of some of the boundary conditions. A well-posed elliptic problem generally leads to an irreducible finite difference matrix A.

Theorem 10.2.1. If the matrix A in (10.2.2) is irreducible and satisfies (for $i = 1, 2, \ldots, P$)

(i) $a_{ii} < 0$, $a_{ij} \geq 0$ for $i \neq j$

(ii) $|a_{ii}| \geq \displaystyle\sum_{\substack{j=1 \\ j \neq i}}^{P} a_{ij}$

with strict inequality in (ii) for some i, then the system (10.2.2) can be solved uniquely by Gaussian elimination with no pivoting. Moreover the Gaussian elimination procedure is stable with respect to growth of rounding errors.

EXAMPLE 10.2.1

Consider the difference system of Example 10.1.2,

$$
\begin{bmatrix}
-4 & 1 & 0 & 1 & 0 & 0 & 0 & 0 & 0 \\
1 & -4 & 1 & 0 & 1 & 0 & 0 & 0 & 0 \\
0 & 1 & -4 & 0 & 0 & 1 & 0 & 0 & 0 \\
1 & 0 & 0 & -4 & 1 & 0 & 1 & 0 & 0 \\
0 & 1 & 0 & 1 & -4 & 1 & 0 & 1 & 0 \\
0 & 0 & 1 & 0 & 1 & -4 & 0 & 0 & 1 \\
0 & 0 & 0 & 1 & 0 & 0 & -4 & 1 & 0 \\
0 & 0 & 0 & 0 & 1 & 0 & 1 & -4 & 1 \\
0 & 0 & 0 & 0 & 0 & 1 & 0 & 1 & -4
\end{bmatrix}
\begin{bmatrix}
U_1 \\ U_2 \\ U_3 \\ U_4 \\ U_5 \\ U_6 \\ U_7 \\ U_8 \\ U_9
\end{bmatrix}
=
\begin{bmatrix}
b_1 \\ b_2 \\ b_3 \\ b_4 \\ b_5 \\ b_6 \\ b_7 \\ b_8 \\ b_9
\end{bmatrix}
$$

Using the fact that the coefficient matrix in this difference system is *block tridiagonal* with A of the form

$$
A = [I; B; I] = \begin{bmatrix}
B & I & 0 \\
I & B & I \\
0 & I & B
\end{bmatrix}
$$

where I denotes the 3×3 identify matrix and

$$
B = \begin{bmatrix}
-4 & 1 & 0 \\
1 & -4 & 1 \\
0 & 1 & -4
\end{bmatrix}
$$

we can show that A is irreducible.

Here, $a_{ii} = -4$ for $i = 1, 2, \ldots, 9$ and (i) and (ii) of Theorem 10.2.1 are seen to hold. Also, strict inequality in (ii) holds for row $i = 1$. In fact strict inequality holds for all rows except row $i = 5$. Theorem 10.2.1 guarantees that the example linear system can be solved by Gaussian elimination with no row interchanges. ■■

The following algorithm implements the Gaussian elimination method with no pivoting. A warning is printed if a zero is encountered on the diagonal during the forward substitution.

ALGORITHM 10.1 Gaussian Elimination

Step 1. Document

This algorithm uses Gaussian elimination to solve a system of P linear equations $AU = b$ in the case that A satisfies the hypotheses of Theorem 10.2.1.

INPUT

Array $A = [a_{ij}]$, $i, j = 1, 2, \ldots, P$, coefficient matrix
Array $b_1, b_2, \ldots, b_P$, right side (stored in augmented
   matrix format, $b_1 = a_{1,P+1}, b_2 = a_{2,P+1}, \ldots, b_P = a_{P,P+1}$)

OUTPUT

U, solution to the linear system
or a message that **A** may be singular

Step 2. Check pivot element for a divide by zero

For $i = 1, 2, \ldots, P - 1$
If $a_{ii} = 0$ then
    print a warning that **A** may be singular and exit
otherwise perform steps 3–5 (forward substitution)

Step 3. Begin forward substitution

For $k = i + 1, i + 2, \ldots, P$
    If $a_{ki} \neq 0$ then perform steps 4 and 5

Step 4. Find multiplier r_{ki} for elimination of a_{ki}

$$r_{ki} = a_{ki}/a_{ii}$$

Step 5. Subtract r_{ki} times equation i from equation k

For $j = i, i + 1, i + 2, \ldots, P + 1$ set
$$a_{kj} = a_{kj} - r_{kj} \cdot a_{ij}$$

Step 6. Check for singular matrix A

If $a_{PP} = 0$ then
    print a message that **A** is singular
otherwise perform steps 7 and 8 (backward substitution)

Step 7. Initialize backward substitution

Set $U_P = a_{P,P+1}/a_{PP}$

Step 8. Back substitute and store solution in array U

For $i = P - 1, P - 2, \ldots, 1$ set
$$U_i = \frac{1}{a_{ii}} \left[a_{i,P+1} - \sum_{j=i+1}^{P} a_{ij} U_j \right]$$

Step 9. Output solution

For $i = 1, 2, \ldots, P$
    Print U_i

EXAMPLE 10.2.2 ————————————————————————————————

To illustrate an application of Algorithm 10.1, consider the linear system of equations of Example 10.2.1 and suppose that all entries in **b** are zero except b_5, which is given by $b_5 = -64$. The result of using Algorithm 10.1 to solve that system is the solution vector.

$$\begin{bmatrix} U_1 \\ U_2 \\ U_3 \\ U_4 \\ U_5 \\ U_6 \\ U_7 \\ U_8 \\ U_9 \end{bmatrix} = \begin{bmatrix} 4.00 \\ 8.00 \\ 4.00 \\ 8.00 \\ 24.00 \\ 8.00 \\ 4.00 \\ 8.00 \\ 4.00 \end{bmatrix}$$

■ ■

An inspection of Algorithm 10.1 reveals a number of inefficiencies. For example, the array **U** is not required since the solution could be written back into the column vector $a_{i,P+1}, i = 1, 2, \ldots, P$. Also, Algorithm 10.1 does not recognize the bandedness of the matrix **A**. As a result, the forward substitution inspects a number of entries that are known to be zero because they lie outside the bandwidth of **A**. The elimination of these inefficiencies is left to the exercises. See Exercise 5.

For most problems of small to moderate size, the basic Gaussian elimination method described in Algorithm 10.1 is a suitable approach for obtaining the solution to a system of finite-difference equations. However, in some applications the solution of systems of the form (10.2.2) may dominate the computation. In these cases it is necessary to develop numerical methods for solving (10.2.2) that are more efficient than Algorithm 10.1.

$\mathscr{L}\mathscr{U}$ Factorizations

A matrix $\mathscr{L} = [l_{ij}]$ is called *lower triangular* if $i < j$ implies $l_{ij} = 0$. Similarly, a matrix $\mathscr{U} = [u_{ij}]$ is called *upper triangular* if $i > j$ implies $u_{ij} = 0$.

EXAMPLE 10.2.3 ————————————————————————————————

A matrix of the form

$$\mathscr{L} = \begin{bmatrix} l_{11} & 0 & 0 & 0 \\ l_{21} & l_{22} & 0 & 0 \\ l_{31} & l_{32} & l_{33} & 0 \\ l_{41} & l_{42} & l_{43} & l_{44} \end{bmatrix}$$

is lower triangular and a matrix of the form

$$\mathscr{U} = \begin{bmatrix} u_{11} & u_{12} & u_{13} & u_{14} \\ 0 & u_{22} & u_{23} & u_{24} \\ 0 & 0 & u_{33} & u_{34} \\ 0 & 0 & 0 & u_{44} \end{bmatrix}$$

is upper triangular.

■ ■

Given a system of linear equations $\mathbf{AU} = \mathbf{b}$, it is often the case that the coefficient matrix $\mathbf{A}$ can be factored into the product of a lower triangular matrix $\mathscr{L}$ times and upper triangular matrix $\mathscr{U}$ with $\mathbf{A} = \mathscr{L}\mathscr{U}$. If $\mathbf{A}$ is nonsingular and $\mathbf{A} = \mathscr{L}\mathscr{U}$, then it follows that l_{ii} and u_{ii} are nonzero for $i = 1, 2, \ldots, P$. If the system (10.2.2) is written in the factored form

$$\mathscr{L}\mathscr{U}\mathbf{U} = \mathbf{b} \tag{10.2.3}$$

then the solution $\mathbf{U}$ can be easily obtained by the following variant of Gaussian elimination. Let $\mathbf{V}$ be defined by

$$\mathbf{V} = \mathscr{U}\mathbf{U}$$

the components of $\mathbf{V}$ can be determined recursively from

$$V_1 = b_1/l_{11}$$
$$V_2 = [b_2 - l_{21}V_1]/l_{22}$$
$$V_3 = [b_3 - l_{31}V_1 - l_{32}V_2]/l_{33}$$
$$\vdots$$

$$V_m = \left[b_m - \sum_{k=1}^{m-1} l_{mk}V_k\right]\bigg/ l_{mm}, \qquad m = 2, 3, \ldots, P \tag{10.2.4}$$

The calculation of $\mathbf{V}$ corresponds to the forward-substitution part of Gaussian elimination.

Once $\mathbf{V}$ is known, the unknown vector $\mathbf{U}$ in (10.2.3) can be found from $\mathscr{U}\mathbf{U} = \mathbf{V}$, which leads to the recursion

$$U_P = V_P/u_{PP}$$
$$U_{P-1} = [V_{P-1} - u_{P-1,P}U_P]/u_{P-1,P-1}$$
$$U_{P-2} = [V_{P-2} - u_{P-2,P-1}U_{P-1} - u_{P-2,P}U_p]/u_{P-2,P-2}$$
$$\vdots$$

$$U_i = \left[V_i - \sum_{k=i+1}^{P} u_{ik}U_k)\right]\bigg/ u_{ii}, \qquad i = P - 1, P - 2, \ldots, 2, 1 \tag{10.2.5}$$

The preceding calculation of $\mathbf{U}$ corresponds to the backward-substitution part of Gaussian elimination.

The following result tells us that whenever a coefficient matrix $\mathbf{A}$ is such that Algorithm 10.1 can be successfully implemented, then an $\mathscr{L}\mathscr{U}$ factorization of $\mathbf{A}$ can be obtained.

Theorem 10.2.2. If Gaussian elimination can be performed on the system $\mathbf{AU} = \mathbf{b}$ without any row interchanges, then the matrix $\mathbf{A}$ can be factored into the product of a lower triangular matrix $\mathscr{L}$ and upper triangular matrix $\mathscr{U}$. Moreover, if r_{kj} are the multipliers constructed in step 3 of Algorithm 10.1 and $\bar{a}_{kj}$ denote the values stored on or above the diagonal of $\mathbf{A}$ at the completion Algorithm 10.1, then one factorization is given by

$$\mathcal{L} = \begin{bmatrix} 1 & 0 & 0 & 0 & \cdots & 0 \\ r_{21} & 1 & 0 & 0 & \cdots & 0 \\ r_{31} & r_{32} & 1 & 0 & \cdots & 0 \\ \vdots & \vdots & \vdots & \vdots & & \vdots \\ r_{P1} & r_{P2} & r_{P3} & r_{P4} & \cdots & 1 \end{bmatrix}$$

and

$$\mathcal{U} = \begin{bmatrix} \tilde{a}_{11} & \tilde{a}_{12} & \tilde{a}_{13} & \tilde{a}_{14} & \cdots & \tilde{a}_{1P} \\ 0 & \tilde{a}_{22} & \tilde{a}_{23} & \tilde{a}_{24} & \cdots & \tilde{a}_{2P} \\ 0 & 0 & \tilde{a}_{33} & \tilde{a}_{34} & \cdots & \tilde{a}_{3P} \\ \vdots & \vdots & \vdots & \vdots & & \vdots \\ 0 & 0 & 0 & 0 & \cdots & \tilde{a}_{PP} \end{bmatrix}$$

Theorem 10.2.2 together with Theorem 10.2.1 assures us that for most systems of the form $\mathbf{AU} = \mathbf{b}$ that are difference equations for an elliptic boundary-value problem an $\mathcal{LU}$ decomposition of $\mathbf{A}$ is available. When the system (10.2.2) must be solved for a fixed $\mathbf{A}$ and several different right sides $\mathbf{b}$, then an $\mathcal{LU}$ decomposition of $\mathbf{A}$ is usually preferred over repeated applications of Algorithm 10.1.

Exercises

1. Apply Gaussian elimination, Algorithm 10.1, to find the solution of each of the following systems

(a) $\begin{bmatrix} -4 & 1 & 0 \\ 1 & -4 & 1 \\ 0 & 1 & -4 \end{bmatrix} \begin{bmatrix} U_1 \\ U_2 \\ U_3 \end{bmatrix} = \begin{bmatrix} -2 \\ -4 \\ -10 \end{bmatrix}$

(b) $\begin{bmatrix} -4 & 2 & 0 \\ 1 & -4 & 1 \\ 0 & 2 & -4 \end{bmatrix} \begin{bmatrix} U_1 \\ U_2 \\ U_3 \end{bmatrix} = \begin{bmatrix} -2 \\ -2 \\ -2 \end{bmatrix}$

(c) $\begin{bmatrix} 4 & -1 & 0 & -1 & 0 & 0 \\ -1 & 4 & -1 & 0 & -1 & 0 \\ 0 & -1 & 4 & 0 & 0 & -1 \\ -1 & 0 & 0 & 4 & -1 & 0 \\ 0 & -1 & 0 & -1 & 4 & -1 \\ 0 & 0 & -1 & 0 & -1 & 4 \end{bmatrix} \begin{bmatrix} U_1 \\ U_2 \\ U_3 \\ U_4 \\ U_5 \\ U_6 \end{bmatrix} = \begin{bmatrix} 1 \\ 1 \\ 1 \\ 0 \\ 0 \\ 0 \end{bmatrix}$

$\begin{bmatrix} -4 & 2 & 0 & 1 & 0 & 0 & 0 & 0 & 0 \\ 1 & -4 & 1 & 0 & 1 & 0 & 0 & 0 & 0 \\ 0 & 2 & -4 & 0 & 0 & 1 & 0 & 0 & 0 \\ 1 & 0 & 0 & -4 & 2 & 0 & 1 & 0 & 0 \\ 0 & 1 & 0 & 1 & -4 & 1 & 0 & 1 & 0 \\ 0 & 0 & 1 & 0 & 2 & -4 & 0 & 0 & 1 \\ 0 & 0 & 0 & 1 & 0 & 0 & -4 & 2 & 0 \\ 0 & 0 & 0 & 0 & 1 & 0 & 1 & -4 & 1 \\ 0 & 0 & 0 & 0 & 0 & 1 & 0 & 2 & -4 \end{bmatrix} \begin{bmatrix} U_1 \\ U_2 \\ U_3 \\ U_4 \\ U_5 \\ U_6 \\ U_7 \\ U_8 \\ U_9 \end{bmatrix} = \begin{bmatrix} 0 \\ 0 \\ 0 \\ 0 \\ 0 \\ 0 \\ 4 \\ 4 \\ 4 \end{bmatrix}$

2. Obtain an $\mathcal{LU}$ factorization of each of the coefficient matrices of Exercise 1.

3. Using the $\mathcal{LU}$ factorizations from Exercise 2, find the solutions to each of the linear systems of Exercise 1.

4. A matrix **A** is said to have bandwidth $2q + 1$ if $a_{ij} = 0$ for $|i - j| > q$. In a difference equation for Laplace's equation, show that q is equal to the maximum of the differences of the subscripts of the U_n that lie on any five-point stencil.

5. Suppose that **A** has bandwidth $2q + 1$.
   (a) Modify Algorithm 10.1 so that subdiagonal entries outside the bandwidth are not involved in the forward substitution.
   (b) Modify Algorithm 10.1 so that superdiagonal entries outside the bandwidth are not involved in the backward substitution.
   (c) Modify Algorithm 10.1 so that the solution is stored in the locations that were allocated to **b** on input.

6. Let Ω be the unit square $0 < x < 1$, $y < 0 < 1$. For each of the following boundary-value problems obtain a system of nine finite difference equations. Introduce ghost points as necessary and provide a sketch of the grid on which your difference equations hold. Then solve the resulting difference equation using Algorithm 10.1.
   (a) $\nabla^2 u = 0$ in Ω
   $$u(0, y) = 0, \quad u(1, y) = 1, \quad 0 < y < 1$$
   $$u_y(x, 0) = 0, \quad u_y(x, 1) = 5, \quad 0 < x < 1$$
   (b) $\nabla^2 u = xy$ in Ω
   $$u_x(0, y) = 0, \quad u_x(1, y) = 1, \quad 0 < y < 1$$
   $$u(x, 0) = 0, \quad u(x, 1) = 5, \quad 0 < x < 1$$
   (c) $\nabla^2 u = 0$ in Ω
   $$u_x(0, y) = 0, \quad u(1, y) = y, \quad 0 < y < 1$$
   $$u_y(x, 0) = 0, \quad u(x, 1) = 0, \quad 0 < x < 1$$
   (d) $\nabla^2 u = 0$ in Ω
   $$u_x(0, y) = 0, \quad u(1, y) = 1, \quad 0 < y < 1$$
   $$u_y(x, 0) = 0, \quad u_y(x, 1) = 0, \quad 0 < x < 1$$

10.3 FOURIER'S METHOD

Eigenvalues and Eigenvectors

The finite-difference systems that were formulated for Laplace's and Poisson's equations in Section 10.1 can be solved exactly by an eigenvector expansion technique similar to the one used in Section 8.4. Because the eigenvector expansion method produces the exact solution to the difference equations in a finite number of operations, it is a direct method for solving the difference equations.

For small systems of difference equations the eigenvector expansion method, or the Fourier method, of this section can be considered as an alternative to the direct methods presented in Section 10.2. For large systems of difference equations, Fourier's method of solution cannot compete with the standard direct methods of Sections 10.2 unless some version of the fast Fourier transform (FFT) is incorporated. See Section 2.4.

To preview the approach to be used in this section, suppose that **A** is a $P \times P$

matrix that results from formulating a system of finite-difference equations in the matrix format

$$AU = b \qquad (10.3.1)$$

For simplicity, let us begin with the assumption that A is symmetric, $A = A^T$. Then A has P mutually orthogonal eigenvectors $V_1, V_2, \ldots, V_P$ satisfying

$$AV_p = \lambda_p V_p, \qquad V_p \cdot V_q = 0 \quad \text{for } p \neq q \qquad (10.3.2)$$

If we know the eigenvalues and eigenvectors of A, then the system (10.3.1) is easily solved as follows.

Write U in the form of an eigenfunction expansion

$$U = \sum_{p=1}^{P} c_p V_p \qquad (10.3.3)$$

in which the coefficients c_p are to be determined. Putting the expansion (10.3.3) into (10.3.1) and using (10.3.2), we find

$$A \left[\sum_{p=1}^{P} c_p V_p \right] = \sum_{p=1}^{P} c_p A V_p = \sum_{p=1}^{P} c_p \lambda_p V_p = b$$

If we multiply both sides of the last equality by V_q and use the orthogonality relationship of (10.3.2), then the coefficients c_p are seen to be

$$c_p = \frac{1}{\lambda_p} \frac{V_p \cdot b}{V_p \cdot V_p}, \qquad \lambda_p \neq 0$$

and the solution of (10.3.1) is complete.

Reviewing the solution procedure outlined in the preceding, it is clear that the crucial step is the determination of the eigenvalues and eigenvectors of the matrix A. The next result allows us to find the eigenvalues and eigenvectors of A for a fairly large class of finite-difference matrices that are block tridiagonal.

Theorem 10.3.1. Let B be an $M \times M$ matrix with M distinct eigenvalues $\gamma_1, \gamma_2, \ldots, \gamma_M$ and corresponding eigenvectors $U_1, U_2, \ldots, U_M$. Suppose A is an $N \times N$ block tridiagonal matrix of the form

$$A = \begin{bmatrix} B & I & & & & 0 \\ I & B & I & & & \\ & I & B & I & & \\ & & I & B & \ddots & \\ & & & \ddots & \ddots & \\ 0 & & & I & B & I \\ & & & & I & B \end{bmatrix}$$

Then the eigenvalues of A are

$$\lambda_{mn} = \gamma_m + 2 \cos \frac{n\pi}{N+1}, \qquad m = 1, 2, \ldots, M, n = 1, 2, \ldots, N$$

and the corresponding eigenvectors are

$$
\mathbf{V}_{mn} =
\begin{bmatrix}
\sin \dfrac{n\pi}{N+1}\,\mathbf{U}_m \\[2mm]
\sin \dfrac{n\pi 2}{N+1}\,\mathbf{U}_m \\[2mm]
\sin \dfrac{n\pi 3}{N+1}\,\mathbf{U}_m \\[2mm]
\vdots \\[2mm]
\sin \dfrac{n\pi N}{N+1}\,\mathbf{U}_m
\end{bmatrix},
\qquad m = 1, 2, \ldots, M,\; n = 1, 2, \ldots, N
$$

Proof. Motivated by the form of **A**, we look for an eigenvector of **A** in the form

$$
\mathbf{V} = [\alpha_1\mathbf{U}, \alpha_2\mathbf{U}, \alpha_3\mathbf{U}, \ldots, \alpha_N\mathbf{U}]^T
$$

where **U** is an eigenvector of **B** with $\mathbf{BU} = \gamma\mathbf{U}$. Putting this vector **V** into $\mathbf{AV} = \lambda\mathbf{V}$, we find that the vector $\boldsymbol{\alpha} = [\alpha_1, \alpha_2, \ldots, \alpha_N]^T$ must satisfy $\mathbf{G}\boldsymbol{\alpha} = \lambda\boldsymbol{\alpha}$ or

$$
\begin{bmatrix}
\gamma & 1 & & & & \\
1 & \gamma & 1 & & \mathbf{0} & \\
 & 1 & \gamma & 1 & & \\
 & & 1 & \gamma & \ddots & \\
 & \mathbf{0} & & \ddots & \gamma & 1 \\
 & & & & 1 & \gamma
\end{bmatrix}
\begin{bmatrix}
\alpha_1 \\ \alpha_2 \\ \alpha_3 \\ \vdots \\ \alpha_{N-1} \\ \alpha_N
\end{bmatrix}
= \lambda
\begin{bmatrix}
\alpha_1 \\ \alpha_2 \\ \alpha_3 \\ \vdots \\ \alpha_{N-1} \\ \alpha_N
\end{bmatrix}
$$

Since $\mathbf{G} = (\gamma + 2)\mathbf{I} - \mathbf{F}_1$, it follows from Theorem 8.4.1 that $\boldsymbol{\alpha}$ has the form

$$
\boldsymbol{\alpha} = \left[\sin \frac{n\pi}{N+1}, \sin \frac{n\pi 2}{N+1}, \sin \frac{n\pi 3}{N+1}, \ldots, \sin \frac{n\pi N}{N+1} \right]^T
$$

and

$$
\lambda = \gamma + 2 \cos \frac{n\pi}{N+1}
$$

which completes the proof.

EXAMPLE 10.3.1

To illustrate Theorem 10.3.1, consider the matrix

$$
\mathbf{A} =
\begin{bmatrix}
-4 & 1 & 0 & 1 & 0 & 0 & 0 & 0 & 0 \\
1 & -4 & 1 & 0 & 1 & 0 & 0 & 0 & 0 \\
0 & 1 & -4 & 0 & 0 & 1 & 0 & 0 & 0 \\
1 & 0 & 0 & -4 & 1 & 0 & 1 & 0 & 0 \\
0 & 1 & 0 & 1 & -4 & 1 & 0 & 1 & 0 \\
0 & 0 & 1 & 0 & 1 & -4 & 0 & 0 & 1 \\
0 & 0 & 0 & 1 & 0 & 0 & -4 & 1 & 0 \\
0 & 0 & 0 & 0 & 1 & 0 & 1 & -4 & 1 \\
0 & 0 & 0 & 0 & 0 & 1 & 0 & 1 & -4
\end{bmatrix}
$$

This matrix in the preceding difference system is *block tridiagonal* with **A** of the form

$$A = [I; B; I] = \begin{bmatrix} B & I & 0 \\ I & B & I \\ 0 & I & B \end{bmatrix}$$

where **I** denotes the 3×3 identity matrix and

$$B = \begin{bmatrix} -4 & 1 & 0 \\ 1 & -4 & 1 \\ 0 & 1 & -4 \end{bmatrix}$$

Since $B = -2I - F_1$, it follows from Theorem 8.4.1 that the eigenvalues of **B** are given by

$$\gamma_m = -4 + 2 \cos \frac{m\pi}{3 + 1}, \qquad m = 1, 2, 3$$

Now, according to Theorem 10.3.1, the eigenvalues of **A** are given by

$$\lambda_{mn} = -4 + 2 \cos \frac{m\pi}{3 + 1} + 2 \cos \frac{n\pi}{3 + 1}, \qquad m, n = 1, 2, 3 \qquad \blacksquare \blacksquare$$

EXAMPLE 10.3.2 _____

To provide an alternate method for finding the eigenvalues and eigenvectors of the matrix **A** of Example 10.3.1, we consider a partial differential equation eigenvalue problem of the form

$$\nabla^2 u = \mu u \quad \text{in } \Omega$$
$$u = 0 \quad \text{on } S$$

where Ω is the unit square, $0 < x < 1$, $0 < y < 1$, and μ is a constant.

If we introduce a square finite-difference grid with $h = \frac{1}{4}$, then there are nine interior grid points on Ω. On this grid, the difference equations for this eigenvalue problem are

$$\delta_x^2 V_{mn} + \delta_y^2 V_{mn} = \lambda_{mn} V_{mn}, \qquad m, n = 1, 2, 3$$

and the homogeneous Dirichlet boundary conditions are

$$V_{0,n} = 0 = V_{4,n}, \qquad n = 1, 2, 3$$
$$V_{m,0} = 0 = V_{m,4}, \qquad m = 1, 2, 3$$

If we write V_{mn} in the form $V_{mn} = \phi_m \psi_n$ to separate the indices, then the difference equation for V_{mn} yields

$$\psi_n \, \delta_x^2 \Phi_m + \Phi_m \, \delta_y^2 \Psi_n = \lambda_{mn} \Phi_m \Psi_n$$

or

$$\frac{\delta_x^2 \Phi_m}{\Phi_m} + \frac{\delta_y^2 \Psi_n}{\Psi_n} = \lambda_{mn}$$

Theorem 10.3.1 implies λ_{mn} has the form $\lambda_{mn} = \lambda_m + \sigma_n$, so

$$\frac{\delta_x^2 \Phi_m}{\Phi_m} - \lambda_m = -\frac{\delta_y^2 \Psi_n}{\Psi_n} + \sigma_n$$

Since m and n can vary independently the ratio $\delta_x^2 \Phi_m/\Phi_m$ must be n-independent, $\delta_x^2 \Phi_m/\Phi_m = \alpha_m$. Similarly, $\delta_y^2 \Psi_n/\Psi_n$ must be m-independent, $\delta_y^2 \Psi_n/\Psi_n = \beta_n$. Incorporating the boundary conditions

$$\Phi_0 = 0 = \Phi_4 \quad \text{and} \quad \Psi_0 = 0 = \Psi_4$$

we obtain

$$
\begin{aligned}
-2\Phi_1 + \Phi_2 &= \alpha_m \Phi_1 \\
\Phi_1 - 2\Phi_2 + \Phi_3 &= \alpha_m \Phi_2 \\
\Phi_2 - 2\Phi_3 &= \alpha_m \Phi_3
\end{aligned}
\quad \text{and} \quad
\begin{aligned}
-2\Psi_1 + \Psi_2 &= \beta_n \Psi_1 \\
\Psi_1 - 2\Psi_2 + \Psi_3 &= \beta_n \Psi_2 \\
\Psi_2 - 2\Psi_3 &= \beta_n \Psi_3
\end{aligned}
$$

or in matrix notation,

$$
\begin{bmatrix} -2 & 1 & 0 \\ 1 & -2 & 1 \\ 0 & 1 & -2 \end{bmatrix}
\begin{bmatrix} \Phi_1 \\ \Phi_2 \\ \Phi_3 \end{bmatrix}
= \alpha_m
\begin{bmatrix} \Phi_1 \\ \Phi_2 \\ \Phi_3 \end{bmatrix},
\qquad
\begin{bmatrix} -2 & 1 & 0 \\ 1 & -2 & 1 \\ 0 & 1 & -2 \end{bmatrix}
\begin{bmatrix} \Psi_1 \\ \Psi_2 \\ \Psi_3 \end{bmatrix}
= \beta_n
\begin{bmatrix} \Psi_1 \\ \Psi_2 \\ \Psi_3 \end{bmatrix}
$$

$$-\mathbf{F}_1 \Phi = \alpha_m \Phi, \qquad\qquad\qquad -\mathbf{F}_1 \Psi = \beta_n \Psi$$

In the preceding eigenvalue problems, the coefficient matrix is seen to be the 3×3 matrix $-\mathbf{F}_1$, where $\mathbf{F}_1$ is given in Theorem 8.4.1. From Theorem 8.4.1 it follows that α must be

$$\alpha_1 = 2 - 2\cos \pi/4 \quad \text{or} \quad \alpha_2 = 2 - 2\cos 2\pi/4 \quad \text{or} \quad \alpha_3 = 2 - 2\cos 3\pi/4$$

and the corresponding eigenvectors must be

$$
\Phi_1 = \begin{bmatrix} \Phi_{11} \\ \Phi_{12} \\ \Phi_{13} \end{bmatrix} = \begin{bmatrix} \sin 1\pi 1/4 \\ \sin 1\pi 2/4 \\ \sin 1\pi 3/4 \end{bmatrix},
\qquad
\Phi_2 = \begin{bmatrix} \Phi_{21} \\ \Phi_{22} \\ \Phi_{23} \end{bmatrix} = \begin{bmatrix} \sin 2\pi 1/4 \\ \sin 2\pi 2/4 \\ \sin 2\pi 3/4 \end{bmatrix}
$$

$$
\Phi_3 = \begin{bmatrix} \Phi_{31} \\ \Phi_{32} \\ \Phi_{33} \end{bmatrix} = \begin{bmatrix} \sin 3\pi 1/4 \\ \sin 3\pi 2/4 \\ \sin 3\pi 3/4 \end{bmatrix}
$$

Identical expressions hold for β_1, β_2, β_3, Ψ_1, Ψ_2, and Ψ_3. Nine linearly independent solutions for the eigenvalue problem

$$
\begin{aligned}
\delta_x^2 V_{mn} + \delta_y^2 V_{mn} &= \lambda_{mn} V_{mn}, & m, n &= 1, 2, 3 \\
V_{0,n} &= 0 = V_{4,n}, & n &= 1, 2, 3 \\
V_{m,0} &= 0 = V_{m,4}, & m &= 1, 2, 3
\end{aligned}
$$

are seen to be

$$V_{mn} = \Phi_{rm} \Psi_{sn} = \sin r\pi m/4 \, \sin s\pi n/4, \qquad r, s = 1, 2, 3$$

If for each fixed choice of r and s the components of V_{mn} are arranged by a FORTRAN ordering with the leftmost subscript varying most rapidly, then the eigenvectors of Theorem 10.3.1 result. ■■

EXAMPLE 10.3.3

To illustrate how a finite-difference system can be solved by Fourier's method, consider difference equations for

$$\nabla^2 u = 0 \quad \text{in } \Omega, \qquad 0 < x < 1, 0 < y < 1$$
$$u(0, y) = 0 = u(1, y), \qquad u(x, 0) = 1, \qquad u(x, 1) = 0$$

Using the grid of Figure 10.3.1, we find the difference equations to be $\mathbf{AU} = \mathbf{b}$, or

$$
\begin{bmatrix}
-4 & 1 & 0 & 1 & 0 & 0 & 0 & 0 & 0 \\
1 & -4 & 1 & 0 & 1 & 0 & 0 & 0 & 0 \\
0 & 1 & -4 & 0 & 0 & 1 & 0 & 0 & 0 \\
1 & 0 & 0 & -4 & 1 & 0 & 1 & 0 & 0 \\
0 & 1 & 0 & 1 & -4 & 1 & 0 & 1 & 0 \\
0 & 0 & 1 & 0 & 1 & -4 & 0 & 0 & 1 \\
0 & 0 & 0 & 1 & 0 & 0 & -4 & 1 & 0 \\
0 & 0 & 0 & 0 & 1 & 0 & 1 & -4 & 1 \\
0 & 0 & 0 & 0 & 0 & 1 & 0 & 1 & -4
\end{bmatrix}
\begin{bmatrix}
U_1 \\ U_2 \\ U_3 \\ U_4 \\ U_5 \\ U_6 \\ U_7 \\ U_8 \\ U_9
\end{bmatrix}
=
\begin{bmatrix}
-1 \\ -1 \\ -1 \\ 0 \\ 0 \\ 0 \\ 0 \\ 0 \\ 0
\end{bmatrix}
$$

From Theorem 8.2.1 and Example 10.3.1, we know the eigenvalues of $\mathbf{A}$ to be

$$\lambda_{mn} = -4 + 2 \cos m\pi/4 + 2 \cos n\pi/4, \qquad m, n = 1, 2, 3$$

and the corresponding eigenvectors to be

$$\mathbf{V}_{mn} = \begin{bmatrix} \sin n\pi 1/4 \ \mathbf{\sin m\pi x} \\ \sin n\pi 2/4 \ \mathbf{\sin m\pi x} \\ \sin n\pi 3/4 \ \mathbf{\sin m\pi x} \end{bmatrix}, \qquad m, n = 1, 2, 3$$

where

$$\mathbf{\sin m\pi x} = \begin{bmatrix} \sin m\pi 1/4 \\ \sin m\pi 2/4 \\ \sin m\pi 3/4 \end{bmatrix}, \qquad m = 1, 2, 3$$

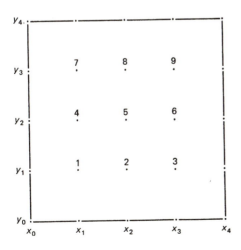

Figure 10.3.1 Difference grid for Example 10.3.3.

Since the matrix $\mathbf{A}$ in this example is symmetric, it follows from Theorem 8.4.2 that the exact solution to the finite-difference system is

$$\mathbf{U} = \sum_{m=1}^{3} \sum_{n=1}^{3} c_{mn} \mathbf{V}_{mn} \quad \text{where } c_{mn} = \frac{1}{\lambda_{mn}} \frac{\mathbf{V}_{mn} \cdot \mathbf{b}}{\mathbf{V}_{mn} \cdot \mathbf{V}_{mn}}$$

Using Theorem 8.4.1, we can show that

$$\mathbf{V}_{mn} \cdot \mathbf{V}_{mn} = 4$$

A direct calculation shows

$$\mathbf{V}_{mn} \cdot \mathbf{b} = \sin \frac{n\pi}{4} \sum_{k=1}^{3} (-1)\sin \frac{m\pi k}{4}$$

$$= -\sin \frac{n\pi}{4} \left\{ \sin \frac{m\pi}{4} + \sin \frac{m\pi}{2} + \sin \frac{m\pi 3}{4} \right\}$$

Now all the coefficients c_{mn} can be calculated and the eigenvector expansion solution is complete.   ■ ■

Nonsymmetric Difference Matrices

If a finite-difference system of the form $\mathbf{AU} = \mathbf{b}$ is nonsymmetric with $\mathbf{A} \neq \mathbf{A}^T$, then the eigenvectors of both $\mathbf{A}$ and $\mathbf{A}^T$ are required to obtain a solution as an eigenvector expansion. See Theorem 8.4.2. In some cases, it is possible to use Theorem 10.3.1 to obtain the eigenvectors of both $\mathbf{A}$ and $\mathbf{A}^T$.

EXAMPLE 10.3.4 _____

Consider the difference system

$$\begin{bmatrix} -4 & 2 & 0 & 1 & 0 & 0 & 0 & 0 & 0 \\ 1 & -4 & 1 & 0 & 1 & 0 & 0 & 0 & 0 \\ 0 & 2 & -4 & 0 & 0 & 1 & 0 & 0 & 0 \\ 1 & 0 & 0 & -4 & 2 & 0 & 1 & 0 & 0 \\ 0 & 1 & 0 & 1 & -4 & 1 & 0 & 1 & 0 \\ 0 & 0 & 1 & 0 & 2 & -4 & 0 & 0 & 1 \\ 0 & 0 & 0 & 1 & 0 & 0 & -4 & 2 & 0 \\ 0 & 0 & 0 & 0 & 1 & 0 & 1 & -4 & 1 \\ 0 & 0 & 0 & 0 & 0 & 1 & 0 & 2 & -4 \end{bmatrix} \begin{bmatrix} U_1 \\ U_2 \\ U_3 \\ U_4 \\ U_5 \\ U_6 \\ U_7 \\ U_8 \\ U_9 \end{bmatrix} = \begin{bmatrix} b_1 \\ b_2 \\ b_3 \\ b_4 \\ b_5 \\ b_6 \\ b_7 \\ b_8 \\ b_9 \end{bmatrix}$$

The preceding coefficient matrix is clearly not symmetric. However, we can write

$$\mathbf{A} = \begin{bmatrix} \mathbf{B} & \mathbf{I} & \mathbf{0} \\ \mathbf{I} & \mathbf{B} & \mathbf{I} \\ \mathbf{O} & \mathbf{I} & \mathbf{B} \end{bmatrix} \quad \text{where } \mathbf{B} = \begin{bmatrix} -4 & 2 & 0 \\ 1 & -4 & 2 \\ 0 & 2 & -4 \end{bmatrix}$$

Notice that $\mathbf{B}$ can be written as

$$\mathbf{B} = -2\mathbf{I} - \mathbf{F}_3$$

where $\mathbf{F}_3$ is as given in Theorem 8.4.4. Since $\mathbf{B}$ and $\mathbf{F}_3$ share the same set of eigenvectors, it follows from Theorems 8.4.4 and 10.3.1 that the eigenvectors of $\mathbf{A}$ are

$$\mathbf{V}_{mn} = \begin{bmatrix} (\sin n\pi 2/4)\mathbf{U}_m \\ (\sin n\pi 2/4)\mathbf{U}_m \\ (\sin n\pi 2/4)\mathbf{U}_m \end{bmatrix} \quad \text{where } \mathbf{U}_m = \begin{bmatrix} 1 \\ \cos(2m-1)\pi/6 \\ \cos(2m-1)\pi/3 \end{bmatrix}$$

where $m, n = 1, 2, 3$.

To find the eigenvectors of $\mathbf{A}^T$, we observe that

$$\mathbf{A}^T = \begin{bmatrix} \mathbf{B}^T & \mathbf{I} & 0 \\ \mathbf{I} & \mathbf{B}^T & \mathbf{I} \\ 0 & \mathbf{I} & \mathbf{B}^T \end{bmatrix} \quad \text{where } \mathbf{B}^T = \begin{bmatrix} -4 & 1 & 0 \\ 2 & -4 & 2 \\ 0 & 1 & -4 \end{bmatrix}$$

Since $\mathbf{B}^T = -2\mathbf{I} - \mathbf{F}_3^T$, it follows from Theorems 8.4.4 and 10.3.1 that the eigenvectors of $\mathbf{A}^T$ are

$$\mathbf{W}_{mn} = \begin{bmatrix} (\sin n\pi 2/4)\mathbf{Z}_m \\ (\sin n\pi 2/4)\mathbf{Z}_m \\ (\sin n\pi 2/4)\mathbf{Z}_m \end{bmatrix} \quad \text{where } \mathbf{Z}_m = \begin{bmatrix} 1 \\ 2\cos(2m-1)\pi/6 \\ 2\cos(2m-1)\pi/3 \end{bmatrix}$$

where $m, n = 1, 2, 3$.

Let λ_{mn} denote the eigenvalues of the matrix $\mathbf{A}$ (or $\mathbf{A}^T$). Theorem 10.3.1 can be used to determine the λ_{mn}. See Exercise 7. Suppose we want to solve the difference system $\mathbf{AU} = \mathbf{b}$ by an eigenvector expansion. Then, according to Theorem 8.4.2, the solution is given by

$$\mathbf{U} = \sum_{m=1}^{3} \sum_{n=1}^{3} c_{mn}\mathbf{V}_{mn} \quad \text{where } c_{mn} = \frac{\mathbf{b} \cdot \mathbf{W}_{mn}}{\lambda_{mn}\mathbf{V}_{mn} \cdot \mathbf{W}_{mn}} \qquad \blacksquare\ \blacksquare$$

Eigenvalues and eigenvectors of other nonsymmetric difference matrices can often be determined from results similar to Theorem 10.3.1. The following theorems can be established using the arguments that led to Theorem 10.3.1.

Theorem 10.3.2. Let $\mathbf{B}$ be an $M \times M$ matrix with M distinct eigenvalues $\gamma_1, \gamma_2, \ldots, \gamma_M$ and corresponding eigenvectors $\mathbf{U}_1, \mathbf{U}_2, \ldots, \mathbf{U}_M$. Suppose $\mathbf{A}$ is an $N \times N$ block tridiagonal matrix of the form

$$\mathbf{A} = \begin{bmatrix} \mathbf{B} & 2\mathbf{I} & & & & \\ \mathbf{I} & \mathbf{B} & \mathbf{I} & & \mathbf{0} & \\ & \mathbf{I} & \mathbf{B} & \mathbf{I} & & \\ & & & \ddots & & \\ & \mathbf{0} & & \mathbf{I} & \mathbf{B} & \mathbf{I} \\ & & & & 2\mathbf{I} & \mathbf{B} \end{bmatrix}$$

Then the eigenvalues of $\mathbf{A}$ are

$$\lambda_{mn} = \gamma_m + 2\cos\frac{(n-1)\pi}{N-1}, \quad m = 1, 2, \ldots, M, n = 1, 2, \ldots, N$$

and the corresponding eigenvectors are

$$
\mathbf{V}_{mn} =
\begin{bmatrix}
\cos\left[(n-1)\dfrac{\pi 0}{N-1}\right]\mathbf{U}_m \\[2mm]
\cos\left[(n-1)\dfrac{\pi 1}{N-1}\right]\mathbf{U}_m \\[2mm]
\cos\left[(n-1)\dfrac{\pi 2}{N-1}\right]\mathbf{U}_m \\[2mm]
\vdots \\[2mm]
\cos\left[(n-1)\dfrac{\pi(N-2)}{N-1}\right]\mathbf{U}_m \\[2mm]
\cos\left[(n-1)\dfrac{\pi(N-1)}{N-1}\right]\mathbf{U}_m
\end{bmatrix}
, \quad m = 1, 2, \ldots, M, \; n = 1, 2, \ldots, N
$$

The matrix $\mathbf{A}^T$ has the same set of eigenvalues and corresponding eigenvectors

$$
\mathbf{W}_{mn} =
\begin{bmatrix}
\cos\left[(n-1)\dfrac{\pi 0}{N-1}\right]\mathbf{U}_m \\[2mm]
2\cos\left[(n-1)\dfrac{\pi 1}{N-1}\right]\mathbf{U}_m \\[2mm]
2\cos\left[(n-1)\dfrac{\pi 2}{N-1}\right]\mathbf{U}_m \\[2mm]
\vdots \\[2mm]
2\cos\left[(n-1)\dfrac{\pi(N-2)}{N-1}\right]\mathbf{U}_m \\[2mm]
\cos\left[(n-1)\dfrac{\pi(N-1)}{N-1}\right]\mathbf{U}_m
\end{bmatrix}
, \quad m = 1, 2, \ldots, M, \; n = 1, 2, \ldots, N
$$

Theorem 10.3.3. Let $\mathbf{B}$ be an $M \times M$ matrix with M distinct eigenvalues $\gamma_1, \gamma_2, \ldots, \gamma_M$ and corresponding eigenvectors $\mathbf{U}_1, \mathbf{U}_2, \ldots, \mathbf{U}_M$. Suppose $\mathbf{A}$ is an $N \times N$ block tridiagonal matrix of the form

$$
\mathbf{A} =
\begin{bmatrix}
\mathbf{B} & 2\mathbf{I} & & & & & \mathbf{0} \\
\mathbf{I} & \mathbf{B} & \mathbf{I} & & & & \\
 & \mathbf{I} & \mathbf{B} & \mathbf{I} & & & \\
 & & & \ddots & & & \\
\mathbf{0} & & & \mathbf{I} & \mathbf{B} & \mathbf{I} \\
 & & & & \mathbf{I} & \mathbf{B}
\end{bmatrix}
$$

Then the eigenvalues of $\mathbf{A}$ are

$$\lambda_{mn} = \gamma_m + 2 \cos \frac{(2n - 1)\pi}{2N}, \qquad m = 1, 2, \ldots, M, \, n = 1, 2, \ldots, N$$

and the corresponding eigenvectors are

$$\mathbf{V}_{mn} = \begin{bmatrix} \cos\left[(2n - 1)\dfrac{\pi 0}{2N}\right]\mathbf{U}_m \\[2mm] \cos\left[(2n - 1)\dfrac{\pi 1}{2N}\right]\mathbf{U}_m \\[2mm] \cos\left[(2n - 1)\dfrac{\pi 2}{2N}\right]\mathbf{U}_m \\[2mm] \vdots \\[2mm] \cos\left[(2n - 1)\dfrac{\pi(N - 2)}{2N}\right]\mathbf{U}_m \\[2mm] \cos\left[(2n - 1)\dfrac{\pi(N - 1)}{2N}\right]\mathbf{U}_m \end{bmatrix}, \qquad m = 1, 2, \ldots, M, \, n = 1, 2, \ldots, N$$

The matrix $\mathbf{A}^T$ has the same set of eigenvalues and corresponding eigenvectors

$$\mathbf{W}_{mn} = \begin{bmatrix} \cos\left[(2n - 1)\dfrac{\pi 0}{2N}\right]\mathbf{U}_m \\[2mm] 2 \cos\left[(2n - 1)\dfrac{\pi 1}{2N}\right]\mathbf{U}_m \\[2mm] 2 \cos\left[(2n - 1)\dfrac{\pi 2}{2N}\right]\mathbf{U}_m \\[2mm] \vdots \\[2mm] 2 \cos\left[(2n - 1)\dfrac{\pi(N - 2)}{2N}\right]\mathbf{U}_m \\[2mm] 2 \cos\left[(2n - 1)\dfrac{\pi(N - 1)}{2N}\right]\mathbf{U}_m \end{bmatrix}, \qquad m = 1, 2, \ldots, M, \, n = 1, 2, \ldots, N$$

EXAMPLE 10.3.5

Consider the boundary-value problem

$$\nabla^2 u = f \quad \text{in } \Omega, \qquad 0 < x < 1, 0 < y < 1$$
$$u_x(0, y) = 0 = u(1, y), \qquad 0 < y < 1$$
$$u_y(x, 0) = 0 = u(x, 1), \qquad 0 < x < 1$$

on the grid of Figure 10.3.2. If we form the finite-difference system from

$$\delta_x^2 U_{mn} + \delta_y^2 U_{mn} = h^2 f_{mn}, \qquad m, n = 1, 2, 3, h = \tfrac{1}{3}$$

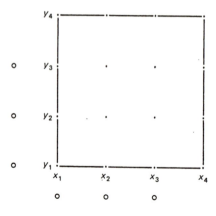

Figure 10.3.2   Finite-difference grid for Example 10.3.5; (○) ghost point.

and incorporate the boundary conditions by setting

$$U_{2,n} - U_{0,n} = 0 = U_{4,n}, \qquad n = 1, 2, 3$$

and

$$U_{m,2} - U_{m,0} = 0 = U_{m,4}, \qquad m = 1, 2, 3$$

then the difference system that results is

$$\begin{bmatrix} -4 & 2 & 0 & 2 & 0 & 0 & 0 & 0 & 0 \\ 1 & -4 & 1 & 0 & 2 & 0 & 0 & 0 & 0 \\ 0 & 1 & -4 & 0 & 0 & 2 & 0 & 0 & 0 \\ 1 & 0 & 0 & -4 & 2 & 0 & 1 & 0 & 0 \\ 0 & 1 & 0 & 1 & -4 & 1 & 0 & 1 & 0 \\ 0 & 0 & 1 & 0 & 1 & -4 & 0 & 0 & 1 \\ 0 & 0 & 0 & 1 & 0 & 0 & -4 & 2 & 0 \\ 0 & 0 & 0 & 0 & 1 & 0 & 1 & -4 & 1 \\ 0 & 0 & 0 & 0 & 0 & 1 & 0 & 1 & -4 \end{bmatrix} \begin{bmatrix} U_1 \\ U_2 \\ U_3 \\ U_4 \\ U_5 \\ U_6 \\ U_7 \\ U_8 \\ U_9 \end{bmatrix} = \begin{bmatrix} b_1 \\ b_2 \\ b_3 \\ b_4 \\ b_5 \\ b_6 \\ b_7 \\ b_8 \\ b_9 \end{bmatrix}$$

where

$$[U_1, U_2, U_3, U_4, U_5, U_6, U_7, U_8, U_9]^T = [U_{11}, U_{21}, U_{31}, U_{12}, U_{22}, U_{32}, U_{13}, U_{23}, U_{33}]^T$$

and

$$[b_1, b_2, b_3, b_4, b_5, b_6, b_7, b_8, b_9]^T = h^2[f_{11}, f_{21}, f_{31}, f_{12}, f_{22}, f_{32}, f_{13}, f_{23}, f_{33}]^T$$

For this example, the coefficient matrix can be written in the form

$$A = \begin{bmatrix} B & 2I & 0 \\ I & B & I \\ 0 & I & B \end{bmatrix} \quad \text{where } B = \begin{bmatrix} -4 & 2 & 0 \\ 1 & -4 & 1 \\ 0 & 1 & -4 \end{bmatrix}$$

From the Theorems 10.3.3 and 8.4.4 we conclude that the eigenvalues of **A** are given by

$$\lambda_{mn} = -4 + 2 \cos \frac{(2n-1)\pi}{6} + 2 \cos \frac{(2m-1)\pi}{6}, \qquad m, n = 1, 2, 3$$

The corresponding eigenvectors of $\mathbf{A}$ are

$$\mathbf{V}_{mn} = \begin{bmatrix} \cos(2n-1)\dfrac{\pi 0}{6}\,\mathbf{U}_m \\[2ex] \cos(2n-1)\dfrac{\pi 1}{6}\,\mathbf{U}_m \\[2ex] \cos(2n-1)\dfrac{\pi 2}{6}\,\mathbf{U}_m \end{bmatrix} \quad \text{where } \mathbf{U}_m = \begin{bmatrix} \cos(2m-1)\dfrac{\pi 0}{6} \\[2ex] \cos(2m-1)\dfrac{\pi 1}{6} \\[2ex] \cos(2m-1)\dfrac{\pi 2}{6} \end{bmatrix}$$

The determination of the eigenvectors of $\mathbf{A}^T$ is left as an exercise.   ■ ■

Fourier's method, or the eigenvector expansion method, can be used to obtain expressions for the exact solution of a number of finite-difference systems. From the Fourier solution, we can calculate the solution to the difference system at only selected values of (x_m, y_n). This contrasts with the situation encountered in Section 10.2 where we saw that to obtain the solution at one point, we must solve the system at all points. Having this ability to selectively evaluate the solution of the difference equations can result in a savings in computer time. However, if the solution is to be computed on the entire grid, then one of the computational methods of Section 10.2 will usually produce the solution more economically than if the eigenvector expansion is evaluated at each node.

The eigenvector expansion solution to the difference equation has the desirable property that it displays the roles of the various problem data in the solution. For example, the right side of the differential equation, the boundary conditions, and the choice of mesh spacing all appear in the eigenvector solution. This allows us to make qualitative as well as quantitative statements about the solution of the difference equations.

Finally, among the results of this section, perhaps the most valuable are the expressions for the eigenvalues of the various finite-difference matrices. As we see in Section 10.5, a knowledge of these eigenvalues allows us to choose the optimal successive overrelaxation (SOR) parameter to obtain a maximum convergence rate when applying point or block SOR. See Sections 10.4 and 10.5.

Exercises

1. Find the eigenvalues of each of the followig matrices:

$$\textbf{(a)} \quad \begin{bmatrix} -4 & 1 & 0 & 1 & 0 & 0 \\ 1 & -4 & 1 & 0 & 1 & 0 \\ 0 & 1 & -4 & 0 & 0 & 1 \\ 1 & 0 & 0 & -4 & 1 & 0 \\ 0 & 1 & 0 & 1 & -4 & 1 \\ 0 & 0 & 1 & 0 & 1 & -4 \end{bmatrix}$$

(b)
$$
\begin{bmatrix}
-4 & 2 & 0 & 1 & 0 & 0 \\
1 & -4 & 1 & 0 & 1 & 0 \\
0 & 2 & -4 & 0 & 0 & 1 \\
1 & 0 & 0 & -4 & 2 & 0 \\
0 & 1 & 0 & 1 & -4 & 1 \\
0 & 0 & 1 & 0 & 2 & -4
\end{bmatrix}
$$

(c)
$$
\begin{bmatrix}
-4 & 2 & 0 & 2 & 0 & 0 \\
1 & -4 & 1 & 0 & 2 & 0 \\
0 & 2 & -4 & 0 & 0 & 2 \\
2 & 0 & 0 & -4 & 2 & 0 \\
0 & 2 & 0 & 1 & -4 & 1 \\
0 & 0 & 2 & 0 & 2 & -4
\end{bmatrix}
$$

(d)
$$
\begin{bmatrix}
-4 & 1 & 0 & 2 & 0 & 0 & 0 & 0 & 0 \\
1 & -4 & 1 & 0 & 2 & 0 & 0 & 0 & 0 \\
0 & 1 & -4 & 0 & 0 & 2 & 0 & 0 & 0 \\
1 & 0 & 0 & -4 & 1 & 0 & 1 & 0 & 0 \\
0 & 1 & 0 & 1 & -4 & 1 & 0 & 1 & 0 \\
0 & 0 & 1 & 0 & 1 & -4 & 0 & 0 & 1 \\
0 & 0 & 0 & 2 & 0 & 0 & -4 & 1 & 0 \\
0 & 0 & 0 & 0 & 2 & 0 & 1 & -4 & 1 \\
0 & 0 & 0 & 0 & 0 & 2 & 0 & 1 & -4
\end{bmatrix}
$$

(e)
$$
\begin{bmatrix}
-4 & 1 & 0 & 2 & 0 & 0 & 0 & 0 & 0 \\
1 & -4 & 1 & 0 & 2 & 0 & 0 & 0 & 0 \\
0 & 1 & -4 & 0 & 0 & 2 & 0 & 0 & 0 \\
1 & 0 & 0 & -4 & 1 & 0 & 1 & 0 & 0 \\
0 & 1 & 0 & 1 & -4 & 1 & 0 & 1 & 0 \\
0 & 0 & 1 & 0 & 1 & -4 & 0 & 0 & 1 \\
0 & 0 & 0 & 1 & 0 & 0 & -4 & 1 & 0 \\
0 & 0 & 0 & 0 & 1 & 0 & 1 & -4 & 1 \\
0 & 0 & 0 & 0 & 0 & 1 & 0 & 1 & -4
\end{bmatrix}
$$

2. Find the eigenvectors for each of the matrices of Exercise 1.

3. Use an eigenvector expansion to represent the solution of the difference system

$$
\begin{bmatrix}
-4 & 1 & 0 & 1 & 0 & 0 & 0 & 0 & 0 \\
1 & -4 & 1 & 0 & 1 & 0 & 0 & 0 & 0 \\
0 & 1 & -4 & 0 & 0 & 1 & 0 & 0 & 0 \\
1 & 0 & 0 & -4 & 1 & 0 & 1 & 0 & 0 \\
0 & 1 & 0 & 1 & -4 & 1 & 0 & 1 & 0 \\
0 & 0 & 1 & 0 & 1 & -4 & 0 & 0 & 1 \\
0 & 0 & 0 & 1 & 0 & 0 & -4 & 1 & 0 \\
0 & 0 & 0 & 0 & 1 & 0 & 1 & -4 & 1 \\
0 & 0 & 0 & 0 & 0 & 1 & 0 & 1 & -4
\end{bmatrix}
\begin{bmatrix}
U_1 \\ U_2 \\ U_3 \\ U_4 \\ U_5 \\ U_6 \\ U_7 \\ U_8 \\ U_9
\end{bmatrix}
=
\begin{bmatrix}
0 \\ 0 \\ 0 \\ 0 \\ -64 \\ 0 \\ 0 \\ 0 \\ 0
\end{bmatrix}
$$

and compare the solution with that of Example 10.2.2.

4. Repeat Exercise 3 using the matrix **A** of Exercise 1,
   (a) part (d)
   (b) part (e)

5. Use separation of variables to show that the eigenvalues and corresponding eigenfunctions for the partial differential equation of Example 10.2.2 are

$$\mu_{mn} = -[(m\pi)^2 + (n\pi)^2], \qquad m, n = 1, 2, \ldots$$

and

$$u_{mn}(x, y) = \sin m\pi x \sin n\pi y, \qquad m, n = 1, 2, \ldots$$

Compare these eigenvalues and eigenfunctions to those for the discrete problem of Example 10.3.2.

6. Determine a boundary-value problem and a finite-difference grid that corresponds to the difference system of Example 10.3.4.

7. (a) Find the eigenvalues of the matrix $\mathbf{A}$ of Example 10.3.4.
   (b) Find the eigenvectors of the matrix in (a).

8. Find the eigenvectors of $\mathbf{A}^T$ in Example 10.3.5.

10.4 ITERATIVE METHODS

Point Interative Methods

Although iterative methods for dealing with difference equations usually do not produce the exact solution to the difference equations, they often are preferred over a direct method. Iterative methods are very easy to implement on a computer and generally require less computer memory than a direct method. We introduce iterative methods for approximating the solution to a system of difference equations by way of a small system of difference equations for Laplace's equation.

EXAMPLE 10.4.1 _____

Consider the boundary-value problem

$$\nabla^2 u = 0, \quad \text{in } \Omega, \qquad 0 < x < 1, 0 < y < 1$$
$$u(0, y) = 0, \qquad u(1, y) = 16y, \qquad 0 < y < 1$$
$$u(x, 0) = 0, \qquad u(x, 1) = 16x, \qquad 0 < x < 1$$

The exact solution to this problem is $u(x, y) = 16xy$. On the grid of Figure 10.4.1, we obtain the following discretization:

$$\delta_x^2 U_{mn} + \delta_y^2 U_{mn} = 0, \qquad m, n = 1, 2, 3$$
$$U_{0,1} = U_{0,2} = U_{0,3} = 0, \qquad U_{4,1} = 4, U_{4,2} = 8, U_{4,3} = 12$$
$$U_{1,0} = U_{2,0} = U_{3,0} = 0, \qquad U_{1,4} = 4, U_{2,4} = 8, U_{3,4} = 12$$

It is easy to verify that the exact solution to this system of difference equations is $U_{mn} = mn$. As we have seen, the difference equation

$$\delta_x^2 U_{mn} + \delta_y^2 U_{mn} = 0$$

is equivalent to the discrete mean-value property that states

$$U_{mn} = \tfrac{1}{4}(U_{m-1,n} + U_{m,n-1} + U_{m+1,n} + U_{m,n+1})$$

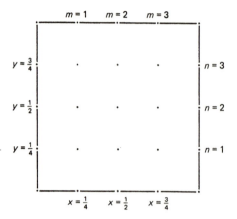

Figure 10.4.1 Difference grid for Example 10.4.1.

This mean-value property motivates a method for solving the system of difference equations. Let us make an initial guess of the values of U_{mn} for $m, n = 1, 2, 3$ and then successively apply the discrete mean-value property with the aim of improving our initial guess. For example, we might begin by guessing U_{mn} to be some value between the maximum and minimum of the boundary values, say,

$$U^0_{mn} = 8, \qquad m, n = 1, 2, 3$$

Then successive estimates of U_{mn} could be computed from

$$U^{i+1}_{mn} = \tfrac{1}{4}(U^i_{m-1,n} + U^i_{m,n-1} + U^i_{m+1,n} + U^i_{m,n+1}), \qquad m, n = 1, 2, 3$$

for $i = 1, 2, \ldots$. This method is known as Jacobi's method.

In Figure 10.4.2 we show the results of two applications of Jacobi's method. Examining Jacobi's method, an improvement that comes to mind is to *use new information about the iteratively generated solution as soon as it is available*. If we scan the grid points in the order $U_{11}, U_{21}, U_{31}, U_{12}, U_{22}, U_{32}, U_{13}, U_{23}, U_{33}$ and apply the discrete mean-value property together with the latest information about the solution, the result is

$$U^{i+1}_{mn} = \tfrac{1}{4}(U^{i+1}_{m-1,n} + U^{i+1}_{m,n-1} + U^i_{m+1,n} + U^i_{m,n+1})$$

which is known as the Gauss–Seidel method. Two iterations of the Gauss–Seidel method are shown in Figure 10.4.3. ■ ■

The methods illustrated in Example 10.4.1 are easy to extend to the Poisson equation boundary-value problem

$$\nabla^2 u = f(x, y) \quad \text{in } \Omega, \qquad 0 < x < 1, 0 < y < 1 \tag{10.4.1}$$
$$u = g(x, y) \quad \text{on } S \tag{10.4.2}$$

where S denotes the boundary of Ω. On the square grid

$$(x_m, y_n) = (mh, nh), \qquad m, n = 0, 1, 2, \ldots, N$$

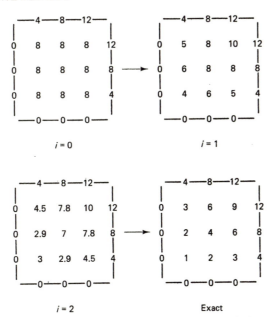

Figure 10.4.2 Two applications of Jacobi's method to Example 10.4.1.

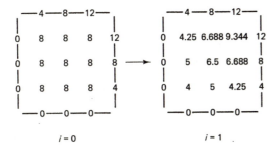

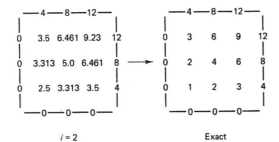

Figure 10.4.3 Two applications of the Gauss–Seidel method to Example 10.4.1.

with $h = 1/(N + 1)$, the difference equations for (10.4.1) read

$$U_{m-1,n} - 2U_{mn} + U_{m+1,n} + U_{m,n-1} - 2U_{mn} + U_{m,n+1} = h^2 f_{mn},$$
$$m, n = 1, 2, \ldots, N \tag{10.4.3}$$

and the boundary conditions are incorporated by setting

$$U_{0,n} = g(0, y_n), \qquad U_{N+1,n} = g(1, y_n), \qquad n = 1, 2, \ldots, N$$
$$U_{m,0} = g(x_m, 0), \qquad U_{m,N+1} = g(x_m, 1), \qquad m = 1, 2, \ldots, N \tag{10.4.4}$$

Solving (10.4.3) for U_{mn} and introducing the iteration parameter i, we obtain *Jacobi's method* for solving the difference equations

$$U_{mn}^{i+1} = \tfrac{1}{4}[U_{m-1,n}^i + U_{m,n-1}^i + U_{m+1,n}^i + U_{m,n+1}^i - h^2 f_{mn}],$$
$$i = 1, 2, \ldots, m, n = 1, 2, \ldots, N \tag{10.4.5}$$

Similarly, if we update the U values as soon as possible, we have the *Gauss–Seidel method*,

$$U_{mn}^{i+1} = \tfrac{1}{4}[U_{m-1,n}^{i+1} + U_{m,n-1}^{i+1} + U_{m+1,n}^i + U_{m,n+1}^i - h^2 f_{mn}],$$
$$i = 1, 2, \ldots, m, n = 1, 2, \ldots, N \tag{10.4.6}$$

In both methods (10.2.5) and (10.4.6), the Dirichlet boundary conditions are incorporated through the boundary nodes, $m = 0$, $m = N + 1$, $n = 0$, and $n = N + 1$. The iterative methods (10.4.5) and (10.4.6) reach to, but never adjust, the U values at the boundary nodes. Constrast this simple boundary treatment with the single-indexed version of Sections 10.1 and 10.22 where we had to store boundary conditions in appropriate locations in the vector **b** to solve the difference equations in the matrix format $\mathbf{AU} = \mathbf{b}$.

The Gauss–Seidel method (10.4.6) is written assuming that the nodes (x_m, y_n) are scanned according to the convention that m varies most rapidly. If some other scanning of the grid points is introduced, then (10.4.6) must be modified.

With only minor changes in the forms of (10.4.5) and (10.4.6), a rectangular grid can be used. This is done in Algorithm 10.2.

We postpone until the next section a formal discussion of the convergence of the iteratively defined sequences $\{U_{mn}^i\}_{i=1}^\infty$. However, from a naive point of view, if the Jacobi iterates (10.4.5) and the Gauss–Seidel iterates (10.4.6) both converge, we expect the Gauss–Seidel method to converge faster since it uses new information as soon as possible. Another technique for accelerating the convergence of an iterative method is known as *successive overrelaxation* (SOR).

To implement SOR, we begin by calculating a Gauss–Seidel update, call it $\overline{U}_{mn}^{i+1}$, at node (x_m, y_n). Then, from $\overline{U}_{mn}^{i+1}$ and the value of U_{mn}^i the direction of convergence can be estimated. Successive overrelaxtion calls for defining U_{mn}^{i+1} to be an extrapolation through U_{mn}^i and $\overline{U}_{mn}^{i+1}$.

Thus, the SOR method for (10.4.3) can be written as

$$\overline{U}_{mn}^{i+1} = \tfrac{1}{4}[U_{m-1,n}^{i+1} + U_{m,n-1}^{i+1} + U_{m+1,n}^i + U_{m,n+1}^i - h^2 f_{mn}]$$
$$U_{mn}^{i+1} = \omega\overline{U}_{mn}^{i+1} + (1 - \omega)U_{mn}^i, \qquad 0 < \omega < 2,$$
$$m, n = 1, 2, \ldots, N, i = 1, 2, \ldots \tag{10.4.7}$$

The constant ω in (10.4.7) is called the *relaxation parameter*. As we show in the next section, the restriction $0 < \omega < 2$ is necessary and sufficient for convergence when SOR is applied to many difference equations. If $\omega > 1$, the method (10.4.7) is called an *overrelaxation method*; if $0 < \omega < 1$, (10.4.7) is called an *underrelaxation method*. When $\omega = 1$, the SOR method reduces to the Gauss–Seidel method.

As with any iterative method, some stopping criteria are required in the application of the Jacobi, Gauss–Seidel, and SOR methods. One method for stopping the iterations that should be built into every computer program is a maximum number of iterations to be allowed. This avoids the embarassment of an infinite loop. Of course, if we have found an answer correct to several digits after, say, 100 iterations, it would be wasteful to continue iterating until the maximum iteration bound, say 1000 iterations, has been reached.

In engineering circles, a stopping criterion, other than a maximum on the iteration count, is often referred to as a *closure criterion*. A number of closure criteria are routinely applied. Most are based on some sort of measurement of a residual associated with the estimate to the solution U_{mn}^i. If the U_{mn}^i values are singly indexed to produce $\mathbf{U}^i$ and the difference system (10.4.3)–(10.4.4) is written in the matrix format $\mathbf{AU} = \mathbf{b}$, then a *residual* associated with $\mathbf{U}^i$ is given by

$$\mathbf{R}^i = \mathbf{b} - \mathbf{AU}^i \tag{10.4.8}$$

Common stopping criteria are (1) that the sum of the squares of the residual vector $\mathbf{R}^i$ or (2) that the maximum of the absolute values of the components of $\mathbf{R}^i$ be below some preset tolerance. Often the application will dictate the closure criterion. In flow problems it is often related to a statement regarding mass balance.

Fortunately, it is not necessary to use the matrix equation (10.4.8) to calculate residuals. Residuals can be calculated *at the end of an iteration* after all nodes (x_m, y_n) have been assigned a value of U_{mn}^i. The components R_{mn}^i of a residual can be calculated as follows:

$$R_{mn}^i = f_{mn} - [U_{m-1,n}^i + U_{m+1,n}^i + U_{m,n-1}^i + U_{m,n+1}^i - 4U_{mn}^i]/h^2 \tag{10.4.8}$$

Algorithm 10.2 illustrates the application of the SOR method to a Poisson equation subject to Dirichlet boundary conditions. ■ ■

ALGORITHM 10.2   SOR/Gauss–Seidel Method

Step 1.

This algorithm applies the SOR method to the boundary-value problem

$$\nabla^2 u = f(x, y), \qquad 0 < x < a, 0 < y < b$$
$$u(0, y) = px(y), \qquad u(X, y) = qx(y), \qquad 0 < x < a$$
$$u(x, 0) = py(x), \qquad u(x, Y) = qy(x), \qquad 0 < y < b$$

INPUT

Integer *mmax*, number of interior x grid points
Integer *nmax*, number of interior y grid points
Integer *itmax*, maximum number of iterations allowed
Real a, x dimension
Real b, y dimension
Real ω, relaxation parameter
Real tol, tolerance for maximum of $|R^i_{mn}|$
Functions $px(y)$, $qx(y)$, boundary conditions on $x = 0$ and $x = a$
Functions $py(x)$, $qy(x)$, boundary conditions on $y = 0$ and $y = b$
Function $f(x, y)$, right side of Poisson equation

OUTPUT

Array U_{mn}, numerical solution for $m, n = 1, 2, \ldots, nmax$
Integer i, number of SOR iterations performed

Step 2. Set boundary conditions and initialize U^i_{mn}

Set $hx = a/(mmax + 1)$
    $hy = b/(nmax + 1)$
    $q = (hx/hy)^2$
    $x_0 = 0$
    $y_0 = 0$
    sum $= 0$
For $m = 1, 2, \ldots, mmax$, set
    $x_m = x_{m-1} + hx$
    $U_{m,0} = py(x_m)$
    $U_{m,nmax+1} = qy(x_m)$
    sum $=$ sum $+ U_{m,0} + U_{m,nmax+1}$
For $n = 1, 2, \ldots, n$ max, set
    $y_n = y_{n-1} + hy$
    $U_{0,n} = px(y_n)$
    $U_{mmax+1,n} = qx(y_n)$
    sum $=$ sum $+ U_{0,n} + U_{mmax+1,n}$
Set ave $=$ sum$/(mmax + nmax)/2$
For $n = 1, 2, \ldots, nmax$, for $m = 1, 2, \ldots, mmax$,
set

        $U_{mn} =$ ave   (initialize U^0_{mn} to average of boundary values)
        $F_{mn} = -(hx)^2 f(x_m, y_n)$

Step 3. Begin SOR iterations

For $i = 1, \ldots, itmax$, perform steps 4–6

Step 4. Calculate ith SOR iterate U_{mn}^i

For $n = 1, 2, \ldots, nmax$, for $m = 1, 2, \ldots, mmax$,
set

$$uold = U_{mn}$$
$$U_{mn} = [U_{m-1,n} + qU_{m,n-1} + U_{m+1,n} + qU_{m,n+1} + F_{mn}]/(2 + 2q)$$
$$U_{mn} = \omega U_{mn} + [1 - \omega]uold$$

Step 5. Calculate maximum residual for ith SOR iterate

Set $rmax = 0$
For $n = 1, 2, \ldots, nmax$, for $m = 1, 2, \ldots, mmax$, set
$$\text{res} = \{F_{mn} + [U_{m-1,n} + U_{m+1,n} - (2 + 2q)U_{mn} + qU_{m,n-1} + qU_{m,n+1}]/hx^2$$
If $|\text{res}| > rmax$, set $rmax = |\text{res}|$

Step 6. Compare maximum residual to tolerance

If $rmax < tol$, then go to step 7.

Step 7. Output solution

Write i, the number of SOR iterations performed
For $n = nmax, nmax - 1, \ldots, 2, 1$, for $m = 1, 2, \ldots, mmax$
    write U_{mn}

EXAMPLE 10.4.2 _____

To illustrate Algorithm 10.2, consider the boundary-value problem

$$\nabla^2 u = 4, \quad 0 < x < 1, 0 < y < 1$$
$$u(0, y) = y^2, \quad u(1, y) = 1 + y^2, \quad 0 < y < 1$$
$$u(x, 0) = x^2, \quad u(x, 1) = x^2 + 1, \quad 0 < x < 1$$

It is easy to verify that $u = x^2 + y^2$ is the exact solution to this boundary-value problem. Let us choose

$$mmax = 9 = nmax \quad \text{and} \quad tol = 0.005$$

For the first application of Algorithm 10.2, we select $\omega = 1$ so that the SOR procedure reduces to the method of Gauss–Seidel. Table 10.4.1 shows the resulting numerical solution. It easy to verify that the numerical solution agrees with the exact solution to three places. Algorithm 10.2 produced a starting guess of $U_{mn}^0 = 0.82$ and with $\omega = 1$ required 66 iterations to drive the maximum residual below the tolerance of 0.005. This example was selected partly to dispel the notion that iterative methods are inferior to direct methods because iterative methods produce only an approximate solution to the difference system. As Table 10.4.1 shows, the approximation can be very good.

TABLE 10.4.1 RESULTS OF GAUSS–SEIDEL ON EXAMPLE 10.4.2[a]

	$x = 0.1$	$x = 0.2$	$x = 0.3$	$x = 0.4$	$x = 0.5$	$x = 0.6$	$x = 0.7$	$x = 0.8$	$x = 0.9$
$y = 0.9$	0.820	0.850	0.900	0.970	1.060	1.170	1.300	1.450	1.620
$y = 0.8$	0.650	0.680	0.730	0.800	0.890	1.000	1.130	1.280	1.450
$y = 0.7$	0.500	0.530	0.580	0.650	0.740	0.850	0.980	1.130	1.300
$y = 0.6$	0.370	0.400	0.450	0.520	0.610	0.720	0.850	1.000	1.170
$y = 0.5$	0.260	0.290	0.340	0.410	0.500	0.610	0.740	0.890	1.060
$y = 0.4$	0.170	0.200	0.250	0.320	0.410	0.520	0.650	0.800	0.970
$y = 0.3$	0.100	0.130	0.180	0.250	0.340	0.450	0.580	0.730	0.900
$y = 0.2$	0.050	0.080	0.130	0.200	0.290	0.400	0.530	0.680	0.850
$y = 0.1$	0.020	0.050	0.100	0.170	0.260	0.370	0.500	0.650	0.820

[a]Tolerance = 0.005; 66 Gauss–Seidel iterations.

In Section 10.5, it is shown that the choice

$$\omega = \frac{2}{1 + \sqrt{1 - \cos^2(\pi/10)}} \approx 1.528$$

provides the optimal convergence rate in the SOR method for Example 10.4.2. Table 10.4.1 also displays the numerical solution for this new choice of ω. However, compare the 66 iterations required by the Gauss–Seidel method with the 22 iterations for the SOR method with the optimal choice of the relaxation parameter ω.

Derivative Boundary Conditions

It is easy to modify Algorithm 10.2 to allow for a Neumann or a mixed type of boundary condition. To illustrate, consider the boundary-value problem

$$\nabla^2 u = f(x, y), \qquad 0 < x < a, \, 0 < y < b \qquad (10.4.9)$$
$$u_x(0, y) = g(y), \qquad u(a, y) = qx(y), \qquad 0 < y < b$$
$$u(x, 0) = py(x), \qquad u(x, b) = qy(x), \qquad 0 < x < a \qquad (10.4.10)$$

To account for the Neumann boundary condition on the boundary $x = 0$, we introduce a grid as shown in Figure 10.4.4. The ghost points above the position $x = x_0$ can be eliminated from the SOR computation by using the derivative boundary condition to write

$$\frac{U_{2,n} - U_{0,n}}{2hx} = g(y_n) = g_n$$

or

$$U_{0,n} = U_{2,n} - 2hxg_n, \qquad n = 1, 2, \ldots, n \text{ max}$$

To incorporate this Neumann boundary condition into Algorithm 10.2, we modify step 4.

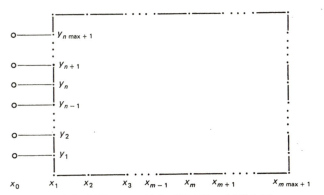

Figure 10.4.4 Difference grid for (10.4.9) and (10.4.10)

Step 4. Calculate *i*th SOR iterate U^i_{mn}

For $n = 1, 2, \ldots, nmax$, set
$$uold = U_{1,n}$$
$$U_{1,n} = [-2hxg_n + qU_{1,n-1} + 2U_{2,n} + qU_{1,n+1} + F_{1,n}]/(2 + 2q)$$
$$U_{1,n} = omega\ U_{1,n} + [1 - omega]\ uold$$
For $m = 2, 3, \ldots, mmax$, set
$$uold = U_{mn}$$
$$U_{mn} = [U_{m-1,n} + qU_{m,n-1} + U_{m+1,n} + qU_{m,n+1} + F_{mn}]/(2 + 2q)$$
$$U_{mn} = omega\ U_{mn} + [1 - omega]\ uold$$

General Five-Point Difference Equations

A large number of finite-difference methods link the value of U_{mn} at a grid point $P = (x_m, y_n)$ to the U values at the four neighbor grid points to the west, east, north, and south. For such methods, we can associate with each computational grid point a *computational stencil* of the form indicated in Figure 10.4.5. If a difference method can be written in the form

$$CP_{mn} \cdot U_{mn} - CW_{mn} \cdot U_{m-1,n} - CE_{mn} \cdot U_{m+1,n} - CS_{mn} \cdot$$
$$U_{m,n-1} - CN_{mn} \cdot U_{m,n+1} = F_{mn}, \qquad m, n = 1, 2, \ldots, N \qquad (10.4.11)$$

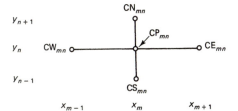

Figure 10.4.5 Computational stencil.

with

$$CP_{mn} > 0, \quad CW_{mn} \geq 0, \quad CE_{mn} \geq 0, \quad CN_{mn} \geq 0, \quad CS_{mn} \geq 0$$

and

$$CP_{mn} \geq CW_{mn} + CE_{mn} + CN_{mn} + CS_{mn} \tag{10.4.12}$$

with strict inequality in (10.4.12) for a least one pair of m, n, then the Jacobi, Gauss–Seidel, and SOR methods can be applied to the difference system (10.4.1). For (10.4.1) the analogues of (10.4.5)–(10.4.7) are as follows.

Jacobi method:

$$U_{mn}^{i+1} = [CW_{mn} \cdot U_{m-1,n}^{i} + CE_{mn} \cdot U_{m+1,n}^{i} + CS_{mn} \cdot U_{m,n-1}^{i} \tag{10.4.13}$$
$$+ CN_{mn} \cdot U_{m,n+1}^{i} + F_{mn}]/CP_{mn}$$

Gauss–Seidel method:

$$U_{mn}^{i+1} = [CW_{mn} \cdot U_{m-1,n}^{i+1} + CE_{mn} \cdot U_{m+1,n}^{i} + CS_{mn} \cdot U_{m,n-1}^{i+1} \tag{10.4.14}$$
$$+ CN_{mn} \cdot U_{m,n+1}^{i} + F_{mn}]/CP_{mn}$$

SOR method:

$$\overline{U}_{mn}^{i+1} = [CW_{mn} \cdot U_{m-1,n}^{i+1} + CE_{mn} \cdot U_{m+1,n}^{i} + CS_{mn} \cdot U_{m,n-1}^{i+1}$$
$$+ CN_{mn} \cdot U_{m,n+1}^{i} + F_{mn}]/CP_{mn}$$
$$U_{mn}^{i+1} = \omega \overline{U}_{mn}^{i+1} + (1 - \omega)U_{mn}^{i} \tag{10.4.15}$$

Clearly, the difference equations (10.4.3) have the form (10.4.11) with $CP_{mn} = 4$ and $CW_{mn} = CE_{mn} = CN_{mn} = CS_{mn} = 1$. Equations (10.1.14) and (10.1.15) provide two examples of difference equations of the form (10.4.11) with nonconstant coefficients C_{mn}^{*}. The next example shows how to apply iterative methods to a difference equation for a parabolic initial-boundary-value problem.

EXAMPLE 10.4.3 _____

Consider the two-dimensional diffusion equation

$$u_t - a^2(u_{xx} + u_{yy}) = S(x, y, t), \qquad 0 < x < 1, 0 < y < 1$$
$$u(x, y, t) = f(x, y), \qquad 0 < x < 1, 0 < y < 1$$
$$u(0, y, t) = 0 = u(1, y, t), \qquad 0 < y < 1$$
$$u(x, 0, t) = 0 = u(x, 1, t), \qquad 0 < x < 1$$

If we choose a square mesh with $h = 1/(N + 1)$ and a time step of k and use the backward-in-time method to discretize this partial differential equation, the resulting difference equation is

$$\frac{U_{mn} - V_{mn}}{k} - a^2 \frac{\delta_x^2 U_{mn}}{h^2} - a^2 \frac{\delta_y^2 U_{mn}}{h^2} = S_{mn}$$

where we have introduced the notation

$$U_{mn} \simeq u(x_m, y_n, t_{j+1}), \qquad V_{mn} \simeq u(x_m, y_n, t_j), \qquad S_{mn} = S(x_m, y_n, t_{j+1})$$

The boundary conditions provide zero values for $U_{0,n}$, $U_{N+1,n}$, $U_{m,0}$, $U_{m,N+1}$ and $V_{0,n}$, $V_{N+1,n}$, $V_{m,0}$, $V_{m,N+1}$. The initial condition provides values for V_{mn} for the first time step.

The problem of advancing the numerical solution from time $t = t_j$ to time t_{j+1} can be reduced to that of determining U_{mn} given values of V_{mn} and S_{mn}. The preceding difference equation can be written in the form

$$(1 + 4r)U_{mn} - rU_{m-1,n} - rU_{m+1,n} - rU_{m,n-1} - rU_{m,n+1} = V_{mn} + kS_{mn},$$
$$m, n = 1, 2, \ldots, N, \qquad r = a^2k/h^2$$

The last equation has the form of (10.4.11). The strict inequality required in (10.4.12) is obtained for all m, n. Now, the methods (10.4.13)–(10.4.15) are available for the solution of the difference equations for the diffusion equation with two space variables. ■ ■

Block Iterative Methods

The iterative method discussed in the preceding are called *point iterative methods* because in the iterative process we deal with one grid point at a time. As we show in the next section, improved convergence rates can be realized if blocks of grid points are updated simultaneously. To obtain this improved convergence rate through a block iterative method, we must deal with systems of linear equations. Most *block iterative methods* are based on simultaneously updating all grid points on a given row or in a given column. By restricting the blocks to be one dimensional, the linear equations that result can be arranged to be tridiagonal and thus easily solved.

If we choose the blocks to be the horizontal line of nodes defined by $n = $ const, then we obtain the *row iterative* counterparts of (10.4.5)–(10.4.7):

Jacobi row iteration:

$$U_{mn}^{i+1} = \tfrac{1}{4}[U_{m-1,n}^{i+1} + U_{m,n-1}^{i} + U_{m+1,n}^{i+1} + U_{m,n+1}^{i} - h^2f_{mn}],$$
$$n = 1, 2, \ldots, N, i = 1, 2, \ldots \qquad (10.4.16)$$

Gauss–Seidel row iteration:

$$U_{mn}^{i+1} = \tfrac{1}{4}[U_{m-1,n}^{i+1} + U_{m,n-1}^{i+1} + U_{m+1,n}^{i+1} + U_{m,n+1}^{i} - h^2f_{mn}],$$
$$n = 1, 2, \ldots, N, i = 1, 2, \ldots \qquad (10.4.17)$$

SOR row iteration:

$$\overline{U}_{mn}^{i+1} = \tfrac{1}{4}[\overline{U}_{m-1,n}^{i+1} + U_{m,n-1}^{i+1} + \overline{U}_{m+1,n}^{i+1} + U_{m,n+1}^{i} - h^2f_{mn}]$$
$$U_{mn}^{i+1} = \omega\overline{U}_{mn}^{i+1} + (1 - \omega)U_{mn}^{i},$$
$$0 < \omega < 2, n = 1, 2, \ldots, N, i = 1, 2, \ldots \qquad (10.4.18)$$

Algorithm 10.3 outlines the implementation of the SOR row iteration method.

ALGORITHM 10.3   Row SOR/Gauss-Seidel Method

Step 1. Document

The following algorithm applies the row SOR method to the boundary value problem

$$\nabla^2 u = f(x, y), \qquad 0 < x < \alpha, 0 < y < \beta$$
$$u(0, y) = px(y), \qquad u(\alpha, y) = qx(y), \qquad 0 < x < \alpha$$
$$u(x, 0) = py(x), \qquad u(x, \beta) = qy(x), \qquad 0 < y < \beta$$

INPUT

Integer *mmax*, number of interior x grid points
Integer *nmax*, number of interior y grid points
Integer *itmax*, maximum number of iterations allowed
Real α, x dimension
Real β, y dimension
Real omega, relaxation parameter
Real tol, tolerance for maximum of $|R^i_{mn}|$
Functions $px(y)$, $qx(y)$, boundary conditions on $x = 0$ and $x = \alpha$
Functions $py(x)$, $qy(x)$, boundary conditions on $y = 0$ and $y = \beta$
Function $f(x, y)$, right side of Poisson equation

OUTPUT

Array U_{mn}, numerical solution for $m = 1, 2, \ldots, mmax$
$$n = 1, 2, \ldots, nmax$$
integer i, number of SOR iterations performed

Step 2. Set boundary conditions and Initialize U^i_{mn}

Set $hx = \alpha/(mmax + 1)$
$\quad hy = \beta/(nmax + 1)$
$\quad q = (hx/hy)^2$
$\quad x_0 = 0$
$\quad y_0 = 0$
$\quad$ sum $= 0$
For $m = 1, 2, \ldots, nmax$, set
$\quad x_m = x_{m-1} + hx$
$\quad U_{m,0} = py(x_m)$
$\quad U_{m,nmax+1} = qy(x_m)$
$\quad$ sum $=$ sum $+ U_{m,0} + U_{m,nmax+1}$
For $n = 1, 2, \ldots, n$ max, set
$\quad y_n = y_{n-1} + hy$
$\quad U_{0,n} = px(y_n)$
$\quad U_{mmax+1,n} = qx(y_n)$
$\quad$ sum $=$ sum $+ U_{0,n} + U_{mmax+1,n}$
Set ave $=$ sum$/(mmax + nmax)/2$
For $n = 1, 2, \ldots, nmax$

For $m = 1, 2, \ldots, mmax$, set
$$U_{mn} = \text{ave} \quad \text{(initialize } U^0_{mn} \text{ to average of boundary values)}$$
$$F_{mn} = -(hx)^2 f(x_m, y_n)$$

Step 3. Begin SOR iterations

For $i = 1$ to $itmax$, perform steps 4–9

Step 4. Begin stepping through horizontal rows

For $n = 1, 2, \ldots, nmax$, perform steps 5–7

Step 5. Define tridiagonal system for row n

For $m = 1, 2, \ldots, mmax$, set
$$a_m = -1$$
$$b_m = 2 + 2q$$
$$c_m = -1$$
$$d_m = F_{mn} + q \cdot U_{m,n-1} + q \cdot U_{m,n+1}$$
Set $d_1 = d_1 + U_{0,n}$
$$d_{mmax} = d_{mmax} + U_{mmax+1,n}$$

Step 6. Solve tridiagonal system for row n

Call TRIDI($mmax, a, b, c, d$)   (Algorithm 8.2)

Step 7. Perform SOR update on row n

For $m = 1, 2, \ldots, mmax$, set

$$U_{mn} = \text{omega} \cdot d_m + [1 - \text{omega}] \cdot U_{mn}$$

Step 8. Calculate maximum residual for ith SOR iterate

Set $rmax = 0$
For $n = 1, 2, \ldots, nmax$
   For $m = 1, 2, \ldots, mmax$, set
      res $= \{F_{mn} + [U_{m-1,n} + U_{m+1,n} - (2 + 2q)U_{mn} + qU_{m,n-1}$
         $+ qU_{m,n+1}]\}/h_x^2$
   If $|\text{res}| > rmax$, then
   Set $rmax = |\text{res}|$

Step 9. Compare maximum residual to tolerance

If $rmax < tol$ then go to step 10

Step 10. Output Solution

Write i, the number of SOR iterations performed

For $n = nmax, nmax - 1, \ldots, 2, 1$
For $m = 1, 2, \ldots, mmax$
     write U_{mn}

Exercises

1. Given the difference equations

$$
\begin{aligned}
4U_{11} - U_{21} - U_{12} &= 6 \\
-U_{11} + 4U_{21} \phantom{- U_{12}} - U_{22} &= 10 \\
-U_{11} \phantom{+ 4U_{21}} + 4U_{12} - U_{22} &= 10 \\
- U_{21} - U_{12} + 4U_{22} &= 14
\end{aligned}
$$

use the initial estimate $[U_{11}^0, U_{21}^0, U_{12}^0, U_{22}^0]^T = [1, 1, 1, 1]^T$ and five applications of
(a) Jacobi method
(b) Gauss–Seidel method
(c) SOR with $\omega = 1.2$, $\omega = 1.5$, and $\omega = 1.8$ to estimate the solution.
(d) Use Algorithm 10.1, Gaussian elimination, to obtain the exact solution.

2. Apply point Jacobi iteration and Gauss–Seidel iteration to estimate the solution of

(a)
$$
\begin{bmatrix}
4 & -1 & 0 & -1 & 0 & 0 \\
-1 & 4 & -1 & 0 & -1 & 0 \\
0 & -1 & 4 & 0 & 0 & -1 \\
-1 & 0 & 0 & 4 & -1 & 0 \\
0 & -1 & 0 & -1 & 4 & -1 \\
0 & 0 & -1 & 0 & -1 & 4
\end{bmatrix}
\begin{bmatrix}
U_{11} \\ U_{21} \\ U_{31} \\ U_{12} \\ U_{22} \\ U_{32}
\end{bmatrix}
=
\begin{bmatrix}
1 \\ 1 \\ 1 \\ 0 \\ 0 \\ 0
\end{bmatrix}
$$

(b)
$$
\begin{bmatrix}
-4 & 2 & 0 & 1 & 0 & 0 & 0 & 0 & 0 \\
1 & -4 & 1 & 0 & 1 & 0 & 0 & 0 & 0 \\
0 & 2 & -4 & 0 & 0 & 1 & 0 & 0 & 0 \\
1 & 0 & 0 & -4 & 2 & 0 & 1 & 0 & 0 \\
0 & 1 & 0 & 1 & -4 & 1 & 0 & 1 & 0 \\
0 & 0 & 1 & 0 & 2 & -4 & 0 & 0 & 1 \\
0 & 0 & 0 & 1 & 0 & 0 & -4 & 2 & 0 \\
0 & 0 & 0 & 0 & 1 & 0 & 1 & -4 & 1 \\
0 & 0 & 0 & 0 & 0 & 1 & 0 & 2 & -4
\end{bmatrix}
\begin{bmatrix}
U_{11} \\ U_{21} \\ U_{31} \\ U_{12} \\ U_{22} \\ U_{32} \\ U_{13} \\ U_{23} \\ U_{33}
\end{bmatrix}
=
\begin{bmatrix}
0 \\ 0 \\ 0 \\ 0 \\ 0 \\ 0 \\ 4 \\ 4 \\ 4
\end{bmatrix}
$$

3. Let Ω be the unit square $0 < x < 1$, $y < 0 < 1$. For each of the following boundary-value problems obtain a system of nine finite-difference equations. Introduce ghost points as necessary and provide a sketch of the grid on which your difference equations hold. Then solve the resulting difference equation using Gauss–Seidel point iteration.

(a) $\nabla^2 u = 0$ in Ω
$u(0, y) = 0$, $u(1, y) = 1$, $0 < y < 1$
$u_y(x, 0) = 0$, $u_y(x, 1) = 5$, $0 < x < 1$

(b) $\nabla^2 u = xy$ in Ω
$u_x(0, y) = 0$, $u_x(1, y) = 1$, $0 < y < 1$
$u(x, 0) = 0$, $u(x, 1) = 5$, $0 < x < 1$

(c) $\nabla^2 u = 0$ in Ω
$u_x(0, y) = 0$, $u(1, y) = y$, $0 < y < 1$
$u_y(x, 0) = 0$, $u(x, 1) = 0$, $0 < x < 1$

(d) $\nabla^2 u = 0$ in Ω
$u_x(0, y) = 0$, $u(1, y) = 1$, $0 < y < 1$
$u_y(x, 0) = 0$, $u_y(x, 1) = 0$, $0 < x < 1$

4. Repeat Exercise 3 using the point SOR method with (a) $\omega = 1.5$; (b) with optimal ω.

5. Repeat Exercise 3 using a finer grid with nine unknowns in each direction.

6. Consider the boundary-value problem

$$u_{xx} + u_{yy} = 0, \qquad 0 < x < \pi, 0 < y < \pi$$
$$u(0, y) = \sin y, \qquad u(\pi, y) = e^{\pi}\sin y$$
$$u(x, 0) = 0, \qquad u(x, \pi) = 0$$

whose exact solution is $u(x, y) = e^x \sin y$. With $hx = hy = \pi/10$, approximate the solution using the

    (a) point Jacobi method

    (b) point Gauss–Seidel method

    (c) point SOR method, $\omega = 1.528$

    (d) row Jacobi method

    (e) row Gauss–Seidel method

    (f) row SOR method (experiment with ω)

7. Rework Example 10.2.4 using the row Gauss–Seidel method of Algorithmn 10.3.

8. Rework parts (b) and (c) of Exercise 6 using $hx = \pi/20$ and $hy = \pi/10$.

9. Consider the boundary-value problem

$$(xu_x)_x + (e^y u_y)_y = e^y(1 + 2xe^y), \qquad 1 < x < 2, 0 < y < 1$$
$$u(1, y) = e^y, \qquad u(2, y) = 2e^y$$
$$u(x, 0) = x, \qquad u(x, 1) = xe^1$$

whose exact solution is $u(x, y) = xe^y$. On a square grid with $h = 0.1$:

    (a) Formulate a set of difference equations.

    (b) Approximate the solution to the difference equations of (a) using the Gauss–Seidel method.

10.5 CONVERGENCE OF ITERATIVE METHODS

Matrix Formulation of Iterative Methods

In Section 10.4, we described the Jacobi, the Gauss–Seidel, and the SOR iterative methods for approximating the solution to a system of linear finite-difference equations. Given an initial approximation to the solution of the difference equations U_{mn}^0, each method defines a sequence of recursively defined approximations U_{mn}^i, $i = 1, 2, \ldots$, to the solution of the difference system.

Section 10.4 was restricted to a discussion of the implementation of the iterative methods. In this section we take up the important topic of the convergence of the sequence of approximations U_{mn}^i, $i = 1, 2, \ldots$. Specific questions of interest are:

    1. How important is the choice of the initial estimate U_{mn}^0? Will an iterative method converge independent of U_{mn}^0, or must U_{mn}^0 be close to the solution of the difference system?

    2. How many iterations of a particular method will be required to obtain a satisfactory approximation to the solution of the difference system? What are the *convergence rates* of the various iterative methods?

    3. In the SOR method, how can we choose the relaxation parameter ω to maximize the convergence rate?

By formulating iterative methods of Section 10.4 in terms of the doubly subscripted array U_{mn}, we are able to obtain easily coded algorithms that use very little computer storage. Since the U_{mn} are in direct correspondence with the physical grid points (x_m, y_n), we refer to the approach based on multiply subscripted unknowns as the *physical approach.*

Although the physical labeling is preferred for computational purposes, in a discussion of the convergence of the various iterative methods, it is more convenient to use a single index to identify the unknowns in the linear system. We refer to a single indexing of the unknowns as the *linear algebra labeling.*

EXAMPLE 10.5.1 ⎯⎯⎯⎯⎯⎯⎯⎯⎯⎯⎯⎯⎯⎯⎯⎯⎯⎯⎯⎯⎯⎯⎯⎯⎯⎯⎯⎯⎯⎯⎯⎯⎯⎯

Consider the grid

$$(x_m, y_n) = (mh, nh), \qquad m = 1, 2, 3, n = 1, 2, h = \tfrac{1}{2}$$

A physically labeled set of unknowns would be

$$(U_{11}, U_{21}, U_{31}, U_{12}, U_{22}, U_{32})$$

The following DO LOOP defines a corresponding linear algebra labeling

$$
\begin{aligned}
&\text{DO 1 N} = 1, 2\\
&\text{DO 1 M} = 1, 3\\
&\quad \text{P} = (\text{N} - 1){*}3 + \text{M}\\
1\quad &\quad \text{V(P)} = \text{U(M, N)}
\end{aligned}
$$

$$(V_1, V_2, V_3, V_4, V_5, V_6) = (U_{11}, U_{21}, U_{31}, U_{12}, U_{22}, U_{32}) \qquad ■\ ■$$

Consider a system of linear difference equations given in linear algebra labeling as

$$\mathbf{AU} = \mathbf{b} \tag{10.5.1}$$

We assume that the connection between the physical and linear algebra labelings is as in Example 10.5.1. That is, the grid points are labeled from left to right from bottom to top. Let us write the coefficient matrix $\mathbf{A}$ in the form

$$\mathbf{A} = -\mathbf{L} + \mathbf{D} - \mathbf{U}$$

where $\mathbf{L}$, $\mathbf{D}$, and $\mathbf{U}$ are, respectively, strictly lower triangular, diagonal, and strictly upper triangular. Assume $\mathbf{D}$ is nonsingular, $\det(\mathbf{D}) \neq 0$.

In matrix notation, the iterative methods (10.4.5)–(10.4.7) can be expressed as follows.

Jacobi point iteration:

$$\mathbf{U}^{i+1} = \mathbf{T}_J \mathbf{U}^i + \mathbf{c}_J, \qquad \mathbf{T}_J = \mathbf{D}^{-1}(\mathbf{L} + \mathbf{U}), \qquad \mathbf{c}_J = \mathbf{D}^{-1}\mathbf{b} \tag{10.5.2}$$

Gauss–Seidel point iteration:

$$\mathbf{U}^{i+1} = \mathbf{T}_G \mathbf{U}^i + \mathbf{c}_G, \qquad \mathbf{T}_G = (\mathbf{D} - \mathbf{L})^{-1}\mathbf{U}, \qquad \mathbf{c}_G = (\mathbf{D} - \mathbf{L})^{-1}\mathbf{b} \tag{10.5.3}$$

SOR point iteration:

$$\mathbf{U}^{i+1} = \mathbf{T}_\omega \mathbf{U}^i + \mathbf{c}_\omega, \qquad \mathbf{T}_\omega = (\mathbf{D} - \omega\mathbf{L})^{-1}[(1 - \omega)\mathbf{D} + \omega\mathbf{U}],$$
$$\mathbf{c}_\omega = \omega(\mathbf{D} - \omega\mathbf{L})^{-1}\mathbf{b} \tag{10.5.4}$$

Convergence

In (10.5.2)–(10.5.4), if $\mathbf{U}^i$ converges to $\mathbf{U}^*$ as $i \to \infty$, then it is easy to show that $\mathbf{U}^*$ satisfies $\mathbf{A}\mathbf{U}^* = \mathbf{b}$. Clearly, each of the iterative methods (10.5.2)–(10.5.4) has the form

$$\mathbf{U}^{i+1} = \mathbf{T}\mathbf{U}^i + \mathbf{c} \qquad (10.5.5)$$

The spectral radius of a matrix $\mathbf{T}$, denoted by $\rho(\mathbf{T})$, is defined by

$$\rho(\mathbf{T}) = \max|\lambda_p|$$

where λ_p is an eigenvalue of $\mathbf{T}$. The following theorem relates the convergence of an iterative method of the form to the spectral radius of the matrix $\mathbf{T}$.

Theorem 10.5.1. The sequence $\mathbf{U}^i$ defined by $\mathbf{U}^{i+1} = \mathbf{T}\mathbf{U}^i + \mathbf{c}$ with $\mathbf{U}^0$ arbitrary converges to a unique vector $\mathbf{U}^*$ if and only if $\rho(\mathbf{T}) < 1$.

EXAMPLE 10.5.2 ——————————————————————————————————

Given the matrix

$$\mathbf{A} = \begin{bmatrix} 4 & -1 & -1 & 0 \\ -1 & 4 & 0 & -1 \\ -1 & 0 & 4 & -1 \\ 0 & -1 & -1 & 4 \end{bmatrix}$$

then using the decomposition $\mathbf{A} = -\mathbf{L} + \mathbf{D} - \mathbf{U}$,

$$\mathbf{D} = \begin{bmatrix} 4 & 0 & 0 & 0 \\ 0 & 4 & 0 & 0 \\ 0 & 0 & 4 & 0 \\ 0 & 0 & 0 & 4 \end{bmatrix}, \quad \mathbf{L} = \begin{bmatrix} 0 & 0 & 0 & 0 \\ 1 & 0 & 0 & 0 \\ 1 & 0 & 0 & 0 \\ 0 & 1 & 1 & 0 \end{bmatrix}, \quad \mathbf{U} = \begin{bmatrix} 0 & 1 & 1 & 0 \\ 0 & 0 & 0 & 1 \\ 0 & 0 & 0 & 1 \\ 0 & 0 & 0 & 0 \end{bmatrix}$$

Thus, $\mathbf{T}_J$ is given by $\mathbf{D}^{-1}(\mathbf{L} + \mathbf{U})$ to be

$$\mathbf{T}_J = \begin{bmatrix} 0 & \frac{1}{4} & \frac{1}{4} & 0 \\ \frac{1}{4} & 0 & 0 & \frac{1}{4} \\ \frac{1}{4} & 0 & 0 & \frac{1}{4} \\ 0 & \frac{1}{4} & \frac{1}{4} & 0 \end{bmatrix}$$

The eigenvalues of $\mathbf{T}_J$ can be related to the eigenvalues of $\mathbf{A}$. Let λ denote an eigenvalue of $\mathbf{A}$ and μ an eigenvalue of $\mathbf{T}_J$.

$$\mathbf{A}\mathbf{V} = \lambda\mathbf{V} \Rightarrow (-\mathbf{L} + \mathbf{D} - \mathbf{U})\mathbf{V} = \lambda\mathbf{V} \Rightarrow \mathbf{D}\mathbf{V} = (\mathbf{L} + \mathbf{U})\mathbf{V} + \lambda\mathbf{V}$$
$$\Rightarrow \mathbf{R}_J\mathbf{V} = (\mathbf{I} - \lambda\mathbf{D}^{-1})\mathbf{V} \Rightarrow \mathbf{T}_J\mathbf{V} = (\mathbf{I} - \tfrac{1}{4}\lambda\mathbf{I})\mathbf{V} \Rightarrow \mathbf{T}_J\mathbf{V} = (1 - \tfrac{1}{4}\lambda)\mathbf{V}$$

which leads us to conclude that

$$\mu = 1 - \tfrac{1}{4}\lambda$$

Since the eigenvalues of $\mathbf{A}$ are given by

$$\lambda_{mn} = 4 - 2\left[\cos\frac{m\pi}{3} + \cos\frac{n\pi}{3}\right], \qquad m, n = 1, 2$$

it follows that the eigenvalues of T_J are

$$\mu_{mn} = \frac{1}{2}\left[\cos\frac{m\pi}{3} + \cos\frac{n\pi}{3}\right], \qquad m, n = 1, 2$$

and the spectral radius of T_J is $\rho(T_J) = \sqrt{3}/2$. ■■

The proof of Theorem 10.5.1 can be found in R. L. Burden and J. D. Faires, *Numerical Analysis*, 3rd, ed., Prindle, Weber & Schmidt, Boston, 1985, pp. 430–431. In many practical applications, the calculation of the spectral radius cannot be easily performed. Theorem 10.5.2 provides conditions on A, rather than $\rho(T)$, that ensure the convergence of the Jacobi method. As we see in what follows Theorem 10.5.2 together with Theorem 10.5.3 also shows that the Gauss–Seidel method converges if the hypotheses of Theorem 10.5.2 are met.

Theorem 10.5.2. Given the system $AU = b$, $A = [a_{ij}]$, if

    (i) $a_{ii} > 0$, $a_{ij} \leq 0$, $i, j = 1, 2, \ldots, P$, $i \neq j$

    (ii) $\displaystyle\sum_{\substack{j=1 \\ j\neq i}}^{P} |a_{ij}| \leq a_{ii}$, $i = 1, 2, \ldots, P$, and

    (iii) A is irreducible,

then the Jacobi method (10.5.2) converges.

The proof of Theorem 10.5.2 is not difficult. First, conditions (i)–(iii) are used to show that A is nonsingular. Then, if we assume that T_J has an eigenvalue λ with $|\lambda| \geq 1$, we are able to conclude that either $\det(A) = 0$ or the eigenvector associated with λ is identically zero. Thus a contradiction is obtained, and we must conclude that all eigenvalues of T_J have magnitude less than 1. See W. F. Ames, *Numerical Methods for Partial Differential Equations*, 2nd ed, Academic Press, New York, 1977, p. 111.

Convergence Rates

To gain a clearer understanding of the role that the spectral radius $\rho(T)$ plays in the convergence of an iterative method, suppose that T has a complete set of eigenvectors V_p with

$$TV_p = \lambda_p V_p, \qquad |V_p| \neq 0, \qquad p = 1, 2, \ldots, P$$

Assume that $\rho(T) < 1$, so that Theorem 10.5.1 guarantees that U^i defined by

$$U^i = TU^{i-1} + c \tag{10.5.6}$$

converges to U^* satisfying

$$U^* = TU^* + c \tag{10.5.7}$$

As a measure of the discrepancy between the ith iterate U^i and the solution of the linear system U^*, we introduce the "error vector"

$$E^i = U^* - U^i$$

Subtracting (10.5.6) from (10.5.7) shows that the error vectors satisfy

$$\mathbf{E}^i = \mathbf{T}\mathbf{E}^{i-1} \tag{10.5.8}$$

If $\mathbf{E}^0$ has the eigenvector expansion

$$\mathbf{E}^0 = \sum_{p=1}^{P} a_p \mathbf{V}_p$$

then

$$\mathbf{E}^1 = \mathbf{T}\mathbf{E}^0 = \mathbf{T} \sum_{p=1}^{P} a_p \mathbf{V}_p = \sum_{p=1}^{P} a_p \mathbf{T}\mathbf{V}_p = \sum_{p=1}^{P} a_p \lambda_p \mathbf{V}_p$$

so repeated applications of (10.5.8) yield

$$\mathbf{E}^i = \sum_{p=1}^{P} a_p (\lambda_p)^i \mathbf{V}_p \quad \text{or} \quad \mathbf{E}^i = \rho(\mathbf{T})^i \sum_{p=1}^{P} a_p \left[\frac{\lambda_p}{\rho(\mathbf{T})} \right]^i \mathbf{V}_p \tag{10.5.9}$$

Equation (10.5.9) shows why all eigenvalues of $\mathbf{T}$ must satisfy

$$|\lambda_p| < 1$$

for (10.5.5) to be convergent.

Using (10.5.9) to represent $\mathbf{E}^i$ and $\mathbf{E}^{i-1}$, it is not difficult to show that

$$|\mathbf{E}^i|/|\mathbf{E}^{i-1}| \simeq \rho(\mathbf{T}) \quad \text{as } i \to \infty \tag{10.5.10}$$

Having completed i iterations of (10.5.5) to obtain $\mathbf{U}^i$, suppose we want to estimate how many additional iterations I are required to reduce the magnitude of the error vector by a factor of 10. That is, what value of I gives

$$|\mathbf{E}^{i+I}| = \tfrac{1}{10}|\mathbf{E}^i| \quad \text{or} \quad |\mathbf{E}^{i+I}|/|\mathbf{E}^i| = \tfrac{1}{10}$$

Using (10.5.10), we find

$$\rho(\mathbf{T})^I \simeq \tfrac{1}{10}$$

and, upon taking the logarithm of both sides of the last equation, that

$$I \simeq -1/\log_{10}\rho(\mathbf{T}) \tag{10.5.11}$$

Because of the role it plays in (10.5.11),

$$-\log_{10}\rho(\mathbf{T})$$

is called the *convergence rate* for the iterative method (10.5.5). For large i, the reciprocal of the convergence rate is roughly the number of iterations of (10.5.5) required to reduce the maximum residual by a factor of 10. A large convergence rate, and thus a rapidly convergent iterative method, corresponds to a small value of $\rho(\mathbf{T})$. In constructing iterative methods of the form (10.5.5), much effort is directed to selecting an iteration matrix $\mathbf{T}$ that has a small spectral radius.

The *asymptotic convergence rate* is the dominant term in the convergence rate as the grid spacing approaches zero. For a square of side length π and a square grid with size h, the convergence rates and asymptotic convergence rates for the Jacobi,

Gauss–Seidel, and SOR are displayed in Table 10.5.1. To obtain the rates for a square of side length L, replace h by $\pi h/L$ in Table 10.5.1.

TABLE 10.5.1 CONVERGENCE RATES FOR T_J, T_G, and T_ω

Method	Convergence rate	Asymptotic convergence rate
Point Jacobi	$-\log(\cos h)$	$h^2/2$
Point Gauss–Seidel	$-\log(\cos^2 h)$	h^2
Optimal point SOR	$-\log \dfrac{1 - \sin h}{1 + \sin h}$	$2h$
Row Jacobi	$-\log \dfrac{\cos h}{2 - \cos h}$	h^2
Row Gauss–Seidel	$-\log \left[\dfrac{\cos h}{2 - \cos h}\right]^2$	$2h^2$
Optimal row SOR	$-\log \left[\dfrac{1 - \sqrt{2}\,\sin(h/2)}{1 + \sqrt{2}\,\sin(h/2)}\right]^2$	$2\sqrt{2}\,h$

Relationships for $\rho(T_J)$, $\rho(T_G)$, and $\rho(T_\omega)$

The next three theorems relate the spectral radii of the Jacobi, Gauss–Seidel, and SOR methods. The first result is due to Stein and Rosenberg. Its proof can be found in D. M. Young and R. T. Gregory, *A Survey of Numerical Mathematics*, Vol. 1, Addison-Wesley, Reading, Mass., 1972, pp. 120–127.

Theorem 10.5.3. Given the system $\mathbf{AU} = \mathbf{b}$, $\mathbf{A} = [a_{ij}]$, if

$$a_{ii} > 0, \quad i = 1, 2, \ldots, P$$
$$a_{ij} \leq 0, \quad j \neq i, i = 1, 2, \ldots, P$$

then one and only one of the following statements holds

(i) $0 < \rho(T_G) < \rho(T_J) < 1$
(ii) $1 < \rho(T_J) < \rho(T_G)$
(iii) $\rho(T_G) = \rho(T_J) = 0$
(iv) $\rho(T_G) = \rho(T_J) = 1$

One consequence of Theorem 10.5.3 is that for elliptic finite-difference equations, the Jacobi and Gauss–Seidel methods converge and diverge together. Moreover, when convergence takes place, we can expect the Gauss–Seidel method to converge more rapidly than the Jacobi method. Theorems 10.5.2 and 10.5.3 together imply that under condiitons (i)–(iii) of Theorem 10.5.3, the Gauss–Seidel method is convergent. To discuss the convergence of the SOR method, we require two more definitions.

We say that the matrix $\mathbf{A}$ is *two-cyclic* if there exists a permutation of the rows and corresponding columns that allows $\mathbf{A}$ to be written in the form

$$\mathbf{A} = \begin{bmatrix} \mathbf{D}_1 & \mathbf{F} \\ \mathbf{G} & \mathbf{D}_2 \end{bmatrix}$$

where $\mathbf{D}_1$ and $\mathbf{D}_2$ are square diagonal matrices and $\mathbf{F}$ and $\mathbf{G}$ are rectangular matrices. In the terminology of linear algebra, we say $\mathbf{A}$ is two-cyclic if there exists a permutation matrix $\mathbf{P}$, a matrix $\mathbf{P}$ having only 0 and 1 entries with exactly one 1 in each column and each row, such that

$$\mathbf{PAP}^T = \mathbf{PAP}^{-1} = \begin{bmatrix} \mathbf{D}_1 & \mathbf{F} \\ \mathbf{G} & \mathbf{D}_2 \end{bmatrix}$$

If a matrix $\mathbf{A} = \mathbf{D} - \mathbf{L} - \mathbf{U}$ is two-cyclic, then it is called *consistently ordered* if all the eigenvalues of the matrix

$$\beta\mathbf{L} + \beta^{-1}\mathbf{U}, \qquad \beta \neq 0$$

are independent of β. In general, it is not an easy matter to show that a matrix is two-cyclic and consistently ordered. For discussions of elliptic finite-difference equations, it is important to know that matrices of the form

$$\begin{bmatrix} \mathbf{D}_1 & \mathbf{F}_1 \\ \mathbf{G}_2 & \mathbf{D}_2 & \mathbf{F}_2 \\ & \mathbf{G}_3 & \mathbf{D}_3 & \mathbf{F}_3 \\ & & & \ddots \\ & & & & \ddots \\ & & & & \mathbf{G}_{q-1} & \mathbf{D}_{q-1} & \mathbf{F}_{q-1} \\ & & & & & \mathbf{G}_q & \mathbf{F}_q \end{bmatrix}$$

where $\mathbf{D}_i$, $i = 1, 2, \ldots, q$, are square diagonal matrices, not necessarily of the same order, are two-cyclic, and consistently ordered. A good discussion of two-cyclic matrices and consistent orderings can be found in G. D. Smith, *Numerical Solutions of Partial Differential Equations: Finite-Difference Methods*, Oxford, Univ. Press, London, 1978, pp. 243–263.

EXAMPLE 10.5.3

Consider a finite-difference equation that arises in the discretization of the Laplacian on a square 3×3 grid of the type indicated in Figure 10.5.1. Namely, $\mathbf{AU} = \mathbf{b}$, or

$$\begin{bmatrix} -4 & 1 & 0 & 1 & 0 & 0 & 0 & 0 & 0 \\ 1 & -4 & 1 & 0 & 1 & 0 & 0 & 0 & 0 \\ 0 & 1 & -4 & 0 & 0 & 1 & 0 & 0 & 0 \\ 1 & 0 & 0 & -4 & 1 & 0 & 1 & 0 & 0 \\ 0 & 1 & 0 & 1 & -4 & 1 & 0 & 1 & 0 \\ 0 & 0 & 1 & 0 & 1 & -4 & 0 & 0 & 1 \\ 0 & 0 & 0 & 1 & 0 & 0 & -4 & 1 & 0 \\ 0 & 0 & 0 & 0 & 1 & 0 & 1 & -4 & 1 \\ 0 & 0 & 0 & 0 & 0 & 1 & 0 & 1 & -4 \end{bmatrix} \begin{bmatrix} U_1 \\ U_2 \\ U_3 \\ U_4 \\ U_5 \\ U_6 \\ U_7 \\ U_8 \\ U_9 \end{bmatrix} = \begin{bmatrix} b_1 \\ b_2 \\ b_3 \\ b_4 \\ b_5 \\ b_6 \\ b_7 \\ b_8 \\ b_9 \end{bmatrix}$$

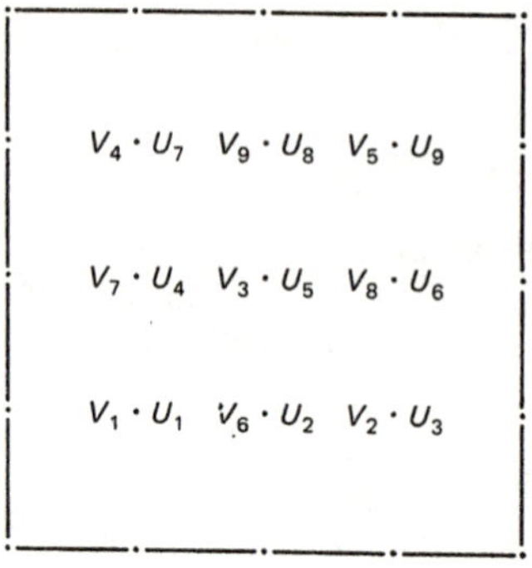

Figure 10.5.1 A 3×3 grid with U values labeled in usual fashion; V values labeled in red–black fashion.

If we write the difference system in terms of the red–black, or checkerboard-labeled, V values indicated in Figure 10.5.1, then the difference system takes the form $\mathcal{A}\mathbf{V} = \mathbf{B}$, or

$$
\begin{bmatrix}
-4 & 0 & 0 & 0 & 0 & 1 & 1 & 0 & 0 \\
0 & -4 & 0 & 0 & 0 & 1 & 0 & 1 & 0 \\
0 & 0 & -4 & 0 & 0 & 1 & 1 & 1 & 1 \\
0 & 0 & 0 & -4 & 0 & 0 & 0 & 1 & 1 \\
0 & 0 & 0 & 0 & -4 & 0 & 0 & 1 & 0 \\
1 & 1 & 1 & 0 & 0 & -4 & 0 & 0 & 0 \\
1 & 0 & 1 & 1 & 0 & 0 & -4 & 0 & 0 \\
0 & 1 & 1 & 0 & 1 & 0 & 0 & -4 & 0 \\
0 & 0 & 1 & 1 & 1 & 0 & 0 & 0 & -4
\end{bmatrix}
\begin{bmatrix}
V_1 \\ V_2 \\ V_3 \\ V_4 \\ V_5 \\ V_6 \\ V_7 \\ V_8 \\ V_9
\end{bmatrix}
=
\begin{bmatrix}
B_1 \\ B_2 \\ B_3 \\ B_4 \\ B_5 \\ B_6 \\ B_7 \\ B_8 \\ B_9
\end{bmatrix}
$$

Using the dotted lines to decompose the coefficient matrix $\mathcal{A}$, we see it to be two-cyclic.

Let $\mathbf{P}$ be the permutation matrix obtained by writing the rows of the 9×9 identity matrix in the order 1, 3, 5, 7, 9, 2, 4, 6, 8. This is the ordering of the U variables relative to the V variables. It is not difficult to show that

$$\mathbf{P}\mathbf{A}\mathbf{P}^T = \mathcal{A}$$

from which we conclude that $\mathbf{A}$ is two-cyclic. ■ ■

The next results allow us to relate the SOR method to the Jacobi method. The proofs of Theorems 10.5.4 and 10.5.5 can be found in Mittchel and Griffiths, *The Finite Difference Method in Partial Differential Equations*, Wiley, New York, 1978, pp. 143–145.

Theorem 10.5.4. If $\mathbf{A}$ is two-cyclic and consistently ordered, then the eigenvalues μ of $\mathbf{T}_J$ and the eigenvalues λ of $\mathbf{T}_\omega$ satisfy

$$(\lambda + \omega - 1)^2 = \lambda\omega^2\mu^2, \qquad \omega \neq 0, \lambda \neq 0 \tag{10.5.12}$$

Since $\mathbf{T}_G = \mathbf{T}_\omega$ when $\omega = 1$, Theorem 10.5.4 relates the eigenvalues of $\mathbf{T}_J$ to the eigenvalues of both $\mathbf{T}_G$ and $\mathbf{T}_\omega$. In particular Theorem 10.5.4 shows that

$$\rho(\mathbf{T}_G) = \rho(\mathbf{T}_J)^2$$

which implies that the convergence rate for the Gauss–Seidel method is twice that of the Jacobi method.

Equation (10.5.12) can be used to determine the optimal relaxation parameter to be used in the SOR method. Let $\overline{\omega}$ denote the value of ω that minimizes the spectral radius of the SOR iteration matrix. Minimizing the maximum of the λ values given in (10.5.12) leads to the following result.

Theorem 10.5.5. If $\mathbf{A}$ is two-cyclic and consistently ordered, the optimal SOR relaxation parameter is

$$\overline{\omega} = \frac{2}{1 + \sqrt{1 - \rho(\mathbf{T}_J)^2}} \tag{10.5.13}$$

Moreover, the spectral radius $\rho(\mathbf{T}_\omega)$ of the SOR method is given by

$$\rho(\mathbf{T}_\omega) = \overline{\omega} - 1$$

EXAMPLE 10.5.4 ───

Let us determine the optimal point SOR relaxation parameter for the difference system $\mathbf{AU} = \mathbf{b}$ of Example 10.5.3. From Theorem 10.3.1 and the two examples that follow it, we know the eigenvalues of $\mathbf{A}$ to be

$$\lambda_{mn} = 4 - 2\left[\cos\frac{m\pi}{4} + \cos\frac{n\pi}{4}\right], \qquad m, n = 1, 2, 3$$

Following Example 10.5.2, we see that the eigenvalues of $\mathbf{T}_J$ are

$$\mu_{mn} = \frac{1}{2}\left[\cos\frac{m\pi}{4} + \cos\frac{n\pi}{4}\right], \qquad m, n = 1, 2, 3$$

and, therefore, that $\rho(\mathbf{T}_J) = 1/\sqrt{2} \cong 0.707$. According to Theorem 10.5.5, the value of ω that minimizes $\rho(\mathbf{T}_\omega)$ is

$$\overline{\omega} = 2/(1 + \sqrt{\tfrac{1}{2}}) \cong 1.172$$

Moreover $\rho(\mathbf{T}_\omega) \cong 0.172$.

Exercises

1. Determine the eigenvalues of the matrix

$$\mathbf{A} = \begin{bmatrix} 4 & -1 & 0 \\ -1 & 4 & -1 \\ 0 & -1 & 4 \end{bmatrix}$$

and find its spectral radius.

2. Let $\mathbf{P}$ be the permutation matrix

$$\mathbf{P} = \begin{bmatrix} 1 & 0 & 0 \\ 0 & 0 & 1 \\ 0 & 1 & 0 \end{bmatrix}$$

(a) Find $\mathbf{PAP}^T$ where $\mathbf{A}$ is as in Exercise 1.

(b) Show that $\mathbf{A}$ is two-cyclic.

(c) Show that $\mathbf{A}$ is consistently ordered.

3. Show that matrix $\mathbf{A}$ of Example 10.5.3 is consistently ordered.

4. Find the spectral radius $\rho(\mathbf{T}_G)$ for the matrix $\mathbf{A}$ of Example 10.5.3.

5. For the matrix $\mathbf{A}$ of Example 10.5.2, find the spectral radius $\rho(\mathbf{T}_G)$ of the point Gauss–Seidel method

6. For the matrix $\mathbf{A}$ of Example 10.5.2, find the optimal point SOR relaxation parameter $\bar{\omega}$.

7. Given the matrix

$$\mathbf{A} = \begin{bmatrix} 4 & -2 & -1 & 0 \\ -2 & 4 & 0 & -1 \\ -1 & 0 & 4 & -2 \\ 0 & -1 & -2 & 4 \end{bmatrix}$$

(a) Find the eigenvalue of $\mathbf{A}$.

(b) Find the eigenvalues of the point Jacobi iteration matrix $\mathbf{T}_J$.

(c) Find the spectral radius $\rho(\mathbf{T}_J)$ of the point Jacobi method.

(d) Find the spectral radius $\rho(\mathbf{T}_G)$ of the point Gauss–Seidel method.

(e) Find the optimal relaxation parameter for point SOR.

8. Repeat Exercise 7 for

$$\mathbf{A} = \begin{bmatrix} -4 & 2 & 0 & 1 & 0 & 0 & 0 & 0 & 0 \\ 1 & -4 & 1 & 0 & 1 & 0 & 0 & 0 & 0 \\ 0 & 2 & -4 & 0 & 0 & 1 & 0 & 0 & 0 \\ 1 & 0 & 0 & -4 & 2 & 0 & 1 & 0 & 0 \\ 0 & 1 & 0 & 1 & -4 & 1 & 0 & 1 & 0 \\ 0 & 0 & 1 & 0 & 2 & -4 & 0 & 0 & 1 \\ 0 & 0 & 0 & 1 & 0 & 0 & -4 & 2 & 0 \\ 0 & 0 & 0 & 0 & 1 & 0 & 1 & -4 & 1 \\ 0 & 0 & 0 & 0 & 0 & 1 & 0 & 2 & -4 \end{bmatrix}$$

9. Repeat Exercise 7 for

$$\mathbf{A} = \begin{bmatrix} -4 & 2 & 0 & 1 & 0 & 0 & 0 & 0 & 0 \\ 1 & -4 & 1 & 0 & 1 & 0 & 0 & 0 & 0 \\ 0 & 1 & -4 & 0 & 0 & 1 & 0 & 0 & 0 \\ 1 & 0 & 0 & -4 & 2 & 0 & 1 & 0 & 0 \\ 0 & 1 & 0 & 1 & -4 & 1 & 0 & 1 & 0 \\ 0 & 0 & 1 & 0 & 1 & -4 & 0 & 0 & 1 \\ 0 & 0 & 0 & 1 & 0 & 0 & -4 & 2 & 0 \\ 0 & 0 & 0 & 0 & 1 & 0 & 1 & -4 & 1 \\ 0 & 0 & 0 & 0 & 0 & 1 & 0 & 1 & -4 \end{bmatrix}$$

10. For the matrix $\mathbf{A}$ of Exercise 8, calculate the convergence rate for

(a) point Jacobi method

(b) point Gauss–Seidel method

(c) point optimal SOR method

11. Repeat Exercise 10 for the matrix $\mathbf{A}$ of Exercise 9.

12. For the matrix $\mathbf{A}$ of Example 10.5.3, find the eigenvalues of the iteration matrix for the row Jacobi method.

13. For the matrix A of Example 10.5.3, find the spectral radius for the row Gauss–Seidel method.

14. For the matrix A of Example 10.5.3, find the relaxation parameter that minimizes the spectral radius of the row SOR method.

15. Show that the matrix A of Example 10.5.2 is
    (a) two-cyclic
    (b) consistently ordered

16. Consider the system of P linear equations $AU = b$, $A = [a_{pq}]$. If $a_{ii} \neq 0$, show that
    (a) the Jacobi method can be written in the form

$$U_n^i = \frac{b_n - \sum_{q=1}^{n-1} a_{nq} U_n^{i-1} - \sum_{q=n+1}^{P} a_{nq} U_n^{i-1}}{a_{nn}}$$

    (b) the Gauss–Seidel method can be written in the form

$$U_n^i = \frac{b_n - \sum_{q=1}^{n-1} a_{nq} U_n^i - \sum_{q=n+1}^{P} a_{nq} U_n^{i-1}}{a_{nn}}$$

17. Apply point Jacobi iteration and Gauss–Seidel iteration and optimal SOR to estimate the solution of

(a)
$$\begin{bmatrix} -4 & 1 & 0 \\ 1 & -4 & 1 \\ 0 & 1 & -4 \end{bmatrix} \begin{bmatrix} U_1 \\ U_2 \\ U_3 \end{bmatrix} = \begin{bmatrix} -2 \\ -4 \\ -10 \end{bmatrix}$$

(b)
$$\begin{bmatrix} -4 & 2 & 0 \\ 1 & -4 & 1 \\ 0 & 2 & -4 \end{bmatrix} \begin{bmatrix} U_1 \\ U_2 \\ U_3 \end{bmatrix} = \begin{bmatrix} -2 \\ -2 \\ -2 \end{bmatrix}$$

(c)
$$\begin{bmatrix} 4 & -1 & 0 & -1 & 0 & 0 \\ -1 & 4 & -1 & 0 & -1 & 0 \\ 0 & -1 & 4 & 0 & 0 & -1 \\ -1 & 0 & 0 & 4 & -1 & 0 \\ 0 & -1 & 0 & -1 & 4 & -1 \\ 0 & 0 & -1 & 0 & -1 & 4 \end{bmatrix} \begin{bmatrix} U_1 \\ U_2 \\ U_3 \\ U_4 \\ U_5 \\ U_6 \end{bmatrix} = \begin{bmatrix} 1 \\ 1 \\ 1 \\ 0 \\ 0 \\ 0 \end{bmatrix}$$

(d)
$$\begin{bmatrix} -4 & 2 & 0 & 1 & 0 & 0 & 0 & 0 & 0 \\ 1 & -4 & 1 & 0 & 1 & 0 & 0 & 0 & 0 \\ 0 & 2 & -4 & 0 & 0 & 1 & 0 & 0 & 0 \\ 1 & 0 & 0 & -4 & 2 & 0 & 1 & 0 & 0 \\ 0 & 1 & 0 & 1 & -4 & 1 & 0 & 1 & 0 \\ 0 & 0 & 1 & 0 & 2 & -4 & 0 & 0 & 1 \\ 0 & 0 & 0 & 1 & 0 & 0 & -4 & 2 & 0 \\ 0 & 0 & 0 & 0 & 1 & 0 & 1 & -4 & 1 \\ 0 & 0 & 0 & 0 & 0 & 1 & 0 & 2 & -4 \end{bmatrix} \begin{bmatrix} U_1 \\ U_2 \\ U_3 \\ U_4 \\ U_5 \\ U_6 \\ U_7 \\ U_8 \\ U_9 \end{bmatrix} = \begin{bmatrix} 0 \\ 0 \\ 0 \\ 0 \\ 0 \\ 0 \\ 4 \\ 4 \\ 4 \end{bmatrix}$$

Linear Algebra

1. ELEMENTARY NOTIONS

Most readers will be familiar with the two-and three-dimensional geometric vector spaces arising in physics and calculus. In that context they will have encountered the notions of vector addition, scalar multiplication, dot and cross products, orthogonality, and possibly even linear independence, subspaces, and bases. This knowledge, unfortunately, hardly constitutes an adequate knowledge of linear algebra for the purposes of studying applied mathematics. A more comprehensive grasp of linear algebra is essential for the presentation of the subjects treated in the second part of this book (the numerical solution of partial differential equations). In addition, linear algebra provides a framework in which the topics in the first part of the book can be understood on more than just a superficial level.

As an aid to those readers whose background in linear algebra may be weak, we have collected in this appendix the definitions and results that are central to our presentation. Explanations are necessarily brief and results are generally given without proof. References where a more complete treatment of the subject of linear algebra can be found are given at the end of this appendix.

The discussion in this appendix will be motivated by the practical problems of solving:

(a) Systems of simultaneous linear algebraic equations

$$Ax = b$$

(b) Algebraic eigenvalue problems

$$Ax = \mu x$$

Therefore we shall always here think of vectors as being N-tuples of numbers. We use the notation

$\mathbf{x}$ for a column vector of numbers and the notation $\mathbf{x}^T$ to represent a row vector; that is,

$$\mathbf{x} = \begin{bmatrix} x_1 \\ \cdot \\ \cdot \\ \cdot \\ x_N \end{bmatrix} \quad \text{and} \quad \mathbf{x}^T = (x_1, \ldots, x_N)$$

This distinction may seem artificial at this point, but its significance will become apparent when we introduce matrix multiplication.

We take the scalars in our vector space to be the real numbers unless otherwise specified. This will be adequate for the treatment of simultaneous equations, but when we begin to consider the algebraic eigenvalue problem, it will be necessary to consider a vector space over the complex numbers. The vector space over the reals will be denoted by R^N. It consists of all possible column or row vectors of N real numbers. Although we shall use different notation for column and row vectors, we think of these as two alternative ways of writing the same vector.

The operations of vector addition and subtraction and scalar multiplication are defined componentwise:

$$\mathbf{x}^T \pm \mathbf{y}^T = (x_1 \pm y_1, \ldots, x_N \pm y_n) \quad \text{for all vectors } \mathbf{x}, \mathbf{y}$$
$$\alpha\mathbf{x}^T = (\alpha x_1, \ldots, \alpha x_N) \quad \text{for all vectors } \mathbf{x} \text{ and scalars } \alpha$$

We give the definitions here in terms of the row vectors since this is more convenient from a typographical standpoint.

An additional operation we shall want to have is the dot, or inner, product, which is defined as

$$\mathbf{x} \cdot \mathbf{y} = (\mathbf{x}, \mathbf{y}) = \sum_{j=1}^{N} x_j y_j \quad \text{for all vectors } \mathbf{x}, \mathbf{y}$$

The notation $\mathbf{x} \cdot \mathbf{y}$ is the one more common in physics or calculus books while $(\mathbf{x}, \mathbf{y})$ usually is used to denote the dot, or inner, product in more advanced mathematics books. Either is acceptable as is another notation we present later.

We also need the notion of a matrix, that is an $m \times n$ array of numbers. When we speak of an $m \times n$ matrix, we mean an array with m rows and n columns. Then, for example, a row vector $\mathbf{x}^T$ is a $1 \times n$ matrix while a column vector $\mathbf{x}$ is an $n \times 1$ matrix.

For matrices, as with vectors, the algebraic operations of addition, subtraction, and multiplication by a scalar are defined entry wise:

$$\mathbf{A} \pm \mathbf{B} = [a_{ij} \pm b_{ij}] \quad \text{for } \mathbf{A} = [a_{ij}], \mathbf{B} = [b_{ij}]$$
$$\alpha\mathbf{A} = [\alpha a_{ij}] \quad \text{for } \mathbf{A} = (a_{ij}) \text{ and scalar } \alpha$$

Note that addition and subtraction are defined only for matrices of the same dimensions. One could not add an $m \times n$ matrix to another matrix that was not $m \times n$.

We also define multiplication of matrices. For an $m \times n$ matrix $\mathbf{A}$ and an $n \times p$ matrix $\mathbf{B}$, we define the product $\mathbf{AB}$ as that matrix whose (i, j) entry is the number

$$[\mathbf{AB}]_{ij} = \sum_{k=1}^{N} a_{ik}b_{kj}, \quad 1 \le i \le m, 1 \le j \le p$$

Note that the product $\mathbf{AB}$ is defined only if the number of columns of $\mathbf{A}$ equals the number of rows of $\mathbf{B}$. Here that number is equal to n. The dimensions of the product matrix $\mathbf{AB}$ are then $m \times p$. In particular, the product of a $1 \times n$ matrix by an $n \times 1$ matrix is defined, and the product is a 1×1 matrix; that is, a number. This is just the dot, or inner, product. In this case we would write it as

$$\mathbf{x}^T\mathbf{y} = \sum_{k=1}^{N} x_{1k}y_{k1} = \sum_{k=1}^{N} x_k y_k$$

The matrix product has the following properties:

  (i) **AB** does not in general equal **BA**
  (ii) **A(B + C) = AB + AC**
  (iii) **A(BC) = (AB)C**

In connection with the treatment of our two motivating problems of linear algebra, we very often consider the product of an $M \times N$ matrix with an $N \times 1$ column vector. This product can be expressed in two equivalent ways,

$$\mathbf{A}\mathbf{x} = \begin{bmatrix} \mathbf{R}_1 \cdot \mathbf{x} \\ \vdots \\ \mathbf{R}_M \cdot \mathbf{x} \end{bmatrix} = x_1\mathbf{C}_1 + \cdots + x_N\mathbf{C}_N$$

Here $\mathbf{R}_1, \ldots, \mathbf{R}_M$ denote the M rows of the matrix **A**. Since **A** is an $M \times N$ matrix, each of these rows is an N vector so that the dot product with **x** is well defined. Each of the columns of **A** is an M vector. Thus we can think of **A** times **x** as the sum of N M vectors or as a single M vector whose entries are the dot products of the M rows of **A** with the N vector **x**. We shall have occasion to refer back to these alternative interpretations of the product **Ax** later.

2. SIMULTANEOUS LINEAR EQUATIONS

Consider the problem of finding an N vector **x** satisfying

$$\mathbf{A}\mathbf{x} = \mathbf{b} \tag{2.1}$$

where **A** denotes (in general) a known $M \times N$ matrix and **b** denotes a given M vector. Although there are many practical applications in which M is different from N, they will not arise in this book, and so we consider just the case $M = N$ here. There are three cases to consider in connection with (2.1),

  (a) Nonsingular case. There exists a unique vector **x** satisfying (2.1).
  (b) Singular indeterminate case. There exist an infinite collection of solutions for (2.1).
  (c) Singular inconsistent case. There is no vector **x** that satisfies (2.1).

In order to find which of these cases applies and to find the unique solution when we are in case (a), the most common procedure is to use some form of Gaussian elimination. This amounts to applying to the $N \times (N + 1)$ augmented matrix [**A** : **b**], the so-called elementary row operations, in order to bring the augmented matrix into row-echelon form.

A matrix is said to be in *row-echelon form* if

  (i) in each row, the first nonzero entry lies on or above the diagonal. This first nonzero entry in each row is called the pivot for that row,
  (ii) below each pivot lies a column of zeroes, and
  (iii) no trivial row (a row of all zeros) lies above any nontrivial row.

A matrix in row-echelon form has no nonzero entries below the diagonal and is then an upper triangular matrix.

The elementary row operations consist of the following:

$C_j(\alpha)$: Replace row j with α times row j.

P_{ij}: Interchange rows i and j.

$E_{ij}(\alpha)$: Replace row i with row i minus α times row j.

Here α denotes a nonzero scalar.

If matrix **A** can be transformed into matrix **B** by finitely many elementary row operations, then we say **A** and **B** are row equivalent. Then we can define the *row rank* of a matrix **A** as the number of nontrivial rows in a row equivalent upper triangular matrix. We denote the row rank of **A** by rank[**A**]. Then we have:

Theorem 2.1. Let **A** be a known $N \times N$ matrix and **b** a given N vector. Then (2.1) is

  (i) nonsingular if rank[**A**] $=$ rank[**A** : **b**] $= N$,
  (ii) singular indeterminate if rank[**A**] $=$ rank[**A** : **b**] $< N$, and
  (iii) singular inconsistent if rank[**A**] $<$ rank[**A** : **b**].

An alternative way of classifying the system (2.1) as singular or nonsingular is by means of determinants. Determinants are defined only for square matrices, and for a square matrix **A** that is upper triangular we have

$$\det[\mathbf{A}] = d_1 d_2 \cdots d_N$$

where d_i denote the diagonal entries. In addition, we have

$$\det[E_{ij}(\alpha)\mathbf{A}] = \det[\mathbf{A}]$$
$$\det[C_j(\alpha)\mathbf{A}] = \alpha \det[\mathbf{A}]$$
$$\det[P_{ij}\mathbf{A}] = -\det[\mathbf{A}] \quad \text{for } i \neq j$$

Then we can compute the determinant of any square matrix and we have the following result.

Theorem 2.2. Let **A** be a given $N \times N$ matrix and **b** a given N vector. Then:

  (i) $\mathbf{Ax} = \mathbf{b}$ has a unique solution if and only if $\det[\mathbf{A}]$ is not equal to zero.
  (ii) $\mathbf{Ax} = \mathbf{0}$ has nontrivial solutions (necessarily nonunique) if and only if $\det[\mathbf{A}] = 0$.

An $N \times N$ matrix **A** for which $\det[\mathbf{A}]$ is not zero is said to be nonsingular. For such a matrix it is possible to find an $N \times N$ matrix **B** such that $\mathbf{AB} = \mathbf{BA} = \mathbf{I}$ (the identity matrix); it is usual to write $\mathbf{A}^{-1}$ for this matrix **B**, and to refer to $\mathbf{A}^{-1}$ as the inverse of **A**.

Theorem 2.3. For an $N \times N$ matrix **A**, the following are equivalent:

  (a) $\mathbf{A}^{-1}$ exists
  (b) rank[**A**] $= N$
  (c) $\det[\mathbf{A}]$ is not zero
  (d) $\mathbf{Ax} = \mathbf{b}$ is nonsingular

3. ADDITIONAL NOTIONS

In order to prove, understand, and extend the results of Section 2, it is necessary to introduce some additional notions.

Definition 3.1. A collection M of vectors is said to form a *subspace* if for all vectors $\mathbf{x}$ and $\mathbf{y}$ in M and all salcars α and β, the linear combination $\alpha\mathbf{x} + \beta\mathbf{y}$ necessarily belongs to M.

Examples of Subspaces

1. Let $\mathbf{A}$ denote an $N \times N$ matrix and consider the following collection of vectors in R^N:

$$M_1 = \text{set of all vectors in } R^N \text{ satisfying } \mathbf{Ax} = \mathbf{0}$$

To see that this is indeed a subspace, we must show that if $\mathbf{x}$ and $\mathbf{y}$ are both in M_1, then the linear combination $\alpha\mathbf{x} + \beta\mathbf{y}$ is also in M_1 for any choice of the scalars α and β. But if $\mathbf{x}$ and $\mathbf{y}$ are both in M_1, then $\mathbf{Ax} = \mathbf{0}$ and $\mathbf{Ay} = \mathbf{0}$ by the definition of M_1. Then it is easy to see that

$$\mathbf{0} = \alpha(\mathbf{Ax}) + \beta(\mathbf{Ay}) = \mathbf{A}(\alpha\mathbf{x} + \beta\mathbf{y})$$

which is to say $\alpha\mathbf{x} + \beta\mathbf{y}$ is in M_1; that is, M_1 is a subspace.

2. Let $\mathbf{v}_1, \ldots, \mathbf{v}_P$ denote a collection of p fixed vectors in R^N and consider the following set of vectors in R^N

$$M_2 = \text{collection of all possible linear combinations}$$
$$\text{of vectors } \mathbf{v}_1, \ldots, \mathbf{v}_P$$

In this case it is clear that M_2 must be a subspace, and we say that M_2 is the subspace *spanned* *by* the vectors $\mathbf{v}_1, \ldots, \mathbf{v}_P$. We denote this as follows,

$$M_2 = \text{span}[\mathbf{v}_1, \ldots, \mathbf{v}_P]$$

and we say the vectors $\mathbf{v}_1, \ldots, \mathbf{v}_P$ are a *spanning set* for the subspace M_2.

The subspace M_1 is an example of a subspace defined by a "membership rule"; that is, $\mathbf{x}$ is in M_1 if $\mathbf{Ax} = \mathbf{0}$. In case of a subspace defined by a membership rule, it is easy to see whether any given vector is or is not in the subspace. One simply tests whether the given vector satisfies the membership rule. On the other hand, one must *prove* that the collection defined by the membership rule is, in fact, a subspace.

The collection M_2 is a subspace defined in terms of a spanning set. Then it is clear from Definition 3.1 that M_2 is a subspace but given an arbitrary vector $\mathbf{w}$ in R^N, it is not so clear in general whether $\mathbf{w}$ does or does not belong to M_2. We have for this purpose the following:

Lemma 3.2. If

$$\text{rank} \begin{bmatrix} \mathbf{v}_1^T \\ \vdots \\ \mathbf{v}_P^T \end{bmatrix} = \text{rank} \begin{bmatrix} \mathbf{v}_1^T \\ \vdots \\ \mathbf{v}_P^T \\ \mathbf{w}^T \end{bmatrix}$$

then $\mathbf{w}$ belongs to $\text{span}[\mathbf{v}_1, \ldots, \mathbf{v}_P]$.

Here we have formed the matrices whose rows are, respectively, the vectors $\mathbf{v}_1^T, \ldots, \mathbf{v}_P^T$ and $\mathbf{v}_1^T, \ldots, \mathbf{v}_P^T, \mathbf{w}^T$. If the rank of the second matrix does not exceed the rank of the first, then w must be equal to some linear combination of the vectors $v_1, \ldots, v_p$; that is, w must belong to span$[v_1, \ldots, v_p]$. Then:

Definition 3.3. The vectors $v_1, \ldots, v_P$ are *linearly dependent* if and only if there exists scalars $\alpha_1, \ldots, \alpha_P$, not all zero, such that

$$\alpha_1 v_1 + \cdots + \alpha_P v_P = 0$$

Lemma 3.2 suggests that we can determine if a set of vectors is dependent by computing the rank of a matrix whose rows are the vectors.

Lemma 3.4. If

$$\text{rank} \begin{bmatrix} v_1^T \\ \vdots \\ v_P^T \end{bmatrix} < p$$

then the vectors $v_1, \ldots, v_P$ are linearly dependent.

Conversely, we can define:

Definition 3.5. The vectors $v_1, \ldots, v_P$ are *linearly independent* if and only if $\alpha_1 v_1 + \cdots + \alpha_P v_P = 0$ implies that $\alpha_1 = \cdots = \alpha_P = 0$.

Then the following analogue of Lemma 3.4 provides a computational recipe for establishing the independence of a set of vectors.

Lemma 3.6. If

$$\text{rank} \begin{bmatrix} v_1^T \\ \vdots \\ v_P^T \end{bmatrix} = P$$

then the vectors $v_1, \ldots, v_P$ are *linearly independent*.

Let $M = \text{span}[v_1, \ldots, v_P]$. Then for any vector x in M there are scalars $\alpha_1, \ldots, \alpha_P$ such that $\alpha_1 v_1 + \cdots + \alpha_P v_P = x$. In general, there may be more than one choice for the set of scalars. However, when the vectors $v_1, \ldots, v_P$ are linearly independent, then the scalars $\alpha_1, \ldots, \alpha_P$ such that

$$\alpha_1 v_1 + \cdots + \alpha_P v_P = x$$

are *uniquely* determined by x. This motivates:

Definition 3.7 The vectors $[v_1, \ldots, v_P]$ are said to form a *basis* for the subspace M if

    (i) $M = \text{span}[v_1, \ldots, v_P]$ and
    (ii) the vectors $[v_1, \ldots, v_P]$ are linearly independent.

In this case we say that the subspace M has *dimension* equal to p.

We list now several properties associated with the notions of basis and dimension:

1. If subspace M had dimension p, then every basis for M must contain exactly p elements.
2. If M is a subspace of R^N, then dim $M \leq N$.
3. dim $R^N = N$ and the vectors $e_1, \ldots, e_N$ form a basis for R^N called the "standard basis." Here, e_j denotes the N-tuple whose jth entry is a 1 and all other entries are zeros, $j = 1, \ldots, N$.
4. If $[v_1, \ldots v_P]$ is a basis for subspace M then each vector in M can be written in a *unique* way as a linear combination of the vectors $[v_1 \ldots v_P]$.

Two vectors in R^N are said to be *orthogonal* if their dot (inner) product is zero. Then given any subspace M in R^N, we can defne a new subspace, denoted M^*, consisting of all the vectors in R^N that are orthogonal to every vector in M. Here, M^* is called the orthogonal complement of M, and since it is defined by membership rule, it must be *shown* to be a subspace. This is left as an exercise, as are the following properties of M^*:

1. $(M^*)^* = M$.
2. If M is a subspace of R^N with dim $M = p$, then dim $M^* = N - p$.
3. The only vector in R^N belonging to *both* M and M^* is the zero vector.
4. Each vector x in R^N can be written in a unique way as

$$x = u + v \quad \text{where } u \in M \text{ and } v \in M^*$$

5. If $[v_1, \ldots, v_P]$ is a basis for subspace M, then there exists also a basis $[u_1, \ldots, u_P]$ for M with the property that

$$(u_j, u_k) = \delta_{jk}, \quad 1 \leq j, k \leq p$$

The basis $[u_1, \ldots, u_P]$ is called an *orthonormal basis* for M. There is a systematic procedure for obtaining the basis $[u_1, \ldots, u_P]$ from the basis $[v_1, \ldots, v_P]$. It is called the Gram–Schmidt orthogonalization procedure.

A function that is defined in R^N with values in R^M and satisfies the condition

$$L(\alpha x + \beta y) = \alpha Lx + \beta Ly$$

for all x, y in R^N and all scalars α, β is said to be a *linear transformation* from R^N into R^M. For example, if A denotes an $M \times N$ matrix, then

$$Lx = Ax \quad \text{for } x \text{ in } R^N$$

defines a linear transformation from R^N into R^M. In fact, this is the *only* example of a linear transformation from R^N into R^M. That is:

Lemma 3.8. If R^N and R^M each carries the standard basis, then for each linear transformation from R^N into R^M there is a unique $M \times N$ matrix A with the property $Ax = Lx$ for every x in R^N.

Note: The linear transformation L is associated with the matrix A only so long as the bases in R^N and R^M are fixed. If the bases are changed, then A will change.

Let $R_1, \ldots, R_M$ and $C_1, \ldots, C_N$ denote, respectively, the M rows and N columns of

A. Then we have:

Subspaces Associated with A
Row Space of A

$$RS(A) = span[\mathbf{R}_1, \ldots, \mathbf{R}_M] \quad \text{(subspace of } R^N\text{)}$$

Column Space of A

$$CS(A) = span[\mathbf{C}_1, \ldots, \mathbf{C}_N] \quad \text{(subspace of } R^M\text{)}$$

Let dim $RS(A) = p$ and dim $CS(A) = q$. Then $p \le N$ and $q \le M$.

Subspaces associated with L
Null Space or Kernel of L

$$N(L) = \text{all } x \text{ in } R^N \text{ such that } Lx = \mathbf{0} \quad \text{(subspace of } R^N\text{)}$$

Range of L

$$Rng(L) = \text{all } \mathbf{y} \text{ in } R^M \text{ such that } Lx = \mathbf{y} \text{ for some } \mathbf{x} \text{ in } R^N$$

Remark. Since $Lx = Ax$ for all x we also write $N(A)$ and $Rng(A)$ for $N(L)$ and $Rng(L)$.

If we denote by $\mathbf{A}^T$ the $N \times M$ matrix whose rows are the columns of $\mathbf{A}$, then $RS(A^T) = CS(A)$ and $CS(A^T) = RS(A)$. Recalling now,

$$\mathbf{Ax} = \begin{bmatrix} \mathbf{R}_1 \cdot \mathbf{x} \\ \vdots \\ \mathbf{R}_m \cdot \mathbf{x} \end{bmatrix} = x_1\mathbf{C}_1 + \cdots + x_N\mathbf{C}_N$$

we can show:

1. $p = \dim RS(A) = \dim CS(A) = q$. Then $rank[A^T] = \dim RS(A^T) = \dim CS(A) = p$.
2. $N(A) = RS(A)^*$ and $\dim N(A) = N - p$, $Rng(A) = CS(A)$ and $\dim Rng(A) = p$.
3. $N(A^T) = RS(A^T)^* = CS(A)^* = Rng(A)^*$, $\dim N(A^T) = M - p$; $Rng(A^T) = CS(A^T) = RS(A) = N(A)^*$, $\dim Rng(A^T) = p$.
4. R^N: $N(A) + Rng(A^T) \xrightarrow{\quad \mathbf{A} \quad} N(A^T) + Rng(A)$: R^M

The meaning of 4 is the following:

Each $\mathbf{v}$ in R^N can be written in a unique way as a sum $\mathbf{w} + \mathbf{z}$ of orthogonal vectors $\mathbf{z}$ in $N(A)$ and $\mathbf{w}$ in $Rng(A^T)$.

Each $\mathbf{v}'$ in R^M can be written in a unique way as a sum $\mathbf{w}' + \mathbf{z}'$ of orthogonal vectors $\mathbf{z}'$ in $N(A^T)$ and $\mathbf{w}'$ in $Rng(A)$.

Result 1 is the only one of these that is not a trivial consequence of the preceding observation about $\mathbf{Ax} = \mathbf{b}$. Using these results, we can prove:

Theorem 3.9 Let $\mathbf{A}$ denote an $N \times N$ matrix. Then, either

(a) $p = N$ and $\mathbf{Ax} = \mathbf{b}$ has a unique solution for every $\mathbf{b}$ in R^N or

(b) $p < N$ and $\mathbf{Ax} = \mathbf{b}$ has a solution if and only if $\mathbf{b}$ is orthogonal to every vector $\mathbf{z}'$ in $N(\mathbf{A}^T)$.

In case (b), if a solution exists, it is not unique. To any solution we can add an arbitrary vector from $N(\mathbf{A})$ and the sum will also be a solution.

Here case (a) is the nonsingular case and case (b) contains the two singular cases; when $\mathbf{b}$ is orthogonal to every $\mathbf{z}'$ in $N(\mathbf{A}^T)$, the problem is singular indeterminate, and when $\mathbf{b}$ is not orthogonal to every $\mathbf{z}'$ in $N(\mathbf{A}^T)$, the problem is inconsistent.

4. THE ALGEBRAIC EIGENVALUE PROBLEM

Throughout this section, $\mathbf{A}$ will denote an $N \times N$ matrix. Then consider the problem of finding nonzero vectors $\mathbf{x}$ and scalars μ such that

$$\mathbf{Ax} = \mu\mathbf{x} \tag{4.1}$$

This is the algebraic eigenvalue problem, and it is clear that $\mathbf{x} = \mathbf{0}$ is a solution for *every* choice of the parameter μ. However, for certain values of μ, called *eigenvalues*, Equation (4.1) has nonzero vector solutions $\mathbf{x}$ called *eigenvectors*. Clearly, if $\mathbf{x}$ is an eigenvector corresponding to the eigenvalue μ, then $\mathbf{x}$ belongs to $N(\mathbf{A} - \mu\mathbf{I})$, the null space of $\mathbf{A} - \mu\mathbf{I}$. Then dim $N(\mathbf{A} - \mu\mathbf{I}) > 0$, and according to Theorem 2.2, this is equivalent to

$$\det[\mathbf{A} - \mu\mathbf{I}] = 0 \tag{4.2}$$

When $\mathbf{A}$ is $N \times N$, Equation (4.2) is a polynomial equation in μ of degree N, and even if $\mathbf{A}$ has all real entries, the equation may have complex roots. Therefore we now consider instead of R^N, the complex vector space C^N, whose vectors are N-tuples of complex numbers and whose scalars are complex numbers. The inner product changes accordingly as follows:

$$\langle\!\langle\mathbf{x}, \mathbf{y}\rangle\!\rangle = \sum_{j=1}^{N} x_j \bar{y}_j = \mathbf{x}^T \bar{\mathbf{y}} \tag{4.3}$$

Here the bar over the y_j indicates the complex conjugate, and we use the angle brackets to emphasize that this is the *complex* inner product. Note that for arbitrary vectors $\mathbf{x}$ and $\mathbf{y}$,

$$\langle\!\langle\mathbf{Ax}, \mathbf{y}\rangle\!\rangle = (\mathbf{Ax})^T \bar{\mathbf{y}} = \mathbf{x}^T \mathbf{A}^T \bar{\mathbf{y}} = \mathbf{x}^T \overline{(\bar{\mathbf{A}}^T \mathbf{y})} = \langle\!\langle\mathbf{x}, \bar{\mathbf{A}}^T \mathbf{y}\rangle\!\rangle \tag{4.4}$$

where we have made use of the facts that, for any $N \times N$ matrices $\mathbf{A}$, $\mathbf{B}$ and all complex numbers z

$$(\mathbf{AB})^T = \mathbf{B}^T \mathbf{A}^T \quad \text{and} \quad \overline{(\bar{z})} = z$$

It follows from (4.4) that for any $N \times N$ matrix $\mathbf{A}$ that satisfies

$$\mathbf{A} = \bar{\mathbf{A}}^T \tag{4.5}$$

we have, $\langle\!\langle\mathbf{Ax}, \mathbf{y}\rangle\!\rangle = \langle\!\langle\mathbf{x}, \mathbf{Ay}\rangle\!\rangle$ for all x, y. A matrix satisfying (4.5) is said to be *Hermitian*; note that if $\mathbf{A}$ has only real entries, then (4.5) reduces to $\mathbf{A} = \mathbf{A}^T$, which means that $\mathbf{A}$ is symmetric about its diagonal.

For a Hermitian matrix we have:

Theorem 4.1.   Suppose A is $N \times N$ and Hermitian. Then:

(a) The eigenvalues of A are all real numbers.
(b) The eigenvectors of A corresponding to distinct eigenvalues are (complex) orthogonal.
(c) The matrix A has N linearly independent eigenvectors; that is, the eigenvectors of A form a basis for C^N.

The meaning of (b) is the following. Suppose $Ax = \mu x$ and $Ay = \theta y$ for $\mu \neq \theta$. Then,

$$\mu \ll x, y \gg = \ll x, \mu y \gg = \ll x, Ay \gg \quad \text{since } \mu \text{ is real}$$

and

$$\theta \ll x, y \gg = \ll x, \theta y \gg = \ll x, Ay \gg \quad \text{since } \theta \text{ is real}$$

Subtraction leads to $(\mu - \theta) \ll x, y \gg = 0$, and since $\mu - \theta$ is not zero by assumption, we conclude that $\ll x, y \gg = 0$.

If A is Hermitian and, in addition, satisfies

$$\ll Ax, x \gg > 0 \quad \text{for all } x \neq 0 \tag{4.6}$$

then we say that A is Hermitian and *positive definite*. Since $Ax = \mu x$ implies

$$\ll Ax, x \gg = \mu \ll x, x \gg$$

we have:

Corollary 4.1   (Continued). If A is Hermitian and positive definite, then in addition to (a), (b), and (c), we have:

. (d) The eigenvalues of A are all positive.

In the case that the $N \times N$ real matrix A is *not* equal to A^T, Theorem 4.1 does not apply. However, we have:

Theorem 4.2.   Let A denote an $N \times N$ real matrix with A not equal to A^T. Then:

(a) μ is an eigenvalue of A if and only if μ is an eigenvalue of A^T.
(b) If μ, θ are distinct eigenvalues for A and

$$Ax = \mu x, \qquad Ay = \theta y$$
$$A^T w = \bar{\mu} w, \qquad A^T z = \bar{\theta} z$$

then $\ll x, z \gg = 0 = \ll y, w \gg$ and $\ll x, w \gg$, $\ll y, z \gg$ are not zero.
(c) If A has N *distinct* eigenvalues, then the corresponding eigenvectors are linearly independent; that is, they form a basis for C^N.

Suppose the eigenvectors of A form a basis for C^N, and let these eigenvectors be denoted by $v_1, \ldots, v_N$. Let the corresponding eigenvalues be denoted by $\mu_1, \ldots, \mu_N$ so that

$$Av_j = \mu_j v_j \quad \text{for } j = 1, \ldots, N$$

Then if we let P denote the $N \times N$ matrix whose columns are just the eigenvectors $v_1, \ldots, v_N$, it is easy to show that

$$AP = A[v_1, \ldots, v_N] = [Av_1, \ldots, Av_N]$$
$$= [\mu_1 v_1, \ldots, \mu_N v_N] = PD$$

where $\mathbf{D}$ denotes the $N \times N$ *diagonal* matrix whose diagonal entries are just the eigenvalues; that is, $(D)_{jj} = \mu_j$. Since the eigenvectors form a basis for C^N, it follows that the rank of P is N, and then Theorem 2.3 implies that $\mathbf{P}^{-1}$ exists. Then,

$$\mathbf{P}^{-1}\mathbf{A}\mathbf{P} = \mathbf{D} \quad \text{and} \quad \mathbf{A} = \mathbf{P}\mathbf{D}\mathbf{P}^{-1} \tag{4.7}$$

In the special case that $\mathbf{A}$ is Hermitian, the eigenvectors are orthogonal, and if we normalize them so that $\|\mathbf{v}_j\| = 1$, $j = 1, \ldots, N$, then it is easy to check that $\mathbf{P}^{-1} = \mathbf{P}^T$.

When (4.7) holds, note that

$$\begin{aligned} \mathbf{A}^2\mathbf{x} &= \mathbf{A}[\mathbf{P}\mathbf{D}\mathbf{P}^{-1}]\mathbf{x} = \mathbf{P}\mathbf{D}\mathbf{P}^{-1}\mathbf{P}\mathbf{D}\mathbf{P}^{-1}\mathbf{x} \\ &= \mathbf{P}\mathbf{D}^2\mathbf{P}^{-1}\mathbf{x} \end{aligned}$$

where $\mathbf{D}^2$ is the diagonal matrix satisfying $(D^2)_{jj} = (\mu_j)^2$, $j = 1, \ldots, N$. More generally, we have

$$\mathbf{A}^q\mathbf{x} = \mathbf{P}\mathbf{D}^q\mathbf{P}^{-1}\mathbf{x} \quad \text{for positive integers } q$$

and for analytic functions $F(x)$,

$$F(\mathbf{A})x = \mathbf{P}F(\mathbf{D})\mathbf{P}^{-1}\mathbf{x}$$

where $F(\mathbf{D})$ is the diagonal matrix satisfying,

$$(F(\mathbf{D}))_{jj} = F(\mu_j), \qquad j = 1, \ldots, N$$

Note that an Nth-degree polynomial equation such as (4.2) *must* have N complex roots. However, these roots need not be distinct. If μ^* is a k-times repeated root of Equation (4.2), we say that $\mu = \mu^*$ is an eigenvalue with algebraic multiplicity equal to k. If there are k linearly independent eigenvectors that correspond to $\mu = \mu^*$ (i.e., if the dimension of the null space of the matrix $[\mathbf{A} - \mu^*\mathbf{I}]$ is equal to k) then we say that μ^* has geometric multiplicity equal to k as well. Clearly the geometric multiplicity of an eigenvalue cannot exceed the algebraic multiplicity, but it can happen that the two are unequal.

This summary of some of the principal results of linear algebra is necessarily very brief. Further elaboration on these and other topics relating to linear algebra can be found in B. Nobel and J. Daniel, *Applied Linear Algebra*, 2nd ed., Prentice-Hall, Englewood Cliffs, NJ, 1977, and G. Strang, *Linear Algebra and Its Applications*, 2nd ed., Academic, New York, 1977.

Solutions to Selected Exercises

CHAPTER 1

Section 1.2 (page 14)

2.

	n = 0	n = 1	n = 2	n = 3	n = 4	n = 5	n = 6	n = 7	n = 8	n = 9	n = 10
j = 0	1	0	0	0	0	0	0	0	0	0	−1
j = 1	1	0.1	0	0	0	0	0	0	0	−0.1	−1
j = 2	1	0.18	0.01	0	0	0	0	0	−0.01	−0.18	−1
j = 3	1	0.245	0.026	0.001	0	0	0	−0.001	−0.026	−0.245	−1
j = 4	1	0.248	0.045	0.003	0.00001	0	−0.00001	−0.003	−0.045	−0.248	−1
j = 5	1	0.343	0.066	0.007	0.00004	0	−0.00004	−0.007	−0.066	−0.343	−1

4.

	n = 0	n = 1	n = 2	n = 3	n = 4	n = 5	n = 6	n = 7	n = 8	n = 9	n = 10
j = 0:	.9	.9	.8	.7	.6	.5	.4	.3	.2	.1	.1
j = 1:	.89	.89	.8	.7	.6	.5	.4	.3	.2	.11	.11
j = 2:	.881	.881	.799	.7	.6	.5	.4	.3	.201	.119	.119
j = 3:	.873	.873	.797	.699	.6	.5	.4	.3001	.203	.127	.127
j = 4:	.865	.865	.795	.70	.60	.5	.400	.3003	.205	.134	.134
j = 5:	.858	.858	.792	.699	.60	.5	.400	.3007	.207	.141	.141

8. u_n^1 is zero for $n = 0, 1, \ldots, 10$ since u_n^0 vanishes for every n. Even though $u(x, 0) = \sin(10x)$ is not the zero function, $\sin(10x_n) = \sin(n\pi)$ vanishes for $n = 0, 1, \ldots, 10$.

9. For every positive integer n, u_1 satisfies a and b, u_2 satisfies c and d, u_3 satisfies c and b, and u_4 satisfies a and d.

10. $u_1(t) = 1 \exp[-2t]$

Section 1.3 (page 20)

1. $u_9 = 2.255$, $u_{10} = 2.005$, $u_{11} = 2.005$

2. $u_9 = 2.36$, $u_{10} = 2.505$, $u_{11} = 2.005$

3. u cannot be determined at any of the interior nodes.

5. (a) with $u_2 = 1$, $u_9 = 1.661$, $u_{10} = 2.63$, $u_{11} = 1.505$

Section 1.4 (page 27)

1.

n:	...	8	9	10	11	12	13	14	15	16	17	
$j = 2$:	...	0	1	0	1	1	1	1	0	1	0	...

2. $k = 11$ is the first k for which $u_0^k \neq 0$. $u_0^k = 0$ for all $k \geq 16$.

CHAPTER 2

Section 2.1 (page 52)

1. (a) $a_0 = 1$; $a_n = 2 \sin(n\pi/2)/n\pi$, $b_n = 0$, $n = 1, 2, \ldots$
(c) $a_0 = L/4$; $a_n = L[1 + (-1)^n - 2 \cos(n\pi/2)]/(n\pi)^2$, $n = 1, 2, \ldots$;
$b_n = -L[1 + (-1)^n]/n\pi$

3. (a) $\tilde{f}(x)$ and $\tilde{f}'(x)$ the $2L$-periodic extensions for $f(x)$ and $f'(x)$, respectively, are each sectionally continuous but not continuous. Then the Fourier series for $f(x)$ converges to $\tilde{f}(x)$ pointwise but not uniformly. The Fourier series for $f'(x)$ does not converge either pointwise or uniformly since $\tilde{f}(x)$ is not continuous. The same answer applies to (b) and (c).

4. (a) $a_n = 0$ for all n; $b_1 = 1$ and $b_n = 0$ for $n > 1$
(b) $a_n = 0$ for all n; $b_n = 8n(-1)^n/[\pi(1 - 4n^2)]$, $n = 1, 2, \ldots$

6. (a) $a_0 = \frac{2}{3}$; $a_n = 4(-1)^n/(n\pi)^2$, $b_n = 0$, $n = 1, 2, \ldots$

10. (a) $a_0 = 2a$ and $a_n = 2 \sin(n\pi a)/n\pi$, $n = 1, 2, \ldots$
(b) $a_0 = 4s$ and $a_n = 4 \sin(n\pi s)\cos(n\pi/2)/n\pi$, $n = 1, 2, \ldots$

In both (a) and (b), $\tilde{f}(x)$ is sectionally continuous but not continuous and $\tilde{f}'(x)$ is sectionally continuous. Then the Fourier series converges pointwise to $\frac{1}{2}[\tilde{f}(x^+) + \tilde{f}(x^-)]$, and we do not consider the convergence of the differentiated series.

14. $\tilde{f}$ is continuously differentiable if (i) f and f' are continuous on $[-\pi, \pi]$ and (ii) $f(\pi) = f(-\pi)$ and $f'(\pi) = f'(-\pi)$.

18. $\sin\dfrac{\pi x}{2} \sim \dfrac{2}{\pi^2} + \displaystyle\sum_{n=1}^{\infty} (2n - 1)^{-2} \cos n\pi x$

Differentiating this series term by term leads to a series that converges (in the pointwise sense) to the odd 2-periodic extension of $\cos(\pi x/2)$, $0 < x < 1$.

Section 2.2 (page 63)

1. Since there is no interval where $\Phi_j(x)\Phi_k(x) \neq 0$ when $j \neq k$, it follows that

$$\int_0^1 \Phi_j(x)\Phi_k(x)\, dx = 0 \quad \text{for } j \neq k$$

Then the family $\{\Phi_1(x), \ldots, \Phi_{10}(x)\}$ is an orthogonal family in $L^2(0, 1)$. The function

$$F(x) = \begin{cases} x - \frac{1}{20} & \text{for } 0 < x < \frac{1}{10} \\ 0 & \text{otherwise} \end{cases}$$

is orthogonal to all of the $\Phi_j(x)$ but $F(x)$ is not the zero function. Thus the family is not complete. The function

$$f\#(x) = \sum_{j=1}^{10} (f, \theta_j)\theta_j(x)$$

is a "staircase" function, a piecewise constant approximation to $f(x)$.

Section 2.3 (page 75)

1. (a) $\mu_n = [(n - \frac{1}{2})\pi]^2$, $u_n(x) = \sin(n - \frac{1}{2})\pi x$, $n = 1, 2, \ldots$
(b) $\mu_n = [n\pi/2]^2$, $u_n(x) = \sin(n\pi/2)x$, $n = 1, 2, \ldots$
(c) $\mu_n = [(n - \frac{1}{2})\pi]^2$, $u_n(x) = \cos(n - \frac{1}{2})\pi x$, $n = 0, 1, 2, \ldots$

4. $\tilde{f}(x) \sim 1 + \dfrac{4}{\pi} \sum_{n=1}^{\infty} (-1)^{n+1}(2n - 1)^{-1}\cos(n - \frac{1}{2})\pi x$. The series converges pointwise to the following extension of $f(x)$:

$$\tilde{f}(x) = \begin{cases} 1 & \text{if } |x| < 1 \\ 0 & \text{if } 1 < |x| < 2 \end{cases}$$

$$\tilde{f}(x) = \tilde{f}(x + 4) \quad \text{for all } x$$

7. If $\alpha < 0$, the eigenvalues and eigenfunctions are

$$\mu_{-1} = -\beta^2 \quad \text{where } e^{2\beta} = (\beta - \alpha)/(\beta + \alpha)$$
$$u_{-1}(x) = e^{\beta x} + [(\beta - \alpha)/(\beta + \alpha)]e^{-\beta x}$$
$$\mu_n = \beta_n^2 \quad \text{where } \alpha \cos \beta_n + \beta_n \sin \beta_n = 0$$
$$u_n(x) = \cos \beta_n x - (\alpha/\beta_n)\sin \beta_n x, \quad n = 1, 2, \ldots$$

If $\alpha = 0$, the eigenvalues and eigenfunctions are $\mu_n = (n\pi)^2$, $u_n(x) = \cos n\pi x$, $n = 0$, $1, \ldots$.

8. For $L > 4$, the Sturm–Liouville problem will have a negative eigenvalue. This is a special case of Example 2.3.3.

Section 2.4 (page 80)

2. Using the result of Exercise 1, we have

$$\sum_{n=0}^{N-1} c_n e^{inx} = c_0 + c_1 e^{ix} + c_{N-1}e^{i(N-1)x} + \cdots$$
$$+ c_2 e^{i2x} + c_{N-2}e^{i(N-2)x} + c_{N/2}e^{ixN/2}$$
$$= c_0 + c_1 e^{ix} + c_{N-1}e^{-ix} + c_2 e^{i2x} + \cdots$$
$$+ c_{N-2}e^{-i2x} + \tfrac{1}{2}c_{N/2}e^{ixN/2} + \tfrac{1}{2}c_{N/2}e^{-ixN/2}$$
$$= c_0 + (c_1 + c_{N-1})\cos x + i(c_1 - c_{N-1})\sin x + \cdots + c_{N/2} \cos Nx/2$$

4. (a) $F(x) = \cos 5x$, $F = [1, 0, -1, 0]$, $C = [0, \frac{1}{2}, 0, \frac{1}{2}]$
(b) $F(x) = \cos 3x$, $F = [1, 0, -1, 0]$, $C = [0, \frac{1}{2}, 0, \frac{1}{2}]$
(c) $F(x) = \cos 4x$, $F = [1, 1, 1, 1]$, $C = [1, 0, 0, 0]$
(d) $F(x) = \cos 5x + 100 \cos 4x$, $F = [101, 100, 99, 100]$

Section 2.5 (page 87)

1. f and g are not in $L^2(0, 1)$; h and p are in $L^2(0, 1)$ with $\|h\|^2 = \sum_{n=1}^{\infty} \dfrac{1}{n^2} < \infty$ and $\|p\|^2 < 1$.

3. For $m = 10$, $k = 360$, $c = 0.1$, $\Omega = 6$, the system is stable (i.e., c is positive). The magnification factor $D_n^{-1/2}$ for $n = 1, \ldots, 5$ is as follows:

n	$D_n^{-1/2}$
1	1.66
2	0.0009
3	0.0003
4	0.0002
5	0.0001

Section 2.6 (page 90)

1. (a) $f_{mn} = \dfrac{4[1 - (-1)^m]\sin(n\pi/2)}{\pi^2 mn}$

(b) $g_{mn} = \dfrac{4[1 - (-1)^n](-1)^m}{\pi mn^2}$

(c) $h_{mn} = \dfrac{4[(-1)^{n+1} - 1]}{\pi^2 n(m^2 - n^2)}\left[\dfrac{(m + n)\sin(m - n)\pi}{2} - \dfrac{(m - n)\sin(m + n)\pi}{2}\right]$

CHAPTER 3

Section 3.1 (page 110)

2. $u(x, y) = B - 4\pi^{-3} \sum_{n=1}^{\infty} \dfrac{1}{(2n - 1)^3} \dfrac{\cosh(2n - 1)\pi(1 - x)}{\sinh(2n - 1)\pi} \cos(2n - 1)\pi y$ where B is an arbitrary constant.

4. $u(x, y) = \dfrac{4}{\pi} \sum_{n=1}^{\infty} (-1)^{n+1} \dfrac{\cosh[(n - \frac{1}{2})\pi(1 - x)]}{2n - 1} \cos(n - \frac{1}{2})\pi y$

8. Eigenvalues: $\mu_{mn} = [(m - \frac{1}{2})^2 + (n - \frac{1}{2})^2]\pi^2$, $m, n = 1, 2, \ldots$
Eigenfunctions: $u_{mn}(x, y) = \sin(n - \frac{1}{2})\pi x \cos(m - \frac{1}{2})\pi y$

$$x^2 + y^2 \sim \sum_{m=1}^{\infty} \sum_{n=1}^{\infty} C_{mn}\sin(n - \tfrac{1}{2})\pi x \cos(m - \tfrac{1}{2})\pi y$$

where

$$C_{mn} = 4(-1)^{m+1}[2(-1)^{n+1}/\alpha_n - 2/\alpha_n^2 - 2/\beta_m^2 + 1]/\alpha_n\beta_m$$
$$\alpha_n = (n - \tfrac{1}{2})\pi, \qquad \beta_m = (m - \tfrac{1}{2})\pi$$

Section 3.1 (page 118)

1. $u(r, \theta) = 2 \sum_{n=1}^{\infty} \cos\left(\dfrac{n\pi}{4}\right) \dfrac{[1 - (-1)^n]2^n(r^{2n} - 1)}{n\pi r^n(4^n - 1)} \sin n\theta$

5. $u(r, \theta) = \dfrac{4}{\pi} \sum\limits_{n=1}^{\infty} f_n \left(\dfrac{r}{2}\right)^{2n} \cos 2n \left(\theta + \dfrac{\pi}{4}\right)$

where

$$f_n = \int_{-\pi/4}^{\pi/4} f(\theta)\cos 2n \left(\theta + \dfrac{\pi}{4}\right) d\theta$$

Section 3.1 (page 119)

4. The method of eigenfunction expansion fails since the region is not bounded in either the x or the y direction; hence neither the x nor the y problem is a Sturm–Liouville problem.

5. The region $\Omega = \{0 < x, y < 1, y < x\}$ is not a coordinate cell. Then the method of eigenfunction expansion fails.

6. Assuming $u(x, y) = X(x)Y(y)$ leads to the Sturm–Liouville problem

$$-Y''(y) = (1 + y^2)^{-1}\mu Y(y), \qquad Y(0) = Y(1) = 0$$

If we can solve this problem, we can solve the boundary-value problem.

Section 3.1 (page 123)

1. $\mu_n = (n - \tfrac{1}{2})\pi, \; n = 1, 2, \ldots$

$$u(x, y) = 2 \sum\limits_{n=1}^{\infty} \mu_n^{-3}\exp[-\mu_n y]\sin \mu_n x$$

3. $u(r, \theta) = \tfrac{1}{2} + \sum\limits_{n=1}^{\infty} \dfrac{r^{-n}}{(n\pi)} [\sin n\theta + \sin n(\pi - \theta)]$

4. $u(r, \theta) = \dfrac{4}{\pi} \sum\limits_{n=1}^{\infty} (2n - 1)^{-1} r^{-(2n-1)} \sin(2n - 1)\theta$

Section 3.2 (page 147)

1. Eigenvalues: $\mu_n = (n - \tfrac{1}{2})\pi, \; n = 1, 2, \ldots$
Eigenfunctions: $X_n(x) = \sin \mu_n x$

$$u(x, t) = \sum\limits_{n=1}^{\infty} C_n \exp[-\mu_n^2 Dt]\sin \mu_n x$$

where $C_n = 4\mu_n^{-2} \sin \mu_n/2 - \mu_n^{-1} \cos \mu_n/2$.

3. Eigenvalues: $\mu_n = n\pi/2, \; n = 1, 2, \ldots$
Eigenfunctions: $X_n(x) = \sin \mu_n x$

$$u(x, t) = \sum\limits_{n=1}^{\infty} C_n \exp[-\mu_n^2 Dt]\sin \mu_n x$$

$$u_x(1, t) = \sum\limits_{n=1}^{\infty} \mu_n C_n \exp[-\mu_n^2 Dt]\cos \mu_n$$

where $C_n = [\sin n\pi/4 + \sin 3n\pi/4]/\mu_n^2 - 2[\cos 3n\pi/4 - \cos n\pi/4]/\mu_n$. Since $\cos \mu_n = 0$ if n is odd and $C_n = 0$ if n is even, it follows that $u_x(1, t) = 0$. Then the solution for

Exercise 3 satisfies all of the conditions of Exercise 1, and we conclude that the two solutions are equal for $0 < x < 1$, $t > 0$.

5. Let $v(x, t) = u(x, t) - (1 - x)$. Then,

$$v(x, t) = \sum_{n=1}^{\infty} C_n \exp[-(n\pi)^2 Dt]\sin n\pi x$$

where $C_n = 2 \int_0^1 (x - 1)\sin n\pi x \, dx$.

7. $u(x, t) = 1 + v(x, t)$, where

$$v(x, t) = \frac{-2}{\pi} \sum_{n=1}^{\infty} \frac{1}{\mu_n} \exp[-\mu_n^2 Dt]\sin \mu_n x$$

and $\mu_n = (n - \frac{1}{2})\pi$.

9. Let $u(x, t) = \exp[\alpha t]v(x, t)$ and let $w(x, t) = v(x, t) - x\exp[\alpha t]$. Then $w(x, t)$ satisfies an inhomogeneous equation with homogeneous boundary conditions. We find

$$w(x, t) = \sum_{n=1}^{\infty} \frac{C_n \alpha}{[\mu_n^2 D - \alpha]} \{\exp[-\alpha t] - \exp[-\mu_n^2 Dt]\}\sin \mu_n x$$

where $\mu_n = (n - \frac{1}{2})\pi$ and $C_n = 2(-1)^{n+1}/\mu_n^2$.

11. Let $u(x, t) = \exp[x\beta/2D - t\beta^2/4D]v(x, t)$. Then

$$v(x, t) = \sum_{n=1}^{\infty} C_n \exp[-\mu_n^2 Dt]\sin \mu_n x$$

where the eigenvalues μ_n satisfy $\tan \mu_n = -2D\mu_n/\beta$, the eigenfunctions are $\sin \mu_n x$, and $C_n = 2 \int_0^1 f(x)\exp\left[-\frac{\beta x}{2D}\right]\sin \mu_n x \, dx$. The solution $u(x, t)$ tends toward the steady state $u(x, \infty) = 0$.

13. **(a)** Steady state for Exercise 7: $u(x, \infty) = 1$. This satisfies $u''(x) = 0$, $u(0) = 1$, $u'(1) = 0$.
   (b) Steady state for Exercise 9: $u(x) = (D/\alpha)^{1/2}\sin[x(\alpha/D)^{1/2}]/\cos[(\alpha/D)^{1/2}]$, that is, $Du''(x) + \alpha u(x) = 0$, $u(0) = 0$, $u'(1) = 1$.

16. $u(x, y, t) = \sum_{m=0}^{\infty} \sum_{n=1}^{\infty} f_{mn}\exp[-(n^2 + m^2)\pi^2 Dt]\sin n\pi x \cos m\pi y$

$u(x, y, t) \to 0$ as $t \to \infty$.

18. $C(x, t) = C_0 e^{\sigma t}\left[1 - \sum_{n=1}^{\infty} \frac{2}{\mu_n} \exp[-\mu_n^2 Dt]\sin \mu_n x]\right]$

If $\sigma = 0$, then $C(L, t) \to C_0$ as $t \to \infty$, and if $\sigma > 0$, then $C(L, t) \to \infty$ as $t \to \infty$.

Section 3.3 (page 171)

1. $u(x, t) = \frac{1}{2}[\tilde{f}(x) + at) + \tilde{f}(x - at)]$, where $\tilde{f}(x)$ denotes the odd periodic extension of period 2 of the function $f(x = x(1 - x)$. Then

$$u(0.3, 5/a) = \frac{1}{2}[\tilde{f}(0.3 + 5) + \tilde{f}(0.3 - 5)]$$

and since $\tilde{f}$ is periodic of period 2 and odd,

$$\tilde{f}(5.3) = \tilde{f}(5.3 - 6) = \tilde{f}(-0.7) = -f(0.7) = -0.21$$
$$\tilde{f}(-4.7) = \tilde{f}(-0.7) = -f(0.7) = -0.21$$

Therefore, $u(0.3, 5/a) = -0.21$. Similarly,

$$u(0.3, 7/a) = \tfrac{1}{2}[\tilde{f}(7.3) + \tilde{f}(-6.7)]$$

and because of the nature of the extension $\tilde{f}$,

$$\tilde{f}(7.3) = \tilde{f}(7.3 - 8) = \tilde{f}(-0.7) = -0.21$$
$$\tilde{f}(-6.7) = \tilde{f}(-6.7 + 6) = \tilde{f}(-0.7) = -0.21$$
$$u(0.3, 7/a) = -0.21$$

Finally,

$$u(0.3, 10/a) = \tfrac{1}{2}[\tilde{f}(0.3 + 10) + \tilde{f}(0.3 - 10)]$$
$$= \tilde{f}(0.3) = f(0.3) = 0.21$$

3. In this problem, the boundary conditions dictate that we use the mixed extension for $f(x)$. That is, we first extend the function $f(x) = x(1 - x)$ to the interval $(1, 2)$ so that it is symmetric about the line $x = 1$ [this ensures that $u_x(1, t) = 0$],

$$\tilde{f}(x) = f(x - 1) \quad \text{for } 1 < x < 2$$

Next we extend this extension to $(-2, 0)$ as an odd function of x so that $u(0, t) = 0$. Finally we extend this function (which is now defined for $-2 < x < 2$) to the whole real line as a periodic function of period 4. Then,

$$u(0.3, 5/a) = \tfrac{1}{2}[\tilde{f}(5.3) + f(-4.7)]$$
$$= \tfrac{1}{2}[\tilde{f}(1.3) + \tilde{f}(-0.7)] = \tfrac{1}{2}[0.21 + (-0.21)] = 0$$
$$u(0.3, 10/a) = \tfrac{1}{2}[\tilde{f}(10.3) + \tilde{f}(-9.7)]$$
$$= \tfrac{1}{2}[\tilde{f}(-1.7) + \tilde{f}(-1.7)] = -0.21$$

5. Here,

$$u(x, t) = \int_{x-at}^{x+at} \bar{g}(s)\, ds$$

where $\bar{g}(x)$ denotes the even periodic extension of period 2 for $g(x)$. Since $\bar{g}(x)$ is even, the integral over one period is not zero as it was for the odd periodic extension in the previous problem. Instead the integral over one period is equal to 1. Then,

$$u\left(0.3, \frac{5}{a}\right) = \int_{-4.7}^{5.3} \bar{g}(s)\, ds = 5$$

That is,

$$\int_{-4.7}^{-0.7} \bar{g}(s)\, ds = \int_{1.3}^{5.3} \bar{g}(s)\, ds = 2 \quad \text{(integral over two periods)}$$

and

$$\int_{-0.7}^{1.3} \bar{g}(s)\, ds = \int_{-1}^{1} \bar{g}(s)\, ds = 1$$

7. Let $\theta_n^2 = a^2 - [\beta^2/(2n\pi)]^2$, $n = 1, 2, \ldots$. Then

$$u(x, t) = \sum_{n=1}^{\infty} D_n \exp\left[-\frac{\beta^2 t}{2}\right] \sin n\pi\theta_n t \, \sin n\pi x$$

or

$$u(x, t) = \sum_{n=1}^{\infty} \tfrac{1}{2}D_n \exp\left[-\frac{\beta^2 t}{2}\right] [\cos n\pi(x - \theta_n t) - \cos n\pi(x + \theta_n t)]$$

where

$$n\pi\theta_n D_n = g_n = 2\int_0^1 g(x)\sin n\pi x \, dx, \qquad n = 1, 2, \ldots$$

9. Let $\theta_n^2 = a^2 + [\beta^4 + \sigma^2 a^2]/(2n\pi)^2$, $n = 1, 2, \ldots$, and $p = \frac{1}{2}(\beta/a)^2$. Then $u(x, t) = \exp[px - \sigma t/2]v(x, t)$, where

$$v(x, t) = \sum_{n=1}^\infty f_n \cos n\pi\theta_n t \sin n\pi x$$

$$v(x, t) = \sum_{n=1}^\infty \tfrac{1}{2}f_n[\sin n\pi(x + \theta nt) + \sin n\pi(x - \theta nt)]$$

and

$$f_n = 2\int_0^1 f(x)\sin n\pi x \, dx, \qquad n = 1, 2, \ldots$$

11. Choose $\phi(x) = 1$. Then, $u(x, t) = A \sin \Omega t - v(x, t)$, where

$$v(x, t) = -\frac{2A\Omega}{a} \sum_{n=1}^\infty \frac{1 - \cos \mu_n}{\mu_n^2} \sin \mu_n at \sin \mu_n x$$

$$+ 2A\Omega^2 \sum_{n=1}^\infty \frac{1 - \cos \mu_n}{\mu_n} \int_0^t \cos \mu_n a(t - \tau) \sin \Omega\tau \, d\tau \sin \mu_n x$$

where $\mu_n = (n - \frac{1}{2})\pi$, $n = 1, 2, \ldots$.

CHAPTER 4

Section 4.2 (page 190)

2. (a) $F(\alpha) = \dfrac{2 \exp[-i\alpha]\sin \alpha}{\pi\alpha} - \dfrac{\exp[3i\alpha/2]\sin(\alpha/2)}{\pi\alpha}$

(b) $G(\alpha) = \dfrac{\sin \pi(\alpha - 1)}{2\pi i(\alpha - 1)} - \dfrac{\sin \pi(\alpha + 1)}{2\pi i(\alpha + 1)}$

(c) $H(\alpha) = \dfrac{\sin \pi(\alpha - 1)/2}{2\pi(\alpha - 1)} - \dfrac{\sin \pi(\alpha + 1)/2}{2\pi(\alpha + 1)}$

4. (a) $F(\alpha) = b[a^2 - (\alpha - b)^2]^{-1}$ (b) $G(\alpha) = [a + i\alpha][a^2 - (\alpha - b)^2]^{-1}$

6. (a) $h(x) = \exp[ix]f(x)$ (b) $h(x) = \frac{1}{2}[f(x + b) + f(x - b)]$

(c) $h(x) = [4\pi t]^{-1/2} \int_{-\infty}^\infty \exp\left[-\dfrac{(x - y)^2}{4t}\right] f(y) \, dy$

(e) $h(x) = (\pi/4)^{1/2}\{\exp[-x^2/4] - x^2/2 \exp[x^2/4]\}$

8. Note that $f*g = 0$ if $F(\alpha)G(\alpha) = 0$. For example,

$$F(\alpha) = I_1(\alpha - 1) \quad \text{and} \quad G(\alpha) = I_1(\alpha + 1)$$
$$f(x) = \exp[ix]\sin x/x, \qquad g(x) = \exp[-ix]\sin x/x$$

Section 4.3 (page 202)

1. (a) $s/(s^2 - a^2)$ (b) $2sa/[(s - a)^2(s + a)^2]$

(c) $[\pi/4]^{1/2}s^{-3/2}$ (d) $2/s^3 + 2/s^2 + 1/s$

(e) $e^2/(s-1)^2 - 2e^2/(s-1)$

(g) $\pi/2 - \arctan(s/3)$

(h) $e^{-s}/s - 3e^{-3s}/s + 2e^{-5s}/s$

3. $\exp[-2ks][1 - e^{-s}]/s$

7. (a) $\mathcal{L}^{-1}[s\hat{f}(s)] = \begin{cases} 1 & \text{for } 0 < t < 1 \\ -1 & \text{for } 1 < t < 2 \\ 0 & \text{for } t > 2 \end{cases}$

(b) $\mathcal{L}^{-1}[s^2\hat{f}(s)] = \mathcal{L}^{-1}[1 - 2e^{-s} + e^{-2s}]$ is not a function

$\mathcal{L}^{-1}[s^2\hat{f}(s)] = \delta(t) - 2\delta(t-1) + \delta(t-2)$ ⠀⠀$\delta \equiv$ Dirac delta

(c) $H(t-2) - 2H(t-3) + H(t-4)$

(d) $\mathcal{L}^{-1}[s\hat{f}(s-1)] = \begin{cases} e^t(t+1) & \text{for } 0 < t < 1 \\ e^t(1-t) & \text{for } 1 < t < 2 \\ 0 & \text{for } \quad t > 2 \end{cases}$

CHAPTER 5

Section 5.1 (page 211)

2. $u_y(x, 0) = 2d/[\pi(x^2 - 1)]$

4. $u(x, y) = [x^2 + y^2]^{1/2} - x$ is not bounded on $H = \{y > 0\}$

6. (a) $\dfrac{1}{\pi}\left[\arctan\left(\dfrac{1+x}{y}\right) - \arctan\left(\dfrac{x}{y}\right)\right] - \dfrac{1-x}{\pi}\left[\arctan\left(\dfrac{x}{y}\right) + \arctan\left(\dfrac{1-x}{y}\right)\right]$

⠀⠀$+ \dfrac{y}{2\pi}\ln\left[\dfrac{x^2 + y^2}{(x-1)^2 + y^2}\right]$

8. (a) $u(x, y) = \dfrac{1}{\pi}\displaystyle\sum_{n=0}^{\infty}\left[\arctan\left(\dfrac{x+2}{2n+y}\right) - \arctan\left(\dfrac{x-2}{2n+y}\right)\right.$

⠀⠀⠀⠀⠀⠀$\left. - \arctan\left(\dfrac{x+2}{2n+2-y}\right) + \arctan\left(\dfrac{x-2}{2n+2-y}\right)\right]$

(b) $v(x, y) = u(x, 1-y)$

10. $u(x, y) = \dfrac{1}{\pi}\displaystyle\int_{-\infty}^{\infty} K(z, y)f(x-z)\,dz$

where

$$K(x, y) = \sum_{n=0}^{\infty}(-1)^n\left[\frac{2n+y}{z^2+(2n+y)^2} + \frac{2n+2-y}{z^2+(2n+2-y)^2}\right]$$

Section 5.1 (page 218)

2. $u(x, t) = \frac{1}{2}\left[\mathrm{erf}\left(\dfrac{x+Vt-0.9}{(4Dt)^{1/2}}\right) - \mathrm{erf}\left(\dfrac{(x+Vt-1.1)}{(4Dt)^{1/2}}\right)\right.$

⠀⠀⠀⠀$\left. + \mathrm{erf}\left(\dfrac{x+Vt+0.9}{(4Dt)^{1/2}}\right) - \mathrm{erf}\left(\dfrac{x+Vt-1.1}{(4Dt)^{1/2}}\right)\right]$

CHAPTER 5

6. $u_x(0, t) = \pi^{-1/2} \int_{-\infty}^{\infty} \exp[-\sigma^2] f'(Vt - \sigma\{4Dt\}^{1/2}) \, d\sigma$

$u_x(0, t) = \pi^{-1/2} \int_{-\infty}^{\infty} \exp[-\sigma^2 + Bt] f'(-\sigma\{4Dt\}^{1/2}) \, d\sigma$

8. (a) $u_t(0, t) = 0$ if $f(x)$ is odd
(b) $u_t(0, t) > 0$ if $f(x)$ is even and

$$\int_0^{\infty} \left(1 - \frac{z^2}{2Dt}\right) \exp\left(-\frac{z^2}{4Dt}\right) f(z) \, dz > 0$$

Section 5.1 (page 220)

1. $u(x, t) = e^{-t} u(x, 0)$
2. If $C = +1$, then $\Sigma(t) = \{x, t : x + t > 0\}$

Section 5.1 (page 226)

1. $u(x, t) = [\cos \pi(x - at) - \cos \pi(x + at)]/2\pi a$
5. If $g(x)$ is odd and periodic with period $2L$, then $u(0, t) = u(L, t) = 0$.
6. $\Sigma(t) = \{x, t : |x - 4t| < 2\} \cup \{x, t : |x + 4t| < 2\}$
7. $u(x, t) = \frac{1}{2}[\tilde{f}(x + at) + \tilde{f}(x - at)] + \frac{1}{2a} \int_{x-at}^{x+at} \tilde{g}(s) \, ds$ where $\tilde{f}$, $\tilde{g}$ denote the odd periodic
extensions of period $2L$ for $f(x)$ and $g(x)$.
9. Choose the extensions $\tilde{f}$ and $\tilde{g}$ to be periodic of period $4L$ and such that

$$\tilde{f}(L + x) = \begin{cases} f(L - x) & \text{for } 0 < x < L \\ -f(-x) & \text{for } x < 0 \end{cases}$$

Section 5.2 (page 233)

1. $u(x, y) = \{\arctan[(x + 1)/y] - \arctan[(x - 1)/y]\}$

3. $u(x, y) = \frac{1}{2\pi} \ln\left[\frac{(x + 1)^2 + y^2}{(x - 1)^2 + y^2}\right] - \frac{2}{\pi}$

$$+ \frac{y}{\pi}\left[\arctan\left(\frac{x + 1}{y}\right) - \arctan\left(\frac{x - 1}{y}\right)\right]$$

5. $u(x, y) = \left\{\frac{1}{\pi}\left[\arctan\left(\frac{x + 2}{y}\right) + \arctan\left(\frac{x - 2}{y}\right)\right.\right.$

$$\left.\left. - \arctan\left(\frac{x + 3}{y}\right) - \arctan\left(\frac{x - 3}{y}\right)\right]\right\}$$

$$+ \left\{\frac{1}{\pi}\left[\arctan\left(\frac{y + 2}{x}\right) + \arctan\left(\frac{y - 2}{x}\right)\right.\right.$$

$$\left.\left. - \arctan\left(\frac{y + 1}{x}\right) - \arctan\left(\frac{y - 1}{x}\right)\right]\right\}$$

Section 5.2 (page 237)

1. $u(x, t) = 1 - \text{erf}[x/(4Dt)^{1/2}]$

3. $u(x, t) = \int_0^\infty [K(x - z, t) + K(x + z, t)]f(z)\, dz - 2D \int_0^t K(x, t - \tau)g(\tau)\, d\tau$

where

$$K(x, t) = [4\pi Dt]^{-1/2}\exp[-x^2/(4Dt)]$$

4. (a) $u(x, t) = \frac{1}{2}\,\text{erf}\left(\dfrac{x + 1}{(4Dt)^{1/2}}\right) - \frac{1}{2}\,\text{erf}\left(\dfrac{x - 1}{(4Dt)^{1/2}}\right)$

(b) $u(x, t) = \begin{cases} -2D \displaystyle\int_0^t K(x, t - \tau) & \text{if } 0 < t < 1 \\[2mm] -2D \displaystyle\int_0^1 K(x, t - \tau)\, d\tau & \text{if } t > 1 \end{cases}$

7. (a) $t_1 = \frac{1}{4}D[1/\text{erf}^{-1}(\frac{1}{2})]^2$ where $x = \text{erf}^{-1}(z)$ if $z = \text{erf}(x)$

(b) $t_2 = 1/4D[2/\text{erf}^{-1}(\frac{1}{2})]^2 = 4t_1$

(c) $u_x(0, t) = [\pi Dt]^{-1/2}$

11. $u(x, t) = \displaystyle\int_0^t \frac{x}{[\pi D(t - \tau)^3]^{1/2}} \exp\left\{-\frac{x - V(t - \tau)^2}{4D(t - \tau)}\right\} g(\tau)\, d\tau$

Section 5.2 (page 246)

1. $u(x, t) = \begin{cases} 0 & \text{if } t < x/4 \text{ or } t > x/4 + \pi \\ \exp[-2t]\sin(t - x/4) & \text{if } 0 < t - x/4 < \pi \end{cases}$

For x between 0 and 10π, $u(x, t) = 0$ for $t > 3.5\pi$.

3. $u(x, t) = \begin{cases} 0 & \text{if } 0 < t < x/a \\ -a(t - x/a) & \text{if } x/a < t < 1 + x/a \\ -a & \text{if } 1 + x/a < t \end{cases}$

5. $u(x, t) = \dfrac{a}{K} \displaystyle\int_0^t \left[1 - \exp\left(\dfrac{K(t - \tau)}{aM}\right)\right] g(\tau)\, d\tau$

Section 5.2 (page 252)

1. $u(x, t) = \dfrac{2A}{\pi^{1/2}} \displaystyle\sum_{n=0}^\infty \left\{ \int_{C_n}^{D_n} \exp[-z^2]\, dz - \int_{E_n}^{F_n} \exp[-z^2]\, dz \right\}$

where

$$C_n^2 = (2n + x)^2/4Dt, \qquad D_n^2 = [(2n + x)^2 + 1]/4Dt$$
$$E_n^2 = (2n + 2 - x)^2/4Dt, \qquad F_n^2 = [(2n + 2 - x)^2 + 1]/4Dt$$

4. $u(x, t) = \displaystyle\int_0^t \sum_{n=0}^\infty \frac{(-1)^n(2n + x)}{[\pi D(t - \tau)^3]^{1/2}} \exp[-(2n + x)^2/4D(t - \tau)]f(\tau)\, d\tau$

$\qquad + \displaystyle\int_0^t \sum_{n=0}^\infty \frac{(-1)^n(2n + 2 - x)}{[\pi D(t - \tau)^3]^{1/2}} \exp[-(2n + 2 - x)^2/4D(t - \tau)]f(\tau)\, d\tau$

9. $u(x, t) = \sum_{n=0}^{\infty} (-1)^n H\left(t - \frac{2n + x}{a}\right) f\left(t - \frac{2n + x}{a}\right)$

$\qquad + \sum_{n=0}^{\infty} (-1)^n H\left(t - \frac{2n + 2 - x}{a}\right) f\left(t - \frac{2n + 2 - x}{a}\right)$

10. A double root occurs if $\Omega = \sigma_n$, where σ_n denotes any of the (infinite number of) roots of the equation $\tan(\sigma/a) = K/(aM\sigma)$.

Section 5.4 (page 262)

1. $u(x, y) = \dfrac{1}{8\pi} \displaystyle\int_0^{\infty} \int_0^{\infty} \ln \dfrac{(x + \sigma)^2 + (y - z)^2}{(x + \sigma)^2 + (y + z)^2} \dfrac{(x - \sigma)^2 + (y - z)^2}{(x - \sigma)^2 + (y + z)^2} f(\sigma, z)\, d\sigma\, dz$

3. $u(x, t) = \displaystyle\int_0^t \left[4\pi D(t - \tau)\right]^{-1/2} \int_{-\infty}^{\infty} \exp\left(-\dfrac{(x - y)^2}{4D(t - \tau)}\right) f(y, \tau)\, dy\, d\tau$

5. $u(x, t) = \begin{cases} \dfrac{1}{2a} \displaystyle\int_0^t \int_{x-a(t-\tau)}^{x+a(t-\tau)} f(z, \tau)\, dz\, d\tau & \text{if } 0 < t < x/a \\[2ex] \dfrac{1}{2a} \displaystyle\int_0^{t-x/a} \int_{a(t-\tau)-x}^{a(t-\tau)+x} f(z, \tau)\, dz\, d\tau & \text{if } x/a < t \\[2ex] + \dfrac{1}{2a} \displaystyle\int_{t-x/a}^t \int_{x-a(t-\tau)}^{x+a(t-\tau)} f(z, \tau)\, dz\, d\tau & \end{cases}$

CHAPTER 7

Section 7.1 (page 293)

1. (a) $u(x, t) = \dfrac{1}{1 - (x - t)^2}$

   (c) $u(x, t) = \begin{cases} 1 - |x - 2t|, & |x - 2t| < 1 \\ 0, & |x - 2t| > 1 \end{cases}$

   (e) $u(x, t) = e^{-(x - 2t)^2}$

2. (a) $u(x, t) = \begin{cases} 1, & 0 < x < t \\ e^{-(x - t)^2}, & x > t \end{cases}$

   (c) $u(x, t) = \begin{cases} 1, & 0 < x < 2t \\ 1 - (x - 2t), & 2t < x < 2t + 1 \\ 0, & x > 2t + 1 \end{cases}$

   (e) $u(x, t) = \begin{cases} 0 & 0 < t < \frac{1}{3}x \\ \frac{1}{3}(3t - x), & \frac{1}{3}x < t < 1 + \frac{1}{3}x \\ 1, & t > 1 + \frac{1}{3}x \end{cases}$

3. $u(x, t) = e^{x - 3t} + 5t$

5. Let $v(x, t) = e^{-ct} u(x, t)$

Section 7.2 (page 300)

1. (a) $u(x, t) = e^t \cos(x - t)$, $-\infty < x, t < \infty$
   (c) $u(x, t) = x - 2t$, $-\infty < x, t < \infty$
   (e) $u(x, t) = te^{2x} + e^x - 1$, $-\infty < x, t < \infty$
   (g) $u(x, t) = \frac{1}{6}t^3 + (x - \frac{1}{2}t^2)(t + x - \frac{1}{2}t^2)$, $-\infty < x, t < \infty$
   (i) $u(x, t) = \dfrac{3xt}{3xt - t + 4x}$, $xt \neq 0$, $3xt \neq t - 4x$
   (k) $u(x, t) = [1 + \cos(x - t)]e^{-t}$, $-\infty < x, t < \infty$
5. $u(x, t) = e^t(x - t^2 - t)$

Section 7.3 (page 306)

1. $u(x, t) = x/(1 + t)$
3. (a) $u = x - e^u t$
   (b) $x = u + e^u t$
5. (b) $u = x - t(1 - u)e^{-u}$
9. (a) $\alpha = -V/U$; $\beta = V$
   (b) $u = \frac{1}{2}U$

Section 7.4 (page 316)

1. (a) $u(x, t) = \begin{cases} 1, & x \le t \\ x/t, & t < x < 3t, \; t > 0 \\ 3, & x \ge 3t \end{cases}$

   (c) $u(x, t) = \begin{cases} -3, & x \le -3t \\ x/t, & -3t < x < -2t, \; t > 0 \\ 3, & x \ge -2t \end{cases}$

   (e) $u(x, t) = \begin{cases} 0, & x \le t \\ \ln(x/t), & t < x < et, \; t > 0 \\ 1, & x \ge et \end{cases}$

3. $u(x, t) = \begin{cases} 0, & x \le 0 \\ x/(1 + t), & 0 < x < 1 + t, \; t > -1 \\ 1, & x \ge 1 + t \end{cases}$

5. (b) No

   (c) $u(x, t) = \begin{cases} 0, & x \le 0 \\ x/t, & 0 < x < 3t \\ 1, & x \ge 3t \end{cases}$

7. $u(x, t) = \begin{cases} 1, & x < 2t \\ \dfrac{1 - x}{1 - 2t}, & 2t < x < 1, \; 0 \le t < \frac{1}{2} \\ 0, & x > 1 \end{cases}$

and

$$u(x, t) = \begin{cases} 1, & x < s(t), \ s(t) = t + \frac{1}{2} \\ 0, & x > s(t) \end{cases}$$

9. $u(x, t) = \begin{cases} 0, & x < -1 \\ \dfrac{1 + x}{1 + 2t}, & -1 < x < 2t, \ 0 < t < \frac{1}{2} \\ \dfrac{1 - x}{1 - 2t}, & 2t < x < 1 \\ 0, & x > 1 \end{cases}$

and

$$u(x, t) = \begin{cases} \dfrac{1 + x}{1 + 2t}, & x < s(t) \\ 0, & x > s(t) \end{cases} \quad t > \frac{1}{2}$$

where

$$s(t) = \frac{1}{3}\left\{ \frac{2t - 1 + 3\sqrt{2}}{\sqrt{1 + 2t}} \right.$$

Section 7.5 (page 325)

1. $u_2(s(t), t) - u_1(s(t), t = \begin{cases} -\frac{1}{2}, & 0 < t < 4 \\ -\sqrt{3}/2(t - 1), & t > 4 \end{cases}$

2. (a) $u(x, t) = \begin{cases} \frac{1}{4}U, & x < \frac{1}{2}Vt \\ (U/2)(1 - x/Vt), & \frac{1}{2}Vt < x < \frac{2}{3}Vt \\ \frac{1}{6}U, & x > \frac{2}{3}Vt \end{cases}$

   (c) $u(x, t) \begin{cases} \frac{1}{4}U, x < 0 \\ \frac{3}{4}U, x > 0 \end{cases}$

   (e) $u(x, t) = \begin{cases} 0, & x < Vt \\ \dfrac{U(x - Vt)}{(L - 2Vt)}, & Vt < x < L - Vt, \ t < L/2V \\ U, & x > L - Vt \end{cases}$

   and

   $$u(x, t) = \begin{cases} 0, & x < \frac{1}{2}L, \\ U, & x > \frac{1}{2}L \end{cases} \quad t > L/2V$$

5. (a) $s(t) = 0$ for $0 < t < 3L/V$
   (b) $t = 6L/V$

6. (a) $u(x, t) = \begin{cases} \frac{1}{4}, & x < \frac{1}{2}t \\ x/2t, & \frac{1}{2}t < x < t; \text{ wetting front} \\ \frac{1}{2}, & x > t \end{cases}$

   (c) $u(x, t) = \begin{cases} \frac{1}{2}, & x > t, \\ x/2t, & x < t \end{cases} \quad 0 < t < 1$

and

$$u(x, t) = \begin{cases} \frac{1}{2}, & x < t - \sqrt{t} \\ x/2t, & t - \sqrt{t} < x < t, t > 1 \\ \frac{1}{2}, & x > t \end{cases}$$

surface drying from $t = 0$ to $t = 1$ followed by wetting front.

Section 7.6 (page 338)

1. (a) Hyperbolic, linear, homoegeneous
   (c) Hyperbolic, linear, nonhomogeneous
   (e) Hyperbolic if $u > \ln (\sqrt{2} - 1)$ and elliptic if $u < \ln (\sqrt{2} - 1)$; quasilinear
   (g) Parabolic, linear

5. (a) $u(x, t) = \frac{1}{5}[6 \sin(x - 3t) - 6 \cos(x - 3t) - \sin(x + 2t) + 6 \cos(x + 2t)]$
      $v(x, t) = \frac{1}{5}[\sin(x - 3t) - \cos(x - 3t) - \sin(x + 2t) + 6 \cos(x + 2t)]$
   (c) $u(x, t) = \frac{1}{2}f(x + t) - \frac{1}{4}g(x + t) + \frac{1}{2}f(x - 3t) + \frac{1}{4}g(x - 3t)$
      $v(x, t) = -f(x + t) + \frac{1}{2}g(x + t) + f(x - 3t) + \frac{1}{2}g(x - 3t)$
   (e) $u(x, t) = [2f(x - t) - g(x - t) + f(x - 4t) + g(x - 4t)]/3$
      $v(x, t) = [-2f(x - t) + g(x - t) + 2f(x - 4t) + 2g(x - 4t)]/3$
   (f) $u(x, t) = [f(x - 3t) + g(x - 5t) - g(x - 3t) + f(x - 5t)]/2$
      $v(x, t) = [f(x - 5t) - f(x - 3t) + g(x - 5t) + g(x - 3t)]/2$

CHAPTER 8

Section 8.1 (page 351)

1. (a) $u_x(\pi/4, 0) = \sqrt{2}/2 = 0.707107$
      $u_t(\pi/4, 0) = -\sqrt{2}/2$
      $u_{xx}(\pi/4, 0) = -\sqrt{2}/2$
      $u_{tt}(\pi/4, 0) = \sqrt{2}/2$

   (c) $h = 0.5$: $\dfrac{1}{2h}\left[u\left(\dfrac{\pi}{4} + h, 0\right) - u\left(\dfrac{\pi}{4} - h, 0\right)\right] = 0.678010$

      $h = 0.1$: $\dfrac{1}{2h}\left[u\left(\dfrac{\pi}{4} + h, 0\right) - u\left(\dfrac{\pi}{4} - h, 0\right)\right] = 0.705929$

      $h = 0.01$: $\dfrac{1}{2h}\left[u\left(\dfrac{\pi}{4} + h, 0\right) - u\left(\dfrac{\pi}{4} - h, 0\right)\right] = 0.707095$

   (e) $k = 0.5$: $\dfrac{1}{2k}\left[u\left(\dfrac{\pi}{4}, 0 + k\right) - u\left(\dfrac{\pi}{4}, 0 - k\right)\right] = -1.042191(\sqrt{2}/2)$

      $k = 0.1$: $\dfrac{1}{2k}\left[u\left(\dfrac{\pi}{4}, 0 + k\right) - u\left(\dfrac{\pi}{4}, 0 - k\right)\right] = -1.001667(\sqrt{2}/2)$

      $k = 0.01$: $\dfrac{1}{2k}\left[u\left(\dfrac{\pi}{4}, 0 + k\right) - u\left(\dfrac{\pi}{4}, 0 - k\right)\right] = -1.000017(\sqrt{2}/2)$

3. $P_1(x, t) = 1 + x + t$, $R_1(x, t) = x^2 + xt + t^2$
   $P_n(x, t) = u(x, t)$, $R_n(x, t) = 0$, $n = 2, 3$

Section 8.2 (page 360)

1. $h = \frac{1}{4}$, $r = k/h^2$, $U_n^0 = \sin nh$, $n = 1, 2, 3$
$U_0^j = U_4^j = 0$, $j = 0, 1, 2, \ldots$
(a) $U_1^{j+1} = U_1^j - 2rU_1^j + rU_2^j$
$U_2^{j+1} = U_2^j + rU_1^j - 2rU_2^j + rU_3^j$
$U_3^{j+1} = U_3^j + rU_2^j - 2rU_3^j$
or $\mathbf{U}^{j+1} = (\mathbf{I} - r\mathbf{F}_1)\mathbf{U}^j$
(c) $(1 + r)U_1^{j+1} - \frac{r}{2}U_2^{j+1} = (1 - r)U_1^j + \frac{r}{2}U_2^j$
$-\frac{r}{2}U_1^{j+1} + (1 + r)U_2^{j+1} - \frac{r}{2}U_3^{j+1} = \frac{r}{2}U_1^j + (1 - r)U_2^j + \frac{r}{2}U_3^j$
$-\frac{r}{2}U_2^{j+1} + (1 + r)U_3^{j+1} = \frac{r}{2}U_2^j + (1 - r)U_3^j$
or $(\mathbf{I} - \frac{r}{2}\mathbf{F}_1)\mathbf{U}^{j+1} = (\mathbf{I} + \frac{r}{2}\mathbf{F}_1)\mathbf{U}^j$

3. $h = \frac{1}{4}$, $r = k/h^2$, $U_n^0 = 0$, $n = 1, 2, 3, 4$
$U_2^j - U_0^j = 2h(1 - e^{-jk}) \equiv 2hp^j$, $U_5^j - U_3^j = 2h \sin(jk) \equiv 2hq^j$
(a) $U_1^{j+1} = U_1^j - 2hrp^j - 2rU_1^j + 2rU_2^j$
$U_2^{j+1} = U_2^j - rU_1^j - 2rU_2^j + rU_3^j$
$U_3^{j+1} = U_3^j - rU_2^j - 2rU_3^j + rU_4^j$
$U_4^{j+1} = U_4^j - 2rU_3^j - 2rU_4^j + 2hrq^j$
or in matrix form,
$\mathbf{U}^{j+1} = (\mathbf{I} - r\mathbf{F}_2)\mathbf{U}^j - 2hr(1 - e^{-jk})\mathbf{e}_1 + 2hr \sin(jk)\mathbf{e}_4$
(c) $(1 + r)U_1^{j+1} - rU_2^{j+1} = (1 - r)U_1^j + rU_2^j - hr(p^j + p^{j+1})$
$-\frac{1}{2}rU_1^{j+1} + (1 + r)U_2^{j+1} - \frac{1}{2}rU_3^{j+1} = \frac{1}{2}rU_1^j + (1 - r)U_2^j + \frac{1}{2}rU_3^j$
$-\frac{1}{2}rU_2^{j+1} + (1 + r)U_3^{j+1} - \frac{1}{2}rU_4^{j+1} = \frac{1}{2}rU_2^j + (1 - r)U_3^j + \frac{1}{2}rU_4^j$
$-rU_3^{j+1} + (1 + r)U_4^{j+1} = rU_3^j + (1 - r)U_4^j + hr(q^j + q^{j+1})$
or, in matrix form,
$(\mathbf{I} + \frac{1}{2}r\mathbf{F}_2)\mathbf{U}^{j+1} = (\mathbf{I} - \frac{1}{2}r\mathbf{F}_2)\mathbf{U}^j - hr(p^{j+1} + p^j)\mathbf{e}_1 + hr(q^{j+1} + q^j)$

5. $h = 1/(N + 1)$, $r = k/h^2$
$$U_n^{j+1} - U_n^j = r\delta_x^2 U_n^j + \frac{3k}{2h}[U_{n+1}^j - U_{n-1}^j] + 5kU_n^j, \quad n = 1, 2, \ldots, N$$

Section 8.3 (page 374)

1. **Algorithm 8.1N** Forward Difference Method: Neumann Initial-Boundary-Value Problem
Step 1. Document
This algorithm uses the forward difference method (8.2.9) to approximate the solution to the initial-boundary-value problem

$$u_t - a^2 u_{xx} = S(x, t), \quad 0 < x < L, t > 0$$
$$u(x, 0) = f(x), \quad 0 < x < L$$
$$u_x(0, t) = p(t), \quad u_x(L, t) = q(t), \quad t > 0$$

INPUT
Real a^2, diffusivity
Real L, endpoint
Real k, time step
Integer $jmax$, number of time steps
Integer $nmax$, number of computational nodes
Function $S(x, t)$, right side of diffusion equation

Function $f(x)$, initial condition
Functions $p(t)$, $q(t)$, boundary derivatives at $x = 0$ and $x = L$
OUTPUT
t, time for $j = 0, 1, \ldots, jmax$
U_n^j, approximate solution for $n = 1, \ldots, nmax$

Step 2. Define a grid
   Set $h = L/(nmax - 1)$
   $r = a^2 k/h^2$
   If $r > \frac{1}{2}$, output message that method is unstable.

Step 3. Initialize numerical solution
   Set $t = 0$
   For $n = 1, 2, \ldots, nmax$ set
   $x_n = (n - 1)h$
   $V_n = f(x_n)$
   Output t
   For $n = 1, 2, \ldots, nmax$
   Output x_n, V_n

Step 4. Begin time stepping
   For $j = 1, 2, \ldots, jmax$
   Do steps 5–7

Step 5. Advance solution one time step
   Set $U_1 = -2hrp(t) + (1 - 2r)V_1 + 2rV_2 + kS(x_1, t)$
   For $n = 2, 3, \ldots, nmax - 1$ set
   $U_n = rV_{n-1} + (1 - 2r)V_n + rV_{n+1} + kS(x_n, t)$
   Set
   $U_{nmax} = 2rV_{nmax-1} + (1 - 2r)V_{nmax} + 2hrq(t) + kS(x_{nmax}, t)$
   Set $t = t + k$

Step 6. Output numerical solution
   Output t
   For $n = 1, 2, \ldots, nmax$
   Output x_n, U_n

Step 7. Prepare for next time step
   For $n = 1, 2, \ldots, nmax$ set
   $V_n = U_n$

3. Change steps 2 and 5 in Algorithm 8.3 to read:
   Step 2. Define a grid
   Set $h = L/(nmax + 1)$
   $r = a^2/k/h^2$
   $z = \beta k/(2h)$

   Step 5. Define tridiagonal system
   Set $t = t + k$
   For $n = 1, 2, \ldots, nmax$ set
   $a_n = -r - z$
   $b_n = (1 + 2r + \gamma)$
   $c_n = -r + z$
   $d_n = U_n + kS(x_n, t)$
   Set $d_1 = d_1 + rp(t)$
   $d_{nmax} = d_{nmax} + rq(t)$

Section 8.4 (page 390)

1. (a) $\lambda_1 = 0$, $\lambda_2 = 2$, $\lambda_3 = 4$

   (b) $\mathbf{V}_1 = \begin{bmatrix} 1 \\ 1 \\ 1 \end{bmatrix}$, $\mathbf{V}_2 = \begin{bmatrix} 1 \\ 0 \\ -1 \end{bmatrix}$, $\mathbf{V}_3 = \begin{bmatrix} 1 \\ -1 \\ 1 \end{bmatrix}$

   (c) $\mathbf{W}_1 = \begin{bmatrix} 1 \\ 2 \\ 1 \end{bmatrix}$, $\mathbf{W}_2 = \begin{bmatrix} 1 \\ 0 \\ -1 \end{bmatrix}$, $\mathbf{W}_3 = \begin{bmatrix} 1 \\ -2 \\ 1 \end{bmatrix}$

3. $\mathbf{f} = \begin{bmatrix} 0 \\ \frac{1}{9} \\ \frac{4}{9} \end{bmatrix} = c_1 \begin{bmatrix} 1 \\ \sqrt{3}/2 \\ 1 \end{bmatrix} + c_2 \begin{bmatrix} 1 \\ 0 \\ -1 \end{bmatrix} + c_3 \begin{bmatrix} 1 \\ -\sqrt{3}/2 \\ 1 \end{bmatrix}$

   where

   $$c_1 = \frac{4 + \sqrt{3}}{27}, \qquad c_2 = \frac{-8}{27}, \qquad c_3 = \frac{4 - \sqrt{3}}{27}$$

5. (a) $\begin{bmatrix} U_1^{j+1} \\ U_2^{j+1} \\ U_3^{j+1} \end{bmatrix} = \begin{bmatrix} U_1^j \\ U_2^j \\ U_3^j \end{bmatrix} - r \begin{bmatrix} 2 & -2 & 0 \\ -1 & 2 & -1 \\ 0 & -2 & 2 \end{bmatrix} \begin{bmatrix} U_1^j \\ U_2^j \\ U_3^j \end{bmatrix}$, $\begin{bmatrix} U_1^0 \\ U_2^0 \\ U_3^0 \end{bmatrix} = \begin{bmatrix} 0 \\ \frac{1}{4} \\ 1 \end{bmatrix}$

   or

   $$\mathbf{U}^{j+1} = (\mathbf{I} - r\mathbf{F}_2)\mathbf{U}^j$$

   (b) $(\mathbf{I} + r\mathbf{F}_2)\mathbf{U}^{j+1} = \mathbf{U}^j$

7. (a) $\begin{bmatrix} U_1^{j+1} \\ U_2^{j+1} \\ U_3^{j+1} \\ U_2^{j+1} \\ U_3^{j+1} \end{bmatrix} = \begin{bmatrix} U_1^j \\ U_2^j \\ U_3^j \\ U_2^j \\ U_3^j \end{bmatrix} - r \begin{bmatrix} 2 & -2 & 0 & 0 & 0 \\ -1 & 2 & -1 & 0 & 0 \\ 0 & -1 & 2 & -1 & 0 \\ 0 & 0 & -1 & 2 & -1 \\ 0 & 0 & 0 & -2 & 2 \end{bmatrix} \begin{bmatrix} U_1^j \\ U_2^j \\ U_3^j \\ U_2^j \\ U_3^j \end{bmatrix}$

   with

   $$[U_1^0, U_2^0, U_3^0, U_4^0, U_5^0]^T = [0, \tfrac{1}{4}, \tfrac{1}{2}, \tfrac{3}{4}, 1]^T$$

   or, in matrix notation,

   $$\mathbf{U}^{j+1} = (\mathbf{I} - r\mathbf{F}_2)\mathbf{U}^j$$

   (b) $(\mathbf{I} + r\mathbf{F}_2)\mathbf{U}^{j+1} = \mathbf{U}^j$
   (c) $(\mathbf{I} + \tfrac{1}{2}r\mathbf{F}_2)\mathbf{U}^{j+1} = (\mathbf{I} - \tfrac{1}{2}r\mathbf{F}_2)\mathbf{U}^j$

Section 8.5 (page 405)

1. $r = ak/h^2$. For the spectral stability condition to hold, h and k must satisfy:

   (a) $\left| 1 - 4r \sin^2 \dfrac{n\pi}{2(N+1)} \right| \le 1$, $h = 1/(N+1)$, $n = 1, 2, \ldots, N$

(b) $\left| 1 - 4r \sin^2 \dfrac{(n-1)\pi}{2(N-1)} \right| \le 1$, $h = 1/(N-1)$, $n = 1, 2, \ldots, N$

(c) $\left| 1 - 4r \sin^2 \dfrac{(2n-1)\pi}{4N} \right| \le 1$, $h = 1/N$, $n = 1, 2, \ldots, N$

3. The spectral stability condition is satisfied for any choice of h and k.

4. (a) $\left| \dfrac{1}{1 + 4r \sin^2 \beta/2} \right| \le 1$, $0 \le \beta \le 2\pi$

5. (a) $u(x, t) = \dfrac{e^{ct}}{2a\sqrt{\pi t}} \displaystyle\int_{-\infty}^{\infty} \exp\left[-\dfrac{(x-\xi)^2}{4a^2 t} \right] f(\xi)\, d\xi$

7. (a) $c_n = \dfrac{2}{N+1} \displaystyle\sum_{n=1}^{N} f_n \sin \dfrac{n\pi}{N+1}$

Section 8.7 (page 430)

1. $c[U_1^{j+1} - U_1^j] = re^{1/8} \cdot 0 - r[e^{1/8} + e^{3/8}]U_1^j + re^{3/8}U_2^j$

$c[U_2^{j+1} - U_2^j] = re^{3/8}U_1^j - r[e^{3/8} + e^{5/8}]U_2^j + re^{5/8}U_3^j$

$c[U_3^{j+1} - U_3^j] = re^{5/8}U_2^j - r[e^{5/8} + e^{7/8}]U_3^j + re^{7/8} \cdot 0$

$U_1^0 = -\frac{3}{16}, \quad U_2^0 = -\frac{1}{4}, \quad U_3^0 = -\frac{3}{16}$

4. $c[U_1^{j+1} - U_1^j] = -re^{3/8}U_1^j + re^{3/8}U_2^j$

$c[U_2^{j+1} - U_2^j] = re^{3/8}U_1^j - r[e^{3/8} + e^{5/8}]U_2^j + re^{5/8}U_3^j$

$c[U_3^{j+1} - U_3^j] = re^{5/8}U_2^j - re^{5/8}U_3^j$

$U_1^0 = -\frac{3}{16}, \quad U_2^0 = -\frac{1}{4}, \quad U_3^0 = -\frac{3}{16}$

7. (a) $c(x_n, \frac{1}{2}(t_j + t_{j+1}))[U_n^{j+1} - U_n^j] = \frac{1}{2}r\{\sigma(\xi_n, t_{j+1})U_{n-1}^{j+1}$

$\qquad\qquad - [\sigma(\xi_n, t_{j+1}) + \sigma(\xi_{n+1}, t_{j+1})]U_n^{j+1}$

$\qquad\qquad + \sigma(\xi_{n+1}, t_{j+1})U_{n+1}^{j+1} + \sigma(\xi_n, t_j)U_{n-1}^j$

$\qquad\qquad - [\sigma(\xi_n, t_j) + \sigma(\xi_{n+1}, t_j)]U_n^j$

$\qquad\qquad + \sigma(\xi_{n+1}, t_j)U_{n+1}^j\}$, $r = k/h^2$

(b) $c(U_n^j)[U_n^{j+1} - U_n^j] = \frac{1}{2}r\{\mu_{n-1}^j U_{n-1}^{j+1} - [\mu_{n-1}^j + \mu_{n+1}^j]U_n^{j+1} + \mu_{n+1}^j U_{n+1}^{j+1}$

$\qquad\qquad + \mu_{n-1}^j U_{n-1}^j - [\mu_{n-1}^j + \mu_{n+1}^j]U_n^j + \mu_{n+1}^j U_{n+1}^j\}$, $r = k/h^2$

9. $x_1 = \frac{1}{2}(h_0 + h_1)$, $x_2 = x_1 + \frac{1}{2}(h_1 + h_2), \ldots, x_n = x_{n-1} + \frac{1}{2}(h_{n-1} + h_n)$

11. $c_n^j[U_n^{j+1} - U_n^j]h_n = \left\{ \dfrac{-2\sigma_n^j \sigma_{n-1}^j}{\sigma_n^j h_{n-1} + \sigma_{n-1}^j h_n}[U_n^j - U_{n-1}^j] \right.$

$\qquad\qquad \left. + \dfrac{2\sigma_n^j \sigma_{n+1}^j}{\sigma_n^j h_{n+1} + \sigma_{n+1}^j h_n}[U_{n+1}^j - U_n^j] \right\} k$

13. (a) $c(x_n, t_j))[U_n^{j+1} - U_n^j] = r\{\sigma(\xi_n, t_j)U_{n-1}^j - [\sigma(\xi_n, t_j) + \sigma(\xi_{n+1}, t_j)]U_n^j$

$\qquad\qquad + \sigma(\xi_{n+1}, t_j)U_{n+1}^j\}$

$\qquad\qquad + \dfrac{k}{2h}\{v(\xi_n, t_j)[U_{n-1}^j + U_n^j]$

$\qquad\qquad - v(\xi_{n+1}, t_j)[U_{n+1}^j + U_n^j]\}$, $r = k/h^2$

Section 8.8 (page 445)

4. $U_{11}^{j+1} - U_{11}^j = re^{1/8} \cdot 0 - r[e^{1/8} + e^{3/8}]U_{11}^j + re^{3/8}U_{21}^j$
$\qquad\qquad + re^{-1/8} \cdot 0 - r[e^{-1/8} + e^{-3/8}]U_{11}^j + re^{-3/8}U_{12}^j$

$U_{21}^{j+1} - U_{21}^j = re^{3/8}U_{11}^j - r[e^{3/8} + e^{5/8}]U_{21}^j + re^{5/8}U_{31}^j$
$\qquad\qquad + re^{-1/8} \cdot 0 - r[e^{-1/8} + e^{-3/8}]U_{21}^j + re^{-3/8}U_{22}^j$

$U_{22}^{j+1} - U_{22}^j = re^{3/8}U_{12}^j - r[e^{3/8} + e^{5/8}]U_{22}^j + re^{5/8}U_{32}^j$
$\qquad\qquad + re^{-3/8} \cdot U_{21}^j - r[e^{-3/8} + e^{-5/8}]U_{22}^j + re^{-5/8}U_{23}^j$

etc.

$$r = 16k$$

CHAPTER 9

Section 9.1 (page 457)

1. (a) Algorithm 9.1 Modified for FTBS

Step 1. Document

This algorithm uses the FTBS method to approximate the solution of the initial-value problem

$$au_x + u_t = c(x, t), \qquad -\infty < x < \infty, t > 0$$
$$u(x, 0) = f(x), \qquad -\infty < x < \infty$$

for $t_j \le t_{jmax}$ and $x_p \le x_n \le x_q$.

INPUT

Integer p, lower index for solution at time t_{jmax}
Integer q, upper index for solution at time t_{jmax}
Integer $jmax$, number of time steps
Real k, time step ($kjmax = t_{jmax}$)
Real h, space step
Real a, coefficient in equation, $a > 0$
Function $f(x)$, initial condition
Function $c(x, t)$, right side of equation

OUTPUT
Time t and approximation U_n^j to $u(x_n, t_j)$ for $n = p, p + 1, \ldots, q$ and $j = 0, 1, \ldots$ $jmax$

Step 2. Define grid ratio

Set $s = k/h$
If $sa > 1$ output message that computation is unstable.

Step 3. Initialize numerical solution

Set $t = 0$
$\quad nmin = p - jmax$
$\quad nmax = q$
For $n = nmin, nmin + 1, \ldots, nmax$ set
$\quad x = nh$
$\quad V_n = f(x)$
Output t
For $n = p, p + 1, \ldots, q$
$\quad$ Output x_n, V_n

Step 4. Begin time stepping
For $j = 1, 2, \ldots, jmax$
    Do steps 5–7
Step 5. Advance solution one time step
Set $nmin = nmin + 1$
For $n = nmin, nmin + 1, \ldots, nmax$ set
    $x = nh$
    $U_n = (1 - sa)V_n + saV_{n-1} + kc(x, t)$
Step 6. Prepare for next time step
Set $t = t + k$
For $n = nmin, nmin + 1, \ldots, nmax$ set
    $V_n = U_n$
Step 7. Output numerical solution
Output t
For $n = p, p + 1, \ldots, q$
    Output x_n, U_n
(b) In Algorithm 9.1, step 3, set $nmax = q$ and modify step 5 to read:
Step 5. Advance solution one time step
Set $nmin = nmin + 1$
For $n = nmin, nmin + 1, \ldots, nmax$ set
    $x = nh$
    $U_n = (1 - sa)V_n + saV_{n-1} + kc(x, t)$
(c) It is only necessary to modify step 5 of Algorithm 9.1 to read:
Step 5. Advance solution one time step
Set $nmin = nmin + 1$
    $nmax = nmax - 1$
For $n = nmin, nmin + 1, \ldots, nmax$ set
    $x = nh$
    $U_n = 0.5(V_{n-1} + V_{n+1}) - 0.5sa(V_{n+1} + V_{n-1}) + kc(x, t)$
5. (a) TE $= -ht_j$
6. (a) $U_n^{j+1} = (1 - sa) + saS_-$

Section 9.2 (page 465)

2. (a) Step 1. Document
This algorithm uses the FTBS method to approximate the solution of the initial-value problem
$$au_x + u_t = c(x, t), \qquad 0 < x < \infty, t > 0$$
$$u(x, 0) = f(x), \qquad 0 < x < \infty$$
$$u(0, t) = g(t), \qquad t > 0$$

for $t_j \leq t_{jmax}$ and $0 \leq x_n \leq x_q$.
INPUT

Integer q, upper index for solution at time t_{jmax}
Integer j max, number of time steps
Real k, time step ($k \cdot jmax = t_{jmax}$)
Real h, space step
Real a, coefficient in equation, $a > 0$
Function $f(x)$, initial condition

Function $g(t)$, boundary condition
Function $c(x, t)$, right side of equation
OUTPUT
Time t and approximation U_n^j to $u(x_n, t_j)$ for $n = 0, 1, \ldots, q$ and $j = 0, 1, \ldots, jmax$

Step 2. Define grid ratio
Set $s = k/h$
If $sa > 1$, output message that computation is unstable.

Step 3. Initialize numerical solution
Set $t = 0$
$\quad V_0 = \frac{1}{2}[g(0) + f(0)]$
For $n = 1, 2, \ldots, q$ set
$\quad x = nh$
$\quad V_n = f(x)$
Output t
For $n = 0, 1, \ldots, q$
$\quad$ Output x_n, V_n

Step 4. Begin time stepping
For $j = 1, 2, \ldots, jmax$
$\quad$ Do steps 5–7

Step 5. Advance solution one time step
For $n = 1, 2, \ldots, q$ set
$\quad x = nh$
$\quad U_n = (1 - sa)V_n + saV_{n-1} + kc(x, t)$

Step 6. Prepare for next time step
Set $t = t + k$
$\quad U_0 = g(t)$
For $n = 0, 1, \ldots, q$ set
$\quad V_n = U_n$

Step 7. Output numerical solution
Output t
For $n = 0, 1, \ldots, q$
$\quad$ Output x_n, U_n

5. (a) $P(S_+, S_-) = 1 + sa + saS_-$ and $Q(S_+, S_-) = 1$; so

$$\left| \frac{\xi^{j+1}}{\xi^j} \right| = \left| \frac{1}{1 + sa + sae^{-i\beta}} \right| \leq 1$$

6. (a) It is only necessary to modify step 5 of Algorithm 9.2 to read:
Step 5. Advance solution one time step
Set $t = t + k$
$\quad U_0 = g(t)$
For $n = 0, 1, \ldots, n \max - 1$ set
$\quad U_{n+1} = (V_{n+1} + saU_n)/(1 + sa)$

Section 9.3 (page 476)

3. For the Lax–Wendroff method (9.1.10),

$$\frac{\partial R}{\partial U_{n-1}} = \frac{sa}{2}(1 + sa) \quad \text{and} \quad \frac{\partial R}{\partial U_{n+1}} = \frac{sa}{2}(-1 + sa)$$

Section 9.4 (page 486)

1. $\omega = \beta - ic$

2. (a) $|\xi| = \left\{1 - 4(sa)^2[1 - (sa)^2]\sin^4\dfrac{mh}{2}\right\}^{1/2}$

   (d) $|\xi| = 1$

3. (d) $\dfrac{\mathcal{R}(\alpha)}{a} = \dfrac{1}{sam\pi/p}\tan^{-1}\left[\dfrac{2sa\,\sin(m\pi/p)}{(1 - s^2a^2) + (1 + s^2a^2)\cos(m\pi/p)}\right]$

8. (a) $v = \frac{1}{24}a^2k^2h^2[1 - s^2a^2]$

9. (a) $\gamma = \frac{1}{24}akh^2[1 - s^2a^2]$

Section 9.5 (page 494)

1. (a) $u_1(x, t) = \frac{1}{3}[2\sin(x - t) - (x - t)^2 + \sin(x - 4t) + (x - 4t)^2]$
     $u_2(x, t) = \frac{1}{3}[-2\sin(x - t) + (x - t)^2 + 2\sin(x - 4t) + 2(x - 4t)^2]$

7. (a) $u_1(x, t) = \frac{1}{2}[2\sin(x - 3t) - \cos(x - 3t) + \sin(x - 5t) + \cos(x - 5t)]$
     $u_2(x, t) = \frac{1}{2}[-\sin(x - 3t) + \cos(x - 3t) + \sin(x - 5t) + \cos(x - 5t)]$

11. (a) FTFS: (i) $\mathbf{G} = \mathbf{I} + s\mathbf{A} - se^{i\beta h}\mathbf{A}$; (ii) if μ is an eigenvalue of $\mathbf{G}$, then $\mu = 1 + s\lambda(1 - e^{i\beta h})$, where λ is an eigenvalue of $\mathbf{A}$, (iii) $|\mu|^2 = 1 + 2s\lambda(1 + s\lambda)(1 - \cos\beta h)$, so $-1 \le s\lambda \le 0$ must hold to ensure $|\mu| \le 1$.

   (c) FTCS: (i) $\mathbf{G} = \mathbf{I} + i(s\,\sin\beta h)\mathbf{A}$; (ii) if μ is an eigenvalue of $\mathbf{G}$, then $\mu = 1 + is\lambda\,\sin\beta h$, where λ is an eigenvalue of $\mathbf{A}$; (iii) $|\mu|^2 = 1 + (s\lambda)^2\sin^2\beta h$, so $|\mu| > 1$ and method is unstable.

   (e) CTCS: (i) $\begin{bmatrix}\xi^j \\ \xi^{j+1}\end{bmatrix}\begin{bmatrix}\mathbf{0} & \mathbf{I} \\ \mathbf{I} & -(2is\,\sin\beta h)\mathbf{A}\end{bmatrix}\begin{bmatrix}\xi^j \\ \xi^{j+1}\end{bmatrix}$ (ii) If μ is an eigenvalue of $\mathbf{G}$, then

$\mu^2 + (2is\lambda\,\sin\beta h)\mu - 1 = 0$ where λ is an eigenvalue of $\mathbf{A}$. (iii) If $|s\lambda| \le 1$ and μ is as in (ii), $|\mu| = 1$ (quadratic formula).

Section 9.6 (page 507)

2. (a) $u(x, t) = \cos 4\pi t\,\sin 2\pi x$

4. (a) $u(x, t) = \frac{1}{2}\{(1 - |x - t|)H(x - t) + (1 - |x + t|)H(x + t)\}$, where

$$H(x) = \begin{cases}1 & \text{if } x \ge 0 \\ 0 & \text{if } x < 0\end{cases}$$

Section 9.7 (page 519)

1. On the characteristic through $(1, 0)$, $x = x(t)$ and $u = u(t)$ are given by
   (a) $x = t + 1$, $u = e^t\cos 1$
   (c) $x = e^{t^2}$, $u = e^t$
   (e) $x = \sqrt{2t + 1}$, $u = [2 - \sqrt{2t + 1}]^{-1}$

3. (a) $u_1(x, t) = \frac{1}{3}[2(x - t) - (x - t)^2 + (x - 4t) + (x - 4t)^2]$
     $u_2(x, t) = \frac{1}{3}[-2(x - t) + (x - t)^2 + 2(x - 4t) + 2(x - 4t)^2]$

4. (a) $u_1(x, t) = \frac{1}{2}[\sin(x - 3t) - (x - 3t)^2 + \sin(x - 5t) + (x - 5t)^2]$

$u_2(x, t) = \frac{1}{2}[-\sin(x - 3t) + (x - 3t)^2 + \sin(x - 5t) + (x - 5t)^2]$

6. (a) $u = \ln x + 2 \ln y$

(b) $y/x = \beta = \text{const}, \; xh = \alpha = \text{const}$

8. (b) $y \exp(-x^2/2) = \beta = \text{const}, \; y \exp(x^2/2) = \alpha = \text{const}, \; P = (\sqrt{\frac{5}{2}}, 2e^{3/4})$

CHAPTER 10

Section 10.1 (page 529)

1. (a) On the vertex-centered grid

$$(x_1, x_2, x_3) = (\tfrac{1}{4}, \tfrac{1}{2}, \tfrac{3}{4}), \qquad (y_1, y_2, y_3) = (0, \tfrac{1}{2}, 1)$$

with $hx = \frac{1}{4}$ and $hy = \frac{1}{2}$, the difference equations are

$$\frac{\delta_x^2 U_{mn}}{hx^2} + \frac{\delta_y^2 U_{mn}}{hy^2} = 0$$

or

$$
\begin{array}{llllll}
-6U_{11} + 2U_{21} & + 2U_{12} & & & & = 0 \\
2U_{11} - 6U_{21} + 2U_{31} & + 2U_{22} & & & & = 0 \\
2U_{21} - 6U_{31} & + 2U_{32} & & & & = -2 \\
U_{11} & - 6U_{21} + 2U_{22} & + U_{13} & & & = 0 \\
U_{21} & + 2U_{12} - 6U_{22} + 2U_{32} & + U_{23} & & & = 0 \\
U_{31} & + 2U_{22} - 6U_{32} & + U_{33} & & & = -2 \\
2U_{12} & - 6U_{13} + 2U_{23} & & & & = -5 \\
2U_{22} & + 2U_{13} - 6U_{23} + 2U_{33} & & & & = -5 \\
2U_{32} & + 2U_{23} - 6U_{33} & & & & = -5 \\
\end{array}
$$

On the cell-centered grid

$$(x_1, x_2, x_3) = (\tfrac{1}{4}, \tfrac{1}{2}, \tfrac{3}{4}), \qquad (y_0, y_1, y_2) = (\tfrac{1}{6}, \tfrac{1}{2}, \tfrac{5}{8})$$

with $hx = \frac{1}{4}$ and $hy = \frac{1}{3}$, the difference equations are

$$-\frac{U_{11} - 0}{\frac{1}{4}} + \frac{U_{21} - U_{11}}{\frac{1}{4}} - 0 + \frac{U_{12} - U_{11}}{\frac{1}{3}} = 0$$

$$-\frac{U_{21} - U_{11}}{\frac{1}{4}} + \frac{U_{31} - U_{21}}{\frac{1}{4}} - 0 + \frac{U_{22} - U_{21}}{\frac{1}{3}} = 0$$

$$-\frac{U_{31} - U_{21}}{\frac{1}{4}} + \frac{1 - U_{31}}{\frac{1}{4}} - 0 + \frac{U_{32} - U_{31}}{\frac{1}{3}} = 0$$

$$-\frac{U_{12} - 0}{\frac{1}{4}} + \frac{U_{22} - U_{12}}{\frac{1}{4}} - \frac{U_{12} - U_{11}}{\frac{1}{3}} + \frac{U_{13} - U_{12}}{\frac{1}{3}} = 0$$

$$-\frac{U_{22} - U_{12}}{\frac{1}{4}} + \frac{U_{32} - U_{22}}{\frac{1}{4}} - \frac{U_{22} - U_{21}}{\frac{1}{3}} + \frac{U_{23} - U_{22}}{\frac{1}{3}} = 0$$

$$-\frac{U_{32} - U_{22}}{\frac{1}{4}} + \frac{1 - U_{32}}{\frac{1}{4}} - \frac{U_{32} - U_{31}}{\frac{1}{3}} + \frac{U_{33} - U_{32}}{\frac{1}{3}} = 0$$

$$-\frac{U_{13} - 0}{\frac{1}{4}} + \frac{U_{23} - U_{13}}{\frac{1}{4}} - \frac{U_{13} - U_{12}}{\frac{1}{3}} + 5 = 0$$

$$-\frac{U_{23} - U_{13}}{\frac{1}{4}} + \frac{U_{33} - U_{23}}{\frac{1}{4}} - \frac{U_{23} - U_{22}}{\frac{1}{3}} + 5 = 0$$

$$-\frac{U_{33} - U_{23}}{\frac{1}{4}} + \frac{1 - U_{33}}{\frac{1}{4}} - \frac{U_{33} - U_{32}}{\frac{1}{3}} + 5 = 0$$

(c) On the vertex-centered grid

$$(x_1, x_2, x_3) = (0, \tfrac{1}{3}, \tfrac{2}{3}), \qquad (y_0, y_1, y_2) = (0, \tfrac{1}{3}, \tfrac{2}{3})$$

with $hx = \tfrac{1}{3} = hy$ and $c_{mn} = \tfrac{1}{9} f_{mn}$,

$$
\begin{aligned}
-4U_{11} + U_{21} \quad\quad + 2U_{12} &= c_{11} \\
U_{11} - 4U_{21} + U_{31} \quad\quad + 2U_{22} &= c_{12} \\
U_{21} - 4U_{31} \quad\quad + 2U_{32} &= c_{13} \\
U_{11} \quad\quad - 4U_{12} + 2U_{22} \quad\quad + U_{13} &= c_{12} \\
U_{21} \quad\quad + U_{12} - 4U_{22} + U_{32} \quad\quad + U_{23} &= c_{22} \\
U_{31} \quad\quad + U_{22} - 4U_{32} \quad\quad + U_{33} &= c_{32} \\
U_{12} \quad\quad - 4U_{13} + 2U_{23} &= c_{13} \\
U_{22} \quad\quad + U_{13} - 4U_{23} + U_{33} &= c_{23} \\
U_{23} \quad\quad + U_{23} - 4U_{33} &= c_{33}
\end{aligned}
$$

Section 10.2 (page 537)

1. (b) $\mathbf{U} = [1, 1, 1]^T$

Section 10.3

1. (a) $\lambda_{mn} = -4 + 2\cos\dfrac{m\pi}{4} + 2\cos\dfrac{n\pi}{3}$, $m = 1, 2, 3, n = 1, 2$

(c) $\lambda_{mn} = -4 + 2\cos\dfrac{(m-1)\pi}{2} + 2\cos(n-1)\pi$, $m = 1, 2, 3, n = 1, 2$

(e) $\lambda_{mn} = -4 + 2\cos\dfrac{m\pi}{4} + 2\cos\dfrac{(2n-1)\pi}{4}$, $m,\ n = 1, 2, 3$

2. (a) $\mathbf{V}_{mn} = \left[\sin\dfrac{n\pi}{3}\sin\dfrac{m\pi}{4}, \sin\dfrac{n\pi}{3}\sin\dfrac{m\pi}{2}, \sin\dfrac{n\pi}{3}\sin\dfrac{3m\pi}{4}, \sin\dfrac{2n\pi}{3}\sin\dfrac{m\pi}{4}, \right.$

$\left. \sin\dfrac{2n\pi}{3}\sin\dfrac{m\pi}{2}, \sin\dfrac{2n\pi}{3}\sin\dfrac{3m\pi}{4} \right]^T$, $m = 1, 2, 3, n = 1, 2$

(c) $\mathbf{V}_{mn} = \left[\sin\dfrac{m\pi}{4}, \sin\dfrac{m\pi}{2}, \sin\dfrac{3m\pi}{4}, \cos((n-1)\pi)\sin\dfrac{m\pi}{4}, \cos((n-1)\pi)\sin\dfrac{m\pi}{2}, \right.$

$\left. \cos((n-1)\pi)\sin\dfrac{3m\pi}{4} \right]^T$, $m = 1, 2, 3, n = 1, 2,$

Section 10.4 (page 564)

1. (a)

	U_{11}	U_{21}	U_{12}	U_{22}
$l = 1$	2.00000000	3.00000000	3.00000000	4.00000000
$l = 2$	3.00000000	4.00000000	4.00000000	5.25000000
$l = 3$	3.50000000	4.56250000	4.56250000	5.81250000
$l = 4$	3.78125000	4.82812500	4.82812500	6.09375000
$l = 5$	3.91406250	4.96875000	4.96875000	6.23046875

(d) $U_{11} = 4$, $U_{21} = 5$, $U_{12} = 5$, $U_{22} = 6$

9. (a) $x_m = 1 + \frac{1}{10}m$, $y_n = \frac{1}{10}n$; $x_{m-1/2} = \frac{1}{2}(x_m + x_{m-1})$;
$x_{m+1/2} = \frac{1}{2}(x_m + x_{m+1})$, $y_{n-1/2} = \frac{1}{2}(y_n + y_{n-1})$, $y_{n+1/2} = \frac{1}{2}(y_n + y_{n+1})$
$- x_{m-1/2}[U_{mn} - U_{m-1,n}] + x_{m+1/2}[U_{m+1,n} - U_{mn}] - \exp(y_{n-1/2})[U_{mn} - U_{m,n-1}]$
$+ \exp(y_{n+1/2})[U_{m,n+1} - U_{mn}] = 100 \exp(y_n)[1 + 2x_n\exp(y_n)]$

Section 10.5 (page 573)

1. $\lambda_n = 4 - 2 \cos \dfrac{n\pi}{4}$, $n = 1, 2, 3$ and $\rho(A) = 4 + \sqrt{2}$

2. (a) $PAP^T = \begin{bmatrix} 4 & 0 & -1 \\ 0 & 4 & -1 \\ -1 & -1 & 4 \end{bmatrix}$

5. $\rho(T_G) = \frac{3}{4}$

6. $\overline{\omega} = \frac{4}{3}$

7. (a) $\lambda_1 = 1$, $\lambda_2 = 3$, $\lambda_3 = 3$, $\lambda_4 = 5$
   (b) $1 - \frac{1}{4}\lambda_n$, $n = 1, 2, 3, 4$, λ_n from (a)
   (c) $\rho(T_J) = \frac{3}{4}$
   (d) $\rho(T_G) = \frac{9}{16}$
   (e) $\overline{\omega} = 2/(1 + \sqrt{7}/4)$.

12. $\lambda_{mn} = \dfrac{\cos(m\pi/4)}{2 - \cos(n\pi/4)}$, $m, n = 1, 2, 3$

13. $2/(4 - \sqrt{2})^2$

Index

A CATALOG OF SELECTED
DOVER BOOKS
IN SCIENCE AND MATHEMATICS

Mathematics

FUNCTIONAL ANALYSIS (Second Corrected Edition), George Bachman and Lawrence Narici. Excellent treatment of subject geared toward students with background in linear algebra, advanced calculus, physics and engineering. Text covers introduction to inner-product spaces, normed, metric spaces, and topological spaces; complete orthonormal sets, the Hahn-Banach Theorem and its consequences, and many other related subjects. 1966 ed. 544pp. 6⅛ x 9¼. 0-486-40251-7

ASYMPTOTIC EXPANSIONS OF INTEGRALS, Norman Bleistein & Richard A. Handelsman. Best introduction to important field with applications in a variety of scientific disciplines. New preface. Problems. Diagrams. Tables. Bibliography. Index. 448pp. 5⅜ x 8½. 0-486-65082-0

VECTOR AND TENSOR ANALYSIS WITH APPLICATIONS, A. I. Borisenko and I. E. Tarapov. Concise introduction. Worked-out problems, solutions, exercises. 257pp. 5⅝ x 8¼. 0-486-63833-2

AN INTRODUCTION TO ORDINARY DIFFERENTIAL EQUATIONS, Earl A. Coddington. A thorough and systematic first course in elementary differential equations for undergraduates in mathematics and science, with many exercises and problems (with answers). Index. 304pp. 5⅜ x 8½. 0-486-65942-9

FOURIER SERIES AND ORTHOGONAL FUNCTIONS, Harry F. Davis. An incisive text combining theory and practical example to introduce Fourier series, orthogonal functions and applications of the Fourier method to boundary-value problems. 570 exercises. Answers and notes. 416pp. 5⅜ x 8½. 0-486-65973-9

COMPUTABILITY AND UNSOLVABILITY, Martin Davis. Classic graduate-level introduction to theory of computability, usually referred to as theory of recurrent functions. New preface and appendix. 288pp. 5⅜ x 8½. 0-486-61471-9

ASYMPTOTIC METHODS IN ANALYSIS, N. G. de Bruijn. An inexpensive, comprehensive guide to asymptotic methods–the pioneering work that teaches by explaining worked examples in detail. Index. 224pp. 5⅜ x 8½ 0-486-64221-6

APPLIED COMPLEX VARIABLES, John W. Dettman. Step-by-step coverage of fundamentals of analytic function theory–plus lucid exposition of five important applications: Potential Theory; Ordinary Differential Equations; Fourier Transforms; Laplace Transforms; Asymptotic Expansions. 66 figures. Exercises at chapter ends. 512pp. 5⅜ x 8½. 0-486-64670-X

INTRODUCTION TO LINEAR ALGEBRA AND DIFFERENTIAL EQUATIONS, John W. Dettman. Excellent text covers complex numbers, determinants, orthonormal bases, Laplace transforms, much more. Exercises with solutions. Undergraduate level. 416pp. 5⅜ x 8½. 0-486-65191-6

RIEMANN'S ZETA FUNCTION, H. M. Edwards. Superb, high-level study of landmark 1859 publication entitled "On the Number of Primes Less Than a Given Magnitude" traces developments in mathematical theory that it inspired. xiv+315pp. 5⅜ x 8½. 0-486-41740-9

CALCULUS OF VARIATIONS WITH APPLICATIONS, George M. Ewing. Applications-oriented introduction to variational theory develops insight and promotes understanding of specialized books, research papers. Suitable for advanced undergraduate/graduate students as primary, supplementary text. 352pp. 5⅜ x 8½.
0-486-64856-7

COMPLEX VARIABLES, Francis J. Flanigan. Unusual approach, delaying complex algebra till harmonic functions have been analyzed from real variable viewpoint. Includes problems with answers. 364pp. 5⅜ x 8½. 0-486-61388-7

AN INTRODUCTION TO THE CALCULUS OF VARIATIONS, Charles Fox. Graduate-level text covers variations of an integral, isoperimetrical problems, least action, special relativity, approximations, more. References. 279pp. 5⅜ x 8½.
0-486-65499-0

COUNTEREXAMPLES IN ANALYSIS, Bernard R. Gelbaum and John M. H. Olmsted. These counterexamples deal mostly with the part of analysis known as "real variables." The first half covers the real number system, and the second half encompasses higher dimensions. 1962 edition. xxiv+198pp. 5⅜ x 8½. 0-486-42875-3

CATASTROPHE THEORY FOR SCIENTISTS AND ENGINEERS, Robert Gilmore. Advanced-level treatment describes mathematics of theory grounded in the work of Poincaré, R. Thom, other mathematicians. Also important applications to problems in mathematics, physics, chemistry and engineering. 1981 edition. References. 28 tables. 397 black-and-white illustrations. xvii + 666pp. 6⅛ x 9¼.
0-486-67539-4

INTRODUCTION TO DIFFERENCE EQUATIONS, Samuel Goldberg. Exceptionally clear exposition of important discipline with applications to sociology, psychology, economics. Many illustrative examples; over 250 problems. 260pp. 5⅜ x 8½.
0-486-65084-7

NUMERICAL METHODS FOR SCIENTISTS AND ENGINEERS, Richard Hamming. Classic text stresses frequency approach in coverage of algorithms, polynomial approximation, Fourier approximation, exponential approximation, other topics. Revised and enlarged 2nd edition. 721pp. 5⅜ x 8½. 0-486-65241-6

INTRODUCTION TO NUMERICAL ANALYSIS (2nd Edition), F. B. Hildebrand. Classic, fundamental treatment covers computation, approximation, interpolation, numerical differentiation and integration, other topics. 150 new problems. 669pp. 5⅜ x 8½. 0-486-65363-3

THREE PEARLS OF NUMBER THEORY, A. Y. Khinchin. Three compelling puzzles require proof of a basic law governing the world of numbers. Challenges concern van der Waerden's theorem, the Landau-Schnirelmann hypothesis and Mann's theorem, and a solution to Waring's problem. Solutions included. 64pp. 5⅜ x 8½.
0-486-40026-3

THE PHILOSOPHY OF MATHEMATICS: AN INTRODUCTORY ESSAY, Stephan Körner. Surveys the views of Plato, Aristotle, Leibniz & Kant concerning propositions and theories of applied and pure mathematics. Introduction. Two appendices. Index. 198pp. 5⅜ x 8½. 0-486-25048-2

INTRODUCTORY REAL ANALYSIS, A.N. Kolmogorov, S. V. Fomin. Translated by Richard A. Silverman. Self-contained, evenly paced introduction to real and functional analysis. Some 350 problems. 403pp. 5⅜ x 8½. 0-486-61226-0

APPLIED ANALYSIS, Cornelius Lanczos. Classic work on analysis and design of finite processes for approximating solution of analytical problems. Algebraic equations, matrices, harmonic analysis, quadrature methods, much more. 559pp. 5⅜ x 8½. 0-486-65656-X

AN INTRODUCTION TO ALGEBRAIC STRUCTURES, Joseph Landin. Superb self-contained text covers "abstract algebra": sets and numbers, theory of groups, theory of rings, much more. Numerous well-chosen examples, exercises. 247pp. 5⅜ x 8½. 0-486-65940-2

QUALITATIVE THEORY OF DIFFERENTIAL EQUATIONS, V. V. Nemytskii and V.V. Stepanov. Classic graduate-level text by two prominent Soviet mathematicians covers classical differential equations as well as topological dynamics and ergodic theory. Bibliographies. 523pp. 5⅜ x 8½. 0-486-65954-2

THEORY OF MATRICES, Sam Perlis. Outstanding text covering rank, nonsingularity and inverses in connection with the development of canonical matrices under the relation of equivalence, and without the intervention of determinants. Includes exercises. 237pp. 5⅜ x 8½. 0-486-66810-X

INTRODUCTION TO ANALYSIS, Maxwell Rosenlicht. Unusually clear, accessible coverage of set theory, real number system, metric spaces, continuous functions, Riemann integration, multiple integrals, more. Wide range of problems. Undergraduate level. Bibliography. 254pp. 5⅜ x 8½. 0-486-65038-3

MODERN NONLINEAR EQUATIONS, Thomas L. Saaty. Emphasizes practical solution of problems; covers seven types of equations. ". . . a welcome contribution to the existing literature...."–*Math Reviews*. 490pp. 5⅜ x 8½. 0-486-64232-1

MATRICES AND LINEAR ALGEBRA, Hans Schneider and George Phillip Barker. Basic textbook covers theory of matrices and its applications to systems of linear equations and related topics such as determinants, eigenvalues and differential equations. Numerous exercises. 432pp. 5⅜ x 8½. 0-486-66014-1

LINEAR ALGEBRA, Georgi E. Shilov. Determinants, linear spaces, matrix algebras, similar topics. For advanced undergraduates, graduates. Silverman translation. 387pp. 5⅜ x 8½. 0-486-63518-X

ELEMENTS OF REAL ANALYSIS, David A. Sprecher. Classic text covers fundamental concepts, real number system, point sets, functions of a real variable, Fourier series, much more. Over 500 exercises. 352pp. 5⅜ x 8½. 0-486-65385-4

SET THEORY AND LOGIC, Robert R. Stoll. Lucid introduction to unified theory of mathematical concepts. Set theory and logic seen as tools for conceptual understanding of real number system. 496pp. 5⅜ x 8¼. 0-486-63829-4

TENSOR CALCULUS, J.L. Synge and A. Schild. Widely used introductory text covers spaces and tensors, basic operations in Riemannian space, non-Riemannian spaces, etc. 324pp. 5⅜ x 8¼. 0-486-63612-7

ORDINARY DIFFERENTIAL EQUATIONS, Morris Tenenbaum and Harry Pollard. Exhaustive survey of ordinary differential equations for undergraduates in mathematics, engineering, science. Thorough analysis of theorems. Diagrams. Bibliography. Index. 818pp. 5⅜ x 8½. 0-486-64940-7

INTEGRAL EQUATIONS, F. G. Tricomi. Authoritative, well-written treatment of extremely useful mathematical tool with wide applications. Volterra Equations, Fredholm Equations, much more. Advanced undergraduate to graduate level. Exercises. Bibliography. 238pp. 5⅜ x 8½. 0-486-64828-1

FOURIER SERIES, Georgi P. Tolstov. Translated by Richard A. Silverman. A valuable addition to the literature on the subject, moving clearly from subject to subject and theorem to theorem. 107 problems, answers. 336pp. 5⅜ x 8½. 0-486-63317-9

INTRODUCTION TO MATHEMATICAL THINKING, Friedrich Waismann. Examinations of arithmetic, geometry, and theory of integers; rational and natural numbers; complete induction; limit and point of accumulation; remarkable curves; complex and hypercomplex numbers, more. 1959 ed. 27 figures. xii+260pp. 5⅜ x 8½. 0-486-63317-9

POPULAR LECTURES ON MATHEMATICAL LOGIC, Hao Wang. Noted logician's lucid treatment of historical developments, set theory, model theory, recursion theory and constructivism, proof theory, more. 3 appendixes. Bibliography. 1981 edition. ix + 283pp. 5⅜ x 8½. 0-486-67632-3

CALCULUS OF VARIATIONS, Robert Weinstock. Basic introduction covering isoperimetric problems, theory of elasticity, quantum mechanics, electrostatics, etc. Exercises throughout. 326pp. 5⅜ x 8½. 0-486-63069-2

THE CONTINUUM: A CRITICAL EXAMINATION OF THE FOUNDATION OF ANALYSIS, Hermann Weyl. Classic of 20th-century foundational research deals with the conceptual problem posed by the continuum. 156pp. 5⅜ x 8½. 0-486-67982-9

CHALLENGING MATHEMATICAL PROBLEMS WITH ELEMENTARY SOLUTIONS, A. M. Yaglom and I. M. Yaglom. Over 170 challenging problems on probability theory, combinatorial analysis, points and lines, topology, convex polygons, many other topics. Solutions. Total of 445pp. 5⅜ x 8½. Two-vol. set.
Vol. I: 0-486-65536-9   Vol. II: 0-486-65537-7

Paperbound unless otherwise indicated. Available at your book dealer, online at **www.doverpublications.com**, or by writing to Dept. GI, Dover Publications, Inc., 31 East 2nd Street, Mineola, NY 11501. For current price information or for free catalogues (please indicate field of interest), write to Dover Publications or log on to **www.doverpublications.com** and see every Dover book in print. Dover publishes more than 500 books each year on science, elementary and advanced mathematics, biology, music, art, literary history, social sciences, and other areas.